T0297856

CAMBRIDGE LIBRARY COLLECTION

*Books of enduring scholarly value*

## Physical Sciences

From ancient times, humans have tried to understand the workings of the world around them. The roots of modern physical science go back to the very earliest mechanical devices such as levers and rollers, the mixing of paints and dyes, and the importance of the heavenly bodies in early religious observance and navigation. The physical sciences as we know them today began to emerge as independent academic subjects during the early modern period, in the work of Newton and other 'natural philosophers', and numerous sub-disciplines developed during the centuries that followed. This part of the Cambridge Library Collection is devoted to landmark publications in this area which will be of interest to historians of science concerned with individual scientists, particular discoveries, and advances in scientific method, or with the establishment and development of scientific institutions around the world.

## A History of the Theory of Elasticity and of the Strength of Materials

A distinguished mathematician and notable university teacher, Isaac Todhunter (1820–84) became known for the successful textbooks he produced as well as for a work ethic that was extraordinary, even by Victorian standards. A scholar who read all the major European languages, Todhunter was an open-minded man who admired George Boole and helped introduce the moral science examination at Cambridge. His many gifts enabled him to produce the histories of mathematical subjects which form his lasting memorial. First published between 1886 and 1893, the present work was the last of these. Edited and completed after Todhunter's death by Karl Pearson (1857–1936), another extraordinary man who pioneered modern statistics, these volumes trace the mathematical understanding of elasticity from the seventeenth to the late nineteenth century. Volume 1 (1886) begins with Galileo Galilei and extends to the researches of Saint-Venant up to 1850.

Cambridge University Press has long been a pioneer in the reissuing of out-of-print titles from its own backlist, producing digital reprints of books that are still sought after by scholars and students but could not be reprinted economically using traditional technology. The Cambridge Library Collection extends this activity to a wider range of books which are still of importance to researchers and professionals, either for the source material they contain, or as landmarks in the history of their academic discipline.

Drawing from the world-renowned collections in the Cambridge University Library and other partner libraries, and guided by the advice of experts in each subject area, Cambridge University Press is using state-of-the-art scanning machines in its own Printing House to capture the content of each book selected for inclusion. The files are processed to give a consistently clear, crisp image, and the books finished to the high quality standard for which the Press is recognised around the world. The latest print-on-demand technology ensures that the books will remain available indefinitely, and that orders for single or multiple copies can quickly be supplied.

The Cambridge Library Collection brings back to life books of enduring scholarly value (including out-of-copyright works originally issued by other publishers) across a wide range of disciplines in the humanities and social sciences and in science and technology.

# A History of the Theory of Elasticity and of the Strength of Materials

Volume 1:
From Galilei to Saint-Venant

Isaac Todhunter
Edited by Karl Pearson

CAMBRIDGE
UNIVERSITY PRESS

University Printing House, Cambridge, CB2 8BS, United Kingdom

Published in the United States of America by Cambridge University Press, New York

Cambridge University Press is part of the University of Cambridge.
It furthers the University's mission by disseminating knowledge in the pursuit of education, learning and research at the highest international levels of excellence.

www.cambridge.org
Information on this title: www.cambridge.org/9781108070423



This edition first published 1886
This digitally printed version 2014

ISBN 978-1-108-07042-3 Paperback

# A HISTORY

OF

# THE THEORY OF ELASTICITY.

London: C. J. CLAY AND SONS,
CAMBRIDGE UNIVERSITY PRESS WAREHOUSE,
Ave Maria Lane.

CAMBRIDGE: DEIGHTON, BELL, AND CO.
LEIPZIG: F. A. BROCKHAUS.

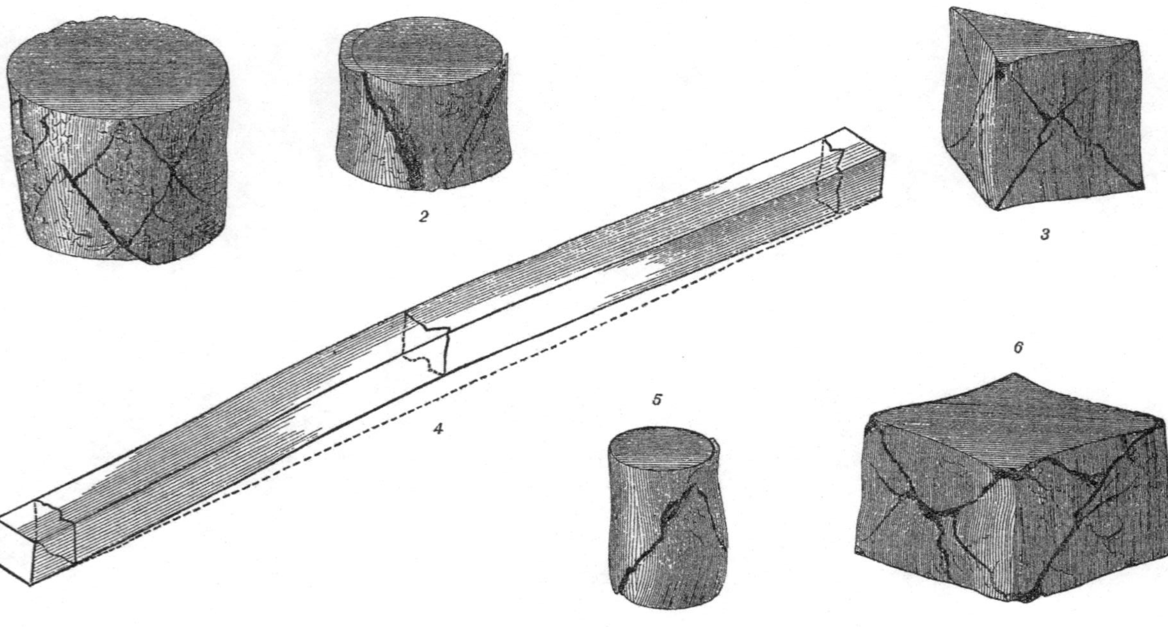


Rupture-Surfaces of Cast-Iron

*Frontispiece*

# A HISTORY OF

# THE THEORY OF ELASTICITY

AND OF

# THE STRENGTH OF MATERIALS

*FROM GALILEI TO THE PRESENT TIME.*

BY THE LATE

ISAAC TODHUNTER, D. Sc., F.R.S.

EDITED AND COMPLETED
FOR THE SYNDICS OF THE UNIVERSITY PRESS
BY

KARL PEARSON, M.A.

PROFESSOR OF APPLIED MATHEMATICS, UNIVERSITY COLLEGE, LONDON.

VOL. I. GALILEI TO SAINT-VENANT
1639—1850.

CAMBRIDGE:
AT THE UNIVERSITY PRESS.
1886

Cambridge:
PRINTED BY C. J. CLAY, M.A. AND SONS,
AT THE UNIVERSITY PRESS.

TO THE MEMORY OF

M. BARRÉ DE SAINT-VENANT

THE FOREMOST OF MODERN ELASTICIANS

THE EDITOR DEDICATES HIS LABOUR

ON THE PRESENT VOLUME.

Si ces imperfections sont malheureusement nombreuses cela vient de ce que la science appliquée est jeune et encore pauvre ; avec ses ressources actuelles, elle peut déjà rendre de grands services, mais ses destinées sont bien plus hautes : elle offre un champ immense au zèle de ceux qui voudront l'enrichir, et beaucoup de parties de son domaine semblent même n'attendre que des efforts légers pour produire des résultats d'une grande utilité.

*Saint-Venant.*

Jedenfalls sieht man aus den angeführten Thatsachen, dass die Theorie der Elasticitat noch durchaus nicht als abgeschlossen zu betrachten ist, und es wäre zu wünschen, dass recht viel Physiker sich mit diesem Gegenstande beschäftigten, um durch vermehrte Beobachtungen die sichere Grundlage zu einer erweiterten Theorie zu schaffen.

*Clausius.*

Ceux qui, les premiers, ont signalé ces nouveaux instruments, n existeront plus et seront complétement oubliés ; à moins que quelque géomètre archéologue ne ressuscite leurs noms. Eh ! qu importe, d'ailleurs, si la science a marché !

*Lamé.*

# PREFACE.

In the summer of 1884 at the suggestion of Dr Routh the Syndics of the University Press placed in my hands the manuscript of the late Dr Todhunter's History of Elasticity, in order that it might be edited and completed for the Press. That the publication might not be indefinitely delayed, it was thought advisable to print off chapter by chapter as the work of revision progressed. That this arrangement has accelerated the publication of the first volume is certain, but at the same time it has introduced some disadvantages to which it is necessary for me to refer. In the first place it was impossible to introduce in the earlier cross-references to later portions of the work; this I have endeavoured to rectify by adding a copious index to the whole volume. In the next place I must mention, that it was not till I had advanced some way into the work that I felt convinced that the reproduction in the analysis of a memoir of the individual writer's terminology and notation must be abandoned and a uniform terminology and notation adopted for the whole book. This was absolutely needful if the book was to be available for easy reference, and not merely of interest to the historical student. The choice, however, of such terminology and notation—considering the enormous diversity, I will even say confusion, on this point to be found in the writings of British and continental elasticians—was an extremely venturesome task. To evolve a really scientific terminology which shall stand any chance of universal adoption from a number of words, which each individual writer has used in his own sense, is no easy matter. If I have in some cases dispensed with such well-worn words as tension, pressure, extension, contraction and so forth, it has been from no desire for novelty, but in order to avoid a conflict of definitions. That the notation and terminology proposed in this work will extend beyond it I hardly venture to hope, I shall be content if they be intelligible to those who may consult this book They will be found fully discussed in Notes B—D of the Appendix, which I would ask the reader to examine before passing to the text. As I have said, it was unfortunately only after I had made some progress in the work, that I became convinced of the need of terminological and notational uniformity. I think, however, consistency in these points will be found after the middle of the chapter devoted to Poisson. The introduction of this uniform system of symbols and terms has itself involved a considerable amount of additional work on the manuscript. The symbols and terms used in the manuscript are occasionally those of the original memoirs, occasionally those of Lamé or of Saint-Venant. The want of uniformity in the first two chapters will perhaps not be considered a disadvantage, the memoirs being of historical rather than scientific interest, and their language often the most characteristic part of their historical value.

The disadvantages which I have pointed out in this first volume will I trust be obviated in the second by the revision and completion of the whole manuscript before the work of printing is commenced. The second volume will contain an analysis of all researches in elasticity from 1850 to the present time From 1850 to 1870 most but not all of the chief mathematical memoirs have been already analysed by Dr Todhunter; there is but little of a later date completed. Considering the amount of work to be done, considering that it is advisable to avoid revision and printing being carried on simultaneously, and finally noting the very limited time, which the teaching duties of my present post allow me to spend in a library where it is possible to carry on historical work of this kind, I fear the publication of the second volume would be much delayed were the task of editing it entrusted to me. I lay stress upon this point as, although I have endeavoured to make the first volume complete in itself, much of its usefulness will be realised only on the appearance of the second. Indeed, in the interests of the reader as well as of the work, I think the Syndics will have to consider the question of appointing another editor, who has more of the needful leisure.

It is proper that I should explain with some detail the manner in which I have performed my task as editor. Dr Todhunter's manuscript consists of two distinct parts, the first contains a purely mathematical treatise on the theory of the 'perfect' elastic solid; the second a history of the theory of elasticity The treatise based principally on the works of Lamé, Saint Venant and Clebsch is yet to a great extent historical, that is to say many paragraphs are composed of analyses of important memoirs. Thus in the History-manuscript after the title of a memoir there is occasionally only a mere reference to the paragraph of the Theory-manuscript, where it will be found discussed. Certain portions also of the manuscript have inscribed upon them in Dr Todhunter's handwriting 'History or Theory?' The Syndics having determined to publish in the first place the History only, it became necessary to determine how the gaps in the 'History' which were covered by mere reference to the 'Theory' should be filled up. With the sanction of the Syndics I have adopted the following principle: the analysis of a memoir wherever possible is to be Dr Todhunter's. Thus certain, on the whole not very considerable, portions of the Theory-manuscript are incorporated in the History, while all portions of the manuscript marked doubtful have been made use of when required.

Dr Todhunter's manuscript contains two versions, a first writing and a revision. The revision has been again read through by the author, but the principal alterations made are notes or suggestions for further consideration; in some cases the note is merely a statement that a criticism must be either modified or entirely reconstructed, in other cases, it involves a valuable cross-reference. One of the most important of these notes is that referred to in my footnote on p. 250; it led to the only considerable excision which I have thought it proper to make before printing Dr Todhunter's manuscript.

The changes I have made in that manuscript are of the following character; the introduction of a uniform terminology and notation, the correction of clerical and other obvious errors, the insertion of cross-references, the occasional introduction of a remark or of a footnote. The remarks are inclosed in square brackets. With this exception any article in this volume the number of which is *not included in square brackets* is due entirely to Dr Todhunter. So far as the arrangement of the memoirs is concerned there was little if anything to guide me in the manuscript. Dr Todhunter had evidently intended to give each of the principal elasticians chapters to themselves, and to group the minor memoirs together into periods. This method although it destroys the strict chronological treatment, and to some extent obscures the order of development, yet possesses such advantages, in that it groups together the researches of one man following his own peculiar lines of thought, that I have followed it without hesitation as the best possible. I even regret that I have not devoted special chapters to such elasticians as Hodgkinson, Wertheim and F. E. Neumann; in the latter case the regret is deepened by the recent publication of his lectures on elasticity.

Turning to my own share in the completing of the work, I fear that at first sight I may appear to have exceeded the duty of an editor. For all the Articles in this volume whose numbers are enclosed in square brackets I am alone responsible, as well as for the corresponding footnotes, and the Appendix with which the volume concludes. The principle which has guided me throughout the additions I have made has been to make the work, so far as it lay in my power, a standard work of reference for its own branch of science. The use of a work of this kind is twofold. It forms on the one hand the history of a peculiar phase of intellectual development, worth studying for the many side lights it throws on general human progress. On the other hand it serves as a guide to the investigator in what has been done, and what ought to be done In this latter respect the individualism of modern science has not infrequently led to a great waste of power; the same bit of work has been repeated in different countries at different times, owing to the absence of such histories as Dr Todhunter set himself to write. It is true that the various *Jahrbücher* and *Fortschritte* now reduce the possibility of this repetition, but besides their frequent insufficiency they are at best but indices to the work of the last few years; an enormous amount of matter is practically stored out of sight in the *Transactions* and *Journals* of the last century and of the first half of the present century. It would be a great aid to science, if, at any rate, the innumerable mathematical journals could be to a great extent specialised, so that we might look to any one of them for a special class of memoir. Perhaps this is too great a collectivist reform to expect in the near future from even the cosmopolitan spirit of modern science. As it is, the would-be researcher either wastes much time in learning the history of his subject, or else works away regardless of earlier investigators. The latter course has been singularly prevalent with even some first-class British and French mathematicians.

Keeping the twofold object of this work in view I have endeavoured to give it completeness (1) as a history of developement, (2) as a guide to what has been accomplished.

Taking the first chapter of this History the author has discussed the important memoirs of James Bernoulli and some of those due to Euler. The whole early history of our subject is however so intimately connected with the names of Galilei, Hooke, Mariotte and Leibniz, that I have introduced some account of their work. The labours of Lagrange and Riccati also required some recognition, so that these early writers form the basis of a chapter, which I believe the reader will not find without interest, whether judged from the special standpoint of the elastician or from the wider footing of insight into the growth of human ideas. With a similar aim I have introduced throughout the volume a number of memoirs having purely histcrical value which had escaped Dr Todhunter's notice.

Another class of memoirs which I have inserted are memoirs of mathematical value, omitted apparently by pure accident. For example all the memoirs of F. E. Neumann, the second memoir of Duhamel, those of Blanchet etc. I cannot hope that the work is complete in this respect even now, but I trust that nothing of equal importance has escaped the author or editor[1]

My greatest difficulty arose with regard to the rigid line which Dr Todhunter had attempted to draw between mathematical and physical memoirs. Thus while including an account of Clausius' memoir of 1849, he had omitted Weber's of 1835, yet the consideration of the former demands the inclusion of the latter, were it not indeed required by the long series of mathematical memoirs which have in recent years treated of elastic after-strain. What seemed to me peculiarly needful at the present time was to place before the mathematician the results of physical investigations, that he might have some distinct guide to the direction in which research is required. There has been far too much invention of solvable problems' by the mathematical elastician; far too much neglect of the physical and technical problems which have been crying out for solution. Much of the ingenuity which has been spent on the ideal body of perfect' elasticity ideally loaded, might I believe have wrought miracles in the fields of physical and technical elasticity, where pressing practical problems remain in abundance unsolved. I have endeavoured, so far as lay in my power, to abrogate this divorce between mathematical elasticity on the one hand, and physical and technical elasticity on the other. With this aim in view I have introduced the general conclusions of a considerable body of physical and technical memoirs, in the hope that by doing so I may bring the mathematician closer to the physicist and both to the practical engineer I trust that in doing so I have rendered this History of value to a wider range of

[1] I should be very glad of a notification of any omissions, so that some reference might be made to them in the second volume.

readers, and so increased the usefulness of Dr Todhunter's many years of patient historical research on the more purely mathematical side of elasticity. In this matter I have kept before me the labours of M. de Saint Venant as a true guide to the functions of the ideal elastician.

It remains for me to thank those friends who have so readily given assistance and sympathy in the labour of editing. Only those, who have undertaken a task of similar dimensions can fully appreciate the value of such help. The aid of two men, strangely alike in character though diverse in pursuit, who exhibited a keen interest in the progress of this work, has been lost to me during its passage through the press. To the late Mr Henry Bradshaw I owe assistance in procuring scarce memoirs, pamphlets, and dissertations, as well as many valuable suggestions on typographical and bibliographical details. To the late M. Barré de Saint Venant I am indebted for the loan of several works, for a variety of references and facts bearing on the history of elasticity, as well as for a revision of the earlier pages of Chapter IX. The later pages of that chapter were revised after the death of M. de Saint-Venant by his friend and pupil M. Flamant, Professeur à l'Ecole Centrale; whom I have likewise to thank for disinterested assistance in the revision of other portions of the work relating to French elasticians.

The assistance of two other friends has left its mark on nearly every article I have contributed to the work. My colleague, Professor A. B. W. Kennedy, has continually placed at my disposal the results not only of special experiments, but of his wide practical experience. The curves figured in the Appendix, as well as a variety of practical and technical remarks scattered throughout the volume I owe entirely to him; beyond this it is difficult for me to fitly acknowledge what I have learnt from mere contact with a mind so thoroughly imbued with the concepts of physical and technical elasticity. Mr W. H. Macaulay, University Lecturer in Applied Mechanics, Cambridge, has given me repeated aid in the discussion of mathematical difficulties, and has saved me from many errors of interpretation, and several of judgment. I have to thank Mr C. Chree of King's College, Cambridge for a very careful revision of the proofs subsequent to Chapter IV., and for a variety of suggestions. Mr T. H. Beare of the Engineering Department, University College, has prepared the copious Index to this volume, upon which much of its usefulness will depend. To Mr R. J. Parker of Lincoln's Inn I owe frequent linguistic assistance and revision. While to Professor Callcott Reilly of Cooper's Hill, to my colleague Professor M. J. M. Hill and to Mr R. Tucker of the London Mathematical Society I am indebted for aid in a variety of ways.

In conclusion I can only hope that this first volume of Dr Todhunter's work will fulfil the object which he had designed for it, that—notwithstanding the want of the author's own revision and the many editorial failings—it may still take its place as a standard work of reference, worthy alike of its author and of the University which publishes it.

KARL PEARSON.

University College, London,
*June* 23, 1886.

## ERRATA.

p. 100, *line* 4 *for* $d^4x/dx^4$ *read* $d^4y/dx^4$.

p. 142, *last line for* Poisson *read* Poinsot.

p. 217, *dele* footnote.

p. 317, *line* 5 *from bottom for* VII. *read* VIII.

p. 327, *footnote for* confusing $f(r)$ *read* confusing f($r$)

p. 359, *line* 1 *for* $xx$ *read* $xx$.

p. 368, *line* 24 *for* Art. 659 *read* Art. 661.

p. 391, *line* 5 *for* horizontal *read* vertical.

p. 438, *line* 14 *for* $\frac{1}{E_2}$ *read* $\frac{1}{E_r}$

p. 446, *first line for* $P\omega$ *read* $P/\omega$.

p. 523, *line* 17 *for* 2·04 *read* 2·047.

p. 705, *line* 15 *for* Art. 366 *read* Art. 365.

p. 855, *line* 16 *for* 1239 *read* 1240.

# CONTENTS.


CHAPTER I.

PAGES

The Seventeenth and Eighteenth Centuries. Galilei to Girard, 1638—1798 . . . . . . . . . . . . 1—79

CHAPTER II.

Miscellaneous Investigations between the Years 1800 and 1822 . . . . . . . . . . . . 80—132

CHAPTER III.

Miscellaneous Researches 1820—1830, Navier, Germain, Savart, Pagani, and others . . . . . . . 133—207

CHAPTER IV.

Poisson . . . . . . . . . . . . 208—318

CHAPTER V.

Cauchy . . . . . . . . . . . . 319—376

CHAPTER VI.

Miscellaneous Researches of the Decade, 1830—1840 . . . 377—543

CHAPTER VII.

Lamé and Clapeyron, Lamé . . . . . . . . . 544—626

CHAPTER VIII.

Miscellaneous Researches of the Decade 1840—1850, including those of Blanchet, Stokes, Wertheim and Haughton . . 627—832

CHAPTER IX.

Saint-Venant's Researches before 1850 . . . . . . 833—872

APPENDIX. NOTES A—E. 873—896

INDEX.

# CHIEF ELASTICIANS BEFORE 1850.

## ARRANGED IN THE ORDER OF THEIR CHIEF MEMOIRS ON ELASTICITY.

| | *Birth. Death.* |
|---|---|
| Galilei | 1564—1642 |
| Hooke | 1635 – 1702 |
| Mariotte | 1620 (?)—1684 |
| James Bernoulli | 1654—1705 |
| Musschenbroek | 1692—1761 |
| Daniel Bernoulli | 1700—1782 |
| Euler | 1707—1783 |
| Coulomb | 1736—1806 |
| Girard | 1765—1836 |
| Young | 1773—1829 |
| Tredgold | 1788—1829 |
| Hodgkinson | 1789—1861 |
| Navier | 1785—1836 |
| Germain | 1776—1831 |
| Savart | 1791—1841 |
| Poisson | 1781—1840 |
| Cauchy | 1789—1857 |
| W. Weber | 1804— * |
| Vicat | 1786—1861 |
| Piola | 1791—1850 |
| F. E. Neumann | 1798— * |
| Gerstner | 1756—1832 |
| Duhamel | 1797—1872 |
| Green | 1793—1841 |
| Poncelet | 1788—1867 |
| Lamé | 1795—1870 |
| Clapeyron | 1799—1864 |
| Stokes | 1819— * |
| Wertheim | 1815—1861 |
| Blanchet | 1813— * |
| Maxwell | 1831—1879 |
| Haughton | (?) — * |
| Jellett | (?) — * |
| Kupffer | 1799—1865 |
| Saint-Venant | 1797—1886 |

* Living scientists.

# CHAPTER I.

## THE SEVENTEENTH AND EIGHTEENTH CENTURIES.

### GALILEI TO GIRARD. 1638—1798.

[1.] THE modern theory of elasticity may be considered to have its birth in 1821, when Navier first gave the equations for the equilibrium and motion of elastic solids, but some of the problems which belong to this theory had previously been solved or discussed on special principles, and to understand the growth of our modern conceptions it is needful to investigate the work of the seventeenth and eighteenth centuries.

[2.] The first memoir that requires notice is by Galileo Galilei and forms the second dialogue of the *Discorsi e Dimostrazioni matematiche*, Leiden 1638[1]. This dialogue both from its contents and form is of great historical interest. It not only gave the impulse but determined the direction of all the inquiries concerning the rupture and strength of beams, with which the physicists and mathematicians for the next century principally busied themselves. Galilei gives 17 propositions with regard to the fracture of rods, beams and hollow cylinders. The noteworthy feature about his method of discussion is that he supposed the fibres of a strained beam to be *inextensible*. There are two

[1] There is an English translation in *Thomas Salusbury's Mathematical Collections and Translations*, London, 1665. Tom. II. p. 89.

problems which Galilei discussed, and which form the starting points of many later memoirs. They are the following:

[3.] A beam ($ABCD$) being built horizontally into a wall (at $AB$) and strained by its own or an applied weight ($E$), to find the breaking force upon a section perpendicular to its axis. This problem is always associated by later writers with Galilei's name, and we shall call it in future *Galilei's Problem.*

(From the *Discorsi*, Leiden 1638.)

The 'base of fracture' being defined as the section of the beam where it is built into the wall; we have the following results:—

(i) The resistances of the bases of fracture of similar prismatic beams are as the squares of their corresponding dimensions.

In this case the beams are supposed loaded at the free end till the base of fracture is ruptured; the weights of the beams are neglected.

(ii) Among an infinite number of homogeneous and similar beams there is only one, of which the weight is exactly in equilibrium with the resistance of the base of fracture. All others, if of a greater length will break,—if of a less length will have a superfluous resistance in their base of fracture.

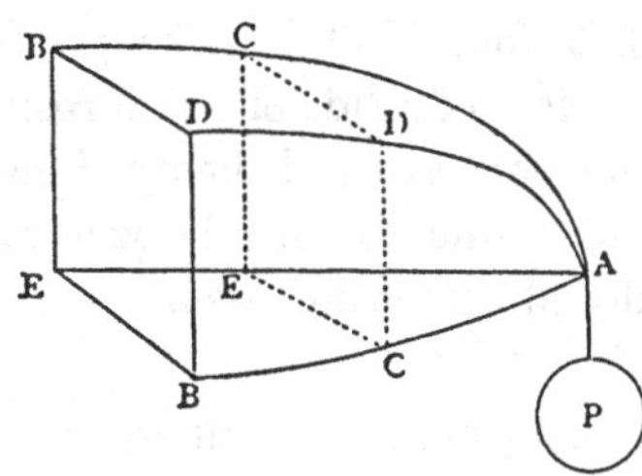


[4.] The second problem with which Galilei particularly busied himself, was the discovery of 'solids of equal resistance.' The problem in its simplest form may be thus stated; $ACB$, $AC'B'$ are two curves in vertical and horizontal planes respectively, a solid is generated by treating $ACB$ and $AC'B'$ as the bases of cylinders with generators perpendicular to the bases. This solid $BEB'DA$ is then treated as a beam built in at the base $BEB'D$ and from $A$ a weight is suspended. The problem is to find the form of the generating curves so that the resistance of a section $CE'C'D$ may be exactly equal to the tendency to rupture at that place Obviously the problem may take a more complex form by supposing any system of forces to act upon the beam As we have stated it, it still remains indeterminate, for we must either be given one of the generating curves or else a relation between them. Galilei supposed the curve $AC'B'$ to be replaced by a line parallel to $AE$, so that all vertical sections of his beam parallel to $ACBE$ were curves equal to $ACB$. In this case he easily determined that the 'solid of equal resistance' must have a parabola for its generating curve.

[5.] This problem of solids of equal resistance led to a memorable controversy in the scientific world. It was discussed by P. Wurtz, François Blondel (*Galilaeus Promotus* 1649 (?); *Sur la résistance des solides, Mém. Acad. Paris, Tom.* I. 1692), Alex. Marchetti (*De resistentia solidorum*, Florence 1669), V. Viviani (*Opere Galilei*, Bologna 1655), Guido Grandi (*La controversia contro dal Sig. A. Marchetti*, and *Risposta apologetica...alle opposizioni dal Sig. A. Marchetti*; both Lucca 1712), and still more fully later, in memoirs to be referred to, of Varignon (1702) and Parent (1710). An interesting account of the controversy and also of writings on the same subject will be found in Girard's work (*cf infra* Art. 124).

Closely as the problem of solids of equal resistance is associated with the growth of the mathematical theory of elasticity, it is nevertheless the problem of the flexure of a horizontal beam which may be said to have produced the entire theory.

[6] While the continental scientists were thus busy with problems, which were treated without any conception of elasticity, and yet were to lead ultimately to the problem of the elastic curve, their English contemporaries seem to have been discussing hypotheses as to the nature of elastic bodies. One of the earliest memoirs in this direction which I have met with is due to Sir William Petty, and is entitled :

*The Discourse made before the Royal Society concerning the use of Duplicate Proportion; together with a new Hypothesis of Springing or Elastique Motions*, London 12*mo*. 1674.

Although absolutely without *scientific* value, this little work throws a flood of light on the state of scientific investigation at the time. On p. 114 we are treated to an 'instance' of duplicate proportion in the "*Compression of Yielding and Elastic Bodies as Wooll, &c.*" There is an appendix (p. 121) on the new hypothesis as to elasticity The writer explains it by a complicated system of atoms to which he gives not only polar properties, but also *sexual* characteristics, remarking in justification that the statement of Genesis i. 27:—"male and female created he them"—must be taken to refer to the very ultimate parts of nature, or, to atoms as well as to mankind! (p. 131.)

Much more scientific value must be granted to the work of the next English writer.

[7] The discovery apparently of the modern conception of elasticity seems due to Robert Hooke, who in his work *De potentiâ restitutiva,* London 1678, states that 18 years before the date of that publication he had first found out the theory of springs, but had omitted to publish it because he was anxious to obtain a patent for a particular application of it. He continues:—

About three years since His Majesty was pleased to see the Experiment that made out this theory tried at *White Hall,* as also my Spring Watch.

About two years since I printed this Theory in an Anagram at the end of my Book of the Descriptions of Helioscopes, *viz. ceiiinosssttuu, id est, Ut Tensio sic vis;* That is, The Power of any spring is in the same proportion with the Tension thereof.

By spring' Hooke does not merely denote a spiral wire, or a bent rod of metal or wood, but any "springy body" whatever. Thus after describing his experiments he writes:

From all which it is very evident that the Rule or Law of Nature in every springing body is, that the force or power thereof to restore it self to its natural position is always proportionate to the Distance or space it is removed therefrom, whether it be by rarefaction, or separation of its parts the one from the other, or by a Condensation, or crowding of those parts nearer together. Nor is it observable in these bodies only, but in all other springy bodies whatsoever, whether Metal, Wood, Stones, baked Earths, Hair, Horns, Silk, Bones, Sinews, Glass and the like. Respect being had to the particular figures of the bodies bended, and to the advantageous or disadvantageous ways of bending them.

[8.] The modern expression of the six components of stress as linear functions of the strain components may perhaps be *physically* regarded as a generalised form of Hooke's Law (See the remark made on this point by Saint-Venant in his *Mémoire sur la Torsion des Prismes,* pp. 256—7, and compare the same physicist's valuable note in his translation of Clebsch's *Theorie der Elasticität fester Korper,* pp. 39—40).

[9.] The principles of the *Congruity* and *Incongruity* of bodies and of the 'fluid subtil matter' or *menstruum* by which all bodies near the earth are incompassed—wherewith Hooke sought to theoretically ground his experimental law will no more satisfy the modern mathematician than the above-mentioned researches of Galilei. They are however very characteristic of the mathematical metaphysics of the period[1].

[10.] Mariotte seems to have been the earliest investigator who applied anything corresponding to the elasticity of Hooke to the fibres of the beam in Galilei's problem. In his *Traité du mouvement des eaux,* Paris 1686 *Partie V. Disc.* 2, pp. 370—400, he publishes the results of experiments made by him in 1680 and shows that Galilei s theory does not accord with experience. He remarks that some of the fibres of the beam extend before rupture, while others again are compressed. He assumes however without the least attempt at proof ("on peut concevoir") that half the fibres are compressed, ha f extended.

[11.] G. W. Leibniz: *Demonstrationes novae de Resistentiâ solidorum. Acta Eruditorum Lipsiae July* 1684. The stir created by Mariotte's experiments and his rejection of the views of the great Italian seem to have brought the German philosopher into the field. He treats the subject in a rather *ex cathedrâ* fashion, as if his opinion would finally settle the matter. He examines the hypotheses of Galilei and Mariotte, and finding that there is always flexure before rupture, he concludes that the fibres are really extensible Their resistance is, he states, in proportion to their extension. In other words he applies "Hooke's Law" to the individual fibres. As to the application of his results to special problems, he will leave that to those who have leisure for such matters. The hypothesis of extensible fibres resisting as their extension is usually termed by the writers of this period the Mariotte-Leibniz theory.

[1] A suggestion which occurs in the tract that one of his newly invented spring scales should be carried to the Pike of Teneriffe to test "whether bodies at a further distance from the centre of the earth do not lose somewhat of their powers or tendency towards it," is of much interest as occurring shortly before Newton's enunciation of the law of gravitation.

[12.] De la Hire: *Traité de Mécanique*, Paris 1695. Proposition CXXVI. of this work is entitled *De la résistance des solides.* The author is acquainted with Mariotte's theory and considers that it approaches the actual state of things closer than that of Galilei. At the same time notwithstanding certain concluding words of his preface, he does little but repeat Galilei's theorems regarding beams and the solid of equal resistance.

[13.] Varignon: *De la Résistance des Solides en général pour tout ce qu'on peut faire d'hypothèses touchant la force ou la ténacité des Fibres des Corps à rompre; Et en particulier pour les hypothèses de Galilée & de M. Mariotte. Mémoires de l'Académie*, Paris 1702.

This author considers that it is possible to state a general formula which will include the hypotheses of both Galilei and Mariotte, but to apply his formula it will in nearly all practical cases which may arise be necessary to assume some definite relation between the extension and resistance of the fibres. As Varignon's method of treating the problem is of some interest, being generally adopted by later writers (although in conjunction with either Galilei's or the Mariotte-Leibniz hypothesis), we shall briefly consider it here, without however retaining his notation.

[14.] Let $ABCNML$ be a beam built into a vertical wall at the section $ABC$, and supposed to consist of a number of parallel

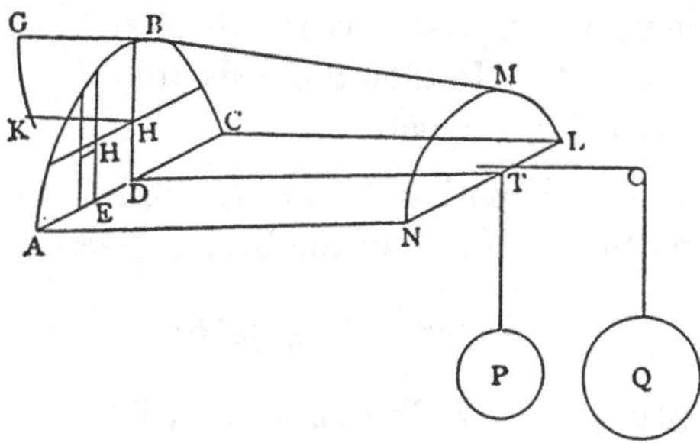


fibres perpendicular to the wall (it is somewhat difficult to see how this is possible in the figure given, which is copied from Varignon) and equal to $AN$ in length. Let $H'$ be a point on the 'base of fracture,' and $H'E$ perpendicular to $AC=y$, $AE=x$. Then if a weight $Q$ be attached by means of a pulley to the extremity of the beam, and be supposed to produce a uniform horizontal force over the whole section $NML$, $Q = r \cdot \int y dx$ where $r$ is the resistance of a fibre of unit sectional area and the integration is to extend over the whole base of fracture. $Q$ is by later writers termed the *absolute resistance* and is given by the above formula. Now suppose the beam to be acted upon at its extremity by a vertical force $P$ instead of the horizontal force $Q$. All the fibres in a horizontal line through $H'$ will have equal resistance, this may be measured by a line $HK$ drawn through $H$ in any fixed direction where $H$ is the point of intersection of the horizontal line through $H'$ and the central vertical $BD$ of the base. As $H$ moves from $B$ to $D$, $K$ will trace out a curve $GK$ which gives the resistance of the corresponding fibres. Take moments for the equilibrium of the beam about $AC$

$$P \cdot l = \iint uy dx dy,$$

where $l =$ length of the beam $DT$ and $u = HK$.

This quantity $\iint uy dx dy$ was termed the *relative resistance* of the beam or the *resistance of the base of fracture.* The meaning of these terms is important for the understanding of these early memoirs. (Varignon speaks of *Résistance absolue* and *Résistance respective,* cf. § XIII.) So far there is little to complain of in Varignon's formulae except that it is necessary to know $u$ before we can make use of it. He then proceeds to apply it to Galilei's and the Mariotte Leibniz hypotheses.

[15.] In Galilei's hypothesis of inextensible fibres $u$ is supposed constant $= r$ and the resistance of the base of fracture becomes

$$= r \int y dx dy = \frac{r}{2} \int y^2 dx$$

On the supposition that the fibres are extensible we ought to consider their extension by finding what is now termed the *neutral line or surface.* Varignon however, and he is followed by later writers *assumes that the fibres in the base ACLN are not extended;* and that the extension of the fibre through $H'$ varies as $DH$, in other words he makes the curve $GK$ a straight line passing through $D$. Hence if $r'$ be the resistance of the fibre at $B$, and $DB=a$, the resistance of the fibre at $H=r'y/a$ or the *resistance of the base of fracture* on this hypothesis becomes

$$\frac{r'}{3a}\int y^3 dx.$$

This resistance in the case of a rectangular beam of breadth $b$ and height $a$ becomes on the two hypotheses

$$\frac{ra^2b}{2} \text{ and } \frac{r'a^2b}{3} \text{ respectively.}$$

Hence in calculating the form of "solids of equal resistance" where the resistance of any section of the beam is taken proportional to the breaking moment at the section, it will be indifferent which hypothesis we make use of (Cf. § XXI. of the memoir.)

[16.] Varignon calculates the forms of various solids of resistance, but it is unnecessary to follow him, his results are practically vitiated when applying the true (Leibniz-Mariotte) theory by his assumption of the position of the neutral surface, but in this error he is followed by so great a mathematician as Euler himself. (See Art. 75.)

[17.] Before entering on the more important work of James Bernoulli we may refer to a memoir by A. Parent entitled *Des points de rupture des figures:...des figures retenués par un de leurs bouts et tirées par telles et tant de puissances qu'on voudra. Mémoires de l'Académie.* Paris 1710, Tom. I. p. 235. I mention this memoir as it practically concludes the theory of solids of equal resistance. The author refers to two of his own earlier memoirs (1704 and 1707) which I have not thought it needful to examine. The point of rupture is deduced from the solid of equal resistance in the following not altogether satisfactory fashion. Consider the case of a beam loaded in any fashion; then retaining the horizontal generating curve of the beam (supposed formed by two cylinders with generators respectively horizontal and vertical in the manner described in our Art. 4,) we may replace the beam by a solid of equal resistance of the same length by changing its vertical generating curve, or again we may invert the process, retaining the vertical and varying the horizontal generating curve. In either case the 'relative resistance' of these two solids at any point (i e. section) will be the same, and a point at which the difference between the relative resistances of the actual solid and of the hypothetical solids of equal resistance is a minimum will be a point of minimum resistance. A point at which this difference vanishes or is negative will be a point of rupture. Parent considers a variety of cases of solids of equal resistance and their points of rupture.

18. The first work of genuine mathematical value on our subject is due to James Bernoulli, who considered the form of a bent elastic lamina in a paper entitled *Curvatura Laminae Elasticae*, printed in the *Acta Eruditorum Lipsiae* for June 1694, p. 262, with *Annotationes et Additiones* thereto in the same *Acta*, Dec. 1695, p. 537. The method of this first examination of the elastic curve did not satisfy Bernoulli, and these memoirs were replaced by another entitled:

*Véritable hypothèse de la résistance des Solides, avec la démonstration de la Courbure des Corps qui font ressort.* This occupies Vol. II. pp. 976—989 of the collected works of the author, published at Geneva, 1744. The date of the memoir is 12th of March 1705; and as the memoirs which follow it in the collected works are entitled *Varia Posthuma*, we may take it to be the last which appeared during the life of this famous mathematician: he died on the 16th of August 1705.

19. The memoir begins by brief notices of what had been already done with respect to the problem by Galilei, Leibniz, and Mariotte; James Bernoulli claims for himself that he first introduced the consideration of the *compression* of parts of the body, whereas previous writers had paid attention to the *extension* alone[1]

20. Three Lemmas which present no difficulty are given and demonstrated:

[1] [As we have seen this remark does not apply to Mariotte.] Ed.

I. Des Fibres de même matière et de même largeur, ou épaisseur, tirées ou pressées par la même force, s'étendent ou se compriment proportionellement à leurs longueurs.

II. Des Fibres homogènes et de même longueur, mais de différentes largeurs ou épaisseurs, s'étendent ou se compriment également par des forces proportionelles à leurs largeurs.

III. Des Fibres homogènes de même longueur et largeur mais chargées de différens poids, ne s'étendent ni se compriment pas proportionellement à ces poids; mais l'extension ou la compression causée par le plus grand poids, est à l'extension ou à la compression causée par le plus petit, en moindre raison que ce poids là n'est à celui-ci.

The third Lemma just stated is strictly true, but is not of great practical importance for our subject, because hitherto in the problems discussed it has been found sufficient to limit the forces so that the extension or contraction of the fibres shall be proportional to the tension or pressure[1]

[21.] The fourth Lemma is more complex in statement, but may be readily understood by reference to Varignon's memoir. We have seen that Varignon supposed the neutral surface to pass through the line $AC$, the so-called 'axis of equilibrium' (see the figure p. 7). James Bernoulli considering there to be both compression and extension does not treat this axis of equilibrium as the horizontal through the lowest point of the base of fracture. He recognises the difficulty of determining the fibres which are neither extended nor compressed, but he comes to the conclusion that the same force applied at the extremity of the same lever will produce the same effect, whether all the fibres are extended, all compressed or part extended and part compressed about the axis of equilibrium. In other words the position of the axis of equilibrium *is indifferent.* This result is expressed by the fourth Lemma and is of course inadmissible. The editor of the collected works has supplied notes recording his dissatisfaction with the reasoning by which this Lemma is supported,

[1] [We shall see reason for somewhat modifying this statement later.] ED.

and Saint Venant remarks in his memoir on the Flexure of Prisms in Liouville's Journal, 1856:

On s'etonne de voir, vingt ans plus tard, un grand géomètre, auteur de la première théorie des courbes élastiques, Jacques Bernoulli tout en admettant aussi les compressions et présentant même leur considération comme étant de lui commettre sous une autre forme, précisement la même méprise du simple au double que Mariotte dans l'évaluation du moment des résistances ce qui le conduit même à affirmer que la position attribuée à l'axe de rotation est tout à fait indifférente.

[22.] In addition to this error of Bernoulli's it must be noted that he rejects the Mariotte-Leibniz hypothesis or the application of Hooke's law to the extension of the fibres. He introduces rather an idle argument against, and quotes an experiment of his own which disagrees with Hooke's *Ut tensio, sic vis.*

23. James Bernoulli next takes a problem which he enunciates thus: "Trouver combien il faut plus de force pour rompre une poutre directement, c'est-à-dire en la tirant suivant sa longueur, que pour la rompre transversalement." The investigation depends on the fourth Lemma, and is consequently not satisfactory.

24. Next we have a second problem entitled: "Trouver la courbure de la Ligne Elastique, c'est-à-dire, celle des lames à ressort qui sont pliées." This problem forms the permanent contribution of the memoir to our subject.

The process is more elaborate than is necessary, because it does not assume the extension or contraction of a fibre to be as the force producing it. But the case in which this assumption is made is especially noticed, and the differential equation to the elastic curve then takes the form $dy = bx^2dx/\sqrt{c^6 - b^2x^4}$.

The investigation considered as the solution of a mechanical problem is imperfect; we know that *three* equations must be satisfied in order to ensure equilibrium among a set of forces in one plane, but here only *one* equation is regarded, namely that of moments[1].

[1] [P. 985 $\int mt\,dt$ is curious notation for $\int_0^t npdp$. Fig. 5 omit the *O* close by *N*.]

25. The method of James Bernoulli with improvements, has been substantially adopted by other writers. The English reader may consult the earlier editions of Whewell's *Mechanics.* Poisson says in his *Traité de Mécanique,* Vol. 1, pages 597 and 600:

Jacques Bernoulli a déterminé, le premier, la figure de la lame élastique en équilibre, d'après des considérations que nous allons développer,...

[26.] Sir Isaac Newton: *Optics or a Treatise of the Reflections Refractions and Colours of Light.* 1717.

The first edition of the Optics (1704) concluded with a series of sixteen *Queries.* To the second edition of 1717 were added fifteen additional queries, making thirty one in all. Of these the XXXI[st], termed 'Elective Attractions,' occupies pp. 242—264 of Vol. IV. of Horsley's edition of the *Opera.* It contains the suggestions upon which were afterwards built up the various hypotheses of the physical nature of elasticity, which seek to explain it by the attractive properties of the ultimate atoms of bodies (see Articles 29. δ. and 36).

The Query commences by suggesting that the attractive powers of small particles of bodies may be capable of producing the great part of the phenomena of nature:—

For it is well known that bodies act one upon another by the attractions of gravity, magnetism and electricity; and these instances shew the tenor and course of nature, and make it not improbable, but that there may be more attractive powers than these. For nature is very consonant and conformable to herself.

We then find certain chemical combinations, fermentations and explosive unions discussed on the ground of attractions between the small particles of bodies.

The parts of all homogeneal hard bodies, which fully touch one another, stick together very strongly. And for explaining how this may be, some have invented hooked atoms, which is begging the question; and others tell us, that bodies are glued together by Rest: that is, by an occult quality, or rather by nothing and others, that they stick together by conspiring motions, that is by relative Rest among themselves. I had rather infer from their cohesion, that their particles attract one another by some force, which in immediate contact is exceeding strong, at small distances performs the chemical operations above-mentioned, and reaches not far from the particles with any sensible effect.

Newton supposes all bodies to be composed of hard particles, and these are heaped up together and scarce touch in more than a few points.

And how such very hard particles, which are only laid together, and touch only in a few points can stick together, and that so firmly as they do, without the assistance of something which causes them to be attracted or pressed towards one another, is very difficult to conceive.

After using arguments from capillarity to confirm these remarks he continues:

Now the small particles of matter may cohere by the strongest attractions, and compose bigger particles of weaker virtue; and many of these may cohere and compose bigger particles, whose virtue is still weaker; and so on for divers successions, until the progression end in the biggest particles, on which the operations in chemistry, and the colours of natural bodies depend; and which by adhering, compose bodies of a sensible magnitude. If the body is compact, and bends or yields inward to pression without any sliding of its parts, it is Hard and Elastick, returning to its figure with a force rising from the mutual attractions of its parts.

The conception of repulsive forces is then introduced to explain the expansion of gases.

Which vast contraction and expansion seems unintelligible, by feigning the particles of air to be springy and ramous, or rolled up like hoops, or by any other means than a Repulsive power. And thus Nature will be very conformable to herself, and very simple; performing all the great motions of the heavenly bodies by the attraction of gravity, which intercedes those bodies and almost all the small ones of their particles, by some other Attractive and Repelling powers.

[27.] The conclusion of the Query and thus also of the Optics is devoted to a semi-theological discussion on the creation of the ultimate hard particles of matter by God. A suggestive paragraph however occurs (p. 261), which is sometimes not sufficiently remembered when gravitation is spoken of as a *cause*:—

These principles—i. e. of attraction and repulsion—I consider not as occult qualities, supposed to result from the specifick forms of things, but as general laws of Nature, by which the things themselves are formed; their truth appearing to us by phenomena, though their causes be not yet discovered[1].

This seems to be Newton's only contribution to the subject of Elasticity, beyond the paragraph of the *Principia* on the collision of elastic bodies. (See Art. 37 and footnote.)

[28.] We may here note that while the mathematicians were beginning to struggle with the problems of elasticity, a number of practical experiments were being made on the flexure and rupture of beams, the results of which were of material assistance to the theorists. Besides the experiments made by Mariotte, we may mention:

*a.* A. Parent: A first memoir by this author dated either 1702 or 1704, I have not been able to discover. It is cited by Girard. *Expériences pour connoître la Résistance des Bois de Chêne et de Sapin. Mém. Acad. Paris,* 1707, *p.* 680. *Les Résistances des Poutres par rapport à leurs Longueurs ou Portées...et des Poutres de plus grande Résistance. Ibid.* 1708, p. 20.

*β.* B. F. de Bélidor: *La Science des ingénieurs dans la conduite des travaux de fortification et d'architecture civile,* La Haye 1729.

*γ.* R. A. F. de Réamur: *Expériences pour connaître, si la force des cordes surpasse la somme des forces des fils qui composent ces mêmes cordes. Mémoires de l'Académie.* Paris 1711. Also, *Expériences et réflexions sur la prodigieuse ductilité de diverses matiéres. Ibid.* 1713. I have referred to these memoirs, because although they can hardly be said to form part of our subject, yet subsequent writers on elasticity draw material from them (e.g. Musschenbroek and Belgrado).

*δ.* Petris van Musschenbroek: *Introductio ad cohaerentiam cor porum firmorum.* This extremely voluminous work commences at p. 423 of the author's *Physicae experimentales et geometricae*

[1] See also the remarkable phrase with which the preface to the second edition concludes; "I do not take gravity for an essential Property of Bodies."

*Dissertationes. Lugduni* 1729. It was held in high repute even to the end of the 18th century.

The author commences with an historical preface, which has been largely drawn upon by Girard. He describes the various theories which have been started to explain cohesion, and rejects successively that of the pressure of the air and that of a subtle medium. Of the latter, he writes:

Quare clarissime conspicimus ex hypothesi aetheris nequaquam fluere phaenomena quae circa cohaerentiam experientia detegit, meri toque hanc Hypothesin ex Physica esse proscribendam (p. 443).

He laughs at Bacon's explanation of elasticity, and another metaphysical hypothesis he terms *abracadabra.* Finally he falls back himself upon Newton's thirty first Query (see Art. 26) and would explain the matter by *vires internae.* These internal forces Musschenbroek assumes to exist, and holds that, without our needing a metaphysical hypothesis as to their cause, we may determine them in each case by experiment.

Haec vis interna â Deo omnibus corporibus indita fuit, voluitque infinite efficax Creator, ut haec in se operantur secundum vim illam: adeoque haec vis est Lex Naturae, cui similis observatur altera, gravitas appellata (p. 451).

The source of elasticity is a *vis interna attrahens.* This theory is obviously drawn directly from Newton's Optics.

Musschenbroek then proceeds to propositions and experiments. He treats of the extension (*cohaerentia vel resistentia absoluta*) and of the flexure (*cohaerentia respectiva aut transversa*) of beams, but does not seem to have considered their compression. His experiments are principally on wood, with a few however on metals. He refers to earlier experiments by Mersenne (*Traité de l'harmonie universelle* 1626. *Lib.* III. *prop.* 7) and Francesco de Lama (*Magisterium naturae et artis* 1684-92. *Lib.* XI. *cap.* 1) and discusses the Galilei and the Mariotte-Leibniz hypotheses. Anything of value in his work is however reproduced by Girard.

Musschenbroek discovered by experiment that the resistance of beams compressed by forces parallel to their length is, all things being equal, in the inverse ratio of the squares of their lengths; a result afterwards deduced theoretically by Euler (see Art. 76).

ε. Buffon, *Moyen facile d'augmenter la solidité, la force et la durée du bois. Mém. Acad. Paris* 1738, p. 241.

*Expériences sur la force du bois.* Two memoirs presented to the French Academy in 1740 (p. 636) and 1741 (p. 394).

Buffon seems to have been the first after Musschenbroek who experimented on bars of iron. See the *Oeuvres complètes,* Tom. VII. p. 61. Paris, 1774—8. Most of the early experimenters deal only with wood.

ζ. M. Perronet. *Oeuvres.* Tom. I. *Sur les pieux et pilotis.*

η. Finally at the end of the century (1798) Girard closed this list of experiments with a remarkable series conducted at Havre referred to later (see Art. 131).

[29.] A few other minor memoirs of the first half of the 18th century must be mentioned before we proceed to the more important work of Jacopo Riccati, Daniel Bernoulli and Euler.

*a.* Père Mazière: *Les Loix du choc des corps à ressort parfait ou imparfait, déduites d'une explication probable de la cause physique du ressort.* Paris, 1727. This was the essay which carried off the prize of the *Académie Royale des Sciences* offered in the year 1726. Père Mazière, *Prêtre de l'Oratoire,* seems to have had remarkable notions on the extreme complexity of the mechanism by which Nature produces her phenomena, and on the slight grounds necessary for the statement of the most elaborate hypotheses. The essay is of purely historical interest, but that interest is considerable; it brings out clearly the union of those theological and metaphysical tendencies of the time, which so checked the true or experimental basis of physical research. It shews us the evil as well as the good which the Cartesian ideas brought to science It is startling to find the French Academy awarding their prize to an essay of this type, almost in the age of the Bernoullis and Euler Finally it more than justifies Riccati's remarks as to the absurdities of these metaphysical mathematicians.

Père Mazière finds a probable explanation of the physical cause of spring in that favorite hypothesis of a 'subtile matter' or *étherée.* In a series of propositions he deduces the following results.

*La cause physique du ressort est un fluide.* This he proves by the method of elimination. Une cause physique n'est pas une intelligence, that is only God or the first cause. Next it cannot be a solid, *ergo* it must be a fluid. This fluid circulates in the imperceptible canals of bodies. *La cause physique du ressort n'est pas l'air, mais la matière subtile.* A single argument from a bladder will hardly strike the modern reader as conclusive.

*La matière subtile a une force infinie, ou comme infinie.*
*La matière subtile est un fluide parfait.*
*La matière subtile est infiniment comprimée.*

The extremely naive proofs of these propositions are well calculated to fill the modern reader with a pharisaical feeling. A remark which occurs under the last heading deserves to be reproduced at length:

Mais comment la matiere subtile ne s'insinuëroit elle pas dans tous les corps créez? C'est elle qui les engendre, pour ainsi dire, & qui les fait croître par des végétations, fermentations, &c. Sans elle que seroit l'Univers Si Dieu qui l'a créée cessoit un instant de la conserver, ou de la comprimer; les Astres n'auroient plus de lumière, ni de mouvement; le feu perdroit sa chaleur, l'eau sa liquidité, & l'aiment toutes ses vertus; l'air que nous respirons se reduiroit à un amas confus de lames spirales sans aucune force; les corps n'auroient plus ni dureté, ni ressort, ni fluidité, ni pesanteur; ils ne tendroient plus vers le centre de la terre; & la terre elle méme que deviendroit-elle? Otez la matiere subtile, l'Univers entier disparoit (p. 16).

The Père Mazière then applies the Cartesian theory of vortices to the aether:

*La matière subtile n'est composée que d'une infinité de tourbillons qui tournent sur leurs centres avec une extréme rapidité.* He attributes this discovery regarding the aether to Malebranche. These vortices are retained in equilibrium by their centrifugal forces, which leads to the corollary: "*la force centrifuge des tourbillons infiniment petits, est infiniment grande.* Finally we have the concluding proposition of this 'explanation': *La matière subtile est la cause physique du ressort par la force centrifuge de ses petits tourbillons.* These little vortices are in the pores of all bodies, etc., etc. Such will sufficiently characterise the method of explanation. Mazière thinks that other solutions might be given by means of the "subtile matter," "mais je me suis arrêté à celle qui m'a paru avoir le plus de vraisemblance."

The rest of the tractate applies the theory to the impact of bodies

*β*. G. B. Bülfinger: *De solidorum Resistentia Specimen, Commentarii Academiae Petropolitanae.* Tomus IV. p. 164.

This is a memoir of August 1729, but first published in the proceedings for that year which appeared in 1735.

The author commences with a reference to the labours of Galilei, Leibniz, Wurtz, Mariotte, Varignon, James Bernoulli and Parent (" Vidimus et Parentium saepius in hoc negotio versatum, virum, cuius longe infra meritum fama est"). He then states as the object of his memoir:

> Nobis id curae est, ut factis repetita vice experimentis, tandem aliquando appareat, num dicta virorum naturae congruant? aut quousque aberrent? Id ut consilio magis, quam casu, fiat, praemitti utique considerationes abstractae debent, sed paucae illae, nec difficiles, nec omnes novae.

The following nine sections (§ 3 to § 12) are concerned with the breaking force on a beam when it is applied longitudinally and transversally. Galilei's and the Mariotte-Leibniz hypotheses are considered. It is shewn that the latter is the more consonant with actual fact, but it is not exact because it neglects the compression (i.e. places the neutral line in the lowest horizontal fibre of the beam).

> Ita *propius* ad naturam accessimus in nova hac hypothesi: sed *absumus* tamen a plena similitudine. Extenduntur fibrae in puncto *D*,

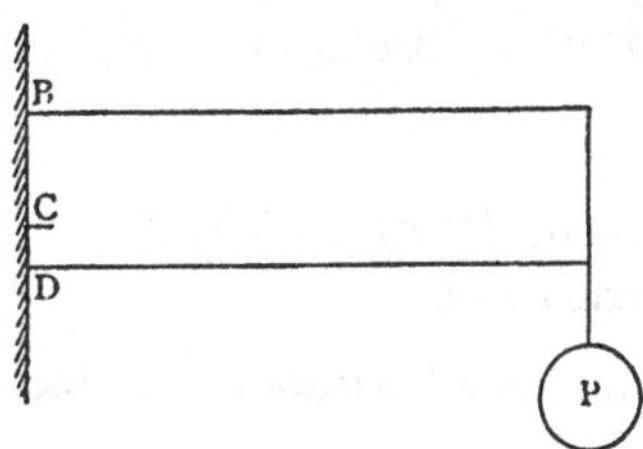

comprimuntur quoque. Nescio, qui factum sit, ut Mariotto maculam inurat, Jac. Bernoullius, quasi compressionem neglexerit, soli extensioni intentus, cum tamen id Mariotto *debeamus*, quod *compressionis primam* ipse *mentionem* fecerit; idemque sit autor propositionis, quam et Bernoullius approbat, quod eaedem conclusiones prodeant, sive solam extensionem fibrarum inde a $B$ ad $D$, sive extensionem a $B$, ad punctum aliquod $C$, et a $C$ ad $D$, compressionem supponas. (§ 12.)

§ 13—§ 21 are concerned with various suppositions as to the relation between extension of a fibre and the extending force. Bülfinger thinks that the *ut tensio sic vis* principle is not consonant with experiment and suggests a parabolic relation of the form

$$\text{tension} \propto (\text{distance from the neutral line})^m,$$

where the power is a constant to be determined by experiment.

In § 22 the writer returns to the question of extension and compression of the fibres of the beam under flexure He cites the two theories; namely that of Mariotte, that the neutral line is the 'middle fibre of the beam, and that of Bernoulli that its position is indifferent. He himself rejects both theories, and gives on the whole sufficient reasons for doing so. Finally, not having accepted Hooke's principle for the fibres of a beam, he holds that till the laws of compression are formulated, the position of the neutral line must be found by experiment.

γ. J. T. Desaguliers: *Thoughts and conceptions concerning the Cause of Elasticity, Phil. Trans.* 1736.

Desaguliers' opinions on elasticity are also expressed in his *Course of Experimental Philosophy.* London 1734—44. Vol. II. pp. 1—11 and p. 38 *et seq.* He supposes elasticity to be due to repulsive and attractive properties in the atoms, but he also endows them with polar properties, so that they are nothing else than "a great number of little loadstones." (Cf. Riccati s statement, Art. 36.)

δ. Jacopo Belgrado: *De corporibus Elasticis Disquisitio Physico-mathematica*, Parma 1748.

This is a quarto pamphlet without any statement as to author, printer or place, but it appears to belong to the above named Italian physicist[1].

There is little that is of value in this production, though the writer is evidently endeavouring to find a dynamical theory of elasticity apart from the hypothesis of some metaphysical mechanism. Here as elsewhere we find a preference among the Italians for the method of Bacon. We may quote the first paragraph as historically interesting:—

Ea seculi nostri indoles est, & ingenium, ut, fictis hypothesibus praetermissis, ex experimentis, observationibus, ac φαινομένοις petenda Philosophia sit, & caussarum scientia ducenda. Propterea desinamus mirari, si ii qui hactenus vis elasticae caussam investigarunt, operam luserint, & nihil invenerint quod vel φαινομένοις congrueret, vel naturae simplicitatem referret. Hypotheses celeberrimae a Cartesio, Bernullio, Mazerio, aliisque institutae, & eloquentissime expositae nodum non solvunt, ac vel *principii petitione* laborant, vel nimis compositae innu meris ambagibus constant. Alii vorticibus, alii viribus centrifugis plus aequo addicti, vel extra ipsa corpora, vel extra solidam partium compagem, & constructionem elaterii caussam quaesiverunt, minus de intimo corporum statu excutiendo solliciti, perinde ac si ad caussae cognitionem ejus investigatio nihil conduceret. Verum, aut ego fallor, aut hujusmodi disquisitio rei extricandae viam quodammodo sternit, ac parat. Non id molior, quod viri ingeniosissimi ante me quicquam tentarunt. Unice animus est nonnulla, quae ad intimum corporum elasticorum statum pertinent, ex observationibus, & experimentis deducta in apricum proferre, quae sin minus vim elasticam constituunt, ac praestant, saltem eam exigunt, ac requirunt, ut, iis positis, elastica corpora sunt, demtis, hujusmodi esse desinant.

The anonymous Italian then states the three principles:

De quibus nemo sanus hac aetate dubitat...Alterum est: nulla vis in natura intercidit, nullo edito effectu; alterum vero est: nihil natura per saltum, ut inquiunt, agit; Tertium vero Leibnitianum nihil unquam accidit, nisi ratio vel caussa sufficiens in promtu sint.

There is little to be learnt from this somewhat diffuse account of elastic bodies. I may note that the author is of opinion that all

[1] A reference of Musschenbroek led me to Belgrado, and Poggendorff places a tractate of the above title among Belgrado's works.

bodies which exist in nature are elastic, and that he proposes the construction of a "scala, seu curva virium elasticarum." Further he recognises that Mariotte had discovered that the fibres of a bent beam are partly compressed and partly extended, and he gives a geometrical method for determining points on the unchanged fibre or neutral line (p. 10). He does not place the neutral line on the surface of a beam, a mistake made by several more important mathematicians of his date. Gravesande's theory of the composition of elastic bodies out of fibres and filaments (see Art. 42) he rejects, remarking the variety of construction in bodies, e.g. the granular. Finally, he states shortly a theory by which the *vis elastica* may be explained by a change of *vis viva* to *vis mortua* followed by the reverse process. This theory is very similar to that of Riccati and was of course suggested by the impact of *elastic* bodies.

ε. H. Manfredi: *De viribus ex elasticorum pulsu ortis, Commentarii Bononienses*, Tom. II. 1748.

(The name of this, together with the fact that I have found no reference to it in later memoirs seemed to render its examination unnecessary for our present purposes.)

[30.] Jacopo Riccati. This author has made two contributions to our subject. The first is a memoir entitled: *Verae et germanae virium elasticarum leges ex phaenomenis demonstratae*, 1731, and printed in the *De Bononiensi scientiarum Academia Commentarii*, Tom. I. p. 523, Bologna 1747.

This memoir is of very considerable interest; it marks the first attempt since Hooke to ascertain by *experiment* the laws which govern elastic bodies. The author commences by laying down the true theory of all physico-mathematical investigations:

Arduum opus aggrediuntur hi, qui quaestionem aliquam enodandam sibi proponunt, in qua cum nihil datum sit, quod ad investigationem perducat; quot data invenire oportet, totidem nova, et difficilia problemata solvenda occurrunt. Et quamquam ea, quae pro notis usurpamus, non ex fictis philosophorum hypothesibus, sed ex ipsa natura, et ab experimentis petenda sunt; saepissime accidit, ut quae propriora videntur, et rem fere attingunt, ea nos a proposito longissime removeant

This latter difficulty Riccati had fallen into with regard to his experiments on elastic bodies:

Verum incassum recidit labor, et rei difficultas qualemcumque industriam meam frustrata est; nullus enim canon observationibus ex omni parte respondens sanciri potuit; quod an materiae defectui, et circumstantiarum varietati, an experimentorum subtilitati tribuam, nescio: parumque abfuit, quin in doctissimi Iacobi Bernoulli sententiam descenderem, unumquodque scilicet corpus, pro varia suarum partium textura, peculiarem, et ab aliis corporibus diversam elasticitatis legem obtinere.

The paragraph expresses very concisely the state of physical investigation with regard to elasticity in Riccati's time. The remark of Bernoulli referred to occurs in the corollary to his third lemma: "Au reste, il est probable que cette courbe" (*ligne de tension et de compression*) "est différente de différens corps, à cause de la différente structure de leurs fibres." (See our article 20.) It struck Riccati however to consider the acoustic properties of bodies. For, he remarks, the harmonic properties of vibrating bodies are well known and must undoubtedly be connected with the elastic properties—("canoni virium elasticarum")

[31.] When however we come to examine the substance of the memoir itself, we find from Riccati's first canon that he has no clear conception of Hooke's Law, nor does the theory he bases upon the known results of acoustic experiments lead him to discover that law In his third canon he states that the 'sounds' of a given length of stretched string are in the sub-duplicate ratios of the stretching weights. The 'sounds' are to be measured by the inverse times of oscillation. Proceeding from this known result he deduces by a not very lucid train of argument that, if $u$ be a weight which stretches a string to length $x$ and $u$ receive a small increment $\delta u$ corresponding to an increment $\delta x$ of $x$, then the law of elastic force is that $\delta u/u$ is proportional to $\delta x/x^2$. Hence according to Riccati we should have instead of Hooke's Law:—
$u = Ce^{-\frac{1}{x}}$, where $C$ is a constant. For compression the law is obtained by changing the sign of $x$

[32.] Riccati points out that James Bernoulli's statements in the memoir of 1705 do not agree with this result or as he expresses it "*fortasse minus veritati consonat*"! He notes that the equation $du/u = \pm dx/x^2$ has been obtained by Taylor and Varignon for the determination of the density of an elastic fluid compressed by its own weight "quod scilicet densitas sit oneri imposito proportionalis et gravitas aeris sit in ratione reciproca duplicata distantiae a centro telluris" (p. 541).

[33.] The second contribution of Riccati is an attempt at a general explanation of the character of elasticity. It occurs in his *Sistema dell' Universo* and must have been written before 1754, which was the year of his death. The *Sistema* was first published in the *Opere del Conte Jacopo Riccati*, Tomus I. Lucca 1761. The third and fourth chapters of the first part of the second book are respectively entitled: *Delle forze elastiche* and *Da quali primi principi derivi la forza elastica.*

These chapters display very clearly the characteristics of the author; dislike namely of any semi metaphysical hypothesis introduced into physics; and desire to discover a purely dynamical theory for physical phenomena.

Chap. III. opens with the statement that the physicists of his time had troubled themselves much with the consideration of elasticity:

> E si può dire, che tante sono le teste, quante le opinioni, fra cui qual sia la vera non si sa, se pure non son tutte false, e quale la più verisimile, tuttavia con calore si disputa.

[34.] Riccati then sketches briefly some of the theories then current. Descartes had supposed elasticity to be produced by a subtle matter (aether) which penetrates the pores of bodies and keeps the particles at due distances; this aether is driven out by a compressing force and rushes in again with great energy on the removal of the compression. (We may compare the conception to a sponge squeezed under water. The Cartesian view was first, I believe, given in the *Principia Philosophiae* published in 1644, or six years later than Hooke in his *De potentiâ restitutiva* (see Art. 9) had also endeavoured to explain elasticity by a 'subtle medium.' There is however a Cartesian character about Hooke's discussion and he may have heard of Descartes' conception. The priority of the idea is only of historical interest and perhaps not worth investigation.)

[35.] The next theorist mentioned is John Bernoulli who in his discourse on motion[1] treats of the cause of elasticity and finds the Cartesian hypothesis insufficient. Bernoulli supposes the aether enclosed in cells in the elastic body and unable to escape. In this captive aether float other larger aether atoms describing orbits. When a compressing force is applied the cells become smaller, and the orbits of these atoms are restricted, hence their centrifugal force is increased; when the compressing force is removed the cells increase and the centrifugal forces diminish. Such is the complicated mechanism invented by Bernoulli to explain (?) how the *forza viva* absorbed by an elastic body can be retained for a time as *forza morta*. (This theory of captive aether was at a later date adopted by Euler although in a slightly more reasonable form, see Art. 94.)

[36.] Finally Riccati gives a characteristic paragraph with regard to the English theorists:

> Escono in campo i matematici Inglesi con una terza assai più delle altre applaudita spiegazione Non ci ha fenomeno in Natura, ch' eglino non ascrivano alle favorite attrazioni, da cui derivano la durezza, la fluidità, ed altre proprietà de' composti, e spezialmente la forza elastica. Se ad una molla si attacca per lungo un grave, che la distenda, viene esso sostenuto, ed equilibrato da una energia attratrice, che rimosso, il peso, accorcia la verga, e la riduce alla sua natural dimensione. All' opposto se l' elastico si comprime, sbuca fuori una forza repulsiva, che coll' azione esterna contrasta, la quale tolta di mezzo, torna prontamente a rimetterlo (p. 154).

[37.] We have quoted so much from Riccati in order to shew exactly the hypotheses as to the nature of elasticity current in his day.

As for Riccati himself he will not enter into these disputes "mercè che il miglior partito di oppugnare le altrui false opinioni

[1] Prize essay of 1724, *Discours sur les loix de la communication du mouvement*, Paris, 1727. Chaps. I. to III.

consiste nel produrre la vera." For his own theory he will not call to his assistance the aether of Descartes or the attractions of Newton: "Il mio giro di raziocinio non uscirà fuori de' confini della Dinamica"—a most excellent principle. We have now to consider how Riccati applied it.

He proceeds first to discuss the results of experiments on elastic bodies, and quotes those of Newton[1] and Rizzetti[2], but he still seems ignorant of Hooke's Law and quotes Gravesande[3] to shew that the relation of extension to force is quite unknown. This is the more curious as he elsewhere cites Hooke for a remark as to the specific gravity of bronze.

[38.] Chap. IV. After again insisting on the importance of the method, which proceeds from the codification of phenomena to the deduction of a principle consistent with experience, Riccati states *la mia novella sentenza*. This principle, so far as I have been able to follow Riccati's not very lucid exposition, is involved in the following statements.

Every deformation is produced by *forza viva* and this force is proportional to the deformation produced. Of this statement Riccati says:

> Io son certo, che non ci sia per essere Fisico, che si opponga ad una verità cosi splendida e dalle allegate sperienze in tante guise comprovata.

[1] I have thought it advisable to omit all consideration of Newton's and other experiments on the collision of *elastic* bodies. The history of this branch of the subject is considerable and there are a number of memoirs from the seventeenth and first half of the eighteenth centuries. I may refer to:

Marcus Marci: *De proportione motus*, Prag, 1639. Historically a most interesting work.

Wren: *Phil. Trans.* Dec. 1668.

Huyghens: *Ibid.* Jan. 1669. *De motu corporum ex percussione*, 1703.

Mariotte: *Traité de la percussion*, Paris, 1676.

Newton: *Principia Naturalis Philosophiae; Scholium to Corol.* VI. p. 23 of the first edition.

A list of further memoirs, *De percussione Corporum*, is given by Reuss, *Repertorium Commentationum*, p. 211, but those I have been able to examine do not seem of much value.

[2] *De Bononiensi Academia Commentarii*, Tomus I. *De corporum collisionibus*, p. 497.

[3] *Physices elementa mathematica experimentis confirmata*, 1720, L. I. c. 26.

The *forza viva* spent in producing a deformation remains in the strained body in the form of *forza morta;* it is stored up in the compressed fibres. Riccati comes to this conclusion after asking whether the *forza viva* so applied could be destroyed? That such a dissipation of energy—to use a modern expression—is possible in the universe he denies, making use strangely enough of the argument from design, a metaphysical conception such as he has told us ought not to be introduced into physics!

La Natura anderebbe successivamente languendo, e la materia diverrebbe col lungo girare de' secoli una massa pigra, ed informe fornita soltanto d' impenetrabilità, e d' inerzia, e spogliata passo passo di quella forza (conciossiachè in ogni tempo una notabil porzione se ne distrugge) la quale in quantità, ed in misura era stata dal sommo Facitore sin dall' origine delle cose ad essa addostata per ridurre il presente Universo ad un ben concertato Sistema.

[39.] This paragraph is singularly interesting as uniting the old theologico-mathematical standpoint, with the first struggling towards the modern conception of the conservation of energy. It is this principle of energy which *la mia novella sentenza* endeavours so vaguely to express, namely that the mechanical work stored up in a state of strain, must be equivalent to the energy spent in producing that state.

[40.] In sections IV. and V. of this chapter Riccati attempts to elucidate, although without much success, his principle by the simple case of a stretched string. He refers to his previous memoir and tells us that the *forza viva* must be measured by the *square* of the velocity. The consideration of the impact of bodies is more suggestive; the *forza viva* existing before impact is converted at the moment into *forza morta* and this re-converted into *forza viva* partly in the motion of either body as a whole, and partly in the vibratory motion of their parts, which we perceive in the sound vibrations they give rise to in the air. With regard to the transition from *forza viva* to *forza morta*, Riccati remarks:

Del perpetuo, e non interroto passaggio delle forze di vive in morte, e di morte in vive fa uso la Natura nel generare con tanta costanza di leggi, e nel tempo stesso con tanta varieta i suoi prodotti, e, quasi direi, per tener equilibrata l' economia del presente Universo (p. 168).

A great example of this is the elastic property of bodies, and to explain it there is no need to fly "alla materia sottile de' Cartesiani, o alle Newtoniane attrazioni." Riccati concludes his discussion with a summary of eight headings, which however contain nothing of additional interest.

[41.] The importance of Riccati's work lies not in his practical results, which are valueless, but in his statement of method, and his desire to replace by a dynamical theory semi-metaphysical hypotheses. In many respects his writings remind us extremely of Bacon, who in like fashion failed to obtain valuable results, although he was capable of discovering a new method. Euler's return to the semi-metaphysical hypothesis (see Art. 95) is a distinct retrogression on Riccati's attempt, which had to wait till George Green's day before it was again broached.

[42.] Of the authorities quoted by Riccati, John Bernoulli, in his *Discours sur les loix de la communication du mouvement*, Paris, 1727, devotes the first three chapters to hardness and elasticity, but without coming to any conclusions worth quoting even for their historical value. Gravesande in his *Physices Elementa Mathematica*, 1720, explains elasticity (Lib. I. Cap. V. p. 6) by Newtonian attractions and repulsions. The 26th chapter of the first book is also entitled *De legibus elasticitatis*. He is of opinion that within the limits of elasticity, the force required to produce any extension is a subject for experiment only. The results are principally experimental, he considers elastic cords, laminae and spheres (supposed built up of laminae), and finds the deflection of the beam in Galilei's problem proportional to the weight. He makes however no attempt to discuss the elastic curve.

[43.] The direct impulse to investigate elastic problems undoubtedly came to Euler from the Bernoullis. Thus in a letter of John Bernoulli to Euler dated March 7, 1739[1], the writer mentions a property of the *Elastica rectangula* (*vel etiam Lintearia; ambae enim eandem faciunt curvam*) which Euler

[1] Fuss: *Correspondance Mathématique et Physique*, St Pétersbourg, 1843, Tom II. p. 23.

had communicated to him[1], but for the full understanding of Euler's work Daniel Bernoulli's letters to Euler are peculiarly instructive.

[44.] On Sept. 23, 1733, he writes[2]:

Unterwegs habe ich einige meditationes mathematicas gemacht de determinandis utique crassitiebus laminae muro horizontaliter infixae, ita ut ubique aequaliter sit rupturae obnoxia lamina, die lamina mag proprio pondere agiren oder noch von einem superincumbente pondere utcunque geladen seyn. Man kann über dieses Thema viele curiose Sachen annotiren, worüber ein sonderbares mémoire abfassen werde.

[45.] In a second letter dated May 4, 1735, we find the following paragraph:

Haben Sie seithero auch gedacht an die vibrationes laminae elasticae muro verticali perpendiculariter infixae. Ich finde pro curva diese Aequation $nd^4y = ydx^4$ allwo $n$ eine quantitas constans, $x$ die abscissae $y$ die applicatae, $dx$ constans. Aber diese Materie ist gar schlüpfrig; und möchte gern Ihre Meinung darüber hören: Obgedachter Aequation satisfacirt die logarithmica wie auch dieser Aequation $n\frac{1}{2}ddy = ydx^2$ Keine aber ist pro presenti negotio general genug[3].

Another letter of the same year (Oct. 26) mentions further results of his researches with regard to the period of vibration[4]

It will be seen at once how Galilei's problem had determined the direction of later researches, how while James Bernoulli solved the problem of the elastic curve his nephew Daniel first obtained a differential equation which really does present itself in the consideration of the transverse vibrations of a bar. Euler when in 1740 he arrived at the same differential equation was already in possession of Bernoulli's results. Neither of these distinguished mathematicians seem at this period, 1735—1740, to have obtained a general solution of their equation. Daniel Bernoulli's results were first published in a memoir printed in the *Commentarii* of the St Petersburg Academy for 1741—1743. I shall return to this memoir later.

[1] His brother James appears to have discovered this curve as early as 1691 See his account of the matter in the *Acta Eruditorum Lipsiae* for 1695, p. 546.

[2] Fuss, II. p. 412. [3] Fuss, II. p. 422. [4] Fuss, II. p. 429.

[46.] A still more important letter is that of Oct. 20, 1742. In this Bernoulli proposes several problems concerning elastic laminae to Euler, and states that several months previously he had sent to him

Eine weitläufige und operose piece de sono laminarum liberarum, darin ich gar viel merkwürdige phaenomena physica explicirt und ausgerechnet habe; hierzu war aber eine neue theoria physica erfordert ehe und bevor ich die mathesin appliciren konnte.

He suggests for Euler's consideration the case of a beam with clamped ends, but states that the only manner in which he has himself found a solution of this "idea generalissima elasticarum" is "per methodum isoperimetricorum." He assumes the "vis viva potentialis laminae elasticae insita" must be a minimum, and thus obtains a differential equation of the fourth order, which he has not solved, and so cannot yet shew that this "aequatio ordinaria elasticae" is general.

Ew. reflectiren ein wenig darauf ob man nicht könne sine interventu vectis die curvaturam immediate ex principiis mechanicis deduciren. Sonsten exprimire ich die vim vivam potentialem laminae elasticae naturaliter rectae et incurvatae durch $\int ds/R^2$, sumendo elementum $ds$ pro constante et indicando radium osculi per $R$. Da Niemand die methodum isoperimetricorum so weit perfectionniret als Sie, werden Sie dieses problema, quo requiritur ut $\int ds/R^2$ faciat minimum, gar leicht solviren[1].

[47.] Bernoulli writes further about the same matter to Euler Feb. 1743 and Sept. 1743. In the latter letter he extends his principle of the 'vis viva potentialis laminae elasticae' to laminae of unequal elasticity, in which case $\int E ds/R^2$ is to be made a minimum. The last letter I have found referring to the subject is written in either April or May 1744, and therein Bernoulli expresses his pleasure that Euler's results on the oscillations of laminae agree with his own[2].

[48.] The memoir of Daniel Bernoulli to which I have referred is published in the *Commentarii Academiae Scientiarum Imperialis Petropolitanae*, Vol. 13, 1751; and is entitled *De vibra-*

[1] Fuss, II. pp. 505—7. [2] *Ibid.* II. pp. 518, 533 and 553.

*tionibus et sono laminarum elasticarum Commentationes Physico-Geometricae*, p. 105. The memoirs in this volume are for the years 1741—1743.

We may note that the volume commences with *Excerpta ex literis a Daniele Bernoulli ad Leonh. Euler*, and on p. 8, after considering several differential equations, Bernoulli takes another example:

Quod praesertim amo, quia pertinet ad argumentum mechanicum iam pridem a me propositum et a nobis ambobus solutum, argumentum intelligo de figura, quam lamina elastica uniformis muro infixa et vibrata affectat...

The equation is $d^4s = f^4 . s . dv^4$,

and the solution given

$$s = ae^{fv} + be^{-fv} + ce^{fv\sqrt{-1}} + de^{-fv\sqrt{-1}},$$

or, since the exponentials may be read as "sinus arcuum circularium,"

$$s = ae^{fv} + be^{-fv} + g \sin \text{arc} (fv + h).$$

[49.] The method of the memoir itself is unsatisfactory because the differential equation is obtained from statical rather than dynamical considerations. For the vibrations of a free beam morticed into a vertical wall Bernoulli finds the differential equation of his earlier letters (see Art. 45), namely,

$$\frac{d^4y}{dx^4} = \frac{y}{f^4},$$

where $x$ is the horizontal distance from the wall and $y$ the corresponding vertical displacement of any point of the beam The solution of this equation is then given, first in the form of four infinite series each with an arbitrary constant, and then in the form of the previous article. Bernoulli states that these arbitrary constants will be in each case determined by the character of the oscillations of that particular case ("cuivis casui proposito accommodari potest, vt et cuivis oscillationum generi"). His method of connecting the "longitudo penduli simplicis isochroni cum vibrationibus laminae" with his differential equation (§§ 5 and

11—12) will hardly satisfy modern desire for accuracy. At the same time he arrives at the equation (stating that he owes it to Euler[1]),

$$l/f = 2 \sin^{-1} \frac{1 + e^{l/f}}{\sqrt{2 + 2e^{2l/f}}},$$

where $l$ is the length of the rod, and deduces that the length of the isochronous pendulum $= gf^4/m^4$ where $m^4$ is an elastic constant. He notes that the equation for $l/f$ gives only numerical values for this ratio;—"igitur longitudo penduli isochroni est in ratione biquadrata longitudinis $l$ et numerus oscillationum laminae in ratione reciproca duplicata ejusdem longitudinis." If we write $l/f = \omega$ the above equation for the periods may be written

$$\cos \omega . (e^{\omega} + e^{-\omega}) + 2 = 0.$$

The memoir concludes with some discussion as to the notes of a beam vibrating transversally.

[50.] A second memoir of Bernoulli in the same volume is entitled *De sonis multifariis quos Laminae Elasticae diversimode edunt disquisitiones Mechanico-Geometricae Experimentis acusticis illustratae et confirmatae*, p. 167. This memoir commences with an enumeration of the four modes of vibration of an elastic lamina which are identical with those afterwards formulated by Euler (see Art. 64). Various forms of the solution of the general differential equation are next considered for these particular modes. Then follows a discussion as to the periods of the vibrations and the position of the nodes. Several experiments in confirmation of the theoretical results are considered in the course of the memoir.

51. Euler, 1740, *De minimis oscillationibus corporum tam rigidorum quam flexibilium. Methodus nova et facilis.* This is published in the *Commentarii Academiae Scientiarum Imperialis Petropolitanae*, Vol. 7, 1740: it occupies pages 99—122.

This memoir is very slightly connected with our subject; it is

[1] This equation appears also in Daniel Bernoulli's letter to Euler of February 1743 mentioned above.

devoted to the exposition of a method of solving problems relating to certain cases of motion, which was useful at an epoch previous to the introduction of D'Alembert's Principle. All that it contains relative to elastic bodies is reproduced in that part of the *Methodus inveniendi lineas curvas* which treats of the vibration of elastic curves; and Euler on page 283 of that work cites the present memoir. A few points may be noticed.

Euler here, as elsewhere, is not very careful with respect to his notation. For instance, in his diagram *g* it will be seen that *a* and *A* denote certain points; in the corresponding text it will be found that *a* denotes also a certain length, and *A* a certain elastic force.

52. Euler assumes that the elasticity along a curve varies inversely as the radius of curvature (see his page 113); he gives only this brief reason, namely that the more the rod is bent the greater is the elastic force. His words are; "Cumque eadem vis elastica sit eo major, quo magis curvatur, erit vis elastica in *M* ut *V* divisum per radium osculi in *M*."

The differential equation $\frac{d^4y}{dx^4} = c^4y$ presents itself, and Euler seems unable to state at once the general form of the solution: (see his pages 116 and 117[1]). In the *Methodus inveniendi*, page 285, he gives the general form, namely,

$$y = Ae^{\frac{x}{c}} + Be^{-\frac{x}{c}} + C\sin\frac{x}{c} + D\cos\frac{x}{c}.$$

53. Euler, 1744. The celebrated work of Euler relating to what we now call the Calculus of Variations appeared in 1744 under the title of *Methodus inveniendi lineas curvas maximi minimive proprietate gaudentes.* This is a quarto volume of 320 pages; of these pages 245—310 form an appendix called *Additamentum I. De Curvis Elasticis,* which we shall now examine.

[1] The following extract from page 122 may interest those who study the history of music: "Ope hujus regulae inveni in instrumentis musicis, quae ad tonum choralem attemperata sunt, chordam infimam littera *C* notatam minuto secundo 118 edere vibrationes; summam vero, quae $\bar{\bar{\bar{c}}}$ signari solet, eodem tempore vibrationes 1888 absolvere (*sic*)."

[54.] The memoir commences with a statement which is of extreme interest as shewing the theologico-metaphysical tendency which is so characteristic of mathematical investigations in the 17th and 18th centuries. It was assumed that the universe was the most perfect conceivable, and hence arose the conception that its processes involved no waste, its 'action' was always the least required to effect a given purpose. That the results obtained by such metaphysical reasoning would differ according to the method in which 'action' was measured, does not seem at first to have occurred to the mathematicians. Thus we find Maupertuis' extremely eccentric attempt at a principle of Least Action. On the whole it is however probable that physicists have to thank this theological tendency in great part for the discovery of the modern principles of Least Action, of Least Constraint, and perhaps even of the Conservation of Energy.

The statement to which we refer is the following:

Cum enim Mundi universi fabrica sit perfectissima atque a Creatore sapientissimo absoluta, nihil omnino in mundo contigit, in quo non maximi minimive ratio quaepiam eluceat, quamobrem dubium prorsus est nullum quin omnes mundi effectus ex causis finalibus, ope methodi maximorum et minimorum, aeque feliciter determinari queant, atque ex ipsis causis efficientibus.

[55.] Euler then cites several examples of this natural principle and mentions the service of the Bernoullis in the same direction. He continues:

Quanquam igitur, ob haec tam multa ac praeclara specimina, dubium nullum relinquitur quin in omnibus lineis curvis, quas Solutio Problematum physico-mathematicorum suppeditat, maximi minimive cujuspiam indoles locum obtineat; tamen saepenumero hoc ipsum maximum vel minimum difficillime perspicitur; etiamsi a priori Solutionem erüere licuisset.

Then stating that Daniel Bernoulli (see Bernoulli's letter of Oct. 1742, Art. 46) had discovered in the course of his investigations that the *vis potentialis* represented by $\int ds/R^2$ was a minimum for the elastic curve, Euler proceeds to discuss the inverse problem, namely:

56. To investigate the equation to a curve which satisfies the following conditions. The curve is to have a given length between two fixed points, to have given tangents at those points, and to render $\int ds/R^2$ a minimum: see pages 247—250 of the book. No attempt is made to shew why this *potential force* should be a minimum in the case of the elastic curve.

By the aid of the principles of his book Euler arrives at the following equations where $a$, $\alpha$, $\beta$, $\gamma$ are constants,

$$dy = \frac{(\alpha + \beta x + \gamma x^2)\,dx}{\sqrt{\{a^4 - (\alpha + \beta x + \gamma x^2)^2\}}};$$

from this we obtain

$$ds = \frac{a^2 dx}{\sqrt{\{a^4 - (\alpha + \beta x + \gamma x^2)^2\}}}.$$

Euler then says

Ex quibus aequationibus consensus hujus curvae inventae cum curva elastica jam pridem eruta manifeste elucet.

Quo autem iste consensus clarius ob oculos ponatur, naturam curvae elasticae a priori quoque investigabo; quod etsi jam a Viro summo Jacobo Bernoullio excellentissime est factum; tamen, hac idonea occasione oblata, nonnulla circa indolem curvarum elasticarum, earumque varias species et figuras adjiciam; quae ab aliis vel praetermissa, vel leviter tantum pertractata esse video.

57. Accordingly Euler gives on his page 250 his investigation of the elastic curve in what he has just called an *a priori* manner. But this method is far inferior to that of James Bernoulli; for Euler does not attempt to estimate the forces of elasticity, but assumes that the moment of them at any point is inversely proportional to the radius of curvature: thus he in fact writes down immediately an equation like (1) on page 606 of Poisson's *Traité de Mécanique*, Vol. I., without giving any of the reasoning by which Poisson obtains the equation[1].

58. Euler starts with the supposition that the elastic curve is fixed at one point, and is bent by the application of a single force

[1] On page 250 observe $P$ is used as elsewhere in two senses, namely for a force, and for the position of a point.

at some other point; and then he considers cases in which instead of his single force we have two or more equivalent forces. On his page 255 he supposes that the forces reduce to a *couple*, and he shews that the elastic curve then becomes a *circle*. Saint-Venant alludes to this in the *Comptes Rendus*, Vol. XIX. page 184, where he remarks:

Déjà M. Wantzel, dans une communication faite le 29 juin à la Société Philomatique, a remarqué que la courbe à double courbure, affectée par une verge primitivement cylindrique, solicitée par un couple, est nécessairement une hélice.

C'est une généralisation du résultat d'Euler, consistant en ce que lorsque la courbe provenant de la verge ainsi solicitée est plane, elle ne peut être qu'un arc de cercle.

59. Euler distinguishes the various species of curves included under the general differential equation of Art. 56; he finds them to be nine in number. The whole discussion is worthy of this great master of analysis; we may notice some of the points of interest which occur.

The third species is that in which the differential equation reduces to $dy = \frac{x^2\,dx}{\sqrt{a^4 - x^4}}$; see his page 261. This is not substantially different from the particular case we have noticed in (Art. 24) our account of James Bernoulli. The curve touches the axis of $x$ at the origin; and Euler calls it the *rectangular elastic curve*. In connection with the discussion of this species Euler introduces two quantities $f$ and $b$ which are thus defined:

$$f = \int_0^a \frac{a^2\,dx}{\sqrt{a^4 - x^4}}, \qquad b = \int_0^a \frac{x^2\,dx}{\sqrt{a^4 - x^4}};$$

and he says:

Quanquam autem hinc, neque $b$, neque $f$ per $a$ accurate assignari potest; tamen alibi insignem relationem inter has quantitates locum habere demonstravi. Scilicet ostendi esse $4bf = \pi aa$.

I do not know where Euler has shewn this; however $b$ and $f$

can now be expressed in terms of the Gamma-function. For put $x = a\sqrt{\sin\theta}$; thus we obtain

$$f = \frac{a}{2}\int_0^{\frac{\pi}{2}} \sqrt{\sin\theta}\, d\theta = \frac{a}{2}\frac{\sqrt{\pi}}{2}\frac{\Gamma(\frac{3}{4})}{\Gamma(\frac{5}{4})},$$

$$b = \frac{a}{2}\int_0^{\frac{\pi}{2}} \frac{d\theta}{\sqrt{\sin\theta}} = \frac{a}{2}\frac{\sqrt{\pi}}{2}\frac{\Gamma(\frac{1}{4})}{\Gamma(\frac{3}{4})}.$$

(See *Integral Calculus*, page 291, Example 24.) Hence we obtain $bf = \frac{\pi a^2}{4}$, since $\Gamma(\frac{5}{4}) = \frac{1}{4}\Gamma(\frac{1}{4})$. Euler says that he finds approximately $f = \frac{5a}{6} \times \frac{\pi}{2}$, and still more closely $f = \frac{\pi a}{2\sqrt{2}} \times 1{\cdot}1803206$; then by a mistake he puts $b = \frac{a}{\sqrt{2}} \times 1{\cdot}1803206$ instead of $b = \frac{a}{\sqrt{2}} \times \frac{1}{1{\cdot}1803206}$. Thus he makes $\frac{b}{a} = {\cdot}834612$, which is too great; in fact $\frac{b}{a}$ is about $\frac{3}{5}$.

60. On his page 270 Euler observes that he has hitherto considered the elasticity *constant*, but he will now suppose that it is variable; he denotes it by $S$, which is supposed a function of the arc $s$; $\rho$ is the radius of curvature. He proceeds to find the curve which makes $\int S ds/\rho^2$ a minimum[1]; and by a complex investigation finds for the differential equation of the required curve

$$\alpha + \beta x - \gamma y = S/\rho,$$

where $\alpha$, $\beta$, $\gamma$ are constants. This he holds to be necessarily the correct result, by the same principle as in Art. 54; and he says on his page 272:

Sic igitur non solum Celeb. Bernoullii observata proprietas Elasticae plenissime est evicta; sed etiam formularum mearum difficiliorum usus summus in hoc Exemplo est declaratus.

[1] [This 'principle' is again due to Daniel Bernoulli: see Art. 74. Ed.]

61. On page 276 the integrals $\int ds \,.\, \sin \frac{ss}{2aa}$ and $\int ds \cos \frac{ss}{2aa}$ present themselves, especially for the case in which the limits are 0 and $\infty$; but Euler cannot assign the values. He says,

Non exiguum ergo analysis incrementum capere existimanda erit, si quis methodum inveniret, cujus ope, saltem vero proxime, valor horum integralium $\int ds \sin \frac{ss}{2aa}$ et $\int ds \cos \frac{ss}{2aa}$ assignari posset, casu quo $s$ ponitur infinitum: quod Problema non indignum videtur, in quo Geometrae vires suas exerceant.

The required integrals can be now expressed; for we have by putting $s^2 = 2a^2 x$,

$$\int_0^\infty ds \sin \frac{s^2}{2a^2} = \frac{a\sqrt{2}}{2} \int_0^\infty \frac{\sin x}{\sqrt{x}}\, dx = \frac{a}{2}\sqrt{\pi},$$

$$\int_0^\infty ds \cos \frac{s^2}{2a^2} = \frac{a\sqrt{2}}{2} \int_0^\infty \frac{\cos x}{\sqrt{x}}\, dx = \frac{a}{2}\sqrt{\pi};$$

see *Integral Calculus*, page 283.

62. On his pages 278 and 279 Euler takes the case in which forces act at every point of the elastic curve; and he obtains an equation like that denoted by ($c$) on page 630 of the first volume of Poisson's *Traité de Mécanique*[1].

63. From page 282 to the end Euler devotes his attention to the *oscillations* of an elastic lamina; the investigation is somewhat obscure for the science of dynamics had not yet been placed on the firm foundation of *D'Alembert's Principle:* nevertheless the results obtained by Euler will be found in substantial agreement with those in Poisson's *Traité de Mécanique*, Vol. II., pages 368—392. The important equations ($a$) and ($a'$) on Poisson's

[1] Page 280, line 7. Go back rather to the first equation of Art. 58, which will reduce to

$$-Ek^2 \frac{d}{dw}\left(\frac{dR}{R^2\, dx}\right) = \sqrt{1+w^2}\,\frac{dv}{dw} - (Q+q);$$

then differentiate again.

pages 377 and 387 respectively agree with corresponding equations on Euler's pages 297 and 287.

64. On the whole Euler examines four cases of vibration of an elastic rod, namely the following, where I use *built in* for the *encastré* of the French writers:

(1) Rod built in at one end, and free at the other; page 285.
(2) Rod free at both ends; page 295.
(3) Rod fixed at both ends; page 305.
(4) Rod built in at both ends; page 308.

[These are identical with the four cases given by Bernoulli in the memoir referred to above, Art. 50.]

[65.] 1757. *Sur la force des colonnes, Mémoires de l'Académie de Berlin*, Tom. XIII. 1759, pages 252—282. This is one of Euler's most important contributions to the theory of elasticity[1].

The problem with which this memoir is concerned, is the discovery of the least force which will suffice to give any the least curvature to a column, when applied at one extremity parallel to its axis, the other extremity being fixed. Euler finds that the force must be at least $=\pi^2.\frac{Ek^2}{a^2}$, where $a$ is the length of the column and $Ek^2$ is the 'moment of the spring or the 'moment of stiffness of the column' (*moment du ressort* or *moment de roideur*). The moment of stiffness multiplied by the curvature at any point of the bent beam is the measure of the moment about that point of the force applied to the beam $\left(\frac{Ek^2}{r} = P.f \text{ in Euler's notation}\right)$.

[66.] If we consider a force $F$ perpendicular to the axis of a beam (or lamina) so as to displace it from the position $AC$ to $AD$, and $\delta$ be the projection of $D$ parallel to $AC$ on a line through $C$ perpendicular to $AC$, Euler finds by easy analysis $D\delta = \frac{F.a^3}{3.Ek^2}$, supposing the displacement to be small. This suggests to him a method of determining the 'moment of stiffness' $Ek^2$ and he makes

[1] Historically, not practically, see footnote, p. 44.

(§§ v.—vii.) various remarks on proposed experimental investigations. He then notes the curious distinction between forces acting parallel and perpendicular to a built-in rod at its free end; the latter, however small, produce a deflection, the former only when they exceed a certain magnitude. It is shewn that the

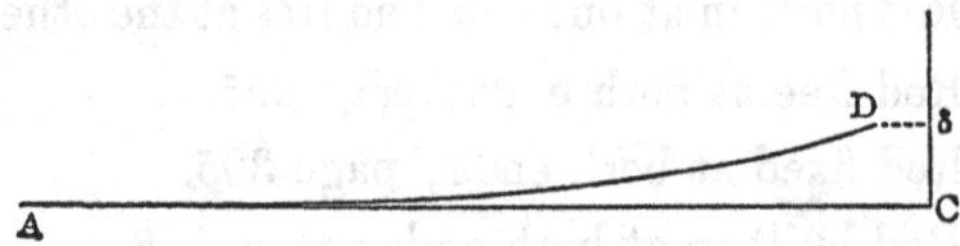


force required to give curvature to a beam acting parallel to its axis would give it an immense deflection if acting perpendicularly (§ xiii.).

[67.] In sections xvi. and xvii. Euler deduces the equation for the curve assumed by the beam $AC$ fixed but not built in at one end $A$ and acted upon by a force $P$ parallel to its axis. If $RM$ be perpendicular to $AC$ and $y = RM$, $x = AM$, he finds

$$\frac{y}{\theta} \cdot \sqrt{\frac{P}{Ek^2}} = \sin\left(x\sqrt{\frac{P}{Ek^2}}\right),$$

where $\theta = \angle RCM$. Hence since $y = 0$, when $x = a$ the length of the

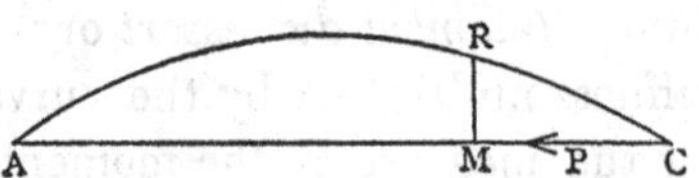


beam, $a\sqrt{\frac{P}{Ek^2}}$ must at least $= \pi$, whence it follows that $P$ must be at least $= \pi^2 \cdot \frac{Ek^2}{a^2}$. This paradox Euler seems unable to explain (see our discussion of Lagrange's Memoir, Art. 108).

[68.] Sections xix.—xxxv. are concerned with beams of varying density or section and are of less interest or importance. In section xxxvi. he returns to the case of a uniform beam, but takes its weight into consideration. If $Q$ be the total weight of the beam the differential equation

$$Ek^2 ad^3y + Pa\,(dx)^2\,dy + Qx\,(dx)^2\,dy = 0$$

is obtained (§ XXXVII). This is reduced by a simple transformation to a special case of Riccati's equation, which is then solved on the supposition that $\frac{Q}{P}$ is small. Euler obtains finally for the force $P$, for which the rod begins to bend, the expression

$$P = \pi^2 . Ek^2/a^2 - Q . (\pi^2 - 8)/2\pi^2;$$

which shews that the minimum force is slightly reduced by taking the weight of the beam into consideration.

69. Euler, 1764. *De motu vibratorio fili flexilis, corpusculis quotcunque onusti.* This is published in the *Novi Commentarii Academiae Scientiarum Imperialis Petropolitanae,* Volume IX. for the years 1762 and 1763; the date of publication is 1764. The memoir occupies pages 215—245 of the volume.

Euler first gives a sketch of the treatment of the general problem, and then discusses the special cases in which the number of weights is respectively one, two, three, and four. The subject of this memoir is considered by Lagrange in the *Mécanique Analytique,* Tome I., Seconde Partie, Section VI.

70. Euler, 1764. *De motu vibratorio cordarum inaequaliter crassarum.* This memoir occupies pages 246—304 of the volume which contains the preceding memoir. Euler implies that this is the first time that the motion of a cord of variable thickness was discussed. He says, on page 247:

> Ne igitur talibus phaenomenis, quae cordis uniformiter crassis sunt propria, nimium tribuatur, haud abs re fore arbitror, si cordarum etiam inaequaliter crassarum motum, quantum quidem Analyseos fines permittunt, examini subjecero, ejusque investigationem latissime complexam instituero. Maxime autem ardua est haec quaestio, atque gravissimis difficultatibus involuta; hancque ob causam etiamsi in ejus enodatione parum profecero, tamen amplissimus nobis aperietur campus, vires nostras in analysi exercendi, hujusque scientiae limites ulterius dilatandi.

These two memoirs by Euler relate to *flexible* bodies, but not to *elastic* bodies; and thus do not really belong to our subject[1]. I was led to examine them in the attempt to verify a reference given by Cauchy in his *Exercices de Mathématiques*, Vol. III. page 312.

71. Euler, 1771. *Genuina Principia Doctrinae de Statu aequilibrii et motu corporum tam perfecte flexibilium quam elasticorum.* This is published in the *Novi Commentarii Academiae Scientiarum Imperialis Petropolitanae*, Volume XV. for 1770; the date of publication is 1771. The memoir occupies pages 381—413 of the volume.

In pages 381—394 general formulae are given for the equilibrium of a flexible string and of an elastic rod. So far as relates

[1] [The accompanying bibliographical note on the vibrations of flexible cords may be of service to the reader, as the subject is intimately connected with our present one.

(1) BROOK TAYLOR. 'De motu nervi tensi.' *Phil. Trans.* 1713, p. 26. 'Methodus incrementorum directa et inversa.' London, 1715, p. 88.

(2) D'ALEMBERT. 'Sur la courbe que forme une corde tendue mise en vibration.' *Mémoires de l'Académie.* Berlin, 1747. Cf. *Opuscules mathématiques*, Tom. I. pp. 1 and 65. Paris, 1761.

(3) EULER. 'Sur les vibrations des cordes.' *Mémoires de l'Académie.* Berlin, 1748.

(4) DANIEL BERNOULLI. 'Réflexions et éclaircissemens sur les nouvelles vibrations des cordes.' *Mémoires de l'Académie.* Berlin, 1753.

(5) EULER. 'Remarques sur les mémoires précédens de M. Bernoulli. *Ibid.*

(6) LAGRANGE. 'Recherches sur la nature et la propagation du son' (a most interesting and important memoir), 1759. *Miscellanea Taurinensia*, Tomus II. Pars 1. 'Addition à la première partie des Recherches, etc., 1762. *Ibid.*

(7) EULER. Mémoires of 1762 and 1764 referred to in the text. 'Sur le mouvement d'une corde qui au commencement n'a été ébranlée que dans une partie.' *Mémoires de l'Académie.* Berlin, 1765. 'Éclaircissemens sur le mouvement des cordes vibrantes.' 'Rech. sur le mouvement des cordes inégalement grosses.' Both in the *Miscellanea Taurinensia*, Tom. III. (1762—65).

(8) DANIEL BERNOULLI. 'Sur les vibrations des cordes d'une épaisseur inégale.' *Mémoires de l'Académie.* Berlin, 1765.

(9) GIORDANO RICCATI. 'Delle corde ovvero fibre elastiche schediasmi.' Bologna, 1767, Schediasma IV.

(10) THOMAS YOUNG. 'Outlines of experiments and enquiries respecting Sound and Light,' § 13 of the 'Vibrations of Chords,' *Phil. Trans.* 1800, pp. 106—150.

There is an interesting historical note by RIEMANN, *Partielle Differentialgleichungen*, § 78, without however references to the original memoirs.] ED.

to the latter the essential part is the hypothesis that if the rod is originally straight the elasticity gives rise to a force along the normal, the moment of which is proportional to the curvature, that is, inversely proportional to the radius of curvature; if the rod is originally curved the moment is taken to be proportional to the difference of the original curvature and the new curvature. The hypothesis agrees with what has since been obtained by considering the nature of the elastic force. See Poisson's *Traité de Mécanique*, Vol. I. pages 603 and 616.

After obtaining his general equations Euler proceeds to particular cases. On his pages 391—400 he discusses the ordinary catenary, and a curve called by the old writers the *Velaria* which he finds to coincide with the catenary. See Whewell's *Mechanics*, Third Edition, page 193.

72. On his pages 400—405 Euler treats the problem of the elastic rod, supposed originally straight; this, as we have seen in Art. 56, he had already discussed in his *Methodus inveniendi*...; the solution given in the present memoir is however more simple than the former. He says on his page 405:

> Plura exempla circa aequilibrium hujusmodi filorum flexibilium et elasticorum, hic subjungere superfluum foret, quoniam hoc argumentum jam passim abunde tractatum reperitur. Hic enim id tantum nobis erat propositum; ut methodum facilem simulque aequabilem, quae ad omnia genera hujusmodi corporum extendatur, traderemus, hocque respectu nullum est dubium, quin haec methodus aliis quibus Geometrae sunt usi, longe sit anteferenda, id quod imprimis ex altera parte hujus dissertationis patebit, ubi ostendemus hanc methodum pari successu adeo ad motus hujusmodi corporum determinandos adhiberi posse.

Accordingly he proceeds to form equations of motion for flexible or elastic cords; the method is in fact coincident with what we should now call an application of D'Alembert's principle; but the name of D'Alembert is not mentioned. Euler does not attempt to integrate these equations.

73. The present memoir is commended by Poisson: see the *Annales de Chimie*, Vol. 38, 1828, page 439; Vol. 39, 1828, page 208; and the *Astronomische Nachrichten*, Vol. 7, 1829, column 353.

[74.] *Determinatio onerum quae columnae gestare valent. Examen insignis paradoxi in theoria columnarum occurrentis. De altitudine columnarum sub proprio pondere corruentium.* These three memoirs occur on pages 121—194 of the *Acta Academiae Petropolitanae* for the year 1778, Pars I. Petersburg, 1780.

The first memoir begins by referring to the memoir of 1757, and points out that vertical columns do not break under vertical pressure by mere crushing, but that flexure of the column will be found to precede rupture. Proceeding in a similar method to that adopted in the Berlin memoir, Euler shews that there will be no flexure, so long as the superincumbent weight is less than $\pi^2 . Ek^2/a^2$, where $Ek^2$ is the 'moment of stiffness' and $a$ is the height of the column. He also notes, § 10, that if a horizontal force displace the top of the column a horizontal distance $\alpha$, $Ek^2 = F . a^3/3\alpha$. So far there is no novelty in this memoir.

[75.] In § 14 however he proposes to deduce a result which is now commonly in use, but which I have not met with before the date of this memoir, namely to find an expression connecting $Ek^2$ with the dimensions of the transverse section of the column. Euler finds $Ek^2 = h . \int x^2 y dx$, where $x$ and $y$ have the following meanings; let a section be taken of the column at any point by a plane perpendicular to the plane of flexure and passing through the centre of curvature of the unaltered fibre (neutral line), then $x$ is the distance of any point on the trace upon this plane of the central plane of flexure from the neutral line, $y$ is the corresponding breadth of the section, and $h$ is the constant now termed the modulus of elasticity. Euler appears however to treat the unaltered fibre or 'neutral line' without remark as the extreme fibre on the concave side of the section of the column made by the central plane of flexure. Thus for a column of rectangular section of dimensions $b$ in, and $c$ perpendicular to the plane of flexure, he finds (§ 21)

$$Ek^2 = \tfrac{1}{3} b^3 ch,$$

and the like method is used in the case of a circular section[1].

[1] [In the case of a beam or column bent by a longitudinal force it may be shewn theoretically that the neutral line does not necessarily lie in the material of the beam, its position and *form* vary with the amount of the deflecting force; in other

[76.] In §§ 26—29 there is a short discussion of the experiments of Musschenbroek (see Art. 28, $\delta$) on the rupture of fir beams; $h$ is evaluated from some of his results and from this value of $h$ the extension of a fir beam under a given tension is calculated theoretically with a result "quod ab experientia non abhorrere videtur."

[77.] In § 31 Euler proceeds to calculate the flexure which may be produced in a column by its own weight. If $y$ be the horizontal displacement of a point on the column at a distance $x$ from its vertex, the equation $Ek^2 . \frac{d^2y}{dx^2} + b^2 \int_0^v x\,dy = 0$ is found, where the weight of unit volume of the column is unity and its section a square of side $b$. This equation Euler solves by a series ascending according to cubes of $x$. Finally, if $a$ be the altitude of the column and $m = Ek^2/b^2$, it is found that the least altitude for which the column will bend from its own weight is the least root of the equation,

$$0 = 1 - \frac{1 . a^3}{4 ! m} + \frac{1 . 4 . a^6}{7 ! m^2} - \frac{1 . 4 . 7 . a^9}{10 ! m^3} + \frac{1 . 4 . 7 . 10 . a^{12}}{13 ! m^4} - \text{etc.}$$

[78.] Euler finds that this equation has no real root, and thus arrives at the paradoxical result, that however high a column may be it cannot be ruptured by its own weight. "Haec autem omnia accuratius examen requirunt, quod in sequente dissertatione instituemus" are the concluding words.

We may note that in this problem of the column bending under its own weight as considered by Euler, the top of the column is during the bending supposed to be in the *same vertical line as the base.*

[79.] In the second memoir Euler returns to the paradox "quod scilicet nulla columna cylindrica, quantumvis fuerit alta,

words $Ek^2$ is not a constant, but a function of the force and of the flexure. The assumption that the 'moment of stiffness' is constant seems to me to vitiate the results not only of Euler and Lagrange but of many later writers on the subject. I have to thank Prof. A. B. W. Kennedy for a series of experiments, the results of which conclusively prove that the position of the neutral line varies with the magnitude of the longitudinal force.] ED.

unquam a proprio pondere frangatur." He starts from the series cited in our Art. 77 and writing $a^3/m = 6v$ states that the paradox depends upon the fact that the series

$$1 - \frac{v}{4.1} + \frac{v^2}{7.1.5} - \frac{v^3}{10.1.5.12} + \frac{v^4}{13.1.5.12.22} - \text{etc.}$$

cannot vanish for real values of $v$.

Instead of considering the roots of this equation by a purely algebraical process as in his first memoir, Euler (§§ 6—16) treats the question graphically and endeavours to sketch the curve which the column would take up when bent by its own weight; this curve has for its equation

$$y = A . \left\{x - \frac{1 . x^4}{4!\, m} + \frac{1\ 4 . x^7}{7!\, m^2} - \text{etc.}\right\},$$

where $A$ is some constant.

The process is extremely complex and leads to no definite results[1].

[80.] In the second part of the memoir Euler inquires whether no solution of the question can be obtained *ex principiis mechanicis*. He starts by placing a load upon the top of a column, which he then converts into an additional length of column, this additional length of column is maintained in a vertical position by means of horizontal forces. Euler argues that, if the weight of this addition be sufficient to bend the column, *a fortiori* it would bend

[1] Euler's equation is $m . \frac{d^2y}{dx^2} + \int_0^x x\,dy = 0$, or differentiating $m\frac{d^3y}{dx^3} + x\frac{dy}{dx} = 0$. Put $x = \sqrt[3]{m} . t$ and $\frac{dy}{dx} = v$, and substitute, we find

$$\frac{d^2v}{dt^2} + vt = 0,$$

which is a case of Riccati's equation. This equation can be solved by the two Bessel's functions (cf. *Messenger of Mathematics*, Vol. IX. p. 129),

$$v = \sqrt{t}\,\{B . J_{\frac{1}{3}}(\tfrac{2}{3} t^{\frac{3}{2}}) + A . J_{-\frac{1}{3}}(\tfrac{2}{3} t^{\frac{3}{2}})\}.$$

The second function only will be found to be admissible, hence

$$v = \sqrt{t} . A \frac{1}{\sqrt{t}}\left(1 - \frac{t^3}{2.3} + \frac{t^6}{2.3.5.6} - \text{etc.}\right),$$

or

$$\frac{dy}{dx} = A\left(1 - \frac{x^3}{2.3.m} + \frac{x^6}{2.3.5.6\,m^2} - \text{etc.}\right),$$

which agrees with Euler's result § 12, or integrating, and remembering that $y$ and $x$ vanish simultaneously, with the value of $y$ given above.

under its own weight, were the column and the addition considered as united and the horizontal forces removed. He obtains the result that a column of height $h = 3 . \sqrt[3]{\frac{\pi^2 Ek^2 . c}{C}}$, where $\frac{C}{c}$ is the weight of unit length of the column, would certainly bend under its own weight (§ 29).

[81.] In the case of cylindrical columns, $Ek^2$ is considered to vary as the fourth power and $C/c$ as the square of the diameter ($b$), hence $h$ varies as the subtriplicate ratio of the area ($\propto b^{\frac{2}{3}}$). Euler thus concludes his second memoir with a *Theorema maxime memorabile* (§ 34):

*Maxima altitudo, qua columnae cylindricae ex eadem materia confectae, proprium pondus etiamnunc sustinere valent, tenet rationem subtriplicatam amplitudinis.*

[82.] The results of these first two memoirs do not appear however to have satisfied Euler. He returns again to the problem in the third. The result of the second memoir, if correct, shews that for some value of $v$ less than the one calculated (Euler finds this limit of $v = \frac{1}{6}$ of 266 in § 32 of the second memoir), the series above given for $v$ ought to vanish. This mechanical result seems to be contradicted by the algebraical investigation of the series, which shewed that it had no real root (see Art. 77). After certain considerations as to the possibility of representing the form of the column by a portion of the curve obtained in the first memoir Euler (§ 6) concludes:

Ut nostram quaestionem rite evolvamus, statum columnae, sive laminae elasticae, initialem aliter constituere debemus atque ante fecimus, scilicet praeter sollicitationes a gravitate oriundas supremo termino $A$ certam quandam vim horizontalem applicatam concipere debemus qua istud punctum $A$ perpetuo in eadem recta verticali contineatur. Facile autem intelligitur, magnitudinem huius vis prius definiri non posse, quam totus calculus ad finem fuerit perductus; quandoquidem tum demum patebit, quanta vi opus sit, ad supremum terminum $A$ in debito situ conservandum[1].

[1] There are *two* paragraphs (6) by some clerical error in the memoir, we refer to that on p. 166.

[83.] It will thus be seen that Euler has quite changed the character of the problem, he has discovered that a horizontal force is necessary to preserve the summit of the column in the same vertical with the base, yet instead of attempting to solve his differential equation

$$m \cdot \frac{d^3y}{dx^3} + x \cdot \frac{dy}{dx} = 0$$

with the true terminal conditions, namely $\frac{d^2y}{dx^2} = 0$ at the summit, $y$ and $\frac{dy}{dx} = 0$ at the base, he turns to the new problem where there is a horizontal force as well as gravity acting on the beam.

[84.] In order to determine this horizontal force $F$, Euler practically takes moments about the base of the column: if $h$ be its height, $G$ the centre of gravity of the displaced column $ACB$ and $M$ the weight of the column,

$$F.h = M.OG = \frac{M}{h} \cdot \int_0^h y dx,$$

with Euler's notation. He then writes

$$g = \left(\int_0^h y dx\right)/h \text{ and } m = Ek^2.h/M.$$

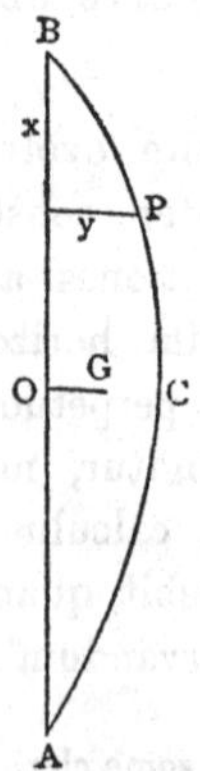

Hence his final equation becomes

$$gx+\int_0^x xdy + m\,\frac{d^2y}{dx^2}=0.$$

This equation (§ 28) is solved by means of two infinite series[1], namely $y=\alpha p+gq$, where $\alpha$ is a constant and

$$p=x-\frac{x^4}{2.3.4.m}+\frac{1.4.x^7}{2.3...7.m^2}-\frac{1.4.7.x^{10}}{2.3...10.m^3}+\text{etc.},$$

$$q=-\frac{x^3}{2.3.m}+\frac{1.3.x^6}{2.3...6.m^2}-\frac{1.3.6.x^9}{2.3...9.m^3}+\text{etc.}$$

At the base we have $x=h$, $y=0$, hence

$$\alpha p+gq=0 \text{ when } x=h,$$

and again $$gh=\int_0^h ydx=\alpha\int_0^h pdx+g\int_0^h qdx,$$

or $$p\left(h-\int_0^h qdx\right)+q\int_0^h pdx=0,$$

an equation to find $h$ the greatest altitude "in qua se tantum non sustinere valebit."

[85.] In §§ 32—37 Euler calculates this value of $h$ and finds that it is nearly equal to $\sqrt[3]{200\,m}$.

Hence we have the following result. Let $b^2$ be the area of a perpendicular section of the column supposed of unit weight per unit volume, then $M=b^2h$, and $m=Ek^2h/M=Ek^2/b^2$. Now let $O$ be the weight which placed upon a column of the same material but of section $d^2$ and height $a$ would suffice to bend it, then

$$O=\frac{Ek'^2\pi^2}{a^2} \text{ and } m=\frac{a^2}{b^2\pi^2}.O.\frac{k^2}{k'^2}.$$

Finally let $\lambda=$ the ratio of the weight $O$ (to be ascertained by experiment) to the weight of the column, which it will just bend, then $O=\lambda ad^2$ and $\dfrac{k^2}{k'^2}=\dfrac{b^4}{d^4}$ Thus

$$h=2{\cdot}7263\,a\sqrt[3]{\frac{\lambda b^2}{d^2}}.$$

This is Euler's final result with regard to the bending of columns under their own weight. Interesting as these three memoirs

[1] Easily obtained in the form of Bessel's Functions.

undoubtedly are, they cannot be said to furnish a complete and satisfactory discussion of the subject.

86. 1782. Euler: *Investigatio motuum quibus laminae et virgae elasticae contremiscunt.* This is published in the *Acta Academiae Scientiarum Imperialis Petropolitanae* for the year 1779, Pars Prior; the date of publication is 1782. The memoir occupies pages 103—161 of the volume.

The memoir commences thus:

Quanquam hoc argumentum jam pridem tam ab Illustriss. D. Bernoulli, quam a me fusius est pertractatum: tamen quia illo tempore neque principia, unde hujusmodi motus determinari oportet, satis erant exculta, neque ea Analyseos pars, quae circa functiones binarum variabilium versatur, satis explorata, actum agere non videbor, si nunc idem argumentum accuratius investigavero. Praeterea vero etiam tot diversa motuum genera in hujusmodi corporibus locum habere possunt, quae accuratiorem enucleationem postulant; quamobrem hic operam dabo, ut universam hujus rei disquisitionem ex primis principiis deducam, et clarius, quam quidem ante est factum, proponam. Imprimis autem omnia diversa motuum genera, quae quidem occurrere possunt, dilucide sum expositurus. Quo igitur omnia fiant magis perspicua, duo praemittam Lemmata, quorum altero status aequilibrii, altero vero motus virgarum utcunque elasticarum et a potentiis quibuscunque sollicitatarum definietur; ubi quidem tam virgam quam potentias perpetuo in eodem plano sitas esse assumo. Demonstrationem autem horum lemmatum non addo, quoniam eam alio loco jam dedi, atque adeo etiam ad eos casus quibus motus non sit in eodem plano, accommodavi.

I do not know to what place Euler alludes in the last sentence which I have quoted[1]. He gives on his pages 104—106 the general equations which he here calls *Lemmas*; they substantially agree with those of Poisson's *Traité de Mécanique*, Vol. I. pages 624—627.

[1] [The following is probably the memoir to which Euler refers. I give a short account of its contents although the memoir of 1782 covers almost exactly the same ground at somewhat greater length.

*De Motu Vibratorio laminarum elasticarum, ubi plures novae vibrationum species hactenus non pertractatae evolvuntur. Novi Commentarii Academiae Petrop.* Tom. XVII. (An. Acad. 1772). St Petersburg, 1773, pp. 449—457. Euler here obtains

87. For the transversal vibrations of a straight elastic rod Euler arrives on his page 109 at an equation which agrees with that in Poisson's Vol. II. page 371; namely,

$$\frac{d^2y}{dt^2} + b^2 \frac{d^4y}{dx^4} = 0.$$

The improvement which we find in comparing the present investigation with that in the *Methodus Inveniendi* consists in the use of what is practically equivalent to D'Alembert's Principle.

88. After obtaining the differential equation Euler adds on his page 109:

Ita ut totum negotium huc sit reductum, quemadmodum ista aequatio differentialis quarti gradus integrari queat; ubi quidem in limine confiteri cogimur ejus integrale nullo adhuc modo inveniri potuisse, ita ut contenti esse debeamus in solutiones particulares inquirere.

by a somewhat cumbrous method the equation $\frac{d^2y}{dt^2}+b^2\frac{d^4y}{dx^4}=0$ for the vibrations of a rod built in at one end and free at the other. He expresses his terminal conditions in analytical form, and obtains the equation

$$2+\cos\omega\,(e^{\omega}+e^{-\omega})=0$$

to determine the periods. He calculates some of the roots of this equation (§§ XVII, XVIII.)

In § XXIII. Euler treats the problem of the vibrations of a free rod (*lamina libera seu incumbat plano horizontali*). In this case he obtains the equation

$$2=\cos\omega\,(e^{\omega}+e^{-\omega}).$$

In § XXIV. we have the case of a rod fixed at one end and free at the other with the equation

$$(e^{\omega}+e^{-\omega})\tan\omega=(e^{\omega}-e^{-\omega}).$$

In § XXVI. the rod is fixed at both ends, and the equation for the periods takes the form $\sin\omega=0$.

In § XXVII. the rod is supposed fixed at one end and built in at the other; the equation for the periods is the same as in the case of § XXIV.

In § XXVII. Problem V, the rod is supposed built in at both ends; the equation for the periods is the same as in the case of § XXIII.

In § XXVII. Problem VI, the rod is free at both ends but fixed (as on a pivot) at some point of its length. In the case where the middle point is fixed, Euler finds the equation

$$2+\cos\omega\,(e^{\omega}+e^{-\omega})=0,$$

which agrees with the result for a rod built in at one end and free at the other.] ED.

In discussing the transversal vibrations for the rod there are six cases, with Euler, as with Poisson, Vol. II. page 371. Each end may be free; each end may be fixed, or as Euler says *simpliciter fixum;* each end may be mortised, or as the French writers say, *encastré*, or as Euler says *infixum*[1].

89. Euler first discusses the case in which both ends are free, which occupies his pages 118—124. He arrives at an equation

$$\cos\omega\,(e^{\omega}+e^{-\omega})=2 \quad\ldots\ldots\ldots\ldots\ldots\ldots(1);$$

this Poisson obtains in his Vol. II. page 372.

This equation is obviously satisfied by $\omega=0$; Euler says that it cannot be satisfied by any other value of $\omega$ which is less than $\frac{\pi}{2}$: see his page 122. To justify this statement he affirms that for such a value of $\omega$ the product $\cos\omega\,(e^{\omega}+e^{-\omega})$ would be less than 2, and he finds by expanding the cosine and the exponentials that

$$\cos\omega\,(e^{\omega}+e^{-\omega})=2\left(1-\frac{\omega^4}{6}\right).$$

But this is not satisfactory, for the equation just given is not exact. If we put for $\cos\omega$ its exponential value and expand, we shall find that

$$\cos\omega\,(e^{\omega}+e^{-\omega})=2\left\{1-\frac{\omega^4}{\underline{|4}}2^2+\frac{\omega^8}{\underline{|8}}2^4-\frac{\omega^{12}}{\underline{|12}}2^6+\ldots\right\},$$

the general term of the series being

$$\frac{\omega^n}{\underline{|n}}2^{\frac{n}{2}}\cos\frac{n\pi}{2}\cos\frac{n\pi}{4}.$$

Now it may be shewn that if $\omega$ lies between 0 and $\frac{\pi}{2}$ the two terms $-\frac{\omega^4}{\underline{|4}}2^2+\frac{\omega^8}{\underline{|8}}2^4$ give a *negative* aggregate; and the same is true for

[1] [For the purposes of this History the terms *clamped*, *mortised* and *built-in* are used as equivalents, to denote not only that the end of a rod is fixed, but its terminal *direction* also. At an end which is only said to be *fixed* the terminal direction is free.] ED.

every succeeding pair: thus the demonstration is completed. Or we may proceed thus: from equation (1) we obtain

$$\sin\omega = \pm \frac{e^{\omega} - e^{-\omega}}{e^{\omega} + e^{-\omega}} \quad \ldots\ldots\ldots\ldots\ldots\ldots(2);$$

and if $\omega$ do not exceed $\frac{\pi}{2}$ we must take the upper sign in (2). Then from (1) and (2) we get

$$\tan\omega = \tfrac{1}{2}(e^{\omega} - e^{-\omega}) \quad \ldots\ldots\ldots\ldots\ldots(3).$$

We shall shew that, except $\omega = 0$, there is no solution of (3) if $\omega$ does not exceed $\frac{\pi}{2}$.

Suppose we trace the curve

$$y_1 = \tan x \ldots\ldots\ldots\ldots\ldots\ldots\ldots \quad (4),$$

and the curve

$$y_2 = \tfrac{1}{2}(e^{x} - e^{-x}) \quad \ldots\ldots\ldots\ldots\ldots\ldots(5).$$

When $x$ is very small we have approximately

$$y_1 = x + \frac{x^3}{3} \text{ and } y_2 = x + \frac{x^3}{6},$$

so that $y_1$ is at first greater than $y_2$. Now $y_2$ cannot become equal to $y_1$, for

$$\frac{dy_1}{dx} = 1 + \tan^2 x, \quad \frac{dy_2}{dx} = \tfrac{1}{2}(e^{x} + e^{-x});$$

and this shews that $y_1$ increases more rapidly than $y_2$. For $y_1$ being at least through some range greater than $y_2$ we have $\tan x$ greater than $\frac{1}{2}(e^{x} - e^{-x})$, and therefore $1 + \tan^2 x$ greater than $1 + \frac{1}{4}(e^{x} - e^{-x})^2$, that is $1 + \tan^2 x$ greater than $\left(\frac{e^{x} + e^{-x}}{2}\right)^2$, and therefore *a fortiori* greater than $\frac{e^{x} + e^{-x}}{2}$: so that through that range $\frac{dy_1}{dx}$ is greater than $\frac{dy_2}{dx}$.

90. We have already stated that the problem involves six cases: see Art. 88. Four of these cases Euler had already discussed in the *Methodus Inveniendi*: see Art. 64. The two cases not discussed there are given in the present memoir, namely that

in which the rod is free at one end and fixed at the other, on pages 126—133; and that in which the rod is fixed at one end and mortised at the other, on pages 143—146. All the cases discussed in the *Methodus Inveniendi* are given again in the memoir, namely that in which the rod is free at both ends, on pages 118—126; that in which the rod is free at one end and mortised at the other, on pages 134—139; that in which it is fixed at both ends on pages 140—143; and that in which it is mortised at both ends on pages 146—150.

91. The mathematical processes consist mainly of discussions of such equations as (1). I have not noticed any points of interest except that presented by page 122, which I have already considered, and something of a similar kind which occurs on page 129. Euler wishes to shew that the equation

$$\tan\omega = \frac{e^{\omega} - e^{-\omega}}{e^{\omega} + e^{-\omega}}$$

has no root, except $\omega = 0$, between 0 and $\frac{\pi}{2}$ He says that

$$\sin\omega\,(e^{\omega} + e^{-\omega}) = 2\omega\,(1 + \tfrac{1}{3}\omega^2 - \tfrac{1}{30}\omega^4),$$

and

$$\cos\omega\,(e^{\omega} - e^{-\omega}) = 2\omega\,(1 - \tfrac{1}{3}\omega^2 - \tfrac{1}{30}\omega^4),$$

and hence it is plain that the former expression is greater than the latter through the whole of the first quadrant. To make this out distinctly it should be observed that the general term in the expansion of $\sin\omega\,(e^{\omega} + e^{-\omega})$ is

$$\frac{2^{\frac{n+2}{2}} \sin\frac{n\pi}{2} \sin\frac{n\pi}{4}}{\lfloor n} \omega^n,$$

and that the general term in the expansion of $\cos\omega\,(e^{\omega} - e^{-\omega})$ is

$$\frac{2^{\frac{n+2}{2}} \sin\frac{n\pi}{2} \cos\frac{n\pi}{4}}{\lfloor n} \omega^n.$$

We may also obtain the result thus: $\sin\omega\,(e^{\omega} + e^{-\omega})$ and $\cos\omega\,(e^{\omega} - e^{-\omega})$ both vanish when $\omega$ vanishes, but the differential coefficient of the former is greater than that of the latter throughout the first quadrant.

92. On his pages 152—157 Euler discusses the case in which the elastic rod has some constraint at a point intermediate between its ends; the constraint is to be of such a kind as to allow of the transmission of motion between the two parts of the rod on the two sides of it: he examines in particular the case in which the constraint is at the middle point.

On his pages 158—162 he notices very briefly the case in which the elastic rod is originally not straight but curved; and especially the case in which the curve is one that returns to itself, as a circle.

Cauchy alludes to the present memoir in his *Exercices*, Vol. III. pages 276 and 312.

[93.] The following additional memoirs or notes by Euler on subjects connected with elasticity occur in the *Opera Posthuma* edited by P. H. and N. Fuss, St Petersburg 1862, Tom. II.

*Page* 126. A note on the problems connected with the vibrations of the Elastic Lamina proposed by D. Bernoulli; Euler's results are found to agree with Bernoulli's.

*Page* 128. A short memoir entitled *De oscillationibus annulorum elasticorum.* This is a discussion on the oscillations of elastic 'annuli' or rings, with a remark at the end which would seem to suggest that the author considered he could apply the results of his memoir to bells. The method and results can hardly be considered of much value, yet it is interesting to note Euler's manner of attacking the problem of an elastic solid. He supposes the annulus to be built up of elastic threads placed transversally: "et a vi horum filorum dependet cohaesio partium materiae ex qua annulus est fabricatus." He then supposes that a certain number of filaments extended to a certain length will support a certain weight; to find the weight or tension which this number of filaments will support for any other extension, a simple proportion is used. In other words Euler applies Hooke's Law[1]

[1] I have found a memoir of Euler's, *Tentamen de sono campanarum*, in the *Novi Commentarii*, Tomus x. (for the academic year 1766) p. 261. The bell is divided into *annuli* by vertical and horizontal sections. The vibrations of these *annuli* are in either case treated as independent, and in both cases a differential equation of the form $\frac{d^2y}{dt^2} + a^2\frac{d^2y}{dx^2} + b^4\frac{d^4y}{dx^4} = 0$ is obtained. The method is unsatisfactory.

[94.] Page 536. *Anleitung zur Naturlehre.* Chapter XVIII. is entitled *Von der Zusammendrückung und Federkraft der Körper.* In this discussion the nature of elasticity is explained by a subtle matter enclosed in the pores of the coarser matter of ordinary bodies. This subtle matter corresponds to the aether, and the magnitude of the elastic force depends upon the quantity, magnitude and figure of the closed pores of a body. "Diese Erklärung der elastischen Kraft, durch die in den verschlossenen Poren befindliche subtile Materie, ist der Natur der elastischen Körper vollkommen gemäss, und wird durch die Art, nach welcher verschiedenen Körpern eine elastische Kraft beigebracht wird, noch mehr bestätigt."

This theory of Euler's to explain the physical nature of Elasticity is interesting, especially as we have noted that Hooke also based the 'springiness' of bodies upon the existence of a like subtle medium. (See Art. 9.)

[95.] Further memoirs by Euler,

*De propagatione pulsuum per medium elasticum*:—*Novi Commentarii Petropol.* Tom. I. 1750.

*Lettre à M. de la Grange contenant des recherches sur la propagation des ébranlemens dans un milieu élastique. Miscellanea Taurinensia.* Tom. II. 1760—1

treat only of the motion of sound waves in an elastic *fluid.*

The memoir entitled:

*De Figura curvae Elasticae contra objectiones quasdam Ill. D'Alembert, Acta Academiae Scientiarum Petropol.* 1779, Pars II. p. 188, is written to defend James Bernoulli's solution of Galilei's problem against an attack of D'Alembert in the 8th volume of his Opuscules. It contains nothing of importance.

A work of Euler's entitled: *Von dem Drucke eines mit einem Gewichte beschwerten Tisches, auf eine Flaeche, A. d. Papieren Eulers gez. von Jak. Bernoulli* 1794, might possibly contain matter relating to our subject, but I have not been able to find a copy.

[96.] The following memoirs are by pupils of Euler, and closely connected with the subject of one or other of the master's papers.

*a.* 1778. Nicolaus v. Fuss: *Varia Problemata circa statum aequilibrii Trabium compactilium oneratarum, earumque vires et pressionem contra anterides. Acta Academiae Scientiarum Petrop.* 1778. Pars I. St Petersburg, 1780. pp. 194—216.

This memoir by Fuss the biographer son-in-law and assistant of Euler applies the results of Euler's papers on the bending of columns to calculate the strength of various simpler cases of framework subjected to pressure. The whole consideration turns on Euler's result quoted in our Art. 74, and contains nothing of any interest or value.

*β.* A. J. Lexell. *Meditationes de formula qua motus Laminarum elasticarum in annulos circulares incurvatarum exprimitur.*

This memoir is on pp. 185—218 of the *Acta Academiae Petrop.* for the year 1781, Pars II. (published 1785).

This memoir by a disciple of Euler's contains a discussion of the equation obtained by Euler in his memoir on the tones of bells (see our footnote, p. 55). In that memoir Euler had obtained an equation of the form

$$\frac{d^2y}{dt^2} + a^2\frac{d^2y}{dx^2} + b^2\frac{d^4y}{dx^4} = 0,$$

for the vibrations of a rod or thin beam in the form of a circular ring, $y$ being the normal displacement at a distance $x$ along the arc from some fixed origin of measurement.

Lexell obtains an equation of the same form, but he differs from Euler in the value he attributes to his coefficients $a$ and $b$.

This paper is hardly of sufficient importance to justify an analysis of its contents. The results obtained are, no more than Euler's, in agreement with the more accurate investigations of Hoppe (Crelle Bd. 63); see also *The Theory of Sound*, Lord Rayleigh, Vol. I. p. 324.

*γ.* In the same volume of the St Petersburg *Acta* immediately preceding Lexell's paper is a short memoir by another disciple M. Golovin (pp. 176—184), which contains an application of Euler's theory of the sounds of bells *ad sonos scyphorum vitreorum, qui sub nomine instrumenti harmonici sunt cogniti.* The notes of the *harmonicum* are calculated from Euler's results, but there is no statement of the amount of agreement the calculated results bear to those of experiment.

[97.] Lagrange: *Sur la Force des Ressorts pliés*, 1770[1]. *Mémoires de l'Académie de Berlin*, XXV. 1771.

Two important memoirs seem to be Lagrange's real contributions to our subject. The above memoir is reprinted pp. 77—110 of Tome III. of Serret's edition of the Oeuvres.

The author commences by telling us that in springs which act by contraction or elongation, the force appears to be proportional to the quantity by which they are contracted or expanded or at least to be a function of these quantities:

Mais ce principe n'a pas lieu dans les lames élastiques inextensibles et pliées en spirale telles que celles qu'on applique aux horloges: le seul principe qu'on puisse employer pour ces sortes de ressorts est que la force avec laquelle le ressort résiste à être courbé est toujours proportionnelle à l'angle même de courbure; et c'est d'après ce principe que de très-grands Géomètres ont déterminé la courbe qu'une lame élastique doit former lorsqu'elle est bandée par des forces quelconques données.

[98.] In order to ascertain the law of force of bent springs, Lagrange proposes the following problem for consideration in his memoir:

Une lame à ressort de longueur donnée et fixe par une de ses extrémités étant bandée par des forces quelconques qui agissent sur l'autre extrémité, et qui la retiennent dans une position donnée, déterminer la quantité et la direction de ces forces.

In order that the equations may not be too complex the lamina is supposed to be of uniform thickness and in its primitive state a straight line.

[1] With regard to the title of Lagrange's work we may refer to three earlier memoirs on the same subject, namely:

Deschamps: *Méthode pour mesurer la force des differens ressorts*. *Mémoires de l'Académie*. Paris, 1723.

C. E. L. Camus: *Du mouvement accéléré par des ressorts et des forces qui résident dans les corps en mouvement*. *Ibid.* 1728.

James Jurin: *On the Action of Springs*. *Phil. Trans.* 1744. The author commences by stating Dr Hooke's principle *Ut tensio, sic vis;* and applies it to a general theorem concerning the compression of a spring struck by a body of weight *M* moving with velocity *V* in the direction of the axis of the spring. The theorem is followed by upwards of 40 corollaries. The paper is of no value.

[99.] Lagrange proceeds in his first section to prove 'd'une manière aussi simple que rigoureuse' that the force of a spring at any point is proportional to the sum of the moments of all the powers which act on the segment marked off by the point. The ingenuity rather than the simplicity or rigour of the proof will strike the modern reader. The exact nature of the 'force of a spring' does not seem to have been quite clear to the writer, or he would have perceived the truth of the proposition from the ordinary statical equations of equilibrium. The force of the spring, we are then told, has been usually held proportional to the curvature.

[100.] Taking the tangent and normal at the extremity at which the forces are applied as axes of $x$ and $y$ respectively, $x$ and $y$ as the co-ordinates of any point of the lamina, $\rho$ as the radius of curvature, and $\phi$ as the tangential angle, Lagrange very easily deduces the relation

$$Py + Qx = \frac{2K^2}{\rho},$$

where $2K^2$ is a coefficient depending on the elasticity of the lamina, and $P$ and $Q$ are the resultant components of the forces along the tangent and normal respectively. Hence by differentiation and integration he finds

$$ds = \frac{Kd\phi}{\sqrt{P - P\cos\phi + Q\sin\phi}},$$

$$dy = \frac{K\sin\phi d\phi}{\sqrt{P - P\cos\phi + Q\sin\phi}},$$

$$dx = \frac{K\cos\phi d\phi}{\sqrt{P - P\cos\phi + Q\sin\phi}}.$$

[101.] If these equations could be integrated, we should have equations to determine $P$, $Q$ and the integral curvature, or the problem would be solved, but Lagrange writes

Il est aisé de voir que l'intégration dont il s'agit dépend en général de la rectification des sections coniques, et qu'ainsi elle échappe à toutes les méthodes connues.

In his third section Lagrange integrates these equations for $Q$

zero and $Q$ very small; in the latter case the results are extremely complex[1].

[102.] A resolution of the applied forces $P$ and $Q$ along and perpendicular to the chord joining the extremities of the spring into components $R$ and $T$ leads to more interesting results. If $T=0$ it is found that

$$R=-\frac{2K^2\mu^2\pi^2}{l^2},$$

where $\mu$ is any integer and $l$ the length of the spring. Hence we deduce that the least force which acting in the direction of the chord will suffice to produce in a spring any the least curvature is $\frac{2K^2\pi^2}{l^2}$. This curious result was first obtained by Euler (see Art. 67). It is also shewn that when the spring is very little bent and $r$ the length of the chord, then the force at the free extremity perpendicular to the chord vanishes, when the angle at the fixed extremity between chord and spring is

$$2\sqrt{\frac{l/r-1}{5}}.$$

[103.] In the seventh section by means of another resolution of the forces at the free end, it is shewn that, if the free end of the spring be compelled to describe a very small circular arc about a centre lying in what would be the unstrained direction of the spring, then the force upon the free end along the tangent to the arc is always proportional to the length of the arc measured from the unstrained position. This elegant property it is suggested might be used to obtain isochronous oscillations in the balance wheel of a watch.

[104.] In the following sections differential equations are obtained connecting the forces $R$ and $T$ with the constants of the problem; they are extremely complex and as Lagrange himself admits throw no light on the nature of these forces. The better method would now be to solve the equations in terms of elliptic functions.

[1] To an error in Lagrange's analysis at this point (on his p. 88) we shall return later when discussing a memoir by Plana (see Art. 153).

[105.] Hitherto the free end has been acted upon by a system of concurrent forces only; in sections X—XII. Lagrange considers the case where the free end is in addition acted upon by a couple. If there is a couple only the curve of the spring will be a circle; if the forces on the free end in addition to the couple are only very small, the figure will be very nearly circular. These results are then in sections XIV—XVI. applied to the case of a spiral spring which produces oscillations in a drum or balance wheel similar to those of a simple pendulum.

Lagrange concludes:

Au reste, il faut toujours se souvenir que ces conclusions sont fondées sur l'hypothèse que la lame du ressort soit naturellement droite et que sa longueur soit très-grande; c'est ce qui fait qu'elles n'ont pas lieu dans les ressorts ordinaires qu'on applique aux horloges, mais il n'est pas impossible qu'elles puissent être d'usage dans quelques occasions.

[106.] *Sur la figure des colonnes. Miscellanea Taurinensia,* Tomus V. This memoir of Lagrange's appears in the mathematical section of the volume of the memoirs of the Royal Society of Turin which embraces the years 1770—1773, p. 123.

The memoir is an important addition to a subject already considered by Euler and Lagrange himself.

On a coutume de donner aux colonnes la figure d'un conoïde qui ait sa plus grande largeur vers le tiers de sa hauteur, et qui aille de là en diminuant vers les deux extrémités; d'où résulte ce qu'on appelle vulgairement le *renflement* et la *diminution* des colonnes.

Lagrange notes that the authors who recommend this seem to have no better argument in its favour than to quote the shape of the human body. A better reasoning from analogy would, Lagrange thinks, have been from the shape suggested by the trunk of a tree. Vitruvius 'le législateur des Architectes modernes' had prescribed the *renflement* and all architects have followed his *dictum*. Since this rule as to the form of a column has such an arbitrary basis, Lagrange in the present memoir proposes to consider the proper figure from the mathematical standpoint.

[107.] Reference is made at the commencement to Euler's memoir of 1757.

Cependant comme le point de vue sous lequel cet illustre Auteur a discuté cette matière est différent de celui dans lequel nous nous proposons de la traiter, nous croyons faire quelque plaisir aux Géomètres en leur communiquant les recherches que nous avons faites sur un sujet, qui interesse également la Mécanique et l'Analyse. (§ 3.)

[108.] Lagrange starts by obtaining in the same fashion as Euler the equation

$$y = f \sin\left(x\sqrt{\frac{P}{K}}\right)$$

for the bent column. Here $y$ is the displacement at vertical distance $x$ from the base; $P$ is the superincumbent weight, $K$ the 'moment of stiffness' (or the $Ek^2$ of Euler's notation); while $f$ is some constant.

Lagrange's treatment of the equation is more satisfactory than Euler's. If $a$ be the height of the column, $y = 0$ for $x = a$, hence

$$a\sqrt{P/K} = m\pi,$$

where $m$ is an integer, or

$$P = m^2\pi^2 K/a^2,$$

and

$$y = f \sin(m\pi x/a),$$

$f$ being an arbitrary constant which is equal to the maximum value of $y$.

[109.] The next paragraph may be cited at length, for it points out results which seem to have escaped Euler.

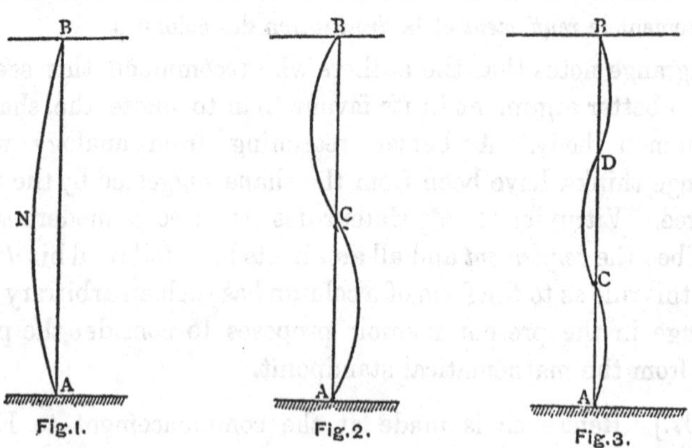

Fig. 1. Fig. 2. Fig. 3.

Si on fait $m=1$, on aura $y=f\sin(\pi x/a)$, d'où l'on voit que la courbe $ANB$ ne coupe l'axe qu'aux deux extrémités $A$ et $B$; et le poids requis pour donner à la colonne cette courbure sera $\pi^2K/a^2$. Si $m=2$, on aura $y=f\sin(2\pi x/a)$, et la courbe coupera l'axe au point où $x=a/2$, c'est-à-dire au point du milieu $C$, en sorte que la colonne prendra la figure 2; mais il faudra pour cela que le poids $P$ soit $4\pi^2K/a^2$, c'est-à-dire quadruple du précédent. Si on faisoit $m=3$, on auroit $y=f\sin(3\pi x/a)$, de sorte que la courbe couperoit l'axe aux points où $x=a/3$ et $x=2a/3$ et seroit semblable à la figure 3, or pour que la colonne soit pliée de cette manière il faudra que le poids $P$ soit $= 9\pi^2K/a^2$, c'est-à-dire neuf fois plus grand que le premier; et ainsi de suite. (§ 6.)

[110.] After remarking that no force less than $\pi^2K/a^2$ will bend the column, Lagrange proceeds to consider what happens when $P$ is not equal to one of the quantities $m^2\pi^2K/a^2$. For this purpose he takes the rigorous equation to the curve and easily deduces for an arc $s$ measured from the base

$$s=\int_0^y \frac{dy}{\sqrt{(f^2-y^2)P/K-(f^2-y^2)^2P^2/4K^2}},$$

and assuming $y=f\sin\phi$, he finds

$$s=\int_0^\phi \frac{d\phi}{\sqrt{P/K-(P^2f^2/4K^2)\cos^2\phi}}.$$

Here $s$ must equal $a$ when $\phi=m\pi$, $m$ being an integer which determines the number of times the bent column cuts the vertical between base and summit $(=\overline{m-1})$ or $m$ is equal the number of bulges (*nombre des ventres*). Lagrange integrates this equation by means of a series, and ultimately obtains

$$a=\frac{m\pi}{\sqrt{P/K}}\left\{1+\frac{Pf^2}{4(4K)}+\frac{9P^2f^4}{4.16(16K^2)}+\frac{9.25P^3.f^6}{4\ \ 16.36(64K^3)}+\text{etc.}\right\},$$

which gives the value of $f$ so soon as $P$ and $a$ are known[1]. A discussion of the roots of this equation for $f$ is then entered upon (§ 10). The quantity $f^2$ can only have a real value when

$$1-\frac{a}{m\pi}\sqrt{\frac{P}{K}}<0,$$

and then only one positive real value. Hence if $P<m^2\pi^2K/a^2$,

[1] Lagrange's result differs slightly from this, but I think his numbers are wrong.

$f$ will have only two real roots equal and opposite; if $P$ be less than this quantity, none at all.

D'où il s'ensuit que tant que $P$ sera $< \pi^2 K/a^2$, la colonne ne pourra pas être courbée; que tant que $P$ sera renfermée entre les limites $\pi^2 K/a^2$ et $4\pi^2 K/a^2$, la colonne sera courbée, mais en ne formant qu'un seul ventre; que tant que $P$ sera entre les limites $4\pi^2 K/a^2$ et $9\pi^2 K/a^2$, la colonne sera nécessairement courbée et *pourra former ou un seul ventre ou deux*, et ainsi de suite. (§ 10.)

[111.] Hitherto the column has been supposed cylindrical, Lagrange now proceeds to treat it as a surface of revolution and accordingly replaces $K$ by a function $X$ of $x$, and obtains the equation $Py + X\dfrac{d^2y}{dx^2} = 0$. He assumes as the solution of this equation $y = \xi \sin \phi$ where $\xi$ and $\phi$ are arbitrary functions of $x$, and finds that the above differential equations will be satisfied if he takes

$$P\xi + X\left\{\frac{d^2\xi}{dx^2} - \xi\left(\frac{d\phi}{dx}\right)^2\right\} = 0, \text{ and } 2\frac{d\phi}{dx}\frac{d\xi}{dx} + \xi\frac{d^2\phi}{dx^2} = 0.$$

The second of these equations is integrable and gives $\phi = h\,(\int dx/\xi^2)$, where $h$ is an arbitrary constant. Putting $\xi^2 = hu$ we have the following equations determining $u$ and then $y$ when $X$ is given:—

$$4Pu^2 + X\left\{2u\frac{d^2u}{dx^2} - \left(\frac{du}{dx}\right)^2 - 4\right\} = 0,$$

$$y = \sqrt{hu}\sin(\int dx/u).$$

With regard to these equations Lagrange makes the following remarks. The expression for $y$ contains two arbitrary constants.

L'une c'est la constante $h$ qui ne $x$ trouve point dans l'équation en $u$; l'autre c'est celle qui est virtuellement renfermée dans l'intégrale $\int dx/u$, c'est pourquoi il suffra d'y substituer une valeur quelconque de $u$ qui satisfasse à l'équation en $u$ sans s'embarasser si elle est une intégrale complette de cette équation ou non.

This expression is also very convenient for determining the weight $P$, for since $y = 0$ when $x = 0$, we can take the limit of the integral $\int dx/u$, so that it vanishes with $x$. Further $y$ must vanish when $x = a$, hence $\int dx/u = m\pi$.

Or comme la quantité $u$ ne doit point contenir de constantes arbitraires, il est visible que cette dernière condition donnera une équation entre les quantités $P$ et $a$, par laquelle on pourra déterminer $P$. Quant au nombre entier $m$ qui demeure indéterminé, il est clair, par ce qu'on a vu plus haut, qu'il sera toujours égal au nombre des ventres que la colonne formera en se courbant par la pression du poids $P$; donc pour avoir la limite des fardeaux que la colonne pourra supporter sans se courber d'une manière quelconque, il faudra toujours prendre pour $m$ le nombre entier qui rendra la valeur de $P$ la plus petite; et cette valeur sera la limite cherchée. (§ 14.)

[112.] We have thus a complete theory for the slight bending of a column of any kind whatever formed by a surface of revolution about its axis. Lagrange proceeds to apply this to various cases, which we shall summarise in the following paragraphs.

I. To the surface generated by the revolution of a conic section about a line in its plane. If $z$ be the distance of any point of the conic from the axis, its equation will be of the form

$$z^2 = \alpha + \beta x + \gamma x^2.$$

Now, 'il paroît que la théorie et l'expérience s'accordent assez à faire $X$ proportionnelle à $z^4 = K(\alpha + \beta x + \gamma x^2)^2$' where $K$ is a constant.

Lagrange easily finds that $u = g(\alpha + \beta x + \gamma x^2)$, where

$$g^{-1} = \sqrt{P/K + \alpha\gamma - \beta^2/4}$$

is a suitable value for $u$.

Hence he deduces from the equation $m\pi = \int dx/u$,

$$P = \{\beta^2/4 - \alpha\gamma + m^2\pi^2/A^2\} \, . \, K,$$

where $A$ is the value of the integral $\int^a \frac{dx}{\alpha + \beta x + \gamma x^2}$, the lower limit being chosen so as to make it vanish for $x = 0$. This value of course depends upon the relations which may exist between $\alpha$, $\beta$, $\gamma$.

To obtain the weight which a column will support without bending, we must put $m = 1$. If we make $P$ a maximum by varying $\alpha$, $\beta$, $\gamma$, and consequently $A$ which is a function of them, we shall have the character of the column which will support a maximum weight. Lagrange remarks however that, if we increase the dimensions of the column, we shall obviously increase the weight it is capable of carrying, and thus, supposing the height given, it is necessary to find a maximum *relative* to the mass of the column. Now the order of $P$ is the fourth power of the linear dimensions of section by the second power of the height (thus, the least weight which will bend a cylindrical column $=\pi^2 Kb^4/a^2$ where $b$ is the radius), hence Lagrange proposes to measure the efficiency of the column by the ratio $P : S^2$ where $S$ is its mass and thus $S^2$ of the fourth order in the linear dimensions of section. This measure of the efficiency of a column frees us from the indeterminateness which would otherwise arise from the possibility of infinitely increasing the magnitude of $P$ by simply increasing the dimensions of the column. Lagrange terms $P : S^2$ 'la force relative d'une colonne' (§ 19).

Case (i). $\frac{\beta^2}{4} - \alpha\gamma = 0$. The column is a right cone; Lagrange finds that its efficiency is greatest when the cone degenerates into a right circular cylinder (§§ 18—20).

Case (ii). $\gamma = 0$. The column is generated by a parabola. Here again the only maximum found for the efficiency is when the surface degenerates into a right circular cylinder (§ 21).

Case (iii). The general equation $z^2 = \alpha + \beta x + \gamma x^2$ is considered. The efficiency is found to be greatest when the conic becomes a straight line or the column is conical. Hence we are thrown back on Case (i), or the column is again a right circular cylinder (§§ 22—26).

Lagrange thus conceives that he has disposed of Vitruvius' dictum as to 'bellied' columns so far as surfaces of revolution of the second order are concerned.

[113.] II. In § 27 the general problem is attacked: to find the curve which by its revolution about an axis in its plane determines the column of greatest efficiency. Lagrange thus expresses the problem analytically:

Il s'agit de trouver une équation entre les ordonnées $z$ et les abscisses $x$, telle que la quantité $P/S^2$ soit la plus grande qu'il est possible, $S$ étant égale à l'intégrale $\pi\int z^2 dx$ prise depuis $x=0$ jusqu'à $x=a$, et $P$ étant une constante qui doit être déterminée par cette condition, que l'intégrale $\int dx/u$ prise ensorte qu'elle soit nulle lorsque $x = 0$, devienne $= \pi$ lorsque $x = a$, en supposant $u$ donnée par l'équation différentielle

$$4Pu^2 + X\left\{2u\frac{d^2u}{dx^2} - \left(\frac{du}{dx}\right)^2 - 4\right\} = 0,$$

où $X$ est une fonction donnée de $z$ que nous avons supposée plus haut $= Kz^4$.

Lagrange's solution of this problem (§§ 28—31) is chiefly interesting as an early application of the Calculus of Variations. His conclusion is that a right circular cylinder is *one* but not the only solution. It is the only solution in the case where the required curve must pass through four points equally distant from the axis, or again where the extreme sections of the column are to be equal and the directions of the tangents to the generating curve at those sections parallel to the axis.

§ 33 treats of a practical case, namely that in which there is a small variation from the cylindrical form. Lagrange by a somewhat laborious calculation finds a class of curves (a curve with variable parameters) for which the efficiency will be a maximum. Causing these parameters to vary (just as in treating of the conic sections in I.), he finds that the cylinder is again the column of greatest efficiency. "D'où l'on doit conclure que la figure cylindrique est celle que donne le *maximum maximorum* de la force" (i.e. efficiency). With these words Lagrange concludes his memoir, which may fairly be said to have shaken the then current architectural fallacies.

[114.] A memoirl of 1777 by Lindquist entitled: *De inflexionibus laminarum elasticarum* Aboae, 1777, which would probably contribute something to our subject, I have searched for in vain.

115. Saint-Venant draws special attention to a few pages which occur in a memoir by Coulomb; see Liouville's *Journal de Mathématiques,* 1856, page 91. Coulomb's memoir is in the volume for 1773 of the *Mémoires...par divers Savans,* published at Paris in 1776. The memoir is entitled *Essai sur une application*

*des règles* de Maximis et Minimis *à quelques Problèmes de Statique, relatifs à l'Architecture.* The memoir occupies pages 343—382 of the volume.

116. The *Introduction* finishes with the following sentences:

Ce Mémoire, composé depuis quelques années, n'étoit d'abord destiné qu'à mon usage particulier, dans les différens travaux dont je suis chargé par mon état; si j'ose le présenter à cette Académie, c'est qu'elle accueille toujours avec bonté le plus foible essai, lorsqu'il a l'utilité pour objet. D'ailleurs les Sciences sont des monumens consacrés au bien public; chaque citoyen leur doit un tribut proportionné à ses talens. Tandis que les grands hommes, portés au sommet de l'édifice, tracent et élèvent les étages supérieurs, les artistes ordinaires répandus dans les étages inférieurs, ou cachés dans l'obscurité des fondemens, doivent seulement chercher à perfectionner ce que des mains plus habiles ont créé.

117. The section of Coulomb's memoir to which Saint-Venant draws attention is entitled *Remarques sur la rupture des Corps;* it occurs on pages 350—354. This substantially amounts to a short theory of the flexure of a beam, and the merit of Coulomb is that he places the *neutral* line at the middle of the transverse section, supposed rectangular; and that he makes an accurate calculation of the moments of the elastic forces over a transverse section[1]. Saint-Venant says:

C'est dans notre siècle seulement que son Mémoire, qui contient sur ce sujet tant de choses dans les trois pages intitulées *Remarques sur la rupture*, a été enfin étudié et compris.

Saint-Venant names Duleau, Barlow, and Tredgold as having fallen into error in the present century in spite of what Coulomb had written.

The section of Coulomb's memoir finishes thus:

M. l'Abbé Bossut, dans un excellent Mémoire sur la figure des digues, ouvrage où l'on trouve réunie, à l'esprit d'invention, la sagacité du Physicien, et l'exactitude du Géomètre, paroît avoir distingué et fixé le

[1] [He considers the problem of flexure by a force *perpendicular* to the beam, so that in this case his position of the neutral line is correct. Belgrado's method was equally correct and earlier.] Ed.

premier la différence qui se trouve entre la rupture des bois et celle des pierres[1].

118. Dr Thomas Young wrote a life of Coulomb for the *Supplement* to the *Encyclopædia Britannica*, and in it he makes a few remarks on the present memoir, which he styles *admirable*. See Young's *Miscellaneous Works*, Vol. II. page 529.

119. 1784. A memoir by Coulomb is contained in the *Histoire de l'Académie*...for 1784, published at Paris in 1787, entitled *Recherches théoriques et expérimentales sur la force de torsion, et sur l'élasticité des fils de métal: Application de cette théorie à l'emploi des métaux dans les Arts et dans différentes expériences de Physique: Construction de différentes balances de torsion, pour mesurer les plus petits degrés de force. Observations sur les loix de l'élasticité et de la cohérence.* The memoir was read in 1784; it occupies pages 229—269 of the volume.

The theory is very simple. Imagine a thread of metal or of silk, fixed at one end; and let the other end be attached to a cylindrical weight so that in equilibrium the string is vertical and its direction in the same straight line with the axis of the cylinder. Let the cylinder be turned through any angle round its axis, and left to itself: it is required to determine the motion. Suppose that a vertical section of the cylinder through its axis makes an angle $\theta$ with its equilibrium position, and *assume* that the force of torsion varies as $\theta$: then the equation of motion is

$$Mk^2 \frac{d^2\theta}{dt^2} = -\mu\theta,$$

where $Mk^2$ is the moment of inertia of the cylinder round its axis, and $\mu$ is some constant. From this equation the time of an oscillation can be found; then by comparing this with the observed time such an agreement is obtained as shews that the assumption made with respect to the force of torsion is correct[2].

[1] [This probably refers to a book published in Paris in 1764 entitled: *Sur la construction la plus avantageuse des digues*. I have been unable to find a copy.] Ed.

[2] [Coulomb first gave his theory of torsion for hairs and silk threads in a *Mémoire sur les boussoles de déclinaison* in the *Mémoires des Savans étrangers*, Tom. IX. 1777. This theory was extended to metal threads in the above-mentioned memoir of 1784,

I presume that this is the memoir which Saint-Venant has in view when he speaks of Coulomb and the ancient theory of torsion, though he does not give any reference: see pages 331, 340, 341 of Saint-Venant's *Torsion.*

Dr Young gives a brief account of Coulomb's memoir on torsion: see Young's *Miscellaneous Works*, Vol. II. pages 532, 533.

[120.] An important conception arrived at by Coulomb is that of *cohesion*, which he defines as follows:

Si l'on suppose un pilier de maçonnerie coupé par un plan incliné à l'horison, ensorte que les deux parties soient unies dans cette section par une cohésion donnée, tandis que tout le reste de la masse est parfaitement solide, ou lié par une adhérence infinie; qu'ensuite on charge ce pilier d'un poids, ce poids tendra à faire couler la partie supérieure du pilier sur le plan incliné par lequel il touche la partie inférieure. Ainsi dans le cas d'équilibre, la portion de la pesanteur qui agit parallèlement à la section, sera exactement égale à la cohérence. Si l'on remarque actuellement dans le cas de l'homogénéité, que l'adhérence du pilier est réellement égale pour toutes les parties, il faut pour que le pilier puisse supporter un fardeau qu'il n'y ait aucune section de ce pilier sur laquelle l'effort décomposé de sa pression puisse faire couler la partie supérieure. Ainsi, pour déterminer le plus grand poids que puisse supporter un pilier, il faut chercher parmi toutes les sections celle dont la cohésion est en équilibre avec un poids qui soit un *minimum*: car, pour lors, toute pression au-dessous de celle déterminée serait insuffisante pour rompre le pilier.

Girard after pointing out the distinction between "substances fibreuses" and "les corps formés de molécules agglutinées" quotes the above with great praise as the ingenious hypothesis of 'citoyen Coulomb.' (Cf. *Traité*, Introduction, p. xxxvi.)

[121.] Giordano Riccati: *Delle vibrazioni sonore dei cilindri.* This memoir occurs on pp. 444—525 of the *Memorie di Matematica e Fisica della Società Italiana*, Tomo I., Verona, 1782.

The basis of this memoir is the appendix to Euler's *Methodus inveniendi;*

and the same subject was still further considered in a paper entitled: *Expériences, destinées à déterminer la cohérence des fluides*, in the *Mémoires de l'Institut National des Sciences*, Tom. III. p. 256 (1806), Paris. Prairial An IX.] Ed.

Quantunque io abbia preso l'essenzial di questa soluzione dall' appendice sopra le curve elastiche aggiunta dal dottissimo Sig. Leonardo Eulero alla sua profonda opera *Methodus inveniendi......*; nulladimeno non isdegnino di leggerla i Matematici, sì perchè ho procurato di rischiarare la materia per sè stessa oscurissima, sì perchè mè riuscito di farci per entro qualche non dispregevole scoperta, e di notare altresì alcun picciolo neo, che anderò opportunamente segnando, nelle speculazioni per altro sublimi del chiarissimo autore.

Riccati still labours under the old error which placed the neutral surface in the lowest fibre of a beam subjected to a perpendicular deflecting force. He also corrects Euler in the same point as Girard (see Art. 130) by stating that the 'absolute elasticity' of a cylindrical beam is proportional to the *cube* of its thickness;

Se col metodo di me tenuto rispettivamente ai cilindri avessi cercato il valore della forza $E$ nelle lamine elastiche, mi farebbe riuscito di trovarlo in ragione composta della rigidità della materia, della larghezza della lamina, e del cubo della sua grossezza. Alla forza $E$ il Signor Eulero dà il titolo di elasticità assoluta della lamina, e senza dimostrarlo, l' asservisce proporzionale al prodotto della rigidità della materia, della larghezza della lamina, e del quadrato della grossezza. Questo sbaglio influisce nella legge dei tempe delle vibrazioni delle lamine elastiche determinata dal nostro Autore, la quale non corrisponde ai fenomeni, salvochè nella circostanza, che le diverse lamine sieno di pari grossezza.

The correction of Riccati, like that of Girard, seems to me to arise from a misunderstanding of Euler. Euler makes his $Ek^2$ (= Riccati's $E$) proportional to the square of the dimension of the beam in the plane of flexure multiplied by the area of its section, and in the case of a beam of circular section this is proportional to the *fourth* power of the diameter. (See Art. 75.)

Riccati arrives at the same result for his cylindrical beams also although he uses (§ III.) a very unsatisfactory proof. In Section X. the equation $d^4y/dx^4 = m^4y$ is found for the vibrations of the beam, by a method which resembles Daniel Bernoulli's or the earlier work of Euler in not introducing D'Alembert's Principle. Equations are obtained similar to those of Euler (see Art. 87 *et seq.*), and Riccati solves them by somewhat lengthy numerical calculations in order to determine the times of vibration. Reference is made on a minor point to the memoir of the father, Jacopo Riccati, which we have considered in Art. 30.

In Section XXVIII. the nodes are determined for any mode of vibration. Finally the results of theory are compared with experiments made on cylinders of steel and bronze.

Viewed as a whole Riccati's memoir contributes little or nothing new to the *mathematical* theory of a vibrating beam, yet his calculations are more complete than those of Daniel Bernoulli (see Art. 50) and the paper has considerable value as a contribution to the theory of sound.

[122.] James Bernoulli: *Essai théorétique sur les vibrations des plaques élastiques rectangulaires et libres.* This memoir was presented to the Academy of St Petersburg in Oct. 1788. It is printed in the *Nova Acta Academiae Scientiarum Petropolitanae*, Tomus V. This is the volume for the academic year 1787, and was published at St Petersburg in 1789. The James Bernoulli in question was the nephew of Daniel and grandson of the John Bernoulli mentioned in Art. 35[1].

The author expresses a desire in § 1 to find a theoretical basis for Chladni's experiments on sound. He notes that the attention of the great 'geometricians' has hitherto been confined to the examination of the vibrations of bodies which can be regarded as having only one dimension, strings and elastic laminae.

> Si M. L. Euler a osé passer plus loin, & traiter du son des cloches, il a reconnu lui-même, que l'hypothèse, qu'il faisoit servir de base à ses calculs, étoit précaire.

(See our Art. 94.)

Even in this case Chladni has shewn that Euler's hypothesis is inadmissible because it leads to results not in accordance with experience.

[1] The following scheme may aid the reader in forming some idea of the relationship.

Nicolaus (b. 1623, d. 1708)
(A merchant of Basel)

James (Fifth Son) (Elastic Curve) — John (Tenth Son) (Discourse on Motion)

John (Tenth Son): Daniel (Second Son) (Vibrations of a beam) — John (Youngest Son)

John (Youngest Son): James (of this memoir)

Bernoulli however, having read Euler's essay on the vibrations of membranes[1], believes that the very plausible hypothesis, therein made use of, might be applied to the case of elastic plates. In the case of membranes no experiments have been made which would enable him to verify Euler's theory, but in the case of vibrating plates, since,

Les sons et les vibrations de ces sortes de corps sont un des principaux objets du livre de M. Chladny, & qu'ainsi il y a une ample provision d'expériences, qu'on pourra confronter avec les résultats de la théorie.

Bernoulli, to obtain an equation for the vibrations of a plate, assumes practically that a curved surface, such as that of a plate in vibration, may be considered as built up of an infinite number of curves of simple curvature. He divides his plate into two series of *annuli* (like Euler divides his bells) *at right angles* to each other, and considers the separate motions of these *annuli* to be the same as those of elastic rods. He then combines these two results to obtain an equation of the form

$$\frac{d^4z}{dx^4} + \frac{d^4z}{dy^4} = c^4\,\frac{d^2z}{dt^2},$$

or, as he writes it, $\qquad = \lambda z.$

It will be seen at once that his method leads him to an equation which wants the term $d^4z/dx^2dy^2$ and is therefore worthless.

In § 31 he proceeds to consider whether the results of his theory are in accordance with the experiments of Chladni, and finds naturally that, with a certain general resemblance, the results of experiment and theory do not accord in any detail. He leaves the question of the validity of his theory to the intelligent reader, who can judge with greater impartiality than the author. (§ 37.)

[123.] P. S. Girard. *Traité Analytique de la Résistance des solides, et des solides d'égale Résistance, Auquel on a joint une*

[1] Euler's paper appears in the *Novi Commentarii* of the St Petersburg Academy for the year 1766 (Tomus x). It is entitled *De motu vibratorio tympanorum*, pp. 243—260. A memoir by Giordano Riccati, entitled *Delle vibrazioni sonore del tamburo*, is contained in the first volume of the *Saggi scientifici dell' Accademia di Padova*. Padua, 1786. To Euler seems due the equation $\frac{d^2z}{dt^2} = e^2\,\frac{d^2z}{dx^2} + f^2\,\frac{d^2z}{dy^2}$ for the vibratory motion of a membrane.

*suite de nouvelles Expériences sur la force, et l'élasticité spécifique des Bois de Chêne et de Sapin.* Paris, 1798. (A German translation appeared at Giessen in 1803.) This work very fitly closes the labours of the 18th century. It is the first practical treatise on Elasticity[1]; and one of the first attempts to make searching experiments on the elastic properties of beams. It is not only valuable as containing the total knowledge of that day on the subject, but also by reason of an admirable historical introduction, which would materially have assisted the compilation of the present chapter had it reached the editor's hands in time. The work appears to have been begun in 1787 and portions of it presented to the Académie in 1792. Its final publication was delayed till the experiments on elastic bodies, the results of which are here tabulated, were concluded at Havre.

[124.] It is interesting to note an *Extrait du rapport fait à la classe des Sciences physiques et mathématiques de l'Institut national des Sciences et Arts.*

*Séance du* 11 *Ventose an* 6.

*Le citoyen Prony lit le rapport suivant:*

*Nous avons été chargés, le citoyen Coulomb et moi, de faire un rapport à la classe sur un ouvrage que lui a présenté le citoyen Girard, Ingénieur des Ponts et Chaussées, etc. etc.*

We are reminded even here that we are considering the period of the French Revolution. In the memoir referred to in Art. 115 Coulomb is 'Ingénieur du Roi.'

[125.] The book is divided into four sections with an introduction. The introduction is occupied with an historical retrospect

[1] The eighteenth century textbooks of Mechanics had usually a chapter on the strength of materials, but they can hardly be said to have been treatises on elasticity. A very popular English book, which may be taken as a sample of the type, was W. Emerson's, *Mechanics or the Doctrine of Motion.* It ran through several editions (2nd Edition, London, 1758); and is quoted with approval as late as 1806. Section VIII. (pp. 93—116) is entitled: 'The strength of beams of timber in all positions; and their stress by any weights acting upon them, or by any forces applied to them.' It labours under the old error as to the position of the neutral line, and discusses the old problems as to solids of equal resistance, etc. There is no sign of the author's acquaintance with the later work of Euler, and the only interesting part of the book is a quaint and characteristic preface.

of the work already accomplished in the field of elasticity. It is written in a clear style and is of great value. The introduction concludes with an analysis of Girard's own work. The original part of this seems to be confined to the fuller consideration of a rod built in at both ends, to a more detailed account than had before been given of the 'solids of equal resistance' and to numerous experiments on the resistance of beams of wood.

[126.] A remark made (p. lii) with regard to these experiments ought to be quoted:—

On a négligé dans la théorie de la résistance des corps la cohérence longitudinale de leurs fibres. Il est évident cependant que cette cohérence doit rendre leur inflexion plus difficile. Aussi avons-nous reconnu dès nos premières expériences que l'élasticité absolue qui, dans l'hypothèse des géomètres sur l'organisation des corps fibreux devrait dépendre uniquement des dimensions de leurs bases de fracture, dépendait encore de la longueur de leurs fibres intégrantes. En conséquence, nous avons recherché la fonction de cette longueur qui représente la cohérence longitudinale dans différentes espèces de bois afin d'en conclure leur *élasticité absolue spécifique.*

This remark shews us exactly how far the 18th century had got; it shews us also where the next steps must be made;—"la cohérence longitudinale" must be recognised in the equations which determine the equilibrium of an elastic body.

[127.] The first section of Girard's treatise is concerned with the resistance of solids according to the hypotheses of Galilei, Leibniz and Mariotte. He notes Bernoulli's objections to the Mariotte-Leibniz theory; but remarks that physicists and geometricians have accepted this theory:

Non-seulement à cause de sa simplicité, mais encore parce qu'elle s'accorde si heureusement avec les observations, et s'éloigne si peu de la vérité, que dans le cas même où la nature nous aurait revélé son secret sur la contexture des corps, elle a tellement multiplié les accidens dans ses productions que des connaissances certaines sur cette contexture ne nous conduiraient pas à des résultats plus avantageux à la pratique des arts que ceux auxquels on parvient à l'aide de cette supposition. (Introduction, p. xxix.)

[128.] At the same time he thinks it probable that Galilei's hypothesis of non-extension of the fibres may hold for some bodies—stones and minerals—while the Mariotte-Leibniz theory is true for sinews, wood and all vegetable matters (cf. p. 6). As to Bernoulli's doubt with regard to the position of the neutral surface, Girard accepts Bernoulli's statement that the position of the axis of equilibrium is indifferent and supposes accordingly that all the fibres extend themselves about the axis $AC$ (see our figure p. 7).

[129.] In this first section Varignon's method is used to obtain a result true for either hypothesis; the only point of interest is an ingenious arrangement of weights and pulleys by which both hypotheses can be represented (pp. 7—9).

[130.] §§ 15—34 discuss various problems on Galilei's hypothesis.

§§ 35—49 treat the same problems on the Mariotte-Leibniz theory.

In §§ 64—68 results are considered which are true on either supposition.

Then follows a discussion on elastic curves, beams and columns, drawn from the various memoirs of Euler and Lagrange. It contains nothing new and is often mere verbal reproduction (§§ 76—124). There is however a remark and footnote as to a supposed error of Euler (see Art. 75, and for the matter of fact of Lagrange too: see Art. 112), in measuring the *absolute elasticity* of a beam (in Euler's notation $Ek^2$) by the *fourth power* of the diameter of the beam supposed cylindrical and not by the *third power*, which James Bernoulli had employed and "celle que nous ont indiquée le raisonnement et l'expérience." Euler was undoubtedly right in his value for the case of a beam of circular or square section, and of these he appears to be treating (see Art. 121).

[131.] The second section of the work treats of solids of equal resistance, it is a reproduction and extension of the results of Galilei, Varignon and Parent, but does not really belong to the theory of elasticity.

The third section considers the long series of experiments undertaken by Girard at Havre on the resistance of oak and deal beams, with an account of his peculiar apparatus.

The fourth section is a not very satisfactory discussion of the oscillations which accompany the flexure of beams, together with some general remarks on subsidence.

[132.] The whole book forms at once a most characteristic picture of the state of mathematical knowledge on the subject of elasticity at the time and marks the arrival of an epoch when science was to free itself from the tendency to introduce theologico-metaphysical theory in the place of the physical axiom deduced from the results of organised experience.

[133.] *General summary.* As the general result of the work of mathematicians and physicists previous to 1800, we find that while a considerable number of particular problems had been solved by means of hypotheses more or less adapted to the individual case, there had as yet been no attempt to form general equations for the motion or equilibrium of an elastic solid. Of these problems the consideration of the elastic lamina by James Bernoulli, of the vibrating rod by Daniel Bernoulli and Euler, and of the equilibrium of springs and columns by Lagrange and Euler are the most important. The problem of a vibrating plate had been attempted, but with results which cannot be considered satisfactory. (See Articles 96 and 122.)

A semi-metaphysical hypothesis as to the nature of Elasticity was started by Descartes and extended by John Bernoulli and Euler. It is extremely unsatisfactory, but the attempt to found a valid dynamical theory by Jacopo Riccati did not lead to any more definite results.

[*Addenda.* References to the following academic dissertations reached me too late for any notices of them to appear in their proper places in this first chapter.

(*a*). *Dissertatio Physica de Corporibus Elasticis, quam .. publicae Eruditorum ventilationi submittit* Heinricus von Sanden, Anno MDCCIV, Regiomonti.

The author classifies the chief tactile properties of bodies under hardness, softness, and elasticity. He defines the latter thus:

Elasticorum denique corporum talis est structura, quod illorum partes tali modo sint locatae, si aliquantulum a se invicem fuerint separatae, distentae aut elongatae violenter ac praeter solitum fuerint, cessante violentiâ hac se reuniant aut in pristinam laxitatem sese restituant.

§ II is devoted to a philological discussion on the derivation of *elasticity* with references to Aristotle, Pliny, Galen, etc.

The sections immediately following divide elastic bodies into natural and artificial; under the latter are described cross-bows and children's spring-guns, etc. Francesco de Lama, Sturm and Mariotte are the chief authorities quoted. Then we find the elasticity of glass, membranes, musical strings, eiderdown, water etc., touched upon. For the compressibility of water experiments of Bacon[1], Boyle and the author himself are cited.

In § XIII—§ XIV the second part of elastic action, viz. restitution after compression, inflexion or extension is discussed. Sanden believes that the hypothesis of an æther is better calculated to explain this restitution than the hypothesis of repulsive and attractive forces inherent in the ultimate particles of matter. He thinks the Newtonian theory rather obscures than elucidates the subject and quotes Halley as of the same opinion. His theory is similar to that of Hooke, but his source seems to be the somewhat later work of Francesco de Lama. It is the extrusion and intrusion of the medium into the pores of bodies which produces the phenomena of elasticity. Sanden's theory however differs from that of Hooke and other writers I have met with, in supposing the medium to be neither air nor æther, but a mixture of the two:

Est itaque corporum elasticorum causa aer gravis, hujus verò aether, cui Deus primus motor immediatè impulsum impressit (p. 19).

The concluding sections apply this theory to particular examples in the usual vague descriptive method characteristic of the

[1] Bacon's experiment with a leaden sphere containing water is described in the *Novum Organum* L. II. Aph. XLV. and in the *Historia Densi et Rari*. (Spedding and Ellis IV. p. 209 and V. p. 395.) Bacon's views on compression will be found at the places indicated. Born three years before Galilei, he is the father of the physical, as Galilei of the mathematical school of elasticians.

period. The air is elastic owing to the presence of æther in it, and any other substance, for example water, is elastic owing to the presence of air in the interstices between its particles.

The dissertation has no scientific but undoubted historical interest.

($\beta$). S. Hannelius: *De causâ elasticitatis.* Aboae, 1746. I have been unable to find a copy of this. It is mentioned in Struve's catalogue of the Pulcova Library.

($\gamma$). *Dissertatio academica de corporum naturalium cohærentia.* Tubingae, 1752. This dissertation seems to have been submitted for public 'wrangling' by G. W. Krafft, Professor of Physics and Mathematics at that time in the University.

It contains nothing of the slightest original value. There is a superficial discussion of the work of Galilei, Mariotte, Leibniz, Bernoulli, Bülfinger and Musschenbroek. It concludes with a suggestion that the results of these writers—regardless of their discordance *inter se* and with experiments—should be applied somehow in the construction of bridges and roofs.] ED.

# CHAPTER II.

## MISCELLANEOUS INVESTIGATIONS BETWEEN THE YEARS 1800 AND 1822.

[134.] 1802. Thomas Young: *A Syllabus of a Course of Lectures on Natural and Experimental Philosophy.* London, 1802.

In Section XI. of the third part of this work, which is devoted to Physics, Young discusses some: *Of the general Properties of Matter.* Without making any very definite statements he would appear to ascribe the properties of cohesion and elasticity to the existence of the ultimate material particles in an ethereal medium, which in some fashion produces between them apparent repulsive and attractive forces. (p. 144.)

Sections XIX. and XX. of the first part of the book, entitled *Of passive strength* and *Of Architecture,* are also more or less concerned with our subject (pp. 39—46). We may note the following paragraphs:

> The strength of the materials employed in mechanics depends on the cohesive and repulsive forces of their particles. When a weight is suspended below a fixed point, the suspending substance is stretched, and retains its form by cohesion; when the weight is supported by a block or pillar placed below it, the block is compressed, and resists primarily by a repulsive force, but secondarily by the cohesion required to prevent the particles from sliding away laterally. When the strain is transverse both cohesion and repulsion are exerted in different parts of the substance (p. 39).

When a body is broken transversely by a force applied close to the place of fracture, it appears to bear more than twice the weight which might be suspended by it.

When a transverse force is applied to a bar at a distance from the place where it is fixed, the parts nearer the one surface resist by their cohesion, and the parts nearer the other surface by their repulsion; and these forces balance each other; but accordingly as the body is more easily extensible or compressible the depth to which each action extends will be different; in general, the neutral point is nearer the concave surface, the incompressibility being the greater (p. 40).

If each end of a beam be firmly fixed, instead of being barely supported, its relative strength will be doubled; for the flexure at each end now adds to the strength a force capable of supporting half as much as the whole weight, and the sum of these additional forces is equal to the whole weight (p. 41).

On the whole there is little to be learnt from these lecture-notes, except that Young supposed that there were different moduli for compression and extension, and that he was not very certain as to the position of the neutral axis. He states, I may add, that there is some difficulty in reconciling the results of various experiments as to the cohesion of materials.

[135.] In 1807 there appeared for the first time the lectures of which the above-mentioned work was the syllabus. They are entitled: *A Course of Lectures on Natural Philosophy and the Mechanical Arts.* This book is in two quarto volumes.

We may first note a very valuable classified list of works on Natural Philosophy and the Mechanical Arts on p. 87 of the second volume. Under the headings *Equilibrium of Elastic Bodies* (p. 136), *Passive Strength* (p. 168), *Columns and Walls: their strongest forms* (p. 173), and *Vibrations from Elasticity* (p. 268), Young gives a nearly complete bibliographical list of memoirs on the subject of elasticity written during the 18th century[1].

[1] The editor did not discover this list till revising Dr Todhunter's account of Young's work. He has endeavoured to supply some omissions discovered by its aid. The list ought to be consulted by all students of the history of mathematics. It is superior in many respects to *Reuss's Repertorium.*

136. On pages 46—50 of the second volume is a section (IX.) entitled: *Of the equilibrium and strength of elastic substances.* This section is reproduced in the *Miscellaneous Works* of Dr Young: it occupies pages 129—140 of the second volume, and the following note by the editor is placed on page 129:

This article has been reprinted in consequence of the originality and importance of some of the propositions which it contains. It was not included in the new edition of Dr Young's lectures which was edited by Professor Kelland.

[137.] Young commences the section by one or two definitions, of which the second relates to that constant of elasticity which has since been named *Young's modulus.* It runs thus:

The modulus of the elasticity of any substance is a column of the same substance, capable of producing a pressure on its base which is to the weight causing a certain degree of compression as the length of the substance is to the diminution of its length.

It will be seen that the modulus as thus defined by Young does not agree with what we now term Young's modulus. In fact the product of the latter and the area of the section of the beam is equal to the *weight* of the modulus adopted by Young[1].

[138.] The definition of the modulus is followed by a series of theorems which in some cases suffer under the old mistake as to the position of the neutral surface. Thus in § 321, when a force acts longitudinally on a beam Young places the neutral surface, or as he here terms it the point of indifference,' in the surface of the beam.

In § 332 Young defines the 'stiffness of a beam' and makes it directly as the breadth and as the cube of the depth, but inversely as the cube of the length. It will thus be seen that his 'stiffness' is not the same as that of Euler, Riccati or Girard.

139. The whole section seems to me very obscure like most of the writings of its distinguished author; among his vast attainments in sciences and languages that of expressing himself clearly in the

[1] That Young originally defined his modulus as a *volume* does not seem quite clearly brought out by Thomson and Tait, *Treatise on Natural Philosophy*, Part II. §§ 686 and 687. See also the Article Elasticity, *Encycl. Brit.* § 46.

ordinary dialect of mathematicians was unfortunately not included. The formulæ of the section were probably mainly new at the time of their appearance, but they were little likely to gain attention in consequence of the unattractive form in which they were presented. They relate to the position of the *neutral* line when a beam is acted on by a given force, to the amount of deflection, and to the form assumed by the beam; also to various points connected with the *strength* of beams. Saint Venant alludes to Dr Young's investigations in Liouville's *Journal de Mathématiques*, 1856, page 92:

...La simple flexion sans rupture, dont l'illustre physicien Th. Young avait en 1807, présenté les formules pour les cas les plus simples,...

and in a note he adds

Ces formules ont été reproduites par le Dr Robison, article *Strength of Materials* de l'*Encyclopédie Britannique*.

140. As a specimen of Dr Young's manner I give one of his Theorems, which can be conveniently extracted as it does not refer to any of the others[1].

Theorem. The force acting on any point of a uniform elastic rod, bent a little from the axis, varies as the second fluxion of the curvature, or as the fourth fluxion of the ordinate.

For if we consider the rod as composed of an infinite number of small inflexible pieces, united by elastic joints, the strain, produced by the elasticity of each joint must be considered as the cause of two effects, a force tending to press the joints towards its concave side, and a force half as great as this, urging the remoter extremities of the pieces in a contrary direction; for it is only by external pressures, applied so as to counteract these three forces, that the pieces can be held in equilibrium. Now when the force acting against the convex side of each joint is equal to the sum of the forces derived from the flexure of the two neighbouring joints the whole will remain in equilibrium; and this will be the case whether the curvature be equal throughout, or vary uniformly, since in either case the curvature at any point is equal to half the sum of the neighbouring curvatures; and it is only the difference of the curvature from this half sum, which is as the second fluxion of the curvature, that determines the accelerating force.

[1] [This appears as an addition on p. 83 of the first edition of the *Course of Lectures*.] ED.

The result which is in view is, I presume, that which forms equation ($f$) of Poisson's *Mécanique*, Vol. I. page 632; but I cannot regard Dr Young's process as offering any intelligible demonstration.

141. Whewell, in the *second* edition of his *Elementary Treatise on Mechanics*, published in 1824, made use of Dr Young's results. He says on page X. of his preface:

I have also added in Chap. X. some important and interesting theorems on the *Elasticity and Compression of Solid Materials*, partly adapted from Dr Young's *Elements of Natural Philosophy*. I would gladly have given a section on the strength and fracture of beams, had there been any mode of considering the subject, which combined simplicity with a correspondence to facts....

The section in the work occurs on pages 195—201; it is entitled *Elasticity and Resistance of Solid Materials*. In this section Whewell gives the essence of Dr Young's results, so far as they relate to the *equilibrium* of elastic substances; and by using ordinary mathematical language and processes he renders the investigations intelligible. But he does not reproduce those parts which relate to the *strength* of elastic substances; and it may be inferred, from the second of the sentences which I have quoted from his preface, that he was not satisfied with them.

[142.] In Lecture L. (Vol. I.) Young treats of *Cohesion* and on p. 628 of elasticity in particular. He writes:

The immediate resistance of a solid to extension or compression is most properly called its elasticity; although this term has sometimes been used to denote a facility of extension or compression, arising from the weakness of this resistance. A practical mode of estimating the force of elasticity has already been explained, and according to the simplest statement of the nature of cohesion and repulsion, the weight of the modulus of elasticity is the measure of the actual magnitude of each of these forces; and it follows that an additional pressure, equal to that of the modulus, would double the force of cohesion and require the particles to be reduced to half their distances in order that the repulsion might balance it; and in the same manner an extending force equal to the weight of half the modulus would reduce the force of cohesion to one half and extend the substance to twice its dimensions. But, if, as there is some reason to suppose the mutual repulsion of the particles of solids varies a little more rapidly than their distance, the modulus of elasticity will be a little greater than the true measure of the whole cohesive and repulsive force; this difference will not, however, affect the truth of our calculations respecting the properties of elastic bodies, founded on the magnitude of the modulus as already determined.

Young then proceeds to discuss stiffness and softness, and at the conclusion of his lecture remarks, that if an ultimate agent for cohesion is to be sought outside a fundamental property of matter, it might perhaps be found in a universal medium of great elasticity. All suppositions founded on analogy in this case (as for example that of the Magdeburg hemispheres) must however be considered as merely conjectural. 'Our knowledge of everything which relates to the intimate constitution of matter, partly from the intricacy of the subject, and partly for want of sufficient experiments, is at present in a state of great uncertainty and imperfection.'

[143.] Besides the semi-mathematical section on the strength of elastic substances (see our Art. 136) and this lecture on cohesion, Young has a separate lecture on *Passive Strength and Friction* (Lecture III. pp. 135—156 of Vol. I.). This contains a purely physical discussion of the elastic properties of bodies. The modulus of elasticity is defined, Hooke's Law is expressly assigned to its discoverer, and references are made to the labours of Coulomb and Musschenbroek. Young also states in ordinary language some of the results of the mathematical theory, for example on p. 139, where he reproduces Lagrange's results for a bent column.

The important subject of 'lateral adhesion' is also discussed, and Young notes that Coulomb makes it nearly equal to the direct cohesion of the same substance or a little greater, while Robison makes it twice as large (p. 146). Young appears to have been among the first, who laid marked stress on the distinction between these two elastic properties of a body. There is as yet however no sign of any attempt to introduce this conception into the mathematical theory of elasticity. Its omission in the consideration of the beam problem before and at this date, does not appear to have arisen from a consciousness that its effects were negligible, but rather from a dogmatic assumption that the curvature of a beam must vary as the moment of the bending force. When this assumption was replaced by considering the beam as made up of individual fibres suffering extension or compression, the question of their lateral adhesion does not seem to have at once occurred to the first investigators.

[144.] Besides the Course of Lectures on Natural Philosophy Young wrote a number of articles connected with the subject of Elasticity. Thus the articles *Bridge* and *Carpentry* were contributed by him to the *Encyclopaedia Britannica.* These articles are equally obscure, and, so far as I have been able to comprehend them, do not treat satisfactorily the position of the neutral line The term 'neutral line' is, I think, due to Young. It certainly owes its general adoption to his constant use of it.

[145.] Closely associated with these *Encyclopaedia* articles of Young are several articles contributed to the same work by Robison. They were written shortly before the death of the author in 1805, and are collected by Brewster in Vol. I. of his edition of Robison's papers entitled: *A System of Mechanical Philosophy.* Edinburgh, 1822. The articles in question are headed: *Strength of Materials* and *Carpentry* (pp. 369—551). They are followed in Brewster's edition by papers on the construction of roofs and arches, which are also connected with our subject.

Robison finds that very few experimental results of importance have yet been obtained. He considers Musschenbroek s investigations to have been made on poor material, and, with the exception of the experiments of Buffon on the strength of timber, holds that nothing can be found from which absolute measures might be obtained that could be employed with confidence. He quotes Hooke's Law and mentions Coulomb's experiments in confirmation of it. James Bernoulli's experiments, which seem to deviate from it, do not do so on closer examination (see Art. 22). Robison does not appear to have had any extended mathematical knowledge and his off-hand treatment of Euler (p. 406) is hardly satisfactory.

On p. 411 he returns to his criticism of Euler:

In the old Memoirs of the Academy of Petersburgh for 1778, there is a dissertation by Euler on the subject, but particularly limited to the strain on columns, in which the bending is taken into account. Mr Fuss has treated the same subject with relation to carpentry in a subsequent volume. But there is little in these papers besides a dry mathematical disquisition, proceeding on assumptions which (to speak favourably) are extremely gratuitous. The most important consequence of the compression is wholly overlooked, as we shall presently see. Our knowledge of the mechanism of cohesion is as yet far too imperfect to entitle us to a confident application of mathematics.

[146.] Such criticism is very idle in a writer, who on the very next page places the neutral line on the concave surface of a beam subjected to flexure! But what we have hitherto quoted is not a tithe of the strong language which Robison applies to Euler. He accuses him with using his great power of analysis regardless of physical truths and merely for the purposes of display:

We are thus severe in our observations, because his theory of the strength of columns is one of the strongest evidences of this wanton kind of proceeding and because his followers in the Academy of St Petersburgh, such as Mr Fuss, Lexell and others, adopt his conclusions, and merely echo his words. We are not a little surprised to see Mr Emerson, a considerable mathematician, and a man of very independent spirit, hastily adopting the same theory, of which we doubt not our readers will easily see the falsity (pp. 465—6).

Now Euler's theory of bent columns does not agree with experiment, and of this Robison was well aware. His consideration of Euler's error (pp. 466—7) is perfectly clear and fair; he shews, in fact, that the neutral line in the case of a bent column does not lie on the concave surface (see our footnote p. 44). But then *this is the very error he had himself fallen into* in the case of a beam subjected to flexure by a force perpendicular to its axis. It is Robison *himself* who perpetuated in the English text-books this error already corrected by Coulomb (see Art. 117). Besides this correction of Euler, I can find nothing of originality in this or any other of Robison's papers.

[147.] The following English and German text-books require a few words of notice.

1803. John Banks: *On the Power of Machines.* Kendal, 1803. This book contains a section entitled: *Experiments on the strength of oak, fir and cast-iron, with many observations respecting the form and dimensions of beams for steam engines etc.* (pp. 73—108). Its author describes himself as a 'lecturer on philosophy.' The section in question contains a set of practical rules for engineers. How far they were of value is doubtful, as the author still makes use of Galilei's hypothesis and also of the erroneous position of the neutral line.

[148.] 1806. Olinthus Gregory: *A Treatise of Mechanics Theoretical, Practical and Descriptive.* London, 1806. Book I. Chapter V. (p. 104) treats of the strength and stress of materials. This author reproduces the whole of Galilei's results, apparently only for the reason 'that they are comparatively simple,'—and this notwithstanding that a century previously they had been recognised as erroneous! Nothing shews more clearly the depth to which English mechanical knowledge had sunk at the commencement of this century. The same section is reproduced in the edition of 1815.

[149.] 1808. Eytelwein: *Handbuch der Statik fester Körper.* Berlin, 1808. The fifteenth chapter of this book (Bd. II. pp. 233—424) is entitled: *Von der Festigkeit der Materialien,* and concludes with a bibliography of the more important works on elasticity published antecedent to 1808. The first section of the chapter, headed: *Von der absoluten Festigkeit,* is principally occupied with experimental results drawn from Musschenbroek, Girard and others.

In the second section of the chapter, headed: *Von der respectiven Festigkeit,* we have (pp. 275—278) a correct consideration of the position of the neutral line for a beam subjected to flexure by a force perpendicular to its axis. There is nothing original in this section, but its author possesses the advantage over Banks and Gregory of being abreast of the mathematical knowledge of his day. In this second section also (p. 302) Eytelwein makes a true statement of the problem of a beam subjected to flexure, but finds the difficulties of analysis too troublesome and ultimately relapses into the old theory which rejected the compression.

Weil alle Körper vor dem Brechen, wenn auch noch so wenig, ausgedehnt werden, so kann hier nicht die Rede davon seyn, die allgemeineren Untersuchungen über die respective Festigkeit auf die Galileische Hypothese zu gründen, sondern da jede Materie als ausdehnbar und compressibel anzunehmen ist, so müssen diese Untersuchungen auch auf diese beiden Eigenschaften der Körper ausgedehnt werden. Allein die nachstehende Auseinandersetzung zeigte, wie weitlauftig die Rechnung wird, wenn man zugleich auf Compressibilität Rücksicht nimmt, und weil diese Eigenschaft denjenigen Körpern, welche hier untersucht werden, nur sehr wenig zukommt, so wird man solche um so mehr bei Seite setzen können.

The results of the second and third sections on prismatic beams and beams of varying sections are in consequence far from satisfactory. The fourth and concluding section of the chapter treats of experiments on the relative strength of beams; Musschenbroek and Girard are the author's chief authorities.

[150.] It may be noted that the subject of compressibility in beams has not even in the present day been reduced to accurate mathematical treatment. Notwithstanding Saint-Venant's classical memoir on the flexure of prisms, the application of the general equations to special cases of the bending of beams under longitudinal pressure still presents difficulties which have not been generally surmounted.

151. 1809. Plana. A memoir by Plana is published in Tom. 18 of the Turin Memoirs for 1809—1810; and is entitled: *Équation de la courbe formée par une lame élastique, quelles que soient les forces qui agissent sur la lame.* The memoir occupies pages 123—155 of the mathematical part of the volume. It was read on the 25th of November, 1809.

152. The memoir, which is confined to the case in which the elastic curve and the forces acting on it are all in one plane, is divided into two parts. The first part occupies pages 123—136; here the investigation of the differential equation to the curve is given, the result being the same as had already been obtained by Euler and Lagrange. Plana says on his page 125:

Les moyens que ces deux auteurs emploient pour parvenir à cette équation ne m'ont pas paru doués de toute la clarté et la simplicité qu'on pourrait souhaiter, et c'est dans l'intention de la démontrer, en suivant une marche précise et naturelle, que je n'ai pas cru inutile d offrir ce mémoire à l'Académie, quoiqu il ait pour but de déterminer une équation déjà connue par les géomètres.

Plana's investigation is simple and intelligible; he states distinctly what forces he supposes to be acting, and uses only common mechanical principles.

153. The second part of the memoir occupies pages 137—155; this is entitled: *Application de la théorie précédente à un cas particulier traité par Lagrange dans les mémoires de l'Académie de Berlin* (année 1769). The memoir by Lagrange to which this refers is the one we have considered in Art. 97. Plana's process amounts to the integration of the differential equation

$$K/\rho = Tx + Ry,$$

where $\rho$ is the radius of curvature at the point $(x, y)$ and $K$, $T$, $R$ are constants. Plana begins by finding an exact integral of the first order, after which he proceeds through a long course of approximations until finally he obtains a result which disagrees with one given by Lagrange in the memoir cited. Plana indicates the point where an error occurs in Lagrange's process.

154. In his approximations Plana follows closely the path traced out by Lagrange, so that in fact the memoir offers nothing new except the correction of Lagrange's mistake. Nevertheless in the new edition of the works of Lagrange published by the French government the mistake remains uncorrected (see pages 85—90 of Vol. III. 1869). It is obvious that there is a mistake, for on the fourth line of page 88 there is an equation in which the left-hand side consists of

$$\frac{r}{l}\left\{1 - \frac{(m-\alpha)^2}{2}\right\};$$

this is less than unity, for $r$ is the chord of an arc $l$; but the right-hand side will be found to be greater than unity in the case under investigation. Afterwards $r/l$ is by mistake changed into $l/r$. The error first enters on page 85, at the sixth line from the foot; here $P^3$ is practically taken as equal to the product of $P^2$ into $+\sqrt{P^2}$, which as $P$ is a negative quantity is incorrect.

155. 1811. Delanges: *Analisi e soluzione sperimentale del problema delle Pressioni.* This occurs in the *Memorie della Società Italiana:* it occupies pages 114—154 of the first part of Vol. XV. published at Verona in 1811. The memoir was received on the 8th of March, 1810.

I allude to this memoir as the title might suggest some relation to our subject, with which however it is totally unconnected. If a heavy body is supported at *three* points in a horizontal plane, but not in the same straight line, the pressure at each point can be found by ordinary statical principles. But if there are *more* than three points the problem becomes indeterminate; works on Statics generally notice this matter. Delanges in fact renders the problem determinate by an arbitrary hypothesis. It will be seen from the *Catalogue of Scientific Papers* published by the Royal Society that Delanges wrote other memoirs on the subject; in the present which is the last he wrote, he alludes to these and to the discussion and controversy to which they had given rise.

156. 1811. The first edition of Lagrange's *Mécanique Analytique* was published in 1788; the first volume of the second edition appeared in 1811, and the second volume in 1815 after the death of Lagrange I shall use the reprint of the work which was edited by Bertrand in 1853.

157. There is nothing to be found here which belongs strictly speaking to our subject; but some of the problems which really form part of the theory of elasticity are here treated by special methods. The pages 128—151 of the first volume contain all that it is necessary to notice; they form three sections of a chapter entitled: *De l'équilibre d'un fil dont tous les points sont tirés par des forces quelconques, et qui est supposé flexible ou inflexible, ou élastique, et en même temps extensible ou non.* The method by which the problems are treated is that for which Lagrange's work is famous, and which amounts to rendering Mechanics a series of inferences drawn from the principle of Virtual Velocities by the aid of the Calculus of Variations.

158. The first problem discussed is the equilibrium of a

string (pages 129—138); the results coincide with those obtained in ordinary works on Statics. Bertrand notices on page 135 that Lagrange considers a part of this problem to be difficult by ordinary methods while it is really easy.

The second problem discussed is the equilibrium of a membrane, called a flexible surface (pages 139—143). The principal result obtained is that, when gravity is the only force and the membrane deviates but little from a horizontal plane, its form is determined by the equation

$$a\left(\frac{d^2z}{dx^2}+\frac{d^2z}{dy^2}\right)=g,$$

where $a$ is a constant. This agrees with later investigations; but the process of Lagrange is unsatisfactory, as is acknowledged by Bertrand, supported by the authority of Poisson, in a good note on page 140.

159. The third problem discussed is the equilibrium of an elastic wire or lamina (pages 143—151). The process is unsatisfactory, as is acknowledged by Bertrand in a note (page 143). In fact one difficulty is that two of Lagrange's indeterminate multipliers are found to be infinite. Here, as in the discussion of the second problem, Lagrange assumes without adequate explanation that the internal force called into action is of a certain kind; and in the present problem it has been found that Lagrange has omitted a part of the force. The error has been corrected by Binet and by Poisson, as we shall see hereafter. Bertrand supplies the necessary correction in a note; and he shews after Binet, that an integration can be effected which Lagrange says *est peut-être impossible en général* (pages 401—405). I do not see why the limitation *la courbe étant primitivement droite* is introduced on page 403. Yet this is involved in Lagrange's page 148; the curvature is supposed to arise entirely from the action of the forces.

160. The discussions of the first and the third problems in the second edition are reproduced, with some additions, from the first edition; the second problem was not discussed in the first edition.

161. The notes of Bertrand throw no light on the error attributed by some writers to Poisson. Thus on page 402 we have the *moment of torsion* held to be constant at this date (1853), while in the *Dublin Calendar* for 1846 this is contradicted[1].

[1812. A memoir by Poisson belonging to this year is noticed in the chapter devoted to his works.]

[162.] 1813. C. Dupin: *Expériences sur la flexibilité, la force et l'élasticité des Bois, avec des applications aux constructions en général, et spécialement à la construction des vaisseaux.* This memoir occurs on pp. 137—211 of the *Journal de l'École Royale Polytechnique,* Tom. x. Paris, 1815. The experiments were made at the arsenal of Corcyre in 1811, and are commended with those of Duleau by Saint-Venant in his memoir on the flexure of beams. They are concerned only with wood, while Duleau's are on iron (see below Art. 226). There is nothing of mathematical value to note in the paper.

163. 1814. G. Belli: *Osservazioni sull' attrazione molecolare* This memoir is published in the *Giornale di Fisica, Chimica ec. di L. V. Brugnatelli,* Vol. VII., Pavia, 1814; it occupies pages 110—126, and 169—202 of the volume.

164. According to Belli two opinions have been held as to molecular force; Newton and his followers maintained that molecular force was quite distinct from the force of gravity or universal attraction; while more recently Laplace had suggested that it was only a modification of gravity. The memoir begins thus:

L' attrazione molecolare, quantunque dall' immortale Newton e dai posteriori Fisici si fosse creduta distinta dalla gravitazione o attrazione universale, nondimeno, in questi ultimi anni dal celebre Sig. Laplace venne sospettata non esserne che una semplice modificazione. Fra queste due opinioni per discernere la vera, e per ispargere qualche debole lume su questa oscura ed importante materia, io ho instituite le presenti ricerche.

165. The memoir opposes the suggestion attributed here to Laplace. Calculations are made on the supposition of the ordinary law of gravity in order to shew that the forces according

[1] Bertrand seems to imply that Binet and Poisson join in correcting Lagrange but Saint-Venant, *Comptes Rendus*, XVII. 953, seems to imply that Poisson went wrong though cautioned by Binet. See our account of Saint-Venant.

to this law are too feeble to produce the results obtained by observation and experiment. The problem respecting the form of the solid of greatest attraction which I called *Silvabelle s Problem* in my *History of the Theories of Attraction* is very well discussed on pages 112—115, and numerical results are obtained respecting it. Other numerical results are also worked out. Thus, suppose a sphere to attract a particle at its surface; let the sphere be changed into a circular cylinder of equal volume, the height of which is equal to the diameter of the sphere, and let the attracted particle be at the centre of the base of the cylinder: then the attraction of the cylinder is about $\frac{63}{64}$ of that of the sphere. If the height of the cylinder instead of being equal to the diameter of the sphere is equal to the product of the radius of the sphere into $\sqrt[3]{4\pi/3}$, then the attraction of the cylinder is about $\frac{166}{165}$ of that of the sphere.

166. The argument on which Belli mainly relies is furnished by the fact that a drop of water will remain in equilibrium hanging from a horizontal surface. Suppose the molecular action of the contents of the drop to follow the law of gravitation, and for a rough approximation take the drop as spherical. The attraction of the drop on its lowest particle must be strong enough to overcome the attraction of the whole earth. Let $r$ be the radius of the drop, $R$ that of the earth; let $\rho$ be the density of the water assumed to be without pores, and $\sigma$ the mean density of the earth: then, according to the ordinary expression for the attraction of a sphere on a particle at its surface we must have $r\rho$ greater than $R\sigma$, and therefore $\rho$ greater than $R\sigma/r$. Suppose $r$ to be a millimetre; then as $R$ is greater than 12000000 metres we must have $\rho$ greater than $12000000000\sigma$. This is altogether inadmissible. Belli says that the density of the earth is really 8 or 9 times that of water; but this is an exaggeration, even according to the state of knowledge at the epoch, for the Schehallien observations and Cavendish's experiment had been made and discussed (see the *History of the Theories of Attraction*, Arts. 730 and 1015).

167. The first part of the memoir concludes on page 126 with the opinion that the true Law of molecular action is yet

unknown, and the writer proposes to attempt to draw from the phenomena some indications respecting it. His words are:

Da tutte queste cose mi sembra che la vera legge dell' attrazione molecolare non sia stata finora interamente conosciuta. Percio io non credetti inopportuno il tentare se non di scoprire qual sia questa legge, almeno di cavare dai fenomeni tutte quelle notizie, che intorno a lei essi ci possono dare.

168. Accordingly in the second part of his memoir Belli examines some phenomena in detail; those of cohesion he treats on his pages 172—188. Suppose we take two equal thin circular plates of metal of the same substance; experiment shews that the attraction between them is insensible so long as the distance between them is sensible, but becomes very great when they are in apparent contact; also it is *independent* of the thickness. Belli calculates the whole attraction between the plates for the cases in which the attraction between particles varies inversely as the second, third, and fourth powers of the distance respectively; he finds that not one of these suppositions will agree with the facts; they do not make the result large enough at apparent contact, and they all assign a very great influence to the thickness. Belli also gives general formulæ for the case in which the attraction between two particles varies inversely as the $n$th power of the distance; he comes to the conclusion that from the facts of cohesion we learn only this, that $n$ must be greater than 4.

169. A note on pages 182 and 183 deserves notice; it resembles, as Belli says, the Prop. 87, Lib. I. of Newton's *Principia*. A single example will be sufficient to illustrate it: Suppose that the attraction between two particles varies inversely as the $n$th power of the distance, we require the limit of $n$ in order that the resultant attraction between two spheres in contact may be *finite*. Imagine the spheres to remain in contact but each sphere to be enlarged in linear dimensions in the ratio of $p$ to 1; then each particle becomes enlarged in the ratio of $p^3$ to 1, so that on this account the resultant attraction becomes enlarged in the ratio of $p^6$ to 1: on the other hand as the distance between any two particles is enlarged the attraction between them is diminished, and becomes $1/p^n$ of what it was originally. Thus on

the whole by this enlargement of dimensions the resultant attraction becomes $p^{6-n}$ times what it was originally. As long as $n$ is less than 6 the resultant attraction then is increased, but when $n=6$ it remains unchanged; thus in this case the additional matter obtained does not augment the attraction of the old: the conclusion is that if $n=6$ the attraction of the spheres in contact must be infinite, and *a fortiori* if $n$ be greater than 6.

170. On pages 188–191 Belli considers the refraction of light on the corpuscular theory, and draws the same inference as from the facts of cohesion, namely that $n$ must be greater than 4 (see Art. 168).

171. On pages 191—202 Belli considers capillary attraction; from the comparison of calculation with observation he draws the conclusion that $n$ must be greater than 5: this extends, without contradicting, the results he had obtained from the consideration of cohesion and refraction. He then says:

Non ci palesa però nemmeno essa quale sia la vera legge se sia quella delle seste potenze o un' altra ancora più rapida. Ma noi ci fermeremo qui, "arrestando la Teorìa ove si ferma l' osservazione," ed aspettando che nuovi fenomeni si manifestino, i quali ci guidino più avanti per questa scabiosa via.

172. All the conclusions obtained by Belli rest, as he allows, on the hypothesis that the law of attraction between particles can be expressed by some *single* power of the distance. The whole memoir forms an interesting exercise on the ordinary theory of attraction though it cannot be considered as an important contribution to molecular mechanics

[1] The following slips may be noted:

Page 112. He does not define $p$; it must be the density.

113. For $\pi$ in the denominator of a fraction read $\sqrt{\pi}$.

115. 'Ad una sfera:' he should say that if the figure is changed *from* a sphere to that of the solid of greatest attraction the increase is only $\frac{28}{1000}$.

123. For cos $KA$ read cos $KAO$.

176. For $AB/AP$ read $AC/AP$.

189. Note line 4; for $x^2$ read $v^2$.

192. In the fig. 12 there is no $hl$; but $KC$ apparently by mistake.

193. In the fig. 14 supply letters $P$, $Q$, $E$ vertically below $C$.

199. 'nella ragione di 1 a $1/R$': he should say, diminishes in the same ratio as $1/R$.

173. 1814. J. Binet. A memoir by this writer is published in the *Journal de l'École Royale Polytechnique*, Vol. x. 1815; it is entitled *Mémoire sur l'expression analytique de l'élasticité et de la raideur des courbes à double courbure:* it occupies pages 418—456 of the volume. It was written in the year preceding publication.

174. As we have seen in Art. 159, Lagrange had treated imperfectly the problem of the equilibrium of an elastic curve of double curvature; in the present memoir the attention of mathematicians was drawn for the first time to this imperfection. The memoir may be conveniently divided into three parts. The first part occupies pages 418—428; this belongs to pure mathematics, and investigates various formulae which now find a place in works on Geometry of Three Dimensions, as for instance an expression for the radius of torsion at any point of a curve of double curvature. The second part occupies pages 428—443; this relates to the equilibrium of forces acting on polygons the sides of which are rigid straight rods. The third part occupies pages 443—456; this applies the results of the former sections to the case of a curve of double curvature.

175. The memoir does not seem to me of much direct service; the author's design apparently was to comment on Lagrange's treatment of the subject, rather than to supply the best independent investigation. He notices the fact that in Lagrange's own process ten of the multipliers are infinite and of different orders: see the note on page 153 of Vol. I. of Bertrand's edition of the *Mécanique Analytique*. The language in which mechanical principles are stated seems to me deficient in precision; and the problems discussed in the second part are not very clearly enunciated. I will give an example of each of these points. On page 419 we read:

> Qu'on se figure, par exemple, un fil métallique plie en forme d'hélice comme le sont les ressorts appelés 'ressorts a boudins'. Si une force agit de maniere à rapprocher ou à éloigner les deux extrémités de ce ressort, on voit assez que le changement de forme qu'il éprouvera aura lieu surtout aux dépens de la torsion du fil métallique.

Here the last few words are too metaphorical. On pages 428 and 429 the first mechanical problem discussed begins thus:

Maintenant prenons un système de trois points 1, 2, 3, auxquels sont appliquées trois forces $P$, $P'$, $P''$, dont nous désignerons les composantes par $X$, $Y$, $Z$, $X'$, $Y'$, &c., dans le sens des trois axes coordonnés. Nous supposons que les verges $a$, $a'$, qui séparent ces points, soient susceptibles d'extension ou de contraction, quoique restant inflexibles et droites; que les forces longitudinales de ces verges, lorsque l'équilibre sera établi, soient $A$, $A'$; et quelle que soit la réaction du troisième point sur le premier, nous la supposerons remplacée par une force intérieure $B$, agissant dans la direction d'une droite $b$ qui joint le point 1 au point 3′, que l'on trouve en prolongeant de toute sa longueur le côté inflexible $a'$.

Here in the first place the body on which the forces act is not well defined; and the part which begins *et quelle que soit,* is very obscure, though as we see afterwards it is the essential part of the process. Binet himself admits that the way in which he supposes his rods connected is not very natural (pp. 420 and 435: see my account of a memoir of Bordoni, Art. 216).

It is remarked by Thomson and Tait in Art. 608 of their *Natural Philosophy* (2nd Ed.):

The fundamental principle that spiral springs act chiefly by torsion seems to have been first discovered by Binet in 1814.

They cite Saint-Venant in the *Comptes Rendus* for September, 1864.

176. 1815. Plana. In the *Journal de l'École Royale Polytechnique,* Vol. x. 1815, we have a *Mémoire sur les oscillations des lames élastiques; par M. Plana ancien élève de l'École polytechnique, et Professeur d'Astronomie à l'université de Turin.* The memoir occupies pages 349—395 of the volume. After a brief introduction on pages 349—351 the memoir is divided into three parts.

177. The first part is entitled *Équations générales du Mouvement d'un Fil ou d'une Lame inextensible et élastique;* it occupies pages 351—360. The method of investigating these equations is that of Lagrange which cannot be considered quite satisfactory: see Art. 159.

178. The second part of the memoir is entitled *Intégration de l'Équation du Mouvement d'une Lame élastique considérée comme*

*non pesante:* it occupies pages 360—391. The differential equation to be integrated is

$$\frac{d^2y}{dt^2} + a^2 \frac{d^4y}{dx^4} = 0 \dots\dots\dots\dots\dots\dots\dots (\delta).$$

By assuming for $y$ a series in ascending whole powers of $t$, where the coefficients are functions of $x$, an integral is obtained involving the arbitrary functions of $x$. Plana however wishes to have the integral in a *finite form,* and this is the essential design of his memoir; accordingly he transforms the two infinite series into double integrals. The process is very laborious, and is unfortunately damaged by important mistakes which are acknowledged on pages 633 and 634 of the volume: I have not verified the whole. The principal point in the investigation is the employment of *Parseval's Theorem:* see Lacroix *Traité du Calcul Différentiel et du Calcul Intégral* Vol. III. 1819, pages 393—395; or Boole's *Differential Equations,* Chapter XVIII.

The value of $y$ obtained by Plana is excessively complex; by using abbreviations he brings it within the compass of sixteen long lines of a quarto page, but if expressed at full it would occupy double the space. He adds somewhat rashly on his page 384:

Telle est, si je ne me trompe, la forme la plus simple dont est susceptible l'expression générale de y, qui satisfait à l équation (δ).

Poisson in the Memoirs of the *Institut* for 1818 puts the value in the following simple form

$$y\sqrt{\pi} = \int_{-\infty}^{+\infty} \cos\left(\frac{\pi}{4} - \zeta^2\right) f\left(x - 2\zeta\sqrt{at}\right) d\zeta$$
$$+ \int_0^t dt \int_{-\infty}^{+\infty} \cos\left(\frac{\pi}{4} - \zeta^2\right) F\left(x - 2\zeta\sqrt{at}\right) d\zeta.$$

179. Plana passes on to the more general differential equation

$$\frac{d^2y}{dt^2} + a^2 \frac{d^4y}{dx^4} = b^2 \frac{d^2y}{dx^2};$$

he proceeds a certain way and then says, on his page 391:

Il ne reste plus maintenant qu'à chercher la partie réelle contenue dans le produit...... ; mais nous n'entreprenons pas d'achever ce calcul, parce qu'il nous paraît que le résultat final doit être excessivement compliqué.

180. The third part of the memoir is entitled *Intégration de l'Équation du Mouvement d'une Lame élastique pesante:* it occupies pages 391—395. The differential equation to be considered now is

$$\frac{d^2y}{dt^2} + a^2\frac{d^4x}{dx^4} = b^2\frac{d^2y}{dx^2} + g\frac{dy}{dx}$$

Plana proceeds to a certain distance in the investigation, and then concludes his memoir thus:

Nous n'entreprendrons pas de pousser plus loin ces recherches qui se compliquent toujours de plus en plus; et nous nous contenterons d'avoir indiqué par-là la possibilité d'exprimer, par des intégrales définies, l'intégrale de l'équation aux différences partielles du quatrième ordre, qui renferme la loi des oscillations d'une lame élastique pesante.

[1816. An article by Poisson belonging to this year is noticed in the chapter devoted to him.]

[181.] 1816. J. B. Biot: *Traité de Physique Expérimentale et Mathématique.* Paris, 1816. This well known work contains a chapter: *De l'Élasticité* (Tom. I. Chap. XXIII. pp. 466—528) which treats of our subject.

[182.] Biot attributes the elastic properties of bodies to their molecular construction and to forces between these molecules.

Nous avons montré les corps comme des assemblages de molécules matérielles extrêmement petites, maintenues en équilibre entre deux forces, savoir une affinité mutuelle, qui tend à les réunir, et un principe répulsif, qui est probablement le même que celui de la chaleur, et qui tend à les écarter. Quoique ces molécules soient si petites que nous ne puissions absolument pas observer leurs formes, nous avons cependant découvert qu'étant placées à de certaines distances les unes des autres, elles exercent des attractions diverses selon les côtés par lesquels elles se présentent.

He thus adopts, what we may term, the Newtonian hypothesis, and the usual description of the difference in construction between gaseous, liquid, and solid bodies follows. A general explanation of elasticity, of the limit of elasticity or permanent distortion, as well as of the elasticity of crystals is then easily deduced from the above molecular hypothesis.

[183.] Biot first considers the elasticity of threads or fibres citing the experiments of 's Gravesande (*que l'on sait avoir été un physicien fort exact*), which lead him to the law of extension. Coulomb's experiments on torsion are then discussed at considerable length. There follows a slight discussion of the action of elastic laminas and further somewhat vague statements as to the molecular condition, and consequent elasticity of bodies acted upon by heat. The chapter concludes with a description of the torsion-balance and the theory of the rotatory oscillations of bodies supported by twisted elastic wires.

[184.] In the portion of his work devoted to sound Biot reproduces some of the known results for the vibrations of elastic bodies. Thus in Chap. IV. Tom. II (p. 72) we have the transverse vibrations of elastic rods with the statement of the six cases of varied terminal conditions (see Art. 88). The following chapter is devoted to the longitudinal vibrations of rods; the sixth to the torsional vibrations (*vibrations circulaires des verges droites*) the seventh to the vibrations of curved rods; the eighth to various forms of vibration in solid bodies including plates. There is little if any mathematical theory and the experimental results are chiefly drawn from Chladni.

The various editions of Biot's *Précis élémentaire de Physique Expérimentale* (1st in 1817, 3rd in 1824) present only an abridgment of the *Traité.*

[185.] 1817. George Rennie: *Account of experiments made on the strength of materials. Phil. Trans.* 1818, p. 118.

This communication to the Royal Society is in the form of a letter to Dr Young.

The author remarks on the somewhat vague and contradictory character of the results obtained by different experimenters. He refers to the work of Emerson, Robison, Banks, Anderson, Beaufoy and Reynolds in this country; of Musschenbroek in Holland and of Buffon, Rondelet, Gauthey, Navier, Aubry and Texier de Norbeck, together with Prony's researches at the École Polytechnique, in France. He remarks also on the discordance between the Euler-Lagrange theory of columns and actual experiment. The first recognition of this discordance he attributes to Coulomb.

[186.] The experiments commence with a set upon the crushing of metal blocks, in particular of cast iron. Rennie finds that vertical cube castings are stronger than horizontal cube castings. The second set of experiments are on the cohesion of metal bars subject to longitudinal tension. In this case the fracture of *cast* bars was attended with very little diminution of section, one, in fact, scarcely perceptible. Then follows a set of experiments on the twist of different materials and another set on the crushing of wood and stone. It would appear that 'hardness is not altogether a characteristic of strength, inasmuch as the limestones, which yield readily to the scratch, have nevertheless a repulsive power approaching to granite itself'.

[187.] The paper concludes with experiments on the transverse strain of cast bars. The paradoxical experiment of Emerson was tried; namely, by cutting off a portion of an equilateral triangle (Emerson's *Mechanics,* p. 114) a bar of this section is made stronger than before, that is, a part stronger than the whole. The experiment was confirmed. In this case the bar was supported at both ends and the weight applied at the centre, the base being uppermost in both cases[1]

In a postscript there is a reference to the then recent experiments of Barlow.

[188.] 1817. Peter Barlow: *An Essay on the strength and stress of timber.* The first edition of this work appeared in 1817, but I have been unable to find a copy of it either in the British Museum or in the Graves Library at University College. My remarks will therefore be based upon the third edition of 1826, the earliest which I have been able to examine.

[189.] Part I. contains an historical sketch of former experiments and theories, which is fairly clear though in the latter part it seems to be rather based upon Girard than on an exhaustive study of the original memoirs. With regard to the labours of Euler and Lagrange, Barlow is of opinion that their instruments of analysis

[1] I have referred to these, as well as to other experimental results, because they state physical facts which seem to me quite unheeded by mathematical elasticians.

have been too delicate to operate successfully upon the materials to which they have been applied; so that while they exhibit, under the strongest point of view, the immense resources of analysis, and the transcendent talents of their authors, they unfortunately furnish but little, very little, useful information. (p. 64.)

The real reason why Euler's and Lagrange's results are not satisfactory is that they integrate the equation for the equilibrium of the beam on the supposition that the 'absolute elasticity' ($Ek^2$ in Euler's notation: see Art. 65) remains constant for all sections of a beam under longitudinal stress, with the additional assumption that it is independent of the length of the beam. The first point, which is intimately associated with the position of the neutral line, Barlow does not seem to have clearly recognised; the second point is considered on p. 75, but is utterly obscured, because the author has not observed that the 'absolute elasticity' for a beam under transverse stress differs from that for a beam under longitudinal stress. This of course follows from the fact that the neutral line is not in the latter case coincident with the line of centres.' This error of Euler's seems nowhere to have been clearly explained, although Coulomb had long since pointed out the true method of finding the neutral line and Robison had indulged in the strongest invective against both Euler's method and its results.

[190.] Barlow also mentions in this historical introduction the discrepancy between Girard's theoretical results and those of experiment in the case of beams subjected to transverse stress. This discrepancy of course arose from Girard's misplacement of the neutral line Barlow however seems to think that Leibniz's erroneous placing of the neutral line in the extreme fibre on the concave side was the accepted view of the theorists in his own day. Robison's, he writes, 'is the only theory of any importance in which the position of the neutral axis, or that line in a beam which suffers neither extension nor compression is introduced as a necessary datum.' Considering that Bülfinger, Riccati and Coulomb had all given this line its true position in the case of transverse strain, Barlow's claim to any merit of discovery in this matter may be at once dismissed.

[191.] In Part III. *On the Deflection of Beams* an obvious error attributed to Girard is corrected; that the deflection varies as the cube and not as the square of the length in the case of a beam subject to transverse strain is of course an obvious result of Bernoulli's theory of the elastic curve[1]

[192.] On p. 164 (§ 117) the writer commences his discussion as to the position of the neutral axis. Although he corrects the misplacement adopted by Leibniz and Girard he falls into an equally serious error himself. He does not make use of the clear principle stated by Coulomb, but assumes that the sum of the moments of the tensions of the extended fibres about the neutral point of any section must be equal to the sum of the moments of the compressed fibres. This is the same erroneous principle as is made use of by Duleau (see Art. 227). There is a footnote to the third edition (p. 167) stating that the above principle is objected to by Hodgkinson (see Art. 233), but the author still thinks his own results correct. He, however, finally accepts Coulomb's principle (or as he terms it Hodgkinson's) in the edition of 1837 (Entitled: *A Treatise on the Strength of Timber and Cast Iron, Malleable Iron and other Materials*, p. 63). The most curious part of the earlier statement however is, that Barlow is led by means of it to deduce from his experimental results that Galilei's theory is more correct than that of Leibniz; he even goes so far as to assert (p. 26), that Leibniz 'rather retrograded than advanced the science.' The discrepancy between the results of Leibniz's theory and his own experiment was of course due to the misplacement of the neutral line, which Barlow had found from experiment practically coincided in the case of transverse stress with the line of sectional centres.

[193.] There is however in this work, as in all the 'practical' English text-books of the period, no clear recognition of what the errors of Leibniz and Euler exactly consist in, although the writers are conscious that the results of these mathematicians do not accord with experience.

[1] Barlow does not state exactly where Girard has made this error. Girard gives the *correct* value of the deflection on p. 125 of his *Traité* and cites that value *correctly* at the end of Table II. of his experimental results.

[194.] In the edition of 1837 (pp. 62, 63), Barlow seems still ignorant that Coulomb had fifty years before correctly considered the position of the neutral axis, and attributed the first determination of the correct position to *his own* experimental researches, and the first accurate theoretical calculation to Hodgkinson in his Manchester memoir!

Of the value of Barlow's experimental results this is not the place to judge; as a theorist he is another striking example of that want of clear thinking, of scientific accuracy, and of knowledge of the work accomplished abroad, which renders the perusal of the English text-books on practical mechanics published in the first half of this century, such a dispiriting, if not hopeless, task to the historian of theory.

[195.] Two short papers, mainly experimental, by Thomas Tredgold in Tillock's Philosophical Magazine, Vol. LI. 1818, entitled: *On the transverse Strength and Resilience of Timber* (p. 214) and *On the Resilience of materials, with Experiments* (p. 276) present nothing of interest or of original value.

[196.] *The Elementary Principles of Carpentry.* London, 1820, by the same author, contains in Section II:—*On the Resistance of Timber, or the Stability of Resistance* (pp. 25—60) some consideration of the flexure and compression of beams. The experimental results are principally drawn from Rumford, Musschenbroek and Girard;—the theory contains nothing original and is often completely confused. Thus (p. 48) the author finds from experience that there is a certain force which will just bend a piece of timber when acting in the direction of its length. He continues:

The strain will be directly as the weight or pressure; and inversely as the strength, which is inversely as the cube of the diameter. The strain will also be directly as the deflexion, which will be directly as the quantity of angular motion, and as the number of parts strained; that is, directly as the square of the length, and inversely as the diameter. Joining these proportions, and retaining the same notation as in the preceding investigations we have $L^2 \times W/D^4$ as the strain.

($L$ = length of column, $D$ its diameter and $W$ superincumbent weight.) On p. 50 Tredgold deduces the following extraordinary result from the above obscure statement:

A column or pillar that shall be equally strong throughout will be generated by the revolution of two parabolas round the axis of the column, the vertices of the curves being at the extremities; for when all other quantities are constant $L^2$ varies as $D^4$ or $L : D^2$, a property of the parabola. But the figure of a column depends on two conditions; the one, that it shall rest firmly on its base, and offer a solid bearing for the load to be supported; the other that it shall be capable of the greatest degree of resistance. To fulfil the first condition, it should be a frustrum of a cone; to fulfil the second, it should be of the form generated by the revolution of two parabolas: and combining the forms which fulfil these conditions, we produce nearly that form which has been adopted for columns; that is, a column with a slight swell in the middle. But where the form of the column is considered rather as the most beautiful than the strongest, one that gradually diminishes from the base to the capital appears preferable.

It is difficult to picture the remarkable scientific ignorance of practical men in England in the first quarter of this century. One can only trust that there may be a closer union of practice and theory in our own day.

[197.] *A Practical Essay on the strength of Cast Iron*, London, 1822. This book contains a number of practical rules founded partly on a not very rigid theory, and partly on experimental results. Tredgold exhibits here as in his *Treatise on Carpentry* the same ignorance of theoretical elasticity, and seems to be acquainted only with the works of Thomas Young. A footnote in his preface is so characteristic that it deserves reproduction:

I have rejected Fluxions in consequence of the very obscure manner in which its (*sic !*) principles have been explained by the writers I have consulted on the subject. I cannot reconcile the idea of one of the terms of a proportion vanishing for the purpose of obtaining a correct result; it is not, it cannot be good reasoning; though, from other principles, I am aware that the conclusions obtained are correct. If the doctrine of Fluxions be freed from the obscure terms, limiting ratios, evanescent increments and decrements, etc., it is in reality not very difficult. If you represent the increase of a variable quantity by a progression, the first term of that progression is the same thing as what is termed a fluxion; and the sum of the progression is the same as a fluent. A fluxion is, therefore the first increase of an increasing variable quantity, and the last decrease of a decreasing one; and the expansion of a variable quantity into a progression is the best and most clear comment that can be added to the Lemmas of Sir Isaac Newton. (p. x.)

Such is the *scientific* capacity of the man whose works remained for years the standard text-books of English engineers!

[198] There is however one paragraph of the work (§ 233, p. 121) which contains something I have not met with elsewhere. Notwithstanding that Robison and others had expressed their dissatisfaction with Euler's theory of columns subjected to *longitudinal* strain, it seems to have been left to Tredgold to attempt the determination of the position of the neutral line[1]. Now although Tredgold's final result for the compressing force in terms of the deflection does not seem to me accurate, for he has neglected the fact that a horizontal force would have to be applied to keep the bent column with its base and summit in the same vertical; although his method of deduction seems to me obscure, if not inexact, still the expression he obtains for the distance between the neutral line and the axis of the column at their middle points agrees with the result I have deduced by a more accurate method. If $d$ be the diameter of the section of the column in the plane of flexure, $x$ the distance from the neutral axis to the axis of the column, and $y$ the deflection of the axis of the column, measured at a point midway between its base and summit, Tredgold finds $x = d^2/12y$ the section of the column being supposed rectangular.

199. 1817. Cisa de Gresy. A memoir by this author is published in the Turin Memoirs, Vol. XXIII. 1818: it is entitled *Considérations sur l'équilibre des surfaces flexibles et inextensibles;* and occupies pages 259—283 of the volume. It was read on the 1st of April, 1817[2].

200. Lagrange had treated the problem of the flexible membrane in a manner which involved the supposition that the tension is the same in all directions; he had employed his

[1] Here, as in the footnote p. 44, I use the term 'neutral line' following Tredgold. The line in question however does not correspond to a real or supposititious unstrained fibre; one or more points of it lie at an infinite distance.

[2] [I have retained this account of Cisa de Gresy's memoir although it hardly belongs to the subject of elasticity.] ED.

peculiar method which depends on the Calculus of Variations: see Art. 158. Poisson had treated the problem on the supposition that the tensions in two directions at right angles to each other are different; he had employed ordinary mechanical principles: see Chap. IV. The main design of the present method is to apply Lagrange's method to Poisson's form of the problem. After some introductory remarks, the memoir is divided into two parts.

201. The introductory remarks occupy pages 259—264. Here we have some valuable criticisms on the solutions of Lagrange and Poisson, shewing that both are deficient in generality. The following is given, on page 261, as the enunciation of the problem treated by Lagrange:

Trouver l'équation d'équilibre d'une surface sollicitée par autant de forces qu'on voudra, la condition du système étant que chaque élément de la surface se trouve également tendu dans tous les sens.

The following is given, on page 262, as the enunciation of the problem treated by Poisson

Trouver l'équation d'équilibre d'une surface sollicitée par autant de forces qu'on voudra, la condition du système étant que chaque élément de la surface soit inextensible suivant deux directions données respectivement perpendiculaires aux côtés adjacents de l'élément.

202. The author enunciates on his page 260 a problem which does not fall within the range of either Lagrange's or Poisson's treatment:

Si on suppose, par exemple, une surface en équilibre, sollicitée uniquement par la gravité, et suspendue à la circonférence d'un cercle fixé horizontalement, il est clair que les éléments de cette surface n'éprouveront qu'une simple tension dans le sens des méridiens ou de la courbe génératrice.

The introduction concludes thus:

L'objet de ce mémoire n'est que de parvenir aux équations données par M. Poisson suivant son hypothèse de deux tensions, mais en y employant le principe des vitesses virtuelles je commencerai cependant par exposer la solution de La-Grange, afin de mieux rapprocher les deux solutions de ces deux illustres géomètres.

Ces sortes de rapprochemens, dit quelque part La-Grange, sont toujours utiles; souvent la véritable métaphysique du problème est renfermée dans ce qu'ont de commun les différentes méthodes que l'on peut employer pour le résoudre.

203. The first part of the memoir is entitled *Hypothèse de Mr. La-Grange;* it occupies pages 264—283. This substantially reproduces Lagrange's process; there are some modifications arising from the fact that a difficulty which once existed in the Calculus of Variations is here tacitly overcome: see my *History of the Calculus of Variations,* Arts. 39, 102, and 304. The point is this: we require correct values of $\delta p$ and $\delta y$, in the notation of this Calculus; such values are given by Cisa de Gresy on his pages 266—269, and in a note on his pages 292—294, he refers to the investigation by Poisson in the *Bulletin Philomatique* 1816, and supplies another investigation due to Plana. The names of Cisa de Gresy and Plana should therefore have found a place in my *History.*

204. On his page 271 Cisa de Gresy arrives at results which coincide with those given by Lagrange on pp. 103 and 149 of the second edition of the *Mécanique Analytique;* these pages correspond respectively to 98 and 140 of Bertrand's edition. He says that there is an error at the second place, for the terms $\left(\frac{dU}{dx}\right)\delta x + \left(\frac{dU}{dy}\right)\delta y$ are omitted in the value of $\delta U$; the mistake is silently corrected in Bertrand's edition, though the appropriate brackets are omitted.

205. We will extract from the memoir the solution which is given, in Lagrange's manner, of the problem enunciated at the beginning of our Art. 202: see pages 280 and 281 of the memoir. In this case we may assume, on account of symmetry, that the surface will be one of revolution, so that we have only to find the equation to the generating curve. Let $x$ and $z$ be the coordinates of any point of it, $z$ being measured vertically downwards. The element of surface, $dS$, may be denoted by $xu\,dx\,d\theta$ where $u$ stands for $\sqrt{1+\left(\frac{dz}{dx}\right)^2}$ and $d\theta$ denotes an infinitesimal angle in the horizontal plane. Now as the surface experiences tension

only in the direction of the meridian the element $dS$ will be invariable provided the surface is inextensible in the direction of the meridian; hence it follows that in taking the variation of $dS$ it is sufficient to vary the element $udx$ of the arc of the generating curve: thus

$$\delta dS = \delta\,(xudxd\theta) = xd\theta\,\delta\,(udx).$$

Multiply this expression by an indeterminate quantity $F$, and add it to the equation of equilibrium: then by Lagrange's well known method we have the following equation as the condition of equilibrium

$$\iint g\,\delta z\,xu\,dx\,d\theta + \iint Fx\,d\theta\,\delta\,(udx) = 0 \text{ ............(1)};$$

this must be developed by the ordinary processes of the Calculus of Variations. Put $p$ for $dz/dx$; we have

$$\delta u = \frac{1}{u} p\delta p = \frac{1}{u} p \left(\frac{d\delta z}{dx} - p\frac{d\delta x}{dx}\right);$$

thus (1) becomes

$$\iint g\delta z\,xu\,dx\,d\theta + \iint Fx\,d\theta\,u\,d\delta x$$
$$+ \iint Fx\,d\theta \frac{p}{u}\left(\frac{d\delta z}{dx} - p\frac{d\delta x}{dx}\right) dx = 0.$$

All the terms except the first are to be integrated by parts, then finally we equate to zero the coefficients of $\delta z$ and $\delta x$ in the expression which remains under the double integral sign: thus

$$gxu - \frac{d}{dx}\left(\frac{Fxp}{u}\right) = 0 \text{ ..................(2)},$$

$$\frac{d}{dx}(Fxu) - \frac{d}{dx}\left(\frac{Fxp^2}{u}\right) = 0 \text{ ..................(3)}.$$

From (3) we obtain

$$Fxu = Fxp^2/u + C,$$

where $C$ is an arbitrary constant;

therefore $$Fx/u = C \text{ ......................(4)}.$$

Substitute from (4) in (2); thus

$$gxu = C\,dp/dx,$$

so that

$$gx = \frac{C}{\sqrt{1+p^2}}\frac{dp}{dx} \text{ ..................(5)}.$$

Hence by integrating we obtain

$$gx^2/2 = C \log \{p + \sqrt{1 + p^2}\} + \text{constant};$$

this is a differential equation of the first order for the required curve.

Cisa de Gresy shews that if we follow Lagrange's own solution we obtain instead of (5) the equation

$$gx(1 + p^2) = (gz + C)\{p + p^3 + x\, dp/dx\}.$$

206. The second part of the memoir is entitled *Hypothèse de Mr Poisson;* it occupies pages 283—291. Here Poisson's results are obtained by the use of Lagrange's method; the process seems to me rather arbitrary, as does also that of Art. 205. Lagrange himself says, without assigning any adequate reason, that, if $F$ denote the elastic force, the sum of the moments of all these forces is expressed by $\iint F\delta dx\, dy \sqrt{1 + p^2 + q^2}$. Now in the present investigation we have instead of his term the two terms

$$\iint T\, dy\, \delta\, dx \sqrt{1 + p^2 + q^2} + \iint T'\, dx\, \delta\, dy \sqrt{1 + p^2 + q^2},$$

where in the first term $q$ is to be considered constant in the variation, and in the second term $p$ is to be considered constant in the variation. But the introduction of these two terms seems to me very inadequately justified

207. 1818. Fourier: *Note relative aux vibrations des surfaces élastiques et au mouvement des ondes.* This is published in the *Bulletin des Sciences par la Société Philomatique de Paris* for 1818; it occupies pages 129—136 of the volume.

The note commences thus:

J'ai présenté à l'Académie des sciences, dans sa séance du 8 juin de cette année, un Mémoire d'analyse qui a pour objet d'intégrer plusieurs équations aux différences partielles, et de déduire des intégrales la connaissance des phénomènes physiques auxquels ces équations se rapportent. Après avoir exposé les principes généraux qui m'ont dirigé dans ces recherches, je les ai appliqués à des questions variées, et j'ai choisi à dessein des équations différentielles dont on ne connaissait point encore les intégrales générales propres à exprimer les phénomènes. Au nombre de ces questions se trouve celle de la propagation du mouvement dans une surface élastique de dimensions infinies.

[1] [P. 269, line 6 for $d\delta y/dx$ read $d\delta y/dy$. P. 272, line 10 for + read − twice.]

The memoir to which he refers does not seem to have been published. Much of Fourier's note is in controversy with Poisson as to the motion of waves and other matters; we shall confine ourselves to what relates to our own subject.

208. Fourier refers to the fact that Poisson had some years before obtained the differential equation for the vibrations of an elastic lamina, which was already known, but that he had not integrated the equation. Fourier quotes the words in which Poisson speaks of Plana's solution for the case of a simple plate as very complicated: see his words in our account of Poisson s memoir of August 1814. Fourier then gives without demonstration his own form of the integral; he confines himself to the case in which the initial velocity is zero, and his form coincides with that obtainea by Poisson at the end of his paper in the *Bulletin* for 1818: see our Chap. IV. The only difference is that Fourier omits the constant factor $1/4\pi a$, or rather $1/4\pi$ as he takes $a$ to be unity. This however is of no consequence for his purpose, as his form satisfies the differential equation.

209. Fourier then proceeds to the equation which relates to the vibrations of a simple lamina; as we have seen in Art. 178, Plana despaired of any solution more simple than his own which expressed at full would occupy more than a quarto page. Fourier also cites from Euler's memoir of 1779 (see our Art. 86) the words:

Ejus integrale nullo adhuc modo inveniri potuisse, ità ut contenti esse debeamus in solutiones particulares inquirere....

The equation is:

$$\frac{d^2z}{dt^2} + a^2 \frac{d^4z}{dx^4} = 0.$$

If we proceed precisely as in Poisson's paper in the *Bulletin* for 1818, and suppose the velocity to be initially zero, but the initial displacement to be given by $z = \phi(x)$, we shall obtain

$$z = \frac{1}{\sqrt{\pi}} \int_{-\infty}^{+\infty} \sin\left(\frac{\pi}{4} + \alpha^2\right) \phi\left(x + 2\alpha\sqrt{at}\right) d\alpha\,;$$

and if we put $x + 2\alpha\sqrt{at} = \xi$ this becomes

$$z = \frac{1}{2\sqrt{\pi a t}} \int_{-\infty}^{+\infty} \sin\left(\frac{\pi}{4} + \frac{(\xi - x)^2}{4at}\right) \phi(\xi)\, d\xi.$$

This coincides with the result which Fourier gives without demonstration, except that he supposes $a$ to be unity, and he omits the factor $\frac{1}{2}$ which is of no consequence for his purpose.

210. If instead of supposing that $dz/dt = 0$ when $t = 0$ we suppose that $dz/dt = F(x)$; then instead of the value first given to $z$ we have

$$z = \frac{1}{\sqrt{\pi}} \int_{-\infty}^{+\infty} \sin\left(\frac{\pi}{4} + \alpha^2\right) \phi\left(x + 2\alpha\sqrt{at}\right) d\alpha$$

$$+ \frac{1}{\sqrt{\pi}} \int_0^t \int_{-\infty}^{+\infty} \sin\left(\frac{\pi}{4} + \alpha^2\right) F\left(x + 2\alpha\sqrt{at}\right) dt\, d\alpha.$$

This is obvious by what has already been shewn; for the second term in the value of $z$ just given satisfies the equation

$$\frac{d^2}{dt^2}\left(\frac{dz}{dt}\right) + a^2 \frac{d^4}{dx^4}\left(\frac{dz}{dt}\right) = c;$$

also it vanishes when $t = 0$, and gives $dz/dt = F(x)$ when $t = 0$. Thus this second term also satisfies the equation

$$\frac{d^2z}{dt^2} + a^2 \frac{d^4z}{dx^4} = 0,$$

and makes $z = 0$, and $dz/dt = F(x)$, when $t = 0$.

211. 1818. Nobili: *Sopra l' identità dell' attrazione molecolare coll' astronomica; Opera del Cavaliere Leopoldo Nobili.* This consists of 84 quarto pages with 4 plates; I have given a notice of it in my *History of the Theories of Attraction*, Art. 1615. The author holds that the law of attraction according to the inverse square of the distance will suffice for the explanation of the phenomena of molecular action as well as for the phenomena of astronomy: but he completely fails in his attempts to maintain his opinion.

[1818. A paper by Poisson belonging to this year is noticed in the Chapter devoted to him.]

[212.] 1818. J. P. S. Voute: *Dissertatio Philosophica inauguralis de Cohaerentia Corporum et de Crystallisatione.* Leyden,

1818. This is a dissertation for the degree of doctor in mathematics and natural philosophy at the University of Leyden. To the dissertation are affixed 15 Theses, which the author was prepared to defend in the greater hall of the University on January 24 of the above year. They contain a number of rather doubtful assertions on a variety of topics, which are singularly characteristic of the time.

The essay itself is divided into three parts; the first treats of that principle of motion which is termed Attraction, and of the nature of particles (those like parts into which a body may be dissolved); the second, of Cohesion, and the last of Crystallisation.

[213.] The first part which might be termed a defence of 'action at a distance' contains nothing touching upon our present subject, nor in fact anything of value.

In the second part on cohesion the writer tells us that in order to understand the *vis cohaerentiae,* which he attributes to attraction, it is needful to compare it with other things:

> Cujus quatuor genera potissimum fuisse adhibita reperio: unum, quo vis cohaerentiae cum attractione generali comparatur; alterum, ubi vis contraria, qualem calor efficit, adversus eam opponitur, et invenitur utrum momentum plus polleat; tertium, cum corpora dura ipsa inter se conferuntur, conflictu aut tritu, ex quo appareat quaenam facilius cedant et penetrentur; quartum, cum partes divelluntur ponderibus vel aliis modis. (p. 15.)

The comparison with gravitation does not lead to any definite results, but the author finds that the *vis cohaerentiae* is exactly opposed to the *vis repulsionis* which arises from heat, but then heat produces other, viz. chemical changes which are totally different from cohesion. A further not very lucid investigation leads Voute to reject this explanation of cohesion.

The following chapters are devoted to hardness and firmness; the discussion of these subjects is of the most general kind; reference is made to the experiments of Musschenbroek, Buffon, Emerson, &c., but there is no theory of any value deduced from them, nor is there any originality of thought. The third part treats of the forms of the various particles under the head of crystallisation. The disputator concludes that the forms of particles are closely connected with the forces which act between them, and that the two questions must not be separated, if we would learn the true nature of bodies:

Attamen nos eas separamus; loquimur enim de viribus ubi formae partium penitus latent; cum vero formae fiunt manifestae, has complectimur, nescimus quae vis partes adducat. Quantum igitur abest ut scientiam illam nacti simus! (p. 88.)

[214.] The whole dissertation is of a negative character, and although occasionally referred to by later writers is practically worthless. It reads more like a mediaeval *disputatio quodlibetaris* than a scientific memoir of the 19th century.

215. 1818. Antonio Bordoni: *Sull' equilibrio delle curve a doppia curvatura rigide, ovvero completamente o solo in parte elastiche.* This memoir is published in the *Memorie...della Societa Italiana...* Vol. XIX., p. 1, Modena, 1821; it was received on the 22nd of May, 1818.

216. We have already noticed that Lagrange treated in an unsatisfactory manner the problem of the equilibrium of a curve of double curvature, and that Binet turned his attention to the same subject: see Arts. 159 and 174. Binet investigated the value of the torsion, but left undetermined two of the elements of the problem, namely the tension and the elasticity, on account of the complexity of the calculation and of the results. Bordoni accepts the challenge thus, as it were, thrown out, and completes the investigation which Binet had left unfinished; but as we shall see in Art. 5, he does not precisely follow the line traced out by Binet. The following is Bordoni's own statement:

Il Sig. Binet Jacques, il primo che dilucidò questo passo della Meccanica di Lagrange, con una Memoria inserita nel tomo decimo del giornale della scuola politecnica di Francia, scrivendo altrimenti la invariabilità della curva, trovò tre altre equazioni indefinite, nelle quali vi sono oltre le forze esteriori e le coordinate della curva, tre nuove quantità atte a misurare, una la vera tensione, l' altra la elasticità, e la terza la torsione; e desunse anche da queste equazioni quella espressione delle forze esteriori, che rappresenta la torsione; ma per rispetto alla tensione ed alla elasticità dichiarò, che "On ne parvient aux valeurs générales de tension et d'élasticité que par des calculs pénibles, dont les résultats paraissent fort compliqués."

In questa breve Memoria, nella quale si parla dell' equilibrio delle curve a doppia curvatura siano rigide, o siano elastiche completamente o solo in parte, si troveranno tre equazioni indefinite fra le coordinate della curva nella posizione di equilibrio, le forze esteriori, la tensione, la elasticità, e la torsione, colle quali si otteranno con facilità, mediante alcuni stratagemmi, queste ultime tre quantità, cioè la tensione, la elasticità, ed anche la torsione, espresse tutte colle forze esteriori e le coordinate della curva nella posizione di equilibrio di esso.

217. Before coming to the mechanical problem, Bordoni investigates the value of what we now call the *angle of torsion;* his method is unsymmetrical and laborious, but he finally gives the correct result in a symmetrical form. It is referred to on page 178 of the work of G. Piola, 1825, hereafter to be noticed.

218. The mechanical problem is treated after the manner of Lagrange. Let $ds$ denote an element of the arc of the curve at the point $(x, y, z)$; let $de$ denote the angle of contingence and $di$ the angle of torsion; then Bordoni says that the sum of the virtual moments is the integral of the following expression,

$$(X\delta x + Y\delta y + Z\delta z)\, dm + \lambda \delta ds + \mu \delta de + \xi \delta di;$$

and that such sum must be zero.

But what are $\lambda$, $\mu$ and $\xi$? It seems to me that the great objection to his method of treating the problem is that we do not start with any clear notion on this point. Bordoni says vaguely that $\lambda$ is the tension, $\mu$ the elasticity, and $\xi$ the torsion; it turns out in the course of the investigation that $\lambda$ is equal to the sum of certain external forces resolved along the tangent at $(x, y, z)$, that $\mu$ is equal to the sum of the *moments* of such forces round the straight line at $(x, y, z)$ which is at right angles to the osculatory plane, and that $\xi$ is the sum of the moments of such forces round the tangent at $(x, y, z)$.

219. In Binet's treatment of the problem the expression for the virtual moment is put in a different form, namely

$$(X\delta x + Y\delta y + Z\delta z)\, dm + a\,(dx\delta dx + dy\delta dy + dz\delta dz)$$
$$+ b\,(d^2x\delta d^2x + d^2y\delta d^2y + d^2z\delta d^2z)$$
$$+ k\,(d^3x\delta d^3x + d^3y\delta d^3y + d^3z\delta d^3z);$$

and the *de* and *di* which Bordoni uses are subsequently introduced.

220. Bordoni's investigation is of a most laborious kind, and forms a good testimony to the patience of the author. There is sufficient detail given to enable the reader to follow the steps without much difficulty, except in one case which occurs on page 16. Here a certain integral has to be found; the author states the result, adding 'come facillissimamente si verifica'... It is only a part of the result which I wish to notice, and expressed in Bordoni's notation it is

$$\int(Cdy - C'dx) = dz/ds:$$

I shall be glad if any reader can be induced to examine this point and shew how this result can be easily obtained; for my own part I find for the integral after a very tedious investigation the value

$$dz/ds - \beta\,(cdy - c'dx),$$

where $$\beta = (dxd^2x + dyd^2y + dzd^2z)/r^2ds^4;$$

this agrees with Bordoni's result if we suppose $s$ to be the *independent variable*, a supposition which he has not stated. The notation is the same as Bordoni's, except that my $c$ and $c'$ are to be obtained by multiplying his $c$ and $c$ by $ds$: it will be observed that at the place where he first introduces this notation, namely near the bottom of page 10, he has omitted $ds$, and this must be restored in order to give correct dimensions in infinitesimals to the expressions.

221. The following are the values which Bordoni obtains for $\lambda$, $\mu$, $\xi$; these values do in effect assign the meanings to these symbols, about which nothing was really known at the beginning of the investigation

$$\lambda = \frac{dx}{ds}\int Xdm + \frac{dy}{ds}\int Ydm + \frac{dz}{ds}\int Zdm,$$

$$\mu = \frac{c}{\omega} L + \frac{c}{\omega} M + \frac{c''}{\omega} N.$$

$$\xi = \frac{dx}{ds} L + \frac{dy}{ds} M + \frac{dz}{ds} N.$$

Here $L = z\int Ydm - y\int Zdm - \int(zY - yZ)\,dm$; and $M$ and $N$ are obtained by symmetrical changes in the variables. Also $c = (dyd^2z - dzd^2y)\,ds$; and $c'$ and $c''$ are obtained by symmetrical changes in the variables; while $\omega^2 = c^2 + c'^2 + c''^2$.

The value of $\mu$ corresponds with that which had already been obtained by Binet.

222. It may be easily shewn that

$$\frac{dx}{ds} dL + \frac{dy}{ds} dM + \frac{dz}{ds} dN = 0,$$

and therefore

$$d\xi = Ld\frac{dx}{ds} + Md\frac{dy}{ds} + Nd\frac{dz}{ds}.$$

[This formula is in fact given in the *Comptes Rendus*, XIX., 182.]

223. In the last three pages of the memoir Bordoni notices the paper by Poisson concerning the problem (see our chapter IV) Bordoni perceives that Poisson has given one result which he himself has not, namely that $d\xi = 0$, so that the moment of torsion is constant; and he tries to explain how it happens that he himself has not obtained this result: but it does not seem to me that he is very successful. I will offer a few remarks on the point. Poisson treats the problem on ordinary mechanical principles, and so it is easy to understand his process. Suppose a normal section of the curve made at any point; the piece between this section and one end must be in equilibrium; the forces acting are the external applied forces and the molecular action at the section, which Poisson *assumes* to be normal to the section. Poisson then forms the three equations of statical equilibrium with respect to moments which we know must hold. Granting these then, we find that $d\xi$ must follow, as Bordoni admits.

Bordoni's account of the absence of the result $d\xi = 0$ from his own investigations amounts in fact to this he does not consider the equation of moments round the principal normal at $(x, y, z)$, holding that motion round such a line would change the nature of the system.

224. It is difficult in the treatment adopted by Bordoni to understand what forces really are supposed to act, but I think Poisson and he really differ as to the nature of the molecular force; Poisson determines his force as altogether *normal* to the transverse section, while Bordoni apparently does not. With Bordoni $\lambda$ denotes that part of the molecular action which is normal, but he does not necessarily assume that this is the total action; there may be other actions in transverse directions, that is, in the plane of the section. It is, I believe, certain that Bordoni's method does not give *all* the conditions of equilibrium, and it is for those who understand and adopt the method to explain and supply the omission. *Transverse* forces will enter into the moment which Bordoni calls $\xi$ and Poisson in his *Mécanique* calls $\tau$; they will not occur in the moments round the two axes in the plane of the section[1].

[225.] C. J. Hill: *De elasticitate torsionis in filis metallicis.* Lundae, 1819. This tract will be considered in the following chapter together with a later work of the same writer.

[226.] A. Duleau: *Essai théorique et expérimental sur la résistance du fer forgé.* Paris, 1820.

This essay arose from experiments made by Duleau on forged iron, when entrusted with the construction of an iron bridge over the Dordogne. It received the approval of the Academy of Sciences after a report upon it by Poisson, Girard and Cauchy.

[227.] The first section only is concerned with theoretical, the remaining three with experimental results.

On p. 2, Duleau defines the neutral line of a beam under flexure; he terms it the *ligne de passage de la tension à la compression,* and makes the following statement concerning it:

[1] Page 27 middle of the page; for $\frac{c}{\omega}\mu$, $\frac{dx}{ds}\xi$ read $-\frac{c}{\omega}\mu$, $-\frac{dx}{ds}\xi$.

Coulomb a supposé que, lorsque la courbure de la pièce élastique est très peu considérable, cette ligne est tellement placée que la somme des momens de tensions des fibres supérieures est égale à celle qu'on obtient en ajoutant les momens des compressions des fibres inférieures. Ce principe, qui n'a pas encore été démontré d'une manière rigoureuse, a été adopté par tous les auteurs qui ont traité ce sujet. Il est inutile d'y avoir recours, s'il agit d'un solide dont la section transversale est divisible par une ligne horizontale en deux portions symétriques; cette droite est alors évidemment celle où existe le passage de la tension à la compression. Il est très rare que dans la pratique on ait à considérer des corps d'une autre forme.

It will be obvious from this quotation that Duleau has misunderstood Coulomb, and adopted a false principle. Coulomb (see Art. 117) placed the neutral line along the axis or middle line of the beam, because he argued that the sum of the longitudinal tensions or resistances of the fibres across any section must be zero if the beam be only acted on by a system of transverse forces. This is a clear result of an elementary principle of statics—the sum of the forces resolved parallel to the axis of the beam must be zero. But Duleau's statement that the moment of the compressions must be equal to the moment of the extensions, is a false principle which only places the neutral line in its right position in the particular case of symmetry mentioned by him at the end of the paragraph cited.

[228.] The following sections are devoted to the problems:

Of a lamina built in at an end and weighted at the other.

Of a lamina supported at both ends and uniformly weighted.

Of a lamina subject to longitudinal pressure.

The treatment of the first two problems contains nothing original, that of the third follows Euler's memoir of 1757, and is liable to the same objections. Duleau does not seem to have been acquainted with Robison's criticism of Euler (see Art. 146) but he has a section entitled: *Application des résultats trouvés pour une lame élastique a un solide prismatique*, which contains his views as to the practical value of Euler's theory. He remarks that the results obtained for the elastic lamina may be extended to the neutral line of an elastic beam, when the flexure of the beam is produced by transverse forces, but when the beam is under longitudinal stress, it cannot with exactitude be treated as an elastic lamina. He observes however that the pieces of iron to which he is about to apply his theory are in practice of great length as compared with their section, and hence their resistance approaches sufficiently that of an elastic lamina. He places the neutral line of such large iron bars at their line of centres, and thus falling into the same error as Euler, is therefore (as a practical engineer) still more deserving of Robison's censure.

[229.] The curious fact remains that he considers Euler's method sufficiently exact for practice, "as experiment confirms" (p. 16).

Note III. p. 79 gives a comparison of the results of experiment and calculation, the mean ratio of the results of experiment to those of calculation equals 1·16. A further section treats of torsion, and contains results of the same character as Coulomb's. The rest of the book is occupied with the discussion of experiments on iron bars.

[230.] 1822. E. Hodgkinson: *On the transverse strain and strength of materials. Memoirs of the Literary and Philosophical Society of Manchester.* Second Series, Vol. IV. London, 1824.

This memoir is contained on pp. 225—289 of the volume, and is divided into two parts: Theoretical and Experimental. The author commences with the following statement of his aims,—

The lateral strength of materials is a subject which has engaged the attention of the greatest mathematicians; but our knowledge of the action of the fibres or particles of bodies during their flexure, chiefly perhaps for want of sufficient experiments, seems still to be imperfect.........The intention of the writer is first to unite in a general formula the commonly received theories, in which all the fibres (with the exception of those on the edge of a bent body) are conceived to be in a state of tension; and next to adapt the investigations to the more general case, where part of the fibres are extended, and part contracted, and to seek experimentally for the laws that regulate both the extensions and the compressions.

[231.] On pp. 226—233 the writer obtains his general formula, and discusses several cases of it. The general formula turns out to be nothing but a very special case of Varignon's formula (see Art. 14), but still sufficiently general to include Galilei's and the Mariotte-Leibniz hypotheses. Throughout this investigation the author supposes the neutral line to be in the lowest fibre of a beam subject to transverse stress,—

And this may probably be not far from the case in such bodies as glass or marble: but (as Dr Robison has shewn in his valuable essay on the Strength of Materials in the *Encyclopaedia Britannica*) it is, when applied to timber, highly erroneous.

[232.] The author cites various experiments in confirmation of this view, and then proceeds to a method for determining the position of the neutral line, when compression as well as extension takes place. He arrives by a somewhat artificial and cumbrous process at Coulomb's result, namely that the sum of the forces of extension must be equal to the sum of the forces of compression across any section. The author attributes this result to Robison, whereas it had long before been stated by Coulomb and something extremely like it by both Bülfinger and Belgrado (Art. 29).

[233.] In a footnote the error of Barlow's method (see Art. 192), is pointed out, it consists in making the moment of the compressing equal to the moment of the extending elastic forces; precisely the same error as Duleau had fallen into. (See Art. 219.) Hodgkinson also criticised Barlow's error in a paper which will be found in Vol. v. of the *Edinburgh Journal of Science.*

[234.] It may be noted that Hodgkinson in the examples he gives of the flexure of beams of various section, supposes the forces necessary to produce the equal extension and compression of a fibre to be unlike; he likewise supposes the relation between tension and extension not to be expressed by Hooke's Law, but the tension to be proportional to some power of the extension. This power, for still greater generality, he treats as having a different value for the relation between pressure and compression. Thus he assumes relations of the form suggested by Bülfinger (Art. 29, $\beta$):

$$\text{Tension} = \mu\,(\text{extension})^n,$$
$$\text{Pressure} = \mu'\,(\text{compression})$$

Very little is practically gained by this generality so far as it concerns the power of the strain, but the adoption of the two moduli is undoubtedly right for some materials, *e.g.* cast iron (see Art. 239); in the case of a prismatic beam, the neutral line would in general not coincide with the axis of figure (see p. 249).

[235.] We may note the following lemma (p. 247):

In the bending of any body this proportion will obtain: as the extension of the outer fibre on one side is to the contraction of that on the other, so is the distance of the former from the neutral line to that of the latter.

This is exactly the same method for finding the position of the neutral line as had been given by Belgrado just seventy-six years earlier. So long does it take for the results of mathematical investigation to find their way into practical text-books or to be rediscovered by the practical engineer!

The rest of the memoir contains a detailed account of experimental investigations on beams of wood (yellow pine, fir, etc.), with special reference to their extension and compression, and to the position of the neutral line. The author found extensions and compressions to be as the forces, till permanent deformation began.

[236.] The merit of Hodgkinson's paper consists in the fact that it led to practical men in England (notably Barlow, who corrected his mistake in later editions, see Art. 185), placing the neutral line in its true position. Hodgkinson was of course only adopting Coulomb's work of half a century earlier, but he may be said to have done for England what Eytelwein did for Germany, namely he gave Coulomb's theory its true place in works on practical mechanics.

[237.] A second paper by Eaton Hodgkinson on the same subject may be perhaps best referred to here. It appears in Vol. 5 (1831) of the *Manchester Literary and Philosophical Society*, pp. 407—544. It is entitled: *Theoretical and Experimental Researches to ascertain the Strength and best Forms of Iron Beams*, and was read April 2nd, 1830.

Hodgkinson begins by recognising the importance of the work of Bernoulli, Euler, Lagrange, Coulomb and Robison.

They have done much, but owing perhaps to a disinclination to the labour and expense of making sufficient experiments, much was left undone for later inquirers.

The author proposes to develop further the results of his previous paper, the theory of which he finds in agreement with those of Robison and Coulomb so far as they have gone.

[238.] The theoretical investigation of the weight which a beam will support and of the position of the neutral line (pp. 410—417) offers nothing of importance. We may note however Hodgkinson's remark that the position of the neutral line may change, if the ratio of the moduli of resistance to compression and of resistance to extension should change owing to the body approaching the limit of elasticity. He quotes (p. 417) an experiment of Tredgold's in confirmation of this.

[239.] On pp. 419—425 he corrects the error of Duleau we have noted in Art. 219, and gives an experiment of his own on the point. These pages are followed by an experimental investigation of the laws which connect the extensions and compressions with the forces in *cast iron*. The first deduction of importance is obtained on p. 432, namely, that deflections from extension are greater than those from compression, the forces being equal, and this whether the elasticity be perfect or not. This result was obtained by measuring the deflection of a T-shaped beam under the same force, first with the more massive head upwards and so subject to extension, and secondly with it downwards and so subject to compression.

Still the difference is not so great, but that the deflections may in most cases, without material error, be assumed as the same. Hence the extensions and compressions from equal forces in cast iron are nearly equal. This is a very interesting fact; it is most likely a property common to tenacious bodies, when not over-strained generally: it has often been assumed by writers, but I have not before seen any proof of it, except in an experiment of M. Duleau, which renders it very probable that it is the case in malleable iron.

[240.] We may note that there are three things to be compared when we are considering the extension and compressions of any substance.

(i) We may note whether the moduli for extension and compression are equal.

Hodgkinson finds that they are not exactly equal for cast iron. This result has been confirmed by later investigators. Hence the neutral line does not exactly pass through the centres of gravity of the sections. The difference however is so small that it does so very approximately.

(ii) We may inquire whether the limits of elasticity are the same in the two cases. This is somewhat difficult to determine in the case of cast iron, because a certain amount of permanent set is found to have arisen from almost any stress we may have applied to the material, although the body of course after removal of the stress returns very nearly to its primitive shape.

Hodgkinson is of opinion that in cast iron a much greater force is required to destroy its elasticity in the case of compression than by extension. It must be noted however that there is an important point to be considered here. Is it the force which will produce any the least permanent set, which is to be taken as that which destroys the elasticity? Or shall we term that the destructive force for which the extension ceases to obey Hooke's Law?

(iii) Lastly we may investigate whether the absolute strength in the two cases is the same. Will the same stress tear a metal by tension and crush it by compression[1]?

[241.] An interesting experiment described on pp. 434—435 gives the reason of Hodgkinson's opinion on our second point. He found in the case of a T-shaped bar of cast iron supported on two props at its ends and weighted at the centre that it required nearly four times as great a weight to break it with the head

[1] The consideration of the limits mentioned in the above remarks is of such importance, that I shall add a note on the subject at the end of this volume and will suggest a terminology.

downwards (⊥) as with the head upwards (T). Hence the absolute strength is different in the cases of extension and compression, being greater in the latter. We must observe however that this rather answers the question we have asked in (iii) than in (ii). It does not *necessarily* follow that because the absolute strength is greater in one case than the other, that the limit of elasticity is greater for compression than for extension. It might happen that the range of imperfect elasticity (partial set) was much greater in one case than the other.

[242.] Hodgkinson next turned his attention to malleable iron, and came to the conclusion that throughout the whole range (by which I suppose he means the ranges of perfect elasticity, and of imperfect elasticity or partial set) the extensions and compressions were nearly equal from the same forces, a result very different from what had been obtained for cast iron. The experiment he gives (p. 437) and the additional one he quotes from Duleau cannot however be considered as conclusive.

[243.] The major portion of the remainder of the paper is devoted to an experimental investigation of the best form for a cast iron beam, in order that for a given mass of material it may, when supported at the ends on two props, best resist transverse stress. The form principally investigated is that composed of two strong ribs united by a thin sheet of metal thus ⌶. Hodgkinson remarks:

As to the comparative strength of these ribs, that appears to me to depend upon the nature of the material, and can only be derived from experiment. Thus, suppose it was found that it required the same force to destroy the elasticity of a piece of metal, whether the force acted by tension or compression. In this case the top rib ought to be equal to the bottom one, supposing it was never intended to strain the beam so as to injure its elasticity. And if it were found that the same weight would be required to tear asunder or to crush a piece of metal according as it acted one way or the other, the beam should have equal ribs to enable it to bear the most without breaking. Now, from the experiments given above, it appears that these qualities are in a great measure possessed by wrought iron; and therefore, whether it was intended to strain a beam of it to the extent of its elasticity or even to the breaking point, there ought to be equal ribs at top and bottom.

If, however, the metal were of such a nature that a force $F$ was needed to destroy its elasticity by stretching it, and another force $G$ to do the same by compressing it, it is evident that the ribs ought to be to one another as $F$ to $G$, in order that the beam might bear the most without injury to its elasticity. And if it took unequal weights $F'$ and $G$ to break the piece by tension and compression, the beam should have ribs as $F'$ and $G'$ to bear the most without fracture.

Our experiments on cast iron were not well adapted to shew what relative forces would be required to destroy the elasticities; but it appears, by the experiments of Mr Rennie[1] that it would take many times the force which would draw it asunder to crush it. The bottom rib must then be several times as large as the top one to resist best an ultimate strain.

[244.] Hodgkinson commences his experiments on beams with equal top and bottom ribs, which he notes had been considered the strongest form so long as the stress did not produce a strain greater than the limit of elasticity. It results however from these experiments that a beam can be found in which the ratio of the top and bottom ribs is such that its *absolute strength* is 2/5 greater than that of a beam of the 'common shape' (an inverted T (⊥) with a somewhat tapering vertical stroke), while the 'common shape' is itself 1/12 stronger than a beam with equal top and bottom ribs. (Cf. Experiments I., IV. and XIX.). The shape of this beam was as follows: top rib 2·33 by ·31, bottom rib 6·67 by ·66, vertical part ·266 by 4·15, the vertical part being slightly spread out where it met the ribs so that the total area of section was 6·4.

The great strength of this section is an indisputable refutation of that theory, which would make the top and bottom ribs of a cast iron beam equal.

[245.] We have referred thus at length to Hodgkinson's second paper because it suggests several points which have not received full treatment at the hands of the mathematician.

[1] See our Art. 185.

Notably the difference in character between cast and malleable iron, the range of imperfect elasticity and the shape of the beam of greatest absolute strength, are all points which seem capable of mathematical treatment with advantage to both theory and practice. The mathematician cannot fail to be struck with the *very small* portion of the phenomena presented by a material subjected to continuously increasing strain, which is covered by the current theory of elasticity.

To other work of Hodgkinson's we shall have occasion to refer later.

[246.] In the *Nouveaux Mémoires de l'Académie Royale des Sciences* (*Acad. An.* 1781), Berlin, 1783, will be found (pp. 347—376) a memoir by John and James Bernoulli[1] entitled: *Mémoire sur l'Usage et la Théorie d'une machine qu'on peut nommer Instrument ballistique.* This paper is not of any value, and would not be mentioned here, had the writers not given a first, although erroneous, theory of a spiral spring in their third section: *Théorie de la Machine ballistique* (pp. 354—358). Their work however led the Italian physicist O. F Mossotti to a more complete consideration of the problem. His memoir: *Sul movimento di un' elice elastica che si scatta,* was presented to the *Società Italiana delle Scienze* in 1817 and published in the mathematical part of Tom. XVIII. pp. 243—268 of their *Memorie di Matematica e di Fisica.* Modena, 1820. Owing to the fact that this and other memoirs of the same volume are unrecorded in the *Royal Society's Catalogue of Scientific Papers,* I did not discover its existence till it was too late to insert any account of it in its proper place in this chapter. The memoir appears to me of considerable interest and remarkably clearly expressed.

[247.] The author describes his method in the following words:

Per risolvere i problemi che mi sono proposti ho assunto due ipotesi, le quali sono però così da vicino verificate dagli sperimenti che, piuttosto che ipotesi, possono risguardarsi come regole di fatto. La

[1] This James Bernoulli is the same as that mentioned in Art. 122. John was his brother.

prima di queste ipotesi è riposta in cio, che l' elice elastica debba in tutto il tempo dello scatto o dilatazione conservare la figura d elice ed un egual numero di rivoluzioni, talmente che nell' allargarsi i passi delle spire, sia soltanto il diametrò dell' elice che venga successivamente a diminuire. Colla seconda ipotesi stabilisco ad imitazione di Daniele e Giacomo Bernoulli che gli accorciamenti o costipazioni che possono farsi soffrire all' elice siano proporzionali alle forze o pesi comprimenti atti a produrle. Allorchè nella soluzione dei problemi mi occorrerà di assumere per la prima volta alcuna di queste ipotesi avrò cura di far conoscere gli esperimenti che mi hanno persuaso ad adotarla, acciò il lettore sia egualmente convinto della legittimità della medesima.

I Bernoulli ed altri autori, che hanno considerato il movimento degli elastri piegati in forma d' elice, hanno per semplicità supposto nei loro calcoli che il movimento oscillatorio di un' elice fissa in un estremo sia eguale a quello di una fibra rettilinea ed omogenea dotata d' una stessa massa e d' una pari elasticità, e la cui lunghezza fosse rappresentata dall' asse stesso dell' elice. Alla fine della presente Memoria farò vedere comme questa supposizione è giusta, e come le equazioni che rappresentano il moto di una fibra rettilinea ed omogenea sono le stesse di quelle appartenenti alle oscillazioni di un' elice elastica. V' è però una notabile differenza fra i miei risultamenti e quelli degli autori che mi hanno preceduto. Secondo questi se si suppone che la fibra elastica sia spogliata in tutta la lunghezza della sua massa, e si imagini che il terzo della medesima sia concentrato nell' estremità mobile, i moti di quest' elastro immaginario devono accompagnare esattamente quelli dell' elastro vero; secondo me non e il terzo della massa dell' elastro che devo supporsi concentrato nell' estremità mobile, ma la metà. (pp. 244–5.)

[248.] Mossotti supposes his spiral wire to be without weight to be placed perpendicular to a rigid plane and compressed by a superincumbent weight; this weight is removed and the motion of the expanding spiral is then required.

Let the axis of the spiral be taken as axis of $z$, and let the axes of $x$ and $y$ be in the rigid plane perpendicular to the axis of the spiral and such that the axis of $x$ passes through one end of the spiral wire. Let $xyz$ be the co-ordinates of a point on the sectional axis of the wire distant $s$ from its extremity; $f, f', f''$ are the 'accelerating forces at the point $xyz$ parallel to the axes;

$\Delta$ is the density and $\pi r^2$ the section of the wire. Mossotti deduces the equations

$$\Delta\pi r^2 . \frac{d^2x}{dt^2} = \frac{df}{ds},$$

$$\Delta\pi r^2 . \frac{d^2y}{dt^2} = \frac{df'}{ds},$$

$$\Delta\pi r^2 . \frac{d^2z}{dt^2} = \frac{df''}{ds}.$$

The equations of the helix formed by the sectional axis of the wire are

$$\left.\begin{aligned} x &= a \cos\theta \\ y &= a \sin\theta \\ z &= a\theta \tan e \end{aligned}\right\} \text{ and again, } \begin{aligned} z &= s \sin e, \\ a\,\theta &= s \cos e, \end{aligned}$$

where $a$ is the radius of the cylinder upon which the helix lies, $e$ the angle the tangent at any point makes with the plane of $xy$, and $\tan\theta = y/x$.

[249.] The experiment made by the author to aid him in the solution of these equations was the following: A steel spiral with one extremity fixed in an immoveable plane was compressed till it formed a continuous cylindrical surface, a white line was then marked upon it parallel to the axis. Being released the wire oscillated so that the series of white marks on the different turns always remained in the same straight line parallel to the axis. Further when the oscillations were slow enough to render each white mark individually visible, they appeared always to be at equal distances from one another. From this experiment Mossotti draws the following conclusions. (1) The wire retains the form of a helix; (2) the number of turns in the helix remains unaltered, in other words, $a$ and $e$ are functions of the time, but not of the arc of the spiral; $\theta$ and $s$ are independent of the time. Since $s/\theta = a/\cos e$, it follows that either of these ratios is a quantity independent alike of the time and of the position of the point $(xyz)$ on the spiral. Terming either of them $1/\lambda$, and substituting for $xyz$ in the equations of motion, we find, after an integration with regard to $s$,

$$\frac{\Delta\pi r^2}{\lambda^2} \frac{d^2 \cos e}{dt^2} \sin\lambda s = f + c$$

$$-\frac{\Delta\pi r^2}{\lambda^2}\frac{d^2\cos e}{dt^2}\cos\lambda s = f' + c',$$

$$\Delta\pi r^2\frac{d^2\sin e}{dt^2}\frac{s^2}{2} = f'' + c'',$$

where $c$, $c'$, $c''$ are constants independent of $s$, but may be functions of $t$. To determine them put $s = 0$; at this point $f = f' = f'' = 0$ thus $c = c'' = 0$ and $c' = -\frac{\Delta\pi r^2}{\lambda^2}\frac{d^2\cos e}{dt^2}$

Let $\sigma$ be the length of the wire supposed to contain a complete number of turns, then putting $s = \sigma$ we deduce for the extremity of the wire

$$f = f' = 0,\quad \Delta\pi r^2\frac{d^2\sin e}{dt^2}\frac{\sigma^2}{2} = f''$$

[250.] In order to continue the discussion we must now make some hypothesis as to the nature of $f''$. Mossotti argues as follows:

È evidente che, se supponiamo l' elastro costipato e posto verticalmente, sovrapponendo un peso che impedisca che piu si allunghi, questo peso misurerà la somma delle forze acceleratrici verticali colle quali l' elastro si distenderebbe in quell' istante essendo in libertà, ossia la forzia $f''$. Questa forza sarà poi diversa anche nello stesso elastro variando la sua lunghezza ossia secondo i diversi stati di compressione, e la sola esperienza può somministrare la legge della variabilità della medesima.

The experience to which the author appeals is that of the above John Bernoulli, of 's Gravesande and of Francesconi. He might also have cited Hooke (see Art. 7).

Let $h$ be the height of the unstrained spiral spring, $z'$ its height at time $t$, and $\mu$ a constant; then, $f = \mu(h - z')$ is the form of the force $f''$ he assumes Here $\mu$ must be determined in each case by experiment. Hence, since $z' = \sigma\sin e$, we have

$$\frac{m}{2}\frac{d^2z'}{dt^2} = \mu(h - z'),$$

where $m = \Delta\pi r^2\sigma$ = the mass of the wire.

[251.] It is obvious that this is the same equation as would hold for the motion of a non-gravitating mass half that of the spiral supported by an elastic string of length equal to the axis of the spiral and modulus of elasticity equal to $\mu$. This is shewn at somewhat unnecessary length in the second part of Mossotti's memoir entitled: *Problema II.* The conclusion of the first part is devoted to the discussion of the simple harmonic motion given by the above equation and its application to the 'ballistic machine.' Although I am not entirely satisfied with Mossotti's method or results, the paper seems to me suggestive

[252.] *Summary.* It will be seen that these miscellaneous investigations of the first quarter of the nineteenth century were principally occupied in extending or correcting the labours of the previous century. We note also how the results of mathematical investigation by a long process were finding their way into practical text-books and being put to the test of every-day experience. From this time onwards *Galilei's Problem* will cease to occupy so much of the energy of the mathematical world. This energy will be directed to the wider question of the equilibrium and motion of elastic solids.

# CHAPTER III.

## MISCELLANEOUS RESEARCHES 1820—1830.

### NAVIER, GERMAIN, SAVART, PAGANI, AND OTHERS[1].

253. NAVIER. Navier more than any other person is to be regarded as the founder of the modern theory of elastic solids. In a memoir presented to the *Institut* on the 14th of May 1821 he gave for the first time the general equations of equilibrium and motion which must hold at every point of the interior of a body, as well as those which must hold at every point of the surface. This memoir is published in Vol. VII. of the Paris Memoirs, dated 1827. Navier had previously presented a memoir to the *Institut*, namely on the 14th of August 1820, in which he treated of the flexure of elastic plates; only an abstract of this seems to have been published. We shall now give an account of the writings of Navier which bear on our subject, taking them in the order of their composition, and not in that of their publication. There are other important memoirs on which we do not touch, as for instance one on the motion of fluids: see Saint-Venant in Moigno's *Statique*, page 695.

254. *Sur la flexion des verges élastiques courbes, par* M. Navier (*Extrait d'un Mémoire présenté à l'Académie des Sciences, le* 25 *novembre* 1819). This is published in the *Bulletin... Philomatique* for 1825, pages 98—100 and 114—118. The abstract consists of two parts; in the first, Navier considers the flexure of rods which are naturally straight, and in the second the flexure of rods which are naturally curved; the memoir was

[1] Memoirs of this period by Poisson, Cauchy and Lamé are considered in the chapters especially devoted to those writers.

written before Navier had commenced the modern theory of elastic bodies, and does not bear on that subject[1].

255. In the first part of the abstract Navier adopts the ordinary principle that the curvature of an elastic lamina at any point is proportional to the moment of the forces round this point. He supposes the rod to be originally horizontal, to be built-in at one end (*encastrée*), and to be acted on by forces at the other end. Take the axis of $x$ horizontal, and that of $y$ vertical, the origin being at the fixed end; and in the expression for the curvature neglect $(dy/dx)^2$ in comparison with unity; then from the assumed principle we have the equation

$$\epsilon\, d^2y/dx^2 = -P(c-x) + Q(f-y) \ldots\ldots\ldots\ldots (1).$$

Here $\epsilon$ is a constant proportional to the force of elasticity; $c$ and $f$ are respectively the horizontal and vertical co-ordinates of the other end; and $P$ and $Q$ are the forces acting there, parallel to the axes of $y$ and $x$ respectively. This equation is well known, allowing for possible differences as to the directions in which $P$ and $Q$ act, and also for the neglect of $(dy/dx)^2$: see for instance, Poisson's *Mécanique*, Vol. I. page 606[2].

256. Put $p^2$ for $P/\epsilon$ and $q^2$ for $Q/\epsilon$; then the integral of (1) is

$$y = f + \frac{p^2}{q^3}\left\{\frac{\sin q(c-x)}{\cos qc} - q(c-x)\right\} \ldots\ldots\ldots\ldots (2),$$

where the constants are determined so that $y$ and $dy/dx$ may vanish when $x = 0$; also the following condition must hold among the quantities

$$\tan qc = qc - q^3 f/p^2 \ldots\ldots\ldots\ldots\ldots\ldots\ldots (3).$$

The length of the curve into which the straight rod is transformed is

$$\int_0^c \sqrt{1 + \left(\frac{dy}{dx}\right)^2}\, dx;$$

[1] [Before 1819 Navier appears to have held an erroneous theory as to the position of the neutral axis of a beam. This he corrected in a lithographed edition of his lectures on applied mechanics given at the *École des ponts et chaussées* 1819—1820. No. 60 is entitled: *De la résistance à la flexion des corps prismatiques etc.* I have not been able to examine this work, but its substance doubtless appears in the later *Leçons:* see Art. 279. Ed.]

[2] [The equation is, I believe, based upon a wrong assumption: see the foot-note to Art. 75. Ed.]

this is approximately $\int_0^c \left\{1 + \frac{1}{2}\left(\frac{dy}{dx}\right)^2\right\} dx$. It will be found that if we neglect the square of $q^3f/p^2$ this gives

$$c + \frac{p^4c^3}{4q^2} + \left(\frac{3p^2}{4q^2} - \frac{p^2c^2}{2}\right)f.$$

It must however be admitted that it is not quite consistent to regard $(dy/dx)^2$ as small in comparison with unity in one part of an investigation, and to retain it in another[1].

257. In the case of rods originally curved, Navier assumes that the moment of the force is proportional to the difference between the curvature of the original rod, and the curvature of the rod as transformed by the action of the forces applied to it. He calls this a *new principle;* it is in fact that adopted by Euler in 1744. For the original rod let $s$ denote the length from the fixed end up to the point $(x, y)$, and let $\phi$ be the inclination to the axis of $x$ of the tangent at this point; let letters with accents have a similar meaning with respect to the transformed rod; then the *new principle* leads to an equation of the form

$$\epsilon\,(d\phi'/ds - d\phi/ds) = -P\,(c - x) + Q\,(f - y)$$
$$= T, \text{ let us suppose,}$$

where the notation is the same as that of the preceding article.

Hence $$\phi' - \phi = \frac{1}{\epsilon}\int T\sqrt{1 + \left(\frac{dy}{dx}\right)^2}\, dx.$$

Now $$\cos\phi' = \cos(\phi + \phi' - \phi) = \cos\phi - (\phi' - \phi)\sin\phi$$

approximately; similarly

$$\sin\phi' = \sin\phi + (\phi' - \phi)\cos\phi.$$

Also $\cos\phi = dx/ds$, $\sin\phi = dy/ds$, $\cos\phi' = dx'/ds$, $\sin\phi' = dy'/ds$.
Hence we have approximately

$$dx' - dx = -\frac{1}{\epsilon}\, dy \int\left\{1 + \tfrac{1}{2}\left(\frac{dy}{dx}\right)^2\right\} T\, dx,$$

$$dy' - dy = \frac{1}{\epsilon}\, dx \int\left\{1 + \tfrac{1}{2}\left(\frac{dy}{dx}\right)^2\right\} T\, dx.$$

Navier makes some interesting applications to the case in which the original curve is the parabola $y = lx^2/a^2$.

[1] On p. 100 in the value of $2P$ for $\sigma'$ read $\sigma_1$.

258. *Extrait des recherches sur la flexion des plans élastiques.* This is published in the *Bulletin...Philomatique* for 1823, pages 95—102. We have here an abstract of the memoir sent to the *Institut* on the 14th of August 1820; it consists of 11 sections.

259. In the first section Navier adverts to the prize essay by Mdlle Germain, to Poisson's memoir of 1814, and to the integration of the partial differential equation by Fourier (see Art. 209); he then states the object of his own memoir thus:

Les recherches dont cet article contient l'exposé avaient pour objet principal les lois suivant lesquelles s'opère la flexion d'un plan élastique, soutenu sur des appuis dans une position horizontale, et chargé par des poids. Elles sont contenues dans un Mémoire présenté à l'Académie des Sciences, le 14 août 1820, et dans une Note manuscrite, remise quelques mois après aux commissaires chargés d'examiner ce Mémoire.

260. In his sections 2 and 3 Navier investigates the general equation which must hold for the equilibrium of an elastic plate that deviates slightly from a plane, and also the conditions which must hold at the boundary; but his process is very obscure. With respect to the elastic forces he appears to adopt an hypothesis like that of Mdlle Germain; but he does not explain it clearly. He seems to present as *exact* the following expression for the virtual moment

$$\epsilon h^3 \left\{ \left( \frac{1}{R'} + \frac{1}{R''} \right) \delta \left( \frac{1}{R'} + \frac{1}{R''} \right) - \delta \frac{2}{3 R' R''} \right\},$$

where $\epsilon$ is a constant coefficient, $h$ is the thickness, and $R'$ and $R''$ are the two principal radii of curvature: but I do not understand how he obtains this. Then he says that this is approximately equal to $\epsilon h^3 \text{E}\delta\text{E}$, where E stands for $\frac{d^2z}{dx^2} + \frac{d^2z}{dy^2}$; but here again I do not follow his reasoning. Starting with this expression and proceeding as in Lagrange's *Mécanique Analytique* (Vol. I. pages 140—142 of Bertrand's edition), Navier obtains the general equation of equilibrium, and also the equations which must hold at the boundary. The general equation coincides in *form* with that given by Poisson on page 219 of his memoir of 1814. Navier puts it thus, where $T$ stands for $\int(Xdx + Ydy)$:

$$Z - X\frac{dz}{dx} - Y\frac{dz}{dy} - T\left(\frac{d^2z}{dx^2} + \frac{d^2z}{dy^2}\right) + \epsilon h^3\left(\frac{d^4z}{dx^4} + 2\frac{d^4z}{dx^2\,dy^2} + \frac{d^4z}{dy^4}\right) = 0.$$

This equation does not agree with that obtained by Poisson in his later researches: see our account of the sixth section of the memoir of 1828.

As to the equations which must hold at the boundary we may observe that Navier has expressed them in a form which is almost unintelligible, and as we shall see hereafter, Poisson would not accept them: when we express them in the notation which Navier ought to have used we find that Poisson's objection does not hold. But these boundary equations do not agree with those obtained by Poisson in his memoir of April 1828, nor with those since proposed by Kirchhoff in opposition to Poisson's. Suppose the boundary of the surface to be entirely free, then Navier has *four* equations which must hold; but two of these apply to forces *in* the plane leaving *two* to apply to the case for which Poisson has *three.* We shall find that the matter was noticed in a controversy which arose between Poisson and Navier.

261. Navier in his sections 4—9 applies his formula to the case of a horizontal rectangular plate, acted on by no force except weights, disposed in any arbitrary manner over the surface. He obtains a general solution in the form of expressions involving sines and cosines of multiple angles, and illustrates it by considering two special cases, namely, first that in which the weight is diffused uniformly over the whole plate, and next, that in which the weight is concentrated on an indefinitely small area close to an assigned point.

262. By suppressing one co-ordinate Navier renders his equations applicable to an elastic *lamina;* and for an example takes the case of a horizontal lamina supported at each end; he thus obtains known results, namely the equation, $f = \frac{5qa^3}{16 \times 24\beta}$, of Poisson's *Mécanique*, Vol. I. page 636, and the equation, $f = \frac{qa^3}{48\beta}$, of the same volume, page 641.

263. Navier in his section 10 draws from his formula some inferences as to the conditions for the *rupture* of the plate, in the two cases mentioned in Art. 261 of the weight uniformly diffused and the same weight concentrated at the middle point of the plate. Suppose that in the former case a weight $W$ will produce rupture, then he says that a weight $W/4$ will produce rupture in the second case; but it seems to me that according to his formula it should be $4W/\pi^2$.

264. In the differential equation given in Art. 260 it will be seen that the *third* power of the thickness occurs as a coefficient; Navier says in his last section that persons who have attended to the subject are not all agreed as to the power of the thickness which should occur; but he seems confident that he is right in using the third power. It is however certain that he is wrong in using $h^3$ instead of $h^2$ if he gives to $X$, $Y$, $Z$ their usual significations: see the equation (10) in our discussion of the sixth section of Poisson's memoir of 1828, confirmed by Clebsch on page 307 of his work when we recur to page 289 for the values of $C'$, $A''$, $B''$. This is quite consistent with the fact noticed by Navier that the expression for the equilibrium of an elastic lamina involves the *cube* of the thickness.

265. *Mémoire sur les lois de l'équilibre et du mouvement des corps solides élastiques.* This memoir occupies pages 375—393 of the *Mémoires...de l'Institut,* Vol. VII. published in 1827; the memoir was read to the Academy on the 14th of May 1821.

266. This memoir is justly famous as being the real foundation of the modern theory of elastic solids. On pages 375—384 an investigation is given of the three equations which hold at any point of the interior of an elastic body; they are obtained in a form equivalent to that given by Lamé, *Leçons,* p. 66, supposing $\lambda = \mu$. Let $\rho$ denote the distance between two particles in the natural state of a body, and $\rho_1$ the distance when forces have been applied to the body; then Navier assumes that the mutual action between the particles is $(\rho_1 - \rho) f(\rho)$, where $f(\rho)$ denotes some function which decreases rapidly as the distance $\rho$ increases; the direction of the force is assumed to be that of the original direction of $\rho$: then the constant $\lambda$ or $\mu$ of Lamé is replaced by $\epsilon$,

where $\epsilon = \frac{2\pi}{15}\int_0^\infty \rho^4 f(\rho)\,d\rho$. The investigation though of great interest as being the first of the kind would not now be accepted as satisfactory. Instead of $(\rho_1 - \rho) f(\rho)$ the expression of the force would now be taken to be $\phi(\rho_1)$, that is $\phi(\rho + \rho_1 - \rho)$, that is approximately $\phi(\rho) + (\rho_1 - \rho)\,\phi'(\rho)$, where $\phi'(\rho)$ would correspond to the $f(\rho)$ of Navier. Also the assumption that the body is continuous, so that summation may be replaced by integration, is not now accepted: see Art. 436 of my account of Poisson's memoir of April 1828.

267. Another investigation is given by Navier on his pages 384—393, which furnishes not only the general equations which must hold at every point of the interior, but also the special equations which must hold at every point of the boundary. The general equations thus obtained agree in *form* with those obtained by the first investigation; the special equations agree practically with those given by Lamé, *Leçons*, p. 20, though in expressing them Navier, by not giving sufficient generality to his symbols, is led to suppose a double statement necessary.

268. I have said that Navier's second investigation leads to general equations of the same *form* as the first; Navier holds that the two forms perfectly agree, and they ought to do so if both processes are sound. But the fact is that if a mistake is corrected by removing $\frac{1}{2}$ which occurs near the foot of page 387, it will be found that in the second form of the equations we get $2\epsilon$ as a coefficient instead of $\epsilon$ in the first form, $\epsilon$ having the value already assigned. Thus one of the two forms must be wrong. The second investigation is conducted by the aid of the Calculus of Variations, but I do not understand the process. The precise point at which the difficulty enters is on page 386, where after saying correctly that a certain force is proportional to $f$, Navier adds:

Le moment de cette force, cette expression étant prise dans le même sens que dans la *Mécanique Analytique*, sera évidemment proportionnel à $f\delta f$, ou à $\frac{1}{2}\delta f^2$.

This seems to me a purely arbitrary statement. I may observe that the letter $f$ is unsuitable, for it has been already appropriated by Navier in the symbol $f(\rho)$.

269. There is a notice of this part of Navier's memoir by Saint-Venant in Moigno's *Statique* page 717: see also page 711. Saint-Venant would seem to agree with me in considering the statement I have quoted as arbitrary; for he says that Navier '*posait* pour le travail ou virtuel moment etc.' Saint-Venant does not notice that if we adopt this expression for the virtual moment the result of the second investigation is inconsistent with that of the first[1].

270. *Sur les lois de l'équilibre et du mouvement des corps solides élastiques. Extrait d'un Mémoire présenté à l'Académie des Sciences, le* 14 *mai* 1821: *par* M. Navier. This occupies pages 177—183 of the *Bulletin...Philomatique* for 1823. The preamble of the memoir of 1821 is here reprinted. Navier then says that there are two ways in which the investigation can be carried on, namely, the two employed in the original memoir; after briefly alluding to the first he fully expounds the second.

271. *Observations communiquées par* M. Navier *à l'occasion du Mémoire de* M. Cauchy. This occurs on pages 36 and 37 of the *Bulletin...Philomatique* for 1823.

Cauchy was one of the persons appointed by the Paris Academy to examine Navier's memoir of August 1820. Cauchy had inserted on pages 9—13 of the *Bulletin* for 1823 an account of some investigations of his own relative to elastic bodies, and in these he mentioned the memoir of Navier, and made some brief remarks upon it. Navier alludes to these remarks, but his main object seems to be to draw attention to his own priority, and he mentions that he had since sent another memoir to the Academy, namely on the 14th of May 1821. We learn from the following sentence the nature of the memoir of August 1820:

La démonstration de l'équation différentielle de la surface élastique ne forme que la moindre partie du travail contenu dans le Mémoire du 14 août 1820, et l'auteur n'y attache aucune importance. L'objet

[1] Lamé on his page 79 is perhaps alluding to such a matter.

spécial de ce travail est la recherche des conditions de la flexion d'un plan chargé par des poids, recherche fondée sur l'intégration de cette équation, comme depuis long-temps.

[272.] *Mémoire sur les ponts suspendus.* Paris, 1823. This memoir is accompanied by a report of Navier to M. Becquey on the results of his examination of the English suspension bridges. There is little in the memoir which belongs properly to our subject. We may note however § XI. (pp. 147—160) entitled: *Des vibrations longitudinales des chaînes, dues à l'élasticité du fer.* This contains a somewhat lengthy discussion of the ordinary equation for the longitudinal vibrations of a rod. The memoir concludes with a long extract from Barlow's *Essay on the strength and stress of timber,* 1817, particularly the experiments contained in that work on the resistance of iron. A few additional experiments due to Brown and to Brunel are cited at the end.

273. *Note sur les effets des secousses imprimées aux poids suspendus à des fils ou à des verges élastiques.* This occurs on pages 73—76 of the *Bulletin...Philomatique* for 1823. Navier quotes the known formula for the extension of an elastic string suspended by one end, and having a weight at the other. He then gives the approximate results of the following easy dynamical problem: Suppose the weight suspended to be very large compared with the weight of the string itself, and let a certain velocity downwards be communicated to the weight; then find the greatest extension and tension at any assigned point of the string. The formula obtained for the tension at the fixed end is illustrated by some numerical examples.

[274.] *Solution de diverses questions relatives aux mouvements de vibration des corps solides. Bulletin des Sciences par la Société Philomatique.* Année 1824, Paris. pp. 178—181. Navier treats here by usual methods problems which presented even at that date no novelty. They are:

(1) Motion of two points connected by an elastic string.

(2) Longitudinal vibrations of an elastic rod with one end fixed.

(3) Special case of the same rod when the other end is free.

(4) Special case when the same rod is treated as infinitely long.

The paper is of no importance.

[275.] *Expériences sur la Résistance de diverses substances à la rupture causée par une tension longitudinale.* Par M. Navier. This occupies pages 225—240 of the *Annales de Chimie,* Vol. XXXIII. 1826. An abstract appeared in the *Quarterly Journal of Science,* Second Series, Vol. I., page 223.

Navier describes his labours as follows:

Les recherches dont je me suis occupé ayant principalement pour objet la résistance des tuyaux et autres vases exposés à des pressions intérieures, j'ai soumis à l'expérience la tôle ou fer laminé, le cuivre laminé, le plomb laminé et le verre, dont on fait quelquefois des vases dans les appareils de physique et de chimie.

The general results of his experiments are given on p. 226, and in a concise tabulated form on p. 240. The experiments themselves seem too few to be of any great value. I may however note experiments 26 and 27 (pp. 238—239) entitled: *Vases sphériques rompus par l'effet d'une pression intérieure.* So far as I am aware, these are the first experiments ever made to determine the internal pressure which will rupture a spherical shell. The spherical iron shells of Navier were not very satisfactorily constructed for comparison with theory, for they consisted of two halves riveted together. In both cases the rupture was first manifested by the formation of a '*très-petite fente,*' through which escaped the water, by means of which the internal pressure was applied.

276. A controversy between Navier and Poisson was carried on during the years 1828 and 1829; an account of this is given in the chapter devoted to Poisson.

277. A memoir by Lamé and Clapeyron entitled: *Mémoire sur l'équilibre intérieur des corps solides homogènes* was sent to the Paris Academy. The memoir was referred to the judgment of Poisson and Navier; a report on it was drawn up by the latter

and communicated to the Academy on the 29th of September 1828: the report is printed in Crelle's *Journal für...Mathematik*, Vol. VII. 1831, pages 145—149.

Navier's report contains nothing of importance; he speaks highly of the memoir, indicating however a certain anxiety which seems to have been habitual in him with respect to his own claims to priority. Two points may be noticed in the report.

> Ils ont reconnu, d'une part qu'un élément de volume, dont la figure serait, dans l'état naturel du corps, une sphère d'un rayon très petit, se changeait toujours en un ellipsoïde.

No doubt this proposition is very closely connected with some which Lamé and Clapeyron give, but I do not see it explicitly stated in the memoir. They do give the *ellipsoid of elasticity*, and the *surface of directions:* see our account of the memoir in the chapter devoted to Lamé.

After alluding to some researches of Cauchy in his *Exercices de Mathématiques* the report says:

> MM. Lamé et Clapeyron remarquent que la théorie exposée dans leur ouvrage diffère essentiellement de celle qu'avait adoptée M. Cauchy.

I see however no allusion to M. Cauchy in the memoir.

[278.] *Note sur la flexion d'une pièce courbe dont la figure naturelle est circulaire. Annales des ponts et chaussées*, 1[re] série, 1[er] semestre. Paris, 1831, pp. 428—436.

In his *Résumé des leçons données à l'École royale des ponts et chaussées* Navier devotes a section to the resistance of prismatic beams whose axes are curved (1st edition, 1826, p. 243).

The equation he makes use of has the following form:

$$\epsilon\,(1/\rho' - 1/\rho) = -P\,(a - x) + Q\,(b - y).$$

In other words, he takes the bending moment at any section proportional to the difference of the imposed and primitive curvatures. He applies his equation to several cases, as where the primitive axis is of parabolic shape. (Cf. Art. 257.)

In the note cited above Navier supposes that the primitive shape is circular. If $r$ be the radius of the circle, before application of force, $\tan\phi = dy/dx$, $\Phi$ = whole angle of arc, he finds equations of the form:

$$\epsilon(x'-x) = -r^3\{P[\sin\Phi(\sin\phi-\phi\cos\phi)+\tfrac{1}{2}\sin^2\phi+\cos\phi-1] + Q[\tfrac{1}{2}\phi-\tfrac{1}{2}\sin\phi\cos\phi-\cos\Phi(\sin\phi-\phi\cos\phi)]\};$$

$$\epsilon(y'-y) = r^3\{P[\sin\Phi(\phi\sin\phi+\cos\phi-1)+\tfrac{1}{2}\sin\phi\cos\phi + \tfrac{1}{2}\phi-\sin\phi]+Q[\tfrac{1}{2}\sin^2\phi-\cos\Phi(\phi\sin\phi+\cos\phi-1)]\},$$

for the displacements of any point determined by $(x, y, \phi)$ parallel respectively to the directions of the forces $Q$ and $P$ applied at the terminal $(\Phi)$. These equations are then applied to various interesting cases on the supposition of $\Phi$ being small, and the displacements at the terminal are calculated to the 5th power of $\Phi$.

Navier afterwards treats the case of a circular arc bent only by its own weight, and determines values for the like displacements. There is no reference in the paper to the work of Lagrange (see Art. 100) or other investigators in the same field. This note is practically reproduced in the second edition of the *Résumé*: see pp. 292—295 and pp. 299—302.

I think that Navier's assumption as to the bending moment is invalid, although the necessary correction does not alter the *form* of his equations. There would in general be a longitudinal force at some section, and hence, the 'neutral axis' not passing through the centres of section, the moment of resistance to flexure ($\epsilon$) would not be an elastic constant.

[279.] *Résumé des Leçons données à l'école des ponts et chaussées sur l'application de la Mécanique à l'établissement des constructions et des machines.* Première Partie. This book was first published in 1826. Navier himself corrected and supplemented it for a second edition in 1833[1], while a third appeared under the superintendence of Saint-Venant in 1863. The second and third parts contain nothing of value for our present purpose. We are concerned here only with the second edition as containing Navier's final revision. It differs considerably from the first

[1] This appears to be the last occasion on which Navier busied himself with elasticity. He died in 1836 and a funeral oration was delivered in the name of the *Institut* by the engineer and elastician Girard.

owing to the progress which had been made in experimental elasticity. The first forty-three pages of the book are devoted principally to experimental results drawn from Rennie, Tredgold, Barlow, Rondelet, Vicat, etc. Article III. (pp. 43—66) contains the usual theory of beams subject only to transverse strain, together with further experimental results from the same sources. Article IV. (pp. 66—99) deals with special cases of the common theory. Articles V. and VI. (pp. 99—112) treat of torsion with reference to the then recent work of Cauchy, Lamé and Clapeyron.

280. An approximate formula relating to the torsion of a rectangular prism is quoted, on page 102 of the first volume, from Cauchy's *Exercices de Mathématiques*, Vol. IV. page 59: see my account of Saint-Venant's memoir on *Torsion*. Navier adds the following note on his page 102:

Les équations différentielles qui expriment les conditions de l'équilibre et du mouvement des corps solides, et qui sont la base des recherches dont il s'agit, ont été données en premier lieu par l'auteur, pour le cas d'un corps homogène, dans un mémoire présenté en 1821 à l'Académie des sciences, et imprimé dans le tome 7[e] de ses Mémoires. Cette matière a été depuis le sujet de recherches très-étendues, qui sont contenues principalement dans un mémoire de MM. Lamé et Clapeyron, présenté à l'Académie des sciences en 1828, et imprimé dans le journal de mathématiques de M. Crelle, dans un mémoire de M. Poisson imprimé dans le tome 8[e] des Mémoires de l'Académie, et dans les Exercices de mathématiques de M. Cauchy.

On page 108 of his first volume Navier gives a formula for the extreme torsion which can be admitted if we are to avoid rupture: this he says is not due to Cauchy, but may be obtained by the analysis which Cauchy used. Saint-Venant on page 413 of his memoir on *Torsion* notices the matter, and explains how Navier's formula is obtained; but he proceeds to demonstrate that it is not trustworthy.

[281.] Art. VII. (pp. 112—120) contains general remarks on the limits of safety for various kinds of stress. This concludes the first section. The second and third sections, devoted to the equilibrium of masses of earth, etc., and to arches, do not directly concern us. The fourth section returns to the subject of elastic beams, under the heading of constructions in carpentry. Art. I. (pp. 227—245) attempts a theory of continuous beams. So far as the results treat of beams *built-in at both ends*, I believe them to be erroneous, for Navier assumes in this case that the *moment de la résistance à la flexion* is a constant depending on the material of the rod, which I think impossible, as longitudinal strain is in such cases introduced. The following article reproduces Euler and Lagrange's theory of columns; both this article and nearly all the following labour under the same error as the first; i.e. the moment of resistance to flexure is treated as an elastic constant although there is longitudinal strain. Various experimental results are quoted from Girard, Lamandé, Aubry and others. A passage on p. 258 may be quoted with which I can hardly agree. After giving several reasons why experiments as to the force which will just bend a column do not agree with theory, Navier continues:

Mais, lorsqu'on prend les précautions convenables pour accorder les circonstances de l'expérience avec les hypothèses sur lesquelles les formules sont fondées, le résultat est alors représenté exactement par cette formule.

The precautions appear to be that the terminal force shall be applied exactly at the axis of figure.

We shall return to this work when considering later the 1864 edition due to Saint-Venant.

[282.] One further remark may be made with regard to Navier. He seems to have been the first to notice that problems relating to reactions, for the determination of which elementary statics does not provide sufficient equations, are perfectly determinate when account is taken of the elasticity of the reacting bodies. The matter is considered by him in a note contributed to the *Bulletin......Philomatique* 1825, p. 35, and entitled: *Sur des questions de statique dans lesquelles on considère un corps supporté par un nombre de points d' appui surpassant trois.* A number of problems of this kind are solved in the *Résumé des Leçons* mentioned above, and were first given by Navier in his lectures for 1824: see Saint-Venant's account in his edition of the *Résumé*, p. cvii.

283. GERMAIN. 1821. *Recherches sur la théorie des surfaces élastiques, par* Mdlle Sophie Germain. Paris, 1821. This is in quarto; the title and preface occupy x. pages, the text occupies 96 pages, and there is one plate. A list of *Errata* is given on the two last pages, but it is far from being complete[1].

284. The preface gives an interesting account of the circumstances which led Mdlle Germain to devote her attention to the subject. As soon as she became acquainted with Chladni's experiments she wished to determine the laws to which the phenomena described by him are subject; she says:

> Mais j'eus occasion d'apprendre d'un grand géomètre, dont les premiers travaux avaient été consacrés à la théorie du son, que cette question contenait des difficultés que je n'avais pas même soupçonnées. Je cessai d'y penser.

The *grand géomètre* was doubtless Lagrange.

The French *Institut* proposed as a subject for a prize: *De donner la théorie mathématique des vibrations des surfaces élastiques, et de la comparer à l'expérience.* October 1st, 1811, was fixed as the date for receiving the essays of candidates. The programme relating to this proposition is reprinted on pages 253—357 of Chladni's *Traité d'Acoustique* 1809: it is stated that the prize was offered by the desire of the Emperor Napoleon. Mdlle Germain was a competitor for the prize; she says respecting her essay:

> J'avais commis des erreurs graves; il ne fallait qu'un simple coup d'œil pour les apercevoir; on aurait donc pu condamner la pièce sans prendre la peine de la lire. Heureusement, un des commissaires, M. de Lagrange, remarqua l'hypothèse; il en déduisit l'équation que j'aurais dû donner moi-même, si je m'étais conformée aux règles du calcul.

It appears from the *Annales de Chimie*, Vol. 39, 1828 (pp. 149 and 207) that the following note was written by Lagrange: *Note*

[1] [A few particulars as to Mdlle Germain's life will be found in the *Biographie universelle*, its *Supplément*, in the *Journal des Débats*, May 18, 1832, reproduced in the prefatory matter to the lady's own *Considérations sur l'état des sciences et des lettres*, published in 1833, two years after her death, and in the *Oeuvres philosophiques*, Paris, 1879.] ED.

*communiquée aux Commissaires pour le prix de la surface élastique* (décembre 1811).

L'équation fondamentale pour le mouvement de la surface vibrante ne me paraît pas exacte, et la manière dont on cherche à la déduire de celle d'une lame élastique, en passant d'une ligne à une surface, me paraît peu juste. Lorsque les $z$ sont très-petits, l'équation se réduit à

$$\frac{d^2z}{dt^2} + gElc\left(\frac{d^6z}{dx^4\,dy^2} + \frac{d^6z}{dy^4\,dx^2}\right) = 0,$$

mais en adoptant, comme l'auteur, $1/r + 1/r'$ pour la mesure de la courbure de la surface, que l'élasticité tend à diminuer, et à laquelle on la suppose proportionnelle, je trouve dans le cas de $z$ très-petit une équation de la forme

$$\frac{d^2z}{dt^2} + k^2\left(\frac{d^4z}{dx^4} + 2\frac{d^4z}{dx^2\,dy^2} + \frac{d^4z}{dy^4}\right),$$

qui est bien différente de la précédente.

Fourier giving a notice of Legendre's second supplement to the *Théorie des Nombres*, 1825, says:

Il cite une proposition remarquable et une démonstration très-ingénieuse que l'on doit à mademoiselle Sophie Germain. On sait que cette dame cultive les branches les plus élevées de l'analyse, et que l'Académie des Sciences de l'Institut lui à décerné en 1825 un de ses grands prix de mathématiques. *Mémoires de l'Acad.* VIII. 1829, page x.

The *Institut* proposed the subject again, fixing Oct. 1st, 1813, as the date for receiving the essays of candidates. Mdlle Germain was again a competitor; she says:

J'envoyai, avant le 1er octobre 1813, un Mémoire dans lequel se trouve l'équation déjà connue, et aussi les conditions des extrémités déterminées à l'aide de l'hypothèse qui avait fourni l'équation. Ce Mémoire est terminé par la comparaison entre les résultats de la théorie et ceux de l'expérience.

The judges made honourable mention of her essay and approved of the comparison between theory and observation.

The *Institut* proposed the subject once more, asking for a demonstration of the equation, and fixing October 1st, 1815, as the date for receiving the essays of candidates. Mdlle Germain competed and gained the prize. The judges must have been far from severe, as they awarded the prize though they were not quite satisfied with her demonstration; and moreover she admits that the agreement between theory and observation was not close, as she had taken without due examination a formula from a memoir by Euler, *De sono campanarum*, which was incorrect (see Art. 93 and footnote). Since that date Mdlle Germain had on various occasions given renewed attention to the subject, and had been assisted by some explanations given to her by Fourier.

There are also various allusions to Poisson, and to his memoir of August 1814, though he is not mentioned by name. In my account of this memoir by Poisson I conjecture that he refers to a memoir by Mdlle Germain (Art. 414); and as he implies that the equation is there given correctly, I suppose that he refers to the *second* memoir which she wrote. In the allusions which Mdlle Germain makes to Poisson there is I think a rather defiant tone, notwithstanding the elaborate praise she confers on him, such as:

Un nom justement célèbre...,...le talent qui caractérise tous ses ouvrages...,...ce savant auteur...,...l'autorité attachée à son nom...,... dont les talens m'inspirent la plus haute estime.

Poisson and Mdlle Germain had both obtained the same equation for the vibration of a plane elastic surface, but by very different methods. The equation is that which we shall presently give, denoted by (B): see Art. 290.

285. Mdlle Germain's work is divided into four sections: The first section is entitled: *Exposition des principes qui peuvent servir de base à la théorie des surfaces élastiques;* it occupies pages 1—12. Mdlle Germain takes the following hypothesis: let $R$ and $R'$ denote the two principal radii of curvature of a surface in its natural condition; let $r$ and $r'$ be the two principal radii of curvature at the same point when the surface has been brought into a new form by any forces, then the action of the forces of elasticity which act on the surface is proportional to $1/r + 1/r' - (1/R + 1/R')$. The authoress tries to justify this hypothesis in two ways; but it seems to me that her statement of the hypothesis is vague, and that the general reasoning by which she tries to support it is quite ineffectual; her pages 2—5 are to me specially unsatis-

factory. She appears to regard her hypothesis as *absolutely* true, and not as a mere *approximation*; and it is not until she arrives at her page 20 that she begins to approximate by supposing that the surface after the action of the forces differs but slightly from its original form. Mdlle Germain seems to say that this hypothesis is not that which she originally adopted. Thus we read on her page viii.:

Dans la pièce que j'adressai à l'Institut avant le 1[er] octobre 1815, je donnai une hypothèse plus générale que celle qui se trouvait dans mes précédens Mémoires. J'essayai de démontrer ma nouvelle hypothèse.

And again on page 27:

L'hypothèse contenue dans le premier de mes Mémoires sur les surfaces élastiques; hypothèse qui, ainsi que je l'ai déjà dit, ne pouvait s'appliquer qu'au seul cas des plaques élastiques.

286. On her page 8 Mdlle Germain adverts to the memoir by Poisson of August 1814; she says with respect to this:

L'auteur borne ses recherches au cas de la surface élastique naturellement plane, et il est facile de voir, en poursuivant la lecture de son Mémoire, que l'hypothèse qu'il admet, mène à regarder les forces d'élasticité qui agissent sur ce genre de surface, comme proportionelles à la quantité $1/r - 1/r'$.

I do not see how this statement with respect to the elastic forces is justified by anything in Poisson's memoir; nor can I assent to the general reasoning by which Mdlle Germain in her next paragraph endeavours to shew that it is practically the same thing in this case whether we take the elastic force to be proportional to $1/r + 1/r'$ or to $1/r - 1/r'$.[1] She says:

Si, comme il me semble permis de le supposer ici, les quantités $1/r + 1/r'$ et $1/r - 1/r'$ sont proportionelles,...

Also on her page vii. she says:

Il résulte d'un théorème dû à l'auteur même du Mémoire dont je viens de parler, que mon hypothèse conduirait également à son équation générale.

The theorem here meant is probably that in the Calculus of Variations which is given at the end of Poisson's memoir; but

[1] [Mdlle Germain is correct in her result if not in her reasoning: see the footnote I have put to Art. 419.] Ed.

there is no need of it for the present purpose, inasmuch as Poisson in the course of his memoir does obtain from his hypothesis the same equation as Mdlle Germain obtains from hers.

287. Poisson as we have just said obtained the general equation which we may call Germain's; but as we shall see in our account of his memoir he postponed the determination of the conditions which must hold at the boundary (Art. 418). Our authoress says on her page 10:

J'ai long temps attendu que l'auteur publiât la détermination dont il s'agit ici; j'aurais désiré, dans l'intérêt de la question, qu'il développât lui-même toutes les conséquences de l'hypothèse qu'il a adoptée.

288. The second section of Mdlle Germain's work is entitled: *Recherche des termes qui doivent conduire à l'équation de la surface élastique;* it occupies pages 12—19. The object of the section is to obtain a general equation for the equilibrium of an elastic surface by imitating the methods used by Lagrange; she however introduces difficulties of her own. The following is the result: let $N^2$ denote a certain constant, and $dm$ an element of mass; then integrate by parts the expression:—

$$-\iint N^2 \left\{\frac{1}{r}+\frac{1}{r'}-\left(\frac{1}{R}+\frac{1}{R'}\right)\right\} \delta\left(\frac{1}{r}+\frac{1}{r'}\right) dm$$

$$+\iint \frac{N^2}{2}\left\{\left(\frac{1}{r}+\frac{1}{r'}\right)^2-\left(\frac{1}{R}+\frac{1}{R'}\right)^2\right\} d\delta m;$$

the terms which remain under the double integral sign must be equated to the terms which express the action of the accelerating forces; the terms which are outside the *double* integral sign will determine the conditions which must hold at the boundary. But the process by which this general conclusion is obtained seems to me very arbitrary and obscure, especially pages 14, 16, 17. It is shewn by Kirchhoff in Crelle's *Journal*, Vol. 40, page 53, that the solution given by Mdlle Germain is untenable.

289. On her pages 13 and 14 Mdlle Germain endeavours to shew by general reasoning that her constant $N^2$ must contain as a factor the *fourth* power of the thickness of the vibrating body.

Poisson in his memoir of April 1828 found only the *second* power of the thickness, and this is confirmed by the researches of Clebsch in his treatise on Elasticity.

She says on page 14:

Dans la suite de ce Mémoire je présenterai encore quelques observations sur la détermination du coefficient $N^2$.

But I cannot perceive that she has kept this promise.

290. The third section of Mdlle Germain's work is entitled: *Equations de la surface cylindrique vibrante et de l'anneau circulaire;* it occupies pages 19—75. This section consists of various parts; the first of these extends to page 27, and in this the differential equations of motion of the problem are definitely obtained. By a circular ring here is meant an indefinitely short cylinder, and the differential equation for this can be found from that for the general circular cylinder, and so the authoress confines herself to the latter; from this she also deduced the equation for the case of a plane lamina. She uses for $1/R+1/R'$ the exact value, namely $-1/a$, where $a$ is the radius of the circle; for $1/r+1/r'$ she uses an approximate value obtained by the supposition that the elastic surface deviates but little from its original form. The process is that of Lagrange, by which the solution of a mechanical problem is made to consist of a process in the Calculus of Variations. The following is the form of the result:

$$N^2\left\{\frac{d^4r}{d\xi^4}+2\frac{d^4r}{d\xi^2\,ds^2}+\frac{d^4r}{ds^4}-\frac{1}{a^2}\left(\frac{d^2r}{d\xi^2}+\frac{d^2r}{ds^2}\right)\right\}+\frac{d^2r}{dt^2}=0\text{.....(A)}.$$

Here $r$ is the difference between the distance of a point from the axis of the cylinder at the time $t$ and its original distance; but from the form of the equation $r$ may also be taken to be the distance itself at the time $t$. The variables $s$ and $\xi$ in conjunction with $r$ determine the position of the point; $s$ is the arc which is intercepted between the generating line of the cylinder corresponding to the point and a fixed generating line; for $\xi$ in the original work we have usually $x''$, which I consider to be inadequately defined, but practically instead of $x''$ we have $x$ which denotes a distance measured along the generating line from one end. This equation is denoted by (A) in the original work. When we have occasion to refer to it we shall suppose that $\xi$ is replaced by $x$.

Suppose $a$ infinite, and change $\xi$ and $s$ into the ordinary rectangular co-ordinates; also put $z$ for $r$; we thus obtain

$$N^2 \left\{\frac{d^4z}{dx^4} + 2\frac{d^4z}{dx^2\,dy^2} + \frac{d^4z}{dy^4}\right\} + \frac{d^2z}{dt^2} = 0 \ldots\ldots\ldots\ldots (B);$$

this equation the authoress gives as that for the vibration of a plane lamina.

Again in (A) efface the terms which involve $\xi$; we thus obtain

$$N^2 \left\{\frac{d^4r}{ds^4} - \frac{1}{a^2}\frac{d^2r}{ds^2}\right\} + \frac{d^2r}{dt^2} = 0 \ldots\ldots\ldots\ldots\ldots (C);$$

this the authoress gives as the equation for the vibrations of a circular ring.

291. I have not found the general equation (A) in any other place, and I cannot understand the demonstration by which it is obtained. The process seems to have been constructed with the express purpose that it should by proper supposition lead to the equation for the vibrations of a plate, that is to equation (B). The latter coincides with that obtained by Poisson on page 221 of his memoir of August 1814, and also on page 533 of his memoir of April 1828; this is also confirmed, at least as approximately true, by Clebsch in his treatise on Elasticity. Mdlle Germain states on her pages vi. and 27 that Lagrange had deduced his equation (B) from her hypothesis; but Poisson says on page 439 of the *Annales de Chimie*, Vol. 38, 1828:

> Je n'ai vu nulle part que Lagrange eût déduit de cette hypothèse l'équation relative aux petites vibrations des plaques élastiques que l'on a trouvée dans ses papiers sans aucune démonstration, et qu'il n'a pas insérée dans la seconde édition de la Mécanique analytique, où il a seulement donné l'équation d'équilibre de la membrane flexible.

See however our Art. 284, page 148.

292. One point connected with the Calculus of Variations must be noticed. Consider the term

$$\iint Q\,\frac{p\delta p + q\delta q}{k}\,dx\,dy,$$

where $p$ stands for $dz/dx$, and $q$ for $dz/dy$; also $k = \sqrt{1 + p^2 + q^2}$. This term occurs on page 25; the notation is the same as Mdlle Germain's, except that she has accents on the variables which are unnecessary for our purpose. $Q$ may be regarded as a function of $x$ and $y$; in her notation it represents $\frac{N^2}{a}\left(\frac{d^2r}{dx''^2} + \frac{d^2r}{ds^2}\right)$. Now the above term is to be transformed; it is sufficient for our purpose to attend to $\delta p$ only. We have then, according to well-known principles,

$$\iint \frac{Qp}{k} \delta p \, dx \, dy = \int \frac{Qp}{k} \delta z \, dy - \iint \frac{d}{dx}\left(\frac{Qp}{k}\right) \delta z \, dx \, dy.$$

Now Mdlle Germain instead of $\frac{d}{dx}(Qp/k)$ uses $Q\frac{d}{dx}(p/k)$; so that she omits $(p/k)\frac{dQ}{dx}$; and for this she offers no justification whatever.

293. I will notice two other points as specimens of the unsatisfactory way in which the process is conducted. On her page 23 she wishes to shew that a certain angle $\omega$ is equal to another angle $\nu$; this she infers from the fact that the straight lines which form $\omega$ are respectively at right angles to those which form $\nu$: the argument would be sufficient if the four straight lines were all in the same plane, but they are not. However in a subsequent publication by the authoress, she seems to admit that pages 21—25 of this section are unsatisfactory[1].

Again, on page 26 she has an equation:

$$\delta z' \, dx' \, dy' = \cos^2 \nu \, \delta r \, dx \, ds;$$

she wishes to change $\cos \nu$ into unity and justifies the step thus:

Mais il est évident que la valeur de $\cos \nu$ dépend uniquement du choix des coordonnées, c'est-à-dire de leur situation autour du centre du cercle qui a été pris pour origine; or, la valeur de $\nu$ ne peut varier qu'entre les limites 0 et $\pi/2$; *la valeur de* $\cos \nu$ *prise entre ces limites, se réduit à l'unité.*

The words which I have put in italics have no meaning as they stand.

[1] Instead of her process on page 23 it would have been better to assume a new variable $x''$ connected with the old $x$ by the relation $x = x'' \cos \nu$. Then on page 24 it is assumed that $dx/dx''$ is constant, which is not in harmony with the meaning of $\nu$.

294. We have seen in Art. 290, that Mdlle Germain gives a certain equation (C) as belonging to the vibrations of a circular ring. According to her this equation was given by Euler in Vol. x. of the Memoirs of St Petersburg in a memoir *De sono campanarum*, but with $+1/a^2$ instead of $-1/a^2$: see the footnote Art. 93. She says on her page 29:

Je tâcherai de prouver, dans les numéros suivans, que l'équation de l'anneau donnée par Euler, dans le Mémoire *De sono campanarum*, n'est affectée que d'une simple erreur de signe, et que cette erreur qui, analytiquement parlant, est peut-être la plus légère qu'un géomètre puisse commettre, suffit cependant pour éloigner entièrement la théorie de l'expérience.

All that she gives in the subsequent articles as to the difference between herself and Euler seems to amount to the consideration that her formula agrees fairly with experiment and therefore Euler's cannot be correct.

But the most curious thing connected with this equation is that Mdlle Germain would agree with Euler had it not been for a mistake in her own work, assuming her process to be otherwise satisfactory. The mistake occurs at the fifth line of her page 25, where she goes wrong in the Calculus of Variations. Using ordinary notation she puts in effect

$$\delta\sqrt{1+p^2+q^2} = -\frac{1}{\sqrt{1+p^2+q^2}}\{p\delta p + q\delta q\};$$

the negative sign should be cancelled.

295. On page 32, Mdlle Germain says:

Revenant à l'objet principal des présentes recherches nous nous bornerons à considérer, parmi les différens mouvemens qui peuvent se manifester, ceux qui intéressent la théorie du son.

Accordingly pages 32—57 are devoted to the integration of the equation (C) of Art. 290, and to numerical deductions from the integral. She assumes for the integral a formula

$$r = M\sin(\zeta + t/\sqrt{k}),$$

where $\zeta$ is a constant and $M$ a function of $s$. Then to determine $M$ we get the equation

$$\frac{d^4M}{ds^4} - \frac{1}{a^2}\frac{d^2M}{ds^2} = \frac{M}{kN^2};$$

this is an ordinary differential equation which can be easily integrated. The result will be of this form:

$$r = \sin(\zeta + t/\sqrt{k}) \{Ae^{\pi\alpha\sigma} + Be^{-\pi\alpha\sigma} + C\sin\pi\beta\sigma + D\cos\pi\beta\sigma\},$$

where $\sigma$ is put for $s/l$, and $l$ is the extreme value of $s$. Here $A, B, C, D$ are arbitrary constants, also $\alpha$ and $\beta$ are constants. This is the equation (G) of our authoress on her page 36. The only difference in notation is that she uses $A$ for our $l$, which is bad, since $A$ has been already used in a different sense; she also puts two accents on $\alpha$ and on $\beta$ which are unnecessary for our purpose. Now to determine the arbitrary constants she has obtained the conditions that $d^2r/ds^2$ and $d^3r/ds^3$ must vanish both when $s=0$ and when $s=l$. These give four equations, and unfortunately she goes quite wrong in them. For example, the two which arise from putting $s=0$ she expresses thus,

$$A + B - D = 0, \; A - B - C = 0,$$

instead of

$$(A+B)\alpha^2 - D\beta^2 = 0, \; (A-B)\alpha^3 - C\beta^3 = 0;$$

and she makes the same mistake in the two equations which arise from putting $s=l$. It is not too much to say that the whole of the rest of the work is ruined by these mistakes, as almost every formula will have to be corrected. For example instead of the equation

$$4 - 2(e^{\pi\alpha} + e^{-\pi\alpha})\cos\pi\beta = 0,$$

which immediately follows, we must have

$$4 - 2(e^{\pi\alpha} + e^{-\pi\alpha})\cos\pi\beta + \frac{\alpha^2 - \beta^2}{\alpha\beta}(e^{\pi\alpha} - e^{-\pi\alpha})\sin\pi\beta = 0.$$

296. I may just notice a small mistake which occurs at the top of page 37. Mdlle Germain says that we have *three* equations; they are of the form

$$H\delta(dr/ds) = 0, \quad J\delta r = 0, \quad K\delta r = 0:$$

but she is wrong, for the principles of the Calculus of Variations would give only *two* equations, namely

$$H\delta(dr/ds) = 0, \; (J+K)\delta r = 0.$$

The lady does not appear to have paid that attention to the

Calculus of Variations which might have been expected from the pupil and friend of its great inventor Lagrange.

297. The serious mistakes which are made in determining the values of the constants $A$, $B$, $C$, $D$ destroy the interest of a reader in the inferences she draws from the value of $r$; and especially I have not examined the numerical results on pages 52—56. Two other mistakes which present themselves may be noted.

On page 39 it is said that $\cos\beta''$ will be positive provided $\beta'' = (4n \pm 1/2 + \delta)\,\pi$, where $n$ is any integer, and $\delta$ is less than $\frac{1}{2}$: this however is insufficient, for we must add the condition that $\delta$ is to be negative if the upper sign is taken, and positive if the lower sign is taken.

Again, on page 36 a formula is given which is said to apply to the case of *two-thirds* of the circumference, and which involves the factor $\sqrt{\beta''^2 + \frac{9}{4}}$; instead of *two-thirds* we must read *three-fourths*; when the formula is employed on page 52, instead of $\frac{9}{4}$ by a mistake $\frac{4}{3}$ has been taken.

298. On her page 57 the authoress proceeds to consider equation (B) of Art. 290; she obtains a particular integral and deduces numerical results as to the nature of the sounds caused by the vibrations. These pages are not affected by the mistakes which I have noted in Art. 295.

On her page 63 the authoress takes a more general form of the integral of equation (A); this is

$$r = \cos\frac{\pi x}{A'} \,.\, M \sin(\zeta + t/\sqrt{k}),$$

where $A'$ is a constant and $M$ is of the same *form* as in Art. 295. For $M$ a value is obtained like that of Art. 295, only $\alpha$ and $\beta$ are not the same as before. Mdlle Germain has to determine the four arbitrary constants which occur in $M$ by the condition that both when $s = 0$ and when $s = l$

$$\frac{d^2r}{ds^2} + \frac{d^2r}{dx^2} = 0, \text{ and } \frac{d}{ds}\left(\frac{d^2r}{ds^2} + \frac{d^2r}{dx^2}\right) = 0.$$

She merely states the values, and I cannot verify them: she seems to me to make the same mistake as I have noticed in Art. 294, and also to neglect altogether the presence of the term $d^2r/dx^2$.

On page vi. of her preface Mdlle Germain says:

J'eus occasion de faire remarquer, contre l'opinion énoncée au programme, qu'il s'en fallait beaucoup que les lignes de repos observées par M. Chladni fussent toujours analogues aux noeuds de vibration des cordes vibrantes.

She seems to return to this matter on pages 31, 64, 65, but I cannot understand exactly what she wishes to establish. She says on page 64:

J'insiste sur cette observation, parce qu'Euler, et après lui plusieurs autres auteurs, ont regardé le mot *libre* appliqué aux extrémités, comme designant un certain état analytique à l'exclusion de tout autre.

299. The fourth section of Mdlle Germain's work is entitled: *Comparaison entre les résultats de la théorie et ceux de l'expérience;* it occupies pages 75—96. I have not examined this very carefully, having little faith in the theoretical formulae. It does not appear that the authoress found any very close agreement between her theory and her experiments. It is curious that the deviations were all in one direction; she states on her page 76 that in a large number of cases the sound obtained was graver than it should have been according to theory, and she never observed the inverse phenomenon.

300. The next production of our authoress which we have to notice is entitled: *Remarques sur la nature, les bornes et l'étendue de la question des surfaces élastiques, et équation générale de ces surfaces.* Paris, 1826. This is in quarto, and consists of 21 pages, besides the title page. We have first some *Observations Préliminaires,* and then the work is divided into two sections. It appears from the preliminary observations that the authoress had no doubt with respect to the accuracy of the formulae which she had already published, though she admits in a vague way some defects in the method she had used:

Il ne me restait aucun doute sur l'exactitude des formules que j'avais publiées; mais je reconnaissais cependant qu'une analyse embarrassée et fautive ôtait à ces formules le caractère d'évidence qui leur est nécessaire.

301. The first section of the work is entitled: *Exposition des conditions qui caractérisent la surface;* it occupies pages 2—10. This consists of general remarks which do not seem to me of any great interest or value. In a note on pages 5 and 6 Mdlle Germain adopts more strongly than in her former work the untenable opinion that a certain coefficient involves the *fourth* power of the thickness: see Art. 289. With respect to her investigations on this subject she says:

Les recherches dont je fais mention ici ont été rassemblées dans un Mémoire que j'ai présenté à l'Académie il y a environ deux ans, et dont MM. de Prony et Poisson, nommés commissaires, n'ont pas encore fait le rapport. Je publierai ce Mémoire lorsque l'examen successif de tout ce qui concerne la théorie des surfaces élastiques en amènera l'occasion.

302. On the last page of this section Mdlle Germain says:

Au reste, je n'aurais rien d'important à ajouter aux deux premiers paragraphes du Mémoire que j'ai déjà publié; ils sont dans un parfait accord avec ce qu'on vient de lire......Le § III du même Mémoire doit être réformé.

The last sentence might suggest that she was dissatisfied with the whole of the third section of the former work; but a note on page 15 limits the part to be given up as that on pages 21—25. After all she does not distinctly admit any *errors* in her former process, but seems to consider she is merely making some *improvements.*

303. The second section is entitled: *Equation générale des surfaces élastiques vibrantes;* it occupies pages 10—21. The result is that an equation of precisely the same *form* as (A) of Art. 290 is now obtained for the vibration of *any surface* whatever, and not merely for a cylindrical surface. It is assumed that the vibrating surface differs very slightly from its original form, and that the direction of motion of any point is along the original normal to the surface at that point.

304. I cannot say that the demonstration convinces me. In the first place the method is liable to the serious objections that have been urged against Lagrange's method, which is imitated:

see Art. 159 of my account of Lagrange. In the second place there are inadmissible steps due to the authoress herself. Thus for instance the error which I have noticed in Art. 292, and that as to signs which I have noticed in Art. 294, are reproduced; also corresponding to the arbitrary step of making $\cos \nu = 1$, which I have noticed in Art. 293, there is an equally arbitrary step, having the same object in view. These three difficulties are all connected with the integration by parts of the second term in the expression given in Art. 288. With respect to the first term she implicitly puts $dsds'$ in her notation for $dm$; she should for consistency have put $\delta dsds'$ for $\delta dm$ in the second term, and then the rest of her process would have been very different from what it is.

305. The work closes with a few remarks as to the possibility of deducing some results from the very general equation which has been presented applicable to any vibrating elastic surface. The authoress says:

Ces considérations, ainsi isolées, perdent sans doute beaucoup de leur vraisemblance; il m'a cependant paru que c'était le lieu d'exposer ce premier aperçu. J'attendrai, pour en développer les conséquences, qu'un travail plus approfondi m'ait mise à même de leur donner le double appui du calcul et de l'expérience.

306. The last publication by Mdlle Germain which we have to notice is entitled: *Examen des Principes qui peuvent conduire à la connaissance des lois de l'équilibre et du mouvement des solides élastiques;* it occupies pages 123—131 of the *Annales de Chimie,* Vol. 38, 1828. This consists of general remarks the object of which is to recommend the method she had adopted of dealing with the problem of elastic surfaces, that is by starting with the hypothesis we have stated in Art. 285; she holds that it is better than the attempt to construct a theory of the action of molecular forces. The article does not contain anything of importance. The authoress seems to have been dissatisfied with the reception given to her memoir of 1821; she says on her page 124:

Je voyais s'établir une opposition redoutable, surtout en ce qu'au lieu de procéder par la discussion elle se réfugiait dans le dédain des généralités que j'ai toujours regardées comme incontestables.

307. Laplace. The fifth volume of the *Mécanique Céleste* was published in 1825. Livre XII. is entitled: *De l'attraction et de la répulsion des sphères, et des lois de l'équilibre et du mouvement des fluides élastiques.* The contents of this book though not very closely connected with our subject may be conveniently noticed here[1]; they occupy pp. 87—144 of the original edition of this volume.

308. The first chapter is entitled: *Notice historique des Recherches des Géomètres sur cet Objet;* it occupies pages 87—99. Part of this had appeared in the *Annales de Chimie,* Vol. 18, 1821, pages 181—187. Laplace adverts to the two remarkable propositions demonstrated by Newton relative to the attraction of spheres, under the ordinary law of attraction; namely that a sphere attracts an external particle in the same manner as if the mass of the sphere were collected at its centre; and that a shell bounded by concentric spherical surfaces, or by similar and similarly situated ellipsoidal surfaces, exerts no attraction on an internal particle. Laplace then passes to his own researches; he had shewn that among all laws of attraction in which the attraction is a function of the distance and vanishes when the distance is infinitely great, the law of nature is the only law which is consistent with Newton's two propositions: see my *History of the Theories of Attraction...* Chapter XXVIII. Laplace's demonstrations have passed into the ordinary text books.

309. Laplace then speaks of researches of his own which had for their object to establish the ordinary laws of elastic fluids on hypotheses of a reasonable nature as to the mutual action of the molecules. The ten pages which are spent on this consist mainly of the substance of his mathematical processes divested of mathematical symbols; they are scarcely intelligible apart from the following two Chapters which contain the mathematical processes, and when these have been mastered they become superfluous.

[1] [I am unable to give any reason for Dr Todhunter's introduction here of these paragraphs relating to elastic *fluids*. There is a long series of memoirs on this subject of an earlier date, to which he has not referred, and notices of which it did not seem to me advisable to introduce into the work. On the other hand I have followed here my almost invariable rule of printing all matter which Dr Todhunter inserted in his manuscript. ED.]

Laplace himself well observes on page 186 of Vol. 18 of the *Annales de Chimie*...after some verbal statements:

Les géomètres saisiront mieux ces rapports traduits en langage algébraique.

310 The velocity of sound in air is an interesting subject which is specially noted by Laplace. He adverts, on his page 95, to the formula obtained by Newton, and says:

Sa théorie, quoique imparfaite, est un monument de son génie.

He gives his own famous correction which, as is well known, brings the theory into agreement with observation. This he first published, without demonstration, in the *Annales de Chimie*... Vol. 3, 1816, pages 238—241. There speaking of Newton's formula he says:

La manière dont il y parvient est un des traits les plus remarquables de son génie.

Laplace corrects a mistake he had made on page 166 of the same volume with respect to the velocity of sound in water, by which he obtained a result $\sqrt{3}$ times too large.

311. Laplace finishes the Chapter with the following sentences relative to forces which are sensible only at imperceptible distances:

Dans ma théorie de l'action capillaire, j'ai ramené à de semblables forces les effets de la capillarité. Tous les phénomènes terrestres dépendent de ce genre de forces, comme les phénomènes célestes dépendent de la gravitation universelle. Leur considération me parait devoir être maintenant le principal objet de la Philosophie mathématique. Il me semble même utile de l'introduire dans les démonstrations de la Mécanique, en abandonnant les considérations abstraites de lignes sans masse flexibles ou inflexibles, et de corps parfaitement durs. Quelques essais m'ont fait voir qu'en se rapprochant ainsi de la nature, on pouvait donner à ces démonstrations, autant de simplicité et beaucoup plus de clarté que par les méthodes usitées jusqu'à ce jour.

312. The second chapter is entitled: *Sur l'attraction des Sphères, et sur la répulsion des fluides élastiques.* It occupies pages

100—118. Laplace begins with quoting his own results for the attraction of a spherical shell on an external particle; from this he derives an expression for the attraction of one sphere on another: this might usefully be introduced into the text-books. The same expression will hold for spheres which *repel* each other. Newton supposed that between two molecules of air a repulsive force is exerted which is inversely as the distance; Laplace examines this hypothesis briefly, and shews that it affords no prospect of agreement with observation. He says, on his page 105:

...aussi ce grand géomètre ne donne-t-il à cette loi de répulsion qu'une sphère d'activité d'une étendue insensible. Mais la manière dont il explique ce défaut de continuité est bien peu satisfaisante. Il faut sans doute admettre entre les molécules de l'air une force répulsive qui ne soit sensible qu'à des distances imperceptibles: la difficulté consiste à en déduire les lois que présentent les fluides élastiques. C'est ce que l'on peut faire par les considérations suivantes.

313. Laplace's hypothesis is that in a gas we have molecules of two kinds, which may be called *matter* and *caloric;* matter attracts matter and caloric, but caloric repels caloric. Also for the permanent gases the attraction of the matter is insensible compared with the repulsion due to caloric. Starting from these principles Laplace obtains the ordinary facts with respect to gas enclosed in an envelope, namely that the pressure is constant throughout, and that the laws of Mariotte and Gay Lussac hold. The mathematical investigation is reasonably satisfactory; it involves a certain quantity $K$ which represents a definite integral $\int_0^\infty \psi(s)\, ds$: this can not be effected because the function $\psi$ is not known. Approximations are freely used in the investigation.

314. Much of the second chapter originally appeared in pages 328—343 of the *Connaissance des Tems* for 1824, published in 1821; the following points of difference may be noted. A passage on pages 103 and 104 of the *Mécanique Céleste* is new, beginning with *Dans les sept intégrations,*...and ending with...*Newton a démontré.* Three short paragraphs from page 336 of the *Connaissance des Tems* are omitted, beginning with *Il résulte*...and

ending with...*développée.* Pages 111—113 of the *Mécanique Céleste* first appeared in the *Connaissance des Tems* for 1825 published in 1822, pages 219—223. Pages 114...118 of the *Mécanique* differ from the pages 339—343 of the *Connaissance* for 1824.

In the catalogue of *Scientific Papers* published by the Royal Society under the head Laplace No. 51 there is a reference to the *Journal de Physique* XCIV. 1822, pages 84—90; so I presume there is some abstract or account of what is contained in the chapter of the *Mécanique Céleste,* which we are considering; but I have not seen the volume referred to.

315. There is also an article in the *Annales de Chimie*...Vol. 18, 1821, which like that just cited bears the title: *Sur l'attraction des corps sphériques, et sur la répulsion des fluides élastiques.* It occupies pages 181—190 of the volume. The same article occurs almost identically on pages 83—87 of the *Bulletin...Société Philomatique,* 1821. This is substantially embodied in the chapter of the *Mécanique Céleste* which we are considering. On pages 273—280 of the same volume of the *Annales* is another article by Laplace entitled: *Éclaircissemens de la théorie des fluides élastiques;* this is only partially reproduced in the *Mécanique Céleste.* Laplace alludes to his article in the *Connaissance des Tems* for 1824, and promises to return to the subject in the volume for 1825.

Some criticisms by Mr Herapath on the Theory of Elastic Fluids contained in Laplace's second Chapter will be found in the *Philosophical Magazine* Vol. 62, 1823, pages 61—66 and 136—139; but they are not connected with our subject, and so we will not investigate them.

316. The third chapter is entitled: *De la vitesse du Son et du mouvement des Fluides élastiques.* It occupies pages 119—144. Here Laplace supplies the mathematical investigation of a result respecting the velocity of sound which he had made known in 1816: see Art. 310. Laplace's formula is now universally received, but it is demonstrated in a more simple manner in the usual works on the subject. Laplace compares his theoretical value of the velocity of sound in air with that given by recent observation, and finds that the difference is only about 3 metres in 340. This

comparison had already appeared in pages 266—268 of the *Annales de Chimie*, Vol. 20, 1822, and pages 371 and 372 of the *Connaissance des Tems* for 1825 published in 1822.

317. Laplace having discussed the velocity of sound passes on to other subjects connected with elastic fluids under the following heads: *Équations générales du mouvement des fluides élastiques; Du mélange de plusieurs gaz; Des atmosphères; De la vapeur aqueuse; Considérations sur la Théorie précédente des gaz.*

318. The third chapter, in substance, first appeared in the *Connaissance des Tems* for 1825, published in 1822: see pages 219—227, 302—323, 386 and 387; the following differences may be noted. In the *Mécanique Céleste* the passage on pages 135 and 136 as to the velocity of sound in a mixture of gases is new; and so also is the section on pages 139 and 140 entitled: *De la vapeur aqueuse.* The passage on pages 386 and 387 of the *Connaissance des Tems* is not reproduced in the *Mécanique Céleste*; here Laplace alludes to the recent curious experiments by Cagniard Latour as to the compression of certain liquids; this passage is also printed in the *Annales de Chimie*...Vol. 21, 1822, pages 22 and 23. An English translation of it is given on pages 430 and 431 of the *Quarterly Journal of Science*...Vol. 14, 1823.

In pages 161—172 of the *Bulletin...Philomatique*, 1821, is an article by Laplace entitled: *Développement de la Théorie des fluides élastiques, et application de cette théorie à la vitesse du son.* This corresponds very closely with pages 219—227, 302—306 of the *Connaissance des Tems* for 1825.

319. The pages of the *Mécanique Céleste* contain several errata which are reproduced in the National edition. On page 111 there is a formula in which we have in succession $u$, $u$, $u''$; here for $u''$ we must read $u$: the National edition reads $u$, $u'$, $u''$, which introduces another misprint. The formula is given correctly in the *Connaissance* for 1825, page 221, and also in the *Bulletin... Philomatique*, 1821. On page 133 we read:

La chaleur spécifique du mélange sous une pression constante, ou sous un volume constant, est visiblement....

Here the words 'ou sous un volume constant' must be omitted. In the *Con. des Tems* for 1825 page 313 the words 'sous une pressure constante, ou sous un volume constant' do not occur; so that the passage is correct, though we have to ascertain from the context which specific heat is meant. Le Verrier drew attention to the inaccuracy of the National edition in the *Comptes Rendus*, Vol. 29, 1879, page 22. A memoir by M. Pouillet entitled: *Mémoire sur la théorie des fluides élastiques et sur la chaleur latente des vapeurs* is published in the *Comptes Rendus*, Vol. 20, pages 915—927. It begins thus:

Le XII$^e$ livre de la *Mécanique céleste* contient une théorie générale des fluides élastiques qui repose uniquement sur les lois de l'attraction des sphères, et sur quelques propriétés primitives attribuées aux éléments de la chaleur; c'est à la fois l'un des derniers et des plus beaux travaux de Laplace. Il eut la satisfaction de voir sa théorie confirmée d'une manière remarquable, d'un côté par les expériences relatives à la vitesse du son, qu'il avait lui-même proposées dans ce but, et d'un autre côté par quelques expériences de dégagement de chaleur, exécutées toutefois comme elles pouvaient l'être, dans des limites assez restreintes de température et de pression.

[320.] Fresnel. The important works upon Light, notably the memoirs on double refraction of this great physicist belong to this period (memoirs of 1821 to 1825). They can hardly however be treated as contributions to the theory of elasticity. So far as Fresnel treats of molecular motions, he understands by the elasticity of his medium a cause producing a force proportional to absolute and not relative molecular displacement; so far as he treats of vibrations his medium possesses properties which we cannot reconcile with those belonging to our theoretical elastic solid.

Premièrement Fresnel admet, sans démonstration suffisante, que les élasticités mises en jeu dans la propagation des ondes planes sont uniquement déterminées par la direction des vibrations et ne dépendent pas de la direction du plan des ondes (see the second memoir on double refraction, *Œuvres complètes*, Tome II. p. 532). Ensuite, il regarde comme négligeable et absolument inefficace, en vertu des propriétés de l'éther, la composante de l'élasticité normale sur le plan des ondes, oubliant qu'après avoir constitué son milieu elastique avec des points matériels disjoints et soumis à leurs actions réciproques, il n'avait plus le droit de recourir à des suppositions auxiliaires du genre de celles sur lesquelles on a coutume de fonder l'hydrostatique et l'hydrodynamique, sans avoir égard à la vraie constitution moléculaire des fluides. Il pouvait sembler singulier que le résultat définitif d'un raisonnement incomplet et inexact en deux points fût une des lois de la nature dont l'expérience a le mieux confirmé la vérité. (Note by E. Verdet to Fresnel's first memoir on double refraction in the *Œuvres complètes*, Tome II. p. 327.)

Saint-Venant remarks with regard to Fresnel's relation to the history of elasticity:

Nous ne pensons pas pour cela qu'on doive attribuer à Fresnel, non plus qu'à Ampère, qui a présenté à ce sujet quelques considérations élevées, l'invention de la théorie de l'élasticité qui, après Navier, doit être regardée comme appartenant à Cauchy. *Historique Abrégé*, p. cl. in the 3rd edition of Navier's *Résumé des Leçons*.

On the contrary it seems to me that but for Cauchy's magnificent molecular researches, it might have been possible for Fresnel to completely sacrifice the infant theory of elasticity to that flimsy superstition, the mechanical dogma, on which he has endeavoured to base his great discoveries in light. Cauchy inspired Green[1], and Green and his followers have done something, if not all, to reconcile Fresnel's results with the now fully developed theory of elasticity, the growth of which his dogma at one time seriously threatened to check.

[321.] SAVART. A long and most valuable series of memoirs by this author is spread over the pages of the *Annales de Chimie et de Physique* from 1819 to 1840. They are principally experimental and belong more especially to that portion of elasticity which falls properly under the theory of sound. They have been largely influenced by Chladni's acoustic experiments, but at the same time present the views of an original, if not very mathematical physicist. The importance they possess for our subject arises from the strong light they occasionally cast on the structure of elastic bodies.

[1] Not Green Cauchy, as Sir William Thomson seems to suggest. *Lectures on Molecular Dynamics*, p. 2.

[322.] *Sur la communication des mouvemens vibratoires entre les corps solides.* This paper was read to the Academy of Sciences, Nov. 15, 1819, and published in the *Annales de Chimie,* Tom. 14, pp. 113—172. Paris, 1820. The experiments to determine the nodal lines of rods subject to longitudinal vibrations are described in § 1 (p. 116). It would be interesting to compare the results with those given by the theory of the *longitudinal* vibrations of a rectangular plate with three free edges, but I am not aware that this problem has been mathematically considered.

[323.] *Mémoire sur les Vibrations des corps solides considérées en général.* This was presented to the Academy of Sciences on April 22, 1822, and it is printed in the *Annales,* Tom. 25, pp. 12—50; pp. 138—178; and pp. 225—269, 1824. The aim of the author in this paper is to consider the most general character of the vibrations which it is possible for solid bodies to perform. He begins by recognising three different kinds of vibratory motions: *longitudinales, transversales et tournantes.* It is not easy to see under which head he would have included vibrations involving contraction and expansion. Such vibrations might be conveniently termed *pulsations;* they are recognised as a fourth distinct class by Poisson: see Art. 428.

Savart believes that all these motions are of the like kind:

Les vibrations transversales d'une verge, par exemple, paraissent avoir été considérées comme un simple mouvement de flexion du corps entier, et non comme un mouvement moléculaire d'où résulterait le mouvement de flexion; il en est de même des vibrations tournantes: les vibrations longitudinales sont les seules pour lesquelles on a admis que les mouvemens généraux sont le résultat de mouvemens plus petits imprimés aux particules mêmes; et il faut avouer qu'on ne pouvait guère se faire une autre idée de cette espèce de vibrations. Mon but, dans ce mémoire, est de montrer qu'il n'existe qu'une seule espèce de mouvement de vibration, et que selon que sa direction est parallèle perpendiculaire, ou oblique aux arêtes ou aux faces d'un corps, d'une verge, par exemple, il en résulte des vibrations longitudinales transversales ou obliques.

As I understand Savart here, *both* his longitudinal and

transverse vibrations are in the face of the wave, and so unaccompanied by dilatation or contraction. His paper would seem to hint that some vibrations had previously not been attributed to molecular motion. If he only means by *une seule espèce de mouvement* a mere motion of molecules, his statement is correct, but he seems to have some idea that the *characteristics* of that motion are the same in all cases.

[324.] In his experiments vibrations of various kinds were induced in the solid body by means of an oscillating cord or string attached to it; by changing the plane of oscillation of the cord, different vibrations were produced in the body.

I confess that I am unable to understand how by means of a string oscillating in a plane it would be possible to give every variety of vibration to a solid body, nor does Savart really appear to have done so in his experiments. The kinds of vibration classified by him have distinct analytical characteristics and I should judge distinct physical characteristics also, especially in the case of fibrous bodies like wooden bars used in these experiments.

[325.] In the first section of the memoir the author treats of the communication of vibrations by means of a cord united to one or more solid bodies. This discussion involves, the author holds, the chief end of the memoir, namely to shew that:

*Il n'existe qu'une seule espèce de mouvement vibratoire qui s'accompagne de circonstances particulières selon le sens dans lequel il a lieu relativement à la forme du corps vibrant.*

In the second and last section of the memoir Savart treats of various phenomena which present themselves in bodies, when the motion takes place in the sense of their length, breadth, height, or in any direction more or less oblique to these.

[326.] The first section contains a most interesting and valuable experimental investigation of the tangential and normal vibrations of circular and rectangular plates. The apparatus and method of experiment are extremely suggestive, but in several places I cannot feel satisfied with the author's deductions. His view that tangential and normal vibrations are the same, would I think involve the absolute elastic isotropy of all bodies, but further than this it does not seem to me to recognise the distinction between a vibration which produces dilatation and one only involving shear. If we take as Savart does a plate with *free* edges, it is quite conceivable that the normal vibrations may be of the same character as the tangential vibration in any direction in its face. This involves only the complete isotropy of the material of the plate, or the capacity of the particles of the plate for vibrating in any direction in parallel lines, but that this complete isotropy can exist in a substance like wood seems to me highly improbable. Still more obvious is it, that, if the plate had not *free* edges, there would be something quite different in the tangential vibrations of such a plate from the normal vibrations of a plate with free edges. I have equally strong objections to the theory that if a string be fastened to one end of a rectangular beam clamped at the other, then the vibrations of the particles of the beam will be parallel to the plane of oscillation of the string; this seems to me at the very least to assume that the beam has a uniform elastic character in the plane perpendicular to its axis. Even for a beam whose sides are parallel to the fibres this can hardly be true; there must be a distinction in the elastic character between directions parallel and perpendicular to the ligneous strata. Unsatisfactory also seems the discussion of the torsional vibrations on pp. 174—177. The first section concludes with the following paragraph, which I leave to the judgment of the reader:

Puisque les vibrations appelées tournantes ne sont qu'une espèce de vibrations normales, il résulte de toutes les recherches qui précèdent, que les vibrations normales, ainsi que celles qui sont obliques ou qui sont tangentielles, soit dans le sens de la longeur, soit dans le sens de la largeur, ne diffèrent entre elles que par les mouvemens de transport ou de flexion qui sont produits par les petites oscillations moléculaires. Il faudrait donc rechercher quelle est la nature de ces mouvemens secondaires dans les différentes espèces de corps, selon que les molécules oscillent dans une direction ou dans une autre. Il est clair que puisque ces diverses espèces de mouvemens généraux sont produits par une même cause, elles doivent avoir un lieu entre elles, et qu'on ne doit pas les isoler en cherchant à en découvrir la nature; c'est pourquoi je les considérerai toutes en même temps dans la section suivante. (p. 177.)

[327.] The second section is occupied with the experimental discovery of the nodal surfaces for vibrations of various kinds in prisms of circular and rectangular section. This again appears to me extremely suggestive. The section concludes with the statement of seven general results. They are briefly the following:

(i). When a body gives a sound, there is always a molecular movement, which is accompanied by particular phenomena according to the direction in which it takes place relative to the faces of the body. This is indisputable.

(ii). That in all cases of vibration the molecules move in straight lines, *en ligne droite*, as 'has been admitted in the case of longitudinal vibration.' This seems to me less obvious.

(iii). That vibrations *tournantes* are only a particular case of normal vibration. I do not feel convinced by Savart's remarks on this point.

(iv). When a body is in vibration there are always faces or sides upon which the nodal lines do not correspond.

(v). Dans les cylindres rigides pleins ou creux, dans les cordes qui exécutent des vibrations longitudinales, il existe une suite de points immobiles dont l'ensemble constitue une ligne de repos continue, qui tourne en rampant autour du corps.

(vi). The laws of normal vibrations have been verified by experiment, when the depth of the body examined is much greater than its breadth.

(vii). Dans un système de corps disposés d'une manière quelconque, toutes les molécules se meuvent suivant des droites parallèles entre elles et à la droite suivant laquelle on promène l'archet (i.e. direction of excitation); ce qui conduit à considérer un tel système comme ne formant qu'un seul corps, puisque les molécules s'y meuvent de la même manière: toutefois il est à remarquer que cela n'est vrai qu'autant que les parties du système sont unies bien intimement entre elles.

For the reasons given above I am disinclined to accept this.

[328.] The mathematical reader of this as well as other of Savart's papers will be struck with the amount of theoretical investigation still wanting in the theory of sound. Particularly I may note the need for a full investigation of every kind of vibration of a rectangular plate or beam of non-homogeneous elastic structure. This of course can only be investigated from the general equations of elasticity.

[329.] *Note sur les Modes de division des corps en vibration. Annales,* Tom. 36, pp. 384—393, 1826. There is an abridged translation of this paper in the *Edinburgh Journal of Science,* Vol. VI. pp. 204—209, 1827.

This note, like the last memoir, is concerned with the modes of vibration possible to a solid body. The author shews that there are an infinite number of nodal systems possible to a body, corresponding either to free or forced vibrations. The memoir is illustrated by series of gradually changing systems of nodal lines for square and circular membranes, for square plates and for flat rods. We may quote the first and last paragraphs as containing the general conclusions which Savart draws from his experiments.

Les diverses recherches qu'on a faites jusqu'ici sur les modes de division des corps qui résonnent, conduisent toutes à ce résultat, que chaque corps d'une forme donnée est susceptible de se diviser en parties vibrantes dont le nombre va toujours croissant suivant une certaine loi; de sorte que chaque corps ne peut produire qu'une série déterminée de sons qui deviennent d'autant plus aigus que le nombre même des parties vibrantes est plus considérable. D'un autre côté, c'est un fait que j'ai établi par une foule d'expériences, que quand deux ou plusieurs corps sont en contact, et qu'ils sont ébranlés l'un par l'autre, ils s'arrangent toujours pour exécuter le même nombre de vibrations; d'où il semble qu'on doive tirer cette conséquence, qu'il n'est pas vrai que les corps ne soient susceptibles que d'une certaine série déterminée de modes de division, entre lesquels il n'y a pas d'intermédiaire, et qu'au contraire ils en peuvent produire qui se transforment graduellement les uns dans les autres: ce qui fait qu'ils sont aptes à exécuter des nombres quelconques de vibrations. J'ai pour but dans cette Note, de faire voir que cette dernière assertion est la seule qui soit conforme à la vérité.

Les divers résultats que contient cette note étant réunis, on peut en déduire cette conséquence générale, que les modes de mouvement des corps qui résonnent sont beaucoup plus variés qu'on ne l'a cru jusqu'ici; et qu'on ne doit admettre l'existence des séries déterminées de sons pour chaque corps d'une forme donnée qu'avec cette restriction importante, que le caractère propre des modes de subdivision doit demeurer le même.

[330.] Savart in this paper is really stating from experimental considerations results which are easily deduced now-a-days from the general theory of small oscillations. He does not in this note however appear to distinguish clearly between the free and forced oscillations of his vibrating body.

[331.] *Mémoire sur un mouvement de rotation dont le système des parties vibrantes de certains corps devient le siège.* This memoir was read before the Academy of Sciences, July 30, 1827. It is printed pp. 257—264 of Vol. 36 of the *Annales,* 1827. It belongs properly to the Theory of Sound, and is concerned with cases of the rotation of a system of nodal lines which has been set up in a plate. The paper is, as one would expect from the author, of considerable interest.

[332.] *Recherches sur la structure des métaux. Annales,* pp. 61—75, Tom. 41, 1829.

This contains an analysis of the structure of metals by means of the nodal systems produced by the vibrations of circular metallic plates. The author has much developed the method in a memoir of 1830 to be referred to later. Although metals are supposed among the most homogeneous of bodies, they act with regard to sound vibrations as if they belonged to fibrous or regularly crystallised substances. Savart deduces this result from the facts that the nodal system composed of two crossed lines at right angles cannot be made to take up any position in a circular metallic plate, and that laminae cut in different directions from a block of metal do not present the same acoustic properties[1]. Thus again there is a great difference between metal plates which have been cast and those which have been cut from a block of metal.

Ces faits et beaucoup d'autres du même genre que je pourrais rapporter montrent nettement que les métaux ne possèdent pas une structure homogène, mais qu'ils ne sont pas non plus cristallisés régulièrement. Il ne reste donc qu'une supposition à faire, c'est qu'ils possèdent une structure semi-régulière, comme si, au moment de la

[1] Sir William Thomson seems to have rediscovered this peculiarity, but attributes the unique position of the quadrantal nodal lines to the plates used by him being only approximately circular and symmetrical. *Lectures on Molecular Dynamics,* pp. 62—64.

solidification, il se formait dans leur intérieur plusieurs cristaux distincts, d'un volume assez considérable, mais dont les faces homologues ne seraient pas tournées vers les mêmes points de l'espace. Dans cette idée, les métaux seraient comme certains cristaux groupés, dont chacun, considéré en particulier, offre une structure régulière, tandis que la masse entière parait tout-à-fait confuse.

This seems to some extent confirmed by the consideration of bright lines in the spectrum of a simple metal.

[333.] *Mémoire sur la réaction de torsion des lames et des verges rigides. Annales* pp. 373—397, Tom. 41, 1829.

This is an endeavour to extend experimentally the results which Coulomb had obtained for the torsion of a wire; Poisson had obtained theoretically like results for cylindrical rods in his *Mémoire sur l'équilibre et le mouvement des corps élastiques,* and Cauchy had extended his laws to rods of rectangular section: see our Chap. V.

Savart proposes to verify the results of Poisson and Cauchy. He refers in a footnote to Duleau's experiments (Art. 229), but holds them not to have been sufficiently general or conclusive.

[334.] After a general description of his apparatus in the first section, the writer proceeds to detail his experiments on rods of circular, square, rectangular and triangular section. On pp. 393 and 394, Savart states three experimental laws of torsion:

(i). Quelque soit le contour de la section transversale des verges les arcs de torsion sont directement proportionnels au moment de la force et à la longeur.

(ii). Lorsque les sections des verges sont semblables entre elles... les arcs de torsion sont en raison inverse de la quatrième puissance des dimensions linéaires de la section.

(iii). Lorsque les sections sont des rectangles et que les verges possèdent une élasticité uniforme dans tous les sens, les arcs de torsion sont en raison inverse du produit des cubes des dimensions transversales, divisé par la somme de leurs carrés; d'où il suit que, si la largeur est très-grande relativement à l'épaisseur, les arcs de torsion seront sensiblement en raison inverse de la largeur et du cube de l'épaisseur, lois qui sont encore vraies dans le cas où l'élasticité n'est pas la même dans toutes les directions.

These laws, Savart holds, are completely in accord with the theory of Poisson and Cauchy:—

On peut même ajouter que le calcul ne s'est jamais mieux accordé avec l'expérience qu'il ne le fait en cette circonstance. (See however Art. 398.)

The memoir concludes with a few experiments and remarks on the relation of heat to the torsional resistance of bodies.

[335.] *Recherches sur l'élasticité des corps qui cristallisent régulièrement. Mémoires de l'Académie de France,* Tom. IX., Paris, 1830. An abridgment of this paper appeared earlier in the *Annales de Chimie,* Tom. 40, Paris, 1829, and a translation in *Taylor's Scientific Memoirs,* Vol. 1, 1837, pp. 139—152 and pp. 255—268. See also the *Edinburgh Journal of Science,* I., 1829, pp. 206—247.

[336.] The author commences his memoir by remarking that precise notions of the inner structure of bodies have hitherto been obtained by two methods, (i) by the cleavage for substances which crystallise regularly whether transparent or opaque, (ii) for transparent bodies only, by the modifications they produce in the propagation of light. Although many new conceptions have been obtained by these methods yet the author considers that that part of physics which treats of the arrangement of the particles of bodies, and of the properties which result from them, such as elasticity, hardness, fragility, malleability, etc., is still in its infancy.

Les travaux de Chladni sur les modes de vibrations des lames de verre ou de métal, et les recherches que j'ai publiées sur le même sujet surtout celles qui se rapportent aux modes de division des disques de substance fibreuse, comme le bois, permettaient de soupçonner qu'on parviendrait, par ce moyen, à acquérir des notions nouvelles sur la distribution de l'élasticité dans les corps solides; mais on ne voyait pas nettement par quel procédé l'on pourrait arriver à ce résultat, quoique la marche qu'il fallait suivre fût d'une grande simplicité.

Toutefois, si ce mode d'expérience, dont nous allons donner la description, est simple en lui-même, il ne laisse pas cependant de s'environner d'une foule de difficultés de détail qui ne pourront être levées qu'après de nombreuses tentatives, et qui, je l'espère, serviront d'excuse à l'imperfection de ces recherches, que je ne donne d'ailleurs que comme les premiers rudiments d'un travail plus étendu.

[337.] It will thus be seen that Savart proposes to investigate the elasticity of crystals, by considering the Chladni figures, which arise from the vibrations of laminae cut from them in various directions of section. His method, which seems to me of very considerable value, is explained in the first section of the memoir and is based upon the following train of argument. If a circular plate of uniform thickness and elasticity uniform in all directions be set vibrating, a certain system of circular and diametral nodal lines will be produced. This nodal system owing to the symmetry in form and structure of the plate will be quite independent of the place of excitation, so long as it remains at the same distance from the centre. The same nodal system is capable of taking up successive positions all round the plate. If now the plate, still remaining circular and of equal thickness, have not the same degree of elasticity in different directions, it will become impossible to shift the same nodal system into a continuous series of positions round the plate. It is found that there are two positions and two only in which the same mode of excitement relative to the centre produces like nodal systems. Intermediate nodal systems vary more or less from these like systems. This immoveability of nodal figures and the double position they can assume are distinctive characters of circular plates, all the diameters of which do not possess a uniform elasticity[1].

[338.] We have thus a method of analysing the character of the structure of a body. Savart remarks that he has not found any body in which the same nodal figure can take up all positions, and this seems to him to indicate that there are very few bodies which possess the same properties in all directions.

The author proposes to commence by analysing a simple body, and having ascertained the laws connecting the nodal lines with the axes of elasticity in such a case to proceed to the more complicated phenomena presented by regular crystals.

[339.] In the second section of his memoir he analyses wood by means of the nodal lines presented by plates cut in different directions. In the case of wood, if we take a small block near the surface of a tree, the ligneous layers may be considered

[1] As to the correctness of this deduction see however the footnote, p. 173.

sensibly plane, hence there are three rectangular directions of varying elasticity, i.e. in the direction of the fibres, in the direction of the ligneous layer and perpendicular to the fibres, and lastly in the direction perpendicular to the fibres and to the ligneous layer. Five series of circular plates are then cut from wood in various directions of section, and the resulting vibrations and nodal lines described and figured at considerable length.

[340.] Savart draws the following conclusions with regard to bodies having three rectangular axes of elasticity (p. 427):

(i). Lorsque l'un des axes d'élasticité se trouve dans le plan de la lame, l'une des figures nodales se compose toujours de deux lignes droites qui se coupent à angle droit, et dont l'une se place constamment sur la direction même de cet axe; l'autre figure est alors formée par deux courbes qui ressemblent aux branches d'une hyperbole.

(ii). Lorsque la lame ne contient aucun des axes dans son plan, les deux figures nodales sont constamment des courbes hyperboliques; jamais il n'entre de lignes droites dans leur composition.

(iii). Les nombres des vibrations qui accompagnent chaque mode de division sont, en général, d'autant plus élevés que l'inclinaison de la lame sur l'axe de plus grande élasticité devient moindre.

(iv). La lame qui donne le son le plus aigu, ou qui est susceptible de produire le plus grand nombre de vibrations, est celle qui contient dans son plan l'axe de plus grande élasticité et celui de moyenne élasticité.

(v). La lame qui est perpendiculaire à l'axe de plus grande élasticité est celle qui fait entendre le son le plus grave, ou qui est susceptible de produire le plus petit nombre de vibrations.

(vi). Quand l'un des axes est dans le plan de la lame, et que l'élasticité dans le sens perpendiculaire à cet axe est égale à celle qu'il possède lui-même, les deux systèmes nodaux sont semblables; ils se composent chacun de deux lignes droites qui se coupent rectangulairement, et ils se placent à 45° l'un de l'autre. Il n'y a, dans un corps qui possède trois axes inégaux d'élasticité, que deux plans qui jouissent de cette propriété.

(vii). Le premier axe des courbes nodales se placent toujours suivant la direction de la moindre résistance à la flexion, il suit de là que, quand dans une série de lames cet axe se place dans la direction occupée d'abord par le second, c'est que, suivant cette dernière direction, l'élasticité est devenue relativement moindre que dans l'autre.

(viii). Dans un corps qui possède trois axes inégaux d'élasticité il y a quatre plans pour lesquels l'élasticité est distribuée de telle manière que les deux sons des lames parallèles à ces plans deviennent égaux, et que les deux modes de division se transforment graduellement l'un dans l'autre, en tournant autour de deux points fixes que pour cette raison j'ai appelés *centres nodaux*.

(ix). Les nombres des vibrations ne sont liés qu'indirectement avec les modes de division, puisque deux figures nodales semblables s'accompagnent de sons très-différents; tandis que, d'un autre côté, les mêmes sons sont produits à l'occasion de figures très-différentes.

(x). Enfin une conséquence plus générale qu'on peut tirer des différents faits que nous venons d'examiner, c'est que quand une lame circulaire ne jouit pas des mêmes propriétés dans toutes les directions, ou en d'autres termes, quand les parties qui la constituent ne sont pas arrangées symétriquement autour de son centre, les modes de division dont elle est susceptible affectent des positions déterminées par la structure même du corps; et que chaque mode de division considéré en particulier, peut toujours, en subissant toutefois des altérations plus ou moins considérables, s'établir dans deux positions également déterminées, de sorte qu'on peut dire que, dans les lames circulaires hétérogènes, tous les modes de division sont doubles.

[341.] In the third section of the memoir the author, starting from these *data*, attempts an analysis of rock crystal. We premise that the rock-crystal is a hexahedral prism terminated by pyramids with six faces; also that its primitive form is a rhombohedron such as would be obtained were the crystal susceptible of cleavage parallel to three non-adjacent pyramidal faces.

[342.] Savart's results again seem of sufficient interest to be cited at length. His *résumé* is given on p. 445:

(i). L'élasticité de toutes les diamétrales d'un plan quelconque perpendiculaire à l'axe d'un prisme de cristal de roche, peut être considérée comme étant sensiblement la même.

(ii). Tous les plans parallèles à l'axe sont loin de posséder le même état élastique; mais si l'on prend trois quelconques de ces plans, en s'astreignant seulement à cette condition, que les angles qu'ils forment entre eux soient égaux, alors leur état élastique est le même.

(iii). Les transformations des lignes nodales d'une série de lames taillées autour de l'une des arêtes de la base du prisme sont tout-à-fait analogues à celles qu'on observe dans une série de lames taillées autour de l'axe intermédiaire dans les corps qui possèdent trois axes inégaux et rectangulaires d'élasticité.

(iv). Les transformations d'une série de lames perpendiculaires à l'un quelconque des trois plans qui passent par deux arêtes opposées de l'hexaèdre sont, en général, analogues à celles d'une série de lames taillées autour d'une ligne qui partage en deux parties égales l'angle plan compris entre deux des trois axes d'élasticité dans les corps où ces axes sont inégaux et rectangulaires.

(v). Au moyen des figures acoustiques d'une lame taillée dans un prisme de cristal de roche, à peu près parallèlement à l'axe, et non parallèlement à deux faces de l'hexaèdre, on peut toujours distinguer quelles sont celles des faces de la pyramide qui sont susceptibles de clivage. L'on peut encore arriver au même résultat par la disposition des modes de division d'une lame prise à peu près parallèlement à l'une des faces de la pyramide.

(vi). Quelle que soit la direction des lames, l'axe optique ou sa projection sur leur plan y occupe toujours une position qui est liée intimement avec l'arrangement des lignes acoustiques: ainsi, par exemple, dans toutes les lames taillées autour de l'une des arêtes de la base du prisme, l'axe optique ou sa projection correspond constamment à l'une des deux droites qui composent le système nodal formé de deux lignes qui se coupent rectangulairement.

[343.] Comparing these results with those derived from bodies having three rectangular axes of elasticity, Savart concludes that the rock-crystal has three axes of elasticity but that they are not rectangular. He deduces that the axis of greatest elasticity (that of greatest resistance to flexure) and that of intermediate elasticity are perpendicular to each other and lie in the lozenge face of the primitive rhombohedron, the smaller diagonal of the face being the direction of the former axis; the axes of least and intermediate elasticity are also perpendicular to each other and lie in the diagonal plane through the shorter diagonal of the lozenge face of the rhombohedron. Thus the angle between the axis of least and greatest elasticity is equal to that between the face and the diagonal plane of the rhombohedron, or to $57^{\circ}\ 40'\ 13''$. (p. 448.)

[344.] This analysis of rock-crystal is followed by a brief discussion of carbonate of lime. The memoir concludes as follows:

Les recherches qui précèdent sont loin, sans doute, de pouvoir être considérées comme un travail complet sur l'état élastique du cristal de roche et de la chaux carbonatée; néanmoins nous espérons qu'elles suffiront pour montrer que le mode d'expérience dont nous avons fait usage pourra devenir, par la suite, un moyen puissant pour étudier la structure des corps solides cristallisés régulièrement ou même confusément. C'est ainsi, par exemple, que les relations qui existent entre les modes de division et la forme primitive des cristaux permettent de présumer qu'on pourra, par les vibrations sonores, déterminer la forme primitive de certaines substances qui ne se prêtent nullement à une simple division mécanique. Il est également naturel de penser que les notions moins imparfaites que celles qu'on possède sur l'état élastique et de cohésion des cristaux, pourront jeter du jour sur beaucoup de particularités de la cristallisation; par exemple, il ne serait pas impossible que les degrés de l'élasticité d'une substance determinée ne fussent pas exactement les mêmes, pour une même direction rapportée à la forme primitive, lorsque d'ailleurs la forme secondaire est différente; et, s'il en était ainsi, comme quelques faits m'induisent à le soupçonner, la détermination de l'état élastique des cristaux conduirait à l'explication des phénomènes les plus compliqués de la structure de ces corps. Enfin, il semble que la comparaison des résultats fournis d'une part, par le moyen de la lumière, touchant la constitution des corps, et de l'autre, par le moyen des vibrations sonores, doit nécessairement concourir aux progrès de la science de la lumière elle-même, ainsi qu'à ceux de l'acoustique.

[345.] I have reproduced so much of this admirable paper because its methods seem to me extremely suggestive. Their fuller development should lead to increased knowledge of the part the ether plays in the transmission of light through crystals, whose elastic character had once been analysed by Savart's method. There ought not to be much difficulty in deducing Savart's results theoretically, yet so far as I am aware the theory of the vibratory motion of a plate of unequal elasticity has not yet been discussed.

[346.] We may here mention two somewhat later memoirs by Savart the date of which is somewhat later than that of our present chapter, but which are included here as belonging to the same mode of investigation.

[347.] *Recherches sur les vibrations longitudinales*, pp. 337—402 of the *Annales de Chimie*, Tom. 65, Paris, 1837.

This memoir is concerned with the nodal surfaces of rods or bars which vibrate longitudinally. These rods or bars differ from those usually considered in the theory of sound, in that their section is of finite dimensions as compared with their length.

[348.] Savart remarks the following peculiarity of the nodal lines:

Si les verges sont carrées ou bien cylindriques, il pourra arriver que les lignes de repos soient disposées en hélice, tournant tantôt de droite à gauche, tantôt de gauche à droite d'un bout à l'autre des verges, ou bien tournant dans un sens pour l'une des moitiés de la longueur, et en sens contraire pour l'autre moitié.

Cette disposition alterne des lignes nodales n'existe pas seulement dans les corps qui sont alongés et qui vibrent dans le sens de leur plus grande dimension : on l'observe aussi dans les corps dont les trois dimensions sont entre elles dans un rapport quelconque, mais seulement pour celle de ces dimensions qui est parallèle à la direction du mouvement. Elle existe également dans les bandes et les cordes tendues ; dans les verges fixées par une ou par deux extrémités, dans les verges ébranlées en travers comme dans celles qui le sont longitudinalement.

[349.] The production of these nodal lines is susceptible of two interpretations, either they are an inherent peculiarity of the longitudinal vibrations of solid bodies, or they belong generally to that kind of vibration which is produced by longitudinal motion. Savart in the first part of his memoir endeavours to establish the latter proposition, namely, he wishes to show that these nodal lines are the result of a normal movement of a particular character which is the product of the alternate contractions and dilatations which accompany longitudinal motion.

In the second part of the memoir this normal movement, established in the first, is considered for rigid bars of different forms and for bands and cords stretched by weights. The third part of the memoir is chiefly busied with the relation of temperature, tension, etc. upon the disposition of the nodal lines.

[350.] Savart distinguishes (pp. 347—353) between transverse vibrations which are the result of flexure and the normal vibrations which accompany longitudinal vibrations; these latter being the outcome of alternate contractions and expansions. He considers that the coexistence of the normal vibrations and of those of contraction and dilatation in a rod vibrating longitudinally must give rise to two sounds and only their isochronism hinders us from distinguishing one from the other.

[351.] In his third section Savart compares the extension produced by a weight and that produced by a longitudinal vibration in a bar, and comes to the following conclusion:

La comparaison des alongemens des verges, par les vibrations longitudinales et par des poids, montre qu'un léger ébranlement moléculaire peut donner lieu à un développement de force qui paraît énorme en égard à la cause qui le produit, et qui est d'autant plus extraordinaire qu'il semble proportionnel à l'aire de la section des verges. (p. 402.)

This memoir, like all Savart's work, is very suggestive for the extension in various directions of the *mathematical* theory of elasticity, especially that branch of it which falls under the Theory of Sound. The author seems to me however to have deserted the standpoint taken up by him in the memoir of 1822, see Articles 323—327.

[352.] *Extrait d'un Mémoire sur les modes de division des plaques vibrantes. Annales,* Tom. 73, pp. 225—273. 1840. There is a footnote to this title as follows:

Tout ce qui a rapport, dans ce travail, aux modes de division des plaques carrées et des plaques circulaires a été donné au cours d'acoustique que j'ai fait au Collége de France en 1838 et 1839. Un extrait de ce travail a été inséré, avec planches, dans le journal intitulé *l'Institut* où il a été rendu compte du cours que j'ai fait en 1839.

Savart commences with an eulogy of Chladni as the discoverer of the nodal figures, but remarks that that distinguished physicist has almost entirely confined himself to rectangular or circular plates. In the present memoir Savart proposes to consider plates whose boundaries are squares, triangles, pentagons, hexagons, heptagons, octagons, dodecagons, circles, ellipses, rectangles and lozenges.

The nodal lines were obtained by scattering a colouring powder on the plate and then, when the system had been formed, pressing a sheet of paper slightly damped with gum-water on the top. Over 1800 figures were obtained in this fashion.

The memoir is accompanied by numerous figures, and is remarkable for its freshness and ingenuity. In conclusion we must again point out the wide field for mathematical investigation which the verification of Savart's experimental results opens out.

[353.] The researches of Chladni and Savart on the nodal figures of vibrating plates were taken up in Germany by F. Strehlke, who contributed several papers to *Poggendorff's Annalen* on the subject. We notice them here only briefly, as they belong rather to the history of acoustics.

[354.] *Beobachtungen über die Klangfiguren auf ebenen nach allen Dimensionen schwingenden homogenen Scheiben. Annalen der Physik*, Bd. 4, pp. 205—318. Leipzig, 1825.

Strehlke states two conclusions as the outcome of his experimental investigations:

(i) Die Klangfiguren, oder die bei schwingenden Scheiben in Ruhe bleibenden Stellen der Oberfläche sind nicht gerade, sondern stets krumme Linien, aber Linien im Sinne der Geometrie, keine Flächen.

(ii) Diese Linien durchschneiden sich nicht.

[355.] These conclusions do not appear to be sufficiently proved, and Chladni at once objected to them. To Chladni's objections Strehlke replied in a memoir entitled:

*Ueber Klangfiguren auf Quadratscheiben. Annalen der Physik*, Bd. 18, pp. 198—225, 1830. In this memoir he makes more accurate measurements of the position of the nodal lines and tries to represent them by means of conic sections. That their forms are not those of the conic sections is now known, and his results are not satisfactory; still less do they conclusively prove that nodal lines never cut each other.

Other later papers by Strehlke may be conveniently referred to here.

[356.] *Ueber die Lage der Schwingungsknoten auf elastischen geraden Stäben, welche transversal schwingen, wenn beide Enden frei sind. Annalen der Physik*, Bd. 27, pp. 505—542. Leipzig, 1833.

This paper is an extension of the results of Daniel Bernoulli and Giordano Riccati (see Arts. 50 and 121). The latter had determined the position of the nodes for the first modes of vibration, the present memoir proposes to give formulae for the distance of the nodes from the nearest end of the rod, whatever their number may be.

The equation for the distance $s$ of the nodes from one end of a rod free at both ends, is

$$0 = \frac{e^{s/c}\cos\alpha + e^{-s/c}(\pm 1 - \sin\alpha)}{\cos\alpha - \sin\alpha \pm 1} + \frac{(\pm 1 - \cos\alpha)\sin(s/c)}{\sin\alpha} + \cos(s/c),$$

where $c$ is a constant depending on the length, material and elasticity of the rod, and $\alpha$ is a root of the equation

$$\cos\alpha = 2/(e^{\alpha} + e^{-\alpha}).$$

It is next shewn that $s$ is approximately a root of the equation

$$e^{-s/c} - \sin(s/c) + \cos(s/c) = 0.$$

Strehlke then calculates the roots of this equation approximately. The values calculated for the position of the nodes are afterwards compared with experimental results, and found in close accordance.

[357.] A *Nachtrag* in Bd. XXVIII., p. 512 of the *Annalen*, 1833, makes a few numerical corrections. It also contains a table of the values of $s$ for the first twelve modes of vibration.

[358.] *Ueber Biot's Behauptung, Galiläi sey der erste Entdecker der Klangfiguren. Annalen der Physik*, Bd. 43, pp. 521—527. Leipzig, 1838.

Biot had asserted in his Experimental Physics (Part I., p. 388), that Galilei was the discoverer of the method of investigating the nodal lines of vibrating bodies by means of sand. He refers only to Galilei's *Dialogi.* Strehlke, examining passages in that work which may possibly refer to the subject, dismisses his claim entirely, and thereby reinstates Chladni as the real discoverer.

[359.] *Ueber die Schwingungen homogener elastischer Scheiben. Annalen der Physik*, Bd. XCV., pp. 577—602. Leipzig, 1855.

This memoir commences with a reference to a paper of Lissajous (to which we shall refer later) confirming Strehlke's results of 1833, and extending them to other cases of the vibratory motion of a rod. The substance however of the memoir is a comparison of the theoretical results of Kirchhoff for the nodal lines of a vibrating circular plate, with experimental measurements; a like comparison is also made for a square plate. References are given to several papers on the same subject, some of which we shall consider in their places, others would carry us too far into the theory of sound: see Articles 512—520.

[360]. We may mention finally a short note by Strehlke on pp. 319 and 320 of the *Annalen der Physik*, Bd. CXLVI. Leipzig, 1872. It is a reprint from the programme of the *Petrisschule* in Dantzig for 1871. It points out the incorrect shapes of the Chladni figures given by recent French and English writers, notably Tyndall in his well-known book on Sound.

361. Paoli. *Ricerche sul moto molecolare de' solidi di D. Paoli.* Pesaro, 1825. This is an octavo volume of XXIII + 350 pages, together with a page of corrections. It is not connected with our subject; there are no mathematical investigations; the author's design seems to be to shew that all bodies, mineral as well as vegetable, possess a life analogous to that of animals. I have not studied the work.

There is a review of the book by G. Belli in the *Giornale di Fisica, Chimica,*...Decade II., Tomo IX., Pavia, 1826; the review occupies pages 167—171 and 322—334 of the volume. The review speaks well of the work on the whole, especially for its collection of facts; but it expresses the hope that if a new edition is issued all the doubtful statements may be collected together, so that the rest of the work may be left unaffected by the incredulity which these are likely to provoke. Belli adverts to his memoir of 1814, and says that he was then ignorant of the experiment of Cavendish and of the observations of Maskelyne: see our Art. 166. He promises to return very soon to the question whether molecular action can be made to depend on universal attraction: but his next contribution to our subject did not appear until 1832. It will be considered later.

362. A work was published at Milan in 1825 entitled: *Sull' applicazione de' principj della Meccanica Analitica del Lagrange ai principali problemi Memoria di Gabrio Piola.* This memoir obtained a prize from the Imperial and Royal Institute of Sciences. Piola alludes to elastic curves on his pages 170—178, but there is nothing really bearing on our subject; he notices on his pages 175 and 178 the correction which Binet had made of an error in Lagrange: see Art. 159 of the account of Lagrange.

The subject of virtual velocities was proposed for a prize by the Academy of Turin in 1809; and an essay written in competition for the prize by J. F. Servois is published on pages 177—244 of the mathematical part of the volume of the Turin Memoirs for 1809—1810.

[363.] 1827. P. Lagerhjelm. This Swedish physicist undertook for the *Bruks-Societet* of Stockholm a long and interesting series of experiments on the density, elasticity, malleability and strength of cast and wrought iron. His results are published in the *Jern-Contorets Annaler; Tionde Årgången. Sednare Häftet,* 1826. They are entitled: *Försök att bestämma valsadt och smidt stångjerns täthet, jemnhet, spänstighet, smidbarhet och styrka,* and dated 1827. They occupy 287 octavo pages. A German translation by Dr F. W. Pfaff appeared in Nürnberg in 1829[1]. Previously a short account of Lagerhjelm's results had been given in *Poggendorff's Annalen der Physik und Chemie,* Bd. 13, p. 404, 1828, and some remarks of Lagerhjelm's upon this account appear as a letter to the editor on p. 348 of Bd. XVII. of the same periodical. A fairly good account will also be found in *Férussac's Bulletin des Sciences Technologiques,* t. 11, p. 41, 1829.

[1] Neither the British Museum nor the Institution of Civil Engineers possess a copy of this translation.

[364.] Lagerhjelm adopted, I believe for the first time, a testing machine involving the hydraulic press and balanced lever. His results are principally of practical value, although he shews an acquaintance with the theoretical work of Young, as well as with such books as those of Tredgold, Duleau and Eytelwein. Some of his statements seem however to bear so closely upon the physical structure of bodies, that it will not be out of place to reproduce them here. I shall make use of the analysis of the more interesting points provided by Poggendorff.

[365.] If a bar be fixed at one end and subject to an extending force at the other, the limit to which it can be extended without permanent set is termed the *limit of elasticity*. If $C$ be the extension of the bar when this limit is reached, and $\Delta$ the extension when the bar breaks, Lagerhjelm finds that $C\sqrt{\Delta}$ is constant.

The quantity $C$ seems to have been measured by *deflection* experiments: see p. 248 of the *Försök att bestämma*. I have not found any later confirmation of this result.

[366.] A further very remarkable result is that all sorts of iron, hard, soft or brittle, appear 'within the limits' of elasticity to possess the same degree of elasticity, i.e. the modulus of elasticity is the same for all of them[1]. Thus the tempering or hardening of steel does not alter its modulus. Two tuning forks of like dimensions which gave the same note, also gave the same note after one had been hardened. Lagerhjelm himself adds to this (Bd. XVII., p. 349) that wrought iron and steel possess the same modulus.

[367.] The experimenter found a slight apparent variation in the modulus as the limits of elasticity were approached. Hence it would seem that in that neighbourhood Hooke's law is not absolutely true.

The limit of elasticity depends on the character of the iron, and is greater for hard iron than soft although the modulus of both is the same. This remark applies also to the absolute strength (breaking weight) of the material which increases with the limit of elasticity, and according to Lagerhjelm is nearly proportional to it.

[1] This result is practically confirmed by more recent experiments.

If a beam or wire be subject to tension, which produces a permanent extension, its limit of elasticity is increased, and in proportion to its diameter it possesses a greater absolute strength.

[368.] An interesting property with regard to specific weight of an iron bar extended to rupture is noticed. Namely that the specific weight of the material at the point of rupture, that is where it has been most extended, is smaller than at other places. Thus a 'permanent set' produces an increase of volume. The writer in Poggendorff here refers to a mathematical investigation by Poisson in the previous volume of the same periodical (p. 516). The note there printed is a translation of a *Note sur l'Extension des Fils et des Plaques élastiques;* by Poisson in Tom. 36 of the *Annales de Chimie et de Physique,* p. 384. Paris, 1827. The following is the important part of this note:

Soit $a$ la longueur d'un fil élastique qui ait partout la même. épaisseur; soit $b$ l'aire de la section normale à sa longueur, et par conséquent $ab$ son volume. Supposons qu'on lui fasse subir une petite extension, de sorte que sa longueur devienne $a(1+\alpha)$, $\alpha$ étant une très-petite fraction; en même temps le fil s'amincira: et si nous désignons par $b(1-\beta)$ ce que deviendra l'aire de la section normale, $\beta$ étant aussi une très-petite fraction, son nouveau volume sera à très-peu près $ab(1+\alpha-\beta)$. Or, d'après la théorie des corps élastiques que j'exposerai dans un prochain mémoire, on doit avoir $\beta=\frac{1}{2}\alpha$, d'où il résulte que par l'extension $\alpha$ d'un fil élastique, son volume se trouve augmenté, suivant le rapport de $1+\frac{1}{2}\alpha$ à l'unité, et sa densité diminuée suivant le rapport inverse.

Poisson quotes an experiment on the point by Cagniard Latour.

[369.] With regard however to this note of Poisson's it must be remarked that he supposes $\beta=\frac{1}{2}\alpha$ from a theory of *elasticity*, but in the case of rupture of a bar, we have long passed the limits of elasticity; in fact the section of the bar does not uniformly diminish but it reaches what is a condition of *flow* or *plasticity*, namely it draws out at some point, often very considerably, before rupture[1].

[1] Some experiments by McFarlane on the augmentation of density by traction are mentioned by Sir William Thomson in his article on Elasticity, § 3. *Encycl. Brit.*

[370.] Finally we may note that Lagerhjelm has attempted a comparison of the modulus of elasticity as found from the velocity of sound in metals with that derived from the extension or flexure of metal bars. The velocity itself is calculated from the note of a bar vibrating longitudinally. The two values of the moduli as calculated for iron, copper and silver from these methods are extremely close. Thus the modulus for iron as obtained by Lagerhjelm from experiments on its extensibility, etc., was about 1070, but as calculated from Savart's experiments on the notes of iron bars it was 1033.

[371.] 1826—29. Benjamin Bevan. There is a series of short experimental papers by this author in the *Philosophical Transactions* and the *Philosophical Magazine.*

[372.] *Account of an experiment on the elasticity of ice. Phil. Trans.*, 1826, pp. 304—6. This is a letter to Dr Young with a note attached by the latter physicist on the modulus of ice. Bevan's experiments to determine the modulus were made upon ice-beams subject to transverse strain. Adopting Young's definition of the modulus, he finds that its value for ice = 2,100,000 feet[1]. He then compares this with the modulus for *water* calculated from Young's account of Canton's experiments on its compression. This modulus he reckons to be 2,178,000 feet, which agrees pretty closely with his own experiments on ice, Dr Young remarks in his footnote that:

It does not appear quite clear from reasoning that the modulus ought to come out different in experiments on solids and fluids; for though the linear compression in a fluid may be only $\frac{1}{3}$ as much as in a solid, yet the number of particles acting in any given section must be greater in the duplicate ratio of this compression, and ought apparently to make up the same resistance. And in a single experiment made hastily some years ago on the sound yielded by a piece of ice, the modulus did appear to be about 800,000 feet only: but the presumption of accuracy is the greater in this case the higher the modulus appears.

[1] Sir William Thomson in his Art. Elasticity, § 77, gives ice a modulus ten times too great. The error is repeated in Thomson and Tait's *Natural Philosophy*, Art. 686. Some interesting experiments as to the bending of ice bars with valuable references are given by Prof. Morgan in *Nature*, May 7th, 1885.

Both Young and Bevan seem to have been quite unconscious of any distinction to be drawn between a *fluid* subject to compression and a *solid* subject to traction. The traction modulus is essentially zero for the former.

[373.] *On the adhesion of glue*, p. 111; *On the strength of bone*, p. 181; *On the strength of cohesion of wood*, p. 269 and p. 343. These four short papers are in the *Philosophical Magazine*, Vol. LXVIII. London, 1826.

[374.] Bevan finds that the actual cohesion of glue is something greater than 715 pounds to the square inch, when a thin coat is placed between two surfaces. This he remarks is greater than the lateral cohesion of fir wood; this cohesion being only 562 pounds to the square inch according to an experiment of his own. He finds from an experiment on *solid glue* that its cohesion is 4000 lbs. to the square inch, 'from which it may be inferred that the application of this substance as a cement is susceptible of improvement.'

[375.] In his experiments on bones, Bevan obtains results which are much in excess of those of Musschenbroek. Thus fresh mutton bone supported 40,000 lbs. to the sq. inch, while the modulus of elasticity for beef-bone was found to be 2,320,000 pounds.

A substance like bone, so universally abounding, possessing such great strength, and considerable flexibility, ought to be restored to its proper place in the scale of bodies, applicable to so many purposes in the arts.

[376.] The paper on bone leads up to those on wood by a criticism of Barlow's apparatus and experiments, which the author thinks liable to objection. The two papers on wood contain only the results of experiment on the cohesion of various kinds, without detail of the individual experiments.

[377.] *Experiments on the cohesion of cast-iron. Philosophical Magazine, New Series.* Vol. I., p. 14. London, 1827. This paper corrects an error in the last paper of the preceding year and notes the irregular results of experiments on cast iron bars.

[378.] *Experiments on the Modulus of Torsion. Phil. Trans.* p. 127. 1829. Bevan remarks that numerous experiments have been made on the strength of wood and other substances as far as regards their cohesion and elasticity, but he knows of no extensive table of the modulus of torsion for different kinds of wood. This he has endeavoured to supply in the present paper.

He states, without proof, the following rule;

To find the deflection $\delta$ of a prismatic shaft of length $l$, when strained by a given force $\omega$ in pounds avoirdupois acting at right angles to the axis of the prism, and by a leverage of given length $= r$; the side of the square shaft $= d$. $T$ being the modulus of torsion from the following table; $l$, $r$, $\delta$ and $d$ being in inches and decimals:—

$$\frac{r^2 l \omega}{d^4 T} = \delta.$$

The modulus of torsion thus appears to be a weight divided by an area, or a *surface pressure* according to Bevan's notation.

[379.] Bevan draws two results from his tables for the modulus.

(1) That the modulus of torsion bears a near relation to the weight of the wood when dry, whatever may be the species. If $s$ be the specific gravity, he finds that for practical purposes we may take

$$\frac{r^2 l \omega}{30000\, d^4 s} = \delta, \text{ or } T = 30000\,.\,s.$$

(2) From some experiments on the modulus of torsion of metals, he finds that for metallic substances the modulus of torsion is about 1/16 of the modulus of elasticity.

It may be noted that the meaning Bevan gives to the term modulus fluctuates from Young's definition to the more modern conception of it as a weight.

380. Pagani. *Mémoire sur l'équilibre des systèmes flexibles*, par M. Pagani. This is published in the *Nouveaux Mémoires de l'Académie...de Bruxelles*, Vol. 4, 1827; it occupies pages 193—244 of the volume. The memoir was read on the 24th of Feb. 1827.

In the preamble the author alludes to the equation of a flexible surface in equilibrium, given by Lagrange in the second edition of the *Mécanique Analytique*, to the memoir of 1814 by

Poisson, and to the memoir of Cisa de Gresy. Pagani then divides his memoir into parts; in the first part he treats of linear flexible systems, and in the second part of superficial flexible systems.

381. The first part of the memoir is substantially a discussion of the well-known mechanical problem of the equilibrium of a funicular polygon; it occupies pages 197—221 of the volume; this presents nothing of importance. Two particular cases are treated separately which may easily be reduced to one. First suppose the system to be composed of rigid straight rods, without weight, hinged together, and let a weight $m$ be suspended at each hinge; next suppose each rod to weigh $m'$. Now Pagani supposes, quite justly, that we may take the weight of each rod to act at its middle point; but instead of this he might suppose the weight $\frac{1}{2}m'$ to be placed at each end, and then the second case becomes practically the same as the first. Then a result which he obtains on pages 217—220, and which he calls *assez remarquable*, is obviously included in what he had previously given.

382. The second part of the memoir occupies pages 221—244; this treats of the equilibrium of a flexible membrane. Two investigations are given; the first is based on ordinary statical principles, and resembles that adopted by Poisson in his memoir of 1814, but is not completely worked out: the second uses the Calculus of Variations, and to this we shall confine ourselves as it presents a little novelty.

383. De Gresy maintained that Poisson's solution of the problem in 1814 was not general, but involved a certain limitation; Pagani holds that the solution was general, and proposes to obtain Poisson's result by the aid of a method resembling Lagrange's; the difference between Lagrange and Pagani we will now state.

Let $dm$ stand for an element of surface, that is for

$$dxdy\sqrt{1+p^2+q^2};$$

then Lagrange takes as the type of virtual moments

$$(X\delta x + Y\delta y + Z\delta z)\,dm + F\delta dm,$$

where $F$ is an undetermined multiplier.

Pagani in effect says that the type should be

$$(X\delta x + Y\delta y + Z\delta z)\, dm + F\frac{d\delta m}{dx}\, dx + G\frac{d\delta m}{dy}\, dy,$$

where $F$ and $G$ are undetermined multipliers.

Then in each case the solution is to be obtained by making the sum of such virtual moments vanish. Pagani's statement as to the proper type for the virtual moments seems to me quite arbitrary.

384. Starting with this assumption, Pagani works out the problem and obtains three general equations referred to rectangular coordinates, which coincide with those in Poisson's memoir of 1814. Then he gives a second investigation in which he uses the ordinary polar coordinates in space instead of the ordinary rectangular coordinates.

385. *Note sur le mouvement vibratoire d'une membrane élastique de forme circulaire;* lue à l'Académie Royale des Sciences de Bruxelles le 1[er] Mai, 1829, par M. Pagani. This is published in Quetelet's *Correspondance Mathématique et Physique,* Vol. v., 1829, pages 227—231, and Vol. vi., 1830, pages 25—31.

386. The object of this memoir is the discussion of those vibrations of the membrane which are performed in the direction of the normal to the plane of the membrane in equilibrium. Pagani starts with the differential equation—

$$\frac{d^2z}{dt^2} = c^2\left(\frac{d^2z}{dr^2} + \frac{1}{r}\frac{dz}{dr} + \frac{1}{r^2}\frac{d^2z}{d\theta^2}\right) \text{..................}(1).$$

He does not say from what source he takes this, but probably it is from Poisson's memoir of April, 1828. In the example which Poisson considered in detail he supposed $z$ a function of $r$ only, so that the term $d^2z/d\theta^2$ vanished; thus the problem as discussed by Pagani is more general than that to which Poisson confined himself; see Art. 472, page 241.

387. We have to find $z$ from equation (1) subject to the following conditions:

$z = 0$ when $r = a$, the radius of the membrane......(2),

$$z = \phi(r, \theta) \text{ and } \frac{dz}{dt} = \psi(r, \theta), \text{ when } t = 0 \text{ ............(3)},$$

where $\phi$ and $\psi$ denote known functions.

Suppose $z = \Pi(r, \theta, t)$ to be a value of $z$ which satisfies (1); we may by Fourier's Theorem develop this in the form

$$z = \tfrac{1}{2}\zeta_0 + \zeta_1 + \zeta_2 + \text{.......................(4)},$$

where $\pi\zeta_n = \cos n\theta \int_0^{2\pi} \Pi(r, \theta, t) \cos n\theta \, d\theta$

$$+ \sin n\theta \int_0^{2\pi} \Pi(r, \theta, t) \sin n\theta \, d\theta \text{...(5)}.$$

Put for $z$ in (1) the series (4); then we obtain a series of differential equations of which the type is

$$\frac{d^2\zeta_n}{dt^2} = c^2\left(\frac{d^2\zeta_n}{dr^2} + \frac{1}{r}\frac{d\zeta_n}{dr} - \frac{n^2\zeta_n}{r^2}\right).$$

We can satisfy this differential equation by supposing

$$\zeta_n = (A \cos c\mu t + B \sin c\mu t)\, u \text{..................(6)},$$

where $A$ and $B$ denote functions of $\theta$, and $\mu$ a constant, all at present undetermined, and $u$ is a function of $r$ which satisfies the differential equation

$$\left(\frac{n^2}{\mu^2 r^2} - 1\right) u = \frac{1}{\mu^2}\frac{d^2u}{dr^2} + \frac{1}{\mu^2 r}\frac{du}{dr} \text{.................(7)}.$$

For the integral of this equation Pagani refers to a memoir by Poisson in Cahier XIX. of the *Journal de l'École Polytechnique.* It will be sufficient for us to cite formulae now to be found in elementary books. If we put $x$ for $\mu r$ the equation (7) coincides with that satisfied by Bessel's Functions: see Art. 370 of my *Laplace's Functions.* Hence by Art. 371 of that work we have as a solution of (7)—

$$u = r^n f(\mu r) \text{..........................(8)},$$

where $$f(\mu r) = \int_0^{\pi} \cos(\mu r \cos\omega) \sin^{2n}\omega \, d\omega \text{ ...........(9)}.$$

In order to satisfy the condition (2) we put

$$\int_0^{\pi} \cos(\mu a \cos\omega) \sin^{2n}\omega \, d\omega = 0 \text{..................(10)},$$

and from this equation we must find the possible values of $\mu$ a constant at present undetermined.

We may substitute in (6) all the different values thus found for $\mu$, combining the values which differ only in sign; then we may write

$$\zeta_n = \Sigma (A_\mu \cos c\mu t + B_\mu \sin c\mu t)\, r^n f(\mu r) \text{.........(11)},$$

where $\Sigma$ denotes a summation extending to all the positive values of $\mu$ found from (10); also $A_\mu$ and $B_\mu$ denote functions of $\mu$ at present undetermined.

388. The value of $\zeta_n$ will be completely determined when we know the value of $A_\mu$ and of $B_\mu$.

Suppose $t = 0$; then by (11), (5) and the first of (3), we have

$$\pi r^n \Sigma A_\mu f(\mu r) = \cos n\theta \int_0^{2\pi} \phi(r, \theta) \cos n\theta \, d\theta$$

$$+ \sin n\theta \int_0^{2\pi} \phi(r, \theta) \sin n\theta \, d\theta \text{......(12)}.$$

Again, differentiate (11) and (5) with respect to $t$, and then put $t = 0$; thus by the second of (3) we get—

$$\pi c\mu r^n \Sigma B_\mu f(\mu r) = \cos n\theta \int_0^{2\pi} \psi(r, \theta) \cos n\theta \, d\theta$$

$$+ \sin n\theta \int_0^{2\pi} \psi(r, \theta) \sin n\theta \, d\theta \text{......(13)}.$$

From these equations (12) and (13) we shall be able to determine separately the quantities of which $A_\mu$ and $B_\mu$ are the types, as we will now shew.

389. Let $\mu_1$ be another root of (10), the square of which differs from $\mu^2$ and $u_1$ the corresponding solution of (7); then it may be shewn that:

$$\int_0^a u u_1 r\, dr = 0,$$

and

$$\int_0^a u^2 r\, dr = \tfrac{1}{2} \left\{ a^{n+1} f'(\mu a) \right\}^2,$$

where $f'$ denotes the derived function of that denoted by $f$. These two formulae are the same as equations (20) and (22) of Chapter XXXV. of my *Laplace's Functions.*

Hence it follows that if we multiply both sides of (12) and (13) by $r^{n+1}f(\mu r)\,dr$ and integrate between the limits 0 and $a$, we get

$$A_\mu = \frac{M}{P},\quad B_\mu = \frac{N}{c\mu P} \quad\text{.....................(14)},$$

where

$$\left.\begin{aligned} M &= \int_0^a f(\mu r)\, r^{n+1}\, dr \left\{\cos n\theta \int_0^{2\pi} \phi(r,\theta)\cos n\theta\, d\theta \right.\\ &\qquad\qquad \left. + \sin n\theta \int_0^{2\pi} \phi(r,\theta)\sin n\theta\, d\theta\right\},\\ N &= \int_0^a f(\mu r)\, r^{n+1}\, dr \left\{\cos n\theta \int_0^{2\pi} \psi(r,\theta)\cos n\theta\, d\theta \right.\\ &\qquad\qquad \left. + \sin n\theta \int_0^{2\pi} \psi(r,\theta)\sin n\theta\, d\theta\right\},\\ P &= \frac{\pi}{2}\left\{a^{n+1} f'(\mu a)\right\}^2 \end{aligned}\right\} \quad (15).$$

Thus $\zeta_n$ is known, and we have for the complete integral of (1), subject to the conditions (2) and (3),

$$z = \tfrac{1}{2}\zeta_0 + S\zeta_n \quad\text{.........................(16)},$$

where $S$ denotes a summation with respect to $n$ from unity to infinity.

Pagani applies his general formulae to special cases, in all of which it is supposed that the membrane is originally plane.

390. Suppose then that $\phi(r,\theta) = 0$, so that the membrane is originally plane. In this case we have from (16):

$$z = \frac{1}{2c}\Sigma \frac{N_0}{\mu P_0} f(\mu r)\sin c\mu t + \frac{1}{c} S r^n \Sigma \frac{N}{\mu P} f(\mu r)\sin c\mu t \ldots (17),$$

when $f(\mu r)$, $N$, and $P$ are to be determined from (9) and (15); also $N_0$ and $P_0$ indicate the values of $N$ and $P$ respectively when $n$ is made zero. $\Sigma$ denotes a summation with respect to the values of $\mu$ furnished by (10), and $S$ denotes a summation with respect to $n$ from unity to infinity.

391. Let us suppose the initial velocity to be a function of $r$ only, so that $\psi(r, \theta)$ may be replaced by $F(r)$. Since $\int_0^{2\pi} \sin n\theta \, d\theta$ vanishes for every value of $n$, and $\int_0^{2\pi} \cos n\theta \, d\theta$ also vanishes, except when $n = 0$, we get

$$z = \frac{2}{a^2 c} \Sigma \frac{f(\mu r) \sin c\mu t}{\mu C^2} \int_0^a F(r) f(\mu r) r dr \text{ .........(18)},$$

where $f(\mu r)$ is given by

$$f(\mu r) = \int_0^\pi \cos(\mu r \cos \omega) \, d\omega \text{ ..................(19)}$$

and $\mu$ is to be determined by

$$0 = \int_0^\pi \cos(\mu a \cos \omega) \, d\omega, \text{ ..................(20)}$$

also $C$ is put for $f'(\mu a)$.

The values of $\mu$ found from (20) will not be commensurable, and then the various terms in (18) will not vanish simultaneously, except for $t = 0$; thus the sound produced by the membrane will not be unique, unless $F(r)$ be such that (18) reduces to a single term. Suppose it does reduce, so that

$$z = \frac{2}{a^2 c} \frac{f(\mu r) \sin c\mu t}{\mu C^2} \int_0^a F(r) f(\mu r) r dr \text{.........(21)}.$$

Differentiate with respect to $t$, and then put $t = 0$; thus we obtain the initial value of the velocity, which by supposition is $F(r)$; so that—

$$F(r) = \frac{2 f(\mu r)}{a^2 C^2} \int_0^a F(r) f(\mu r) r dr.$$

If we suppose $F(r) = K f(\mu r)$, where $K$ is any constant, this equation is satisfied; for we know, by Art. 389, that—

$$\int_0^a \{f(\mu r)\}^2 r dr = \frac{a^2}{2} \{f'(\mu a)\}^2 = \frac{a^2 C^2}{2}.$$

Substitute in (21); then we obtain for determining the motion on the supposition that initially $z = 0$, and $\frac{dz}{dt} = F(r)$,

$$z = \frac{K}{c\mu} f(\mu r) \sin c\mu t \text{ ....................(22)}.$$

We see from (22) that $z$ will vanish when $f(\mu r) = 0$ whatever $t$ may be. Let $\rho_1, \rho_2, \ldots\ldots$ denote values of $\mu a$ found from (20); then there will be as many *nodal circles* as there are values of $r$ less than $a$ in the series $\rho_1/\mu_1, \rho_2/\mu_2, \rho_3/\mu_3, \ldots\ldots$

392. Suppose that the initial velocity is equal to $b$ for all points of the membrane from the centre to a very small distance $h$, and is zero for all other points. From (18) we have

$$z = \frac{2b}{a^2 c} \Sigma \frac{f(\mu r) \sin c\mu t}{\mu C^2} \int_0^h f(\mu r)\, r dr.$$

Develope $f(\mu r)$ in powers of $r$; then by (19) we have

$$f(\mu r) = \pi \left\{1 - \frac{\mu^2 r^2}{2^2} + \frac{\mu^4 r^4}{2^2 . 4^2} - \frac{\mu^6 r^6}{2^2 . 4^2 . 6^2} + \ldots\ldots \right\},$$

so that

$$\int_0^h f(\mu r)\, r dr = \pi \left\{\frac{h^2}{2} - \frac{\mu^2 h^4}{4 . 2^2} + \frac{\mu^4 h^6}{6 . 2^2\, 4^2} \ldots\ldots \right\}.$$

Suppose $h$ to be so small that we may neglect all the terms of this series after the first; then we have

$$z = \frac{\pi b h^2}{a^2 c} \Sigma \frac{f(\mu r) \sin c\mu t}{\mu C^2} \ldots\ldots\ldots\ldots\ldots\ldots (23).$$

It follows then from the analysis that if the membrane is struck at the centre and within a small extent round the centre, it will give out several sounds simultaneously. The gravest sound will be that which corresponds to the least value of $\mu$, and only two or three of the sounds will be appreciable.

393. We will now suppose that the function $\psi(r, \theta)$ is equal to the constant $b$ for all values of $r$ comprised between the limits $r'$ and $r''$, and for all values of $\theta$ comprised between the limits 0 and $\theta'$; and that it is zero for all points of the membrane not within these limits; we will also assume that $r'' - r'$ is very small.

By (15) we have:

$$N_n = b \int_{r'}^{r''} f(\mu r)\, r^{n+1}\, dr \left\{\cos n\theta \int_0^{\theta'} \cos n\theta d\theta + \sin n\theta \int_0^{\theta'} \sin n\theta d\theta\right\}.$$

Since $r'' - r'$ is very small, we have approximately:

$$\int_{r'}^{r''} f(\mu r)\, r^{n+1}\, dr = (r'' - r') f(\mu\rho)\, \rho^{n+1},$$

where $\rho$ is some value intermediate between $r'$ and $r''$, and may be supposed equal to $(r'+r'')/2=r'$, if $r'$ be not zero. Thus

$$N_n=\frac{b}{n}(r''-r')f(\mu\rho)\,\rho^{n+1}\{\sin n\theta-\sin n(\theta-\theta')\},$$

and again $$N_0=b\,(r''-r')f(\mu\rho)\,\rho\theta'.$$

Also $P_n=\frac{\pi}{2}a^{2n+2}\,C_n^{\,2}$, and therefore $P_0=\frac{\pi}{2}a^2\,C_0^2$.

Substitute in the formula (17); then the result may be expressed thus:

$$z=\frac{2b\,(r''-r')\,\rho}{\pi a^2 c}\left[\frac{\theta'}{2}\Sigma\frac{f(\mu_0\rho)\,f(\mu_0 r)}{\mu_0 C^2_0}\sin c\mu_0 t\right.$$

$$+\frac{\rho}{a^2}\{\sin\theta-\sin(\theta-\theta')\}\,\Sigma\frac{f(\mu_1\rho)\,f(\mu_1 r)}{\mu_1\,C_1^2}\sin c\mu_1 t$$

$$\left.+\frac{\rho^2}{a^4}\{\sin 2\theta-\sin 2(\theta-\theta')\}\,\Sigma\frac{f(\mu_2\rho)\,f(\mu_2 r)}{\mu_2\,C_2^2}\sin c\mu_2 t+\ \ldots\ldots\ldots\ldots\right],$$

where $\mu_n$ denotes a value of $\mu$ found from (10), and $C$ denotes the corresponding value of $f'(\mu a)$; also $\Sigma$ denotes the sum of all the values found from (10).

If $\rho/a^2$ is so small that we may reject all powers of it above the first we get:

$$z=\frac{b\,(r''-r')\,\rho\theta'}{\pi a^2 c}\Sigma\frac{f(\mu_0\rho)\,f(\mu_0 r)\,\sin c\mu_0 t}{\mu_0 C_0^2}.$$

Pagani remarks:

En comparant cette valeur de $\zeta$ avec celle de la formule (23), on voit que les sons qu'elle rendrait seraient les mêmes dans les deux cas, quoique les circonférences nodales aient des rayons différens.

I do not see how these radii are different; they are determined with respect to (23) by the equation $f(\mu r)=0$, and in the present case by the equation $f(\mu_0 r)=0$, which means precisely the same thing. In fact if we suppose $\theta'=2\pi$, $\rho=(r''+r')/2$, and $r'=0$, the present case coincides with that in (23).

394. Pagani finishes his memoir thus:

Nous terminerons cette note par la remarque que la série donnée par l'équation (23), ainsi que les séries que nous fournissent les deux dernières formules, ne pouvant être réduites à un seul terme, qu'en négligeant ceux qui ont des valeurs comparables à celle du premier terme dont la valeur est la plus grande; la membrane fera entendre, dans tous ces cas, outre le son fondamental qui est le même, plusieurs autres sons appréciables, lesquels n'étant pas harmoniques avec le premier, causent cette singulière sensation que l'on éprouve lorsqu'on frappe d'un coup de baguette la caisse d'un tambour. Ceci nous explique ainsi pourquoi la corde d'un *piano* fait entendre un son lorsqu'on la frappe d'un coup de marteau, tandis que la membrane du tambour ne fait entendre qu'un bourdonnement, et enfin, pourquoi ce bourdonnement est sensiblement le même, soit que l'on frappe la membrane au centre ou dans un autre endroit quelconque peu éloigné de ce point.

395. *Considérations sur les principes qui servent de fondement à la théorie mathématique de l'équilibre et du mouvement vibratoire des corps solides élastiques;* par M. Pagani (Extrait d'un mémoire lu le 5 décembre 1829, à l'Académie Royale des sciences). This is published in Quetelet's *Correspondance Mathématique et Physique*, Vol. VI. 1830, pp. 87—91.

The opening paragraphs explain the object of the Article.

M. *Navier* a donné, le premier, les équations fondamentales de l'équilibre et du mouvement des corps solides élastiques. M. Poisson est parvenu ensuite aux mêmes équations dans un mémoire fort étendu, où l'on trouve plusieurs applications des formules générales (tom. VIII. des mémoires de l'Institut de France). Il s'est pourtant élevé une contestation entre ces deux illustres académiciens au sujet du principe qui leur a servi de base, et du mode que l'on a employé pour le traduire en langage algébrique.

Nous examinerons d'abord les principes et la marche adoptés par les deux savans géomètres, afin d'arriver, s'il est possible, à expliquer la contradiction apparente des hypothèses et la coïncidence remarquable des résultats.

In this article Pagani shews that by adopting certain special and arbitrary suppositions as to the nature of the molecular force he can bring the hypotheses of Navier and Poisson into agreement. I do not attach any importance to this article, and I presume that the memoir of which it is an abstract has not been published.

396. *Note sur l'équilibre d'un système dont une partie est supposée inflexible et dont l'autre partie est flexible et extensible.* This is published in the memoirs of the Academy of Brussels, Vol. VIII. 1834, pp. 1—14. It does not relate to our subject but to the well-known indeterminate problem of Statics, a simple example of which occurs when a body is on a horizontal plane, and in contact with it at more than three points. The subject was much discussed in the early half of the century by various Italian mathematicians: see for example Fusinieri, *Ann. Sci. Lomb. Venet.* II. 1832, pp. 298—304.

397. *Mémoire sur l'équilibre des colonnes.* This is published in the *Memorie della reale Accademia...di Torino*, Vol. I. 1839, pp. 355—371. The article is stated by the author to form a supplement to the *Note* which we have mentioned in the preceding Article. It has scarcely any connection with our subject, as the author adopts nothing with respect to elasticity except a portion of the ancient assumptions, such as will be found for instance in the section on the equilibrium of an elastic rod in Poisson's *Mécanique*, Vol. I. [He assumes for example the neutral line of a beam under longitudinal stress to coincide with the mean fibre: see p. 357 of the memoir.]

[398.] C. J. Hill. *Disputatio Physica de elasticitate torsionis in filis metallicis*, Lundae, 1819. This tract was submitted by by C. J. Hill and G. Lagergren. We have referred to it in Art. 225. Pp. 1—3 contain dedications, 4—22 text, and 23—29 tables of experimental results.

The tract commences with a few remarks as to compression, flexure and torsion, referring on these points respectively to the experiments of 's Gravesande, Bernoulli and Coulomb. The authors propose to experiment on the laws of torsion by a statical, as distinguished from Coulomb's kinetic method. The torsion balance adopted for this purpose is described in Art. 6. If we lay any stress upon these experiments it would appear that Coulomb's rules are not so completely in agreement with the phenomena of torsion as Coulomb himself and Savart (see Art. 334) seem to have supposed.

[399.] *Tractatus Geometricus de curvarum quae ab elasticitate nomen habent, theoria, aptissimoque construendarum modo...submittit* Carolus Joh. Ds. Hill. Londini Gothorum, 1829. This is a university dissertation of the Swedish type, in which the dissertation is the work of a Professor or some such person as *Praeses*. The candidates for the degree or the *Respondentes* were probably examined in the subject matter of the dissertation. Every eight pages has a fresh title page with a different *Respondens* and date. These title-pages do not mark any chapters or sections of the work, but are placed in the middle of sentences. The several *Respondentes* had the privilege of dedicating their respective eight pages. The tract concludes abruptly with the end of the third respondent's eight pages. Poggendorff has no reference either to this tract or to the one considered in the previous article.

[400.] The author proposes to treat those curves which derive their name from elasticity with that fulness which has been reached in the case of cycloids, catenaries, caustics, etc. He lays down the scheme of his work as follows:

Primum igitur quaestionis de curvis elasticis historiolam breviter exponamus; deinde, cum omnes fere curvas, si placuerit, ut elasticas spectare liceat, accuratius, quaenam praecipue hoc nomine sint insigniendae, definiamus; tum vero, priusquam ad harum theoriam atque constructionem nos propius accingamus, aliqua, istis mox applicanda, generalia de curvis analytice considerandis praecepta, itemque de functionibus Ellipticis, quae quidem nostris curvis arctissime necti constat, praemonere lubet.

[401.] The *historiola* contains an interesting remark on the first statement of the elastic curve for a *heavy* rod, which I had not met with before:

Primus igitur, quoad resciverimus, hujusmodi problematis mentionem fecit anonymus quidam, qui Parisiis degens, doctis quidem omnibus, praesertim vero iis qui illustrissimis, quae Parisiis atque Londoni florent, Societatibus adscripti erant, celeberrimum illud proposuit his fere verbis[a]: "*catenulae* mediocriter flexilis, attamen realis (ideoque etiam elasticae), utroque in limite clavis retentae, figuram, quam pondere innato induat, curvam indagare." Cui quaestioni cum statim haud responsum fuisset idem eandem denuo proposuit, atque ad collectores

**Actorum Eruditorum transmisit**[b]**. Attamen quadriennium responsio expectanda fuit: tum vero, idque eodem propemodum tempore, Clar. tum D. Bernoulli tum L. Eulerus hujus problematis dederunt solutiones**[c]**, easque egregie inter se convenientes.**

[a] ***Journ. des Sav.*, 1723 p. 366.** [b] ***Act. Erud. Lips.*, 1724, p. 366.**
[c] ***Comm. Ac. Petro p.*, T. III. p. 84. (1728).**

[402.] The writer refers (p. 5) to Lagrange's treatment of the elastic lamina in the *Mécanique Analytique* (see our Art. 159), and to his error with regard to the torsion. His equations are thus only true for plane curves, and, when reduced to two co-ordinates, agree with Euler's. Binet's correction of Lagrange (see our Art. 174) is thus alluded to:

**Cel. vero Binet, formulas generales emendaturus, torsionis effectum considerat, formulamque hujus momenti exhibet, eas vero, quae ad tensionem fili atque flexionem attinent, haud calculavit, dicens: "on ne parvient aux valeurs générales (de celles-ci) que par des calcules pénibles, dont les résultats paraissent fort compliqués." Omne igitur haud tulit punctum.**

[403.] The *historiola* is followed by a discussion on the *notio curvarum elasticarum.* The *curva elastica genuina* is that assumed by what we term the neutral axis of a beam, built in at one end and subject to transverse force at the other. It is thus a plane curve. The writer uses the following notation, $y$ is the vertical height of any point of the neutral line above the loaded end, $x$ is the corresponding horizontal distance, $x_0$ the horizontal distance between the ends, $s$ the arc from the loaded end to the point $(xy)$, $\phi$ is the angle between the tangent at $(xy)$ and the horizontal, $\rho$ the radius of curvature, and $a, b, c$ certain constants.

[404.] He obtains the following equations, which agree with the results of Euler and Lagrange:

$$\rho = \sqrt{a/(\sin\phi - c)}$$

$$ds = d\phi\sqrt{a/(\sin\phi - c)},$$

$$dx = d\phi \,.\cos\phi\,\sqrt{a/(\sin\phi - c)} \text{ or } x - x_0 = 2\sqrt{a\,(\sin\phi - c)},$$

$$dy = d\phi\sin\phi\,\sqrt{a/(\sin\phi - c)},$$

$$dy = (bc + x'^2)\,dx'/\sqrt{b^2 - (bc + x'^2)^2} \text{ where } x' = x - x_0.$$

This last is the differential equation to the *genuina curva elastica*. If the built-in end of the rod be horizontal $c=0$, and this equation reduces to $dy = x'^2\,dx'/\sqrt{b^2 - x'^4}$ which is termed the *rectangula*, in every other case the equation represents an *obliquangula* (p. 10).

[405.] Proceeding from the differential equation of the *obliquangula* the author defines the *familia elastica* as the series of curves whose co-ordinates $x$ and $y$ are related by an equation of the form

$$y = \int P\,dx/\sqrt{Q},$$

where $P$ is any rational algebraical function of $x$ and $Q$ an integral function of the fourth degree. We thus see that the relation between the co-ordinates can always be expressed by elliptic functions.

[406.] Pp. 11—20 contain a discussion on the forms into which the equations to curves can be thrown, Möbius and Ampère being the authorities chiefly made use of. Simple formulae for the osculating parabola of a curve are here obtained, and are claimed by the author as original. The one of which use is afterwards made is the now well-known

$$\tan\epsilon = 1/3\ d\rho/ds,$$

where $\epsilon$ is the angle between the diameter of the parabola, to the point of osculation, and the normal to the osculated curve at that point. The relation between $\rho$ and $\epsilon$ for any curve is (p. 19) termed its *caracteristica*.

[407.] On p. 21 the *caracteristica* for the genuine elastic curve is investigated. From the equations cited above the author easily finds

$$36a^2\tan^2\epsilon = (1-c^2)\rho^4 - 2ac\rho^2 - a^2.$$

What remains of the memoir (pp. 22—24) is occupied with showing that the *caracteristica* as deduced from Lagrange's equations in the *Mécanique Analytique* (ed. 1811, p. 156) is really identical with the above. It appears from Marklin's catalogue of Swedish Dissertations (Upsal, 1856) that nothing was published after page 24, where the essay breaks off abruptly.

[408.] 1830. William Ritchie. *On the elasticity of threads of glass with some of the most useful applications of this property to torsion balances.* *Phil. Trans.* 1830, pp. 215—222.

The author commences his memoir with the following statement:

From facts connected with crystallization and elasticity, it seems extremely probable, that the atoms of matter do not attract each other indifferently on all sides. There appear to be peculiar points on their surfaces which have a more powerful attraction for each other than for other points on the same molecule. This property is not peculiar to the atoms of ponderable matter, but seems also to belong to those of light and heat. It is as impossible to prove directly the existence of this property, as it is to prove the existence of atoms themselves; but on account of the satisfactory manner in which it enables us to explain the phenomena of crystallization and elasticity, it is now generally adopted.

To this polar property of atoms the author attributes that peculiar elastic effect termed torsion. He has noticed the very large amount of twist which can be given to such a brittle substance as glass thread before it obtains a permanent set or breaks. He thinks the resistance of glass threads to torsion is due to vitreous molecules being held together by the attractions of their poles or points of greatest affinity. These points are displaced by torsion along the whole line of communication, and as they endeavour to regain their former state of stable equilibrium, the thread will of course untwist itself. He remarks that if a thread could be drawn so fine as to consist of a single line of vitreous molecules, torsion would have no tendency to displace the points of greatest attraction, and this elementary thread might be twisted for ever without breaking,—the compound molecules of glass would only turn round their points of greatest attraction like bodies revolving on a pivot. This theory is exemplified by the statement that the author has drawn threads of glass of such extreme tenuity, that one of them, not more than a foot long, may be twisted nearly a hundred times without breaking.

It can hardly be said that the above theory contains any very lucid atomic explanation of the phenomenon of torsion.

[409.] In § 4, Ritchie states the difficulty which arises in proving for glass threads the laws of torsion determined by Coulomb for metallic wires. This is due to the impossibility of obtaining glass threads of uniform diameter, and hence the force of torsion cannot be shown to vary as the fourth power of the diameter. The property however required for torsion-balances,—namely, that 'the force of torsion or that force with which a thread tends to untwist itself is directly proportional to the number of degrees through which it has been twisted,—is easily deduced experimentally. Experiments for this purpose are described[1].

[410.] § 5 explains the best method of drawing a fine glass-thread.

§ 6 is devoted to the description and use of a torsion galvanometer with glass-thread.

The concluding paragraphs (§ 7, and § 8) contain an account of an ingenious torsion balance for the weighing of very minute portions of matter.

[411.] *Summary.* Although the most important work of that period to which this chapter has been devoted has still to be considered in the following chapters, yet the reader cannot fail to remark the great stride which the theory of elasticity made in these ten years. Within this decade the theory may be said to have been discovered and in broad outline completed. It is entirely to French scientists that we owe this great contribution to a wider knowledge of the physical universe, and however we may regard the relative merits of Navier, Poisson, and Cauchy, there cannot be the least doubt as to their dividing between them the entire merit of the discovery. Even if we

[1] The torsional *imperfection* of glass fibres is however emphasised by Sir William Thomson in his paper on Elasticity in the *Encycl. Brit.* Art. 4.

put on one side the more important work of these leaders of scientific investigation we cannot fail to be struck with the essentially modern character of the minor memoirs. The methods of Lagrange and Fourier had become general, and the more complex forms of analysis were wielded, not without success, by lesser mathematicians. Sophie Germain with all her vagaries at least succeeded in finally establishing the equation for the normal vibrations of a plate; while Pagani following Poisson gave some very general results for the vibrations of a circular membrane. Lastly we may note that amidst this wealth of theoretical power, France possessed in Savart a physical elastician of an extremely thoughtful and suggestive kind.

# CHAPTER IV.

## POISSON.

412. The contributions of Poisson to our subject begin with his *Mémoire sur les surfaces élastiques;* this was read to the French *Institut* on the 1st of August, 1814, and is contained in the volume of the *Mémoires* for 1812, published in 1814; it occupies pages 167—225 of the volume.

413. The introductory remarks occupy pages 167—172; these supply some historical information. Poisson says

> Jacques Bernouilli est, comme on sait, le premier qui a donné l'équation d'équilibre de la lame élastique, en se fondant sur cette hypothèse, que l'élasticité, en chaque point, est une force normale à la courbe, dont le moment est proportionnel à l'angle de contingence, ou en raison inverse du rayon de courbure en ce point. Depuis ce grand géomètre, plusieurs autres, et principalement Euler et Daniel Bernouilli, ont publié un grand nombre de Mémoires sur les conditions d'équilibre des lignes élastiques et sur les lois de leurs vibrations; mais il n'a paru que quelques Essais infructueux qui aient pour objet les surfaces élastiques, pliées à-la-fois en deux sens différents.

Poisson also refers to *un autre Jacques Bernouilli* who considered the problem of the vibration of an elastic lamina in the St Petersburg Memoirs for 1788, but the equation he obtained was deficient in a term: see Art. 122.

414. The *Institut* about five years before the date of Poisson's memoir had proposed the vibrations of sonorous plates as a prize

subject; only one essay had been received worthy of attention, which was from an anonymous author; in this essay an equation was given without demonstration, which contained the term omitted in the St Petersburg memoir of 1788. Probably this anonymous writer was Mademoiselle Sophie Germain. The equation is that which will be found in Art. 290.

415. The introductory remarks conclude thus:

Dans un autre Mémoire, j'appliquerai les mêmes considérations aux lignes élastiques, à simple ou à double courbure, d'une épaisseur constante ou variable suivant une loi donnée; ce qui me conduira d'une manière directe et exempte d'hypothèse, non-seulement à leurs équations d'équilibre, mais aussi à l'expression des forces qu'on doit appliquer à leurs extrémités, pour les tenir fixes et balancer l'effet de l'élasticité.

I do not think that this design was carried out, though, as we shall see, a note on the subject appeared in the third volume of the *Correspondance sur l'École...Polytechnique*. The present memoir is divided into two sections, which are devoted respectively to inelastic and elastic surfaces.

416. The first section occupies pages 173—192; it is entitled: *Equation d'équilibre de la surface flexible et non-élastique*. The problem had been considered by Lagrange, but not in a satisfactory manner: see Art. 158 of my account of Lagrange. Poisson works it out in an intelligible and accurate manner; he assumes that a rectangular element of the membrane is acted on by tensions at right angles to the edges, but he does not assume that the tension on one pair of parallel edges is the same as the tension on the other pair. As a particular case he shews that if certain conditions are satisfied the two tensions may be taken equal, and then his equation for determining the figure of the membrane coincides with Lagrange's. The equation in question is

$$Z - pX - qY + \frac{T}{k^2}\left\{(1+q^2)\frac{d^2z}{dx^2} - 2pq\frac{d^2z}{dxdy} + (1+p^2)\frac{d^2z}{dy^2}\right\} = 0,$$

where the axis of $z$ is perpendicular to the plane of the membrane, $X$, $Y$, $Z$ are the components of applied force per unit area at the point $(x, y, z)$, $p = dz/dx$, $q = dz/dy$, $T$ is the tension and $k$ an elastic constant. Poisson justly says that Lagrange's equation cannot be considered as the general equation to flexible surfaces in equilibrium.

417. The second section occupies pages 192—225; it is entitled: *Equation de la surface élastique en équilibre.* The method is peculiar. On each rectangular element tensions are supposed to act, as in the first section, but now they are assumed equal; and besides these tensions repulsive forces are supposed to act on each particle arising from elasticity; the repulsive forces are supposed to be sensible only so long as the distance is insensible. The surface is supposed to be of uniform thickness. The repulsive force between two particles at the distance $r$ is denoted by the product of $\rho(r)$ into the mass of the particles. In the course of the investigation the powers of the distance beyond the fourth are neglected; and finally a complicated differential equation of the fourth order is obtained to determine the form of the surface in *equilibrium*: see page 215 of the Memoir. In this equation the repulsive force enters only in the values of two constants, namely $a^2$ which stands for $\int r^2\rho(r)\,dr$, and $b^2$ which stands for $\int r^4\rho(r)\,dr$; the integrals are supposed taken from $r=0$ to the extreme value of $r$ for which $\rho(r)$ is sensible.

418. With respect to *motion* Poisson confines himself to the case of a surface nearly plane, with no applied forces; and then he arrives at the equation

$$\frac{d^2z}{dt^2}+n^2\left(\frac{d^4z}{dx^4}+2\,\frac{d^4z}{dx^2dy^2}+\frac{d^4z}{dy^4}\right)=0;$$

this agrees with the result he obtained by a later investigation: see Art. 485. The second James Bernoulli had omitted the term $d^4z/(dxdy)^2$ (see Art. 122). With respect to the solution of the equations Poisson says on his page 170:

Malheureusement cette équation ne peut s'intégrer sous forme finie, que par des intégrales définies qui renferment des imaginaires; et si on les fait disparaître, ainsi que M. Plana y est parvenu dans le cas des simples lames, on tombe sur une équation si compliquée, qu'il paraît impossible d'en faire aucun usage. (See Art. 178.)

A form of solution has been since obtained, which is not complicated. See the account of a paper of 1818 by Fourier, Art. 207.

419. In the course of the investigation Poisson has to effect certain integrations in order to determine the force which acts on an element of the elastic lamina. If the element is at a sensible distance from the boundary of the surface no difficulty occurs with respect to the limits of the integrations; but if the element is extremely near to the boundary, a difficulty does occur, because the particle considered is not completely surrounded by other particles up to the limit of molecular activity. Poisson notices this point on his page 202, and says:

Mais pour trouver l'équation différentielle de la surface élastique en équilibre, il suffit de considérer les points intérieurs, situés à une distance quelconque de son contour; et l'on n'a besoin d'examiner ce qui arrive aux points extrèmes, que pour déterminer les forces particulières que l'on doit appliquer aux limites de la surface pour la tenir en équilibre; détermination très-délicate sur laquelle je me propose de revenir par la suite, mais dont il ne sera pas question dans ce mémoire.

The intention here expressed does not seem to have been carried into effect.

420. A curious property of the elastic surface in equilibrium is noticed on pages 221—225. Poisson determines by the Calculus of Variations the differential equation to the surface of constant area for which $\iint\left(\frac{1}{\rho}+\frac{1}{\rho'}\right)^2 \sqrt{(1+p^2+q^2)}\, dxdy$ is a maximum or minimum, where $\rho$ and $\rho'$ are the two principal radii of curvature at the point $(x, y, z)$; he finds that the differential equation is the same as that which he had already obtained for the elastic surface in equilibrium: see my *History of the Calculus of Variations*, page 333[1].

[1] [In the place referred to, Poisson states that the differential equation obtained is the same as that which arises from making $\iint\left(\frac{1}{\rho}-\frac{1}{\rho'}\right)^2 \sqrt{1+p^2+q^2}\, dxdy$ a

421. The Memoir considered as an exercise in mathematics is a fine specimen of Poisson's analytical skill, but it adds little to the discussion of the physical problem. In his later writings Poisson objects strongly to the use of integrals instead of finite summations, in questions relating to molecular force, which here he adopts. I have quoted in Art. 435, of my account of his memoir of April 1828 the opinion which he there expresses of the present memoir.

422. A very good abstract of the memoir by Poisson himself was published in the *Bulletin de la Société Philomatique*, 1814, and afterwards in the *Correspondance de l'École Polytechnique*, Vol. III. 1816. A note in the later work, on page 154, states that the *Institut* has again offered as a prize subject the theory of the vibrations of elastic plates; and on page 410, it is said that the prize has been awarded to Mademoiselle Sophie Germain.

423. 1816. An article by Poisson entitled: *Sur les lignes élastiques à double courbure*, occurs on pages 355—360 of the *Correspondance sur l'École Polytechnique*, Vol. III. 1816. The object of the article is to prove that the *moment of torsion* round the tangent to the curve is constant throughout the curve when there is equilibrium. The mechanical problem assumed is that at any point of the curve the elasticity tends to produce two effects,

minimum, for $\iint \delta \frac{\sqrt{1+p^2+q^2}}{\rho\rho'} dxdy$ vanishes so far as the terms under the sign of double integration are concerned. Hence the same differential equation must arise from making $\iint \left\{A\left(\frac{1}{\rho}+\frac{1}{\rho'}\right)^2 + B\left(\frac{1}{\rho}-\frac{1}{\rho'}\right)^2\right\} \sqrt{1+p^2+q^2}\, dxdy$ a minimum, or from making the surface integral of $C\left(\frac{1}{\rho^2}+\frac{1}{\rho'^2}+\frac{2\mu}{\rho\rho'}\right)$ a minimum. But this is the form usually adopted for the potential energy per unit of area due to the bending of a thin plate of uniform thickness and isotropic material. (Cf. Lord Rayleigh, *Theory of Sound*, I. p. 293.) I must remark here that I am not quite satisfied with this expression for the potential energy of the bending of a plate. My doubt arises exactly as in the case of a bent rod, for which the potential energy of bending does not seem to be proportional to $\frac{1}{\rho^2}$ per unit of length, if the bending is accompanied in any degree by longitudinal stress. In the same manner I am inclined to think the above expression only true for a plate when the applied forces are wholly normal to the plane of the plate.] ED.

namely one to change the value of the angle of contingence in the osculating plane, and the other to twist the curve round the tangent at the point considered[1].

As we have seen in Arts. 159 and 173, the problem of the equilibrium of such an elastic curve had been considered by Lagrange and by Binet. Poisson says

M. Binet a eu regard le premier à la torsion dont les courbes élastiques sont susceptibles; mais on n'avait point encore expliqué la nature de cette force, et montré que son moment est constant dans l'état d'équilibre. Lagrange a donné, dans la Mécanique analytique, des équations de la ligne élastique à double courbure, qu'il a trouvées par une analyse très-différente de la nôtre, et qui reviennent cependant à nos équations (1), en y supposant $\theta=0$.

Here $\theta$ denotes the moment of torsion. Poisson's article is reproduced substantially in his *Mécanique*, 1833, Vol. I. pages 622—627.

[424.] 1818 *Remarques sur les rapports qui existent entre la propagation des ondes à la surface de l'eau, et leur propagation dans une plaque élastique. Bulletin des Sciences par la Société Philomatique*, Année 1818, Paris, pp. 97—99. This is a short note by Poisson on Fourier's memoir on the vibrations of elastic plates (see Art. 207). He notes how relations between waves at the surface of water and those in an infinite elastic plate arise from both being determined by linear partial differential equations with constant coefficients.

Ces rapports singuliers tiennent à ce que les lois de ces deux mouvemens sont renfermées dans des équations aux différences partielles de même nature, savoir, des équations linéaires à coëfficients constans, qui ne sont pas du même ordre par rapport au temps et par rapport aux distances des points mobiles au lieu de l'ébranlement primitif, mais avec cette différence, que l'équation du problême des ondes est du quatrième ordre par rapport au temps, et du second par rapport aux coordonnées; tandis que dans l'autre problême elle est

[1] [Poisson's equations are not the most general conceivable, as there would usually be a couple round the radius of curvature of the 'mean fibre' of the wire. His result as to the moment of torsion is not generally true. We shall return to this point in the Chapter devoted to Saint-Venant.] ED.

au contraire du second ordre par rapport au temps, et du quatrième par rapport aux coordonnées. De là vient que tout ce qui se dit du temps ou des distances dans le premier problême, doit s'appliquer aux distances ou au temps dans le second, et *vice versâ*.

Poisson remarks in conclusion:

Au reste, cette propagation des sillons dans les plaques élastiques infinies est une question de pure curiosité, qu'il ne faut pas confondre avec la propagation du son dans ces mêmes plaques; celle-ci se fait toujours d'un mouvement uniforme; la vitesse ne dépend ni de l'ébranlement primitif ni de l'épaisseur de la plaque.

425. 1818. *Sur l'intégrale de l'équation relative aux vibrations des plaques élastiques.* This is published in the *Bulletin des Sciences par la Société Philomatique*, Paris, 1818; it occupies pages 125—128 of the volume.

The object of the paper is to give the integral of the differential equation for the vibration of an elastic plate: see Art. 418.

The equation is

$$\frac{d^2z}{dt^2} + a^2\left(\frac{d^4z}{dx^4} + 2\,\frac{d^4z}{dx^2dy^2} + \frac{d^4z}{dy^4}\right) = 0 \ldots\ldots\ldots\ldots\ldots(1).$$

Let $\zeta$ be another function of $x$, $y$ and $t$, which satisfies the equation

$$\frac{d\zeta}{dt} = m\left(\frac{d^2\zeta}{dx^2} + \frac{d^2\zeta}{dy^2}\right) \ldots\ldots\ldots\ldots\ldots\ldots\ldots (2),$$

where $m$ is a constant. Differentiate with respect to $t$; then

$$\frac{d^2\zeta}{dt^2} = m\left(\frac{d^3\zeta}{dx^2dt} + \frac{d^3\zeta}{dy^2dt}\right);$$

put for $d\zeta/dt$ on the right-hand side its value derived from (2); thus we get

$$\frac{d^2\zeta}{dt^2} = m^2\left(\frac{d^4\zeta}{dx^4} + 2\,\frac{d^4\zeta}{dx^2dy^2} + \frac{d^4\zeta}{dy^4}\right).$$

Hence if we put $m^2 = -a^2$ we shall satisfy (1) by taking $z = \zeta$. In this way we shall obtain only a particular integral of the equation; but if we take in succession $m = +a\iota$ and $-a\iota$, where $\iota$ is put for $\sqrt{-1}$, the equation (2) will give two values of $\zeta$, the sum of which will express the complete integral of the equation (1). The question then is reduced to the integration of the equation (2).

Now Laplace has given the integral of the equation

$$\frac{d\zeta}{dt} = m\frac{d^2\zeta}{dx^2},$$

in the form

$$\zeta = \int_{-\infty}^{+\infty} e^{-\alpha^2} \phi\,(x + 2\alpha\sqrt{mt})\, d\alpha,$$

where $\phi$ denotes an arbitrary function. It is easy to extend this form of integral to equation (2); with respect to this we shall have

$$\zeta = \int_{-\infty}^{+\infty}\int_{-\infty}^{+\infty} e^{-\alpha^2-\beta^2}\phi\,(x + 2\alpha\sqrt{mt},\, y + 2\beta\sqrt{mt})\, d\alpha d\beta.$$

If we put successively in this formula $+a\iota$ and $-a\iota$ for $m$, and take the sum of the two results, we shall have for the complete integral of the equation (1)

$$z = \int_{-\infty}^{+\infty}\int_{-\infty}^{+\infty} e^{-\alpha^2-\beta^2}\phi\,(x + 2\alpha\sqrt{a\iota t},\, y + 2\beta\sqrt{a\iota t})\, d\alpha d\beta$$

$$+ \int_{-\infty}^{+\infty}\int_{-\infty}^{+\infty} e^{-\alpha^2-\beta^2}\psi\,(x + 2\alpha\sqrt{-a\iota t},\, y + 2\beta\sqrt{-a\iota t})\, d\alpha d\beta,$$

where $\phi$ and $\psi$ denote two arbitrary functions.

To shew how these arbitrary functions are to be determined from the initial conditions of the plate, suppose that the origin of the motion corresponds to $t = 0$, and that at that instant the equation to the plate is $z = \chi\,(x, y)$; then we have

$$\chi\,(x, y) = \{\phi\,(x, y) + \psi\,(x, y)\}\int_{-\infty}^{+\infty}\int_{-\infty}^{+\infty} e^{-\alpha^2-\beta^2}\, d\alpha d\beta \ldots (3).$$

Let us also suppose that initially the velocity of every point is zero; we must then have $dz/dt = 0$, when $t = 0$, for all values of $x$ and $y$: this condition will be satisfied by supposing the functions $\phi$ and $\psi$ to be equal, so that from (3) we have

$$\chi\,(x, y) = 2\phi\,(x, y)\int_{-\infty}^{+\infty}\int_{-\infty}^{+\infty} e^{-\alpha^2-\beta^2}\, d\alpha d\beta = 2\pi\phi\,(x, y).$$

Thus

$$\phi\,(x, y) = \psi\,(x, y) = \frac{1}{2\pi}\chi\,(x, y).$$

It is easy to get rid of the imaginary quantities which occur in the general value of $z$, by putting instead of $\alpha$ and $\beta$ respectively $\frac{\alpha}{\sqrt{\iota}}$ and $\frac{\beta}{\sqrt{\iota}}$ in the first integral, and $\frac{\alpha}{\sqrt{-\iota}}$ and $\frac{\beta}{\sqrt{-\iota}}$ in the second integral, which will make no change in the limits. Let us also put $\chi$ in place of the arbitrary functions $\phi$ and $\psi$, and change the imaginary exponentials into sines and cosines; thus we obtain

$$z=\frac{1}{\pi}\int_{-\infty}^{+\infty}\int_{-\infty}^{+\infty}\sin(\alpha^2+\beta^2)\,\chi(x+2\alpha\sqrt{at},\,y+2\beta\sqrt{at})\,d\alpha d\beta.$$

Another form may be given to this expression by making $x+2\alpha\sqrt{at}=p$, and $y+2\beta\sqrt{at}=q$; thus it becomes

$$z=\frac{1}{4a\pi t}\int_{-\infty}^{+\infty}\int_{-\infty}^{+\infty}\chi(p,q)\sin\left\{\frac{(x-p)^2+(y-q)^2}{4at}\right\}dpdq.$$

Poisson adds with respect to the last formula:

Sous cette dernière forme, l'intégrale de l'équation (1) coincide avec celle que l'on trouve en résolvant d'abord cette équation par une série infinie d'exponentielles réelles ou imaginaires, et sommant ensuite cette série par des intégrales définies, ainsi que l'a fait M. Fourier dans son Mémoire sur les vibrations des plaques élastiques.

I presume therefore that the integral was first obtained by Fourier in the form here given; but the memoir by Fourier to which Poisson here refers seems not to have been published: see Art. 207 for an account of a note published by Fourier in 1818.

426. 1823. A memoir by Poisson entitled: *Sur la distribution de la chaleur dans les corps solides* was published in the *Journal de l'École Polytechnique*, XIX$^e$ Cahier, 1823. This does not belong to our subject, but I mention it because of a reference made to it by Saint-Venant in Moigno's *Statique*, page 619. After speaking of two different definitions of *pressure* in the theory of elasticity Saint-Venant says:

Au reste, M. Poisson a montré en 1821 (*Journal de l'École Polytechnique*, XIX$^e$ Cahier, 1823, p. 272) que les deux définitions analogues, relatives au *flux de chaleur*, donnent dans les calculs les mêmes résultats quand on néglige certains ordres de quantités.

427. 1827[1]. An article by Poisson entitled: *Note sur les vibrations des corps sonores* was published in the *Annales de Chimie*, Vol. 36, 1827. It occupies pages 86—93 of the volume. It is also printed in *Ferussac's Bulletin*, Vol. IX. 1828, pages 27—31. The first paragraph explains the nature of the article:

Je m'occupe actuellement d'un travail fort étendu sur les lois de l'équilibre et du mouvement des corps élastiques, et particulièrement sur les vibrations des corps sonores. En attendant que j'aie pu en terminer la rédaction définitive, je demande à l'Académie la permission de lui faire connaître le principe de mon analyse et plusieurs des conséquences qui s'en déduisent.

The article states some results obtained in the memoir as to the vibrations of an elastic rod; and for the case of longitudinal vibrations seven experiments made by Savart are compared with calculations made by Poisson, and a satisfactory agreement obtained. Two points of interest will now be noticed.

428. In the *Annales* Poisson says on page 87:

Une même verge élastique peut vibrer de quatre manières différentes. Elle exécutera, 1° des vibrations *longitudinales*, lorsqu'on l'étendra ou qu'on la comprimera suivant sa longueur; 2° des vibrations *normales* quand on la dilatera ou qu'on la condensera perpendiculairement a sa plus grande dimension; 3° des vibrations que Chladni a nommées *tournantes*, qui auront lieu en vertu de la torsion autour de son axe; 4° enfin des vibrations *transversales*, dues aux flexions qu'on lui fera éprouver.

He seems to imply on the next page that he has determined the laws of the normal vibrations, but that they are too complex to be indicated in the present note. In the memoir which this note announces all that is said about *normal* vibrations seems contained in a brief passage on pages 452 and 453.

Mais il faut ajouter que, dans tous les cas, les vibrations longitudinales seront accompagnées de vibrations normales de la même durée....

Finally in his *Mécanique*, 1833, Vol. II. page 368, Poisson says:

Les vibrations normales consistent en des dilatations et condensations alternatives des sections de la verge, perpendiculaires à sa longueur; elles n'ont pas encore été déterminées par la théorie.

[1] [A letter of Poisson's in the *Annales* for this year will be noticed when we consider his memoir of October, 1829.] ED.

There seems to me an inconsistency between these various statements, but I cannot explain it[1].

429. In the *Annales* Poisson says that he has determined the ratio of the transversal to the longitudinal vibrations for the case of a cylindrical rod and for the case of a rod in the form of a parallelepiped. The former case however alone appears in the memoir. Of the seven experiments made by Savart, three refer to the former case and four to the latter; the former three are alone given in the memoir.

The following are Poisson's words on this point:

Le rapport des vibrations transversales aux longitudinales dépend de la forme de la verge; je l'ai déterminé dans deux cas différens: dans le cas des verges cylindriques et dans celui des verges parallélepipédiques. S'il s'agit, par exemple, d'une verge libre par les deux bouts, rendant le ton le plus grave dont elle est susceptible; que l'on représente par $l$ sa longueur, par $n$ le nombre de ses vibrations longitudinales, et par $n'$ celui des transversales; et que l'on désigne par $e$ son épaisseur dans le cas des verges parallélepipédiques, ou son diamètre dans le cas des verges cylindriques, on aura,

$$n' = (2 \cdot 05610)\, ne/l$$

dans le premier cas, et

$$n' = (1 \cdot 78063)\, ne/l$$

dans le dernier; le second nombre compris entre parenthèses se déduisant du premier en le multipliant par $\frac{1}{2}\sqrt{3}$.

See pp. 88 and 89.

430. Another article by Poisson occurs on pages 384—387 of the volume of the *Annales* cited in Art. 427; it is entitled: *Note sur l'extension des fils et des plaques élastiques.* We have referred to this article in the previous chapter (Art. 368).

At the end of the article Poisson cites another result which he had obtained from theory, which he says would be less easy to verify by experiment: see Art. 483. Poisson's article is translated into German in Poggendorff's *Annalen,* Vol. XII. 1828, pages 516—519.

[1] [This note of Poisson's should be read in conjunction with Savart's memoir of 1822: see Articles 323—327.] ED.

431. 1828. An article by Poisson entitled: *Mémoire sur l'équilibre et le mouvement des corps élastiques* was published in the *Annales de Chimie*, Vol. 37, 1828; it occupies pages 337—355 of the volume. The object of the article is to give an account of the memoir which Poisson read to the Academy on the 14th of April, 1828; the pages 337—348 of the article coincide with the introduction to the memoir, that is with the pages 357—368 of it. A note in the memoir relative to a point in the Theory of Equations is not in the article; on the other hand a brief note is given in the article which is not in the memoir; this note in the article consists of simple reasoning to illustrate in a particular case the general result obtained by Poisson, and denoted by $K=0$: see Art. 442 of my account of the memoir.

432. After this introduction Poisson enumerates various results obtained in his memoir which may interest students of physics; these he distributes under eleven heads.

There are a few lines in the *Annales* with respect to the vibrations of plates which are not in the memoir. Poisson obtained by calculation in the memoir for the ratio of the lowest two sounds in the case of a free circular plate the number 4·316; he says in the *Annales*:

> M. Savart a obtenu pour ce rapport, un nombre qui surpasse un peu 4, mais d'une fraction sensiblement moindre que nous ne le trouvons. Il pense que la différence entre le calcul et l'expérience n'est pas hors des limites des erreurs dont est susceptible ce genre d'observations.

433. The article in the *Annales* concludes with the following paragraph in which Poisson puts forth the just claims of his memoir to consideration:

> La discussion qui s'est élevée à l'Académie après la lecture de mon Mémoire, m'oblige de faire observer qu'il se compose de deux parties: l'une, toute spéciale, est relative à des questions d'acoustique, dont ce qui précède est un résumé, et que personne jusqu'à présent n'avait traitées; l'autre renferme des considérations générales sur l'action moléculaire, et sur l'expression des forces qui en résultent. On y fait voir que pour parvenir à cette expression, il est nécessaire d'apporter quelque restriction à la fonction de la distance qui exprime l'action mutuelle de deux molécules, et qu'il ne suffit pas, comme

on l'avait supposé jusqu'ici, que cette fonction soit une de celles qui deviennent insensibles dès que la variable a acquis une grandeur sensible. On y prouve aussi que la somme qui exprime la résultante totale des actions moléculaires n'est pas de nature à pouvoir se convertir en une intégrale; ce qui n'avait pas non plus été remarqué, et ce qui est cependant essentiel, puisque la représentation de cette résultante par une intégrale définie rendrait nul son coefficient après le changement de forme du corps produit par des forces données, et par conséquent impossible la formation de ses équations d'équilibre. Enfin, après avoir établi les équations générales de l'équilibre d'un corps élastique de forme quelconque, en ayant égard aux diverses circonstances que je viens d'indiquer, j'en ai déduit celles qui appartiennent aux cordes, aux verges, aux membranes et aux plaques élastiques; déduction que personne, à ma connaissance, n'avait cherché à effectuer, et qui exige des transformations d'analyse par lesquelles j'ai été longtemps arrêté même dans le cas le plus simple, celui de la corde élastique.

On this paragraph three brief remarks may be made. The discussion in the Academy probably consisted mainly of criticisms by Navier which we shall notice hereafter. The restrictions which must be imposed on the function representing the mutual action between two molecules do not seem to be very decisively stated in the memoir, though there are certainly hints bearing on the point: see Art. 439 of my account of the memoir. It is not obvious what great difficulties Poisson can have found in his discussion of the simple problem of the elastic cord.

434. 1829. *Mémoire sur l'équilibre et le mouvement des corps élastiques.* This memoir was read to the Paris Academy on the 14th of April, 1828; it is published in the Memoirs of the Academy, Vol. VIII. 1829, where it occupies pages 357—570; on pages 623—627 of the volume there is an *Addition* to the memoir which was read on the 24th of November, 1828. The memoir is one of the most famous written by this great mathematician.

435. The introduction to the memoir occupies pages 359—368. We have here a rapid sketch of the history of questions connected with the theory of elasticity. Leibniz and the Bernoullis solved the problem of the form of the catenary, as to which Galileo had

erred; James Bernoulli investigated the form of an elastic lamina in equilibrium. After these problems relating to equilibrium some relating to motion were discussed. D'Alembert was the first who solved the problem of vibrating cords, and Lagrange some years later gave another solution. Euler and Daniel Bernoulli determined the vibrations of an elastic lamina for all the circumstances in which the ends of the lamina could be placed. Then Poisson proceeds thus on his page 360:

Tels sont, en peu de mots, les principaux résultats relatifs à l'équilibre et au mouvement des corps élastiques, qui étaient connus lorsque j'essayai, d'aller plus loin dans un *Mémoire sur les surfaces élastiques*, lu à l'Institut en 1814. J'ai supposé que les points d'une plaque élastique, courbée d'une manière quelconque, se repoussent mutuellement suivant une fonction de la distance qui décroît très-rapidement et devient insensible dès que la variable a acquis une grandeur sensible; hypothèse qui m'a conduit à une équation d'équilibre des surfaces élastiques, laquelle prend la même forme que celle de la simple lame courbée en un seul sens, quand on l'applique à ce cas particulier. Mais cette manière d'envisager la question ne convient rigoureusement qu'à une surface sans épaisseur, sur laquelle sont placés des points matériels, contigus ou très-peu distants les uns des autres; et quand, au contraire, on a égard à l'épaisseur de la plaque courbée, ses particules se distinguent en deux sortes: les unes se repoussent effectivement en vertu de la contraction qui a lieu du côté de la concavité, et les autres s'attirent en vertu de la dilatation produite du côté opposé. Il était donc nécessaire de reprendre de nouveau cette question; et pour qu'elle soit complètement résolue il faudra trouver, relativement à une plaque élastique d'une épaisseur donnée, les conditions qui doivent être satisfaites, soit en tous ses points, soit à ses bords en particulier, pour l'équilibre des forces qui lui sont appliquées et des actions mutuelles de ses molécules.

436. In his introductory remarks Poisson draws attention to a point which he considers very important; see his pages 365 and 366. It amounts to this: in all cases hitherto in which molecular action has been considered, such as questions of capillary attraction and heat, the forces which arise from this action have been expressed by definite integrals, but this mode of expression is inapplicable. For in the natural state of the body no force would

be called into action, and the definite integral would vanish; it would follow that after the body has suffered deformation the definite integral would still vanish, and so no force would be called out, which is absurd. The mathematical process is given at page 398 of the memoir, and we shall recur to the point: see Art. 443.

437. A remark is made by Poisson in a note on his page 367 relating to the Theory of Equations. Fourier on page 616 of the volume contradicts Poisson, and goes fully into the matter in the Memoirs of the Paris Academy, Vol. x. pages 119—146.

438. The first section of the memoir is entitled: *Expression des forces résultantes de l'action moléculaires;* it occupies pages 368—405. The object of this section may be said to be generally the investigation of the equations of equilibrium of elasticity; and the results coincide with those of the ordinary text-books on elasticity provided we take $\lambda = \mu$. Thus where Lamé writes:

$$N_1 = \lambda\theta + 2\mu \frac{du}{dx},$$

the corresponding equation with Poisson is, if $k = \mu$,

$$N_1 = k\left(3\frac{du}{dx} + \frac{dv}{dy} + \frac{dw}{dz}\right).$$

We proceed to notice some special points in this section.

439. On his page 369 Poisson gives an example which may illustrate the law of molecular force[1]. Let $r$ be the distance between two particles; then the molecular force may be the product of some constant into the expression

$$b^{-(r/\overline{na})^m}$$

Here $b$ denotes a constant greater than unity, $m$ is a large positive quantity, $\alpha$ is the distance between two consecutive particles, $n$ is a large integer but yet such that $n\alpha$ is a line of imperceptible length. This expression will remain nearly constant so long as $r$ is not a considerable multiple of $\alpha$; but when $r$ becomes greater than $n\alpha$ the expression diminishes very rapidly, and soon becomes

[1] [This example is not wholly satisfactory as the molecular force cannot in this case become repulsive; it is necessary to consider the *difference* of two expressions of this kind.] ED.

insensible. We may denote the product of this expression into a constant by $\rho(r)$.

440. On page 375 Poisson uses without any formal demonstration a theorem as to the tension produced by molecular action across an infinitesimal plane area taken in the substance of an elastic body.

[The following interesting historical note relative to this theorem is given by Saint-Venant:

> Nous l'avons démontré pour la première fois *pour cette définition de la pression* à la Société philomatique en mars 1844, et aux *Comptes rendus*, 7 juillet 1845 (t. XXI, p. 24). Cauchy l'a demontré de même dans un Mémoire inédit (*Comptes rendus*, 23 juin et 14 juillet, t. XX, p. 1765, et t. XXI, p. 125). Il est, au reste, presque évident, et il a été admis aussi par Poisson (t. VIII. des *Mémoires de l'Institut*, p. 375, et *Journal de l'École Polytechnique*, XX[e] cahier, art. 16 ou p. 31), par MM. Lamé et Clapeyron (*Savants étrangers*, t. IV, p. 485 ou Art. 20), par Cauchy (*Exercices*, troisième année, p. 316), pour une autre définition de la pression; et il n'a été contesté que depuis qu'on a prétendu établir les formules des pressions par un simple raisonnement mathématique, sans se baser sur la loi physique des actions moléculaires à distance.

The theorem in question assumed by Poisson is thus stated by Saint-Venant:

> La pression sur une petite face, ou la résultante de toutes les actions moléculaires sensibles qui s'exercent à travers sa superficie $\omega$, peut être remplacée par la résultante des actions qui seraient exercées sur chaque molécule $m$ d'un des côtés de son plan, par une masse concentrée à son centre égale, pour chaque molécule $m$, à celle d'un cylindre de la matière du côté opposé, ayant $\omega$ pour base, et une hauteur égale à la distance de $m$ au plan de $\omega$. Moigno's *Leçons de Mécanique Analytique: Statique*, pp. 673—675.]

441. Poisson as we saw in Art. 421 condemns the mode of expressing the elastic forces by definite integrals; but he allows himself to do this to some extent: see Art. 542 of my account of the memoir of 1829. In the case of a triple summation Poisson transforms two summations into integrations leaving only one summation untransformed: see his page 378.

**442.** As we have stated in Art. 438 if we put $\lambda = \mu = k$ in the well-known formula we obtain such an expression as

$$N_1 = k\left(3\frac{du}{dx} + \frac{dv}{dy} + \frac{dw}{dz}\right).$$

The corresponding form for $N_1$ which Poisson first obtains is

$$N_1 = K\left(1 + \frac{du}{dx}\right) + k\left(3\frac{du}{dx} + \frac{dv}{dy} + \frac{dw}{dz}\right),$$

where $K$ stands for $\frac{2\pi}{3}\Sigma\frac{r^3}{a^5}f(r)$, and $k$ for $\frac{2\pi}{15}\Sigma\frac{r^5}{a^5}\frac{d}{dr}\left\{\frac{1}{r}f(r)\right\}$.

Then Poisson shews that as the body was originally in equilibrium $K$ must be zero; and this reduces his expressions for the elastic tensions to the ordinary forms in which we take $\lambda = \mu$.

Two matters occur for notice which we shall consider in the account of the memoir of 1829: see Articles 542 and 554.

**443.** We can now give the mathematical process alluded to in Art. 436. Suppose that we could replace the summations by integrations. Then in the summation denoted by $K$ multiply by $dr/\alpha$; thus we obtain

$$0 = \frac{2\pi}{3.}\int_0^\infty \frac{r^3}{\alpha^6}f(r)\,dr.$$

And $k$ becomes

$$\frac{2\pi}{15}\int_0^\infty \frac{r^5}{\alpha^6}\frac{d}{dr}\left\{\frac{1}{r}f(r)\right\}dr,$$

and this by integration by parts is

$$-\frac{2\pi}{3}\int_0^\infty \frac{r^4}{\alpha^6}\frac{1}{r}f(r)\,dr,$$

that is $-K$; so that $k$ vanishes. But this is absurd for then the elastic tensions all vanish. Hence we see that the summation with respect to $r$ cannot be transformed into an integral[1].

In the preceding operation it will be seen that we assume $r^4 f(r)$ to vanish both when $r = 0$ and when $r = \infty$. Poisson says that $f(r)$ is zero at the two limits; but $f(r)$ does not necessarily vanish with $r$, as we see by the example which he suggests for $f(r)$ in Art. 439; and it is not enough that $f(r)$ should vanish when $r$ is infinite, we must have $r^4 f(r)$ vanishing.

[1] [I shall return to this point in so far as it involves a criticism of Navier when considering a paper by Clausius. The legitimacy of the molecular hypothesis which leads to $\lambda = \mu$ will be best discussed after the chapter devoted to Cauchy.] Ed.

444. Poisson's process seems to me sometimes deficient in rigour; the following will serve as an example, taken from his page 378. He arrives at a certain result and then adds:

Ce résultat exige à la vérité, que $r$ soit un multiple très-considérable de $z$; mais d'après la supposition que nous avons faite sur le mode de décroissement de l'action moléculaire (No. 1), on peut, sans erreur sensible, négliger dans la somme $\Sigma$ relative à $r$, la partie où cette condition n'est pas remplie par rapport à l'autre partie.

This statement as to what may be neglected in comparison with the rest seems to me quite arbitrary.

445. Poisson in his pages 392—395 obtains the general equations for the equilibrium of a body subject to applied forces by transforming his equations by the aid of a process like that adopted by Lamé in his *Elasticité*, p. 21.

446. Poisson objects on his page 400 to a use which had been made of the Calculus of Variations, following the example of Lagrange; he says that the method is not applicable to the case in which we regard a body as made up of molecules separated by intervals, however small the intervals may be; but he is very brief and does not unfold his objection. Compare Saint-Venant in Moigno's *Statique*, page 718.

447. Suppose an elastic body under the action of no applied forces except a constant uniform normal pressure over the surface; Poisson states that all the conditions of equilibrium will be fulfilled if we suppose that the body is every where and in all directions equally compressed or expanded: see his pages 400—402. In fact the fundamental equations of elasticity will be satisfied by supposing that the three normal tensions are equal and constant, and the three tangential tensions zero.

448. After establishing the equations for the equilibrium of an elastic body, namely the three which hold at every point of the interior and the three which hold at every point of the surface, Poisson says that they agree with those given by Navier in the seventh volume of the Paris memoirs, but that Navier's own

process would really have made all the forces vanish, according to what we have seen in Art. 442. As Poisson gives no reference to Cauchy, or any other writer except Navier, we are I presume to infer that the equations were first established by Navier. Saint-Venant considers that the step in Navier's investigation to which Poisson objects might easily have been avoided: see Moigno's *Statique*, page 719.

449. Poisson's second section is entitled: *Vibrations d'une sphère élastique*; it occupies pages 405—421. We reproduce this investigation with some change of notation, not at first putting with Poisson $\lambda = \mu$.

We assume that all points of the sphere at the same distance from the centre are moving with the same velocity along their respective radii. Thus we must have

$$u = \frac{x}{r}\psi,\ v = \frac{y}{r}\psi,\ w = \frac{z}{r}\psi;$$

where $\psi$ is some function of $r$, the distance of the point considered from the centre of the sphere.

We may denote $\psi$ by $d\phi/dr$, where $\phi$ is a function of $r$; thus

$$u = \frac{d\phi}{dx},\ v = \frac{d\phi}{dy},\ w = \frac{d\phi}{dz}.$$

Hence for the cubical dilatation we have

$$\theta = \frac{d^2\phi}{dx^2} + \frac{d^2\phi}{dy^2} + \frac{d^2\phi}{dz^2} = \frac{d^2\phi}{dr^2} + \frac{2}{r}\frac{d\phi}{dr} = \nabla^2\phi.$$

Now we assume that there are no applied internal or external forces. Then the first of the usual equations of an elastic solid becomes

$$(\lambda + \mu)\frac{d\theta}{dx} + \mu\,\nabla^2 u = \rho\frac{d^2u}{dt^2},$$

that is—

$$(\lambda + \mu)\frac{d\theta}{dx} + \mu\nabla^2\left(\frac{d\phi}{dx}\right) = \rho\,\frac{d}{dx}\left(\frac{d^2\phi}{dt^2}\right).$$

Hence by integrating with respect to $x$ we get

$$(\lambda + \mu)\,\theta + \mu\nabla^2\phi = \rho\,\frac{d^2\phi}{dt^2} + C \ldots\ldots\ldots\ldots\ldots\ldots(1),$$

where $C$ is constant with respect to $x$, but so far as we see at present may involve $y$, $z$, and $t$. But by taking the other two equations we arrive at results exactly like (1); and in this way we find that $C$ in (1) cannot involve $y$ or $z$; thus it cannot involve any variable except $t$.

Thus (1) may be written

$$(\lambda + 2\mu)\,\theta = \rho \frac{d^2\phi}{dt^2} + C,$$

that is
$$\frac{\lambda + 2\mu}{r} \frac{d^2 r\phi}{dr^2} = \rho \frac{d^2\phi}{dt^2} + C,$$

that is
$$a^2 \frac{d^2 r\phi}{dr^2} = \frac{d^2 r\phi}{dt^2} + \frac{r}{\rho} C, \qquad \text{(2)},$$

where $a^2$ is put for $(\lambda + 2\mu)/\rho$.

Poisson himself assumes that $\lambda = \mu$, so that with him $a^2 = 3\lambda/\rho$.

450. Thus we have to discuss equation (2). It is unnecessary to trouble ourselves with the term $rC/\rho$; for suppose that we have obtained a general solution of (2) without this term, and denote it by $\phi = \Phi$. Then for the solution of (2) as it stands we have

$$\phi = \Phi - \frac{1}{\rho} \iint C dt\, dt.$$

We shall therefore confine ourselves to the equation

$$\frac{d^2 r\phi}{dt^2} = a^2 \frac{d^2 r\phi}{dr^2} \qquad \text{(3)}.$$

Now it is obvious that the following is *a* solution

$$\phi = (A \cos\nu at + B \sin\nu at) \frac{\sin\nu r}{r} + (A' \cos\nu at + B' \sin\nu at) \frac{\cos\nu r}{r};$$

where $A$, $B$, $A'$, $B'$ and $\nu$ are independent of $r$ and $t$; and the aggregate of any number of such expressions would also be a solution. But $(\cos\nu r)/r$ would be infinite at the centre of the sphere, and to avoid this we must suppress the term which involves $(\cos\nu r)/r$; hence we may take for the solution

$$\phi = \Sigma\,(A \cos\nu at + B \sin\nu at) \frac{\sin\nu r}{r} \qquad \text{(4)},$$

where $\Sigma$ denotes the summation of terms in which different values are given to $A$, $B$, and $\nu$.

451. We must seek to determine the quantities at present unknown by means of the equations to be satisfied at the surface of the sphere; we will denote the radius of the sphere by $l$. The first equation of these equations takes the form:

$$\frac{x}{l} N_1 + \frac{y}{l} T_3 + \frac{z}{l} T_2 = 0;$$

this gives the following result in which we are to put finally $l$ for $r$,

$$\frac{x}{r}\left(\lambda\theta + 2\mu\frac{du}{dx}\right) + \frac{\mu y}{r}\left(\frac{du}{dy} + \frac{dv}{dx}\right) + \frac{\mu z}{r}\left(\frac{du}{dz} + \frac{dw}{dx}\right) = 0,$$

that is $$\frac{x}{r}\left(\frac{\lambda}{r}\frac{d^2\phi r}{dr^2} + 2\mu\frac{d^2\phi}{dx^2}\right) + \frac{2\mu y}{r}\frac{d^2\phi}{dxdy} + \frac{2\mu z}{r}\frac{d^2\phi}{dzdy} = 0,$$

or $$\frac{\lambda x}{r^2}\frac{d^2 r\phi}{dr^2} + \frac{2\mu}{r}\left(x\frac{d^2\phi}{dx^2} + y\frac{d^2\phi}{dxdy} + z\frac{d^2\phi}{dxdz}\right) = 0,$$

that is $$\frac{\lambda x}{r^2}\frac{d^2 r\phi}{dr^2} + \frac{2\mu x}{r}\frac{d^2\phi}{dr^2} = 0,$$

that is $$(\lambda + 2\mu)\frac{d^2\phi}{dr^2} + \frac{2\lambda}{r}\frac{d\phi}{dr} = 0.$$

Put $b$ for $(\lambda + 2\mu)/\lambda$, then we obtain the following equation which is to hold when $r = l$

$$b\frac{d^2\phi}{dr^2} + \frac{2}{r}\frac{d\phi}{dr} = 0 \;\ldots\ldots\ldots\ldots\ldots\ldots\ldots\ldots (5).$$

The other two surface equations also lead to (5).

Substitute in (5) the value of $\phi$ from (4); then we shall find that the following relation must hold:

$$\left(-\frac{\nu^2 b}{l} + \frac{2b}{l^3} - \frac{2}{l^3}\right)\sin\nu l + \frac{2\nu}{l^2}(1-b)\cos\nu l = 0 \;\ldots\ldots (6).$$

452. According to Poisson $\lambda = \mu$, so that $b = 3$; thus (6) simplifies to

$$(4 - 3\nu^2 l^2)\sin\nu l = 4\nu l\cos\nu l \ldots\ldots\ldots\ldots\ldots\ldots (7).$$

We shall keep henceforward to his case, as it sufficiently illustrates all that is interesting in the discussion. We have now to shew how $A$ and $B$ can be found.

453. From (4) we have

$$r\frac{d\phi}{dr} = \Sigma\,(A\cos\nu at + B\sin\nu at)\,\frac{\nu r\cos\nu r - \sin\nu r}{r}$$

$$= \Sigma\,(A\cos\nu at + B\sin\nu at)\,R \text{ say}\ldots\ldots\ldots\ldots\ldots(8).$$

Now equation (3) gives

$$\frac{d^2\phi}{dt^2} = \frac{a^2}{r}\left(r\frac{d^2\phi}{dr^2} + 2\frac{d\phi}{dr}\right);$$

differentiate with respect to $r$, thus

$$\frac{d^2}{dt^2}\frac{d\phi}{dr} = a^2\left(\frac{d^3\phi}{dr^3} + \frac{2}{r}\frac{d^2\phi}{dr^2} - \frac{2}{r^2}\frac{d\phi}{dr}\right) = \frac{a^2}{r}\frac{d^2}{dr^2}\left(r\frac{d\phi}{dr}\right) - \frac{2a^2}{r^2}\frac{d\phi}{dr};$$

thus by (8)
$$-\frac{R\nu^2a^2}{r} = \frac{a^2}{r}\frac{d^2R}{dr^2} - \frac{2a^2}{r^3}R,$$

so that
$$\left(\nu^2 - \frac{2}{r^2}\right)R + \frac{d^2R}{dr^2} = 0\ldots\ldots\ldots\ldots\ldots\ldots\ldots(9).$$

And from (5) the following must hold at the surface

$$b\frac{dR}{dr} + (2-b)\frac{R}{r} = 0$$

that is
$$3\frac{dR}{dr} - \frac{R}{r} = 0\ldots\ldots\ldots\ldots\ldots\ldots\ldots(10).$$

This is easily verified by means of (7).

454. Now the initial circumstances must be supposed known, so that $d\phi/dr$ and $d^2\phi/(dr\,dt)$ must be known initially in terms of $r$. Let us suppose that initially

$$\frac{d\phi}{dr} = \chi(r) \text{ and } \frac{d^2\phi}{drdt} = \chi_1(r);$$

then putting $t=0$ in (8) and in the differential coefficient of it with respect to $t$ we get

$$\Sigma AR = r\chi(r)\ldots\ldots\ldots\ldots\ldots\ldots\ldots(11).$$

$$\Sigma a\nu BR = r\chi_1(r)\ldots\ldots\ldots\ldots\ldots\ldots\ldots(12).$$

We have now by means of (11) to isolate and thus to determine the value of each coefficient of which $A$ is the general type; and by means of (12) to do the same with respect to $B$. Let $\nu_1$ stand for a specific value of the general symbol $\nu$; so that $\nu_1$ is a root of (7): let $\nu_2$ stand for another specific value. Let $A_1$

and $B_1$ correspond to the former value of $\nu_1$ and $A$ and $B$ to the latter. Then we shall obtain our end by means of the following formula

$$\int_0^l R_1 R_2 dr = 0 \ldots\ldots\ldots\ldots\ldots\ldots\ldots(13),$$

where $R_1$ and $R_2$ are any two values of $R$ which correspond to two different values of $\nu$ satisfying (7).

455. To establish (13) we observe that the integral may be written in two ways;

$$\int_0^l R_1 R_2 dr = \int_0^l (\nu_1 r \cos \nu_1 r - \sin \nu_1 r) \frac{d}{dr} \frac{\sin \nu_2 r}{r} dr,$$

$$\int_0^l \int R_1 R_2 dr = \int_0^l (\nu_2 r \cos \nu_2 r - \sin \nu_2 r) \frac{d}{dr} \frac{\sin \nu_1 r}{r} dr.$$

Integrate by parts; thus we get

$$\int_0^l R_1 R_2 dr = (\nu_1 l \cos \nu_1 l - \sin \nu_1 l) \frac{\sin \nu_2 l}{l} + \nu_1^2 \int_0^l \sin \nu_1 r \sin \nu_2 r \, dr,$$

$$\int_0^l R_1 R_2 dr = (\nu_2 l \cos \nu_2 l - \sin \nu_2 l) \frac{\sin \nu_1 l}{l} + \nu_2^2 \int_0^l \sin \nu_1 r \sin \nu_2 r \, dr.$$

Now by (7) of which both $\nu_1$ and $\nu_2$ are roots

$$\nu_1 l \cos \nu_1 l - \sin \nu_1 l = -\frac{3\nu_1^2 l^2}{4} \sin \nu_1 l,$$

$$\nu_2 l \cos \nu_2 l - \sin \nu_2 l = -\frac{3\nu_2^2 l^2}{4} \sin \nu_2 l.$$

Hence we have

$$\int_0^l R_1 R_2 dr = \nu_1^2 \left\{ \int_0^l \sin \nu_1 r \sin \nu_2 r dr - \frac{3l}{4} \sin \nu_1 l \sin \nu_2 l \right\},$$

$$\int_0^l R_1 R_2 dr = \nu_2^2 \left\{ \int_0^l \sin \nu_1 r \sin \nu_2 r dr - \frac{3l}{4} \sin \nu_1 l \sin \nu_2 l \right\}.$$

Hence it follows that

$$(\nu_2^2 - \nu_1^2) \int_0^l R_1 R_2 dr = 0,$$

and therefore $$\int_0^l R_1 R_2 dr = 0.$$

456. Now to apply (13). Multiply both sides of (11) by $R_1$,

and integrate with respect to $r$ from 0 to $l$; then by (13) all the terms vanish except one, and we are left with

$$A_1 \int_0^l R_1^2 dr = \int_0^l r\chi(r) R_1 dr:$$

this determines $A_1$, and similarly all the other coefficients of which $A$ is the general type may be determined.

In like manner from (12) we get

$$av_1 B_1 \int_0^l R_1^2 dr = \int_0^l r\chi_1(r) R_1 dr.$$

457. The value of $\int_0^l R_1^2 dr$ which occurs in the preceding Article may easily be found. For

$$\int_0^l R^2 dr = \int_0^l (\nu r \cos \nu r - \sin \nu r) \frac{d}{dr} \frac{\sin \nu r}{r} dr.$$

Integrate by parts; thus

$$\int_0^l R^2 dr = (\nu l \cos \nu l - \sin \nu l) \frac{\sin \nu l}{l} + \nu^2 \int_0^l \sin^2 \nu r \, dr,$$

that is $$\int_0^l R^2 dr = \frac{1}{2l} \{\nu^2 l^2 + \nu l \sin \nu l \cos \nu l - 2 \sin^2 \nu l\}.$$

458. It may be shewn that the equation (7) has no imaginary roots. For if possible suppose that there is a root $\nu_1 = p + q\sqrt{-1}$; then there must be also a root $p - q\sqrt{-1}$, which we will denote by $\nu_2$. Let $R_1$ take the form $P + Q\sqrt{-1}$; then $R_2$ must take the form $P - Q\sqrt{-1}$. Hence (13) becomes

$$\int_0^l (P^2 + Q^2)\, dr = 0;$$

but this is obviously impossible, as every element of the integral is positive.

459. Let $m$ be put for $\nu l$, so that the equation (7) becomes

$$(4 - 3m^2) \sin m = 4m \cos m \quad \text{..............(14)}.$$

According to Poisson we have for approximate values of the least two roots of his equation

$$m = 2{\cdot}56334, \quad m = 6{\cdot}05973 \quad \text{................(15)}.$$

The larger roots are found approximately from $m = i\pi$, where $i$ is an integer. Thus if we put $m = i\pi - x$ we shall have $x$ small; and by (14) the value of $x$ is approximately to be found from

$$\tan x = \frac{i\pi}{\frac{3}{4}i^2\pi^2 - 1}\left\{1 + \frac{\frac{3}{4}i^2\pi^2 + 1}{(\frac{3}{4}i^2\pi^2 - 1)^2}\right\}.$$

If we put $i = 2$ we shall find from these formulæ that $m = 6{\cdot}05917$, which differs very little from the value given above.

460. Suppose that in the interior of the sphere there are one or more surfaces concentric with that of the sphere, all the points of which remain at rest during the isochronous vibrations which correspond to a specific value of $\nu$ found from (7); then the radii of such surfaces will be determined by the equation $\dfrac{d\phi}{dr} = 0$; this by (8) leads to

$$\nu r \cos \nu r - \sin \nu r = 0 \text{.....................(16).}$$

The sound which corresponds with the specific value of $\nu$ will be accompanied by as many nodal surfaces as (16) will give values of $r$ less than $l$. The least two roots of (16) are found to be

$$\nu r = 4{\cdot}49331, \quad \nu r = 7{\cdot}73747.$$

On comparing these with the values of $m$ or $\nu l$ given in (15), we see that for the least value of $m$ there can be no node; but for the next value of $m$ there is one, and the radius of this is determined by

$$r = \frac{4{\cdot}49331}{6{\cdot}05973}\, l = {\cdot}74150\, l;$$

so that the radius is about three-quarters of that of the sphere. Poisson adds:

Il est inutile de dire que l'existence des surfaces nodales intérieures ne pourrait pas se vérifier par l'expérience; et il paraît même difficile que l'observation puisse faire connaître les différents sons d'une sphère élastique qui sont déterminés par le calcul.

461. We have hitherto supposed that no force acts on the surface. Let us now however suppose that there is such a force, namely a normal pressure which is the same at every point of the surface, but is a function of the time denoted by $Ne^{-ht}$, where $N$ and $h$ are positive constants.

We can satisfy the conditions of the problem by adding to the value of $\phi$ as given by (4) the term

$$-(e^{\frac{hr}{a}} - e^{\frac{-hr}{a}}) \frac{N}{\gamma r} e^{-ht},$$

provided we determine $\gamma$ properly, as we will now shew. In the first place with this additional term the fundamental equation (3) is still satisfied, as is obvious on trial.

In the next place consider the surface equations as in Art. 451; instead of zero we have now on the right-hand side $X_0$, $Y_0$, $Z_0$, where

$$X_0 = -\frac{x}{r} Ne^{-ht}, \quad Y_0 = -\frac{y}{r} Ne^{-ht}, \quad Z_0 = -\frac{z}{r} Ne^{-ht}$$

Then proceeding as in Art. 451' we shall now obtain

$$Ne^{-ht} + (\lambda + 2\mu) \frac{d^2\phi}{dr^2} + \frac{2\lambda}{r} \frac{d\phi}{dr} = 0;$$

and putting, as sufficient for our purpose, $\lambda = \mu$, we get

$$Ne^{-ht} + 3\mu \frac{d^2\phi}{dr^2} + \frac{2\mu}{r} \frac{d\phi}{dr} = 0 \quad \text{..................(17)}.$$

This is to hold when $r = l$; it replaces the more simple equation (5) which we formerly used. The value of $\phi$ as hitherto used will contribute nothing to (17). We have now to regard the terms added in the present Article; and using these we find that (17) leads to

$$\frac{\gamma l}{\mu} = \left(\frac{4}{l^2} + \frac{3h^2}{a^2}\right)\left(e^{\frac{hl}{a}} - e^{\frac{-hl}{a}}\right) - \frac{4h}{al}\left(e^{\frac{hl}{a}} + e^{\frac{-hl}{a}}\right)$$

this determines $\gamma$. Thus the problem with the specified condition as to the force acting at the surface is satisfied.

462. Poisson says that one of his objects in discussing the example of the vibrations of a sphere was to elucidate a certain difficulty: see his page 405. The difficulty seems to be this, if I understand Poisson correctly. Imagine that the sphere has been subjected to a uniform normal compression, and that this compression is *instantaneously* removed, and the sphere left to pass into motion; then we cannot obtain formulæ which will satisfy these conditions. That is we cannot obtain formulæ which

shall correspond to the state of compression of the body when $t=0$, and when $t$ has any value different from zero shall exhibit that motion which has been developed in the preceding articles. Or we may state the matter generally thus: we have found in equation (5) that we must have at the surface

$$3\frac{d^2\phi}{dr^2}+\frac{2}{r}\frac{d\phi}{dr}=0;$$

hence the $\chi(r)$ which is introduced in (11) cannot be taken arbitrarily, but must be subject to the condition

$$3\frac{d}{dr}\chi(r)+\frac{2}{r}\chi(r)=0.$$

463. Poisson takes an example as sufficient to illustrate the difficulty and the explanation. Suppose that the sphere is originally compressed uniformly; thus a radius of natural length $r$ becomes $r(1-\tau)$ where $\tau$ is a constant quantity Then

$$\frac{d\phi}{dr}=\chi(r)=-\tau r;$$

hence $$3\frac{d}{dr}\chi(r)+\frac{2}{r}\chi(r)$$

instead of being zero is $-5\tau$. Poisson says that however rapidly the compressing force may be removed the removal cannot be really instantaneous; so we may represent such a compression by $Ne^{-ht}$ where $h$ is very large; thus when $t=0$ this compression is denoted by $N$, and it is practically zero as soon as $t$ has any sensible value. With the value of $\chi(r)$ just considered we have initially $N-5\mu\tau=0$ (p. 416). This is quite consistent with the case of uniform compression; for then $u=-\tau x$, $v=-\tau y$, $w=-\tau z$. Thus $\theta=-3\tau$ and each normal stress $=-(3\lambda+2\mu)\tau=-5\mu\tau$ on Poisson's hypothesis of the equality of $\lambda$ and $\mu$.

This problem is considered by Lamé in his pages 202—210; his treatment is in some respects more general than Poisson's, for he does not assume his expressions to be functions of the radius only, but allows them to involve angles. On the other hand Lamé does not regard any condition which must hold at the surface, but assumes that a fluid medium surrounding the body will furnish what normal pressure may be required.

Clebsch on pages 55—61 of his work gives a solution which substantially agrees with Poisson's. Put $\nu$ for the $k/a$ on Clebsch's page 58; then as $h=4$ the function on the right-hand side of Clebsch's equation (45) will be found to be $\frac{3}{\nu^2}\frac{\nu r\cos\nu r-\sin\nu r}{r^3}$.

And Clebsch's equation (48) on his page 59 will agree with Poisson's equation (4) on his page 409, that is with my equation (7). Observe that the expression $\overline{1+\mu}/\overline{1-\mu}$ of Clebsch is 5/3 according to Poisson.

464. Poisson's third section is entitled: *Equations de l'équilibre et du mouvement d'une corde élastique,* and occupies pages 422—442. The conditions of equilibrium of an elastic cord had been long known and given in ordinary works on Statics; it is the object of this section to deduce these conditions from the theory of elasticity, and thus to explain the nature of the force which is called the *tension* of the cord in the ordinary investigation. Poisson's process is very simple and satisfactory; it is reproduced substantially in pages 93—106 of Lamé's work.

[465.] If $s$ be the length of the cord measured from some fixed point to the point $xyz$, $\omega$ its section, $\rho$ its density and $XYZ$ the applied forces on the element $ds$, Poisson finds from the equations of equilibrium of an elastic solid the following equations:

$$X\rho\omega=\frac{d\left(T\omega\frac{dx}{ds}\right)}{ds},\quad Y\rho\omega=\frac{d\left(T\omega\frac{dy}{ds}\right)}{ds},\quad Z\rho\omega=\frac{d\left(T\omega\frac{dz}{ds}\right)}{ds},$$

where $T$ is *une nouvelle inconnue qui reste indéterminée.*

The elimination of $T$ gives the equations to the curve assumed by the cord.

If $\theta$ be the dilatation of the element $ds$ of the cord, Poisson easily deduces $T=-5\mu/2\,.\,\theta$.

When there are no applied forces and therefore $T$ is a constant $\theta$ will be uniform along the cord and equal therefore to the total extension divided by the length. The above equation then agrees with Hooke's Law, for if $W$ be a weight suspended at an extremity $T\omega+W=0$, or (see his p. 430)

$$W=\frac{5\mu\omega}{2}\cdot\frac{\text{extension}}{\text{length}}.$$

The following pages are devoted to the discussion of the longitudinal and transverse vibrations of a cord. We may note one point.

Suppose $n$ the frequency of *longitudinal* vibrations of a cord, and $n'$ the frequency of transversal vibrations; then Poisson obtains the result $n'/n = \sqrt{\alpha/l}$, where $l$ is the length of the cord, and $\alpha$ the elongation of the cord produced by the tension it experiences. Poisson says that this result had not been noticed before; it is now usually given in treatises on this subject: see Lamé, page 106. Poisson on his page 438 states that the result had been confirmed by an experiment made by M. Cagniard-Latour. He adds that the cord was 14·8 metres long, and that observation gave $n/n' = \frac{188}{7}$: he asserts that by calculation from the formula we deduce $\alpha = {\cdot}052$ metres which differs but little from the observed value ·05 of a metre. But there is some mistake here; for with the value which he assigns to $l$ and to $n/n'$ the formula gives for $\alpha$ a value very different from that which he obtains. He himself quotes his own result wrongly here; he quotes it correctly in his *Mécanique*, Vol. II. page 316, but there he does not give the figures of the experiment. In the *Annales de Chimie*, Vol. XXXVII. page 349, instead of $n/n' = \frac{188}{7}$ we have $n/n' = 16{\cdot}87$, which is consistent with the rest.

466. Poisson's fourth section is entitled: *Equations de l'équilibre et du mouvement d'une verge élastique.* It occupies pages 442—488. The section begins thus:

> Dans ce paragraphe, nous allons considérer une verge élastique proprement dite, qui tend à revenir à sa forme naturelle quand on l'en a écartée en la faisant fléchir, et capable de vibrer transversalement sans avoir besoin, comme une simple corde, d'être tendue suivant sa longueur. Cette verge sera homogène et partout à la même température; dans son état naturel, nous supposerons qu'elle soit un cylindre droit à base circulaire; et quoique le rayon de sa base soit très-petit, nous aurons maintenant égard, dans chaque section perpendiculaire à l'axe, aux petites variations des forces moléculaires et du mouvement des points de la verge, circonstances dont on fait abstraction dans le cas d'une corde élastique.

467. The investigation soon becomes a process of approxima-

tion, and is rather complex. With respect to the equations of motion the most important case is that of transversal vibration; and then, supposing there are no applied forces, the following are the principal results:

$$\frac{d^2y}{dt^2} = -\frac{5k\epsilon^2}{8\rho}\frac{d^4y}{dx^4} + \frac{\epsilon^2}{4}\frac{d^4y}{dx^2dt^2},$$

$$\frac{d^2z}{dt^2} = -\frac{5k\epsilon^2}{8\rho}\frac{d^4z}{dx^4} + \frac{\epsilon^2}{4}\frac{d^4z}{dx^2dt^2}.$$

The axis of $x$ coincides with the axis of the rod originally, $y$ and $z$ are transverse coordinates of a point which was originally on the axis of the rod at the distance $x$ from one end; $\epsilon$ is the radius of the rod, $\rho$ is the density, and $k$ is a constant which corresponds to the $\lambda$ of Lamé, supposing $\lambda = \mu$. These equations correspond with one in Clebsch's work, namely (17) on page 253; observing that the $M$ of Clebsch denotes a force acting at one end and is zero with Poisson.

468. Taking the equations as just given Poisson observes that if in the second term on the right-hand side we substitute from the left-hand side we introduce $\epsilon^4$, which he neglects; thus he reduces his equations to

$$\frac{d^2y}{dt^2} + \frac{5k\epsilon^2}{8\rho}\frac{d^4y}{dx^4} = 0, \quad \frac{d^2z}{dt^2} + \frac{5k\epsilon^2}{8\rho}\frac{d^4z}{dx^4} = 0.$$

It is sufficient to discuss one of these as they are precisely similar; and Poisson shews how to integrate the first. This process of integration is reproduced by Poisson in his *Mécanique* 1833, Vol. II. pages 371—392. Some numerical values on page 485 of the memoir are not quite the same as on the corresponding page, namely 389, of the *Mécanique*. The value of $\lambda'$ on the second line of page 390 of the *Mécanique* does not follow from what he has given before it should be 1·87511.

469. In the course of his process Poisson arrives at the following result: suppose that $l$ is the original length of the cylinder, and $\epsilon$ the original radius, and let the former become $l(1+\delta)$ by the deformation; then the latter will become $\epsilon(1-\frac{1}{4}\delta)$: see his pages 449, 451. Thus the volume was originally $\pi l\epsilon^2$; and it becomes by the deformation $\pi l\epsilon^2(1+\delta)(1-\frac{1}{4}\delta)^2$, that is approxi-

mately $\pi l \epsilon^2 (1 + \frac{1}{2}\delta)$: so that the volume is augmented in the ratio of $1 + \frac{1}{2}\delta$ to unity. Poisson adds

Ce fait intéressant de l'augmentation de volume des fils élastiques, par l'effet de leur allongement, a été observé par M. Cagniard-Latour; et sur ce point, le calcul et l'observation s'accordent d'une manière remarquable comme on peut le voir dans la note où j'ai rendu compte de son expérience.

A note refers to the *Annales de Physique et de Chimie*, Tome XXXVI. page 384. See Art. 368.

[470.] Poisson obtains a proportion between the number of longitudinal vibrations which a cylinder will perform in a given time, and the number of torsion vibrations which it will perform in the same time: this he makes to be $\sqrt{10}/2$. By experiments Chladni put it at $\frac{3}{2}$, and Savart more recently at 1 6668; the mean of the experimental values differs little from Poisson's theoretical value: see page 456 of the memoir, and also page 369 of Vol. II. of the *Mécanique.*

Poisson's method is as follows: he obtains equations for the torsional and longitudinal vibrations of the forms

$$\frac{d^2\psi}{dt^2} = \frac{k}{\rho}\frac{d^2\psi}{dx^2} \text{ and } \frac{d^2u}{dt^2} = \frac{5k}{2\rho}\frac{d^2u}{dx^2}$$

respectively.

These agree with those usually given in treatises on sound, if we remark the relation supposed by Poisson to exist between the elastic constants. Thus Lord Rayleigh (*The Theory of Sound*, Vol. I. pp. 191 and 199) uses the same equations, if we remember that his $q$ is Poisson's $k$, and that his $\mu$ is taken by Poisson to be equal to 1/4[1].

If $n$ be the number of longitudinal, $n'$ of torsional vibrations in unit time, $l$ the length of the rod, Poisson finds[2]

$$n = \frac{1}{2l}\sqrt{\frac{5k}{2\rho}}, \qquad n' = \frac{1}{2l}\sqrt{\frac{k}{\rho}},$$

or

$$\frac{n}{n'} = \frac{\sqrt{10}}{2};$$

[1] Lord Rayleigh's $\mu$ is equal to the ratio of lateral contraction to longitudinal extension, and this Poisson makes equal to 1/4.

[2] Poisson has dropped the coefficient 1/2 in the values of both $n$ and $n'$ see p. 456 of the memoir.

Ce qui montre que les sons d'une même verge cylindrique qui exécute successivement des vibrations tournantes et des vibrations longitudinales, sont dans un rapport indépendant de la longueur, du diamètre et la matière de la verge. C'est ce que Chladni avait reconnu par l'expérience.

It will be seen that Poisson's remark as to the ratio being independent of the material of the rod is not indisputably true. In Lamé's notation the ratio equals $\sqrt{\frac{2\mu + 3\lambda}{\mu + \lambda}}$, which will only be a constant if the ratio $\mu : \lambda$ be admitted to be the same for all materials.

471. Again, Poisson obtains another proportion; namely between the number of vibrations of a cylindrical rod which vibrates longitudinally, and the number of vibrations of the same rod in the same time when it vibrates transversely: see his page 486. Take the case of a rod free at its ends and giving forth its gravest sound; then if $n_1$ represent the number of transversal vibrations in a unit of time he finds that $n_1 = 3{\cdot}5608 \frac{\epsilon}{2l^2} \sqrt{\frac{5k}{2\rho}}$, the notation being the same as in Arts. 467 and 468. Let $n$ denote the corresponding number for the case of longitudinal vibrations; then he finds that $n = \frac{1}{2l} \sqrt{\frac{5k}{2\rho}}$ Thus $\frac{n_1}{n} = 3{\cdot}5608 \frac{\epsilon}{l}$; Poisson records some observations communicated to him by Savart, which agree well with this theoretical result.

Later in the memoir Poisson obtains another such proportion, namely between the number of longitudinal vibrations of a rod, and the number of transversal vibrations of a plate; and he expresses a wish to have this result tested by experiment: see his page 567.

[472.] Poisson's fifth section is entitled: *Equations de l'équilibre et du mouvement d'une membrane élastique.* It occupies pp. 488—523.

This section is occupied with the equilibrium and vibrational motion of a membrane. It contains much the same matter as the ninth and tenth of Lamé's *Leçons* with of course Poisson's usual supposition as to the equality of the elastic constants.

After the general equations of equilibrium (p. 491) we may note the following results:

(i) For a plane membrane subject to no normal force

$$\rho X+\frac{k}{3}\left(3\frac{d^2u}{dy^2}+5\frac{d^2v}{dxdy}+8\frac{d^2u}{dx^2}\right)=0,$$

$$\rho Y+\frac{k}{3}\left(3\frac{d^2v}{dx^2}+5\frac{d^2u}{dxdy}+8\frac{d^2v}{dy^2}\right)=0$$

The notation being the same as in the earlier part of the memoir, p. 494.

Special cases are considered, thus that of a membrane subject only to tension at its boundary. On p. 498 Poisson makes the remark:

On voit que la même force appliquée successivement à la superficie d'un corps, au contour d'une membrane, et aux extrémités d'une corde, produit des dilatations linéaires qui sont entre elles comme les nombres 2, 3, 4; la quantité $k$ dépendante de la matière étant supposée la même dans les trois cas. (Cf. Lamé, p. 114.)

The case of longitudinal vibrations follows with special treat ment of a circular membrane, whose boundary is subject to uniform tension, and while it vibrates radially (pp. 499—508).

(ii) For a plane membrane subjected to force which may be partially normal and so rendered slightly curved.

$$\rho(Z-Xp-Yq)+Q_2\frac{d^2z}{dy^2}+2P_2\frac{d^2z}{dxdy}+P_3\frac{d^2z}{dx^2}=0,$$

where the plane of $xy$ is the primitive plane of the membrane, $Z, X, Y$ are the components of applied force, and

$$P_2=-k\left(\frac{du}{dy}+\frac{dv}{dx}\right),\qquad P_3=-\frac{2k}{3}\left(4\frac{du}{dx}+\frac{dv}{dy}\right),$$

$$Q_2=-\frac{2k}{3}\left(4\frac{dv}{dy}+\frac{du}{dx}\right),$$

values not involving $z$. These results are given by Lamé: see page 110 equation (6) and page 112 equation (8).

Poisson applies the equation to the case of transverse vibrations of a membrane subject to uniform tension. He deduces the equation $\frac{d^2z}{dt^2}=c^2\left(\frac{d^2z}{dx^2}+\frac{d^2z}{dy^2}\right)$, which he applies at considerable length to the case of a square membrane (pages 510—519) and to the case of a circular membrane (pages 519—523) after transformation to the form $\frac{d^2z}{dt^2} = c^2\left(\frac{d^2z}{dr^2} + \frac{1}{r}\frac{dz}{dr}\right)$.

In his treatment of the first case he has been closely followed by Lamé and later writers, while the analysis he presents of the second is by no means without interesting points. For the integration of the equation for the transverse as well as for the radial vibrations of a circular membrane Poisson refers to a memoir of his own in the *Journal de l'École polytechnique*, 19e Cahier, page 239[1].

473. A few additional remarks may be made. Poisson refers on his pages 491 and 493 to his former memoir on elastic surfaces. After stating that he will confine himself to the case in which the membrane deviates but little from a plane, he says on his page 493:

L'équation différentielle de la surface flexible que Lagrange a donnée, et qui se trouve aussi dans le Mémoire que je viens de citer, n'est pas soumise à cette restriction; elle est seulement fondée sur la supposition particulière qu'en chaque point de cette surface, la tension est la même dans toutes les directions: on la déduirait sans difficulté des équations (2) en y introduisant cette hypothèse.

The equations (2) to which Poisson refers are substantially equivalent to equations (5) on Lamé's page 110. On the same page Lamé expresses emphatically his surprise that Poisson adopted a method so long and complex; but it seems to me that Lamé gives an exaggerated notion of the length and complexity of the method, which is in fact connected in a natural manner with formulæ already given by Poisson in the present and in his former memoirs.

474. Poisson's sixth section is entitled: *Equations de l'équilibre et du mouvement d'une plaque élastique.* It occupies pages 523—545.

This is the first investigation of the problem of the elastic plate from the general equations of elasticity. Owing to the importance of the subject and the considerable controversy which has arisen over Poisson's contour-conditions, we substantially reproduce this investigation. The reader will remember Poisson's hypothesis as to the equality of the elastic constants.

[1 The very important *Mémoire sur l'intégration des équations linéaires aux différences partielles*, read before the Academy Dec. 31st, 1821.] ED.

The lamina will be supposed in its natural condition to be plane and of constant thickness; that is to say it will be comprised between two parallel planes which constitute its *faces*, and the distance between these will express its thickness. The *boundary* of the lamina will consist of planes or of portions of cylindrical surfaces perpendicular to the faces. We will denote the thickness by $2\epsilon$, and we will suppose it very small with regard to the other dimensions of the lamina; but the thickness will be sufficiently great for the lamina to tend to recover its plane figure when it has been disturbed from that by given forces, and for the lamina to execute transversal vibrations when these forces have been withdrawn. It will now be necessary to pay attention to the variation of the molecular forces in the direction of the thickness; this is the essential difference between the case we are about to consider and that of a flexible membrane, where the thickness is taken indefinitely small, and these forces constant.

475. Let a plane be taken in the natural state of the lamina parallel to its faces, and equidistant from them: this we will call the *mean section* of the lamina, and we will take the plane of it for that of $(xy)$. Let $M$ be any point of this section; let $x$ and $y$ be its original co-ordinates; let $x+u$, $y+v$, $z$ be the co-ordinates of this point after the change of form of the lamina, so that $u$, $v$, $z$ are unknown functions of $x$ and $y$. Let $M'$ be another point situated originally on the straight line through $M$ at right angles to the lamina; and let its original co-ordinates be $x$, $y$, $\zeta$; and let these co-ordinates become $x+u'$, $y+v'$, $\zeta+w'$ after the change of form: then $u'$, $v'$, $w'$ are unknown functions of $x$, $y$, $\zeta$ which coincide with $u$, $v$, $z$ respectively in the case of $\zeta=0$. These displacements of $M$ and $M'$ will be taken to be very small, and it will be then assumed that the lamina does not deviate much from the plane of $(xy)$ when its form is changed.

476. The expressions for the elastic tensions are (see Art. 438)

$$\left.\begin{aligned}
N_1 &= k\left(3\frac{du'}{dx}+\frac{dv'}{dy}+\frac{dw'}{d\zeta}\right), & T_1 &= k\left(\frac{dv'}{d\zeta}+\frac{dw'}{dy}\right)\\
N_2 &= k\left(\frac{du'}{dx}+3\frac{dv'}{dy}+\frac{dw'}{d\zeta}\right), & T_2 &= k\left(\frac{dw'}{dx}+\frac{du'}{d\zeta}\right)\\
N_3 &= k\left(\frac{du'}{dx}+\frac{dv'}{dy}+3\frac{dw'}{d\zeta}\right), & T_3 &= k\left(\frac{du'}{dy}+\frac{dv'}{dx}\right)
\end{aligned}\right\}\ldots(1).$$

The variable $\zeta$ being very small, if we develop $u'$, $v'$, $w'$ in powers of $\zeta$ we shall have series which are in general very convergent. We shall exclude the case of exception which might exist if these quantities varied very rapidly in the direction of the thickness, as for example would happen if their values depended on the ratio $\zeta/\epsilon$. The following analysis and the consequences deduced from it are all founded on the possibility of this development. We shall then have the following convergent series:

$$u' = u + \left[\frac{du'}{d\zeta}\right]\zeta + \tfrac{1}{2}\left[\frac{d^2u'}{d\zeta^2}\right]\zeta^2 + \ldots,$$

$$v' = v + \left[\frac{dv'}{d\zeta}\right]\zeta + \tfrac{1}{2}\left[\frac{d^2v'}{d\zeta^2}\right]\zeta^2 + \ldots,$$

$$w' = z + \left[\frac{dw'}{d\zeta}\right]\zeta + \tfrac{1}{2}\left[\frac{d^2w'}{d\zeta^2}\right]\zeta^2 + \ldots,$$

the square brackets indicating that $\zeta$ is to be made zero after the differentiation.

477. By means of the equations of equilibrium we must determine in succession the coefficients of the preceding series, and then we shall know the condition of the plate to as close a degree of approximation as we please. We will stop at the first power of $\zeta$, and thus we have simply

$$u' = u + \left[\frac{du'}{d\zeta}\right]\zeta, \quad v' = v + \left[\frac{dv'}{d\zeta}\right]\zeta, \quad w' = z + \left[\frac{dw'}{d\zeta}\right]\zeta \ldots\ldots(2).$$

These formulæ contain only six unknown quantities which must be expressed in terms of $x$ and $y$. (They are, I suppose, $u$, $v$, $z$, $\left[\frac{du'}{d\zeta}\right]$, $\left[\frac{dv'}{d\zeta}\right]$, and $\left[\frac{dw'}{d\zeta}\right]$.) The expression for $z$ when obtained will give the form which the mean section assumes; the values of $u$ and $v$ will be the displacements of the points of the mean section in the original direction of the plane; the difference of the values of $w'$ which correspond to $\zeta = \pm\,\epsilon$ will give the thickness of the lamina, which becomes variable from point to point after the change of form.

478. Let $X'$, $Y'$, $Z'$ denote the components parallel to the axes of $x$, $y$, $z$ respectively of the inner applied forces which act at the

point $M'$ Then by the general equations of body-equilibrium we have

$$\left.\begin{aligned} \frac{dN_1}{dx}+\frac{dT_3}{dy}+\frac{dT_2}{d\zeta}+\rho X'=0 \\ \frac{dT_3}{dx}+\frac{dN_2}{dy}+\frac{dT_1}{d\zeta}+\rho Y'=0 \\ \frac{dT_2}{dx}+\frac{dT_1}{dy}+\frac{dN_3}{d\zeta}+\rho Z'=0 \end{aligned}\right\}\text{..................(3).}$$

479. Let us now consider the *faces* of the lamina. We will suppose that there are no outer applied forces at these faces: hence $N_3$, $T_1$ and $T_2$ must vanish at every point of these faces, that is both when $\zeta=\epsilon$ and when $\zeta=-\epsilon$. Hence if we expand, and neglect powers of $\epsilon$ beyond $\epsilon^3$, we shall obtain

$$\left.\begin{aligned} [N_3]+\tfrac{1}{2}\left[\frac{d^2N_3}{d\zeta^2}\right]\epsilon^2=0, \quad \left[\frac{dN_3}{d\zeta}\right]+\tfrac{1}{6}\left[\frac{d^3N_3}{d\zeta^3}\right]\epsilon^2=0 \\ [T_1]+\tfrac{1}{2}\left[\frac{d^2T_1}{d\zeta^2}\right]\epsilon^2=0, \quad \left[\frac{dT_1}{d\zeta}\right]+\tfrac{1}{6}\left[\frac{d^3T_1}{d\zeta^3}\right]\epsilon^2=0 \\ [T_2]+\tfrac{1}{2}\left[\frac{d^2T_2}{d\zeta^2}\right]\epsilon^2=0, \quad \left[\frac{dT_2}{d\zeta}\right]+\tfrac{1}{6}\left[\frac{d^3T_2}{d\zeta^3}\right]\epsilon^2=0 \end{aligned}\right\}\text{....(4).}$$

The equations which belong to the boundary will be given further on. The investigation to which we now proceed is to eliminate between equations (3) and (4) the differential coefficients of $u'$, $v'$, $w'$ with respect to $\zeta$ which are of the second order, or of a higher order, and to deduce the values of the six unknown quantities in (2).

480. Let $X$, $Y$, $Z$ denote the inner applied forces at the point $M$ of the mean section; that is what $X'$, $Y'$, $Z'$ respectively become when $\zeta=0$. Give to $\zeta$ this value in the first two equations of (3); then substitute for $\left[\frac{dT_2}{d\zeta}\right]$ and $\left[\frac{dT_1}{d\zeta}\right]$ their values from (4), and neglect the terms which have $\epsilon^2$ for a factor, and which are very small compared with the terms independent of $\epsilon$: we have thus simply

$$\left.\begin{aligned} X\rho+\left[\frac{dN_1}{dx}\right]+\left[\frac{dT_3}{dy}\right]=0 \\ Y\rho+\left[\frac{dT_3}{dx}\right]+\left[\frac{dN_2}{dy}\right]=0 \end{aligned}\right\}\text{..................(5).}$$

Make $\zeta=0$ in the third equation of (3), and substitute for the three terms involving elastic tensions from (4),—as all these terms involve $\epsilon^2$ we must preserve them all—thus:

$$Z\rho=\frac{\epsilon^2}{6}\left\{\left[\frac{d^3N_3}{d\zeta^3}\right]+3\left[\frac{d^3T_1}{d\zeta^2dy}\right]+3\left[\frac{d^3T_2}{d\zeta^2dx}\right]\right\}.$$

481. Differentiate the last equation of (3) twice with respect to $\zeta$; thus:

$$\frac{d^2Z\rho}{d\zeta^2}=-\left\{\frac{d^3N_3}{d\zeta^3}+\frac{d^3T_1}{d\zeta^2dy}+\frac{d^3T_2}{d\zeta^2dx}\right\}.$$

Make $\zeta=0$ in this formula, multiply by $\epsilon^2/6$, and add to the last result in Art. 480; thus

$$Z\rho+\frac{\epsilon^2}{6}\left[\frac{d^2Z'\rho}{d\zeta^2}\right]=\frac{\epsilon^2}{3}\left\{\left[\frac{d^3T_1}{d\zeta^2dy}\right]+\left[\frac{d^3T_2}{d\zeta^2dx}\right]\right\}.$$

The first two equations of (3) give in like manner

$$\frac{d^2X'\rho}{d\zeta dx}=-\left\{\frac{d^3N_1}{d\zeta dx^2}+\frac{d^3T_3}{d\zeta dxdy}+\frac{d^3T_2}{d\zeta^2dx}\right\},$$

$$\frac{d^2Y'\rho}{d\zeta dy}=-\left\{\frac{d^3T_3}{d\zeta dxdy}+\frac{d^3N_2}{d\zeta dy^2}+\frac{d^3T_1}{d\zeta^2dy}\right\}.$$

Make $\zeta=0$ in these formulæ, multiply by $\epsilon^2/3$ and add to the preceding formula: thus

$$Z\rho+\frac{\epsilon^2}{6}\left\{\left[\frac{d^2Z'\rho}{d\zeta^2}\right]+2\left[\frac{d^2X'\rho}{d\zeta dx}\right]+2\left[\frac{d^2Y'\rho}{d\zeta dy}\right]\right\}$$
$$=-\frac{\epsilon^2}{3}\left\{\left[\frac{d^3N_1}{d\zeta dx^2}\right]+2\left[\frac{d^3T_3}{d\zeta dxdy}\right]+\left[\frac{d^3N_2}{d\zeta dy^2}\right]\right\}\text{......(6)}.$$

In the equations (5) and (6) there remain to be eliminated only $N_1$, $N_2$, $T_3$, and their differential coefficients with respect to $\zeta$; we cannot simplify them further without substituting the values of these quantities which correspond to $\zeta=0$; but in calculating these values we may neglect the terms dependent on $\epsilon^2$, which will allow us to reduce the equations (4) to

$$[N_3]=0,\quad [T_1]=0,\quad [T_2]=0,\quad \left[\frac{dN_3}{d\zeta}\right]=0,\quad \left[\frac{dT_1}{d\zeta}\right]=0,\quad \left[\frac{dT_2}{d\zeta}\right]=0.$$

482. We suppose the lamina homogeneous; the density $\rho$

and the coefficient $k$ will then be constant; and by means of (1) the six equations just given become

$$\frac{du}{dx}+\frac{dv}{dy}+3\left[\frac{dw'}{d\zeta}\right]=0, \qquad \frac{dz}{dx}+\left[\frac{du'}{d\zeta}\right]=0,$$

$$\left[\frac{d^2u'}{d\zeta dx}\right]+\left[\frac{d^2v'}{d\zeta dy}\right]+3\left[\frac{d^2w'}{d\zeta^2}\right]=0, \qquad \frac{dz}{dy}+\left[\frac{dv'}{d\zeta}\right]=0,$$

$$\left[\frac{d^2w'}{d\zeta dy}\right]+\left[\frac{d^2v'}{d\zeta^2}\right]=0, \qquad \left[\frac{d^2w'}{d\zeta dx}\right]+\left[\frac{d^2u'}{d\zeta^2}\right]=0.$$

These six equations give

$$\left.\begin{aligned}\left[\frac{du'}{d\zeta}\right]&=-\frac{dz}{dx}, & \left[\frac{d^2u'}{d\zeta^2}\right]&=\frac{1}{3}\left(\frac{d^2u}{dx^2}+\frac{d^2v}{dxdy}\right)\\ \left[\frac{dv'}{d\zeta}\right]&=-\frac{dz}{dy}, & \left[\frac{d^2v'}{d\zeta^2}\right]&=\frac{1}{3}\left(\frac{d^2u}{dxdy}+\frac{d^2v}{dy^2}\right)\\ \left[\frac{dw'}{d\zeta}\right]&=-\frac{1}{3}\left(\frac{du}{dx}+\frac{dv}{dy}\right), & \left[\frac{d^2w'}{d\zeta^2}\right]&=\frac{1}{3}\left(\frac{d^2z}{dx^2}+\frac{d^2z}{dy^2}\right)\end{aligned}\right\}\ldots\ldots(7).$$

The values of $N_1$, $N_2$, $T_3$ and their differential coefficients with respect to $\zeta$, when $\zeta$ is zero, are

$$\left.\begin{aligned}[][N_1]&=\frac{2k}{3}\left(4\frac{du}{dx}+\frac{dv}{dy}\right), & \left[\frac{dN_1}{d\zeta}\right]&=-\frac{2k}{3}\left(4\frac{d^2z}{dx^2}+\frac{d^2z}{dy^2}\right)\\ [N_2]&=\frac{2k}{3}\left(4\frac{dv}{dy}+\frac{du}{dx}\right), & \left[\frac{dN_2}{d\zeta}\right]&=-\frac{2k}{3}\left(4\frac{d^2z}{dy^2}+\frac{d^2z}{dx^2}\right)\\ [T_3]&=k\left(\frac{du}{dy}+\frac{dv}{dx}\right), & \left[\frac{dT_3}{d\zeta}\right]&=-2k\frac{d^2z}{dxdy}\end{aligned}\right\}\ldots\ldots(8).$$

483. From these formulæ we see that the unknown quantity $z$ will enter singly into equation (6), and that the equations (5) will contain only the unknown quantities $u$ and $v$. Thus the displacements of the points of the mean section in the direction of the section and the form which this mean section will take are independent, and can be determined separately. We see also that the equations (5) are the same as in the case of a flexible *membrane*; so that with respect to the values of $u$ and $v$ we have nothing to add to what has been mentioned in Art. 472. But it appears from the third of equations (7) that these displacements will be always accompanied by a normal dilatation $\left[\frac{dw'}{d\zeta}\right]$ equal and of contrary sign to a third

of the sum of the dilatations $du/dx$ and $dv/dy$ which occur in the directions of $x$ and $y$ respectively. Suppose that a lamina has experienced parallel to its faces a linear dilatation equal in all directions and throughout its extent, and represent this by $\delta$; then there will at the same time be a normal condensation equal to $\frac{2\delta}{3}$. By the dilatation the volume of the lamina will be increased in the ratio of $(1+\delta)^2$ to 1; by the condensation it will be diminished in the ratio of $1-\frac{2\delta}{3}$ to 1: the total variation will be in the ratio of $(1+\delta)^2\left(1-\frac{2\delta}{3}\right)$ to 1, that is approximately in the ratio of $1+\frac{4\delta}{3}$ to 1. Thus the extension of an elastic lamina in the direction of its largest dimensions gives rise to an increase of volume, and consequently to a diminution of density; this result is similar to what occurs in the case of an elastic string, but would be more difficult to verify by experiment. (Cf. Art. 368.)

484. In what follows we may omit the consideration of the displacement of the points of the mean section parallel to $x$ and $y$, and suppose that $u$ and $v$ are zero. By reason of the first three equations (7) the formulæ (2) will then become

$$u' = -\frac{dz}{dx}\zeta, \qquad v' = -\frac{dz}{dy}\zeta, \qquad w' = z.$$

These values of $u'$ and $v'$ shew that the points of the lamina which were originally on one perpendicular to the mean section will after the change of form lie on a common normal to the curved surface into which the mean section is transformed. At every point there will be on one side of this section dilatation parallel to its direction, and on the other side condensation. It is this difference of state of the faces of the lamina which produces its elasticity by flexion, or its tendency to recover its natural form. In the directions of the principal curvatures of the mean section the dilatations and condensations will be proportional to the distances from this surface, and inversely proportional to the corresponding radius of curvature. Whether there is equilibrium or motion the condition of the lamina will be known throughout its thickness when the form of the mean section is determined, that is when we know the value of $z$ in terms of $x$ and $y$.

By means of the formula (8) equation (6) becomes

$$Z+\frac{\epsilon^2}{6}\left\{\left[\frac{d^2Z'}{d\zeta^2}\right]+2\left[\frac{d^2X'}{d\zeta dx}\right]+2\left[\frac{d^2Y'}{d\zeta dy}\right]\right\}$$
$$=\frac{8k\epsilon^2}{9\rho}\left(\frac{d^4z}{dx^4}+2\frac{d^4z}{dx^2dy^2}+\frac{d^4z}{dy^4}\right)\text{...........(9)}.$$

485. In the case of equilibrium this will be the partial differential equation to the surface required; the forces $Z, X', Y', Z'$ being then given as functions of $x, y, \zeta$. In the case of motion, and supposing that there is no inner applied force, we must put $Z=-\frac{d^2z}{dt^2}$, $X'=-\frac{d^2u'}{dt^2}$, $Y'=-\frac{d^2v'}{dt^2}$, $Z'=-\frac{d^2w'}{dt^2}$ If we make $\zeta=0$ after the differentiations we have by the aid of (7)—

$$\left[\frac{dX'}{d\zeta}\right]=\frac{d^3z}{dxdt^2},\qquad \left[\frac{dY'}{d\zeta}\right]=\frac{d^3z}{dydt^2},$$
$$\left[\frac{d^2Z'}{d\zeta^2}\right]=-\frac{1}{3}\left(\frac{d^4z}{dx^2dt^2}+\frac{d^4z}{dy^2dt^2}\right);$$

and consequently the first member of equation (9) will become

$$-\frac{d^2z}{dt^2}+\frac{5\epsilon^2}{18}\left(\frac{d^4z}{dx^2dt^2}+\frac{d^4z}{dy^2dt^2}\right)$$

But the term which has $\epsilon^2$ for a factor is evidently very small and may be neglected with respect to the first term; this equation will then be simply

$$\frac{d^2z}{dt^2}+a^2\left(\frac{d^4z}{dx^4}+2\frac{d^4z}{dx^2dy^2}+\frac{d^4z}{dy^4}\right)=0\text{...........(10)},$$

where for abbreviation we put $a^2=\frac{8k\epsilon^2}{9\rho}$.

To the equations (9) and (10) we must add those which relate to the boundary of the plate, and which will serve with the initial state of the plate in the case of motion, to determine the arbitrary quantities involved in the integrals. These special equations we now propose to investigate.

486. Let $X_0$, $Y_0$, $Z_0$ be the outer applied forces at any point of the boundary, $\theta$ the angle which the normal to the boundary there

makes with the axis of $x$. Then (cf. Lamé, page 20, equations (8)),

$$X_0 = N_1 \cos\theta + T_3 \sin\theta, \qquad Y_0 = T_3 \cos\theta + N_2 \sin\theta,$$
$$Z_0 = T_2 \cos\theta + T_1 \sin\theta \quad \ldots\ldots\ldots\ldots (11).$$

Since the quantities $u$ and $v$ have been supposed zero the values of $N_1$, $N_2$ and $T_3$ which correspond to $\zeta = 0$ are also zero. Neglecting the powers of $\zeta$ above the first the foregoing values of $X_0$ and $Y_0$ become

$$X_0 = \zeta \left[\frac{dT_3}{d\zeta}\right] \sin\theta + \zeta \left[\frac{dN_1}{d\zeta}\right] \cos\theta,$$
$$Y_0 = \zeta \left[\frac{dN_2}{d\zeta}\right] \sin\theta + \zeta \left[\frac{dT_3}{d\zeta}\right] \cos\theta.$$

Now it seems to me that according to the ordinary principles of the theory of elasticity the three equations (11) must hold at *every point* of the boundary; and thus the last two must hold approximately at every point. Hence we may if we please multiply by $\zeta$ and integrate between the limits $-\epsilon$ and $\epsilon$: thus

$$\int_{-\epsilon}^{+\epsilon} X_0 \zeta d\zeta = \frac{2\epsilon^3}{3} \left\{ \left[\frac{dT_3}{d\zeta}\right] \sin\theta + \left[\frac{dN_1}{d\zeta}\right] \cos\theta \right\},$$
$$\int_{-\epsilon}^{+\epsilon} Y_0 \zeta d\zeta = \frac{2\epsilon^3}{3} \left\{ \left[\frac{dN_2}{d\zeta}\right] \sin\theta + \left[\frac{dT_3}{d\zeta}\right] \cos\theta \right\}.$$

Therefore by the aid of equations (8) we have

$$\left.\begin{aligned} \int_{-\epsilon}^{+\epsilon} X_0 \zeta d\zeta &= -\frac{4k\epsilon^3}{9} \left\{ 3\frac{d^2z}{dxdy} \sin\theta + \left(4\frac{d^2z}{dx^2} + \frac{d^2z}{dy^2}\right) \cos\theta \right\} \\ \int_{-\epsilon}^{+\epsilon} Y_0 \zeta d\zeta &= -\frac{4k\epsilon^3}{9} \left\{ 3\frac{d^2z}{dxdy} \cos\theta + \left(4\frac{d^2z}{dy^2} + \frac{d^2z}{dx^2}\right) \sin\theta \right\} \end{aligned}\right\} \ldots (12).$$

Again, take the last of the three equations (11); suppose $T_1$ and $T_2$ to be expanded in powers of $\zeta$, eliminate by means of equations (4) the values of $[T_1]$, $[T_2]$, $\left[\frac{dT_1}{d\zeta}\right]$ and $\left[\frac{dT_2}{d\zeta}\right]$, and neglect terms of the third order with respect to $\epsilon$ and $\zeta$: thus

$$Z_0 = \tfrac{1}{2}(\zeta^2 - \epsilon^2) \left\{ \left[\frac{d^2T_1}{d\zeta^2}\right] \sin\theta + \left[\frac{d^2T_2}{d\zeta^2}\right] \cos\theta \right\}.$$

Then as this is true at every point of the boundary we are at liberty to integrate with respect to $\zeta$, and thus we get

$$\int_{-\epsilon}^{+\epsilon} Z_0 \, d\zeta = -\frac{2\epsilon^3}{3} \left\{ \left[\frac{d^2T_1}{d\zeta^2}\right] \sin\theta + \left[\frac{d^2T_2}{d\zeta^2}\right] \cos\theta \right\}.$$

Now the first two equations (3) give by the aid of (8)

$$\left[\frac{d^2T_2}{d\zeta^2}\right] = -\rho\left[\frac{dX'}{d\zeta}\right] - \left[\frac{d^2N_1}{dxd\zeta}\right] - \left[\frac{d^2T_3}{dyd\zeta}\right]$$

$$= -\rho\left[\frac{dX'}{d\zeta}\right] + \frac{8k}{3}\left\{\frac{d^3z}{dx^3} + \frac{d^3z}{dxdy^2}\right\}.$$

Similarly, $\left[\frac{d^2T_1}{d\zeta^2}\right] = -\rho\left[\frac{dY'}{d\zeta}\right] + \frac{8k}{3}\left\{\frac{d^3z}{dy^3} + \frac{d^3z}{dx^2dy}\right\}.$

Hence by substitution we get

$$-\int_{-\epsilon}^{+\epsilon} Z_0\, d\zeta = \frac{2\epsilon^3}{3}\left\{\frac{8k}{3}\left(\frac{d^3z}{dx^3} + \frac{d^3z}{dxdy^2}\right) - \rho'\left[\frac{dX'}{d\zeta}\right]\right\}\cos\theta$$

$$+ \frac{2\epsilon^3}{3}\left\{\frac{8k}{3}\left(\frac{d^3z}{dy^3} + \frac{d^3z}{dx^2dy}\right) - \rho'\left[\frac{dY'}{d\zeta}\right]\right\}\sin\theta \text{......(13)}.$$

487. The equations (12) and (13) seem to me to coincide with those which Poisson maintains to hold at the boundary; they correspond to the equations (13) on his page 538: but the point is one of great difficulty. Poisson's mode of obtaining them is circuitous and not very clear; I have put them in what seems the most natural connexion with the ordinary theory of elasticity. But at the same time that theory requires that (11) should be true at *every point* of the boundary, and this condition Poisson does not attempt to satisfy; but deduces certain aggregate or average results by integration. It may be asked too why Poisson chose the two equations (12) instead of $\int_{-\epsilon}^{+\epsilon} X_0\, d\zeta = \int_{-\epsilon}^{+\epsilon} (N_1 \cos\theta + T_3 \sin\theta)\, d\zeta$, and $\int_{-\epsilon}^{+\epsilon} Y_0\, d\zeta = \int_{-\epsilon}^{+\epsilon} (T_3 \cos\theta + N_2 \sin\theta)\, d\zeta$, which it would seem must also hold. Kirchhoff first, followed by Clebsch, objected to Poisson's boundary equations; according to them (13) is approximately true; but they replace the two equations (12) by a single equation: see the end of Art. 531.

[488.] These boundary equations of Poisson[1] express the conditions that the total applied force perpendicular to the plate at

[1] [Dr Todhunter had included in his manuscript at this point a long discussion of the relative merits of the Poisson and Kirchhoff contour-conditions. His general conclusion seems to be that Poisson's three conditions are not only sufficient, but *necessary*. I have felt justified in replacing this discussion by the above article, as Dr Todhunter had added a note at a later date that it would be necessary for him to reconsider the whole subject.] ED.

any element of the boundary shall be equal to the shearing force produced by the strain at that point (13), and again that the stress couples with axes parallel to the axes of $x$ and $y$ shall be in equilibrium with the applied couples at the same element of the boundary (12).

Now the couples applied to the edge of the plate at any element may be reduced to two, having respectively the tangent and normal to the contour in the plane of the plate for axes. The forces of the latter couple lie in the same plane and may be taken parallel to the shearing force. Now it has been argued that it is not needful that the three conditions, equality of shearing force and shearing stress, equality of applied couple and stress couple about the tangent, and equality of applied couple and stress couple about the normal, should hold. For the latter couple may be combined with the shear to give a single condition of equilibrium without affecting the state of the plate at distances from the edge sensibly greater than the thickness. Hence instead of Poisson's *three* boundary-conditions we should have only *two*. Such appears to be Thomson and Tait's explanation of the discrepancy between the Poisson and Kirchhoff boundary conditions[1]. (See their *Treatise on Natural Philosophy*, Part II. pages 190—193.) The literature of the controversy will be treated more fully when we consider Kirchhoff's discussion of the problem. We may remark here however that Poisson's work is not in the least invalidated supposing Thomson and Tait's view to be the correct one Poisson finds that the stress consists of a certain shear and two couples about the axes of $x$ and $y$. These must certainly be kept in equilibrium by the applied system, say the force $F$ and the couples $H$ and $H$ That is Poisson's real statement as to the boundary conditions (page 538). Thomson and Tait now add that if the couples $H$, $H'$ be replaced by others $M$, $N$ about the tangent and normal respectively, then the couple distribution $N$ may be replaced by a shear distribution $dN/ds$ where $ds$ is an element of arc of the contour. It seems to me that if we are given the distribution of force upon the edge of a plate, Poisson's boundary

[1] This reconciliation of Poisson and Kirchhoff has been attributed by French writers to M. Boussinesq. His memoirs however are of considerably later date than the first edition of Thomson and Tait's *Treatise*.

conditions ought to give exactly the same result as Kirchhoff's, for we should make $F$, $M$ and $N$ in equilibrium with the given distribution and in doing so find that $F$, $M$ and $N$ were not separately determinate. Poisson's error appears first, I think on page 537, when in considering a special case, he adds: *mais en général, ces trois quantités $F$, $H$, $H'$ seront indépendantes l'une de l'autre.* On page 547, when treating a circular plate with a free edge, he supposes $F$, $H$, $H'$ zero, but in this case the symmetry of the plate preserves him from error. I am inclined then to think that Poisson's so-called error has been much exaggerated. It is one of stating the results of analysis and not of analysis itself. Further, in the most general case of a *discontinuous* distribution of shearing force and normal couple, it would seem more convenient to take Poisson's calculation of the shear and couples, and we should have at every element to make them in equilibrium with the discontinuous applied system of force I feel also some doubt as to whether Poisson's method of treating the whole problem is not really more satisfactory and suggestive than Kirchhoff's when the plate has a definite although small thickness[1].

489. Among equations (7) we have

$$\left[\frac{du'}{d\zeta}\right] = -\frac{dz}{dx}, \qquad \left[\frac{dv'}{d\zeta}\right] = -\frac{dz}{dy};$$

we shall now give the equations which correspond to these when the approximation is carried to a higher order.

By (1) and (4) we have

$$[T_1] = k\left[\frac{dv'}{d\zeta}\right] + k\frac{dz}{dy}, \quad [T_1] = -\frac{\epsilon^2}{2}\left[\frac{d^2T_1}{d\zeta^2}\right];$$

hence from the value of $\left[\dfrac{d^2T_1}{d\zeta^2}\right]$ found in Art. 486 we get

$$\left[\frac{dv'}{d\zeta}\right] = -\frac{dz}{dy} + \frac{\epsilon^2}{2}\frac{\rho}{k}\left[\frac{dY'}{d\zeta}\right] - \frac{4\epsilon^2}{3}\left(\frac{d^3z}{dy^3} + \frac{d^3z}{dx^2dy}\right).$$

[1] Saint-Venant remarks: Ce sujet est délicat. Nous ne doutons pas que les équations aux limites de M. Kirchhoff ne soient les véritables; mais quoi celles de Poisson sont-elles fausses? C'est ce que nous n'avons pas encore eu le loisir d'étudier à fond....La matière demande donc un examen approfondi. *Historique Abrégé* prefixed to Navier's *Leçons*, 3rd Ed. p. cclxx. This 'examen approfondi' Saint-Venant has given in his edition of Clebsch, pp. 689—733. He adopts Thomson and Tait's reconciliation, attributing it however to Boussinesq. I shall return to the subject later.

Similarly

$$\left[\frac{du'}{d\zeta}\right] = -\frac{dz}{dx} + \frac{\epsilon^2}{2}\frac{\rho}{k}\left[\frac{dX'}{d\zeta}\right] - \frac{4\epsilon^2}{3}\left(\frac{d^3z}{dx^3} + \frac{d^3z}{dxdy^2}\right).$$

[Poisson seems here as in other places in this memoir to have wrong signs which I have tacitly corrected.]

490. When the parts of the boundary are supported in such a manner that it cannot slide parallel to the axis of $z$ the equation (13) will not hold, and it must be replaced by the condition $z = 0$. The right-hand member of (13) will then express the pressure parallel to the axis of $z$, and referred to the unit of length which the points of support will have to bear.

491. But suppose that at some point or points the plate is so constrained that it cannot slide, cannot turn on itself, and cannot turn round the tangent to the mean section; the plate is then said by Poisson to be *encastrée*, or as we call it *built-in*. Then the conditions (12) and (13) do not hold, but in addition to $z = 0$ it will be necessary that the displacements $u'$ and $v$ should be zero at the boundary throughout the thickness of the plate. Hence by the formulæ (2) we must have

$$\left[\frac{du'}{d\zeta}\right] = 0, \quad \left[\frac{dv'}{d\zeta}\right] = 0;$$

and in these equations we shall employ the values obtained in Art. 489. In special applications instead of these equations it will be more convenient to use

$$\left[\frac{du'}{d\zeta}\right]\frac{dy}{ds} - \left[\frac{dv'}{d\zeta}\right]\frac{dx}{ds} = 0, \quad \left[\frac{du'}{d\zeta}\right]\frac{dx}{ds} + \left[\frac{dv'}{d\zeta}\right]\frac{dy}{ds} = 0,$$

which are equivalent to the preceding, where $ds$ denotes an element of the perimeter of the mean section. In the first of these we may neglect the terms multiplied by $\epsilon^2$, which are small with respect to the terms independent of the thickness; this reduces it to[1]

$$\frac{dz}{dx}\frac{dy}{ds} - \frac{dz}{dy}\frac{dx}{ds} = 0.$$

[1] [This result has been criticised by Cauchy, but Mathieu makes Poisson and Cauchy agree. We shall return to the point later.] Ed.

The case is different with respect to the second equation, which becomes

$$-\left(\frac{dz}{dx}\frac{dx}{ds}+\frac{dz}{dy}\frac{dy}{ds}\right)+\frac{\epsilon^2}{2}\frac{\rho}{k}\left\{\left[\frac{dX'}{d\zeta}\right]\frac{dx}{ds}+\left[\frac{dY'}{d\zeta}\right]\frac{dy}{ds}\right\}$$
$$-\frac{8}{3}\frac{\epsilon^2}{2}\left(\frac{d^3z}{dx^3}+\frac{d^3z}{dxdy^2}\right)\frac{dx}{ds}-\frac{8}{3}\frac{\epsilon^2}{2}\left(\frac{d^3z}{dy^3}+\frac{d^3z}{dx^2dy}\right)\frac{dy}{ds}=0.$$

But if we differentiate the equation $z=0$ with respect to $x$ and $y$, considering $x$ and $y$ to be functions of $s$ given by the equation to the perimeter of the mean section, we have

$$\frac{dz}{dx}\frac{dx}{ds}+\frac{dz}{dy}\frac{dy}{ds}=0;$$

thus the term independent of $\epsilon$ will disappear from the preceding equation, and reduce it to the term dependent on $\epsilon^2$, which has been preserved for this reason. In this manner the three equations relative to a part of the boundary which is *built-in* will be

$$\left.\begin{aligned}&z=0,\quad \frac{dz}{dx}\frac{dy}{ds}-\frac{dz}{dy}\frac{dx}{ds}=0,\\&\frac{\rho}{k}\left\{\left[\frac{dX'}{d\zeta}\right]\frac{dx}{ds}+\left[\frac{dY'}{d\zeta}\right]\frac{dy}{ds}\right\}-\frac{8}{3}\frac{d}{ds}\left(\frac{d^2z}{dx^2}+\frac{d^2z}{dy^2}\right)=0\end{aligned}\right\}\ldots(14).$$

492. These formulæ relative to the different points of the perimeter of the mean section apply to the two cases of equilibrium and motion. But in the case of motion, and supposing that there are no inner applied forces, we put, as in Art. 485,

$$\left[\frac{dX'}{d\zeta}\right]=\frac{d^3z}{dxdt^2},\qquad \left[\frac{dY'}{d\zeta}\right]=\frac{d^3z}{dydt^2}.$$

Now having regard to equations (10), and to the equation $a^2=\frac{8k\epsilon^2}{9\rho}$, we see that the terms $\rho\left[\frac{dX'}{d\zeta}\right]$ and $\rho\left[\frac{dY'}{d\zeta}\right]$ of equation (13) and of the last equation (14) will be very small and may be neglected with respect to the other terms.

493. Poisson closes this section of his memoir by using, as he expresses it, equations (12) and (13) to verify the ordinary equations of equilibrium of all the given forces which act on an elastic plate having its boundary free. He restricts himself however to *three*

equations out of the *six* which must hold for a rigid body. This he says will serve to confirm the analysis. It is perhaps meant to throw some light on what appears arbitrary in his process: see Art. 487.

494. Poisson's seventh section is entitled: *Application des formules précédents à l'équilibre et au mouvement d'une plaque circulaire*, and occupies pages 545—570. As the first attempt at this problem and as an excellent example of the application of the equations of the preceding section, it is reproduced here. It will enable the reader to judge of the comparatively minor importance of later additions to the solution, and give him a most valuable example of the clearness and power of this great master of analysis[1].

495. In the case of equilibrium we will suppose the plate horizontal and heavy, and its boundary entirely free or constrained everywhere in the same manner. Let us apply to its upper face a normal pressure of equal intensity at equal distances from the centre; take the centre for the origin of co-ordinates, and denote by $R$ the pressure at the distance $r$ from the centre, referred to the unit of surface, so that $R$ is a given function of $r$. If we neglect the squares of $dz/dx$ and $dz/dy$ the components of this force parallel to the axes of $x$, $y$, $z$ respectively will be $-R\dfrac{dz}{dx}$, $-R\dfrac{dz}{dy}$, $R$; the first two being very small with respect to the third we will neglect the small horizontal displacements which they produce, and consider only the curvature of the plate or of its mean section. Now it will be the same if instead of applying the pressure $R$ to the *face* of the plate, we suppose all the points of the plate solicited by constant forces in the direction of the thickness, and represented in intensity by $R/(2\rho\epsilon)$ per unit mass. We will put then in equation (9) of Art. 484,

$$Z' = Z = g + \frac{R}{2\rho\epsilon}, \quad Y' = -\frac{R}{2\rho\epsilon}\frac{dz}{dy}, \quad X' = -\frac{R}{2\rho\epsilon}\frac{dz}{dx};$$

[1] [I have thought it better to print this section as it was left by Dr Todhunter, though for the sake of proportion it would perhaps have been fitter to abridge. Readers who take a genuine historical interest in the method by which a great master like Poisson attacked an unsolved problem, will not regret the space devoted to these two sections of his greatest memoir.] Ed.

where $g$ denotes gravity, and the axis of $z$ is supposed to be directed vertically downwards. Thus the equation will become

$$g + \frac{R}{2\rho\epsilon} = \frac{8k\epsilon^2}{9\rho}\left(\frac{d^2\phi}{dx^2} + \frac{d^2\phi}{dy^2}\right),$$

where $\phi$ stands for $\frac{d^2z}{dx^2} + \frac{d^2z}{dy^2}$. The same equation would hold if the plate were drawn by weights suspended from its lower face, and represented at the distance $r$ from the centre by $R$ for every unit of surface.

496. Since we suppose that everything is symmetrical round the centre of the plate, the ordinate $z$ of any point of the mean section will be a function of $r$. Hence we have

$$\phi \text{ or } \frac{d^2z}{dx^2} + \frac{d^2z}{dy^2} = \frac{d^2z}{dr^2} + \frac{1}{r}\frac{dz}{dr},$$

and

$$\frac{d^2\phi}{dx^2} + \frac{d^2\phi}{dy^2} = \frac{d^2\phi}{dr^2} + \frac{1}{r}\frac{d\phi}{dr}.$$

Let $l$ denote the radius of the plate, and $p$ its weight; then

$$p = 2\pi l^2 \epsilon \rho g,$$

so that if we put for brevity $\frac{9}{16k\epsilon^3} = k'$, the equation of equilibrium may be written

$$k'\left(\frac{p}{\pi l^2} + R\right) = \frac{d^2\phi}{dr^2} + \frac{1}{r}\frac{d\phi}{dr} \quad \text{..................(1).}$$

497. In the equations which relate to the boundary of the plate we put

$$\cos\theta = x/r, \ \sin\theta = y/r,$$

because the normal to the perimeter of the mean section coincides with the production of the radius vector. We will suppose that there are no outer applied forces; then the left-hand members of equations (12) and (13) of Art. 486 vanish.

Thus in the case of a perfectly free boundary we have the following two equations which must hold when $r = l$:

$$\frac{d\phi}{dr} = 0, \quad \frac{d^2z}{dr^2} + \frac{1}{4r}\frac{dz}{dr} = 0 \text{.................. (2).}$$

For the two equations (12) reduce to one, namely the first of those just given, and equation (13) reduces to the second of those just given, when we neglect $\left[\frac{dX'}{d\zeta}\right]$ and $\left[\frac{dY'}{d\zeta}\right]$

If the boundary of the plate is supported and cannot ascend or descend vertically, we shall have, as the conditions which must hold when $r = l$,

$$z = 0, \qquad \frac{d^2z}{dr^2} + \tfrac{1}{4}\frac{1}{r}\frac{dz}{dr} = 0 \quad\text{..................(3).}$$

If the plate is *built-in* round the boundary we have the equations (14) of Art. 491 which must hold round the boundary. The hypothesis that $z$ is a function of $r$ makes the third equation of (14) identical; and we are left with the following to hold when $r = l$:

$$z = 0, \quad \frac{dz}{dr} = 0 \quad\text{..........................(4).}$$

In the last two cases the vertical pressure at every point of the perimeter will be equal to the expression on the right-hand side of equation (13) of Art. 486. This reduces to

$$\frac{16k\epsilon^3}{9}\left(\frac{d\phi}{dx}\frac{x}{r} + \frac{d\phi}{dy}\frac{y}{r}\right),$$

that is to $\frac{1}{k'}\frac{d\phi}{dr}$. Hence if we denote by $P$ the vertical pressure on the whole perimeter we have $P = \frac{2\pi l}{k'}\frac{d\phi}{dr}$

498. Putting $C$ and $C'$ for arbitrary constants the complete integral of (1) is

$$\phi = C + C' \log r + k' \left\{\frac{pr^2}{4\pi l^2} + \int (\int Rrdr)\frac{dr}{r}\right\}.$$

Integrating by parts we get

$$\int (\int Rrdr)\frac{dr}{r} = \log r \int Rrdr - \int Rr \log r\, dr;$$

we may suppose that these simple integrals vanish with $r$, and we may write if we please $\log r/l$ instead of $\log r$. But $\phi$, which represents $\frac{d^2z}{dx^2} + \frac{d^2z}{dy^2}$, cannot become very great, and therefore cannot become infinite at any point of the plate; thus the term

$C' \log r$ must disappear, and therefore $C'$ must be zero. Thus the formula for $\phi$ becomes

$$\phi = C + k' \left\{ \frac{pr^2}{4\pi l^2} + \log\frac{r}{l} \int Rrdr - \int Rr \log\frac{r}{l}\, dr \right\},$$

the integrals being supposed to vanish when $r = 0$. Hence we get

$$\frac{d\phi}{dr} = k' \left\{ \frac{pr}{2\pi l^2} + \frac{1}{r} \int Rrdr \right\}.$$

If we put $2\pi \int_0^l Rrdr = \varpi$, then $\varpi$ will be the total pressure exerted on the plate, and the first equation of (2) will reduce to $p + \varpi = 0$; this manifestly ought to hold for the equilibrium of the plate when it is entirely free. In the case of the plate supported or built-in we have from the last result of Art. 497 the equation $P = p + \varpi$; a relation which is also obvious *a priori*.

499. Put $\frac{d^2z}{dr^2} + \frac{1}{r}\frac{dz}{dr}$ for $\phi$, and integrate again; thus

$$\frac{dz}{dr} = \frac{B}{r} + \frac{Cr}{2} + k' \left\{ \frac{pr^3}{16\pi l^2} + \frac{1}{r} \int (\int Rrdr)\, r \log\frac{r}{l}\, dr - \frac{1}{r} \int \left( \int Rr \log\frac{r}{l}\, dr \right) rdr \right\},$$

where $B$ is an arbitrary constant.

Integrating by parts we get

$$\int (\int Rrdr)\, r \log\frac{r}{l}\, dr = -\frac{r^2}{4}\left(1 - 2\log\frac{r}{l}\right) \int Rrdr + \tfrac{1}{4} \int Rr^3 \left(1 - 2\log\frac{r}{l}\right) dr,$$

$$\int \left( \int Rr \log\frac{r}{l}\, dr \right) rdr = \frac{r^2}{2} \int Rr \log\frac{r}{l}\, dr - \tfrac{1}{2} \int Rr^3 \log\frac{r}{l}\, dr.$$

We will suppose that these simple integrals vanish with $r$; and as $dz/dr$ must not be infinite for any value of $r$ we will suppress the term $B/r$. Thus we shall have

$$\frac{dz}{dr} = \frac{Cr}{2} + k \left\{ \frac{pr^3}{16\pi l^2} - \frac{r}{4}\left(1 - 2\log\frac{r}{l}\right) \int Rrdr - \frac{r}{2} \int Rr \log\frac{r}{l}\, dr + \frac{1}{4r} \int Rr^3 dr \right\}.$$

From this we can deduce $d^2z/dr^2$; then in both put $l$ for $r$ and substitute in the second equation of (2) or (3): thus we get

$$C = -k'\left\{\frac{13p}{40\pi} + \frac{3\varpi}{20\pi} - \int_0^l Rr\log\frac{r}{l}\,dr - \frac{3}{10l^2}\int_0^l Rr^3dr\right\}:$$

this value of $C$ will hold for the two cases of the plate entirely free, and of the plate supported at its perimeter. In the third case, that in which the plate is built-in, we deduce from the expression for $dz/dr$ and the second equation (4)

$$C = -k'\left\{\frac{p}{8\pi} - \frac{\varpi}{4\pi} - \int_0^l Rr\log\frac{r}{l}\,dr + \frac{1}{2l^2}\int_0^l Rr^3dr\right\}$$

Hence in all cases the constant $C$ is determined.

500. Integrate the value of $dz/dr$ found in the preceding Article, and put $\gamma$ for the arbitrary constant: thus

$$z = \gamma + \frac{Cr^2}{4} + k'\left\{\frac{pr^4}{64\pi l^2} - \tfrac{1}{4}\int(\textstyle\int Rrdr)\left(1 - 2\log\frac{r}{l}\right)rdr\right.$$
$$\left. - \tfrac{1}{2}\int\left(\textstyle\int Rr\log\frac{r}{l}\,dr\right)rdr + \tfrac{1}{4}\int(\textstyle\int Rr^3dr)\frac{dr}{r}\right\}.$$

Integrating by parts we get

$$\int(\textstyle\int Rrdr)\left(1 - 2\log\frac{r}{l}\right)rdr$$
$$= r^2\left(1 - \log\frac{r}{l}\right)\int Rrdr - \int R\left(1 - \log\frac{r}{l}\right)r^3dr,$$

$$\int\left(\textstyle\int Rr\log\frac{r}{l}\,dr\right)rdr = \frac{r^2}{2}\int Rr\log\frac{r}{l}\,dr - \tfrac{1}{2}\int Rr^3\log\frac{r}{l}\,dr,$$

$$\int(\textstyle\int Rr^3dr)\frac{dr}{r} = \log\frac{r}{l}\int Rr^3dr - \int Rr^3\log\frac{r}{l}\,dr.$$

We shall suppose that the simple integrals vanish with $r$, and we shall have finally

$$z = \gamma + \frac{Cr^2}{4} + \frac{k'}{4}\left\{\frac{pr^4}{16\pi l^2} - r^2\left(1 - \log\frac{r}{l}\right)\int Rrdr\right.$$
$$\left. + \left(1 + \log\frac{r}{l}\right)\int Rr^3dr - r^2\int Rr\log\frac{r}{l}\,dr - \int Rr^3\log\frac{r}{l}\,dr\right\} \ldots(5).$$

This is the equation to the surface formed by the plate in equilibrium.

501. If the plate is entirely free the constant $\gamma$ remains undetermined; and in fact it is indifferent then whether the plate occupies one position or another in space provided it is horizontal. Leaving this constant out of consideration the equation of the plate will be the same in this case as in the case of the plate supported vertically, since in these two cases the same value must be ascribed to the constant $C$. When the plate is supported round the boundary, or when it is built-in, the condition $z = 0$ for $r = l$ will determine the value of $\gamma$. We must put for $C$ the first or the second of the values previously found for this constant, according as the plate is supported or built-in at the boundary. In the two cases the constant $\gamma$ will be the sagitta (*flèche*) of the plate curved by its weight and the pressure which it experiences. We will continue to denote by $\gamma$ the value in the case where the boundary is supported, and by $\gamma_1$ the value when the boundary is built-in. We shall have then

$$\gamma = \frac{k'l^2}{4}\left\{\frac{21p}{80\pi} + \frac{13\varpi}{20\pi} - \frac{13}{10l^2}\int_0^l Rr^3 dr + \frac{1}{l^2}\int_0^l Rr^3 \log\frac{r}{l}\, dr\right\},$$

$$\gamma_1 = \frac{k'l^2}{4}\left\{\frac{p}{16\pi} + \frac{\varpi}{4\pi} - \frac{1}{2l^2}\int_0^l Rr^3 dr + \frac{1}{l^2}\int_0^l Rr^3 \log\frac{r}{l}\, dr\right\}$$

502. If the pressure $R$ is everywhere the same, and therefore equal to $\varpi/\pi l^2$, the integrations indicated may be easily effected. It will be found that the quantity $\log r/l$ disappears from the equation (5), which becomes

$$z = \gamma + \frac{Cr^2}{4} + \frac{k'}{64\pi l^2}(p + \varpi)\, r^4$$

Poisson says that the equation represents a paraboloid of revolution: we see that it is not the common paraboloid.

If we put for $k'$ its value we find that in this case

$$\gamma = \frac{21hl^2}{\epsilon^3}(p + \varpi), \qquad \gamma_1 = \frac{5hl^2}{\epsilon^3}(p + \varpi),$$

where $h$ stands for $\frac{9}{5120} \times \frac{1}{\pi} \times \frac{1}{k}$: thus $h$ depends solely on th quantity $k$, and is the greater the less this quantity is; Poisson expresses this result by saying that $h$ is greater the less the elasticity is: see Art. 522.

503. If the plate is drawn by a weight suspended in its centre, it will be necessary in order to apply the general formulae to this particular case to suppose that the function $R$ has sensible values only when the values of $r$ are insensible. By the nature of this kind of function the integrals $\int_0^l Rr^3 dr$ and $\int_0^l Rr^3 \log \frac{r}{l}\, dr$ will then be suppressed, as being insensible compared with $l^2 \int_0^l Rrdr$, that is with $\frac{l^2 \varpi}{2\pi}$

In this manner we shall have

$$\gamma = \frac{21hl^2}{\epsilon^3}\left(p + \frac{52\varpi}{21}\right), \qquad \gamma_1 = \frac{5hl^2}{\epsilon^3}(p + 4\varpi).$$

By comparing these formulae with those of the preceding Article we see that the weight $\varpi$ produces now a greater sagitta (*flèche*) than when it was spread over the entire surface of the plate. We see also that other things being equal the values of $\gamma_1$ are less than those of $\gamma$; such a result might have been anticipated, but the exact measure of the excess could be found only by calculation.

504. The last case includes that in which the centre of the plate is supported and maintained at the level of the perimeter. We must then consider $\varpi$ as an unknown force which is exerted in the direction opposite to gravity, and represents the resistance of the central point of support. The sagitta of the plate must be zero in this case; this condition will serve to determine $\varpi$, and we shall have

$$\varpi = -\frac{21p}{52}, \text{ or } \varpi = -\frac{p}{4},$$

according as the boundary is supported or built in. These values of $\varpi$, taken with the contrary sign, will express the pressures which are exerted at the centre; the corresponding values of the pressures at the boundary will be

$$P = \frac{31p}{52}, \qquad P = \frac{3p}{4}$$

Thus the pressure exerted by the weight $p$ of the plate is divided between the centre and the boundary in the ratio of 21 to 31 when the plate is simply supported, and in the ratio of 1 to 3 when it is built-in. These ratios then depend only on the manner in which the boundary of the plate is treated. and not at all on the radius, the thickness, or the degree of elasticity. But the elasticity must not be absolutely zero, for if the matter of the plate is supposed to be absolutely rigid, and deprived of all elasticity, which is never the case in nature, the distribution of the pressure $p$ between the centre and the boundary, and even between the different points of the boundary, would be quite indeterminate.

505. Let us now pass to the consideration of the *vibrations* of a circular plate. We omit gravity and every applied force, and we suppose that at any instant the points at the same distance from the centre have the same ordinate normal to the plate; so that the ordinate $z$ is a function of $t$ and $r$, where $t$ denotes the time, and $r$ is the same variable as before. Hence the equation (10) of Art. 485 applied to this particular case will be

$$\frac{d^2z}{dt^2} + a^2\left(\frac{d^2\phi}{dr^2} + \frac{1}{r}\frac{d\phi}{dr}\right) = 0 \text{.................. (6)},$$

where $\phi$ stands for $\frac{d^2z}{dr^2} + \frac{1}{r}\frac{dz}{dr}$

Let $z'$ and $z''$ be two other functions of $r$ and $t$, such that

$$\frac{dz'}{dt} + a\sqrt{-1}\left(\frac{d^2z'}{dr^2} + \frac{1}{r}\frac{dz'}{dr}\right) = 0,$$

$$\frac{dz''}{dt} - a\sqrt{-1}\left(\frac{d^2z''}{dr^2} + \frac{1}{r}\frac{dz''}{dr}\right) = 0;$$

then (6) will be satisfied by $z = z'$ and by $z = z''$, and if we take for $z'$ and $z''$ their most general values the complete integral of (6) will be $z = z' + z''$.

Now the following are values of $z'$ and $z''$, as may be easily verified,

$$z' = \int_{-\infty}^{+\infty}\left\{\int_0^\pi f(r\cos\omega + 2h\alpha\sqrt{at})\,d\omega\right.$$
$$\left.+\int_0^\pi F(r\cos\omega + 2h\alpha\sqrt{at})\log(r\sin^2\omega)\,d\omega\right\}e^{-\alpha^2}d\alpha,$$

$$z'' = \int_{-\infty}^{+\infty} \left\{ \int_0^\pi f_1 (r \cos \omega + 2h_1 \alpha \sqrt{at})\, d\omega \right.$$

$$\left. + \int_0^\pi F_1 (2 \cos \omega + 2h_1 \alpha \sqrt{at}) \log (r \sin^2 \omega)\, d\omega \right\} e^{-\alpha^2} d\alpha,$$

where $f$, $F$, $f_1$, $F_1$ denote arbitrary functions, $e$ is the base of the Napierian logarithms, and $h$ and $h_1$ are $\sqrt{\mp\sqrt{-1}}$; that is

$$h = \frac{1 - \sqrt{-1}}{\sqrt{2}}, \qquad h_1 = \frac{1 + \sqrt{-1}}{\sqrt{2}}.$$

These values of $z'$ and $z''$ Poisson says that he has found in another memoir; they may be easily verified.

506. But in order to give to these expressions a form which is more convenient for the calculation of vibrations, let us suppose that

$$f(x) = \Sigma\, (C \cos \nu x + D \sin \nu x), \qquad F(x) = \Sigma\, (E \cos \nu x + G \sin \nu x),$$
$$f_1(x) = \Sigma\, (C_1 \cos \nu x + D_1 \sin \nu x), \quad F_1(x) = \Sigma\, (E_1 \cos \nu x + G_1 \sin \nu x)$$

Here $C, D, E, G, C_1, D_1, E_1, G_1, \nu$ are quantities independent of the variable $x$, and the sums denoted by $\Sigma$ extend to all their possible values real and imaginary. Now we have

$$\int_{-\infty}^{+\infty} e^{-\alpha^2} \sin (2\nu h \alpha \sqrt{at})\, d\alpha = 0,$$

$$\int_{-\infty}^{+\infty} e^{-\alpha^2} \cos (2\nu h \alpha \sqrt{at})\, d\alpha = \sqrt{\pi}\, e^{-\nu^2 a h^2 t}$$

$$= \sqrt{\pi}\, \{\cos \nu^2 at + \sqrt{-1} \sin \nu^2 at\}.$$

Similar formulae hold when we put $h_1$ for $h$. Hence it will follow that the values of $z'$ and $z''$, and therefore of $z$, will be expressed in a series of quantities of the form

$$R \cos \nu^2 at + R_1 \sin \nu^2 at,$$

where $R$ and $R_1$ are functions of $r$. The terms which involve the *cosine* will depend on the initial values of $z$, and those which involve the *sine* will depend on the initial value of $dz/dt$. The treatment of the two classes of terms is of the same kind, and thus in order to abridge the formulae we will consider only the former; this amounts to supposing the velocity of all parts of the plate to be zero initially, so that the plate is made to take the form of a surface of revolution, and is then abandoned to itself.

507. Since we have

$$\int_0^\pi \sin(\nu r \cos\omega)\, d\omega = 0, \text{ and } \int_0^\pi \sin(\nu r \cos\omega)\log(r\sin^2\omega)\, d\omega = 0,$$

the value of $z$ derived from $z'$ and $z''$ will be

$$z = \Sigma\left\{A\int_0^\pi \cos(\nu r\cos\omega)\, d\omega + A_1\int_0^\pi \cos(\nu r\cos\omega)\log(r\sin^2\omega)\, d\omega\right\}\cos\nu^2 at,$$

where $A$ and $A_1$ are, like $\nu$, quantities independent of $r$ and $t$; and the summation denoted by $\Sigma$ has reference to these quantities. Let $m$ denote another constant; put in succession $m$ and $m\sqrt{-1}$ instead of $\nu$: in the case of $m$ let the coefficients $A$ and $A_1$ be used, and in the case of $\sqrt{-1}$ let them be replaced by $B$ and $B_1$. We shall get

$$z = \Sigma\left\{A\int_0^\pi \cos(mr\cos\omega)\, d\omega + \tfrac{1}{2}B\int_0^\pi (e^{mr\cos\omega} + e^{-mr\cos\omega})\, d\omega + A_1\int_0^\pi \cos(mr\cos\omega)\log(r\sin^2\omega)\, d\omega + \tfrac{1}{2}B_1\int_0^\pi (e^{mr\cos\omega} + e^{-mr\cos\omega})\log(r\sin^2\omega)\, d\omega\right\}\cos m^2 at \;\ldots\; (7),$$

and this is the form we shall employ for $z$.

508. The expression for $\phi$ which follows from that for $z$ just given may be simplified by considering the following differential equations:

$$\frac{d^2u}{dr^2} + \frac{1}{r}\frac{du}{dr} = -m^2u, \qquad \frac{d^2u'}{dr^2} + \frac{1}{r}\frac{du'}{dr} = m^2u'.$$

The complete integrals of these are

$$u = A\int_0^\pi \cos(mr\cos\omega)\, d\omega + A_1\int_0^\pi \cos(mr\cos\omega)\log(r\sin^2\omega)\, d\omega,$$

$$u' = \tfrac{1}{2}B\int_0^\pi (e^{mr\cos\omega} + e^{-mr\cos\omega})\, d\omega + \tfrac{1}{2}B_1\int_0^\pi (e^{mr\cos\omega} + e^{-mr\cos\omega})\log(r\sin^2\omega)\, d\omega:$$

where $A$, $A_1$, $B$, $B_1$ are arbitrary constants. Poisson refers for these to the *Journal de l'École Polytechnique*, 19[e] Cahier, page 475; it is however easy to verify them.

Hence from the formula (7) we infer that

$$\phi = \Sigma m^2 \left\{\tfrac{1}{2} B_1 \int_0^\pi (e^{mr\cos\omega} + e^{-mr\cos\omega})\, d\omega - A \int_0^\pi \cos(mr\cos\omega)\, d\omega \right.$$
$$+ \tfrac{1}{2} B_1 \int_0^\pi (e^{mr\cos\omega} + e^{-mr\cos\omega}) \log(r\sin^2\omega)\, d\omega$$
$$\left. - A_1 \int_0^\pi \cos(mr\cos\omega) \log(r\sin^2\omega)\, d\omega \right\} \cos m^2 at.$$

509. We suppose in what follows that there is no part fixed at the centre of the plate; the value of $z$ must then apply when $r = 0$; and as $z$ cannot be infinite the coefficients $A_1$ and $B_1$ of the terms which involve $\log r$ must be zero. Then the expressions for $z$ and $\phi$ will reduce to

$$\left.\begin{aligned} z &= \Sigma \left\{ A \int_0^\pi \cos(mr\cos\omega)\, d\omega \right. \\ &\qquad \left. + \tfrac{1}{2} B \int_0^\pi (e^{mr\cos\omega} + e^{-mr\cos\omega})\, d\omega \right\} \cos m^2 at, \\ \phi &= \Sigma m^2 \left\{ \tfrac{1}{2} B \int_0^\pi (e^{mr\cos\omega} + e^{-mr\cos\omega})\, d\omega \right. \\ &\qquad \left. - A \int_0^\pi \cos(mr\cos\omega)\, d\omega \right\} \cos m^2 at \end{aligned}\right\} \ldots\ldots(8).$$

The conditions relative to the boundary will be the same as in the case of equilibrium, and will be expressed by the equations (2), (3), or (4) of Art. 497 according as the boundary is free, or is fixed so that it cannot slide normally to the plate, or is built in. These conditions must hold when $r = l$, where $l$ is the radius of the plate, and they must hold for all values of $t$. We will examine these cases in succession: Arts. 510—513 refer to the built in plate, Arts. 514—516 to the plate which is supported, and Arts. 517—520 to the free plate.

510. In order to satisfy the condition $z = 0$ when $r = l$, whatever $t$ may be, we must take

$$A = H \int_0^\pi (e^{ml\cos\omega} + e^{-ml\cos\omega})\, d\omega, \quad B = -2H \int_0^\pi \cos(ml\cos\omega)\, d\omega;$$

$H$ being an unknown coefficient. Put for brevity

$$R = \int_0^\pi (e^{ml\cos\omega} + e^{-ml\cos\omega})\, d\omega \int_0^\pi \cos(mr\cos\omega)\, d\omega$$
$$- \int_0^\pi \cos(ml\cos\omega)\, d\omega \int_0^\pi (e^{mr\cos\omega} + e^{-mr\cos\omega})\, d\omega;$$

then the formula (8) becomes

$$z = \Sigma HR \cos m^2at \ldots\ldots\ldots\ldots\ldots\ldots\ldots(9);$$

and this applies to the two cases of the plate built-in, and the plate simply fixed.

In the former case the condition $\dfrac{dz}{dr} = 0$, or the second of equations (4) of Art. 497, will require that $\dfrac{dR}{dr} = 0$ when $r = l$; this gives

$$\int_0^\pi (e^{ml\cos\omega} + e^{-ml\cos\omega})\, d\omega \int_0^\pi \sin(ml\cos\omega)\cos\omega\, d\omega$$
$$+ \int_0^\pi (e^{ml\cos\omega} - e^{-ml\cos\omega})\cos\omega\, d\omega \int_0^\pi \cos(ml\cos\omega)\, d\omega = 0 \ldots (10).$$

This equation will serve to determine the values of $m$. We may prove as in Art. 458 of this chapter that the equation has no root which is partly real and partly imaginary; and we can determine the value of the coefficient denoted by $H$ as a function of $m$: these processes have been already sufficiently exemplified, and so we will omit them with respect to the present case, and to the cases to be discussed hereafter, and we will proceed to consider the different sounds of a circular plate, which is the essential object of the problem.

511. The roots of (10) being incommensurable it is necessary that the formula (9) should reduce to a single term in order that the plate should perform isochronous vibrations. Let $\lambda$ denote one of the values of $ml$ derived from (10); let $\tau$ denote the corresponding duration of an entire vibration; then $\tau = 2\pi l^2/(\lambda^2 a)$. The number of oscillations in a unit of time will be $1/\tau$; denote this by $n$; then putting for $a$ its value as given in Art. 485 we have $n = \dfrac{\lambda^2\epsilon}{3\pi l^2}\sqrt{\dfrac{2k}{\rho}}$ Hence we see that, other things being equal, the sound measured by this number $n$ will be directly as the thickness of the plate, and inversely as the square of the radius.

512. Poisson gives some numerical results[1] respecting the

[1] [These results are not calculated to any very great degree of accuracy Thus in Art. 518 we find $x_1 = \sqrt{92}$, but in Art. 519, $x_1 = \sqrt{91 \cdot 75}$, and corresponding variations appear in the other quantities.] Ed.

first two solutions of (10). Develop the left-hand member of (10) according to the powers of $ml$ or of $\lambda$, effect the integrations with respect to $\omega$, and make $\lambda^2 = 4x$; thus we get

$$\left\{1 + x + \frac{x^2}{(1.2)^2} + \frac{x^3}{(1.2.3)^2} + \ldots\ldots\right\} \times \left\{1 - \frac{2x}{(1.2)^2} + \frac{3x^2}{(1.2.3)^2} - \frac{4x^3}{(1.2.3.4)^2} + \ldots\ldots\right\}$$

$$+ \left\{1 - x + \frac{x^2}{(1.2)^2} - \frac{x^3}{(1.2.3)^2} + \ldots\ldots\right\} \times \left\{1 + \frac{2x}{(1.2)^2} + \frac{3x^2}{(1.2.3)^2} + \frac{4x^3}{(1.2.3.4)^2} + \ldots\ldots\right\} = 0.$$

This becomes when we multiply out

$$1 - \frac{x^2}{6} + \frac{x^4}{480} - \frac{x^6}{181440} + \frac{x^8}{209018880} - \ldots = 0.$$

By resolving this equation with respect to $x^2$ we obtain for the least two roots

$$x^2 = 6{\cdot}5227, \quad x^2 = 98;$$

the corresponding values of $\lambda^2$ are

$$\lambda^2 = 10{\cdot}2156, \quad \lambda^2 = 39{\cdot}59;$$

and the two gravest sounds of the built-in plate, or the numbers of vibrations which measure these sounds, are in the ratio of the numbers just given, that is nearly as 1 to 4.

513. If we wish to determine the radii of the nodal circles which accompany these sounds we must put $R = 0$; this equation, putting $m^2l^2 = 4x$ and $m^2r^2 = 4y$, becomes when developed

$$\left.\begin{aligned} &\left\{1 + x + \frac{x^2}{(1.2)^2} + \frac{x^3}{(1.2.3)^2} + \ldots\ldots\right\} \times \left\{1 - \frac{y}{(1.2)^2} + \frac{y^2}{(1.2.3)^2} - \frac{y^3}{(1.2.3.4)^2} + \ldots\ldots\right\} \\ &- \left\{1 - x + \frac{x^2}{(1.2)^2} - \frac{x^3}{(1.2.3)^2} + \ldots\ldots\right\} \times \left\{1 + \frac{y}{(1.2)^2} + \frac{y^2}{(1.2.3)^2} + \frac{y^3}{(1.2.3.4)^2} + \ldots\ldots\right\} = 0 \end{aligned}\right\} \ldots\ldots(11).$$

Moreover we must have $r$ less than $l$, that is $y$ less than $x$. If we employ the least value of $x$ we obtain no values of $y$ which are less; this shews that corresponding to the gravest sound there is no nodal line except the perimeter of the plate. If we employ the next value of $x$ we obtain one value of $y$ which is less than $x$, namely $y = 1{\cdot}424$. Thus we obtain $r = {\cdot}381l$ for the radius of the nodal circle in the case of the sound which is next to the gravest.

514. In the case of the plate which is supported at its boundary we substitute the formula (9) in the second equation (3); and as this must hold for all values of $t$ it will follow that

$$\frac{d^2R}{dr^2} + \frac{1}{4r}\frac{dR}{dr} = 0,$$

in which $r$ must be put equal to $l$. This gives after certain reductions[1]

$$2ml \int_0^\pi (e^{ml\cos\omega} + e^{-ml\cos\omega})\, d\omega \int_0^\pi \cos(ml\cos\omega)\, d\omega$$

$$-\tfrac{3}{4}\int_0^\pi (e^{ml\cos\omega} + e^{-ml\cos\omega})\, d\omega \int_0^\pi \sin(ml\cos\omega)\cos\omega\, d\omega$$

$$-\tfrac{3}{4}\int_0^\pi (e^{ml\cos\omega} - e^{-ml\cos\omega})\cos\omega\, d\omega \int_0^\pi \cos(ml\cos\omega)\, d\omega = 0 \ldots (12).$$

This equation will serve to determine the values of $m$, and consequently the different sounds which the plate will produce. If we denote by $\lambda'$ one of the values of $ml$ which are obtained from the equation, and by $n'$ the number of vibrations in the unit of time, which serves to measure the corresponding sound, we shall have $n' = \dfrac{\lambda'^2\epsilon}{3\pi l^2}\sqrt{\dfrac{2k}{\rho}}$, as in Art. 511.

515. By developing the first member of (12) according to powers of $ml$ or $\lambda'$, and putting $\lambda'^2 = 4x'$ Poisson obtains

$$1 - \frac{x'^2}{2} + \frac{x'^4}{96} - \frac{x'^6}{25920} + \frac{x'^8}{23224320} - \ldots\ldots$$

$$-\tfrac{3}{8}\left\{1 - \frac{x'^2}{6} + \frac{x'^4}{480} - \frac{x'^6}{181440} + \frac{x'^8}{209018880}\ldots\ldots\right\} = 0.$$

For the approximate values of the least two roots of this equation in $x'^2$ he gives

$$x'^2 = 1{\cdot}4761, \quad x'^2 = 55;$$

[1] [Poisson's equation (p. 563) seems to have an error in the second integral.] ED.

the corresponding values of $\lambda'^2$ are

$$\lambda'^2 = 4{\cdot}8591, \quad \lambda'^2 = 29{\cdot}67,$$

and the gravest two sounds for the plate, the boundary of which is supported, are in the ratio of these two numbers.

516. If we make $m^2l^2 = 4x'$, and $m^2r^2 = 4y'$, the radii of the nodal circles will be determined by the equation (11) in which $x'$ and $y'$ must be put for $x$ and $y$ respectively. It will be necessary that $y'$ should be less than $x'$; and if we take for $x$ its least value there exists no value of $y'$ which satisfies this condition: if we take for $x'$ its second value there exists one value of $y'$ less than $x'$ namely $y' = 1{\cdot}447$. Hence it follows that corresponding to the gravest sound there is no nodal circle except the perimeter of the plate, and that corresponding to the next sound there is one nodal circle the radius of which is $r = {\cdot}441l$.

517. Consider now the case of a free plate. In order that the second formula (8) may satisfy the first of equations (2) we must take

$$A = H' \int_0^\pi (e^{ml\cos\omega} - e^{-ml\cos\omega}) \cos\omega\, d\omega,$$

$$B = -2H' \int_0^\pi \sin(ml\cos\omega) \cos\omega\, d\omega,$$

$H'$ being a new constant. We shall then have

$$z = \Sigma H'R' \cos m^2at \quad \text{.....................(13)},$$

where for abridgement we put

$$R' = \int_0^\pi (e^{ml\cos\omega} - e^{-ml\cos\omega}) \cos\omega\, d\omega \int_0^\pi \cos(mr\cos\omega)\, d\omega$$

$$- \int_0^\pi \sin(ml\cos\omega) \cos\omega\, d\omega \int_0^\pi (e^{mr\cos\omega} + e^{-mr\cos\omega})\, d\omega.$$

The second equation (2) is the same thing as

$$\phi - \frac{3}{4r}\frac{dz}{dr} = 0;$$

if then we substitute in this from the second formula (8) and the formula (13) we shall obtain

$$\int_0^\pi (e^{ml\cos\omega} + e^{-ml\cos\omega})\, d\omega \int_0^\pi \sin(ml\cos\omega)\cos\omega\, d\omega$$
$$+ \int_0^\pi (e^{ml\cos\omega} - e^{-ml\cos\omega})\cos\omega\, d\omega \int_0^\pi \cos(ml\cos\omega)\, d\omega$$
$$- \frac{3}{2ml}\int_0^\pi (e^{ml\cos\omega} - e^{-ml\cos\omega})\cos\omega\, d\omega \int_0^\pi \sin(ml\cos\omega)\cos\omega\, d\omega = 0 \quad \ldots\ldots(14);$$

this equation will serve to determine the values of $m$.

Let $\lambda_1$ denote a value of $ml$ obtained from this equation, and let $x_1$ denote the number of vibrations in the unit of time which serves to measure the corresponding sound; then we shall have from the formula (13) reduced to a single term

$$n_1 = \frac{\lambda_1^2 a}{2\pi l^2} = \frac{\lambda_1^2 \epsilon}{3\pi l^2}\sqrt{\frac{2k}{\rho}}$$

518. Put $\lambda_1^2 = 4x_1$, and develop the first member of the equation (14): it becomes

$$1 - \frac{x_1^2}{6} + \frac{x_1^4}{480} - \frac{x_1^6}{181440} + \frac{x_1^8}{209018880} - \ldots$$
$$- \tfrac{3}{8}\left\{1 - \frac{x_1^2}{12} + \frac{x_1^4}{1440} - \frac{x_1^6}{725760} + \frac{x_1^8}{1045094400} - \ldots\right\} = 0.$$

For the approximate values of the least two roots of this equation in $x_1^2$ Poisson gives

$$x_1^2 = 4{\cdot}9392, \quad x_1^2 = 92;$$

the corresponding values of $\lambda_1^2$ are

$$\lambda_1^2 = 8{\cdot}8897, \quad \lambda_1^2 = 38{\cdot}36.$$

The ratio of the second value of $\lambda_1^2$ to the first, this ratio being that of the frequency of the gravest two sounds of the free plate, is thus equal to 4·316.

The number of vibrations which serves for the measure of the gravest sound is $n_1 = \frac{1{\cdot}3339}{l^2}\epsilon\sqrt{\frac{k}{\rho}}$. Let us denote by $n$ the number of longitudinal vibrations of a cylindrical rod of length $2l$ and radius $\epsilon$, formed of the same material as the plate, supposing the rod to give forth its gravest sound; then Poisson has shown (see

Art. 470) that $n = \frac{1}{4l}\sqrt{\frac{5k}{2\rho}}$. Hence we get $\frac{n_1}{n} = 3\cdot3746\,\frac{\epsilon}{l}$. This relation Poisson suggests as deserving of being tested by experiment.

519. In order to determine the radii of the nodal circles, which may correspond to the different sounds of the free plate, we shall have to solve the equation $R' = 0$; put $m^2l^2 = 4x_1$ and $m^2r^2 = 4y_1$, then this equation becomes

$$\left\{1 + \frac{2x_1}{(1\,.\,2)^2} + \frac{3x_1^{\,2}}{(1\,.\,2\,.\,3)^2} + \frac{4x_1^{\,3}}{(1\,.\,2\,.\,3\,.\,4)^2} + \dots\right\} \times \left\{1 - y_1 + \frac{y_1^{\,2}}{(1\,.\,2)^2} - \frac{y_1^{\,3}}{(1\,.\,2\,.\,3)^2} + \dots\right\}$$

$$- \left\{1 - \frac{2x_1}{(1\,.\,2)^2} + \frac{3x_1^{\,2}}{(1\,.\,2\,.\,3)^2} - \frac{4x_1^{\,3}}{(1\,.\,2\,.\,3\,.\,4)^2} + \dots\right\} \times \left\{1 + y_1 + \frac{y_1^{\,2}}{(1\,.\,2)^2} + \frac{y_1^{\,3}}{(1\,.\,2\,.\,3)^2} + \dots\right\} = 0.$$

We must take only those values of $y_1$ which are less than $x_1$. If we use the lowest value of $x_1$ we find only one value of $y_1$ which satisfies this condition; if we use the next value of $x_1$ we find two values of $y_1$. Thus in the case of the gravest sound there is only one nodal circle; and in the case of the next sound there are two nodal circles. With respect to the former sound we have

$$x_1 = \sqrt{4\cdot9392}, \quad y_1 = 1\cdot0295;$$

hence we obtain for the radius of the single nodal circle $r = \cdot6806l$. With respect to the latter sound we have

$$x_1 = \sqrt{91\cdot75}, \; y_1 = 1\cdot468, \; y_1 = 6\cdot674;$$

hence we obtain for the radii of the nodal circles $r = \cdot3915l$, $r = \cdot835l$.

520. The radii of the nodal circles which form themselves on circular plates are independent of the matter and of the thickness of the plate; they are proportional to its diameter, and besides this they depend only on the way in which the centre and the boundary are constrained. M. Savart measured them

with care on three plates of copper of different dimensions with the centre and the boundary entirely free. Poisson does not however say in what way the plates were supported. In the case of the gravest sound Savart found for the ratio of the radius of the nodal circle to the radius of the plate on these three plates ·6819, ·6798, ·6812. The slight differences between these values may be attributed to the unavoidable errors of observation, and the mean of them, namely ·6810, agrees remarkably well with the theoretical value. In the case of the next sound Savart found for the inner nodal circle ·3855, ·3876, 3836; and for the outer nodal circle ·8410, ·8427, ·8406. The differences here also are small and fall within the limits of the errors of observation. The mean of the first three numbers is ·3856, and the mean of the other three is ·8414; these numbers agree well with the theoretical values ·3915 and ·835: see Art. 359.

521. In all the cases of vibration which we have examined, the centre of the plate is in motion; for if we make $r=0$ in the formulae (9) and (13) we obtain for the ordinate $z$ of this point a function of $t$ which is not zero. If we suppose, on the contrary, that a circular portion of the plate having the same centre and a radius which we will denote by $\alpha$ is rendered fixed, its perimeter ought to be considered as if it were built-in; and we shall accordingly have the conditions $z=0, \frac{dz}{dr}=0$ for $r=\alpha$; besides the conditions which hold for the boundary of the plate, that is for $r=l$. In like manner if the central part were hollow, and its perimeter entirely free, we should have $\frac{d\phi}{dr}=0, \phi-\frac{3}{4r}\frac{dz}{dr}=0$ for $r=\alpha$. In these two cases the ordinate $z$ would correspond only to values of $r$ between $\alpha$ and $l$, and so it will not be necessary to suppress that part of its expression which would become infinite for $r=0$: thus instead of the formula (8), which we have hitherto employed, we must take the formula (7) and the corresponding expression for $\phi$. The calculations will then be similar to those Poisson has gone through, but the formulae will be longer; Poisson refrains from giving them.

522. A remark may be made here as to what we are to understand by *greater* and *less* elasticity. Poisson obtains on his page 552 an expression for the amount of depression of the centre of a horizontal elastic plate below the plane of the boundary; the expression has in the denominator $k$ in Poisson's notation: see Art. 502. Then Poisson says in effect that this is greater the smaller $k$ is, that is the smaller the elasticity is; so he takes $k$ to be a measure of the elasticity. But this seems contrary to common notions, for one would expect that the greater the elasticity is the greater will be the depression. And on his page 554 Poisson seems to consider that if a body is absolutely rigid it may be said to have *no* elasticity, and then surely there would be no depression; so that contrary to what we have on his page 552 small elasticity would lead to small depression.

523. In the volume of the Paris Memoirs which contains this memoir by Poisson there is an *Addition* to it on pages 623—627. The object of this *Addition* is to give the complete integral of the equations which correspond to the vibrations of an elastic body supposing that there are no applied forces. The equations according to Poisson's memoir are

$$\left.\begin{aligned}\frac{d^2u}{dt^2}&=\frac{a^2}{3}\left(3\,\frac{d^2u}{dx^2}+2\,\frac{d^2v}{dxdy}+2\,\frac{d^2w}{dxdz}+\frac{d^2u}{dy^2}+\frac{d^2u}{dz^2}\right)\\ \frac{d^2v}{dt^2}&=\frac{a^2}{3}\left(3\,\frac{d^2v}{dy^2}+2\,\frac{d^2u}{dydx}+2\,\frac{d^2w}{dydz}+\frac{d^2v}{dz^2}+\frac{d^2v}{dx^2}\right)\\ \frac{d^2w}{dt^2}&=\frac{a^2}{3}\left(3\,\frac{d^2w}{dz^2}+2\,\frac{d^2u}{dzdx}+2\,\frac{d^2v}{dzdy}+\frac{d^2w}{dx^2}+\frac{d^2w}{dy^2}\right)\end{aligned}\right\}\ldots(1).$$

These agree in form with those usually given in the text books if we suppose only one elastic constant.

Put $$\frac{du}{dx}+\frac{dv}{dy}+\frac{dw}{dz}=\frac{d^2\phi}{dt^2}\ldots\ldots\ldots\ldots\ldots\ldots\ldots(2);$$

then equations (1) may be written

$$\left.\begin{aligned}\frac{d^2u}{dt^2}&=\frac{a^2}{3}\left(\frac{d^2u}{dx^2}+\frac{d^2u}{dy^2}+\frac{d^2u}{dz^2}\right)+\frac{2a^2}{3}\frac{d^3\phi}{dx\,dt^2}\\ \frac{d^2v}{dt^2}&=\frac{a^2}{3}\left(\frac{d^2v}{dx^2}+\frac{d^2v}{dy^2}+\frac{d^2v}{dz^2}\right)+\frac{2a^2}{3}\frac{d^3\phi}{dy\,dt^2}\\ \frac{d^2w}{dt^2}&=\frac{a^2}{3}\left(\frac{d^2w}{dx^2}+\frac{d^2w}{dy^2}+\frac{d^2w}{dz^2}\right)+\frac{2a^2}{3}\frac{d^3\phi}{dz\,dt^2}\end{aligned}\right\}\ldots\ldots(3).$$

Differentiate equations (3) with respect to $x$, $y$, $z$ respectively and add; thus

$$\frac{d^4\phi}{dt^4} = a^2\left(\frac{d^4\phi}{dx^2\,dt^2} + \frac{d^4\phi}{dy^2\,dt^2} + \frac{d^4\phi}{dz^2\,dt^2}\right).$$

Integrate twice with respect to $t$, and we obtain

$$\frac{d^2\phi}{dt^2} = a^2\left(\frac{d^2\phi}{dx^2} + \frac{d^2\phi}{dy^2} + \frac{d^2\phi}{dz^2} + Pt + Q\right),$$

where $P$ and $Q$ are arbitrary functions of $x$, $y$, and $z$. Let us denote by $p$ and $q$ two other functions of these variables, and make

$$\phi = \phi' + pt + q;$$

then we can reduce the preceding equation to this:

$$\frac{d^2\phi'}{dt^2} = a^2\left(\frac{d^2\phi'}{dx^2} + \frac{d^2\phi'}{dy^2} + \frac{d^2\phi'}{dz^2}\right) \quad\text{..................(4)},$$

provided we establish between $p$ and $P$, and between $q$ and $Q$ the following relations,

$$\frac{d^2p}{dx^2} + \frac{d^2p}{dy^2} + \frac{d^2p}{dz^2} + P = 0, \quad \frac{d^2q}{dx^2} + \frac{d^2q}{dy^2} + \frac{d^2q}{dz^2} + Q = 0.$$

Now let

$$u = u' + a^2\frac{d\phi'}{dx}, \quad v = v' + a^2\frac{d\phi'}{dy}, \quad w = w' + a^2\frac{d\phi'}{dz} \quad\text{......(5)};$$

substitute these values of $u$, $v$, $w$, and that of $\phi$ in equations (3), then taking into account the results of differentiating (4) with respect to $x$, $y$, and $z$, and reducing, we shall have

$$\left.\begin{aligned}\frac{d^2u'}{dt^2} &= \frac{a^2}{3}\left(\frac{d^2u'}{dx^2} + \frac{d^2u'}{dy^2} + \frac{d^2u'}{dz^2}\right)\\ \frac{d^2v'}{dt^2} &= \frac{a^2}{3}\left(\frac{d^2v'}{dx^2} + \frac{d^2v'}{dy^2} + \frac{d^2v'}{dz^2}\right)\\ \frac{d^2w'}{dt^2} &= \frac{a^2}{3}\left(\frac{d^2w'}{dx^2} + \frac{d^2w'}{dy^2} + \frac{d^2w'}{dz^2}\right)\end{aligned}\right\} \quad\text{...........(6)}.$$

If we make the same substitution in (?) we obtain

$$\frac{du'}{dx} + \frac{dv'}{dy} + \frac{dw'}{dz} = 0 \quad\text{..................(7)}.$$

Now according to a formula obtained by Poisson in the Memoirs of the Paris Academy, Vol. III. 1818, the complete integral of (4) in a finite form is

$$\phi' = \int_0^\pi \int_0^{2\pi} f(x + at\cos\alpha,\ y + at\sin\alpha\sin\beta,$$
$$z + at\sin\alpha\cos\beta)\, t\sin\alpha\, d\alpha d\beta,$$
$$+ \frac{d}{dt}\int_0^\pi \int_0^{2\pi} F(x + at\cos\alpha,\ y + at\sin\alpha\sin\beta,$$
$$z + at\sin\alpha\cos\beta)\, t\sin\alpha\, d\alpha d\beta,$$

where $f$ and $F$ denote arbitrary functions.

The integrals of (6) can be deduced from this by putting $a/\sqrt{3}$ in the place of $a$, and changing the arbitrary functions. Let us denote by $\frac{df_1}{dx}$ and $\frac{dF_1}{dx}$ the arbitrary functions which occur in $v'$, and by $\frac{df_2}{dx}$ and $\frac{dF_2}{dx}$ the arbitrary functions which occur in $w'$; then we shall have

$$v' = \frac{d}{dx}\int_0^\pi\int_0^{2\pi} f_1\left(x + \frac{at}{\sqrt{3}}\cos\alpha,\ y + \frac{at}{\sqrt{3}}\sin\alpha\sin\beta,\right.$$
$$\left. z + \frac{at}{\sqrt{3}}\sin\alpha\cos\beta\right) t\sin\alpha\, d\alpha d\beta$$
$$+ \frac{d^2}{dxdt}\int_0^\pi\int_0^{2\pi} F_1\left(x + \frac{at}{\sqrt{3}}\cos\alpha,\ y + \frac{at}{\sqrt{3}}\sin\alpha\sin\beta,\right.$$
$$\left. z + \frac{at}{\sqrt{3}}\sin\alpha\cos\beta\right) t\sin\alpha\, d\alpha d\beta,$$

$$w' = \frac{d}{dx}\int_0^\pi\int_0^{2\pi} f_2\left(x + \frac{at}{\sqrt{3}}\cos\alpha,\ y + \frac{at}{\sqrt{3}}\sin\alpha\sin\beta,\right.$$
$$\left. z + \frac{at}{\sqrt{3}}\sin\alpha\cos\beta\right) t\sin\alpha\, d\alpha d\beta$$
$$+ \frac{d^2}{dxdt}\int_0^\pi\int_0^{2\pi} F_2\left(x + \frac{at}{\sqrt{3}}\cos\alpha,\ y + \frac{at}{\sqrt{3}}\sin\alpha\sin\beta,\right.$$
$$\left. z + \frac{at}{\sqrt{3}}\sin\alpha\cos\beta\right) t\sin\alpha\, d\alpha d\beta.$$

Then in order to satisfy (7) in the most general manner we must take

$$u' = \psi - \frac{d}{dy}\int_0^\pi\int_0^{2\pi} f_1\left(x + \frac{at}{\sqrt{3}}\cos\alpha,\ y + \frac{at}{\sqrt{3}}\sin\alpha\sin\beta,\right.$$
$$\left. z + \frac{at}{\sqrt{3}}\sin\alpha\cos\beta\right) t\sin\alpha\, d\alpha d\beta$$
$$- \frac{d}{dz}\int_0^\pi\int_0^{2\pi} f_2\left(x + \frac{at}{\sqrt{3}}\cos\alpha,\ y + \frac{at}{\sqrt{3}}\sin\alpha\sin\beta,\right.$$
$$\left. z + \frac{at}{\sqrt{3}}\sin\alpha\cos\beta\right) t\sin\alpha\, d\alpha d\beta$$
$$- \frac{d^2}{dydt}\int_0^\pi\int_0^{2\pi} F_1\left(x + \frac{at}{\sqrt{3}}\cos\alpha,\ y + \frac{at}{\sqrt{3}}\sin\alpha\sin\beta,\right.$$
$$\left. z + \frac{at}{\sqrt{3}}\sin\alpha\cos\beta\right) t\sin\alpha\, d\alpha d\beta$$
$$- \frac{d^2}{dzdt}\int_0^\pi\int_0^{2\pi} F_2\left(x + \frac{at}{\sqrt{3}}\cos\alpha,\ y + \frac{at}{\sqrt{3}}\sin\alpha\sin\beta,\right.$$
$$\left. z + \frac{at}{\sqrt{3}}\sin\alpha\cos\beta\right) t\sin\alpha\, d\alpha d\beta,$$

where $\psi$ denotes an arbitrary function of $y$ and $z$.

524. With respect to the history of the important formula which Poisson gives as the general integral of (4) the reader should consult a paper by Liouville in his *Journal de Mathématiques*, Vol. I. of the New Series, 1856, and a note by the same writer which we have reproduced in a foot-note to Art. 562.

525. The last sentence of the memoir is

Nous reviendrons dans la suite sur les applications des formules precédentes à des problèmes particuliers.

We shall see as we proceed that Poisson in another memoir put the integrals of the general equations in another form; but he does not seem to have applied his formulae to special problems.

526. The values of $u$, $v$, $w$ found by (5) will involve the time $t$ in two forms; in one form we shall have $a$ as the coefficient of $t$, and in the other form we shall have $a/\sqrt{3}$ as the coefficient of $t$. Thus we have two waves, one propagated with the velocity $a$ and the other with the velocity $a/\sqrt{3}$. This is I presume the first appearance of this result in the history of our subject.

527. 1828—29. The publication of Poisson's memoir of April, 1828, gave rise to a controversy between Navier and him which was carried on in the *Annales de Chimie*, Vols. 38 and 39, 1828, and Vol. 40, 1829, and Férussac's *Bulletin*, Vol. XI., 1829. Saint-Venant in Moigno's *Statique*, page 695, states that the controversy appears also in Vols. 36 and 37 of the same series; but these two volumes contain no article by Navier, and the articles in them by Poisson do not mention Navier's name, though it is possible there may be some oblique reference to Navier[1].

The following are the articles which form this controversy, numbered for convenience of reference.

I. *Note relative à l'article intitulé: Mémoire sur l'équilibre et le mouvement des Corps élastiques, inséré page* 337 *du tome précédent;* par M. Navier. Vol. 38, pages 304—314.

II. *Réponse à une Note de M. Navier insérée dans le dernier Cahier de ce Journal;* par M. Poisson. Vol. 38, pages 435—440.

III. *Remarques sur l'Article de M. Poisson inséré dans le Cahier d'août;* par M. Navier. Vol. 39, pages 145—151.

IV. *Lettre de M. Poisson à M. Arago.* Vol. 39, pages 204—211.

V. *Lettre de M. Navier à M. Arago.* Vol. 40, pages 99—107.

VI. Navier. *Note relative à la question de l'équilibre et du mouvement des corps solides élastiques.* Férussac, *Bulletin des Sciences Mathématiques.* Vol. XI. 1829, pages 243—253.

We will now notice briefly the main points of the controversy.

528. In I. the chief complaint of Navier is that his labours on the subject, as shewn by his memoir published in 1827 in the seventh volume of the Memoirs of the Academy, were not adequately appreciated by Poisson. Navier thinks that he is entitled to consideration as having led the way in the right discussion of

[1] [See however Art. 433. On p. 86 of Vol. 36 and on p. 347 of Vol. 37, Poisson lays it down that molecular action cannot be represented by definite integrals, and that therefore the method of Lagrange is not applicable to the very problems in which Navier had used it. Saint-Venant criticises Poisson's view in the *Historique Abrégé*, p. clxiii., and Moigno's *Statique*, p. 695.] ED.

problems concerning elasticity; and this seems to be a reasonable claim: see a note on page 243 of the memoir on *Torsion* by Saint-Venant[1].

Navier makes with emphasis some curious remarks on one point. We have already stated that in the equation for the vibration of a lamina Poisson obtains a coefficient which involves the *square* of the thickness; see Arts. 260 and 289. Now Navier says that Poisson's coefficient ought to vary as the *cube* of the thickness, otherwise his expression will not hold when we suppress one of the coordinates and reduce the elastic plate to an elastic lamina. Poisson seems not to have condescended to notice this remark; it is certain that Navier is wrong here: compare the equation for the transversal vibration of a rod or lamina given in Poisson's *Mécanique*, Vol. II., page 371, and it will be found that the coefficient does involve the square of the thickness, and it is obtained in a manner to which Navier could not have objected. This is quite consistent with the fact that a certain equation of *equilibrium* presents the *cube* of the thickness: see the *Mécanique*, Vol. I., page 606

Navier distinguishes on his page 305 between membranes and elastic surfaces thus:

Les recherches qui ont été faites jusqu'a ces derniers temps sur les lois de l'équilibre ou des mouvemens de vibration des corps, s'appliquent principalement, d'une part, aux cordes et aux membranes ou tissus, supposés parfaitement flexibles, mais susceptibles de résister à l'extension et à la contraction; et, d'autre part, aux plans et surfaces courbes, élastiques, auxquels, outre la même résistance à l'extension et a la contraction, on attribue encore la faculté de résister à la flexion.

529. In II. Poisson states that he had cited Navier's formulae in the place of his memoir which seemed most convenient, and had shewn the passage to Navier in his manuscript: the passage occurs on pages 403 and 404 of the memoir. Considering the

[1] [Saint-Venant has a short paragraph on this polemic in his *Historique Abrégé* (p. clxv) in which he sums up Poisson's attack from the molecular side with the words: 'Tous ces reproches étaient ou sans fondement ou exagérés I cannot quite agree with this. Navier made a distinct mistake, and was only saved from its consequences, because he did not evaluate his integral.] Ed.

habit of the French writers to be extremely sparing in references, it does not appear to me that Poisson could have been expected to do more; but it is obvious that Navier considered this single citation quite insufficient: see page 151 of III. Poisson says that there is an important difference between his own process and that of Navier, for the latter had not considered the natural state of the body, and also by expressing his coefficients in the form of integrals had involved himself in a serious difficulty see Arts. 436 and 443 of the account of the memoir of April 1828.

Poisson alludes to his memoir of 1814, and makes the same admission respecting it as I have quoted in Art. 435 of my account of the memoir of April 1828. He says moreover:

Il en résulte donc qu'en 1814, je n'avais pas trouvé l'équation de la plaque élastique en équilibre; je l'avoue tres volontiers; mais qu'il me soit aussi permis de dire que personne encore ne l'a obtenue par des raisonnemens exacts, et que ce sera dans mon Mémoire sur les Corps élastiques qu'elle se trouvera pour la première fois sans aucune hypothèse et déduite de l'action moléculaire considérée dans toute l'épaisseur de la plaque.

The correct equation of equilibrium to which Poisson here alludes must be that numbered (9) in Art. 484.

Navier in I. spoke of the principle adopted by Mdlle Germain as ingenious and true. Poisson says that this is inadmissible; and he implies that there is not the analogy between this and the hypothesis which James Bernoulli used for the elastic lamina which was apparently claimed for it, since according to Bernoulli it is the *moment* and not the *normal* force which varies inversely as the radius of curvature.

Poisson says that he obtained the equations relative to the boundary of an elastic plate, which had not been given before; at least they did not agree with those of Navier in the *Bulletin de la Société Philomatique*, 1823, page 92. With respect to Navier's equations he says:

Pour s'assurer de l'inexactitude de celles-ci, il suffit de les appliquer à un cas fort simple, au cas d'une plaque circulaire dont tous les points du contour sont soumis à une force constante et normale à la plaque. Il est évident qu'alors la figure d'équilibre sera celle d'une surface de revolution; or, cette figure serait impossible d'après l'équation (7) du Mémoire cité. En effet, cette équation est:

$$Z + k.\left\{\left(\frac{d^3z}{dx^3} + \frac{d^3z}{dy^2dx}\right)\frac{dy}{ds} + \left(\frac{d^3z}{dy^3} + \frac{d^3z}{dx^2dy}\right)\frac{dx}{ds}\right\} = 0,$$

$Z$ désignant la force normale, $k$ un coefficient qui dépend de la matière et de l'épaisseur de la plaque, $x$, $y$, et $z$ les coordonnées d'un point du contour, et $ds$ l'élément de cette courbe. Si l'on appelle $r$ le rayon vecteur du même point, et $\theta$ l'angle qu'il fait avec l'axe de $x$; que l'on place l'origine des coordonnées au centre de la plaque, et qu'on regarde l'ordonnée $z$ comme une fonction de $r$, ce qui exprimera que la figure de la plaque est une surface de révolution, cette équation deviendra:

$$Z + \frac{d}{dr}\left(\frac{d^2z}{dr^2} + \frac{1}{r}\frac{dz}{dr}\right) k\,(\cos^2\theta - \sin^2\theta) = 0\,;$$

résultat impossible, lorsque la force $Z$ est supposée constante, et par conséquent indépendante de l'angle $\theta$.

530. In III. Navier asserts that he had considered the natural state of the elastic body, and that he had obtained seven years before Poisson the equations of equilibrium of such a body. With respect to a claim made by Poisson Navier says:

M. Poisson demande qu'il lui soit permis de dire que personne encore n'a obtenu, par des raisonnemens exacts, l'équation dont il s'agit. Je ne sais si d'autres lui accorderaient cette demande; quant à moi, cela ne m'est pas possible, parce que la démonstration que j'ai indiquée, page 93 du *Bulletin de la Société Philomatique* pour 1823, et qui est contenue dans le Mémoire et dans la Note manuscrite mentionnée dans le même article, est fondée *sur des raisonnements exacts.* Ce travail sera publié dans peu de temps.

Navier contradicts the statement that the hypothesis of Mdlle Germain is inadmissible. He prints the note found among the papers of Lagrange which gives without demonstration the correct form of the equation for the vibration of an elastic lamina; see my account of Sophie Germain, p. 148. As to the objection recorded at the end of Art. 529, Navier says briefly that if the proper values of $dy/ds$ and $dx/ds$ be used, as he has defined these terms, then $\theta$ will disappear, for instead of $\cos^2\theta - \sin^2\theta$ we get $\cos^2\theta + \sin^2\theta$

531. In IV. Poisson repeats his objection against Navier's method of representing the resultants of the mutual actions of

disconnected molecules by definite integrals; he believes he is the first who has called the attention of mathematicians to this point: but Saint-Venant in Moigno's *Statique*, page 694, says that Cauchy did this "en même temps que Poisson."

Poisson also objects that Navier omits some of the forces which are of the same order of magnitude as those which he retains: this I think refers to the point noticed by Saint-Venant in Moigno's *Statique*, pages 696 and 729; namely that instead of putting $f(r_1) = f(r) + (r_1 - r) f'(r)$, he put only $f(r_1) = (r_1 - r) f'(r)$.

Poisson holds that Lagrange could not have been satisfied with the mode of obtaining the equation for the vibration of an elastic lamina to which Navier drew attention, for he did not give it in the second edition of the *Mécanique Analytique.* Poisson says:

Mais je ne veux pas ici reculer devant la difficulté: lors même que l'opinion contraire à la mienne serait actuellement appuyée de l'autorité de Lagrange, ce qu'à la vérité je suis loin de penser, je me croirais toujours fondé à soutenir que dans la lame élastique ordinaire, c'est le moment et non pas la force d'élasticité qui est en raison inverse du rayon de courbure, et que, dans la plaque élastique courbée en différens sens, ni les momens ni les forces ne sont exprimés par la somme des deux rayons de courbure renversés. Je renverrai, sur ce point, à la page 182 de mon Mémoire, où les expressions des momens et de la force normale, c'est-à-dire, de la force qui s'oppose à la flexion, sont données pour le cas de la plaque courbée en tout sens, et au beau Mémoire d'Euler qui fait partie du tome xv des *Novi Commentarii* pour le cas de la lame ordinaire.

The page 182 of Poisson's memoir corresponds to page 538 of the volume of which it forms part.

Poisson gives the calculation by which he found that a certain equation involved $\theta$, while Navier asserted that it was independent of $\theta$. The equation is

$$Z + k\left(\frac{d\phi}{dx}\frac{dy}{ds} + \frac{d\phi}{dy}\frac{dx}{ds}\right) = 0,$$

where $\phi$ stands for $\frac{d^2z}{dx^2} + \frac{d^2z}{dy^2}$.

Poisson applies this to a circular plate; the force $Z$ is constant,

$z$ is a function of the radius vector $r$, the origin being at the centre. Thus $x = r\cos\theta$, $y = r\sin\theta$.

Since $z$ is a function of $r$ so also is $\phi$; thus $\frac{d\phi}{dx} = \frac{x}{r}\frac{d\phi}{dr}$, and $\frac{d\phi}{dy} = \frac{y}{r}\frac{d\phi}{dr}$; also $ds = rd\theta$. Hence the equation becomes

$$Z + k\frac{d\phi}{dr}\frac{d(xy)}{r^2 d\theta} = 0,$$

that is
$$Z + k\frac{d\phi}{dr}(\cos^2\theta - \sin^2\theta) = 0.$$

Poisson adds,

Maintenant l'auteur dit que, pour faire usage de son équation, il faudra changer quelque chose à la signification naturelle et ordinaire des différentielles qu'elle contient; il me semble que ce serait alors changer l'équation elle-même; mais, sur ce point, je n'ai pas assez bien saisi le sens de ses expressions pour essayer d'y répondre.

Finally Poisson adverts to the difference between himself and Navier as to the equations which must hold round the boundary of an elastic plate; the difference related both to the number and form of these equations: Poisson gave *three* and Navier only *two*. With respect to the *number* of these equations Poisson supports his own opinion by the following remarks:

Appelons $\mu$ une portion de la plaque appartenant à son contour, et d'une grandeur insensible. Si l'on tient compte des forces moléculaires qui attachent $\mu$ au reste de la plaque, et aux autres forces données qui lui sont appliquées, on pourra considérer ensuite $\mu$ comme entièrement libre. Or, en faisant abstraction des mouvemens parallèles à la plaque qui donneraient lieu aux conditions d'équilibre dont il n'est pas maintenant question, il restera trois mouvemens que $\mu$ pourra prendre. En effet, supposons la plaque horizontale et menons à l'endroit de son contour où $\mu$ est situé, une tangente et une normale horizontales; il est évident que $\mu$ pourra s'élever ou s'abaisser verticalement, qu'il pourra tourner autour de la tangente, et qu'enfin il pourra tourner autour de la normale. De plus, ces trois mouvemens étant indépendans entre eux, ils donneront lieu à trois conditions d'équilibre, qui ne pourront être exprimées à moins de trois équations distinctes.

Au lieu d'une plaque aussi mince que l'on voudra, s'il s'agissait d'une surface élastique, absolument sans épaisseur, les équations d'équilibre se réduiraient à deux, parce qu'alors il n'y aurait pas lieu de considérer le mouvement de $\mu$ autour d'une tangente dont il ferait partie. C'est encore une différence essentielle entre la plaque élastique, et la surface qui résiste à la flexion en vertu des répulsions mutuelles de ses différens points : la détermination de la forme que celle-ci doit prendre, quand elle est en outre sollicitée par des forces données, n'est plus qu'un simple problème de curiosité, dont la solution exige, ainsi que je l'ai pratiqué dans mon Mémoire sur les surfaces élastiques, que l'on pousse le développement des forces moléculaires plus loin que dans le cas de la plaque un tant soit peu épaisse; mais je conviens de nouveau que j'avais confondu mal à propos l'une avec l'autre dans cet ancien Mémoire.

532. In V. Navier says that he had given in his memoir the expression for his coefficient $\epsilon$, in terms of the molecular action, thus contradicting the statement made by Poisson to the opposite effect: the expression is $\epsilon = \frac{2\pi}{15}\int_0^\infty \rho^4 f(\rho)\, d\rho$. Navier has now read Poisson's memoir of April, 1828, and he criticises that as well as defends his own method. He objects to Poisson's results as resting on the equation $r^4 f(r) = 0$ when $r = 0$; he says that there are many forms of $f(r)$ for which $r^4 f(r)$ does not vanish with $r$. He attempts to defend his use of $(r_1 - r) f'(r)$ where he omits $f(r)$: see Art. 531. He points out that the first volume of the second edition of the *Mécanique Analytique* was issued in September 1811, and that Lagrange's note which we have quoted in Art. 284 was dated December 1811; the second volume of the second edition of the *Mécanique Analytique* did not appear until long after the death of Lagrange; thus the inference which Poisson had drawn in Art. 531 could not be sustained. We see from page 110 of the *Annales*, Vol. 40, that Poisson had discovered and acknowledged his mistake.

Navier defends the equation relative to the boundary of an elastic plate to which Poisson had objected[1]. He says that

[1] [The expression for the potential energy of a plane elastic plate bent to curvatures $1/R'$, $1/R''$, is per unit area equal to a constant $\times$ $\{1/R'^2 + 1/R''^2 + 2\gamma/(R'R'')\}$, where $\gamma$ is the ratio of lateral contraction to longitudinal extension, and therefore in the French theory is put equal to 1/4. Now Navier obtains (see Art. 260) the erroneous expression: constant $\times$ $\{1/R'^2 + 1/R''^2 + 2/(3R'R'')\}$ for this potential energy,

Poisson in applying this equation to the figure he considered neglected a preliminary operation.

Cette opération consiste à remplacer respectivement les rapports $dx/ds$ et $dy/ds$ par $\cos \alpha$ et $\cos \beta$, en désignant par $\alpha$ et $\beta$ les angles formés par l'élément $ds$ du contour avec les axes des $x$ et des $y$. Quant à la nécessité de cette opération, qui se représente dans tous les résultats analogues obtenus par l'application du calcul des variations aux questions de mécanique, et que M. Poisson paraît avoir perdu de vue, il me suffira de renvoyer le lecteur à la page 205 du tome 1[er] de la *Mécanique Analytique* où on la trouve expliquée et démontrée en détail. L'équation dont il s'agit ne peut donc véritablement donner lieu à aucune objection, et elle s'accorde d'ailleurs avec les résultats que MM. Cauchy et Poisson ont donnés dans ces derniers temps.

I do not understand this; the page 205 seems to correspond to the page 200 of Bertrand's edition, and to relate to the process by which Lagrange condenses into one expression the difference of two integrals. I do not know to what results of Cauchy and Poisson allusion is made by Navier.

With respect to Poisson's *third* equation for the boundary Navier says that Cauchy thought it involved some difficulties: see the *Exercices de mathématiques*, Vol. III., page 346.

533. In VI. we have no new point of importance. The controversy in the *Annales* had been finished by the remarks of Arago, one of the editors of the publication; and this article by Navier in Férussac's *Bulletin* is mainly a repetition of what he had said before, addressed now to a fresh audience. We may, I think, fairly sum up the whole controversy thus: the special points which Poisson noticed have been decided generally in his favour by the subsequent history of the subject; the great merit of Navier in commencing a new method of treatment might well have been more warmly commended by Poisson,

having $2/(3R'R'')$ instead of $1/(2R'R'')$. This term only affects the contour-conditions but the error in it naturally leads to wrong expressions. It must however be noted that Navier's method of treating the problem by the Calculus of Variations leads to only *two* contour-conditions, and Kirchhoff's work on this point ought to be considered in the light of Navier's. It might even be more just to speak of *Navier's* two contour-conditions: see Art. 260.] ED.

while on the other hand Navier might have recognised the improvements which had been effected by Poisson's memoir of April 1828.

534. After Navier's letter a note by the editor is given on pages 107—110; the general drift is unfavourable to Navier, both as regards his defence of his own memoir, and his objections to Poisson's. Navier seems to misunderstand the difference between Cauchy and Poisson; Cauchy does not object to the important conditions which according to Poisson must hold at a *free* part of the boundary, but to the less important case of the conditions at a fixed part: see my remarks on this point in my account of the pages 328—355 of the *Exercices*, Vol. III.

535. 1828. *Note sur la Compression d'une sphère.* This occupies pages 330—335 of the *Annales de Chimie*, Vol. 38, 1828. The note was written in consequence of Poisson being consulted as to some opinions adopted by Oersted founded on experiment. The first sentence enunciates the problem which is discussed:

> Une sphère creuse, homogène et d'une épaisseur constante, est soumise en dehors et en dedans à des pressions données; on demande de déterminer le changement qu'éprouvent son rayon extérieur et son rayon intérieur.

I need not go over the process as it is given with greater generality, by the use of two constants of elasticity instead of one, in Lamé's work on Elasticity, pages 214—219. From the *Annales de Chimie*, Vol. 39, page 213, we may infer that the problem had already been discussed in a memoir as yet unpublished by Lamé and Clapeyron. As Poisson says, the equations which he uses had been given in his memoir of April 1828, but he does not supply any exact references. His equation (1) is obtained by using $rd\phi/dr$ instead of $d\phi/dr$ in the equation (1) of page 406, and suppressing the terms involving $t$; his equation (2) coincides with a statement on page 431 of the memoir; and the proposition with which he finishes the article seems derived from a comparison of pages 402 and 430 of the memoir.

536. An article by Poisson occurs on columns 353 and 354 of Schumacher's *Astronomische Nachrichten*, Vol. 7, 1829; it takes the form of a letter to the editor, and is entitled in the Royal Society Catalogue of Scientific Papers: *Note sur l'équilibre d'un fil élastique.*

A criticism by Professor de Schultén on a passage in Lagrange's *Mécanique Analytique* appeared on columns 185—188 of the same volume of the Journal; the editor sent a copy to Poisson requesting his opinion, and accordingly Poisson replied. The passage in Lagrange is comprised in Articles 48—52 of the fifth section of the part on Statics, pages 145—151 of Bertrand's edition; it relates to the equilibrium of an elastic wire.

Suppose the wire to be inextensible; there are three equations of equilibrium given on page 145; the first will serve as a type:

$$Xdm - d\,.\,\frac{\lambda dx}{ds} + d^2\,.\,(I\,d^2x) = 0 \;\ldots\ldots\ldots\ldots\ldots(1).$$

Suppose the wire to be extensible; there are three equations of equilibrium as we see from page 151; the first will serve as a type:

$$Xdm - d\,.\left\{\left(F + d\,.\frac{E\,d^2s}{e\,ds^2} - \frac{Ee}{ds}\right)\frac{dx}{ds}\right\} + d^2\,.(Id^2x) = 0 \;\ldots(2).$$

According to Lagrange $\lambda$ denotes the tension in the first case and $F$ denotes the tension in the second case.

Schultén quotes the formulae (2) with $\lambda$ instead of $F$; and he says that the formulae (1) are wrong, and that (2) are the proper formulae for both cases; and he offers some general reasons to shew that the expression for the tension must be the *same* in both cases, but I cannot say that I understand this. Poisson treats the matter very briefly. He says that Schultén should retain $F$ in the formulae (2), for it is not the same thing as $\lambda$ of the formulae (1); and he says that the tension ought to be the same in the two cases, but that the true tension is not expressed either by $\lambda$ in the first case, or by $F$ in the second. Poisson uses $\int pds$ for the sum of the tangential forces; he puts $a$ for the $K$ of Lagrange, and $ds/d\phi$ for the $\rho$ of Lagrange. I shall assume that the reader has the *Mécanique Analytique* before him in the remarks I make.

Poisson says that from (1) we get $\lambda = \int pds - \frac{K}{\rho^2}$, and from (2) we get $F = \int pds$; these results he says are easily obtained by making $s$ the independent variable. He holds that the true value of the tension is $\int pds - \frac{K}{2\rho^2}$, for which he cites Euler, *Novi Commentarii*, Vol. xv. page 390.

It seems to me however that from (1) we get $\lambda = \int pds - \frac{3K}{2\rho^2}$, and from (2) we get $F - \frac{Ee}{ds} = \int pds - \frac{3K}{2\rho^2}$, so that finally $F = \int pds - \frac{K}{2\rho^2}$. Hence $F$ really gives what Poisson holds to be the true tension.

Schultén obtains from (2) another form of $F$, namely this:

$$F = \frac{dx}{ds}\int Xdm + \frac{dy}{ds}\int Ydm + \frac{dz}{ds}\int Zdm,$$

where arbitrary constants may be considered to occur in all the three integrals; but this does not invalidate the form already given.

537. On the whole it seems to me that Schultén has not shewn that there is any real ground of objection to Lagrange's result, and that Poisson contributes nothing to the question. As Bertrand does not allude to the matter in his edition of the *Mécanique Analytique* I presume that he sees no error here in Lagrange's process.

538. Poisson however proceeds to some general remarks on Lagrange's method which seem to me quite just. He says that Lagrange's manner of applying the principle of virtual velocities to forces, the effect of which is to vary a differential expression, has always appeared to him unsatisfactory: we have no clear idea of the meaning of the undetermined coefficients. Poisson says also that Lagrange had taken the elastic force as represented by the inverse radius of curvature, whereas James Bernoulli had so represented the moment of the force and not the force itself.

539. The subject is resumed in Vol. 8 of the *Astronomische Nachrichten* in a number which appeared in December 1829; the complete volume is dated 1831. On the columns 21—24 there is a note by Schultén entitled: *Note sur la tension des fils élastiques.* He shews that in the case of an elastic thread in one plane, which is the case discussed by Euler, his formula for the tension agrees with Euler's; but he does not say distinctly, as I have done, where Poisson's statements are wrong. Schultén concludes that, as he had originally maintained, Lagrange is in error. A letter from Poisson to the editor follows. Poisson had seen the second article by Schultén, admits its accuracy, but says nothing about the mistakes into which he had himself fallen. Apparently he now agrees with Schultén in attributing to Lagrange an error; but as I have said I do not concur in this.

540. 1829. *Mémoire sur les Équations générales de l'équilibre et du Mouvement des Corps solides élastiques et des Fluides.* This memoir was read to the Paris Academy on the 12th of October, 1829; it is published in the *Journal de l'École Polytechnique,* 20th Cahier, 1831, where it occupies pages 1—174.

541. The first paragraph of the memoir indicates briefly the nature of its contents:

Dans les deux Mémoires que j'ai lus à l'Académie, l'un sur l'équilibre et le mouvement des corps élastiques, l'autre sur l'équilibre des fluides, j'ai supposé ces corps formés de molécules disjointes, séparées les unes des autres par des espaces vides de matière pondérable, ainsi que cela a effectivement lieu dans la nature. Jusque-là, dans ce genre de questions, on s'était contenté de considérer les mobiles comme des masses continues, que l'on décomposait en élémens différentiels, et dont on exprimait les attractions et les répulsions par des intégrales définies. Mais ce n'était qu'une approximation, à laquelle il n'est plus permis de s'arrêter lorsqu'on veut appliquer l'analyse mathématique aux phénomènes qui dépendent de la constitution des corps, et fonder sur la réalité les lois de leur équilibre et de leur mouvement. En même temps, on doit s'attacher à simplifier cette analyse autant qu'il est possible, en conservant au calcul toute la rigueur dont il est susceptible dans ses diverses applications. C'est ce motif qui m'a engagé à reprendre en entier les questions que j'avais déjà traitées dans les mémoires précédens. Mes nouvelles recherches ne m'ont conduit à aucun changement dans les résultats; mais je n'ai rien négligé pour rendre plus simples, et quelquefois plus exactes, les considérations sur lesquelles je m'étais appuyé, surtout dans la partie relative à l'équilibre des fluides.

542. Poisson thus claims as the distinctive character of his own investigations that instead of the definite integrals of the earlier writers he used finite summations; however, as Saint-Venant remarks in Moigno's *Statique*, page 695, definite integrals occur in Poisson's memoir of April, 1828, on pages 378—381. With respect to Poisson's views on this matter Saint-Venant refers to pages 366 and 369 of the memoir of April, 1828, to pages 31 and 378 (rather 278) of Poisson's *Nouvelle théorie de l'action capillaire,* and to Poisson's controversy with Navier in Vols. 36, 37, 38, 39 of the *Annales de Chimie et de Physique.* Saint-Venant proceeds thus:

Cauchy exprime constamment (comme Poisson l'a fait ensuite complétement aussi) ses résultantes de forces, non par des intégrales, mais par des sommes $S$ ou $\Sigma$ d'un nombre fini quoique très-grand d'actions individuelles; et, cela, sans se servir, comme Poisson, de considérations peu rigoureuses relatives à la grandeur moyenne de l'espacement des molécules, et sans avoir besoin de supposer avec lui que "le rayon d'activité comprend un nombre immense de fois l'intervalle moléculaire," de sorte "que les actions entre les molécules les plus voisines puissent être négligées devant les actions moindres mais plus nombreuses qui s'exercent entre les autres," ce qui, comme le remarque Cauchy, conduirait aux mêmes conséquences fausses que la substitution d'un nombre infini de particules contiguës aux molécules isolées et espacées.

Saint-Venant adds references to various parts of this sentence: after *complétement aussi* to pages 41—46 of the memoir of October 1829; after *des molécules* to pages 32 and 42 of the same memoir; after *les autres* to pages 370 and 378 of the memoir of 1828, and to pages 7, 8, 13, 25, and 26 of the memoir of 1829; and at the end of the sentence to various pages of Cauchy's *Exercices, troisième année.* See also a note on pages 261, 262 of the memoir on *Torsion* by Saint-Venant[1].

[1] Or more recently in the *Historique Abrégé*, pp. clxi.—clxv.

543. The first section of Poisson's memoir is entitled: *Notions préliminaires* and occupies pages 4—8. Here some mention is made of *caloric* as supplying a *repulsive* force in addition to the *attractive* force which may be supposed to arise from the action of particles of matter on each other. Thus on the whole what may be called the molecular force between two particles at an assigned distance apart may be positive or negative; and this consideration is kept constantly in view, and constitutes one of the main differences between the present memoir and Poisson's previous writings on the subject of elasticity.

544. The second section is entitled: *Calcul des Actions moléculaires et Équations d'équilibre relativement à des Molécules rangées en ligne droite;* it occupies pages 9—28. The main result is an equation on page 20, namely $\rho X = dp/dx$; here $p$ is what would be the *pressure* if the investigation were relative to a slender column of fluid, and what would be the elastic *tension* if the investigation were relative to a straight solid rod. $X$ denotes the *applied* force along the column or rod. The special part of Poisson's process is that he finds an expression for $p$, which denotes the molecular force, involving undetermined integrals, namely

$$p = \frac{k}{\epsilon^2} - a_1 f(0) + 3a_2\epsilon^2 f''(0) - 5a_3\epsilon^4 f''''(0) + \dots.$$

Here $f(s)$ denotes the molecular force between two particles, each of the unit of mass, at the distance $s$, and $k$ is put for $\int_0^\infty s f(s)\,ds$; also $\epsilon$ denotes the mean distance between two adjacent particles; $a_1, a_2, a_3, \dots$ are numerical constants given by the general formula

$$a_n = \frac{2}{(2\pi)^{2n}} \left\{\frac{1}{1^{2n}} + \frac{1}{2^{2n}} + \frac{1}{3^{2n}} + \dots\right\},$$

so that $$a_1 = \tfrac{1}{12},\quad a_2 = \tfrac{1}{720},\quad a_3 = \tfrac{1}{30240}.$$

This value of $p$ is obtained by the aid of various steps of general reasoning, which are not very convincing I think. The principal mathematical theorem used is that which is called *Euler's Theorem,* made accurate by an expression for the remainder due to Poisson himself, for which he refers to a memoir on Definite Integrals in Vol. VI. of the Paris Memoirs.

I do not know that any application has ever been made of the value of $p$ which Poisson here obtains.

545. The third section is entitled: *Calcul des Pressions moléculaires dans les Corps élastiques; Équations différentielles de l'équilibre et du mouvement de ces Corps;* it occupies pages 28—68. I notice some points of interest which present themselves.

546. On page 29 a definition is given of the term *pressure* (stress) as used in the theory of elasticity. This corresponds with that adopted by Lamé and Cauchy. The action exerted on a certain cylinder of a body by the matters below the plane of the base is estimated. This is perhaps the first introduction of this cylinder.

[547.] Poisson in his memoir expressly considers the solid body first in the state in which there are no applied forces internal or external. He finds that there is no stress within the body, and that round any point $\Sigma rf(r) = 0$: see his page 34. Here $f$ has the same meaning as in Art. 443; and $r$ is the distance of a second particle from the particle considered as the origin: the summation is to extend over all the particles round the origin. On page 37 of the memoir Poisson uses a formula equivalent to the following:

$$s_r = \frac{du}{dx}\cos^2\alpha + \frac{dv}{dy}\cos^2\beta + \frac{dw}{dz}\cos^2\gamma + \left(\frac{dw}{dy} + \frac{dv}{dz}\right)\cos\beta\cos\gamma$$
$$+ \left(\frac{du}{dz} + \frac{dw}{dx}\right)\cos\gamma\cos\alpha + \left(\frac{dv}{dx} + \frac{du}{dy}\right)\cos\alpha\cos\beta,$$

where $s_r$ is the stretch in direction $r$ determined by the angles $(\alpha\beta\gamma)$.

The formula had been already given by Navier, but did not occur in Poisson's memoir of April 1828: see Saint-Venant's memoir on Torsion, page 243.

[548.] On p. 45 a result is given of the following kind; the six stress components are expressed in terms of the strain by equations of the form

$$\widehat{xx} = p_0\left(1 + \frac{du}{dx} - \frac{dv}{dy} - \frac{dw}{dz}\right) + G\frac{du}{dx} + F\left(\frac{dv}{dy} + \frac{dw}{dz}\right),$$

$$\widehat{yz} = (p_0 + F)\left(\frac{dv}{dz} + \frac{dw}{dy}\right).$$

Poisson shews by a somewhat lengthy process and with a different notation that,

$$G - F + 2p_0 = 2(p_0 + F),$$

or

$$G = 3F.$$

This relation appears with our notation[1], namely,

$$\widehat{xx} = \lambda\theta + 2\mu\frac{du}{dx}, \qquad \widehat{yz} = \nu\left(\frac{dv}{dz} + \frac{dw}{dy}\right),$$

under the form $\nu = \mu$.

Some remarks by Saint-Venant on Poisson's procedure at this point will be found in Moigno's *Statique*, p. 684. Poisson in fact here improves upon the sixth section of the Memoir of 1828.

549. As a simple example, Poisson considers the case in which the only applied force is a constant normal pressure at the surface of the body. He shews that the equations are satisfied then by taking

$$\frac{du}{dx} = \frac{dv}{dy} = \frac{dw}{dz} = -s, \text{ a constant},$$

and by supposing the other first fluxions of the shifts to be zero. In this example the stretch will be uniform and equal to $-s$.

550. A simple proposition which is given on pages 61—63 may be noticed. If the stress on a plane at a point is always normal to the plane and of the same value, then the stretch is the same all round the point; and conversely.

For taking the equations for stress across a plane surface $(\alpha, \beta, \gamma)$, and applying them at any point in the interior of a body, we are by supposition to have relations of the form,

$$Q\cos\alpha = \widehat{xx}\cos\alpha + \widehat{xy}\cos\beta + \widehat{xz}\cos\gamma,$$

where $Q$ is a constant. These are to hold for all values of $\alpha, \beta, \gamma$, and so they lead to

$$\widehat{xx} = \widehat{yy} = \widehat{zz} = Q,$$

and

$$\widehat{yz} = 0, \quad \widehat{zx} = 0, \quad \widehat{xy} = 0:$$

[1] [See the note at the end of this volume for the terminology, and the footnote p. 321 for the notation. Ed.]

then by Art. 548 we get from these

$$\frac{du}{dx} = \frac{dv}{dy} = \frac{dw}{dz} = s \text{ say}, \quad \ldots\ldots\ldots\ldots\ldots(1),$$

$$\frac{dv}{dz} + \frac{dw}{dy} = 0, \ \frac{dw}{dx} + \frac{du}{dz} = 0, \ \frac{du}{dy} + \frac{dv}{dx} = 0 \quad \ldots\ldots\ldots(2).$$

The equations (1) and (2) lead to the required result. The converse may be established in like manner.

From equations (1) and (2) we can infer that $s$ is a *linear* function of $x$, $y$, $z$; for we can eliminate $u$, $v$, $w$ by suitable differentiations, and thus obtain

$$\frac{d^2s}{dx^2} = \frac{d^2s}{dy^2} = \frac{d^2s}{dz^2} = \frac{d^2s}{dzdy} = \frac{d^2s}{dxdz} = \frac{d^2s}{dydx} = 0.$$

These lead to

$$s = \alpha + \beta x + \beta' y + \beta'' z,$$

where $\alpha$, $\beta$, $\beta'$, $\beta''$ are constants.

551. Poisson transforms his equations by the aid of a process which I think we ought to ascribe to Lagrange. This he had given before: see Art. 445 of my account of the memoir of April, 1828. At the close of the section Poisson arrives at the same equations as he had previously obtained in the memoir of April 1828. There is an important mistake on page 68: see Stokes's memoir, *Camb. Phil. Trans.* Vol. VIII. Part III. p. 31; or, *Math. and Phys. Papers,* Vol. I. page 125.

552. The fourth section is entitled: *Calcul des pressions moléculaires dans les corps cristallisés; réflexions générales sur ces pressions dans les fluides et dans les solides,* and occupies pages 69—90. This is not so much a general theory as a particular example. Poisson himself says on page 70:

> Mais, pour donner un exemple du calcul des pressions moléculaires dans l'intérieur des cristaux, nous allons faire une hypothèse particulière sur l'action mutuelle et la distribution des molécules, qui sera très-propre à éclaircir la question et à en montrer les difficultés.

The investigation resembles that in the second section; expressions of a complicated character occur involving an unknown function under the form of a definite integral. I am not aware that any application has ever been made of the result.

553. We may notice a remark on pages 82 and 83. In the ordinary mode of treating the subject of elasticity it is laid down that each stress-component is a linear function of *six* quantities, namely the *three* stretches and the *three* slidings. Poisson takes the following view: each stress may be assumed to be a linear function of the nine first fluxions of the shifts, so we may take as the expression for a stress

$$A\frac{du}{dx}+B\frac{dv}{dy}+C\frac{dw}{dz}+D\frac{du}{dy}+E\frac{du}{dz}+F\frac{dv}{dz}$$
$$+D'\frac{dv}{dx}+E'\frac{dw}{dx}+F'\frac{dw}{dy}$$

Then these *nine* coefficients may be immediately reduced to *six*. For suppose the body as a whole to be turned through a small angle $\alpha$ round the axis of $z$; thus we shall have

$$u=-\alpha y,\ v=\alpha x,\ w=0;$$

therefore $\frac{du}{dy}=-\alpha$, $\frac{dv}{dx}=\alpha$, and the other seven first fluxions vanish. Thus the expression for the stress reduces to $(D'-D)\,\alpha$; but no *relative* displacement has taken place, and therefore no stress is exerted: therefore $D'-D=0$, so that $D'=D$. In the same manner we obtain $E'=E$, and $F'=F$ Thus the proposed reduction is effected.

Poisson then, by appealing to Cauchy's Theorem (Art. 606 *infra*), shews that the six shear components of stress reduce to *three;* thus on the whole there are six different stresses which will involve 36 coefficients.

554. Saint-Venant, in Moigno's *Statique*, page 627, makes the following remark with respect to Cauchy's theorem just mentioned:

...déjà trouvé et appliqué par lui, [Cauchy] aussi dès 1822, et dont Poisson a reconnu, en 1829 (12 octobre, Mémoire inséré au xx[e] Cahier du *Journal de l'École Polytechnique*, Art. 38, p. 83), la grande généralité d'abord méconnue (t. VIII. des *Mémoires de l'Institut*).

Thus Saint-Venant holds that Poisson did not fully appreciate Cauchy's theorem; the page of the memoir of April, 1828, which Saint-Venant has in view is perhaps 385, where Poisson seems to say that the theorem, which we call Cauchy's, holds if a certain quantity $K$ vanishes.

555. With respect to the number of independent coefficients, Saint-Venant remarks on page 261 of his memoir on *Torsion:*

On peut remarquer que M. Poisson, après avoir, le premier, présenté les formules avec les 36 coefficients indépendants (*Journal de l'École Polytechnique* 20e cahier, p. 83), en a réduit le nombre à 15, même pour les corps cristallisés, dans son dernier mémoire relatif à ces sortes de corps (*Mémoires (nouv.) de l'Institut,* t. XVIII., Art. 36, 37).

[It is however to be noted that Poisson did not make this reduction till ten years later; the date of the last memoir being 1839.]

556. Poisson finishes his fourth section by some reflections with regard to solid bodies and fluid bodies. They do not seem to me very important. Among other things he is led to conclude that in uncrystallised solids, in liquids, and in gases the pressure $p$ and the density $\rho$ are connected, at least approximately, by the law

$$p = a\rho^2 + b\rho^{\frac{2}{3}};$$

$a$ and $b$ either are constants, or vary, when the temperature varies, according to some law which is to us unknown.

557. The pages 90—174 of the memoir relate to the equilibrium and motion of fluids, and are not sufficiently connected with our subject to require notice here. Saint-Venant, in Moigno's *Statique,* refers twice to this part of the memoir: see his pages 619 and 694. The part of the memoir which we have examined contains numerous misprints, so that a reader must be on his guard. Important criticisms on the memoir of Poisson by Professor Stokes will be found in the *Camb. Phil. Trans.* Vol. VIII. p. 287, or *Math. and Phys. Papers,* Vol. I. p. 116.

558. An account of this memoir of Poisson's is given in Férussac's *Bulletin des Sciences Mathématiques,* Vol. XIII. 1830, pages 394—412. It offers nothing of importance. Another account of the memoir by Poisson himself is given in the *Annales*

*de Chimie et de Physique*, Vol. XLII. 1829, pages 145—171. Here in a note on pages 160 and 161 we have a mathematical investigation of which in the memoir Poisson had given only the result. It is connected with the law stated in Art. 556; Poisson combines this, he says, with the laws of Mariotte and Gay-Lussac which are established by observation, and obtains the result $\gamma = 2 - \frac{4b}{3k^{\frac{2}{3}}(1+\omega\theta)^{\frac{2}{3}} p^{\frac{1}{3}}}$, where $\gamma$ is the ratio of specific heat under constant pressure to specific heat under constant volume, $\theta$ is the temperature, $\omega$ is the coefficient of dilatation of gases, $k$ the ratio of the pressure to the density when $\theta = 0$.

For by the laws of Mariotte and Guy-Lussac we have

$$p = k\rho\,(1+\omega\theta) \quad\text{..........................(1).}$$

Let $q$ be the quantity of heat contained in a gramme of the gas, and consider $q$ as an unknown function of $\rho$ and $p$. Let $i$ be the increment of the temperature, either when $q$ becomes $q+c$ the pressure $p$ not changing, or when $q$ becomes $q+c'$ the density $\rho$ not changing. Suppose $i$ very small, then we shall have $c = \frac{dq}{d\rho}\frac{d\rho}{d\theta}\,i$, and $c' = \frac{dq}{dp}\frac{dp}{d\theta}\,i$; and by reason of (1) we have $\frac{d\rho}{d\theta} = -\frac{\rho\omega}{1+\omega\theta}$, and $\frac{dp}{d\theta} = \frac{p\omega}{1+\omega\theta}$; so that, as $c = c'\gamma$, we get

$$\rho\frac{dq}{d\rho} + p\gamma\frac{dq}{dp} = 0 \quad\text{......................(2).}$$

Now suppose that $p$ and $\rho$ become $p+p'$ and $\rho+\rho'$ respectively, without any change in $q$ the quantity of heat; $p'$ and $\rho'$ being infinitesimal we shall have $\rho'\frac{dq}{d\rho} + p'\frac{dq}{dp} = 0$; and by differentiating $p = a\rho^2 + b\rho^{\frac{5}{3}}$, which holds in this case, we get $p' = 2\left(a\rho + \frac{5}{6}b\rho^{\frac{2}{3}}\right)\rho' = 2\left(p - \frac{1}{6}b\rho^{\frac{5}{3}}\right)\rho'/\rho$. Thus

$$\rho\frac{dq}{d\rho} + 2\left(p - \tfrac{1}{6}b\rho^{\frac{5}{3}}\right)\frac{dq}{dp} = 0.$$

Hence by (2) we get

$$\gamma = 2\left(1 - \frac{2b\rho^{\frac{5}{3}}}{3p}\right) = 2 - \frac{4b}{3k^{\frac{5}{3}}(1+\omega\theta)^{\frac{5}{3}}\,p^{\frac{1}{3}}}.$$

559. In the *Mémoires de l'Académie...de France*, Vol. x., we have a memoir by Poisson entitled: *Mémoire sur le mouvement de deux fluides élastiques superposés;* it occupies pages 317—404. A note at the foot of page 317 says:

Ce Mémoire est une partie de celui que j'ai lu à l'Académie le 24 mars 1823, sous le titre de *Mémoire sur la propagation du mouvement dans les fluides élastiques.*

This memoir does not concern us, but I quote a few words from a note to pages 387 and 388 as they allude to the history of our subject:

...équations d'où dépendent les petits mouvements des corps élastiques, qui sont connues depuis la lecture de ce Mémoire...

560. In the same volume of the Paris Memoirs we have another memoir by Poisson entitled: *Mémoire sur la propagation du mouvement dans les milieux élastiques.* This memoir was read to the Academy on the 11th of October, 1830: it occupies pages 549—605 of the volume. After a short introduction the memoir consists of two parts.

561. The first part of the memoir is entitled: *Propagation du mouvement dans un fluide,* and occupies pages 550—577.

Poisson starts with the ordinary equations of fluid motion:

$$X - \frac{du}{dt} - u\frac{du}{dx} - v\frac{du}{dy} - w\frac{du}{dz} = \frac{1}{\rho}\frac{dp}{dx},$$

$$Y - \frac{dv}{dt} - u\frac{dv}{dx} - v\frac{dv}{dy} - w\frac{dv}{dz} = \frac{1}{\rho}\frac{dp}{dy},$$

$$Z - \frac{dw}{dt} - u\frac{dw}{dx} - v\frac{dw}{dy} - w\frac{dw}{dz} = \frac{1}{\rho}\frac{dp}{dz},$$

$$\frac{d\rho}{dt} + \frac{d\rho u}{dx} + \frac{d\rho v}{dy} + \frac{d\rho w}{dz} = 0.$$

Let $D$ denote the natural density of the fluid, $gh$ the measure of the elastic force there, $g$ being gravity and $h$ the height of a given liquid of which the density is taken to be unity; so that in the state of equilibrium we have

$$\rho = D, \; p = gh.$$

During the motion we shall have

$$\rho = D(1-s), \quad p = gh(1-\gamma s),$$

where $s$ denotes the dilatation of the fluid, and $\gamma$ is a constant greater than unity which represents the ratio of the specific heat under constant pressure to the specific heat under constant volume. Put $a^2$ for $\frac{gh\gamma}{D}$, and neglect quantities of the second order compared with $s, u, v, w$; then supposing that there are no applied forces the equations of motion become approximately

$$\frac{du}{dt} = a^2 \frac{ds}{dx}, \quad \frac{dv}{dt} = a^2 \frac{ds}{dy}, \quad \frac{dw}{dt} = a^2 \frac{ds}{dz},$$

$$\frac{ds}{dt} = \frac{du}{dx} + \frac{dv}{dy} + \frac{dw}{dz}.$$

562. The equations thus obtained are integrated exactly by Poisson; he assumes that the fluid extends to infinity in all directions, so that there are no boundary conditions to be regarded. The process of integration is a fine piece of analysis depending mainly on two important formulae. One of these [(10) of the memoir,] is Poisson's own integral of a certain partial differential equation: see Art. 523 of my account of the memoir of April 1828. The other formula [p. 555 of the memoir] may be expressed thus:

$$\phi(x, y, z) = \frac{1}{8\pi^3} \iiiiii \phi(x', y', z')\, U\, d\alpha\, d\beta\, d\gamma\, dx'\, dy'\, dz',$$

where for $U$ we may put

either $\qquad \cos\{\alpha(x-x') + \beta(y-y') + \gamma(z-z')\}$,

or $\qquad \cos\alpha(x-x')\cos\beta(y-y')\cos\gamma(z-z')$;

the limits of all the six integrals being $\pm\infty$.

Respecting a formula precisely of this kind, with four integrals instead of six, Poisson remarks in a note on page 322 of the volume we are noticing:

M. Fourier a donné le premier cet important théorème pour des fonctions d'une seule variable, qui sont égales et de même signe, ou égales et de signe contraire, quand on y change le signe de la variable. Il était facile de l'étendre à des fonctions quelconques, de deux ou d'un plus grand nombre de variables. On en peut voir la démonstration dans mes précédents Mémoires.

[Poisson's fundamental integral is that marked ($d$) in the footnote on this page[1]. It has been obtained more concisely by M. Liouville.]

[1] M. Liouville's method is contained in the following:

Note sur l'intégration d'une équation aux différentielles partielles qui se présente dans la théorie du son. *Comptes Rendus*, VII. 1838, pages 247, 248.

Dans les *Nouveaux Mémoires de l'Académie des Sciences* (année 1818), M. Poisson a donné l'intégrale de l'équation

$$\frac{d^2\lambda}{dt^2} = a^2\left(\frac{d^2\lambda}{dx^2}+\frac{d^2\lambda}{dy^2}+\frac{d^2\lambda}{dz^2}\right) \quad\ldots\ldots\ldots\ldots\ldots\ldots\ldots\ldots (a).$$

En désignant par $F(x, y, z)$, $a^2\psi(x, y, z)$ les valeurs de $\lambda$ et $\frac{d\lambda}{dt}$ pour $t=0$, il a trouvé

$$\left.\begin{aligned}\lambda=\frac{a^2}{4\pi}\int_0^\pi\int_0^{2\pi}\psi(x+at\cos\theta,\ y+at\sin\theta\sin\omega,\ z+at\sin\theta\cos\omega)\,t\sin\theta\,d\theta d\omega\\ +\frac{1}{4\pi}\frac{d}{dt}\int_0^\pi\int_0^{2\pi}F(x+at\cos\theta,\ y+at\sin\theta\sin\omega,\ z+at\sin\theta\cos\omega)\,t\sin\theta\,d\theta d\omega.\end{aligned}\right\}\ldots\ldots(b)$$

Les deux méthodes qui le conduisent à ce résultat sont assez simples, surtout la seconde, d'ailleurs; il montre que l'on peut aisément en vérifier *à posteriori* l'exactitude.

Mais, dans un autre Mémoire *sur la propagation du mouvement dans les milieux élastiques* (*Nouveaux Mémoires de l'Académie des Sciences*, tome X.), l'illustre géomètre considère, au lieu de l'équation ($a$), l'équation suivante :

$$\frac{d^2\phi}{dt^2} = a^2\left[\frac{d^2\phi}{dx^2}+\frac{d^2\phi}{dy^2}+\frac{d^2\phi}{dz^2}+\psi(x, y, z)\right] \quad\ldots\ldots\ldots\ldots\ldots\ldots (c),$$

à laquelle on doit joindre les conditions définies que voici :

$$\phi=0,\ \frac{d\phi}{dt}=F(x, y, x) \text{ pour } t=0,$$

$\psi(x, y, z)$, $F(x, y, z)$ étant deux fonctions connues de $x$, $y$, $z$. Et le procédé qu'il emploie pour ramener l'intégration de l'équation ($c$) à celle de l'équation ($a$), ou plutôt pour simplifier l'intégrale de l'équation ($c$), exige d'assez longs calculs. On peut éviter ces calculs en adoptant la marche que je vais indiquer.

Je différencie l'équation ($c$) par rapport à $t$, et je pose $\frac{d\phi}{dt}=\lambda$; je trouve ainsi que $\lambda$ doit satisfaire précisement à l'équation ($a$); de plus pour $t=0$, il vient

$$\lambda=\frac{d\phi}{dt}=F(x, y, z),$$

puis
$$\frac{d\lambda}{dt}=\frac{d^2\phi}{dt^2}=a^2\left[\frac{d^2\phi}{dx^2}+\frac{d^2\phi}{dy^2}+\frac{d^2\phi}{dz^2}+\psi(x, y, z)\right],$$

ou simplement
$$\frac{d\lambda}{dt}=a^2\psi(x, y, z),$$

563. Having thus integrated the equation of motion Poisson proceeds to interpret the formulae. This is an approximate investigation of a rough kind; the following is the main result: suppose that originally a certain finite portion of the fluid is disturbed, then the disturbance spreads in every direction about this portion, and at an extremely great distance the wave will be very approximately spherical, and the motion of a particle at any point will be at right angles to the tangent plane to the wave at that point.

564. The second part of the memoir is entitled: *Propagation du mouvement dans un corps solide élastique*: and occupies pages 578—605. The equations to be integrated are (1) of Art. 523. The following is the beginning of the process.

Put for brevity

$$\delta = \alpha (x - x') + \beta (y - y') + \gamma (z - z'),$$

then the equations will be satisfied if we take

$$u = \left( A \cos \rho\lambda a t + A' \frac{\sin \rho\lambda a t}{\rho\lambda a} \right) \cos \rho\delta,$$

$$v = \left( B \cos \rho\lambda a t + B' \frac{\sin \rho\lambda a t}{\rho\lambda a} \right) \cos \rho\delta,$$

$$w = \left( C \cos \rho\lambda a t + C' \frac{\sin \rho\lambda a t}{\rho\lambda a} \right) \cos \rho\delta;$$

$A, B, C, A', B', C', \alpha, \beta, \gamma, \rho, x', y', z'$ being constants, the last four of which are perfectly arbitrary, while the nine others are connected by the equations

puisque $\phi$ s'évanouit en même temps que $t$. La valeur de $\lambda$ ou $\frac{d\phi}{dt}$ est donc celle écrite ci-dessus et fournie par la formule ($b$); pour en déduire $\phi$ il suffit d'intégrer à partir de $t=0$, ce qui donne

$$\left.\begin{aligned} \phi &= \frac{1}{4\pi}\int_0^{\pi}\int_0^{2\pi} F(x + at\sin\theta,\ y + at\sin\theta\sin\omega,\ z + at\sin\theta\cos\omega)\, t\sin\theta\, d\theta d\omega \\ &+ \frac{1}{4\pi}\int_0^{at}\int_0^{\pi}\int_0^{2\pi} \psi(x + \rho\cos\theta,\ y + \rho\sin\theta\sin\omega,\ z + \rho\sin\theta\cos\omega)\, \rho\sin\theta\, d\rho\, d\theta\, d\omega \end{aligned}\right\} \dots (d).$$

C'est la formule de M. Poisson, telle qu'on la lit au No. 5 (p. 561) de son Mémoire.

$$3A\lambda^2 = A\,(3\alpha^2 + \beta^2 + \gamma^2) + 2B\alpha\beta + 2C\alpha\gamma,$$
$$3B\lambda^2 = B\,(3\beta^2 + \alpha^2 + \gamma^2) + 2A\beta\alpha + 2C\beta\gamma,$$
$$3C\lambda^2 = C\,(3\gamma^2 + \alpha^2 + \beta^2) + 2A\gamma\alpha + 2B\gamma\beta,$$

and the three which may be deduced from these by changing $A$, $B$, $C$ into $A'$, $B'$, $C'$ respectively.

Put $D$ for $A\alpha + B\beta + C\gamma$ and $D'$ for $A'\alpha + B'\beta + C'\gamma$; then as we may without loss of generality suppose that

$$\alpha^2 + \beta^2 + \gamma^2 = 1,$$

our equations become

$$A\,(3\lambda^2 - 1) = 2\alpha D, \quad A'\,(3\lambda^2 - 1) = 2\alpha D',$$
$$B(3\lambda^2 - 1) = 2\beta D, \quad B'\,(3\lambda^2 - 1) = 2\beta D',$$
$$C\,(3\lambda^2 - 1) = 2\gamma D, \quad C'\,(3\lambda^2 - 1) = 2\gamma D'.$$

These equations may be satisfied in two ways; we may take

$$\lambda = \pm \frac{1}{\sqrt{3}}, \qquad A = -\frac{B\beta}{\alpha} - \frac{C\gamma}{\alpha}, \qquad A' = -\frac{B'\beta}{\alpha} - \frac{C'\gamma}{\alpha},$$

or we may take

$$\lambda = \pm 1, \quad B = \frac{A\beta}{\alpha}, \qquad C = \frac{A\gamma}{\alpha}, \qquad B' = \frac{A'\beta}{\alpha}, \qquad C' = \frac{A'\gamma}{\alpha}.$$

Hence we obtain two different solutions of the original equations (1), and as these equations are linear the aggregate of the two solutions will constitute a solution. Thus we take for a solution

$$u = \left\{ A \cos \rho at + A' \frac{\sin \rho at}{\rho a} - \left(\frac{B\beta}{\alpha} + \frac{C\gamma}{\alpha}\right) \cos \rho bt \right.$$
$$\left. - \left(\frac{B'\beta}{\alpha} + \frac{C'\gamma}{\alpha}\right) \frac{\sin \rho\, bt}{\rho b} \right\} \cos \rho\delta,$$

$$v = \left\{\frac{A\beta}{\alpha} \cos \rho at + \frac{A'\beta}{\alpha} \frac{\sin \rho at}{\rho a} + B \cos \rho bt + B' \frac{\sin \rho bt}{\rho b}\right\} \cos \rho\delta,$$

$$w = \left\{\frac{A\gamma}{\alpha} \cos \rho at + \frac{A'\gamma}{\alpha} \frac{\sin \rho at}{\rho a} + C \cos \rho bt + C' \frac{\sin \rho bt}{\rho b}\right\} \cos \rho\delta,$$

where $b$ is put for $\dfrac{a}{\sqrt{3}}$.

We have thus gone far enough to obtain a glimpse of the two forms in which the time $t$ occurs in the expressions: see Art. 526 of my account of the memoir of April 1828. Poisson by a most elaborate analysis, starting from the particular solution here given, arrives at complete integrals of the equations; the process depends chiefly on the extension of Fourier's theorem, to which we have adverted in Art. 562. Poisson says that the integrals he now gives are less simple but more symmetrical than those in his former memoir: see Art. 523 of my account of the memoir of April 1828. Poisson adds in a note another form of the integrals communicated to him by Ostrogradsky since his own memoir was written; we shall notice these hereafter.

565. Poisson then proceeds to interpret the formulae obtained; he supposes that the original disturbance is restricted to a small portion of the body, and examines the nature of the motion to which this gives rise at a great distance from the origin. The process is an approximation of a rough kind but the results are very interesting; namely: at a great distance, where the waves have become sensibly plane in a part which is small compared with the whole surface, these waves are of two kinds; in the wave which moves most rapidly the motion of each particle is *normal* to the surface of the wave and is accompanied by a proportional dilatation; in the other wave the motion of each particle is parallel to the surface of the wave, and there is no dilatation; the velocity of the first wave is $\sqrt{3}$ times that of the second.

566. An account of the memoir by Poisson himself is given in the *Annales de Chimie et de Physique*, Vol. XLIV., 1830, pages 423—433. This is very interesting; but it relates not so much to our subject as to fluid motion, and to the controversies round the cradle of the wave theory of light. I will extract a few words which relate to the results mentioned in the previous article: they occur on pages 429—431.

Les intégrales des équations relatives aux vibrations des corps solides, que j'ai donnée, dans l'*Addition* à mon Mémoire sur l'équilibre et le mouvement de ces corps, montrent que le mouvement imprimé à une portion limitée d'un semblable milieu donnera naissance, en général, à deux ondes mobiles, qui s'y propageront uniformément, avec des vitesses différentes dont le rapport sera celui de la racine carrée de trois à l'unité. Ainsi, par exemple, si un ébranlement quelconque avait lieu dans l'intérieur de la terre, nous éprouverions à sa surface deux secousses séparées l'une de l'autre par un intervalle de temps qui dépendrait de la profondeur de l'ébranlement et de la matière de la terre, regardée comme homogène dans toute cette profondeur...... Quelles qu'aient été les directions initiales des vitesses imprimées aux molécules dans l'étendue de cet ébranlement, il ne subsiste finalement que des vitesses dirigées suivant les rayons des ondes mobiles et des vitesses perpendiculaires à ces rayons. Les premières ont lieu exclusivement dans les ondes qui se propagent le plus rapidement, et elles y sont accompagnées de dilatations qui leur sont proportionnelles, en sorte que ces ondes sont constituées comme celles qui se répandent dans les fluides. Les vitesses perpendiculaires aux rayons, ou parallèles aux surfaces, existent, aussi exclusivement, dans les autres ondes dont la vitesse de propagation est à celle des premières comme l'unité est à la racine carrée de trois: elles n'y sont accompagnées d'aucune augmentation ou diminution de la densité du milieu; circonstance digne de remarque, qui ne s'était point encore présentée dans les mouvemens d'ondulation, que les géomètres avaient examinés jusqu'à présent.

567. Poisson published in 1831 his *Nouvelle Théorie de l'action capillaire.* Saint-Venant in Moigno's *Statique*, page 695, refers to pages 31 and 378 of this work as repeating Poisson's objection to the replacing of certain sums by integrals; see Art. 542. Instead of page 378 we must read 278.

568. 1833. In the *Traité de Mécanique* by Poisson, second edition, 1833, there are portions which bear on our subject. In the first volume pages 551—653 form a chapter which is entitled: *Exemples de l'équilibre d'un corps flexible*; this consists of three sections. The first section is on the equilibrium of a funicular polygon; it occupies pages 561—565. The second section is on the equilibrium of a flexible cord; it occupies pages 565—598, and gives the ordinary theory of the catenary and other flexible curves, such as we find now in the ordinary books on statics. The third section is on the equilibrium of an elastic rod, and occupies pages 598—653; this section requires some notice.

569. On his page 599 Poisson makes a few remarks as to the

forces which are called into action when an elastic rod is changed from its natural form into any other. He adds

Le calcul des forces totales qui en résultent et doivent faire équilibre aux forces données, appartient à la Physique mathématique: je renverrai, pour cet objet, à mon *Mémoire sur l'équilibre et le mouvement des Corps élastiques.* Dans ce Traité, on formera les équations d'équilibre d'une verge élastique, en partant de principes secondaires qui sont généralement admis.

The memoir to which Poisson here refers is that of April 1828.

570. Poisson first works out the problem of the equilibrium of an elastic lamina, following the method of James Bernoulli. He obtains for the equation to the curve

$$C\frac{d^2y}{dx^2} = [Q(a-x) - P(b-y)]\left[1 + \left(\frac{dy}{dx}\right)^2\right]^{\frac{3}{2}}.$$

The elastic lamina is supposed fixed at the origin, and its original direction is taken to be that of the axis of $x$. In obtaining the result it is assumed that the action between two parts of the lamina at an imaginary section transverse to the original direction is at *right angles* to the section[1].

571. On his page 620 Poisson proceeds to a more general problem, which he introduces thus:

Formons maintenant les équations d'équilibre d'une verge élastique quelconque, dont tous les points sont sollicités par des forces données.

The problem had been considered by Lagrange, Binet, and Bordoni, as well as by Poisson himself: see Arts. 159, 174, 216, and 423. Poisson's treatment of the problem in his *Mécanique* agrees with that which he gave in 1816: this is now I believe admitted to be unsatisfactory, and Saint-Venant, following Bordoni and Bellavitis, objected to it, but I have not yet found the place

[1] A better solution of the question, taking into account the transversal action, is given by Mr Besant in the *Quarterly Journal of Mathematics*, Vol. IV. pages 12—18. He arrives at the same equation as we have just quoted from Poisson. [The problem has been most thoroughly discussed in Germany by Heim, Klein, Grashoff etc. in works to be considered later. ED.]

where Saint-Venant published his criticism[1]. Kirchhoff states that Saint-Venant has shewn that the suppositions on which Poisson proceeded are partly wrong but gives no reference: *Crelle*, Vol. 56, p. 285. A memoir by Saint-Venant on curves which are not plane curves is published in the 30th Cahier of the *Journal de l'École Polytechnique.* He objects to the phrase *angle of torsion* to denote the angle between two consecutive osculating planes of a curve; and says that it may lead to considerable errors such as have already been committed more than once: see page 55 of the memoir. In a note Saint-Venant cites Poisson's *Mécanique*, Arts. 317 and 318; and also the *Comptes Rendus*, XVII. 953 and 1027, XIX. 41 and 47. See also my account of Bellavitis's memoir of 1839.

572. The general investigation which Poisson gives on his pages 620—629 is very simple and seems correct on the principles which he assumes; so that it is interesting to compare it with the criticisms which have shewn it to be wrong. I presume that the result $d\tau = 0$ which is obtained on page 627 may perhaps be one which is attacked[2].

573. On page 621 Poisson makes an allusion to the memoir of April 1828; the passage to which he alludes is on page 451 of the memoir: it is Art. 469 of my account of the memoir. On page 629 Poisson makes another allusion to the same memoir: the passage to which he alludes is on page 454 of the memoir.

574. On his page 629 Poisson takes the particular case in which the mean thread of the elastic rod forms a plane curve; and in page 631 he limits his process still further by taking the rod homogeneous and naturally prismatic or cylindrical. This leads to an interesting discussion extending to page 643, in which various problems are solved. The treatment however is not very satisfactory; see later my account of Kirchhoff's *Vorlesungen*, namely of page 435 of that book. We may notice especially the problem in which the rod is supported in a horizontal position

[1] [See however the *Historique Abrégé* pp. cxxx. *et seq.* ED.]

[2] Poisson's *Mécanique* was translated into English by the Rev. H. H. Harte, 2 vols., Dublin, 1842; but there is no note on this matter.

and a weight is hung at its middle point; an analogous problem is discussed by M. Chevilliet, who refers to Poisson: see page 6 of his Thesis, 1869[1].

575. Poisson's pages 643—653 are devoted to the investigation of formulae in pure mathematics to which he has referred in the immediately preceding pages; the formulae are those which give the expansion of a function in a series of sines or cosines of multiple angles.

576. In Poisson's second volume the portions with which we are concerned are comprised between pages 292 and 392; but some of these pages are very slightly connected with our subject: the whole constitutes a chapter entitled: *Exemples du mouvement d'un corps flexible.* The first section of this chapter extends over pages 292—316; it is entitled: *Vibrations d'une corde flexible;* here we have the ordinary theory of the vibrations transversal and longitudinal of a stretched cord. Suppose that the tension of the cord is such as to produce the extension $\gamma$ in a cord of length $l$; let $n$ denote the number of transversal vibrations, and $n'$ the number of longitudinal vibrations per second, the vibrations corresponding in both cases to the lowest notes: then the theory shews that

$$\left(\frac{n}{n'}\right)^2 = \frac{\gamma}{l}.$$

Poisson says,

> Ce rapport très simple du nombre des vibrations longitudinales à celui des vibrations transversales d'une même corde, a été vérifié par une expérience que M. Cagniard-Latour a faite sur une corde très longue, dont les vibrations transversales étaient visibles et assez lentes pour qu'on pût les compter.

[1] [The whole of Poisson's analysis for the general case of a rod subject to any system of forces is practically vitiated because he has really assumed the bending moment to be proportional to the curvature. He has fallen into the same error as Euler and Lagrange in applying Bernoulli's theory without modification to the case when there is any force other than transverse applied to the rod. As I have had frequently to point out, if there be any longitudinal stress the so-called neutral axis does not run through the line of centres, and the bending moment is not necessarily proportional to the curvature, *e.g.* a vertical pole bent by its own weight. See the footnote also to Art. 570. Ed.]

The solution to which Poisson here refers has been simplified and corrected by M. Bourget in the *Annales...l'École Normale Supérieure*, Vol. 4, 1867.

577. The second section of Poisson's chapter is entitled: *Vibrations longitudinales d'une verge élastique;* it occupies pages 316—331. The matter is very simple and Poisson had contented himself with a brief notice of it in his memoir of April, 1828: see page 452 of the memoir. At the end of the section Poisson compares the propagation of sound in a solid bar with that of air in a straight tube, and refers to a memoir of his own in the *Mémoires de l'Académie*, Vol. II.

578. The third section of Poisson's chapter is entitled: *Choc longitudinal des verges élastiques*; it occupies pages 331—347: this is simple and interesting. A curious mistake occurs on page 340. Poisson has found an expression in terms of sines of multiple angles which from $x = 0$ to $x = c$, excluding the last value of $x$, is equal to $h$, and from $x = c$ to $x = c + c'$, excluding the first value of $x$, is equal to $h'$; then he professes to shew that for $x = c$ the expression is equal to $h'$: but we know by the general theory of such expressions that the expression must be equal to $\frac{1}{2}(h + h')$. Poisson by mistake has put $\sin \frac{i\pi c}{l} \cos \frac{i\pi c}{l}$, where $l$ stands for $c + c'$, equal to $(-1)^i \sin \frac{i\pi c}{l}$: this is wrong: he should put it equal to

$$\tfrac{1}{2} \sin \frac{2\pi i c}{l} = \tfrac{1}{2} \sin \frac{i\pi (c + l - c')}{l} = \frac{(-1)^i}{2} \sin \frac{i\pi (c - c')}{l}.$$

579. Another mistake occurs in this section, which may be illustrated thus. Suppose we have $n'$ elastic balls, all exactly equal in contact in a row; let there be also $n$ others, exactly equal to the former, in contact in a row; let the second set be in the same straight line as the first, and let them be started with a common velocity to impinge on the first set at rest. Then in analogy with theory and experiment we conclude that out of the $n + n'$ balls the foremost $n$ will go off with the common velocity, and the hindmost $n'$ will remain at rest. Poisson then substantially holds that to ensure this result $n'$ must be greater than $n$; but

this seems to be unnecessary. The mistake, if such it be, appears to be introduced at the first line of page 342; where it is assumed that $c'$ is greater than $c$, apparently without any reason.

580. The fourth section of Poisson's chapter is entitled: *Digression sur les intégrales des équations aux différences partielles;* it occupies pages 347—368. This is a discussion in pure mathematics, and does not fall within our range.

581. The fifth section of Poisson's chapter is entitled: *Vibrations transversales d'une verge élastique;* it occupies pages 368—392. This section is taken substantially from Poisson's memoir of April, 1828, to which he refers for developments. The problem reduces to the solution of the differential equation

$$\frac{d^2y}{dt^2} + b^2 \frac{d^4y}{dx^4} = 0;$$

the corresponding pages of the memoir are 475—488. The pages 382—384 of the *Mécanique* consist of a simple example which was not given in the memoir; the motion of *rotation* of which it treats must be supposed to hold through only an infinitesimal time. On the last page of the chapter we have this note, "pour la comparaison de ces formules à l'observation, voyez les *Annales de Chimie et de Physique*, tome XXXVI, page 86."

Between the dates of the memoir and of the *Mécanique* Cauchy, in his *Mémoire sur l'application du Calcul des Résidus à la solution des problèmes de physique mathématique*, considered the differential equation. This adds to Poisson's solution a fact which amounts to giving the simple value of his $\int X^2 dx$ between the limits. The whole formula is stated by Cauchy on page 35 of his memoir to have been given by Brisson in 1823. Again Cauchy on page 44 of the memoir has a more simple form than that of Poisson. I think Poisson should have noticed these matters.

582. The next memoir by Poisson to be noticed is entitled: *Mémoire sur l'équilibre et le mouvement des corps cristallisés.* This was read to the Academy on the 28th of October, 1839; it is published in the memoirs of the Academy, Vol. XVIII., 1842, where it occupies pages 3—152. Poisson died on the 25th of April, 1840: the memoir as we shall see was left unfinished at his death.

583. A few introductory remarks occupy pages 3—6; these are also printed in the *Comptes Rendus*, Vol. IX. pages 517—519. After these the memoir is divided into three sections; the first section entitled: *Notions préliminaires* occupies pages 6—46; the second section, entitled: *Calcul des pressions moléculaires qui ont lieu dans l'intérieur des corps cristallisés; équations de l'équilibre et des petits mouvements de ces corps* occupies 47—134; the third section entitled: *Lois de la propagation du mouvement, dans un corps cristallisé* occupies pages 134—151, and is only a fragment, unfinished by reason of Poisson's illness and death.

584. In the introductory remarks Poisson states very briefly some of the results of his previous memoirs, and then speaks of the present, and of another which was to follow; from this part I extract some sentences:

Dans ce nouveau Mémoire, je considérerai le cas beaucoup plus compliqué des corps cristallisés. Les équations de leur équilibre, et par suite celles de leur mouvement, sont au nombre de six qui renferment un pareil nombre d'inconnues. Dans le cas du mouvement, trois de ces inconnues se rapportent aux petites vibrations des molécules, et les trois autres à leurs petites oscillations sur elles-mêmes dont ces vibrations sont toujours accompagnées. On peut facilement éliminer les trois dernières inconnues, et l'on parvient ainsi à trois équations aux différences partielles du second ordre, d'où dépendent, à un instant quelconque, les distances suivant trois axes rectangulaires, des molécules à leurs positions d'équilibre dont elles ont été un tant soit peu écartées...

Je présenterai à l'Académie, le plus tôt qu'il me sera possible, un autre Mémoire où se trouveront les lois des petites vibrations des fluides, déterminées d'après le principe fondamental qui distingue ces corps des solides, que j'ai exposé en plusieurs occasions, et dont il est indispensable de tenir compte, lorsque le mouvement se propage avec une extrême rapidité, ce qui rapproche en général les lois de cette propagation, de celles qui ont lieu dans les corps solides. J'appliquerai ensuite les résultats de ce second Mémoire à la théorie des ondes lumineuses, c'est-à-dire, aux petites vibrations d'un éther impondérable, répandu dans l'espace ou contenu dans une matière pondérable, telle que l'air ou un corps solide cristallisé ou non; question d'une grande étendue, mais qui n'a été résolue jusqu'à présent, malgré toute son importance, en aucune de ses parties, ni par moi dans les essais que j'ai tentés à ce sujet, ni selon moi par les autres géomètres qui s'en sont aussi occupés.

A note to the word *occasions* gives a reference to Poisson's *Traité de Mécanique*, Art. 645.

585. The first section of the memoir is devoted to *preliminary* notions; I will notice a few points of interest.

Two kinds of motion with respect to the molecules are contemplated in this memoir. Each molecule may execute vibrations parallel to fixed axes, and, as is usual in this subject, the shifts of a molecule from its mean position parallel to fixed axes are denoted by $u$, $v$, $w$ respectively. Also each molecule may turn on itself; thus a set of rectangular axes is supposed to be fixed in each molecule, and equations are obtained for expressing the change in direction which these axes undergo; this is one of the specialities of the memoir: see pages 16—18 of the memoir. The molecules are not assumed *spherical* in general, and thus the resultant action of one molecule on another is not necessarily a *single* force acting along the straight line which joins what we may call the *centres* of the molecules.

586. The ordinary expressions for stretch and dilatation which involve $u$, $v$, $w$ and their differential coefficients are investigated on pages 18—27 of the memoir; to these are added on pages 28—30 some formulae relative to the change of direction of a plane section of a body, which we will now give.

Let $M$ denote a point the coordinates of which are $x$, $y$, $z$; suppose a plane section passing through $M$, and let $MP$ denote a straight line drawn from $M$ at right angles to this section; let $\lambda$, $\mu$, $\nu$ be the direction angles of $MP$.

By reason of a deformation of the body the original plane section will take a new position, though still remaining plane; suppose $M'$ the new position of $M$, and let $M'P'$ denote the straight line drawn from $M'$ at right angles to the new position of the plane section; let $\lambda'$, $\mu'$, $\nu'$ be the direction angles of $M'P'$: it is required to obtain expressions for $\cos\lambda' - \cos\lambda$, $\cos\mu' - \cos\mu$, and $\cos\nu' - \cos\nu$.

Let $x+x_1$, $y+y_1$, $z+z_1$ be the original coordinates of a point $N$ in the original position of the plane section, and let $N'$ be the position of $N$ after deformation; let $x+u$, $y+v$, $z+w$ be the coordinates of $M'$, and let $x+x'+u$, $y+y'+v$, $z+z'+w$ be the coordinates of $N'$. Then since $MP$ is at right angles to $MN$ we have

$$x_1 \cos\lambda + y_1 \cos\mu + z_1 \cos\nu = 0 \ldots\ldots\ldots\ldots\ldots (1).$$

And since $M'P'$ is at right angles to $M'N'$ we have

$$x' \cos\lambda' + y' \cos\mu' + z' \cos\nu' = 0 \ldots\ldots\ldots\ldots\ldots (2).$$

Also if we suppose $N$ very close to $M$ we have

$$\left.\begin{aligned} x' - x_1 &= x_1 \frac{du}{dx} + y_1 \frac{du}{dy} + z_1 \frac{du}{dz} \\ y' - y_1 &= x_1 \frac{dv}{dx} + y_1 \frac{dv}{dy} + z_1 \frac{dv}{dz} \\ z' - z_1 &= x_1 \frac{dw}{dx} + y_1 \frac{dw}{dy} + z_1 \frac{dw}{dz} \end{aligned}\right\} \ldots\ldots\ldots\ldots\ldots (3).$$

From (1), (2), and (3) we must obtain the required result. From (1) and (2) when we reject the product of $x'-x_1$ into $\cos\lambda' - \cos\lambda$ and similar terms, we have

$$(x'-x_1)\cos\lambda + (y'-y_1)\cos\mu + (z'-z_1)\cos\nu$$
$$+ x_1(\cos\lambda' - \cos\lambda) + y_1(\cos\mu' - \cos\mu) + z_1(\cos\nu' - \cos\nu) = 0;$$

and by (3) this becomes

$$x_1\left(\frac{du}{dx}\cos\lambda + \frac{dv}{dx}\cos\mu + \frac{dw}{dx}\cos\nu + \cos\lambda' - \cos\lambda\right)$$
$$+ y_1\left(\frac{du}{dy}\cos\lambda + \frac{dv}{dy}\cos\mu + \frac{dw}{dy}\cos\nu + \cos\mu' - \cos\mu\right)$$
$$+ z_1\left(\frac{du}{dz}\cos\lambda + \frac{dv}{dz}\cos\mu + \frac{dw}{dz}\cos\nu + \cos\nu' - \cos\nu\right) = 0;$$

we will denote this for brevity thus

$$Ax_1 + By_1 + Cz_1 = 0.$$

Now the last equation must be identical with (1) as the point $N$ is subject only to the condition of lying in a certain plane; therefore

$$\frac{A}{\cos\lambda} = \frac{B}{\cos\mu} = \frac{C}{\cos\nu} \ldots\ldots\ldots\ldots\ldots\ldots\ldots (4).$$

Again we have

$$\cos^2\lambda + \cos^2\mu + \cos^2\nu = 1,$$
$$\cos^2\lambda' + \cos^2\mu' + \cos^2\nu' = 1;$$

subtract, neglecting the square of $\cos\lambda' - \cos\lambda$, and the two similar squares; thus

$$(\cos\lambda' - \cos\lambda)\cos\lambda + (\cos\mu' - \cos\mu)\cos\mu + (\cos\nu' - \cos\nu)\cos\nu = 0 \ldots\ldots(5).$$

The equations (4) and (5) supply three linear equations for finding $\cos\lambda' - \cos\lambda$, $\cos\mu' - \cos\mu$, and $\cos\nu' - \cos\nu$: thus we get

$$\cos\lambda' - \cos\lambda = \left(\frac{du}{dy}\cos\lambda + \frac{dv}{dy}\cos\mu + \frac{dw}{dy}\cos\nu\right)\cos\lambda\cos\mu$$
$$+ \left(\frac{du}{dz}\cos\lambda + \frac{dv}{dz}\cos\mu + \frac{dw}{dz}\cos\nu\right)\cos\lambda\cos\nu$$
$$- \left(\frac{du}{dx}\cos\lambda + \frac{dv}{dx}\cos\mu + \frac{dw}{dx}\cos\nu\right)\sin^2\lambda.$$

The values of $\cos\mu' - \cos\mu$ and $\cos\nu' - \cos\nu$ can be written down by symmetry.

587. On his page 32 Poisson says that it is well to verify a statement which is evident of itself, that a movement of the body as a whole has no influence whatever on the stretch or on the dilatation.

First suppose the motion to be one of translation; then $u$, $v$, $w$ are functions of the time which are independent of $x$, $y$, $z$; so that the differential coefficients of $u$, $v$, $w$ with respect to $x$, $y$, $z$ are zero.

Next suppose the motion to be one of rotation round an axis; then by known formulae we have

$$\frac{du}{dt} = (y\cos\eta'' - z\cos\eta')\,\omega, \qquad \frac{dv}{dt} = (z\cos\eta - x\cos\eta'')\,\omega,$$
$$\frac{dw}{dt} = (x\cos\eta' - y\cos\eta)\,\omega,$$

where $\omega$ is the angular velocity; and $\eta$, $\eta'$, $\eta''$ are the angles which the instantaneous axis of rotation makes with the axes of

$x$, $y$, $z$ respectively. The four quantities $\omega$, $\eta$, $\eta'$, $\eta''$ will be functions only of the time; let

$$\int \omega \cos \eta dt = \zeta, \quad \int \omega \cos \eta' dt = \zeta', \quad \int \omega \cos \eta'' dt = \zeta'',$$

the integrals being taken from $t = 0$ as the lower limit: then

$$u = y\zeta'' - z\zeta', \quad v = z\zeta - x\zeta'', \quad w = x\zeta' - y\zeta;$$

and these values make $\frac{du}{dx} = 0$, $\frac{dv}{dy} = 0$, $\frac{dw}{dz} = 0$; also

$$\frac{dv}{dz} + \frac{dw}{dy} = 0, \quad \frac{dw}{dx} + \frac{du}{dz} = 0, \quad \frac{du}{dy} + \frac{dv}{dx} = 0.$$

Thus the stretch and the dilatation vanish.

588. On page 45 the following sentence occurs, the correctness of which is probably now generally admitted:

Dans la réalité, cet équilibre n'a pas lieu rigoureusement, et ce que nous prenons dans la nature pour l'état de repos d'un corps, n'est autre chose qu'un état dans lequel ses molécules exécutent incessamment des vibrations d'une étendue insensible, et des oscillations sur elles-mêmes, également imperceptibles;......

589. The object of the second section of the memoir is to calculate the stresses at any point of an elastic body, and thence to form the equations for the equilibrium and motion of the body. The first and second sections taken together constitute in fact a treatise on the theory of elasticity so far as concerns the general equations of the subject, without any applications.

590. On page 47 we have a definition of the term *pressure* (stress) as used in this subject; it coincides with that which Poisson had formerly adopted: see Art. 546 of my account of the memoir of October 1829. Saint-Venant objects to the definition as leading to inconveniences which Poisson himself perceived: see Moigno's *Statique*, page 619, and the memoir on *Torsion*, page 249.

591. In his second section Poisson proposes to consider two successive states of a body under the action of different forces; the second state may be a state of equilibrium differing but very little

from the first, or it may be a state of motion. Accordingly pages 47—70 are devoted to the consideration of the first state of the body. Poisson obtains on his pages 48—52 expressions for the stresses; these are left in the form of summations which are indicated, though they cannot really be performed. Then on pages 53—56 he forms the three equations of equilibrium which must hold at every point of the interior of the body, by resolving the forces parallel to the three axes; on pages 57 and 58 he considers the three equations derived from the principle of moments, by virtue of which the nine stresses hitherto used are reduced to six; and on pages 59—61 he obtains the three equations which must hold at every point of the bounding surface, by a method which presents a little novelty.

592. On page 65 Poisson adopts a special hypothesis, which he does not state very distinctly, but which amounts to assuming a symmetrical arrangement of the molecules round any arc; in virtue of this he comes to the conclusion that all the *shears* must vanish. Then on page 69 he says that if the first state of the body is its *natural* state the three *tractions* must also vanish.

593. On page 70 Poisson proceeds to consider the body in its second state. He says

Occupons-nous maintenant du second état du corps, dans lequel ses molécules ont été très-peu déplacées des positions qu'elles avaient dans le premier, soit par de nouvelles forces extérieures ou intérieures qui se font encore équilibre, soit par des causes quelconques qui les ont mises en mouvement.

The pages 70—122 form the most important part of the memoir; the investigations are rather complex, but they are exhibited very fully, so that they may be followed without difficulty. In consequence of the change of the body from its first state to its second the symbols denoting distances and angles receive slight increments, and we have to find the consequent changes produced in the expression for the stresses and the equations of equilibrium and motion.

594. Poisson obtains on his pages 71—83 expressions for the stresses in the second state of the body; these involve three quantities $G_1$, $H_2$, $K_3$ which occur in the expressions relative to the first state, and which are in fact the tractions in the special hypothesis of Art. 592; the expressions involve also fifteen other quantities which take the form of summations, and which will be constants if the body is supposed homogeneous and of the same temperature throughout. On his pages 84—87 Poisson forms the first three equations of equilibrium which must hold at every point of the body, and on his pages 88—91 the second three; on his pages 92—95 he forms the equations which must hold at every point of the surface. In these investigations squares and products of small quantities are neglected, and expansions by Taylor's Theorem are limited to terms involving first differential coefficients; under these limitations the process is satisfactory. Some points of interest which may be considered as digressions from the main investigations will now be noticed.

595. We know that $\frac{du}{dx}+\frac{dv}{dy}+\frac{dw}{dz}$ expresses the dilatation at the point $(x, y, z)$; also $\frac{du}{dx}+\frac{dv}{dy}$ expresses what we may term the *spread* or *areal* dilatation in the plane $xy$ at this point: that is if $A$ denote the original area of any small figure in this plane near the point, its area after deformation will be $A\left(1+\frac{du}{dx}+\frac{dv}{dy}\right)$. Now

$$\frac{du}{dx}+\frac{dv}{dy}=\left(\frac{du}{dx}+\frac{dv}{dy}+\frac{dw}{dz}\right)-\frac{dw}{dz},$$

that is: the spread is equal to the dilatation diminished by the stretch in the direction at right angles to the plane considered. This proposition is general, for the directions of the axes of $x, y, z$ may be any whatever which form a rectangular system. Now the stretch in the direction determined by the angles $\alpha, \beta, \gamma$ is (see Art. 547)

$$\frac{du}{dx}\cos^2\alpha+\frac{dv}{dy}\cos^2\beta+\frac{dw}{dz}\cos^2\gamma+\left(\frac{du}{dy}+\frac{dv}{dx}\right)\cos\alpha\cos\beta$$
$$+\left(\frac{dv}{dz}+\frac{dw}{dy}\right)\cos\gamma\cos\beta+\left(\frac{dw}{dx}+\frac{du}{dz}\right)\cos\gamma\cos\alpha.$$

Subtract this from $\frac{du}{dx}+\frac{dv}{dy}+\frac{dw}{dz}$, and we obtain for the spread in the plane of which the normal has the direction angles $\alpha, \beta, \gamma$—

$$\frac{du}{dx}\sin^2\alpha+\frac{dv}{dy}\sin^2\beta+\frac{dw}{dz}\sin^2\gamma-\left(\frac{du}{dy}+\frac{dv}{dx}\right)\cos\alpha\cos\beta,$$
$$-\left(\frac{dv}{dz}+\frac{dw}{dy}\right)\cos\beta\cos\gamma-\left(\frac{dw}{dx}+\frac{du}{dz}\right)\cos\gamma\cos\alpha.$$

This is given by Poisson on his pages 96 and 97.

596. In order to verify by an example the system of equations which he has obtained, Poisson applies them to the case of a homogeneous non-crystallised body; see his pages 101—109. His results are in agreement with those of his earlier memoirs, when the proper limitations are introduced.

597. As in a former memoir Poisson holds that molecular force is really the difference of an attraction exerted by the molecules themselves and a repulsion exerted by the caloric round them: see Art. 543 of my account of the memoir of October 1829. He says on pages 113 and 114 of the present memoir:

Dans un corps solide, on est obligé d'exercer une très-grande pression à la surface pour produire une très-petite condensation; l'augmentation aussi très-grande de la pression moléculaire qui en résulte dans l'intérieur provient donc alors d'un très-petit rapprochement des molécules; or, cela ne peut avoir lieu à moins que l'action mutuelle de deux molécules voisines ne soit la différence de deux forces contraires, dont chacune est extrêmement grande eu égard à cette différence ou à la force apparente; de manière que pour ce très-petit rapprochement, chacune des deux forces contraires varie d'une très-petite fraction de sa propre grandeur, et qu'il s'ensuive néanmoins dans leur différence une variation comparable à sa valeur primitive, ou même bien plus considérable, qui rende, par exemple, la force apparente décuple ou centuple de ce qu'elle était d'abord entre les deux mêmes molécules. C'est tout ce que nous pouvons savoir sur la répulsion et l'attraction dont nous n'observons jamais les effets séparés, et dont l'excès de l'une sur l'autre produit tous les phénomènes que nous pouvons connaître. J'ai déjà eu plusieurs fois l'occasion de faire cette remarque conforme à ce qui a été avancé au commencement de ce mémoire.

598. On his pages 115—119 Poisson notices the case in which the molecules of the crystal are supposed spherical, so that crystallisation consists merely in the regular distribution of the molecules round each other. In this case the *fifteen* quantities mentioned in Art. 594 reduce to *nine*, including $G_1$, $H_2$, $K_3$; if the first state of the body be the natural state, the last three vanish, and the quantities reduce to *six*.

599. Poisson returns on his page 119 to the general equations which he had obtained for a crystal, and simplifies them by the supposition that the first state of the body is the natural state; also the body is supposed homogeneous. He thus finds that the differential equations for determining $u$, $v$, and $w$ involve *twelve* constants. But *three* more constants occur in the equations by which we determine the changes in the position of the axes supposed to be fixed in each molecule: see Art. 585. Saint-Venant refers to the number of constants: see Moigno's *Statique* page 706, and the memoir on *Torsion* page 261.

600. The third section relates to the propagation of motion in a crystallised body. This is only a fragment. Poisson here shews how to integrate the equations of motion, supposed to be in the simple state noticed in Art. 599; he follows the method of integration given in his memoir of October, 1830: see my Art. 564. With respect to the equations to be integrated Poisson says on his pages 135 and 136:

> Elles sont comprises, comme cas particulier, parmi celles que M. Blanchet, professeur de physique au collége de Henri IV, a intégrées sous forme finie, dans un mémoire lu à l'Académie, il y a environ un an, où il est parvenu à exprimer les valeurs de $u$, $v$, $w$, par des intégrales définies doubles et triples.

The memoir by Blanchet was published in Liouville's *Journal de mathématiques*, Vol. V. 1840, and is considered in our Chapter VII.

601. The following sentences occur after the memoir:

> M. Poisson n'a pas achevé d'écrire le troisième paragraphe de ce premier mémoire, à la suite duquel, ainsi qu'il le dit au préambule de celui-ci il se proposait encore de présenter à l'Académie un second

mémoire sur la lumière. Pendant la maladie longue et douloureuse qui l'a enlevé aux sciences, il a bien souffert du regret d'emporter avec lui les découvertes dont son imagination infatigable était pleine. Quand le mal moins avancé lui permettait encore de causer science avec ses amis, il a dit qu'il avait trouvé comment il pouvait se faire, qu'un ébranlement ne se propageât dans un milieu élastique que suivant une seule direction; le mouvement propagé suivant les directions latérales étant insensible aussitôt que l'angle de ces directions avec celle de la propagation était appréciable. Il arrivait ainsi à la propagation de la lumière en ligne droite. Plus tard, cédant au mal, et se décidant enfin à interrompre l'impression de son mémoire: c'était pourtant, a-t-il dit, la partie originale, c'était décisif pour la lumière; et cherchant avec peine le mot pour exprimer son idée, il a répété plusieurs fois: c'était un *filet* de lumière. Puissent ces paroles, religieusement conservées par les amis de M. Poisson, les dernières paroles de science qui soient sorties de sa bouche, mettre les savants sur la trace de sa pensée, et inspirer un achèvement de son œuvre digne du commencement[1].

On the death of Malus in 1812 Delambre said:

Si Malus eût vécu, c'est lui qui nous eût complété la théorie de la lumière;

and the words might be applied to Poisson, who succeeded to the place of Malus at the *Institut:* see *Mém. de l'Institut* 1812, page xxxiii. A brief, but very good notice of Poisson will be found in the *Monthly Notices of the Royal Astronomical Society,* Vol. v., pages 84—86; it says, he "was placed, by common consent, at the head of European analysts on the death of Laplace."

[His labours as an elastician are only second to those of Saint-Venant, scarcely excelled by those of Cauchy. There is hardly a problem in our subject to which he has not contributed, and many owe their very existence to his initiative.]

[1] Some attempt has been made to reveal the meaning of Poisson's dying words: see *Comptes Rendus* xx. 561.

## CHAPTER V.

### Cauchy.

602. 1823. *Recherches sur l'équilibre et le mouvement intérieur des corps solides ou fluides, élastiques ou non élastiques.* This is published in the *Bulletin...Philomatique*, 1823, pages 9—13; it is the first of the numerous writings of Cauchy on the subject of elasticity; it consists of an abstract of a memoir presented to the Paris Academy on the 30th of September, 1822.

603. Cauchy was one of the commissioners appointed to examine the memoir sent to the Paris Academy by Navier on the 14th of August, 1820; and this led him to turn his own attention to the investigation of the subject. He here states, without the use of any mathematical symbols, the results at which he had arrived. We see that he must at this date have constructed a complete elementary theory including the following particulars: the existence of the *six* stress-components which have to be considered at any point; the representation of the stress on a plane by the reciprocal of the radius vector of a certain ellipsoid; the existence of *principal* tractions; and the representation of the resolved part of the stress at right angles to an assigned plane by the reciprocal of the square of the radius of a surface of the second order[1]. Also he had obtained, I presume, the general equations for the internal equilibrium of a solid body. He speaks of these as *four* in number, one of which deter-

[1] [The reader will do well to consult the note on elastic terminology at the end of this volume. Ed.]

mines separately the dilatation; but it should have been stated that there are only *three* independent equations.

604. The name of Fresnel is introduced after mention of the stresses. Cauchy says:

J'en étais à ce point, lorsque M. Fresnel, venant à me parler des travaux auxquels il se livrait sur la lumière, et dont il n'avait encore présenté qu'une partie à l'Institut, m'apprit que, de son côté, il avait obtenu sur les lois, suivant lesquelles l'élasticité varie dans les diverses directions qui émanent d'un point unique, un théorème analogue au mien.

605. I am not certain what property Cauchy has in view in the following sentence: [It probably refers to what may be termed the *strain-ellipsoid:* see Clebsch, *Theorie der Elasticität,* p. 41; Weyrauch, *Theorie elastischer Körper*, p. 72, erroneously attributes it to Clebsch. Compare our Arts. 612, 617.]

De plus, je démontre que les diverses condensations ou dilatations autour d'un point, diminuées ou augmentées de l'unité, deviennent égales, au signe près, aux rayons vecteurs d'un ellipsoïde.

[606.] The paper is of importance in the history of the subject, as we have here the origin of the theory of stress. We may especially notice the following theorem which may be termed *Cauchy's Theorem.* The stress on any infinitesimal face in the interior of a solid or fluid body at rest is the resultant of the stresses on the three projections of this face on planes through its centre. The projections may be right or oblique: see Saint-Venant on *Torsion,* pages 249 and 250; also Moigno's *Statique,* pages 627, 657, 693. Resal on page 4 of his *Thèse de Mécanique* cites the paper, but ascribes to it the date 1825 instead of 1823.

607. *Sur la théorie des pressions.* This is published in Férussac's *Bulletin,* Vol. IX. 1828, pages 10—22. It does not relate to our subject but to the well-known indeterminate problem of Statics, a simple example of which occurs when a body is on a horizontal plane and in contact with it at more than three points.

608. We have now to notice various writings by Cauchy published in his *Exercices de mathématiques;* the second volume

of this collection is dated 1827, and contains memoirs relating to our subject of which we will give an account in the following five articles.

609. On pages 23 and 24 there is an article entitled: *De la pression dans les fluides.* The object is to demonstrate the *equality of pressure in all directions round a point:* it appears to me unsatisfactory from not explaining what is meant by a fluid, so that it is not very clear what is the foundation of the demonstration. It is referred to in Moigno's *Statique*, page 620.

[610.] An article entitled: *De la pression ou tension dans un corps solide,* occupies pages 41—56, and is followed by an *Addition* on pages 57—59. The article may be described as an investigation of the fundamental equations with respect to elastic stresses; Cauchy refers to the *Bulletin...Philomatique* for January 1823, in which he had enunciated his main results. The following propositions are here substantially investigated by Cauchy:

(i) The stresses exerted at a given point of a solid body against the two faces of any plane whatever placed at the point are equal and opposite forces (p. 46).

(ii) Suppose two infinitesimal faces of equal area to have the same centre at any point of a solid body at rest; then the stress on the first face resolved along the normal to the second face is equal to the stress on the second face resolved along the normal to the first face. Cauchy only treats the case of the two planes being perpendicular. In our notation this is represented by $\widehat{rs} = \widehat{sr}$[1].

[1] [For the purposes of this history I have settled with some hesitation to adopt a slightly modified form of the double-suffix notation originally introduced by Coriolis and afterwards adopted by Cauchy. In some cases where this notation however luminous would be still too cumbrous, the convenient but not very suggestive notation of Lamé has been followed.

The modification adopted consists in printing $\widehat{xy}$ for $p_{xy}$, and so avoiding the troublesome subscripts by suppressing the unnecessary letter $p$. Thus $\widehat{rs}$ denotes either the stress component at a point on a plane whose normal is $r$ in the direction $s$, or on a plane whose normal is $s$ in the direction $r$.

(iii) The properties of the *stress-quadric*, whose equation is

$$\widehat{xx}x^2 + \widehat{yy}y^2 + \widehat{zz}z^2 + 2\,\widehat{yz}yz + 2\,\widehat{zx}zx + 2\,\widehat{xy}xy = \pm 1,$$

namely, that the stress on any plane is normal to that plane which is diametral to the perpendicular to the original plane, and is inversely proportional to the product of the radius vector and central perpendicular to the tangent plane parallel to the given plane (pp. 48—51).

(iv) The property of what we may call *Cauchy's stress-ellipsoid:*

$$(\widehat{xx}x + \widehat{xy}y + \widehat{xz}z)^2 + (\widehat{yx}x + \widehat{yy}y + \widehat{yz}z)^2 + (\widehat{zx}x + \widehat{zy}y + \widehat{zz}z)^2 = 1,$$

namely, that the reciprocal of the radius vector gives the value at the point of the stress on a plane perpendicular to the radius vector, p. 54. These propositions all seem to be due to Cauchy, and are demonstrated here for the first time.

**The following table will serve to connect the various notations for the system of stress components:**

| | Poisson | | | Cauchy (earlier) | | | Coriolis, Cauchy (later), followed by Saint-Venant, Maxwell, and Castigliano | | | Lamé (Winkler, Minchin, etc.) | | | Klein | | | Beer | | |
|---|---|---|---|---|---|---|---|---|---|---|---|---|---|---|---|---|---|---|
| | $x$ | $y$ | $z$ | $x$ | $y$ | $z$ | $x$ | $y$ | $z$ | $x$ | $y$ | $z$ | $x$ | $y$ | $z$ | $x$ | $y$ | $z$ |
| $x$ | $P_3$ | $Q_3$ | $R_3$ | $A$ | $F$ | $E$ | $p_{xx}$ | $p_{xy}$ | $p_{xz}$ | $N_1$ | $T_3$ | $T_2$ | $N_1$ | $T_{xy}$ | $T_{xz}$ | $N_x$ | $T_z$ | $T_y$ |
| $y$ | $P_2$ | $Q_2$ | $R_2$ | $F$ | $B$ | $D$ | $p_{yx}$ | $p_{yy}$ | $p_{yz}$ | $T_3$ | $N_2$ | $T_1$ | $T_{yx}$ | $N_2$ | $T_{yz}$ | $T_z$ | $N_y$ | $T_x$ |
| $z$ | $P_1$ | $Q_1$ | $R_1$ | $E$ | $D$ | $C$ | $p_{zx}$ | $p_{zy}$ | $p_{zz}$ | $T_2$ | $T_1$ | $N_3$ | $T_{zx}$ | $T_{zy}$ | $N_3$ | $T_y$ | $T_x$ | $N_z$ |

| | Kirchhoff, followed by Riemann and Weyrauch | | | Thomson | | | Grashof | | | Clebsch | | | Notation adopted | | |
|---|---|---|---|---|---|---|---|---|---|---|---|---|---|---|---|
| | $x$ | $y$ | $z$ | $x$ | $y$ | $z$ | $x$ | $y$ | $z$ | $x$ | $y$ | $z$ | $x$ | $y$ | $z$ |
| $x$ | $X_x$ | $X_y$ | $X_z$ | $P$ | $V$ | $T$ | $\sigma_x$ | $\tau_z$ | $\tau_y$ | $t_{11}$ | $t_{12}$ | $t_{13}$ | $\widehat{xx}$ | $\widehat{xy}$ | $\widehat{xz}$ |
| $y$ | $Y_x$ | $Y_y$ | $Y_z$ | $V$ | $Q$ | $S$ | $\tau_z$ | $\sigma_y$ | $\tau_x$ | $t_{21}$ | $t_{22}$ | $t_{23}$ | $\widehat{yx}$ | $\widehat{yy}$ | $\widehat{yz}$ |
| $z$ | $Z_x$ | $Z_y$ | $Z_z$ | $T$ | $S$ | $R$ | $\tau_y$ | $\tau_x$ | $\sigma_z$ | $t_{31}$ | $t_{32}$ | $t_{33}$ | $\widehat{zx}$ | $\widehat{zy}$ | $\widehat{zz}$ |

**Of these notations Poisson's, Cauchy's (earlier) and Thomson's, are not very suggestive, Lamé's has obvious advantages, but for a single-suffix notation is inferior to Kirchhoff's. Klein and Beer do not much improve on Lamé, nor Grashof on Kirchhoff. Clebsch's notation has all the disadvantages of the double-suffix notation without its generality or luminosity. Wand follows Coriolis replacing the latter's $p_{xx}$... by $K_{xx}$.... I have, after balancing the claims of these various notations, adopted Coriolis's, which has the use of great authorities in its favour; at the same time to avoid subscripts I use an umbral notation. Ed.]**

611. In the *Addition* Cauchy demonstrates some formulae used by Fresnel in his theory of double refraction. They are equivalent to the well-known relations between the component stresses at the centres of the four faces of a tetrahedron: see Saint-Venant on *Torsion*, page 250; Moigno's *Statique*, page 627[1].

612. [An article entitled: *Sur la condensation et la dilatation des corps solides* occupies pages 60—69 of the volume. The *stretch*, or *linear* dilatation close to an assigned point, in any direction, is shewn to be related to the radius vector of a certain ellipsoid. This is the first formal appearance of the *strain-ellipsoid*: see Arts. 605 and 617. I use the word *stretch* generally for *linear expansion* or *contraction*. Cauchy also finds an expression for the *cubical* dilatation or as I term it simply: the dilatation. He then proceeds to simplify his formulae by the supposition that the displacements are very small, and thus he obtains the usual expressions for the stretches and dilatation. He also gives another surface of the second degree by which the stretch is geometrically represented. The equation to this quadric is

$$s_x x^2 + s_y y^2 + s_z z^2 + \sigma_{yz} yz + \sigma_{zx} zx + \sigma_{xy} xy = \pm 1,$$

where $s_x$, $s_y$, $s_z$ are the three stretches (*dilatations, Dehnungen*) parallel to the three axes and $\sigma_{yz}$, $\sigma_{zx}$, $\sigma_{xy}$ the corresponding slides (*glissements, Schiebungen, Gleitungen*). The property of this *stretch-quadric* is that the inverse square of its radius-vector measures numerically the stretch in the direction of the radius-vector. See Saint-Venant on *Torsion*, pages 243, 281, 283; Moigno's *Statique*, pages 644, 650.]

It will be convenient to shew the way in which Cauchy begins his investigations.

Let $x$, $y$, $z$ be the coordinates of a molecule $m$ in the second state of the body; let $x-\xi$, $y-\eta$, $z-\zeta$ be the coordinates of the same molecule in the original state. Let $x+\Delta x$, $y+\Delta y$, $z+\Delta z$ be the coordinates, in the second state of an adjacent molecule $m'$; let $r$ be the distance between the molecules; and $\alpha$, $\beta$, $\gamma$ the

[1] Pages 43 and 44 for $\cos\gamma\, dxdy$ read $\sec\gamma\, dxdy$.

direction angles of $r$. The coordinates of $m'$ in the original state will be:

$$x+\Delta x-\left(\xi+\frac{d\xi}{dx}\Delta x+\frac{d\xi}{dy}\Delta y+\frac{d\xi}{dz}\Delta z+\ldots\ldots\right),$$

$$y+\Delta y-\left(\eta+\frac{d\eta}{dx}\Delta x+\frac{d\eta}{dy}\Delta y+\frac{d\eta}{dz}\Delta z+\ldots\ldots\right),$$

$$z+\Delta z-\left(\zeta+\frac{d\zeta}{dx}\Delta x+\frac{d\zeta}{dy}\Delta y+\frac{d\zeta}{dz}\Delta z+\ldots\ldots\right),$$

Let $\dfrac{r}{1+s_r}$ denote the original distance of $m$ and $m'$; then we have approximately

$$\left(\frac{r}{1+s_r}\right)^2=\left(\Delta x-\frac{d\xi}{dx}\Delta x-\frac{d\xi}{dy}\Delta y-\frac{d\xi}{dz}\Delta z\right)^2,$$
$$+\left(\Delta y-\frac{d\eta}{dx}\Delta x-\frac{d\eta}{dy}\Delta y-\frac{d\eta}{dz}\Delta z\right)^2,$$
$$+\left(\Delta z-\frac{d\zeta}{dx}\Delta x-\frac{d\zeta}{dy}\Delta y-\frac{d\zeta}{dz}\Delta z\right)^2.$$

Also $\Delta x=r\cos\alpha$, $\Delta y=r\cos\beta$, $\Delta z=r\cos\gamma$: thus

$$\left(\frac{1}{1+s_r}\right)^2=\left(\cos\alpha-\frac{d\xi}{dx}\cos\alpha-\frac{d\xi}{dy}\cos\beta-\frac{d\xi}{dz}\cos\gamma\right)^2,$$
$$+\left(\cos\beta-\frac{d\eta}{dx}\cos\alpha-\frac{d\eta}{dy}\cos\beta-\frac{d\eta}{dz}\cos\gamma\right)^2,$$
$$+\left(\cos\gamma-\frac{d\zeta}{dx}\cos\alpha-\frac{d\zeta}{dy}\cos\beta-\frac{d\zeta}{dz}\cos\gamma\right)^2$$

613. An article entitled: *Sur les relations qui existent, dans l'état d'équilibre d'un corps solide ou fluide, entre les pressions ou tensions et les forces accélératrices* occupies pages 108—111 of the volume. The differential equations between the stresses and applied forces for the equilibrium of elementary volumes, namely the body stress-equations of the type

$$\frac{d\widehat{xx}}{dx}+\frac{d\widehat{xy}}{dy}+\frac{d\widehat{xz}}{dz}+\rho X=0,$$

are here demonstrated for the first time. See Saint-Venant on *Torsion*, page 274, and Moigno's *Statique*, page 638.

We pass to the third volume of the *Exercices* which is dated 1828; to this we shall devote the following twenty-one Articles.

614. An article entitled: *Sur les équations qui expriment les conditions d'équilibre, ou les lois du mouvement intérieur d'un corps solide, élastique, ou non élastique* occupies pages 160—187 of the volume. We have first some repetition of formulae already given in the preceding volume, and then equations are obtained which we may describe thus: take the usual body-equations for the shifts, i.e. those of the type

$$(\lambda+\mu)\frac{d\theta}{dx}+\mu\nabla^2 u+\rho X=0,$$

and suppose $\lambda=0$, but let $\mu$ be variable or constant. The equations are obtained by Cauchy on the *assumption* that the 'stress-quadric' (see Art. 610, iii.) is similar and similarly situated to the 'stretch-quadric' (see Art. 612). We may state the assumption verbally thus: the directions of the principal tractions coincide with the directions of the principal stretches, and the ratio of the traction to the stretch is the same for all the three principal tractions. This assumption is entirely arbitrary. Cauchy soon after proceeds to another hypothesis, namely that each of the principal tractions consists of two terms, one term being $k$ times the stretch in the corresponding direction, and the other term depending only on the position of the point considered; then he specialises the second term by assuming it to represent the dilatation, so as to render it equivalent to our $\theta$, and thus he obtains formulae with two constants[1] equivalent to those of the usual type,

$$(\lambda+\mu)\frac{d\theta}{dx}+\mu\nabla^2 u+\rho X=0.$$

See Moigno's *Statique*, page 657.

[1] [This would seem to be the first appearance in the history of our subject of the equations of isotropic elasticity with *two* constants. Cauchy thus anticipated both Green and Stokes. Remembering this and also remarking that Saint-Venant is and

On his pages 173 and 174 Cauchy gives the equations which must hold at every point of the surface where there is equilibrium; i.e. the surface stress-equations of the type

$$\widehat{xx}\cos\alpha + \widehat{xy}\cos\beta + \widehat{xz}\cos\gamma = X_0.$$

There are some references in the article to memoirs presented to the French Academy which may be noted; on pages 177 and 185 to a memoir by Cauchy of September, 1822; on page 182 to a memoir by Poisson not then published, which must be that of April, 1828; to the memoir by Navier of May, 1821; and to a memoir by Cauchy of October, 1827.

On the whole we may say that in this article Cauchy obtains the fundamental equations of the subject, involving *two* constants, but by methods inferior to those now adopted.

Saint-Venant on *Torsion*, p. 255, Moigno's *Statique*, 658, 702.

[615.] An article entitled: *Sur l'équilibre et le mouvement d'un système de points matériels sollicités par des forces d'attraction ou de répulsion mutuelle* occupies pages 188—212 of the volume. This had been presented to the Paris Academy on the 1st of October, 1827: see Moigno's *Statique*, page 678 and 690. In this important memoir Cauchy obtains the body shift-equations from considering the equilibrium of a single molecule of the body,—just as Navier in 1821 had done for a simpler case. We have here in fact the *first* consideration of the elastic equations for a non-isotropic body. The expressions, really differing only by constant terms from those of the stresses, given by Cauchy in his formulae (56) and (57) and the shift-equations for body-equilibrium and motion in (67) and (68) include *nine* constants.

Les formules (67) et (68) paraissent spécialement applicables au cas où l'élasticité n'étant pas la même, dans les diverses directions, le corps offre trois axes d'élasticité rectangulaires entre eux.

The case is in fact that of a crystal with three rectangular axes of elasticity.

has been the staunchest supporter of an uni-constant isotropy (*Navier's Leçons*, pp. 746—762, and *Théorie de l'élasticité de Clebsch*, pp. 65—67), the following remark of Professor Tait is perhaps not quite satisfactory:

'The erroneousness of this conclusion was first pointed out by Stokes, and his paper, followed by the investigations of de St-Venant and Sir W. Thomson, has put the whole subject in a new light.' *Properties of Matter*, p. 199. ED.]

For the case of isotropy Cauchy proves on pp. 201 an equation like that on page 703 of Moigno's *Statique* expressed by $a_{yyyy} = 3a_{yyzz}$: see our Art. 548. He thus reduces his body-equations to the type

$$(R + G)\nabla^2 u + 2R\frac{d\theta}{dx} + X = 0,$$

involving *two* constants determined by the summations

$$G = \Sigma\left[\pm\frac{mr}{2}\cos^2\alpha\,\chi(r)\right], \qquad R = \Sigma\left[\frac{mr}{2}\cos^2\alpha\cos^2\beta f(r)\right],$$

where $m$ is the mass of a molecule situated at distance $r$ in direction $(\alpha, \beta, \gamma)$ from the molecule whose equilibrium is being considered, and if $\chi(r)$[1] be the law of force between molecules,

$$f(r) = \pm[r\chi'(r) - \chi(r)].$$

We are thus led to biconstant formulae, when we make no assumption as to molecular action. But Cauchy continues, if without altering sensibly the sums $G$ and $R$ *we can neglect such of the molecules as are most near to the molecule in question we may replace the summations by integrals.* He then obtains (p. 204) a formula also given with the same assumptions by Poisson,

$$\int_0^\infty r^4\chi'(r)\,dr = -4\int_0^\infty r^3\chi(r)\,dr:$$

which leads to the absurd result $G + R = 0$. Compare Poisson's memoir of April, 1828, and our Arts. 442 and 443. Now the words in italics are exactly the assumption made by Poisson (see Art. 144), or Cauchy here proves that Poisson's own method is hardly more accurate than that of Navier. Compare the remarks of Saint-Venant on this point in the *Historique Abrégé*, page clxiii.

On his page 211, Cauchy shews that if we put with Navier $\lambda = \mu$ we must consider $G$ such a quantity that the ratio $G/R$ is negligible. He thus recognises uni-constant isotropy as a very special case of bi-constant isotropy.

On voit au reste que, si l'on considère un corps élastique comme un système de points matériels qui agissent les uns sur les autres à de très-petites distances, les lois de l'équilibre ou du mouvement intérieur

[1] Cauchy uses for $\chi(r)$ the very confusing $f(r)$, which can hardly be distinguished from $f(r)$.

de ce corps seront exprimées dans beaucoup de cas par des équations différentes de celles qu'a données M. Navier (page 212).

In fact Navier's equations apply to a particular case and not to the problem in all its generality.

See Moigno's *Statique*, pages 692—695, 702.

[616.] An article entitled: *De la pression ou tension dans un système de points matériels* occupies pages 213—236 of the volume. In the preceding article Cauchy had obtained the ordinary equations of the equilibrium of elasticity without formally introducing the stresses; in this article he explicitly introduces the stresses. The method of summation is used as in the preceding article. See Moigno, p. 675.

The importance of this memoir lies in the step made from a molecular to a continuous state, i.e. the transition from the consideration of the force upon a molecule exerted by its neighbours to the consideration of the stress, or what Cauchy terms pressure, upon a small plane at a point. It therefore involves a definition of such stress in terms of molecular forces. What is the stress across a plane in terms of the action upon each other of the molecules on either side of it? Cauchy's definition is not perhaps entirely satisfactory: see Art. 678. It runs as follows, $OO'O''$ being a small plane in the medium:

> Le produit $p_1 s$ de la pression ou tension $p_1$ par la surface élémentaire $s$ ne sera autre chose que la résultante des actions exercées par les molécules $m_1$, $m_2$,...sur les molécules comprises dans le plan $OO'O''$ et sur celles des molécules $m$, $m'$, $m''$, qui seront situées tout près de la surface $s$.

The investigation is somewhat similar to that of Poisson's, but more general in that it involves nine constants, and makes no assumption as to 'neglecting the irregular part of the action of the molecules in the immediate neighbourhood of the one considered.' On his page 226 Cauchy says:

> On voit par les détails dans lesquels nous venons d'entrer que, pour obtenir l'égalité de pression en tous sens, dans un système de molécules qui se repoussent, on n'a pas besoin d'admettre, comme l'a fait M. Poisson, une distribution particulière des molécules autour de l'une quelconque d'entre elles (voyez dans les Annales de physique et de

chimie un extrait du Mémoire présenté par M. Poisson à l'Académie des Sciences, le 1[er] octobre 1827).

The paper by Poisson to which Cauchy alludes is that of our Art. 431; he alludes to it again on his page 230.

The values of the stresses obtained are of the type

$$\widehat{xx} = G\left(1 + 2\frac{du}{dx}\right) + L\frac{du}{dx} + R\frac{dv}{dy} + Q\frac{dw}{dz},$$

$$\widehat{yz} = (P + I)\frac{dv}{dz} + (P + H)\frac{dw}{dy}.$$

Here $G$, $H$, $I$ are the initial components of stress at $(x, y, z)$ and may be represented by $\widehat{xx}_0$, $\widehat{yy}_0$, $\widehat{zz}_0$.

In the case of isotropy $\widehat{xx}_0 = \widehat{yy}_0 = \widehat{zz}_0$ and the formulae reduce to

$$\widehat{xx} = \widehat{xx}_0\left(1 + 2\frac{du}{dx}\right) + 2R\frac{du}{dx} + R\theta,$$

$$\widehat{yz} = (R + \widehat{xx}_0)\left(\frac{dv}{dz} + \frac{dw}{dy}\right).$$

Hence if the initial stress vanishes or $\widehat{xx}_0 = 0$, we obtain the usual formulae for *uni-constant* isotropy. That is to say, without Poisson's assumptions we come exactly to his result. Cauchy appears to consider that because he has two constants here, $R$ and $xx_0$, and he had two in the last memoir considered that these results confirm each other and point to bi-constant isotropy (page 236). With this end in view he uses rather a misleading notation which has been censured by Saint-Venant as likely to lead to confusion by introducing an apparent similarity between two cases really different: see Moigno's *Statique*, pp. 705—706, and Saint-Venant's edition of *Navier's Leçons*, page 655. The value of $\widehat{xx}_0$ as given by Cauchy page 222 is

$$\Sigma\left[\pm\frac{mr}{2}\cos^2\alpha\,\chi(r)\right],$$

and this is identical in *form* with the value of $G$ in the preceding memoir, the *form* of $R$ is also identical. The condition that the molecules should be initially in a state of equilibrium is expressed on page 191 by equations of the type

$$\Sigma\left[\pm m\cos\alpha\,\chi(r)\right] = 0 \quad \text{..................(i).}$$

The condition that there shall be no initial state of stress is given in the second memoir by equations of the type

$$\widehat{xx}_0 = \Sigma\left[\pm\frac{mr}{2}\cos^2\alpha\,\chi(r)\right] = 0 \ldots\ldots\ldots\ldots\ldots\text{(ii)}.$$

It seems to me that (i) must hold for the molecules of the second memoir always; and (ii) must hold for the molecules of the first memoir if there be no initial stress. This is not distinctly pointed out by Cauchy.

If we put $G=0$ in the first memoir and $\widehat{xx}_0 = 0$ in the second we get precisely the same body shift-equations for equilibrium, or the two methods agree in giving the same uni-constant formulae for isotropy. On the other hand if we do not put $G=0$ in the first set, but $\widehat{xx}_0 = 0$ in the second, we get two different forms for the equations in the case of no initial stress. In the former case bi-constant, in the latter uni-constant isotropy is the result. The stresses not having been introduced in the first memoir, we appear to arrive always at bi-constant isotropy; the fact being that one constant, $G$, depends on an initial stress. How to express the condition for no initial stress, when the stresses are not introduced, is not obvious, but the second memoir shews us that it must be $G=0$. We thus reach uni-constant isotropy in both cases. Otherwise we should have to conclude that the two methods, the one of considering the equilibrium of a single molecule, the second of finding expressions for the stresses lead to different results—the first to bi-constant, the second to uni-constant isotropy. I cannot find that Cauchy had grasped this difficulty.

Saint-Venant, pp. 261, 266. Further, Moigno, pp. 618, 673, 684, 702, 704.

617. An article entitled: *Sur quelques théorêmes relatifs à la condensation ou à la dilatation des corps* occupies pages 237—244 of the volume. This contains theorems which are easy deductions from the *strain-ellipsoid* (*Verschiebungsellipsoid*, Clebsch, p. 43); but it may be interesting to give some account of Cauchy's own method: see Arts. 605, 612.

Let $x$, $y$, $z$ be the coordinates of a point of the body in its second state, that is the state into which it has been brought by the operation of certain forces; suppose that the coordinates of the same point in the first state of the body are $x-\xi$, $y-\eta$, $z-\zeta$ respectively. Let the coordinates of an adjacent point in the second state be $x+\Delta x$, $y+\Delta y$, $z+\Delta z$. Let $r_0$ denote the distance between the points in the first state; $\alpha_0$, $\beta_0$, $\gamma_0$ the direction angles of the straight line $r_0$; in the second state of the body let $r$ denote the distance, and $\alpha$, $\beta$, $\gamma$ the direction angles.

Then the original coordinate of the first point being $x-\xi$, that of the second point is

$$x+\Delta x-\left\{\xi+\frac{d\xi}{dx}\Delta x+\frac{d\xi}{dy}\Delta y+\frac{d\xi}{dz}\Delta z+\ldots\ldots\right\}:$$

and similar expressions hold for the other coordinates. Hence, neglecting squares and products of $\Delta x$, $\Delta y$, and $\Delta z$, we have three equations of the type

$$r_0\cos\alpha_0=\Delta x-\left(\frac{d\xi}{dx}\Delta x+\frac{d\xi}{dy}\Delta y+\frac{d\xi}{dz}\Delta z\right)\ldots\ldots\ldots\ldots\ldots(1).$$

Also

$$r\cos\alpha=\Delta x,\quad r\cos\beta=\Delta y,\quad r\cos\gamma=\Delta z,\ldots\ldots\ldots\ldots\ldots(2).$$

Substitute from (2) in (1), squaring and adding, we obtain

$$\begin{aligned}\left(\frac{r_0}{r}\right)^2=&\left(\cos\alpha-\frac{d\xi}{dx}\cos\alpha-\frac{d\xi}{dy}\cos\beta-\frac{d\xi}{dz}\cos\gamma\right)^2\\&+\left(\cos\beta-\frac{d\eta}{dx}\cos\alpha-\frac{d\eta}{dy}\cos\beta-\frac{d\eta}{dz}\cos\gamma\right)^2\\&+\left(\cos\gamma-\frac{d\zeta}{dx}\cos\alpha-\frac{d\zeta}{dy}\cos\beta-\frac{d\zeta}{dz}\cos\gamma\right)^2\end{aligned}$$

This may obviously be put in the form

$$\begin{aligned}\left(\frac{r_0}{r}\right)^2=A\cos^2\alpha+B\cos^2\beta+C\cos^2\gamma+2D\cos\beta\cos\gamma+2E\cos\gamma\cos\alpha&\\+2F\cos\alpha\cos\beta\ldots\ldots\ldots(3).&\end{aligned}$$

Put $1+s_r$ for $\dfrac{r}{r_0}$ and suppose that from the point $(x, y, z)$ we measure in the direction corresponding to $(\alpha, \beta, \gamma)$ a length $1+s_r$; let the coordinates of the end of the straight line measured from $(x, y, z)$ be $x_1$, $y_1$, $z_1$: then

$$\frac{x_1}{\cos\alpha}=\frac{y_1}{\cos\beta}=\frac{z_1}{\cos\gamma}=1+s_r.$$

Thus (3) becomes

$$Ax_1^2+By_1^2+Cz_1^2+2Dy_1z_1+2Ez_1x_1+2Fx_1y_1=1\ldots\ldots\ldots\ldots(4).$$

Thus a small portion of the body which is spherical in the first state becomes in the second state an ellipsoid similar and similarly situated to (4). From this Cauchy easily deduces a result which he states thus in words:

Supposons qu'un corps se condense ou se dilate par l'effet d'une cause quelconque. Concevons d'ailleurs que l'on construise, dans l'état primitif du corps, une sphère infiniment petite, qui ait pour centre la molécule $m$, et qui renferme en outre un grand nombre de molécules voisines, puis, dans le second état du corps, l'ellipsoïde dans lequel cette sphère s'est transformée. Les molécules primitivement situées près de la molécule $m$ 1°. sur un diamètre de la sphère, 2°. dans un plan perpendiculaire à ce diamètre, se trouveront transportées, après le changement d'état du corps 1°. sur le diamètre de l'ellipsoïde correspondant au diamètre donné de la sphère, 2°. dans le plan diamétral parallèle aux plans tangents menés à l'ellipsoïde par les extrémités du nouveau diamètre.

Other simple theorems are also stated. See Moigno, p. 650.

618. An article entitled: *Sur l'équilibre et le mouvement d'une lame solide* occupies pages 245—325 of the volume, and is followed by an *Addition* on page 327. This long memoir forms by far the most elaborate discussion based on the received theory of elasticity which the problem of the elastic lamina has received, except the researches of Saint-Venant on flexure; the formulae which are used are those which involve *two* elastic constants. The memoir is divided into four sections.

The method which Cauchy adopts in this memoir and in others which relate to an elastic plate and to the problem of torsion in this and the next volume of the *Exercices* is peculiar; we shall give later a portion of the memoir relating to an elastic plate as a specimen. Cauchy himself afterwards adopted another method due to Saint-Venant[1].

[1] [Moigno's *Statique* which is based upon Cauchy's *Exercices*, revised and brought up to date by Saint-Venant for that part which relates to Elasticity has the following footnote on p. 616: Ces Leçons comprennent la *statique* de l'élasticité dans ce qu'elle a de plus général. Nous ne rapportons pas les applications que Cauchy en a faites (*Exercices*, troisième et quatrième années) à la théorie de la flexion, et surtout à celle de la torsion pour laquelle il a ouvert une voie nouvelle, parce qu'il a adopté (*Comptes rendus des séances de l'Académie des Sciences*, 20 février 1854 t. XXXVIII.

619. The first section of the memoir is entitled: *Considérations générales;* it occupies pages 245—247. The main point here is the definition of the body with which the memoir is concerned. Imagine a thin shell bounded by cylindrical surfaces; the cylinders are not to be limited to the circular form, but the generating lines of both surfaces are to be parallel. Take a portion of the shell bounded by two planes parallel to the generating lines, and at any distance apart: such a portion is called a *plate* (*plaque*). Cut the *plate* by two planes at right angles to the generating lines, and very close together; the intercepted part is called a *lamina* (*lame*). The problem to be discussed then is the equilibrium and motion of an elastic lamina; this is simpler than the general problem of the equilibrium and motion of an elastic *body*, for we need only regard forces in one plane, namely a plane at right angles to the generating lines of the cylindrical surfaces: this plane may be called the *plane of the lamina.* We shall have only *two* coordinates of any point to deal with.

620. The second section of the memoir is entitled: *Equations d'équilibre ou de mouvement d'une lame naturellement droite et d'une épaisseur constante;* it occupies pages 247—277. The cylindrical surfaces of Art. 619, bounding the plate from which the lamina is obtained, now become two parallel planes. The section is almost entirely a process of approximation, but is of a reasonably satisfactory kind.

A few remarks may be made on special parts.

621. At a certain point of his investigations Cauchy finds that he has three constants which remain undetermined; such a fact has been noticed and explained in other parts of our subject. The remark which Cauchy makes respecting it on his page 261 does not seem very satisfactory:

p. 329) une autre manière de les traiter, due à M. de Saint-Venant, et qui donne des résultats sensiblement différents, confirmés par diverses expériences, ainsi que par des recherches analytiques entreprises par M. Kirchhoff, l'éminent professeur de Heidelberg (*Ueber das Gleichgewicht und die Bewegung eines unendlich dünnen elastischen Stabes*, t. LVI. du *Journal de Crelle*, p. 285).

These remarks may serve to explain the frequent references to Moigno's *Statique* and Saint-Venant in these pages devoted to Cauchy. ED.]

Il était facile, au reste, de prévoir ces résultats, attendu qu'on ne trouble pas l'équilibre d'une lame élastique dont les extrémités sont libres, lorsqu'on la déplace très-peu, en faisant tourner cette lame sur elle-même, ou en transportant l'une de ses extrémités sur une droite parallèle soit à l'axe des $x$, soit à l'axe des $y$.

622. Cauchy shews on his page 262 that the moment of the elastic tensions on a transverse section of the lamina, taken with respect to an axis at right angles to the plane of the lamina through the mean line, varies as the product of the breadth of the lamina, into the cube of the thickness, into the curvature. This, as he remarks, verifies the hypothesis of James Bernoulli: see too, Poisson's *Mécanique*, Vol. I. page 606. But the solution of James Bernoulli is unsatisfactory, because he assumed the tension at every point of a transverse section of the lamina to be at right angles to the section.

623. On pages 268 and 269 Cauchy cites the integral of a certain differential equation which he had given in his *Mémoire sur l'application du calcul des résidus aux questions de physique mathématique;* I consider the integral incorrect, or at least not demonstrated, and have discussed the question elsewhere.

624. Cauchy finishes the second section of his memoir by the following paragraph:

Parmi les formules obtenues dans ce paragraphe, celles qui sont relatives à l'équilibre ou au mouvement d'une lame élastique sollicitée par une force accélératrice constante et constamment parallèle à elle-même coïncident avec des formules déjà connues, et particulièrement avec celles que renferme le Mémoire d'Euler, intitulé: *Investigatio motuum quibus laminæ et virgæ elasticæ contremiscunt* [voy. les *Acta Academiæ petropolitanæ* pour l'année 1779]. Elles doivent donc s'accorder aussi avec celles que renferme le Mémoire présenté par M. Poisson à l'Académie des Sciences, le 14 avril dernier. En effet, après avoir annoncé, dans les Annales de chimie, qu'il déduit de la considération des forces moléculaires les équations relatives soit à tous les points, soit aux extrémités des cordes et des verges, des membranes et des plaques élastiques, M. Poisson ajoute: *Parmi ces équations, celles qui répondent au contour d'une plaque élastique pliée d'une manière quelconque, et celles qui appartiennent à tous les points d'une plaque ou d'une membrane qui est restée plane n'avaient pas encore été données; les autres coïncident avec les équations trouvées par différents moyens.* Il paraît d'ailleurs par ce passage que M. Poisson s'est occupé seulement des lames et des plaques élastiques d'une épaisseur constante, qui, étant naturellement planes, ne cessent de l'être qu'autant qu'elles sont pliées et courbées par l'action d'une cause extérieure. Lorsqu'une lame ou plaque est dénuée d'élasticité, ou naturellement courbe, ou d'une épaisseur variable, on parvient à des équations d'équilibre ou de mouvement qui sont très-distinctes des équations déjà connues, et ne sont pas indiquées dans le passage cité. C'est ce que montrent les calculs ci-dessus effectués, et ceux que nous développerons ci-après ou dans de nouveaux articles.

In a note Cauchy says with respect to Poisson's memoir of April, 1828.

Ce beau Mémoire, dans lequel M. Poisson a déduit le premier de la considération des forces moléculaires les équations relatives à l'équilibre ou au mouvement des cordes, des verges, des membranes et des plaques élastiques, s'imprime en ce moment, et doit paraître dans le tome VIII. des Mémoires de l'Académie des Sciences.

625. The third section of the memoir is entitled: *Equations d'équilibre ou de mouvement d'une lame naturellement droite, mais d'une épaisseur variable,* and occupies pages 277—284. This is throughout an approximate process; the thickness though not assumed constant is taken to be always small.

626. The fourth section of the Memoir is entitled: *Equations d'équilibre ou de mouvement d'une lame naturellement courbe et d'une épaisseur constante;* it occupies pages 285—326. This form of the problem is rather complex; but the process of approximation is very clearly developed so that it can be readily followed. Cauchy obtains on his page 312 the result that the moment of elasticity is proportional to the product of the breadth of the lamina, into the cube of the thickness, into the change of the curvature; this result he says agrees with the hypothesis admitted by Euler in the *Novi Commentarii* and the *Acta Academiæ Petropolitanæ* for the years 1764 and 1779. I do not know what is meant by the date 1764. In the volume of the *Novi Comm.* for 1764 there are memoirs by Euler, but I cannot find any thing to which Cauchy's words are

applicable[1]. For an application of his general formulae Cauchy supposes that the natural form of the lamina is a circle; he obtains numerical results as to the vibrations in this case, and he says on his page 326 that these are quite in accordance with experiments made by Savart.

627. The formulae of the theory of elasticity which Cauchy uses are those with *two* constants; in the *Addition* to the memoir on page 327 he shews how his results will be modified if he uses formulae slightly more general: this is of service in a later article of the volume beginning on page 356.

628. An article entitled: *Sur l'équilibre et le mouvement d'une plaque solide* occupies pages 328—355 of the volume; it is divided into three sections.

The first section is entitled: *Considérations générales* and occupies pages 328—330. This explains the notation to be used, and gives the three equations which must hold at every point in the interior of an elastic body in equilibrium, and the three which must hold at every point of the surface.

629. The second section is entitled: *Équations d'équilibre ou de mouvement d'une plaque naturellement plane et d'une épaisseur constante;* it occupies pages 330—348. This constitutes a discussion of what I call *Poisson's Problem.* Cauchy's process is somewhat more systematic than Poisson's, and he uses *two* constants of elasticity whereas Poisson used only one; by supposing finally one of his constants double the other he brings his equations into precise agreement with those of Poisson, except as to one unimportant point which will presently be noticed. I reproduce the whole as a confirmation of Poisson's investigation, and as a specimen of the method used by Cauchy throughout pages 245—368 of this volume of the *Exercices,* and pages 1—29, and 47—64 of the next.

630. Let us consider a solid plate which in its natural state is comprised between two curved surfaces very near to each other. Let us suppose moreover that after a change of form in the plate we apply to the molecules of which it consists given accelerating

1 [Cauchy possibly refers to the memoir of 1766 cited in my footnote, p. 55. Ed.]

forces, and also to the surfaces which bound it external pressures normal to these surfaces. Let us refer all points of space to three rectangular axes, and in the state of equilibrium or motion of the plate, let

$m$ denote any molecule whatever,

$x, y, z$ the co-ordinates of this molecule,

$\rho$ the density of the plate at the point $(x, y, z)$,

$\phi$ the accelerating force applied at the molecule $m$,

$p', p'', p'''$ the stresses exerted at the point $(x, y, z)$ on planes perpendicular to the axis of $x$, the axis of $y$, and the axis of $z$,

$A, F, E$ the algebraic projections of the stress $p'$ on the axes of co-ordinates[1],

$F, B, D$ the algebraic projections of the stress $p''$,

$E, D, C$ the algebraic projections of the stress $p'''$.

Then if there is equilibrium we shall have, by Art. 613, three equations of the type:

$$\frac{dA}{dx} + \frac{dF}{dy} + \frac{dE}{dz} + \rho X = 0 \quad \text{..................(1).}$$

But if the plate is in motion, and we denote by $\chi', \chi'', \chi'''$, the resolved parts of the effective accelerating force at $m$ parallel to the axes we shall have three equations of the type:

$$\frac{dA}{dx} + \frac{dF}{dy} + \frac{dE}{dz} + \rho (X - \chi') = 0 \quad \text{..............(2).}$$

Also, both in equilibrium and in motion, let $\alpha$, $\beta$, $\gamma$ denote the angles comprised between the semi-axes of positive co-ordinates and another semi-axis $OO'$ drawn arbitrarily from the point $(x, y, z)$; let $p$ be the pressure or tension exerted at the point $(x, y, z)$ on the plane perpendicular to this semi-axis and on the side which corresponds to it; let $\lambda$, $\mu$, $\nu$ be the angles formed by the direction of the force $p$ with the semi-axes of positive co-ordinates; we shall have three equations of the type:

$$p \cos\lambda = A \cos\alpha + F \cos\beta + E \cos\gamma \quad \text{............(3).}$$

Finally if we suppose the point $(x, y, z)$ situated on one of the curved surfaces which bound the plate, and if we make the semi-axis $OO'$ coincide with the normal to this surface, the preceding

[1] [I have thought it better throughout this investigation to preserve Cauchy's original notation, as the subscripts (see Art. 632) would have rendered the double-suffix notation too cumbrous. Ed.]

values of $p\cos\lambda$, $p\cos\mu$, $p\cos\nu$ will coincide, disregarding the sign, with the algebraic projections of the external pressure applied at this surface in a normal direction. Thus, if we denote by $P$ the external pressure which corresponds to the point $(x, y, z)$ we shall have three equations of the type:

$$A\cos\alpha + F\cos\beta + E\cos\gamma = -P\cos\alpha, \ldots\ldots\ldots\ldots(4).$$

We must remember that these last formulæ hold only for the surfaces mentioned above.

631. It remains to shew how from the equations (1), (2) and (4), we can deduce those which determine at any instant whatever, in the state of equilibrium or in that of motion, the form of the plate, and in particular the various changes in the form of the curved surface which originally divided the thickness of the plate into two equal parts. However, as the determination of this surface, which we will call the mean surface, is effected in different ways, and involves calculations more or less extensive, according as we consider a plate elastic or non-elastic, of constant thickness or variable thickness, we will refer the development of these calculations to the following paragraphs.

Suppose that in the natural state of the plate the two curved surfaces which bound it reduce to two parallel planes separated from each other by a very small distance. Denote by $2i$ this distance, or the natural thickness of the plate; and take for the plane of $x$, $y$ that which originally divided this thickness into two equal parts. Suppose moreover that in the transition from the natural state to the state of equilibrium or of motion, the displacements of the molecules are very small. The mean surface which coincided in the natural state with the plane of $x$, $y$ will become curved by reason of the change in the form of the plate; but its ordinate will remain very small: denote this ordinate by $f(x, y)$. Let $x$, $y$, $z$ be the co-ordinates of any molecule $m$ of the plate; let $s$ be the difference between the ordinates $z$ and $f(x, y)$, measured on the same straight line perpendicular to the plane of $x, y$, so that we have generally

$$z = f(x, y) + s \ldots\ldots\ldots\ldots\ldots\ldots\ldots (5).$$

Finally let

$$x - \xi,\ y - \eta,\ z - \zeta \ldots\ldots\ldots\ldots\ldots\ldots\ldots(6)$$

be the primitive co-ordinates of the molecule $m$; then $\xi$, $\eta$, $\zeta$ will

when we carry the approximation with respect to $A$, $F$, $B$ as far as terms of the same order as $i$, and with respect to $E$, $D$, $C$ as far as terms which are of the order $i^2$. In fact the approximate values in question are respectively

$$A = A_0 + A_1 s, \quad F = F_0 + F_1 s, \quad B = B_0 + B_1 s \ldots\ldots\ldots (28)$$

and

$$E = E_0 + E_2 \frac{s^2}{2} = E_0\left(1 - \frac{s^2}{i^2}\right), \quad D = D_0 + D_2 \frac{s^2}{2} = D_0\left(1 - \frac{s^2}{i^2}\right) \ldots (29),$$

$$C = C_0 + C_2 \frac{s^2}{2} \ldots\ldots\ldots\ldots\ldots\ldots\ldots\ldots (30),$$

or, which is the same thing,

$$\left.\begin{aligned} E &= \tfrac{1}{2}\left(\frac{dA_1}{dx} + \frac{dF_1}{dy} + \rho X_1\right)(i^2 - s^2) \\ D &= \tfrac{1}{2}\left(\frac{dF_1}{dx} + \frac{dB_1}{dy} + \rho Y_1\right)(i^2 - s^2) \end{aligned}\right\} \ldots\ldots\ldots\ldots (31),$$

$$C = -P + \tfrac{1}{2}\rho Z_1 (i^2 - s^2) \ldots\ldots\ldots\ldots\ldots\ldots (32).$$

635. The equations (20) and (27) are the only equations, which in the case of the equilibrium of the solid plate, hold for all the points of the mean surface. Suppose now that the plate is bounded laterally, in its natural state, by planes perpendicular to the plane of $xy$, or by a cylindrical surface having its generating lines parallel to the axis of $z$. Suppose that this cylindrical surface is subjected to a normal pressure $P'$ different from $P$; then denoting by

$$\alpha, \ \beta, \text{ and } \gamma = \frac{\pi}{2},$$

the angles which the normal to the cylindrical surface produced outwards makes with the semi-axes of $x$, $y$, $z$, we replace in equations (4) $\cos\gamma$ by zero and $P$ by $P'$: thus we have for all points situated on the boundary of the plate and so for all values of $s$

$$\left.\begin{aligned} A\cos\alpha + F\cos\beta = -P\cos\alpha, \quad F\cos\alpha + B\cos\beta = -P'\cos\beta, \\ E\cos\alpha + D\cos\beta = 0 \end{aligned}\right\} (33).$$

Then combining equations (33) with the formulae (28) and (29) we deduce

$$(A_0 + P')\cos\alpha + F_0\cos\beta = 0, \quad F_0\cos\alpha + (B_0 + P')\cos\beta = 0 \ldots (34),$$

$$A_1\cos\alpha + F_1\cos\beta = 0, \quad F_1\cos\alpha + B_1\cos\beta = 0 \ldots\ldots (35),$$

and $$E_0 \cos\alpha + D_0 \cos\beta = 0 \;\ldots\ldots\ldots\ldots\ldots\ldots \;(36),$$

or, which comes to the same thing,

$$\left(\frac{dA_1}{dx} + \frac{dF_1}{dy} + \rho X_1\right)\cos\alpha + \left(\frac{dF_1}{dx} + \frac{dB_1}{dy} + \rho Y_1\right)\cos\beta = 0 \ldots (37).$$

The five conditions expressed by (34), (35) and (37) must be satisfied for all points situated on *free* portions of the cylindrical surface which forms the lateral boundary of the plate. As to the points situated on *fixed* portions of the latter surface they must satisfy other conditions which we will now investigate.

636. Let

$$\left.\begin{array}{c} \xi = \xi_0 + \xi_1 s + \xi_2 \dfrac{s^2}{2} + \ldots, \quad \eta = \eta_0 + \eta_1 s + \eta_2 \dfrac{s^2}{2} + \ldots, \\ \zeta = \zeta_0 + \zeta_1 s + \zeta_2 \dfrac{s^2}{2} + \ldots \end{array}\right\} \ldots (38),$$

be the developments of $\xi$, $\eta$, $\zeta$ considered as functions of $x$, $y$, $s$ according to ascending powers of the variable $s$; then $\xi_0$, $\eta_0$, $\zeta_0$ will represent the displacements of the point $(x, y)$ of the mean surface, measured parallel to the axes of co-ordinates. If we neglect in the values of $\xi$, $\eta$, $\zeta$ the terms proportional to the square of $s$ we shall have simply

$$\xi = \xi_0 + \xi_1 s, \quad \eta = \eta_0 + \eta_1 s, \quad \zeta = \zeta_0 + \zeta_1 s \ldots\ldots\ldots (39).$$

Granting this, let us now suppose that a portion of the cylindrical surface which forms the lateral boundary of the plate in its natural state becomes fixed; or rather, that among the points situated on a portion of this surface those which belong to the boundary of the mean surface become fixed, the others being so restricted that every one of them remains constantly on the same generating line of the cylindrical surface. We shall have then for all the points situated on a fixed portion of the last surface, not only

$$\xi_0 = 0, \quad \eta_0 = 0, \quad \zeta_0 = 0 \;\ldots\ldots\ldots\ldots\ldots\ldots \;(40),$$

but also $\xi = 0$ and $\eta = 0$ whatever $s$ may be. Hence we shall obtain from the formulae (39)

$$\xi_1 = 0, \quad \eta_1 = 0 \ldots\ldots\ldots\ldots\ldots\ldots\ldots\ldots (41).$$

In the particular case where the solid plate is rectangular, and bounded laterally by planes perpendicular to the axes of $x$ and $y$, the formulae (34), (35) and (37) give

1°. for the points situated on the surface, supposed free, of one of the planes perpendicular to the axis of $x$,

$$A_0 + P' = 0, \quad F_0 = 0 \quad\ldots\ldots (42),$$

$$A_1 = 0, \quad F_1 = 0 \quad\ldots\ldots (43),$$

$$\frac{dA_1}{dx} + \frac{dF_1}{dy} + \rho X_1 = 0 \quad\ldots\ldots (44),$$

2°. for the points situated on the surface, supposed free, of one of the planes perpendicular to the axis of $y$

$$B_0 + P' = 0, \quad F_0 = 0 \quad\ldots\ldots (45),$$

$$B_1 = 0, \quad F_1 = 0 \quad\ldots\ldots (46),$$

$$\frac{dF_1}{dx} + \frac{dB_1}{dy} + \rho Y_1 = 0 \quad\ldots\ldots (47).$$

637. If we consider the solid plate, no longer in the state of equilibrium, but in the state of motion, we must in the equations (20) and (27), and in the formulae (37) replace the quantities

$$X_0,\ X_1;\quad Y_0,\ Y_1;\quad Z_0,\ Z_1,\ Z_2 \quad\ldots\ldots (48)$$

by the differences

$$\left.\begin{matrix} X_0 - \dfrac{d^2\xi_0}{dt^2},\ X_1 - \dfrac{d^2\xi_1}{dt^2};\quad Y_0 - \dfrac{d^2\eta_0}{dt^2},\ Y_1 - \dfrac{d^2\eta_1}{dt^2}; \\ Z_0 - \dfrac{d^2\zeta_0}{dt^2},\ Z_1 - \dfrac{d^2\zeta_1}{dt^2},\ Z_2 - \dfrac{d^2\zeta_2}{dt^2} \end{matrix}\right\} \ldots\ldots (49).$$

Hence we shall obtain

1°. from the equations (20)

$$\frac{dA_0}{dx} + \frac{dF_0}{dy} + \rho X_0 = \rho\,\frac{d^2\xi_0}{dt^2}, \quad \frac{dF_0}{dx} + \frac{dB_0}{dy} + \rho Y_0 = \rho\,\frac{d^2\eta_0}{dt^2} \ldots\ldots (50),$$

2°. from the equation (27) when we reduce the polynomial

$$\zeta_0 + \frac{i^2}{6}\left(\zeta_2 + 2\,\frac{d\xi_1}{dx} + 2\,\frac{d\eta_1}{dy}\right)$$

to the single term $\zeta_0$,

$$\frac{i^2}{3}\left(\frac{d^2A_1}{dx^2} + 2\,\frac{d^2F_1}{dxdy} + \frac{d^2B_1}{dy^2}\right) + \rho\frac{i^2}{6}\left(Z_2 + 2\,\frac{dX_1}{dx} + 2\,\frac{dY_1}{dy}\right) = \rho\,\frac{d^2\zeta_0}{dt^2} \quad (51),$$

3°. from the formula (37)

$$\left(\frac{dA_1}{dx} + \frac{dF_1}{dy} + \rho X_1\right)\cos\alpha + \left(\frac{dF_1}{dx} + \frac{dB_1}{dy} + \rho Y_1\right)\cos\beta =$$

$$\rho\left(\frac{d^2\xi_1}{dt^2}\cos\alpha + \frac{d^2\eta_1}{dt^2}\cos\beta\right) \ldots (52).$$

We may add that in the case of motion the values of $E, D, C$ furnished by the equations (31) and (32) will become

$$\left.\begin{aligned} E&=\tfrac{1}{2}\left(\frac{dA_1}{dx}+\frac{dF_1}{dy}+\rho X_1-\rho\frac{d^2\xi_1}{dt^2}\right)(i^2-s^2) \\ D&=\tfrac{1}{2}\left(\frac{dF_1}{dx}+\frac{dB_1}{dy}+\rho Y_1-\rho\frac{d^2\eta_1}{dt^2}\right)(i^2-s^2) \end{aligned}\right\}\ldots(53),$$

$$C=-P+\tfrac{1}{2}\rho\left(Z_1-\frac{d^2\zeta_1}{dt^2}\right)(i^2-s^2)\ \ldots\ldots\ldots\ldots (54).$$

638. Suppose now that the given plate becomes elastic, and that the elasticity is the same in all directions. Then by the principles developed by Cauchy in his earlier papers, and taking $x, y, z$ for independent variables, we shall have—

$$\left.\begin{aligned} &A=k\frac{d\xi}{dx}+K\theta,\quad B=k\frac{d\eta}{dy}+K\theta,\quad C=k\frac{d\zeta}{dz}+K\theta \\ &D=\tfrac{1}{2}k\left(\frac{d\eta}{dz}+\frac{d\zeta}{dy}\right),\quad E=\tfrac{1}{2}k\left(\frac{d\zeta}{dx}+\frac{d\xi}{dy}\right), \\ &\qquad\qquad F=\tfrac{1}{2}k\left(\frac{d\xi}{dy}+\frac{d\eta}{dx}\right) \end{aligned}\right\}\ldots(55),$$

where $k$ and $K$ are two constants, and $\theta$ is the cubical dilatation, given by the equation

$$\theta=\frac{d\xi}{dx}+\frac{d\eta}{dy}+\frac{d\zeta}{dz}\ \ldots\ldots\ldots\ldots\ldots\ldots (56).$$

Hence if we put for convenience

$$k+K=\kappa K\ \ldots\ldots\ldots\ldots\ldots\ldots (57),$$

and if we take for independent variables $x, y, s$ instead of $x, y, z$ we shall have

$$\left.\begin{aligned} &\frac{A}{K}=\kappa\frac{d\xi}{dx}+\frac{d\eta}{dy}+\frac{d\zeta}{ds},\quad \frac{B}{K}=\frac{d\xi}{dx}+\kappa\frac{d\eta}{dy}+\frac{d\zeta}{ds}, \\ &\qquad\frac{C}{K}=\frac{d\xi}{dx}+\frac{d\eta}{dy}+\kappa\frac{d\zeta}{ds} \\ &\frac{D}{K}=\frac{\kappa-1}{2}\left(\frac{d\eta}{ds}+\frac{d\zeta}{dy}\right),\quad \frac{E}{K}=\frac{\kappa-1}{2}\left(\frac{d\zeta}{dx}+\frac{d\xi}{ds}\right), \\ &\qquad\frac{F}{K}=\frac{\kappa-1}{2}\left(\frac{d\xi}{dy}+\frac{d\eta}{dx}\right) \end{aligned}\right\}\ldots(58).$$

639. Now let us substitute in the formulae (58) for the functions $A$, $B$, $C$, $D$, $E$, $F$, $\xi$, $\eta$, $\zeta$ their developments arranged according to ascending powers of the variable $s$; then, equating the coefficients of like powers of $s$, we have

$$\left.\begin{aligned}
&\frac{A_0}{K}=\kappa\frac{d\xi_0}{dx}+\frac{d\eta_0}{dy}+\zeta_1,\quad \frac{B_0}{K}=\frac{d\xi_0}{dx}+\kappa\frac{d\eta_0}{dy}+\zeta_1,\\
&\qquad\frac{C_0}{K}=\frac{d\xi_0}{dx}+\frac{d\eta_0}{dy}+\kappa\zeta_1\\
&\frac{D_0}{K}=\frac{\kappa-1}{2}\left(\eta_1+\frac{d\zeta_0}{dy}\right),\quad \frac{E_0}{K}=\frac{\kappa-1}{2}\left(\xi_1+\frac{d\zeta_0}{dx}\right),\\
&\qquad\frac{F_0}{K}=\frac{\kappa-1}{2}\left(\frac{d\xi_0}{dy}+\frac{d\eta_0}{dx}\right)
\end{aligned}\right\}\ \dots (59),$$

$$\left.\begin{aligned}
&\frac{A_1}{K}=\kappa\frac{d\xi_1}{dx}+\frac{d\eta_1}{dy}+\zeta_2,\quad \frac{B_1}{K}=\frac{d\xi_1}{dx}+\kappa\frac{d\eta_1}{dy}+\zeta_2,\\
&\qquad\frac{C_1}{K}=\frac{d\xi_1}{dx}+\frac{d\eta_1}{dy}+\kappa\zeta_2\\
&\frac{D_1}{K}=\frac{\kappa-1}{2}\left(\eta_2+\frac{d\zeta_1}{dy}\right),\quad \frac{E_1}{K}=\frac{\kappa-1}{2}\left(\xi_2+\frac{d\zeta_1}{dx}\right),\\
&\qquad\frac{F_1}{K}=\frac{\kappa-1}{2}\left(\frac{d\xi_1}{dy}+\frac{d\eta_1}{dx}\right)
\end{aligned}\right\}\ \dots (60).$$

By combining the formulae (59) and (60) with equations (19) we obtain

$$\xi_1=-\frac{d\zeta_0}{dx},\quad \eta_1=-\frac{d\zeta_0}{dy},\quad \zeta_1=-\frac{1}{\kappa}\left(\frac{d\xi_0}{dx}+\frac{d\eta_0}{dy}+\frac{P}{K}\right)\dots (61),$$

$$\xi_2=-\frac{d\zeta_1}{dx},\quad \eta_2=-\frac{d\zeta_1}{dy},\quad \zeta_2=-\frac{1}{\kappa}\left(\frac{d\xi_1}{dx}+\frac{d\eta_1}{dy}\right)\ \dots (62),$$

$$\left.\begin{aligned}
&A_0=(\kappa-1/\kappa)\,K\frac{d\xi_0}{dx}+(1-1/\kappa)\,K\frac{d\eta_0}{dy}-\frac{P}{\kappa}\\
&B_0=(1-1/\kappa)\,K\frac{d\xi_0}{dx}+(\kappa-1/\kappa)\,K\frac{d\eta_0}{dy}-\frac{P}{\kappa}\\
&F_0=\frac{\kappa-1}{2}K\left(\frac{d\xi_0}{dy}+\frac{d\eta_0}{dx}\right)
\end{aligned}\right\}\ \dots\dots (63).$$

$$\left.\begin{aligned}
&A_1=(\kappa-1/\kappa)\,K\frac{d\xi_1}{dx}+(1-1/\kappa)\,K\frac{d\eta_1}{dy}\\
&B_1=(1-1/\kappa)\,K\frac{d\xi_1}{dx}+(\kappa-1/\kappa)\,K\frac{d\eta_1}{dy}\\
&F_1=\frac{\kappa-1}{2}K\left(\frac{d\xi_1}{dy}+\frac{d\eta_1}{dx}\right)
\end{aligned}\right\}\ \dots\dots\dots (64).$$

Thus, if we put for brevity

$$(\kappa - 1/\kappa)\,K = \rho\Omega^2 \qquad (65)$$

we shall have

$$\left.\begin{aligned} A_0 &= \rho\Omega^2\left(\frac{d\xi_0}{dx} + \frac{1}{\kappa+1}\frac{d\eta_0}{dy}\right) - \frac{P}{\kappa} \\ B_0 &= \rho\Omega^2\left(\frac{d\eta_0}{dy} + \frac{1}{\kappa+1}\frac{d\xi_0}{dx}\right) - \frac{P}{\kappa} \\ F_0 &= \tfrac{1}{2}\rho\Omega^2\frac{\kappa}{\kappa+1}\left(\frac{d\xi_0}{dy} + \frac{d\eta_0}{dx}\right) \end{aligned}\right\} \qquad (66)$$

and

$$\left.\begin{aligned} A_1 &= \rho\Omega^2\left(\frac{d\xi_1}{dx} + \frac{1}{\kappa+1}\frac{d\eta_1}{dy}\right) \\ B_1 &= \rho\Omega^2\left(\frac{d\eta_1}{dy} + \frac{1}{\kappa+1}\frac{d\xi_1}{dx}\right) \\ F_1 &= \tfrac{1}{2}\rho\Omega^2\frac{\kappa}{\kappa+1}\left(\frac{d\xi_1}{dy} + \frac{d\eta_1}{dx}\right) \end{aligned}\right\} \qquad (67),$$

or, which is the same thing,

$$\left.\begin{aligned} A_1 &= -\rho\Omega^2\left(\frac{d^2\zeta_0}{dx^2} + \frac{1}{\kappa+1}\frac{d^2\zeta_0}{dy^2}\right) \\ B_1 &= -\rho\Omega^2\left(\frac{d^2\zeta_0}{dy^2} + \frac{1}{\kappa+1}\frac{d^2\zeta_0}{dx^2}\right) \\ F_1 &= -\rho\Omega^2\frac{\kappa}{\kappa+1}\frac{d^2\zeta_0}{dx\,dy} \end{aligned}\right\} \qquad (68).$$

640. Granting this the equations (20) and (27) which relate to the equilibrium of a solid plate will give for an elastic plate

$$\left.\begin{aligned} \Omega^2\left(\frac{d^2\xi_0}{dx^2} + \frac{1}{2}\frac{\kappa}{\kappa+1}\frac{d^2\xi_0}{dy^2} + \frac{1}{2}\frac{\kappa+2}{\kappa+1}\frac{d^2\eta_0}{dx\,dy}\right) + X_0 &= 0 \\ \Omega^2\left(\frac{d^2\eta_0}{dy^2} + \frac{1}{2}\frac{\kappa}{\kappa+1}\frac{d^2\eta_0}{dx^2} + \frac{1}{2}\frac{\kappa+2}{\kappa+1}\frac{d^2\xi_0}{dx\,dy}\right) + Y_0 &= 0 \end{aligned}\right\} \quad \ldots (69)$$

and

$$\Omega^2\frac{i^2}{3}\left(\frac{d^4\zeta_0}{dx^4} + 2\frac{d^4\zeta_0}{dx^2dy^2} + \frac{d^4\zeta_0}{dy^4}\right) = Z_0 + \frac{i^2}{6}\left(Z_2 + 2\frac{dX_1}{dx} + 2\frac{dY_1}{dy}\right) \ldots (70).$$

641. Moreover by reason of the formulæ (34), (35) and (37) combined with the equations (66) and (68) we shall have for all the points situated on a free portion of the boundary of the elastic plate

$$\left.\begin{aligned}\Omega^2\left\{\left(\frac{d\xi_0}{dx}+\frac{1}{\kappa+1}\frac{d\eta_0}{dy}\right)\cos\alpha+\tfrac{1}{2}\frac{\kappa}{\kappa+1}\left(\frac{d\xi_0}{dy}+\frac{d\eta_0}{dx}\right)\cos\beta\right\}\\=\frac{1}{\rho}\left(\frac{P}{\kappa}-P'\right)\cos\alpha\\\Omega^2\left\{\left(\frac{d\eta_0}{dy}+\frac{1}{\kappa+1}\frac{d\xi_0}{dx}\right)\cos\beta+\tfrac{1}{2}\frac{\kappa}{\kappa+1}\left(\frac{d\xi_0}{dy}+\frac{d\eta_0}{dx}\right)\cos\alpha\right\}\\=\frac{1}{\rho}\left(\frac{P}{\kappa}-P'\right)\cos\beta\end{aligned}\right\}\dots(71),$$

$$\left.\begin{aligned}\left(\frac{d^2\zeta_0}{dx^2}+\frac{1}{\kappa+1}\frac{d^2\zeta_0}{dy^2}\right)\cos\alpha+\frac{\kappa}{\kappa+1}\frac{d^2\zeta_0}{dx\,dy}\cos\beta=0\\\left(\frac{d^2\zeta_0}{dy^2}+\frac{1}{\kappa+1}\frac{d^2\zeta_0}{dx^2}\right)\cos\beta+\frac{\kappa}{\kappa+1}\frac{d^2\zeta_0}{dx\,dy}\cos\alpha=0\end{aligned}\right\}\dots(72),$$

$$\Omega^2\left\{\left(\frac{d^3\zeta_0}{dx^3}+\frac{d^3\zeta_0}{dx\,dy^2}\right)\cos\alpha+\left(\frac{d^3\zeta_0}{dx^2\,dy}+\frac{d^3\zeta_0}{dy^3}\right)\cos\beta\right\}$$
$$=X_1\cos\alpha+Y_1\cos\beta\;\dots\dots(73).$$

642. On the other hand for all the points situated on a fixed portion of the boundary of the plate the values of the unknown quantities $\xi_0$, $\eta_0$, $\zeta_0$ must satisfy not only the conditions (40) but also the formulae (41), or, which comes to the same thing, see (61), the two following

$$\frac{d\zeta_0}{dx}=0,\quad\frac{d\zeta_0}{dy}=0\;\dots\dots\dots\dots\dots\dots\;(74).$$

643. In the particular case where the solid plate is rectangular and bounded laterally by planes perpendicular to the axes of $x$ and $y$, the formulae (71), (72), (73) give

1°. for the points situated on the surface, supposed free, of one of the planes perpendicular to the axis of $x$,

$$\Omega^2\left(\frac{d\xi_0}{dx}+\frac{1}{\kappa+1}\frac{d\eta_0}{dy}\right)=\frac{1}{\rho}\left(\frac{P}{\kappa}-P'\right),\quad\frac{d\xi_0}{dy}+\frac{d\eta_0}{dy}=0\dots(75),$$

$$\frac{d^2\zeta_0}{dx^2}+\frac{1}{\kappa+1}\frac{d^2\zeta_0}{dy^2}=0,\quad\frac{d^2\zeta_0}{dx\,dy}=0\;\dots\dots\dots\dots(76),$$

$$\Omega^2\left(\frac{d^3\zeta_0}{dx^3}+\frac{d^3\zeta_0}{dx\,dy^2}\right)=X_1\dots\dots\dots\dots\dots\dots(77),$$

2°. for the points situated on the surface, supposed free, of one of the planes perpendicular to the axis of $y$,

$$\Omega^2\left(\frac{d\eta_0}{dy}+\frac{1}{\kappa+1}\frac{d\xi_0}{dx}\right)=\frac{1}{\rho}\left(\frac{P}{\kappa}-P'\right),\quad \frac{d\xi_0}{dy}+\frac{d\eta_0}{dx}=0 \ldots (78),$$

$$\frac{d^2\zeta_0}{dy^2}+\frac{1}{\kappa+1}\frac{d^2\zeta_0}{dx^2}=0,\quad \frac{d^2\zeta_0}{dx\,dy}=0 \ldots\ldots\ldots\ldots (79),$$

$$\Omega^2\left(\frac{d^3\zeta_0}{dx^2dy}+\frac{d^3\zeta_0}{dy^3}\right)=Y_1 \ldots\ldots\ldots\ldots (80).$$

644. If we denote by $\omega$ the sum of the first two terms comprised in the value of $\theta$, in the formula (56), and consequently put

$$\omega=\frac{d\xi}{dx}+\frac{d\eta}{dy} \ldots\ldots\ldots\ldots (81),$$

then $\omega$ will evidently express the superficial dilatation experienced in consequence of the change in the form of the plate by a plane drawn through the point $(x-\xi,\ y-\eta)$ parallel to the plane of $x$, $y$. Moreover if we represent by

$$\omega=\omega_0+\omega_1 s+\omega_2\frac{s^2}{2}+ \ldots\ldots\ldots\ldots (82)$$

the development of $\omega$, considered as a function of the variables $x$, $y$, $s$, according to ascending powers of $s$, we shall have

$$\omega_0=\frac{d\xi_0}{dx}+\frac{d\eta_0}{dy} \ldots\ldots\ldots\ldots (83),$$

$$\omega_1=\frac{d\xi_1}{dx}+\frac{d\eta_1}{dy}=-\left(\frac{d^2\zeta_0}{dx^2}+\frac{d^2\zeta_0}{dy^2}\right) \ldots\ldots\ldots\ldots (84),$$

and the formulae (66), (67) will give

$$\left.\begin{aligned}A_0&=\frac{\rho\Omega^2}{\kappa+1}\left(\kappa\frac{d\xi_0}{dx}+\omega_0\right)-\frac{P}{\kappa}\\ B_0&=\frac{\rho\Omega^2}{\kappa+1}\left(\kappa\frac{d\eta_0}{dy}+\omega_0\right)-\frac{P}{\kappa}\\ F_0&=\tfrac{1}{2}\rho\Omega^2\frac{\kappa}{\kappa+1}\left(\frac{d\xi_0}{dy}+\frac{d\eta_0}{dx}\right)\end{aligned}\right\} \ldots\ldots\ldots\ldots (85),$$

$$\left.\begin{aligned}A_1&=\frac{\rho\Omega^2}{\kappa+1}\left(\kappa\frac{d\xi_1}{dx}+\omega_1\right)\\ B_1&=\frac{\rho\Omega^2}{\kappa+1}\left(\kappa\frac{d\eta_1}{dy}+\omega_1\right)\\ F_1&=\tfrac{1}{2}\rho\Omega^2\frac{\kappa}{\kappa+1}\left(\frac{d\xi_1}{dy}+\frac{d\eta_1}{dx}\right)\end{aligned}\right\} \ldots\ldots\ldots\ldots (86).$$

Moreover the equations (69) will become

$$\left.\begin{array}{l}\dfrac{\Omega^2}{2(\kappa+1)}\left\{\kappa\left(\dfrac{d^2\xi_0}{dx^2}+\dfrac{d^2\xi_0}{dy^2}\right)+(\kappa+2)\dfrac{d\omega_0}{dx}\right\}+X_0=0\\ \dfrac{\Omega^2}{2(\kappa+1)}\left\{\kappa\left(\dfrac{d^2\eta_0}{dx^2}+\dfrac{d^2\eta_0}{dy^2}\right)+(\kappa+2)\dfrac{d\omega_0}{dy}\right\}+Y_0=0\end{array}\right\} \dots (87).$$

Differentiate the first of (87) with respect to $x$, and the second with respect to $y$, and add; thus

$$\Omega^2\left(\frac{d^2\omega_0}{dx^2}+\frac{d^2\omega_0}{dy^2}\right)+\frac{dX_0}{dx}+\frac{dY_0}{dy}=0 \dots\dots\dots\dots (88).$$

Finally the equation (27) will give

$$\Omega^2\frac{i^2}{3}\left(\frac{d^2\omega_1}{dx^2}+\frac{d^2\omega_1}{dy^2}\right)+Z_0+\frac{i^2}{6}\left(Z_2+2\frac{dX_1}{dx}+2\frac{dY_1}{dy}\right)=0 \dots (89).$$

When in the formula (89) we put for $\omega_1$ its value drawn from equation (84), we are immediately brought back to the formula (70).

645. If we wish to consider the elastic plate, no longer in the state of equilibrium but in the state of motion, we must, in equations (69), (70), (87), (88), and in the formulae (73), (77), (80) replace the expressions (48) by the expressions (49). Admitting this we shall obtain

1°. from the equations (87) and (88)

$$\left.\begin{array}{l}\dfrac{\Omega^2}{2(\kappa+1)}\left\{\kappa\left(\dfrac{d^2\xi_0}{dx^2}+\dfrac{d^2\xi_0}{dy^2}\right)+(\kappa+2)\dfrac{d\omega_0}{dx}\right\}+X_0=\dfrac{d^2\xi_0}{dt^2}\\ \dfrac{\Omega^2}{2(\kappa+1)}\left\{\kappa\left(\dfrac{d^2\eta_0}{dy^2}+\dfrac{d^2\eta_0}{dx^2}\right)+(\kappa+2)\dfrac{d\omega_0}{dy}\right\}+Y_0=\dfrac{d^2\eta_0}{dt^2}\end{array}\right\} \dots (90),$$

$$\Omega^2\left(\frac{d^2\omega_0}{dx^2}+\frac{d^2\omega_0}{dy^2}\right)+\frac{dX_0}{dx}+\frac{dY_0}{dy}=\frac{d^2\omega_0}{dt^2} \dots\dots\dots (91).$$

2°. from the equation (70), when we reduce the polynomial

$$\zeta+\frac{i^2}{6}\left(\zeta_2+2\frac{d\xi_1}{dx}+2\frac{d\eta_1}{dy}\right)=\zeta_0-\frac{i^2}{3}(1-1/2\kappa)\left(\frac{d^2\zeta_0}{dx^2}+\frac{d^2\zeta_0}{dy^2}\right)$$

to the single term $\zeta_0$,

$$\Omega^2\frac{i^2}{3}\left(\frac{d^4\zeta_0}{dx^4}+2\frac{d^4\zeta_0}{dx^2dy^2}+\frac{d^4\zeta_0}{dy^4}\right)+\frac{d^2\zeta_0}{dt^2}$$
$$=Z_0+\frac{i^2}{6}\left(Z_2+2\frac{dX_1}{dx}+2\frac{dY_1}{dy}\right) \dots\dots\dots (92).$$

Moreover if we substitute in the formulae (73), (77), (80) for $Z_1$, $Y_1$ the binomials

$$X_1 - \frac{d^2\xi_1}{dt^2} = X_1 + \frac{d^3\zeta_0}{dx\,dt^2},\quad Y_1 - \frac{d^2\eta_1}{dt^2} = Y_1 + \frac{d^3\zeta_0}{dy\,dt^2},$$

or rather the approximate values of these binomials obtained by the aid of equation (92), namely

$$X_1 + \frac{dZ_0}{dx},\quad Y_1 + \frac{dZ_0}{dy},$$

we shall thus obtain

$$\Omega^2\left\{\left(\frac{d^3\zeta_0}{dx^3} + \frac{d^3\zeta_0}{dx\,dy^2}\right)\cos\alpha + \left(\frac{d^3\zeta_0}{dx^2\,dy} + \frac{d^3\zeta_0}{dy^3}\right)\cos\beta\right\}$$
$$= \left(X_1 + \frac{dZ_0}{dx}\right)\cos\alpha + \left(Y_1 + \frac{dZ_0}{dy}\right)\cos\beta \quad\ldots(93),$$

$$\Omega^2\left(\frac{d^3\zeta_0}{dx^3} + \frac{d^3\zeta_0}{dx\,dy^2}\right) = X_1 + \frac{dZ_0}{dx} \quad\ldots\ldots\ldots\ldots\ldots(94),$$

$$\Omega^2\left(\frac{d^3\zeta_0}{dx^2\,dy} + \frac{d^3\zeta_0}{dy^3}\right) = Y_1 + \frac{dZ_0}{dy} \quad\ldots\ldots\ldots\ldots\ldots(95).$$

In the particular case where the accelerating force $\phi$ is the same for all points of the plate the formulae (91) and (92) reduce to

$$\Omega^2\left(\frac{d^2\omega_0}{dx^2} + \frac{d^2\omega_0}{dy^2}\right) = \frac{d^2\omega_0}{dt^2} \quad\ldots\ldots\ldots\ldots\ldots\ldots\ldots\ldots(96),$$

$$\Omega^2\frac{i^2}{3}\left(\frac{d^4\zeta_0}{dx^4} + 2\frac{d^4\zeta_0}{dx^2\,dy^2} + \frac{d^4\zeta_0}{dy^4}\right) + \frac{d^2\zeta_0}{dt^2} = Z \quad\ldots\ldots(97).$$

If this accelerating force vanishes the formula (97) will give simply

$$\Omega^2\frac{i^2}{3}\left(\frac{d^4\zeta_0}{dx^4} + 2\frac{d^4\zeta_0}{dx^2\,dy^2} + \frac{d^4\zeta_0}{dy^4}\right) + \frac{d^2\zeta_0}{dt^2} = 0 \ldots\ldots\ldots(98).$$

646. If now we consider an elastic body as a system of molecules which act upon one another at very small distances, then supposing that the elasticity remains the same in all directions, and that the pressures supported by the free surface of the body in its natural state reduce to zero, we shall obtain between the constants denoted by $k$ and $K$ in the formula (57) the relation

$$k = 2K \quad\ldots\ldots\ldots\ldots\ldots\ldots\ldots\ldots\ldots(99).$$

We shall therefore have $\kappa=3$, and we shall deduce from equations (90), supposing the forces $X_0$ and $Y_0$ to be zero,

$$\left.\begin{aligned}\Omega^2\left\{\tfrac{3}{8}\left(\frac{d^2\xi_0}{dx^2}+\frac{d^2\xi_0}{dy^2}\right)+\tfrac{5}{8}\frac{d\omega_0}{dx}\right\}=\frac{d^2\xi_0}{dt^2},\\ \Omega^2\left\{\tfrac{3}{8}\left(\frac{d^2\eta_0}{dx^2}+\frac{d^2\eta_0}{dy^2}\right)+\tfrac{5}{8}\frac{d\omega_0}{dy}\right\}=\frac{d^2\eta_0}{dt^2}\end{aligned}\right\}\ \ldots\ldots(100).$$

On the same hypothesis, if the pressures $P$, $P'$ vanish with the force $\phi$, the formulae (71), (72), (73) will give for the points situated on free portions of the surface which forms the lateral boundary of the elastic plate

$$\left.\begin{aligned}\left(\frac{d\xi_0}{dx}+\tfrac{1}{4}\frac{d\eta_0}{dy}\right)\cos\alpha+\tfrac{3}{8}\left(\frac{d\xi_0}{dy}+\frac{d\eta_0}{dx}\right)\cos\beta=0,\\ \left(\frac{d\eta_0}{dy}+\tfrac{1}{4}\frac{d\xi_0}{dx}\right)\cos\beta+\tfrac{3}{8}\left(\frac{d\xi_0}{dy}+\frac{d\eta_0}{dx}\right)\cos\alpha=0\end{aligned}\right\}\ \ldots(101).$$

$$\left.\begin{aligned}\left(\frac{d^2\zeta_0}{dx^2}+\tfrac{1}{4}\frac{d^2\zeta_0}{dy^2}\right)\cos\alpha+\tfrac{3}{4}\frac{d^2\zeta_0}{dx\,dy}\cos\beta=0,\\ \left(\frac{d^2\zeta_0}{dy^2}+\tfrac{1}{4}\frac{d^2\zeta_0}{dx^2}\right)\cos\beta+\tfrac{3}{4}\frac{d^2\zeta_0}{dx\,dy}\cos\alpha=0,\end{aligned}\right\}\ \ldots\ldots(102),$$

$$\left(\frac{d^3\zeta_0}{dx^3}+\frac{d^3\zeta_0}{dx\,dy^2}\right)\cos\alpha+\left(\frac{d^3\zeta_0}{dx^2dy}+\frac{d^3\zeta_0}{dy^3}\right)\cos\beta=0\ \ldots(103).$$

647. If the elastic plate were rectangular and bounded laterally by planes perpendicular to the axes of $x$ and $y$, the formulae (101), (102), (103) would give

1°. for the surface, supposed free, of one of the planes perpendicular to the axis of $x$,

$$\frac{d\xi_0}{dx}+\tfrac{1}{4}\frac{d\eta_0}{dy}=0,\quad \frac{d\xi_0}{dy}+\frac{d\eta_0}{dx}=0\ \ldots\ldots\ldots(104),$$

$$\frac{d^2\zeta_0}{dx^2}+\tfrac{1}{4}\frac{d^2\zeta_0}{dy^2}=0,\quad \frac{d^2\zeta_0}{dx\,dy}=0\ \ldots\ldots\ldots(105),$$

$$\frac{d^3\zeta_0}{dx^3}+\frac{d^3\zeta_0}{dx\,dy^2}=0\ \ldots\ldots\ldots\ldots\ldots\ldots(106).$$

2°. for the surface, supposed free, of one of the planes perpendicular to the axis of $y$,

$$\frac{d\eta_0}{dy}+\tfrac{1}{4}\frac{d\xi_0}{dx}=0,\quad \frac{d\xi_0}{dy}+\frac{d\eta_0}{dx}=0\ \ldots\ldots\ldots(107),$$

$$\frac{d^2\zeta_0}{dy^2}+\frac{1}{4}\frac{d^2\zeta_0}{dx^2}=0, \quad \frac{d^2\zeta_0}{dx\,dy}=0 \quad \text{.........(108)},$$

$$\frac{d^3\zeta_0}{dx^2dy}+\frac{d^3\zeta_0}{dy^3}=0 \quad \text{.................(109)}.$$

It is well to notice that by reason of the second of the conditions (105) or (108) the condition (106) can be reduced to

$$\frac{d^3\zeta_0}{dx^3}=0 \quad \text{.......................(110)},$$

and the condition (109) to

$$\frac{d^3\zeta_0}{dy^3}=0 \quad \text{.....................(111)}.$$

648. Most of the equations established here, and particularly the formulae (20), (27), (34), (35), (36), (50), (51), (90), (91), (92), are taken from a memoir which Cauchy presented to the Academy of Sciences on the 6th of October, 1828. These formulae, or at least those which we can deduce from them by putting $\kappa=3$, are found to agree with the formulae contained in Poisson's memoir which was in the press at the time, and which soon afterwards appeared. However to the conditions (74), the first of which involves the second when we have regard to the last of the formulae of (40), Poisson added a third condition which disappears of itself when the plate is circular, and the admission of which in other cases appeared to Cauchy to be subject to some difficulties.

649. We can easily infer from equation (96) that the velocity of sound in an elastic plate of an indefinite extent is precisely the value of $\Omega$ determined by the formula (65). If moreover we suppose $\kappa=3$ we can prove that the velocity in question is to the velocity of sound in an elastic body of which all three dimensions are infinite in the ratio of $\sqrt{8}$ to $\sqrt{9}$ or 3. If we ascribe to $\kappa$ a value different from 3 the ratio of the two velocities will be that of $\sqrt{\kappa^2-1}$ to $\sqrt{\kappa^2}$ or $\kappa$.

The equation (98) has the same form as that which was found without demonstration among the papers of Lagrange, and which served as the foundation of the researches published by Mdlle Sophie Germain in a memoir on elastic plates crowned by the Institut in 1815; see Arts. 284 and 290.

650. Cauchy then gives a page and a half to the case in which the plate is supposed to be without elasticity; and afterwards seven pages to the case in which the plate is supposed to be of variable thickness, but we have given sufficient to illustrate his method[1].

651. Thus we are led to make some remarks on a point which has given rise to controversy, namely Poisson's view of the conditions which must hold at the boundary of the plate.

Cauchy says nothing with respect to Poisson's conditions at a *free* part of the boundary, from which we may infer that he saw no serious objection to them. Cauchy obtains five equations which must hold at a *free* part of the boundary; two of these correspond with those which Poisson obtained in the case of a *membrane* and which he does not formally repeat in the case of a *plate;* there remain *three*, so that as to the *number* of conditions Cauchy exactly agrees with Poisson. Moreover as to the *form* of the conditions there is substantial agreement: Cauchy puts them in the form without integration, see (102) and (103), but as the $X_0$ and $Y_0$ of my Art. 486 are taken by Cauchy to be constant, and the $Z_0$ to be zero, the result obtained by Poisson after integration coincides with Cauchy's.

652. But with respect to the conditions which must hold at a *fixed* part of the boundary there is some difference between the two mathematicians, as we see from Cauchy's opinion given in Art. 648. The part to which Cauchy objects must be involved in my Art. 491; but he does not specify it. Cauchy however should not say that Poisson *has joined a third condition*, for the facts would be more accurately expressed thus: "M. Poisson starts with two equations equivalent to my (41) and therefore to my (74); he transforms these into two others, one of which seems to me subject to certain difficulties."

[1] [The articles 630—647 are a translation of pp. 328—346 of the third year of the *Exercices*. I have retained them here because it seemed to be Dr Todhunter's intention, and because while offering a good example of Cauchy's style, they will for the first time appear side by side in the same work with Poisson's solution, and enable us more easily to form an estimate of later work. ED.]

The matter is of small consequence and seems to have attracted scarcely any attention; what is important to notice is that as to the objection raised by Kirchhoff, and implicitly adopted by Clebsch and others, against Poisson, Cauchy agrees with Poisson[1]

653. An article entitled: *Sur l'équilibre et le mouvement d'une verge rectangulaire* occupies pages 356—368 of the volume. By a *verge rectangulaire* Cauchy means a rod which in its natural state has for its axis a straight line or a plane curve, and any section at right angles to the axis is a rectangle of small dimensions. The equations for this body are deduced very ingeniously from those given in the two preceding memoirs for the elastic lamina and the elastic plate respectively.

654. On page 365 Cauchy finds that the velocity of sound along an elastic rod is to the velocity of sound in an elastic body as $\sqrt{5}$ is to $\sqrt{6}$, and he observes that Poisson had already enunciated this proposition.

655. On his page 366 Cauchy quotes a result which he had obtained on page 271 respecting the ratio between the numbers of longitudinal and of transversal vibrations for an elastic rectangular rod corresponding to the fundamental notes. Let $N$ be the number of longitudinal vibrations (ends free), and $N'$ the number of transversal vibrations (ends fixed) in a unit of time; then

$$N' = (2{\cdot}055838\ldots)\,\frac{2h}{a}\,N;$$

where $a$ is the length and $2h$ the breadth of the rod in the transverse direction considered. Cauchy says:

Cette dernière formule s'accorde parfaitement avec les expériences de M. Savart rapportées dans le Bulletin des Sciences de janvier 1828, et diffère très-peu d'une formule que M. Poisson a présentée sans démonstration dans ce même Bulletin, mais que l'on ne retrouve pas dans le Mémoire publié par ce géomètre sur l'équilibre et le mouvement des corps élastiques.

Cauchy gives on pages 367 and 368 the results of some experiments which at his request Savart had made.

[1] [See my remarks, p. 251. Ed.]

The article in the *Bulletin des Sciences* to which Cauchy alludes is also printed in the *Annales de Chimie*, Vol. 36, 1827, pages 86—93: see my Art. 429 on Poisson. The formula to which Cauchy alludes is that which is there given thus,

$$n' = (2\cdot05610)\,\frac{ne}{l}:$$

Poisson did not reproduce this in his memoir. It agrees closely with that given by Cauchy above.

We proceed to the fourth volume of the *Exercices*, which is dated 1829; to this we shall devote the following thirteen articles.

656. An article entitled: *Sur l'équilibre et le mouvement d'une plaque élastique dont l'élasticité n'est pas la même dans tous les sens* occupies pages 1—14 of the volume. In the memoir which we have noticed in Arts. 638 and 646 Cauchy used *two* constants of elasticity; in the present memoir he shews how the problem is to be treated if we use *fifteen* constants: see Kirchhoff in *Crelle*, Vol. 56, and Thomson, *Quarterly Journal of Mathematics*, 1857. The formulae become complicated but there is nothing of special interest to be noticed. Cauchy calls a body homogeneous when the fifteen coefficients of elasticity are, as we suppose them here to be, really constant. See Saint-Venant on *Torsion*, 261, 263.

657. An article entitled: *Sur l'équilibre et le mouvement d'une verge rectangulaire extraite d'un corps solide dont l'élasticité n'est pas la même en tous sens* occupies pages 15—29 of the volume. The axis of the rod in its natural state is supposed to be a straight line; the constants of elasticity are taken to be *fifteen* in number, as in the memoir immediately preceding; the investigation is a process of approximation not very complex. Cauchy arrives at a theorem which he enunciates thus on his page 28:

Théorème. Une verge élastique étant extraite d'un corps solide homogène qui n'offre pas la même élasticité dans tous les sens, pour obtenir le carré de la vitesse du son dans cette verge indéfiniment prolongée, il suffit de chercher ce que deviennent, en un point quelconque du corps solide, la dilatation ou condensation linéaire $\pm\epsilon$, mesurée parallèlement à l'axe de la verge, et la pression ou tension $p'$

supportée par un plan perpendiculaire à cette axe, tandis que les pressions ou tensions principales se réduisent l'une à $p'$, les deux autres à zéro, puis de diviser la dilatation ou condensation $\pm\epsilon$ par le facteur $p'$ et par la densité $\rho$.

If the section of the elastic rod is a square Cauchy finds that the transverse vibrations produce the same sound whether they are parallel to one or to the other of the sides of the square. He adds on his page 29:

Il était important de voir si cette conclusion, qui peut paraître singulière quand on suppose la verge extraite d'un corps solide dont l'élasticité n'est pas la même dans tous les sens, serait confirmée par l'observation. Ayant consulté, à ce sujet, M. Savart, j'ai eu la satisfaction d'apprendre que des expériences qu'il avait entreprises, sans connaître mes formules, l'avaient précisément conduit au même résultat.

See Saint-Venant on *Torsion*, 297.

658. An article entitled: *Sur les pressions ou tensions supportées en un point donné d'un corps solide par trois plans perpendiculaires entre eux* occupies pages 30—40 of the volume. Here we have formulae expressing the stresses for one set of rectangular axes in terms of the stresses for another set, and some applications of them, the nature of which we will indicate.

Suppose that of the stress-components $\widehat{xx}, \widehat{xy}, \widehat{xz}, \ldots$ all vanish except $\widehat{xx}$; we have six formulae which we may write thus,

$$\widehat{x'x'} = \widehat{xx}\cos^2\alpha, \quad \widehat{y'y'} = \widehat{xx}\cos^2\beta, \quad \widehat{z'z'} = \widehat{xx}\cos^2\gamma,$$

$$\widehat{y'z'} = \widehat{xx}\cos\beta\cos\gamma, \quad \widehat{z'x'} = \widehat{xx}\cos\gamma\cos\alpha, \quad \widehat{x'y'} = \widehat{xx}\cos\alpha\cos\beta \ldots \quad (1).$$

Again as in Art. 547 we have

$$s_r = \frac{du}{dx}\cos^2\alpha + \frac{dv}{dy}\cos^2\beta + \frac{dw}{dz}\cos^2\gamma + \left(\frac{dv}{dz} + \frac{dw}{dy}\right)\cos\beta\cos\gamma$$
$$+ \left(\frac{dw}{dx} + \frac{du}{dz}\right)\cos\gamma\cos\alpha + \left(\frac{du}{dy} + \frac{dv}{dx}\right)\cos\alpha\cos\beta \quad \ldots\ldots(2).$$

Now we can express $\widehat{xx}, \ldots$ as linear functions of $\frac{du}{dx}, \frac{du}{dy}, \ldots$: see Art. 553. Eliminate these shift-fluxions between (1) and (2), and put $\rho s_r \Omega^2$ for $xx$; we thus obtain an equation which may be written thus,

$$\frac{1}{\rho\Omega^2} = A\cos^4\alpha + B\cos^4\beta + C\cos^4\gamma + \ldots\ldots\ldots\ldots(3),$$

where the terms not expressed involve $\cos^3\alpha\cos\beta$, $\cos^2\alpha\cos\beta\cos\gamma$, and like expressions.

Then (3) may be represented by a surface of the fourth degree if we take

$$\frac{\xi}{\cos\alpha} = \frac{\eta}{\cos\beta} = \frac{\zeta}{\cos\gamma} = \rho^{\frac{1}{4}}\,\Omega^{\frac{1}{2}};$$

the equation to this surface being

$$1 = A\xi^4 + B\eta^4 + C\zeta^4 + \ldots\ldots$$

the maxima or minima values of the radius vector of this surface will correspond to the maxima or minima values of $\Omega$.

The process is connected by Cauchy with the theorem which reduces the nine stress-components to six.

See Saint-Venant on *Torsion*, p. 253.

659. An article entitled: *Sur la relation qui existe entre les pressions ou tensions supportées par deux plans quelconques en un point donné d'un corps solide* occupies pages 41 and 42 of the volume. Cauchy had proved on page 48 of the second volume of the *Exercices* a particular case of the theorem which we have called *Cauchy's Theorem*, namely that in which the axes are at right angles: see Art. 610 (ii) of this Chapter. He now demonstrates the theorem generally, and we will reproduce his process. Saint-Venant, on page 250 of his memoir on *Torsion*, seems to say that this generalisation had been given in the preceding year by Lamé and Clapeyron.

Let $OL$, $OM$ be two semi-axes drawn arbitrarily from a given point of a solid body. Refer all the points of the body to three rectangular axes $x, y, z$; let $\alpha_1, \beta_1, \gamma_1$ be the direction angles of $OL$, and $\alpha_2, \beta_2, \gamma_2$ those of $OM$. Let $p_1$ and $p_2$ be the stresses supported at $O$, on the side of $OL$ and $OM$, by planes at right angles to these semi-axes respectively. Let $\lambda_1, \mu_1, \nu_1$ be the direction angles of $p_1$, and $\lambda_2, \mu_2, \nu_2$ those of $p_2$. Let $\varpi_1$ be the angle between $OM$ and the direction of $p_1$, and $\varpi_2$ the angle between $OL$ and the direction of $p_2$.

Let $\widehat{xx}$, $\widehat{xy}$, $\widehat{xz}$ be the algebraic projections of the stress supported at $O$, on the side of the positive direction of the axis of $x$, by a plane at right angles to this axis; let $\widehat{yx}$, $\widehat{yy}$, $\widehat{yz}$ be similar quantities with respect to the axis of $y$; and $\widehat{zx}$, $\widehat{zy}$, $\widehat{zz}$, with respect to the axis of $z$. Then we shall have

$$\left.\begin{aligned} p_1\cos\lambda_1 &= \widehat{xx}\cos\alpha_1 + \widehat{xy}\cos\beta_1 + \widehat{xz}\cos\gamma_1 \\ p_1\cos\mu_1 &= \widehat{yx}\cos\alpha_1 + \widehat{yy}\cos\beta_1 + \widehat{yz}\cos\gamma_1 \\ p_1\cos\nu_1 &= \widehat{zx}\cos\alpha_1 + \widehat{zy}\cos\beta_1 + \widehat{zz}\cos\gamma_1 \end{aligned}\right\} \text{......(1)};$$

$$\cos\varpi_1 = \cos\lambda_1\cos\alpha_2 + \cos\mu_1\cos\beta_2 + \cos\nu_1\cos\gamma_2 \text{......(2)};$$

and therefore

$$\begin{aligned} p_1\cos\varpi_1 = \widehat{xx}\cos\alpha_1\cos\alpha_2 &+ \widehat{yy}\cos\beta_1\cos\beta_2 + \widehat{zz}\cos\gamma_1\cos\gamma_2 \\ + \widehat{yz}(\cos\beta_1\cos\gamma_2 + \cos\beta_2\cos\gamma_1) &+ \widehat{zx}(\cos\gamma_1\cos\alpha_2 + \cos\gamma_2\cos\alpha_1) \\ &+ \widehat{xy}(\cos\alpha_1\cos\beta_2 + \cos\alpha_2\cos\beta_1) \text{.........(3)}. \end{aligned}$$

Now suppose we interchange the semi-axes $OL$ and $OM$; by reason of this the first member of (3) is changed into $p_2\cos\varpi_2$, while the second member remains unchanged: therefore

$$p_1\cos\varpi_1 = p_2\cos\varpi_2 \text{ .......................(4)}.$$

This is the theorem which was to be proved.

Moigno's *Statique*, p. 627.

660. An article entitled: *Sur les vibrations longitudinales d'une verge cylindrique ou prismatique à base quelconque* occupies pages 43—46 of the volume. Here the equation

$$\Omega^2\frac{d^2\xi}{dx^2} + X = \frac{d^2\xi}{dt^2}$$

is obtained for the longitudinal vibrations of a rod whatever may be the form of the transverse section. The article concludes thus:

> Les résultats que nous venons d'exposer subsistent, de quelque manière que l'élasticité du corps, d'où on suppose la verge extraite, varie quand on passe d'une direction à une autre. Ils coïncident d'ailleurs avec ceux que M. Poisson a obtenus, en considérant une verge extraite d'un corps solide dont l'élasticité reste la même en tous sens. Seulement, dans ce cas particulier, le coefficient $\Omega$ devient indépendant de la direction que présentait, avant l'extraction, l'axe de la verge élastique.

The reference to Poisson applies to page 452 of his memoir of April, 1828: see also Poisson's *Mécanique*, Vol. II., page 316.

661. An article entitled: *Sur la torsion et les vibrations tournantes d'une verge rectangulaire* occupies pages 47—64 of the volume; there is however a mistake in the numbering of the pages, and the article really occupies 24 pages. The investigation is a process of approximation and may be considered to be now superseded by the more rigid treatment of the problem by Saint-Venant; the latter however ascribes great merit to this memoir which formed the starting point of his own researches: see Saint-Venant on *Torsion,* page 361. In particular the expression obtained by Cauchy for the moment of torsion corresponds nearly in a certain case with that of Saint-Venant.

[The latter illustrious physicist writes as follows concerning this memoir on pp. clxxv—clxxvii of the *Historique Abrégé:*

Jusqu'au mémoire publié en 1829 par M. Cauchy, on avait cru que les résistances opposées par les diverses fibres d'un prisme à cette sorte de déformation étaient, comme pour le cylindre à base circulaire, proportionnelles aux inclinaisons que ces fibres prennent sur l'axe de torsion en devenant des hélices, et par conséquent à leurs distances à cet axe......L'analyse de Cauchy......est fondée, comme celle qu'il a employée ainsi que Poisson pour la flexion, sur la supposition gratuite que les pressions intérieures sont exprimables en séries convergentes suivant les puissances entières des deux coordonnées transversales, et sur des suppressions, non justifiées, de termes dont on ne connaît pas le rapport de grandeur avec ceux que l'on conserve....... Son mémoire de 1829 a néanmoins fait faire un grand pas à la théorie de la torsion, car, outre que la formule à laquelle il arrive est exacte pour les prismes plats, l'analyse qu'il y développe fait apercevoir un rapport nécessaire entre la torsion $\theta$ et la dérivée seconde du déplacement longitudinal des points de chaque section par rapport aux deux coordonnées transversales; dérivée dont l'existence annonce que les divers points des sections d'un prisme rectangle tordu se déplacent inégalement dans un sens parallèle aux arêtes, ou que ces sections ne restent pas planes.]

Cauchy on his page 60 says that his results are similar to those obtained by Poisson in considering the torsion of a cylindrical rod on a circular base, and that they hold for a cylindrical or prismatic rod on any base; the allusion must be to page 454 of Poisson's memoir of April, 1828. On pages 62 and 64 Cauchy appeals to experiments made by Savart as confirming his theoretical conclusions: see Art. 333.

[Savart's confirmation of this erroneous theory of torsion is remarkable, and it may not be out of place to call attention again to the adverse experimental results of C. J. Hill: see Art. 398.]

662. An article entitled: *Sur les équations différentielles d'équilibre ou de mouvement pour un système de points matériels sollicités par des forces d'attraction ou de répulsion mutuelle* occupies pages 129—139 of the volume; this is connected with that on pages 188—212 of the third volume of the *Exercices*, noticed in my Art. 615. It is difficult to say what is the design of this article, but it seems to be intended to carry on the comparison between the two modes of treating the problems connected with a rigid body; namely that in which the body is regarded as an aggregate of isolated particles, and that in which it is regarded as continuous.

See Saint-Venant *on Torsion*, page 261; Moigno, page 683.

663. An article entitled: *Sur les corps solides ou fluides dans lesquels la condensation ou dilatation linéaire est la même en tous sens autour de chaque point* occupies pages 214—216 of the volume. This we will reproduce.

Suppose that a solid or fluid body changes its form, and by the effect of any cause whatever it passes from one natural or artificial state to a second different from the former. Let us refer all points of space to three rectangular axes, and suppose that the material point corresponding to the coordinates $x$, $y$, $z$ in the second state of the body is exactly that which in the first state of the body had for coordinates the three differences

$$x-\xi,\ y-\eta,\ z-\zeta.$$

If we take $x$, $y$, $z$ for independent variables $\xi$, $\eta$, $\zeta$ will be functions of $x$, $y$, $z$ which will serve to measure the shifts of the point we are considering parallel to the axes of coordinates. Let $r$ be the radius vector drawn, in the second state of the body, from the molecule $m$ to any adjacent molecule $m'$; and let $(\alpha, \beta, \gamma)$ be the direction angles of $r$. If we denote by

$$\frac{r}{1+s_r}$$

the original distance of the molecules $m$ and $m'$, the numerical value of $s_r$ will be the measure of what we call the stretch of the body along the direction of $r$; that is to say the linear dilatation if $s_r$ is a positive quantity, and the linear condensation or contraction if $s_r$ is a negative quantity.

Then we shall have, by Art. 612 of this series,

$$\left(\frac{1}{1+s_r}\right)^2 = \left(\cos\alpha - \frac{d\xi}{dx}\cos\alpha - \frac{d\xi}{dy}\cos\beta - \frac{d\xi}{dz}\cos\gamma\right)^2$$
$$+\left(\cos\beta - \frac{d\eta}{dx}\cos\alpha - \frac{d\eta}{dy}\cos\beta - \frac{d\eta}{dz}\cos\gamma\right)^2$$
$$+\left(\cos\gamma - \frac{d\zeta}{dx}\cos\alpha - \frac{d\zeta}{dy}\cos\beta - \frac{d\zeta}{dz}\cos\gamma\right)^2 \text{......(1)};$$

and hence, supposing that the displacements $\xi$, $\eta$, $\zeta$ are very small, we obtain

$$s_r = \frac{d\xi}{dx}\cos^2\alpha + \frac{d\eta}{dy}\cos^2\beta + \frac{d\zeta}{dz}\cos^2\gamma + \left(\frac{d\eta}{dz} + \frac{d\zeta}{dy}\right)\cos\beta\cos\gamma$$
$$+\left(\frac{d\zeta}{dx} + \frac{d\xi}{dz}\right)\cos\gamma\cos\alpha + \left(\frac{d\xi}{dy} + \frac{d\eta}{dx}\right)\cos\alpha\cos\beta \text{......(2)}.$$

Now we may ask what conditions must be fulfilled by $\xi$, $\eta$, $\zeta$, considered as functions of $x$, $y$, $z$, in order that the stretch may remain the same in all directions round every point: this is the question we propose now to consider.

Let $s_x$, $s_y$, $s_z$ be the stretches measured parallel to the axes of $x$, $y$, $z$: we have by the formula (2)

$$s_x = \frac{d\xi}{dx}, \quad s_y = \frac{d\eta}{dy}, \quad s_z = \frac{d\zeta}{dz}.$$

Thus if we suppose these stretches all equal we shall obtain the condition

$$\frac{d\xi}{dx} = \frac{d\eta}{dy} = \frac{d\zeta}{dz} \text{..........................(3)},$$

and hence equation (2) will give

$$s_r = s_x + \left(\frac{d\eta}{dz} + \frac{d\zeta}{dy}\right)\cos\beta\cos\gamma + \left(\frac{d\zeta}{dx} + \frac{d\xi}{dz}\right)\cos\gamma\cos\alpha$$
$$+\left(\frac{d\xi}{dy} + \frac{d\eta}{dx}\right)\cos\alpha\cos\beta \text{......(4)}.$$

Hence if the linear dilatation $s_r$ remains constantly equal to $s_x$ we shall have for all values of $\alpha, \beta, \gamma$

$$\left(\frac{d\eta}{dz}+\frac{d\zeta}{dy}\right)\cos\beta\cos\gamma+\left(\frac{d\zeta}{dx}+\frac{d\xi}{dz}\right)\cos\gamma\cos\alpha$$
$$+\left(\frac{d\xi}{dy}+\frac{d\eta}{dx}\right)\cos\alpha\cos\beta=0\;\ldots\ldots(5).$$

In (5) put successively $\alpha=\frac{\pi}{2}, \beta=\frac{\pi}{2}, \gamma=\frac{\pi}{2}$; then we deduce

$$\frac{d\eta}{dz}+\frac{d\zeta}{dy}=0,\quad \frac{d\zeta}{dx}+\frac{d\xi}{dz}=0,\quad \frac{d\xi}{dy}+\frac{d\eta}{dx}=0\ldots\ldots\ldots\ldots(6).$$

Thus in order that the value of $s_r$ may become independent of the angles $\alpha, \beta, \gamma$ it is necessary that the shifts $\xi, \eta, \zeta$, considered as functions of $x, y, z$ should satisfy the conditions (3) and (6). Conversely if these conditions are satisfied $s_r$ will be independent of the angles $\alpha, \beta, \gamma$, and we shall obtain from the formula (2)

$$s_r=\frac{d\xi}{dx}=\frac{d\eta}{dy}=\frac{d\zeta}{dz}\ldots\ldots\ldots\ldots\ldots\ldots\ldots\ldots(7).$$

664. It is easy to shew that when the conditions (3) and (6) are satisfied the stretch $s_r$ reduces to a *linear* function of $x, y, z$. In fact suppose we differentiate the first of equations (6) with respect to $x$, the second with respect to $y$, and the third with respect to $z$; we shall have

$$\frac{d^2\eta}{dz\,dx}+\frac{d^2\zeta}{dx\,dy}=0,\quad \frac{d^2\zeta}{dx\,dy}+\frac{d^2\xi}{dy\,dz}=0,\quad \frac{d^2\xi}{dy\,dz}+\frac{d^2\eta}{dz\,dx}=0\ldots.(8)$$

and consequently

$$\frac{d^2s_r}{dy\,dz}=0,\quad \frac{d^2s_r}{dz\,dx}=0,\quad \frac{d^2s_r}{dx\,dy}=0\ldots\ldots\ldots\ldots(9).$$

Differentiate the first of equations (9) with respect to $x$, the second with respect to $y$, and the third with respect to $z$; then having regard to (7) we obtain

$$\frac{d^2s_r}{dy\,dz}=0,\quad \frac{d^2s_r}{dz\,dx}=0,\quad \frac{d^2s_r}{dx\,dy}=0\;\ldots\ldots\ldots(10).$$

Again, differentiate successively the first of (6) with respect to $y$ and $z$, the second of (6) with respect to $z$ and $x$, and the third of (6) with respect to $x$ and $y$; then having regard to (7) we obtain

$$\frac{d^2s_r}{dz^2}+\frac{d^2s_r}{dy^2}=0,\quad \frac{d^2s_r}{dx^2}+\frac{d^2s_r}{dz^2}=0,\quad \frac{d^2s_r}{dy^2}+\frac{d^2s_r}{dx^2}=0\ldots\ldots(11);$$

and consequently

$$\frac{d^2s_r}{dx^2}=0,\quad \frac{d^2s_r}{dy^2}=0,\quad \frac{d^2s_r}{dz^2}=0\ldots\ldots\ldots\ldots\ldots\ldots(12).$$

From (10) and (12) we deduce

$$d\left(\frac{ds_r}{dx}\right)=0,\quad d\left(\frac{ds_r}{dy}\right)=0,\quad d\left(\frac{ds_r}{dz}\right)=0\ldots\ldots\ldots\ldots(13)$$

and consequently

$$\frac{ds_r}{dx}=a,\quad \frac{ds_r}{dy}=b,\quad \frac{ds_r}{dz}=c\ldots\ldots\ldots\ldots\ldots(14);$$

hence

$$ds_r=adx+bdy+cdz\ldots\ldots\ldots\ldots\ldots\ldots(15),$$

$$s_r=ax+by+cz+k\ldots\ldots\ldots\ldots\ldots\ldots(16),$$

where $a$, $b$, $c$, $k$ are constant quantities.

Hence we obtain the following theorem: If a solid or fluid body changes its form so that the stretch remains very small and is the same in all directions round every point, then this stretch must be a linear function of the coordinates $x$, $y$, $z$.

665. The value of the stretch $s_r$ being determined by (16) we can easily deduce the values of the shifts $\xi$, $\eta$, $\zeta$ by means of (6) and (7); these will give

$$\frac{d^2\xi}{dy^2}=\frac{d^2\xi}{dz^2}=-\frac{ds_r}{dx}=-a,\quad \frac{d^2\xi}{dydz}=0,\ \&c.\ldots\ldots\ldots(17).$$

and we shall obtain

$$\left.\begin{aligned}\xi&=(ax+by+cz+k)\,x-\tfrac{1}{2}a\,(x^2+y^2+z^2)+hy-gz+l\\ \eta&=(ax+by+cz+k)\,y-\tfrac{1}{2}b\,(x^2+y^2+z^2)+fz-hx+m\\ \zeta&=(ax+by+cz+k)\,z-\tfrac{1}{2}c\,(x^2+y^2+z^2)+gx-fy+n\end{aligned}\right\}\ldots(18),$$

where $f$, $g$, $h$, $l$, $m$, $n$ denote constant quantities.

666. An article entitled: *Sur l'équilibre et le mouvement intérieur des corps considérés comme des masses continues* occupies pages 293—319 of the volume.

If we assume that the six stress-components are *linear* functions of the nine first shift-fluxions with respect to $x$, $y$, $z$, then on the most general supposition there will be *nine* coefficients in the expression for each stress-component, or *ten* if we include a term independent of the shift-fluxions. Poisson, starting with *nine* coefficients for each stress had shewn that they must reduce to *six:* see Art. 553 of my account of Poisson's memoir of October 1829. Cauchy alludes to this process given by Poisson, and it serves as the foundation of the present memoir. Cauchy starting with *ten* coefficients for each stress reduces them to *seven;* thus on the whole he has 42 coefficients, namely the ordinary 36, and one in each tension independent of differential coefficients. Cauchy observes that if the body had been considered as an aggregate of isolated particles these 42 coefficients would reduce to 21 by virtue of relations between them; the 21 coefficients would be the ordinary 15, and the additional 6 which, as we have already noticed, Cauchy introduces: see Moigno, p. 660. Cauchy then proceeds to consider what relations must hold among the 42 coefficients in order that the body may exhibit the same elasticity in every direction round a fixed axis. He finds that his formulae for the stresses finally involve *five* coefficients; they are

$$\widehat{xx} = a + (3f + a)\frac{d\xi}{dx} + (f - a)\frac{d\eta}{dy} + (d - a)\frac{d\zeta}{dz},$$

$$\widehat{yy} = a + (f - a)\frac{d\xi}{dx} + (3f + a)\frac{d\eta}{dy} + (d - a)\frac{d\zeta}{dz},$$

$$\widehat{zz} = c + (d - c)\frac{d\xi}{dx} + (d - c)\frac{d\eta}{dy} + (k + c)\frac{d\zeta}{dz},$$

$$\widehat{yz} = (d + c)\frac{d\eta}{dz} + (d + a)\frac{d\zeta}{dy},$$

$$\widehat{zx} = (d + a)\frac{d\zeta}{dx} + (d + c)\frac{d\xi}{dz},$$

$$\widehat{xy} = (f + a)\left(\frac{d\xi}{dy} + \frac{d\eta}{dx}\right);$$

the elasticity being supposed the same in every direction which is at right angles to the axis of $z$. If we suppose $a$ and $c$ to vanish these formulae reduce to those usual in uniaxial symmetry. Moigno, page 667.

667. Finally if the elasticity is the same in *all* directions Cauchy's formulae become of the form

$$\widehat{xx} = 2\mu \frac{d\xi}{dx} + \lambda\theta + \tau_0,$$

$$\widehat{yy} = 2\mu \frac{d\eta}{dy} + \lambda\theta + \tau_0,$$

$$\widehat{zz} = 2\mu \frac{d\zeta}{dz} + \lambda\theta + \tau_0,$$

$$\widehat{yz} = \mu \left(\frac{d\eta}{dz} + \frac{d\zeta}{dy}\right),$$

$$\widehat{zx} = \mu \left(\frac{d\zeta}{dx} + \frac{d\xi}{dz}\right),$$

$$\widehat{xy} = \mu \left(\frac{d\xi}{dy} + \frac{d\eta}{dx}\right),$$

where $\tau_0$ represents an initial traction and $\theta$ stands for

$$\frac{d\xi}{dx} + \frac{d\eta}{dy} + \frac{d\zeta}{dz}.$$

Moigno, page 667.

668. 1829. A paper entitled: *Démonstration analytique d'une loi découverte par M. Savart et relative aux vibrations des corps solides ou fluides,* occupies pages 117 and 118 of the *Mémoires* of the Paris Academy, Vol. IX. 1830. It was read on the 12th of January 1829. We will give it.

J'ai donné dans les Exercices de mathématiques les équations générales qui représentent le mouvement d'un corps élastique dont les molécules sont très-peu écartées des positions qu'elles occupaient dans l'état naturel du corps, de quelque manière que l'élasticité varie dans les diverses directions. Ces équations qui servent à déterminer, en fonction du temps $t$ et des coordonnées $x$, $y$, $z$, les déplacements $\xi$, $\eta$, $\zeta$, d'un point quelconque mesurés dans le sens de ces coordonnées, sont de deux espèces. Les unes se rapportent à tous les points du corps élastique, les autres, aux points renfermés dans sa surface extérieure. Or, à l'inspection seule des équations dont il s'agit, on reconnaît immédiatement qu'elles continuent de subsister, lorsqu'on y remplace $x$ par $kx$, $y$ par $ky$, $z$ par $kz$, $\xi$ par $k\xi$, $\eta$ par $k\eta$, $\zeta$ par $k\zeta$, $k$ désignant une constante choisie arbitrairement, et lorsqu'en même temps on fait varier les forces accélératrices appliquées aux diverses molécules dans le rapport de 1 à $1/k$. Donc, si ces forces accélératrices sont nulles, il suffira de faire croître ou diminuer les dimensions du corps solide, et les valeurs initiales des déplacements dans le rapport de 1 à $k$, pour que les valeurs générales de $\xi$, $\eta$, $\zeta$ et les durées des vibrations varient dans le même rapport. Donc, si l'on prend pour mesure du son rendu par un corps, par une plaque, ou par une verge élastique, le nombre des vibrations produites pendant l'unité de temps, ce son variera en raison inverse des dimensions du corps, de la plaque ou de la verge, tandis que ces dimensions croîtront ou décroîtront dans un rapport donné. Cette loi, découverte par M. Savart, s'étend aux sons rendus par une masse fluide contenue dans un espace fini, et se démontre alors de la même manière.

On prouverait encore de même que, si les dimensions d'un corps venant à croître ou à diminuer dans un certain rapport, sa température initiale croît ou diminue dans le même rapport, la durée de la propagation de la chaleur variera comme le carré de ce rapport.

669. A paper entitled: *Mémoire sur la torsion et les vibrations tournantes d'une verge rectangulaire* occupies pages 119—124 of the *Mémoires* of the Paris Academy Vol. IX. 1830. It was read on the 9th of February 1829. This is a summary of the results obtained by Cauchy in his memoir on pages 47—64 of the fourth volume of his *Exercices de mathématiques*. See our account of that memoir in Art. 659.

670. *Sur les diverses méthodes à l'aide desquelles on peut établir les équations qui représentent les lois d'équilibre, ou le mouvement intérieur des corps solides ou fluides.* (Lu à l'Académie royale des Sciences, le 8 mai 1830.) This is published in Férussac's *Bulletin*, Vol. XIII. 1830 pages 169—176. The *diverses méthodes* are really *two* which I have mentioned at the end of Art. 662 of my notice of Cauchy. In the present article Cauchy confines himself to the case in which the body is continuous. The article states briefly results equivalent to those obtained in the memoir which occupies pages 293—319 of the fourth volume of the *Exercices de mathématiques:* see my Arts. 666 and 667.

671. *Allgemeine Sätze über die Ausdehnung und Zusammenziehung fester Körper.* This is published in the *Journal für Chemie*

*und Physik......*vom Dr Fr. W. Schweigger-Seidel. Vol. LXIV. 1832, pages 44—49. It states results obtained by Cauchy in his *Exercices de mathématiques*, Vol. II. pages 60—69, and Vol. III. pages 237—244. This article in German is not by Cauchy himself, but by G. Th. Fechner, as will be seen from page 28 of the volume in which it appears; however, in the *Royal Society Catalogue of Scientific Papers* it is entered under the name of Cauchy as number 68.

672. 1839. In the *Comptes Rendus*, Vol. IX. pages 588—590 there is an article by Cauchy entitled: *Mémoire sur les pressions et tensions dans un double système de molécules sollicitées par des forces d'attraction ou de répulsion mutuelle.* The object of the article is to give an account of the contents of the new memoir.

Cauchy begins by alluding to his own investigations respecting stress in the *Bulletin de la Société Philomatique* (see Art. 602), and in the second volume of his *Exercices de mathématiques*, especially to the theorem that at every point of a body a system of *principal tractions* exists: see Art. 603. He then notices a memoir by Poisson presented to the Academy on the 1st of October 1827, of which an abstract had been published in the *Annales de physique et de chimie;* and he refers to his own investigations in the third volume of his *Exercices de mathématiques*, and in his *Exercices d'analyse et de physique mathématique.* Then he proceeds to speak of his new memoir. Here two systems of molecules are supposed to be included within the same space, and to be under the influence of mutual attractions and repulsions. Cauchy says:

Alors les pressions supportées par un plan quelconque, ou plutôt leurs composantes parallèles aux axes coordonnés, se composent chacune de trois termes qui sont sensiblement proportionnels l'un au carré de la densité du premier système de molécules, l'autre au carré de la densité du second système, l'autre au produit de ces deux densités.

673. Cauchy alludes to the principle with respect to fluids which is usually held to be characteristic of them, namely that of the equality of pressure in all directions; he says that his researches lead him to the conclusion that the principle holds with respect to

the *equilibrium* of fluids, but not with respect to their *motion.* He says:

On se trouve ainsi conduit à révoquer en doute, avec M. Poisson, l'exactitude du principe d'égalité de pression appliqué au mouvement des liquides. Ne serait-ce pas à ce défaut d'exactitude que tiendraient les modifications que l'on a été obligé d'apporter aux formules de l'hydrodynamique pour les rendre propres à représenter les résultats des observations?

I do not know whether the memoir to which this brief notice relates was ever published.

674. During the years extending from about 1840 to 1847 Cauchy published four volumes under the title of: *Exercices d'analyse et de physique mathématique.* The first volume is important with respect to the theory of light, but contains nothing that strictly forms part of our subject; in Moigno's *Statique* there are some references to this volume, namely on pages 656, 677, 690, 701, but the matters are not important for our purpose. The second volume of this series is dated 1841; pages 302—330 are occupied by a memoir entitled: *Mémoire sur les dilatations, les condensations et les rotations produites par un changement de forme dans un système de points matériels.* The memoir begins thus:

Pour être en état d'appliquer facilement la Géométrie à la Mécanique, il ne suffit pas de connaître les diverses formes que les lignes ou surfaces peuvent présenter, et les propriétés de ces lignes ou de ces surfaces, mais il importe encore de savoir quels sont les changements de forme que peuvent subir les corps considérés comme des systèmes de points matériels, et à quelles lois générales ces changements de forme se trouvent assujettis. Ces lois ne paraissent pas moins dignes d'être étudiées que celles qui expriment les propriétés générales des lignes courbes ou des surfaces courbes; et aux théorèmes d'Euler ou de Meunier sur la courbure des surfaces qui limitent les corps, on peut ajouter d'autres théorèmes qui aient pour objet les condensations ou les dilatations linéaires, et les autres modifications éprouvées en chaque point par un corps qui vient à changer de forme. Déjà, dans un Mémoire qui a été présenté à l'Académie des Sciences le 30 septembre 1822, et publié par extrait dans le *Bulletin de la Société Philomatique*, j'ai donné la théorie des condensations ou dilatations linéaires, et les lois de leurs variations dans un système de points matériels. A cette théorie, fondée sur une analyse que j'ai développée dans le second volume des *Exercices de mathématiques*, et que je vais reproduire avec quelques légères modifications, je me propose de joindre ici la théorie des rotations qu'exécutent, en se déformant, des axes menés par un point quelconque du système.

The memoir contains various theorems demonstrated with clearness and simplicity; but with regard to our subject of elasticity they may be considered as analytical superfluities.

675. Three articles published by Cauchy in the *Comptes Rendus*, Vol. XVI. 1843, will now be noticed.

The first article is entitled: *Mémoire sur les dilatations, les condensations et les rotations produites par un changement de forme dans un système de points matériels;* it occupies pages 12—22. This is an abstract of the memoir having the same title published in the second volume of the *Exercices d'analyse et de physique mathématique*, of which we have just spoken.

676. The second article is entitled: *Note sur les pressions supportées, dans un corps solide ou fluide, par deux portions de surface, très-voisines, l'une extérieure, l'autre intérieure à ce même corps;* it occupies pages 151—155.

Cauchy gives some general reasoning to shew that at any point *within* a solid body the stresses on the two faces of a plane passing through the point are equal; or, as it may be expressed, the two stresses on the two faces of an indefinitely thin shell passing through the point are equal: see Art. 610 (i). Then the question occurs whether the extension can be made to the case when the point is very close to the surface of the body, and one face of the shell coincides with that surface, while the thickness of the shell is equal to the radius of the sphere of activity of a molecule. Cauchy says that he himself and Poisson had substantially held that this extension is true. Cauchy adds:

Mais avons-nous raison de le faire, et cette manière d'opérer est-elle légitime? C'est un point sur lequel s'était élevé dans mon esprit quelques doutes, que j'ai cru devoir loyalement exposer aux géomètres, non-seulement dans le Mémoire lithographié sur la théorie de la lumière, mais aussi dans le Mémoire présenté à l'Académie le 18 mars 1839. Aujourd'hui ces doutes sont heureusement dissipés, ainsi que je vais l'expliquer en peu de mots.

Then Cauchy offers some general remarks to establish the truth of the required extension of the proposition, at least under a certain condition. He adds:

Dans le tome VIII des *Mémoires de l'Académie* (page 390) et dans le XX[e] cahier du *Journal de l'École Polytechnique* (page 56), M. Poisson avait déjà cherché à démontrer l'égalité des pressions extérieure et intérieure correspondantes à deux points situés, l'un sur la surface d'un corps, l'autre près de cette surface. Mais la démonstration qu'il a donnée dans les *Mémoires de l'Institut*, et modifiée dans le *Journal de l'École Polytechnique*, en comparant l'une à l'autre les pressions supportées par les bases, tantôt d'un très-petit segment de volume, tantôt d'un cylindre dont la hauteur et les bases sont très-petites, me paraît sujette à quelques difficultés qu'il serait trop long de développer ici...

677. The third article is entitled: *Mémoire sur les pressions ou tensions intérieures, mesurées dans un ou plusieurs systèmes de points matériels que sollicitent des forces d'attraction ou de répulsion mutuelle;* it occupies pages 299—308, 954—967, 1035—1039[1]. So far as we are concerned with this article it contains a process for obtaining the body shift-equations of an elastic solid by the consideration of the equilibrium of a molecule without the introduction of the stress-components; the rest of it consists mainly of generalities with respect to the solution of certain differential equations relating to the motion of particles.

678. *Notes relatives à la mécanique rationnelle;* these occur in the *Comptes Rendus,* Vol. XX. 1845, pages 1760—1766. The part of this communication which concerns us is entitled: *Note relative à la pression totale supportée par une surface finie dans un corps solide ou fluide.* Here Cauchy explicitly adopts the definition of *pressure,* or as we term it *stress,* given by Saint-Venant: i.e. the

[1] [This would appear to be two memoirs and an addition to the second, rather than one continuous paper. The titles of the first two are somewhat different. ED.]

stress on one side of an indefinitely small plane face in the interior of a solid or fluid body is the resultant of all the actions of the molecules on this side of the plane on the molecules on the other side, the directions of which cross the face: see our Art. 616 and Moigno's *Statique*, pages 619, 675. Cauchy makes some remarks on the phrase *moment d'une force d'élasticité* which is used in the *Mécanique analytique*; Lagrange to obtain this moment multiplies the force by the differential of the angle which it tends to diminish. Cauchy says:

Il est clair que, pour obtenir le véritable sens des formules de Lagrange, on ne doit pas attribuer ici aux expressions qu'il a employées leur signification ordinaire.

After some explanations Cauchy concludes thus:

En conséquence, dans la *Mécanique analytique* de Lagrange, par ces mots *force d'élasticité tendant à diminuer un angle*, on doit toujours entendre le moment d'un couple appliqué à l'un des côtés de cet angle, c'est-à-dire la surface du parallélogramme construit sur les deux forces du couple.

679. *Observations sur la pression que supporte un élément de surface plane dans un corps solide ou fluide. Comptes Rendus*, Vol. XXI. 1845, pages 125—133.

Cauchy as we saw in the preceding article adopted the definition of stress given by Saint-Venant, and in this article he investigates expressions for the resolved stresses on the definition. The results are of the same nature as those in Moigno's *Statique* pages 674 and 675. Cauchy's method is somewhat obscure, and does not present any obvious advantage. See Moigno 619.

680. *Mécanique moléculaire. Comptes Rendus*, Vol. XXVIII. 1849, pages 2—6. All that concerns us here is the announcement of a design which it is to be regretted was never accomplished.

Des savants illustres, dont plusieurs sont membres de cette Académie, m'ayant m'engagé à réunir en un corps de doctrines les recherches que j'ai entreprises et poursuivies depuis une trentaine d'années, sur la mécanique moléculaire et sur la physique mathématique, j'ai cru qu'il était de mon devoir de répondre, autant que je le pouvais, à leur attente, et de réaliser prochainement le vœu qu'ils m'avaient exprimé.

Il m'était d'autant moins permis de résister à leur désir, qu'en y accédant je remplis, en quelque sorte, un acte de piété filiale, puisque ce désir était aussi le vœu d'un tendre père, qui joignant, jusqu'en ses derniers jours, l'amour de l'étude et la culture des lettres à la pratique de toutes les vertus, s'est endormi du sommeil des justes, et s'est envolé vers une meilleure patrie. Pressé par tous ces motifs, je me propose de publier bientôt un Traité de mécanique moléculaire où, après avoir établi les principes généraux sur lesquels cette science me paraît devoir s'appuyer, j'appliquerai successivement ces principes aux diverses branches de la physique mathématique, surtout à la théorie de la lumière, à la théorie du son, des corps élastiques, de la chaleur, &c.

This passage is condensed in the *Répertoire d'optique moderne*...par l'Abbé Moigno, page 1741; but the allusion to the father of M. Cauchy is rendered unintelligible, and even ungrammatical, by the omission of its last clause.

681. *Note sur l'équilibre et les mouvements vibratoires des corps solides. Comptes Rendus*, Vol. XXXII. 1851, pages 323—326. This consists merely of generalities, and is apparently of no importance.

[One point in this memoir seems to me suggestive. Cauchy remarks that if we consider a homogeneous body as built up of a system of molecules, and each molecule be in itself a system of atoms, then

les coefficients renfermés dans les équations des mouvements vibratoires de ce corps cesseront d'être des quantités constantes.

The conception as Cauchy points out is by no means without a possible application in the Theory of Light.]

682. *Rapport sur divers Mémoires de* M. Wertheim (Commissaires, MM. Regnault, Duhamel, Despretz, Cauchy rapporteur). *Comptes Rendus*, Vol. XXXII. pages 326—330, 1851.

This report speaks very highly of the investigations of Wertheim. The principal point noticed is the value assigned by Wertheim as the ratio of one constant to another; Wertheim holds, using our notation, that $\lambda = 2\mu$. Cauchy cites experiments in favour of Wertheim's view, and holds that there is no valid theoretical objection against it.

683. Saint-Venant alludes to the papers noticed in Arts. 540 and 682; see his *Torsion* page 262, and Moigno's *Statique* page 706: but the object of the allusions is not very clear to me. For instance at the last cited page we have,

> La possibilité que Cauchy s'efforce d'y établir, contrairement à ses beaux travaux de 1828 à 1845, de plus de quinze coefficients...:

but I do not see to what passage of Cauchy's these words refer[1].

684. *Sur la torsion des prismes. Comptes Rendus*, Vol. XXXVIII. 1854, pages 326—332.

These pages contain a brief introduction to the subject of the torsion of prisms. Cauchy alludes to his own researches on it in the fourth volume of his *Exercices de mathématiques;* these he admits were only approximative, and holding under certain conditions. He speaks very highly of the recent researches of Saint-Venant with respect to the subject, and says that a careful perusal of them had led him to some new reflections. There are only two points which require notice in this article.

Cauchy establishes in a simple way expressions for the six stress-component, involving twelve constants; Saint-Venant reproduces this in his *Torsion* in establishing the equations (18) on page 265 of that work. Compare also equations (33) on page 655 of Moigno's *Statique,* where however only nine constants are preserved.

Again Saint-Venant assumes that the angle of torsion $\tau$, corresponding to a unit of length is *constant;* Cauchy proposes to generalise this by assuming $\tau$ to be a function of the distance of the point from the axis. Saint-Venant himself pursues this suggestion in a note on pages 341—343 of his *Torsion,* and shews that it does not lead to any results of practical value.

Cauchy's article concludes thus:

[1] [Saint-Venant sees in these papers an abandonment by Cauchy of what he himself holds to be the true basis of elasticity, namely that molecular theory which reduces the 36 constants to 15, and not merely to 21, in the case of an aeolotropic elastic solid. Ed.]

Il reste à montrer comment, à l'aide du calcul des résidus, on pourra obtenir immédiatement l'intégrale donnée par M. de Saint-Venant, et l'intégrale du même genre relative au cas où $\tau$ est facteur de $r$. C'est ce que je me propose d'expliquer dans un prochain article.

The intention here expressed seems not to have been carried out[1].

Moigno's *Statique* 616, 625, 640, 664, 665. Saint-Venant on *Torsion* 266, 340.

685. The remarks with which an enthusiastic pupil and friend of Cauchy closes a survey of the contributions of this great mathematician to the theory of light are equally applicable in relation to the subject of elasticity.

See Epiphonème. Moigno's *Répertoire d'optique moderne*, pages 1748—1749.

[1] [It may be remarked that Cauchy in this memoir first employs Coriolis's double suffix notation. Ed.]

# CHAPTER VI.

## MISCELLANEOUS RESEARCHES OF THE DECADE 1830 TO 1840[1].

[686.] WE must here note an historical controversy, which arose on the effect of a uniform tractive load on the inside and outside surfaces of a glass or metal vessel in changing its volume. The problem as to this change of contents sprung from the renewed experiments on the compressibility of water and other liquids which were made in various countries during this and the preceding decade. The controversy is interesting as giving a practical example of the need for a theory of elasticity. We may note the following memoirs:

[687.] I. Oersted. *Sur la compressibilité de l'eau. Annales de Chimie*, Tome 22, pp. 192—198, 1823. This is an account of Oersted's first apparatus and experiments. He appears to have neglected the compressibility of his containing vessel. A remark made by him (p. 196) seems to shew that the material of that vessel obtained a set, not that the water lost its compressibility after several trials:

> Je dois encore signaler une autre circonstance qu'on devrait peut-être prendre en considération ici: c'est que l'eau semble perdre un peu de sa compressibilité après quelques compressions. Je n'oserai cependant assurer ce fait, ne l'ayant pas soumis à des épreuves rigoureuses.

[688.] II. Colladon and Sturm. *Mémoire sur la compression des liquides. Mémoires...par divers savants, sciences mathématiques et physiques.* Tome v. pp. 267—347, Paris, 1838. The paper was read on June 11, 1827.

The authors commence that portion of their subject which concerns us with the following statement:

> ...ce principe assez évident, qu'un corps solide homogène, plongé

[1] This chapter contains a few experimental researches falling outside this decade.

dans un fluide et soumis à une pression uniforme, éprouve, selon chacune de ses dimensions, une diminution proportionelle à leur grandeur, et se contracte en conservant une forme semblable à sa forme primitive.

Supposons en effet que ce corps solide est un prisme parallélépipède dont les trois dimensions ont une mesure commune, et divisons par la pensée ce prisme par des plans parallèles à chacune de ses bases en un grand nombre de petits cubes tous égaux entre eux.

Lorsque la compression sera opérée et l'équilibre établi, ces molécules cubiques supporteront nécessairement sur leurs faces opposées des pressions égales, et cette pression sera la même pour toutes. Ainsi chacune de ces molécules se contractera également selon ses trois dimensions, et le corps, après avoir diminué de volume conservera une forme semblable à celle qu'il avait avant la compression. (p. 283.)

The question then arises as to how this principle is to be applied to their piezometer. They remark:

Que dans l'emploi de ce piézometre le volume intérieur occupé par le liquide diminue pendant la compression de la même quantité dont diminuerait sous une pression égale une masse solide de même matière que l'enveloppe et d'un volume équivalent à celui du liquide comprimé. (p. 285.)

Colladon and Sturm next proceed to obtain the diminution of volume produced by uniform tractive load, from the measurements of the stretch in a prismatic bar subject to uniform terminal tractive loads. There is a description of their apparatus for this purpose on pp. 285—286, and it is figured on Plate II. figs. 1, 2.

In the original memoir presented to the Institut in 1827 (*Annales de Chimie*, Tome 36), the authors really supposed that when the terminals of a prismatic bar were subjected to uniform tractive load, the sectional area did not diminish, and denoting this stretch by $s$, they assumed $3s$ to be the dilatation which would be produced by an uniform tractive load upon the whole surface of a mass of this material. In the memoir as it is printed in the volume at the head of this paragraph there is a footnote saying that Poisson had corrected this error and shewn that the stretch produced by an uniform tractive load at all points of the surface $= s/2$, and therefore that $3s/2$ is the dilatation. Poisson's results suppose uni-constant isotropy and no set in the material.

[689.] III. Oersted. *Sur la compression de l'eau dans les vases de matières différentes. Annales de Chimie,* Tome 38, pp. 326—329, 1828. In this note Oersted attempts from experimental results for lead to prove that, when a vessel is compressed inside and outside by the same uniform tractive load, the effect on the vessel is to render its sides thinner (*ses parois plus minces*) or to increase its capacity a little, although it be very little[1]. The experiments on lead, as the Editor of the *Annales* remarks, are not very satisfactory, and he placed the matter in Poisson's hands, who then wrote the note which immediately follows Oersted's: see our Art. 535.

[690.] IV. Oersted. *On the Compressibility of Water. Report of the Third (or Cambridge) Meeting of the British Association in* 1833, pp. 353—360, London, 1834. This is in the form of a letter to Dr Whewell. The only point which concerns us is a remark on p. 358—the English is apparently Oersted's:

Messrs Colladon and Sturm have in the calculation of their experiments introduced a correction founded upon the supposition that the glass of the bottle in which the water is compressed should suffer a compression so great as to have an influence upon their results. Their supposition is that the diminution of volume produced by a pressure on all sides can be calculated by the change of length which takes place in a rod during longitudinal traction or pression[2]. Thus, a rod of glass, lengthened by a traction equal to the weight of the atmosphere as much as 1·1 millionth, should by an equal pression on all sides lose 3·3 millionths, or, according to a calculation by the illustrious Poisson, by 1·65 millionth. As a mathematical calculation here is founded upon physical suppositions, it is not only allowable, but necessary, to try its results by experiment. Were the hypothesis of this calculation just, the result would be that most solids were more compressible than mercury.

[1] Professor J. D. Forbes in a paper printed in the *Edin. New Phil. Journal*, Vol. xix. 1835, p. 38, surpassing Oersted, appears to hold that a glass vessel equally pressed within and without is unaffected by the pressure.

[2] Cf. *Nature*, August 25, 1885!

With reference to his experiments Oersted adds:

I have not yet exactly discussed all the experiments on this subject, but the numbers obtained are such as to show that the results are widely different from those calculated after the supposition named.

If Oersted's assertion were correct, it would cut at the very root of the mathematical theory of elasticity. The error, if any, in Poisson's correction lies: (1) in the possibly erroneous assumption of uni-constant isotropy, (2) in the fact that a glass bottle is probably anything but isotropic, (3) in a possible set.

[691.] V. In repeating the experiments of Canton, Perkins, Colladon and Sturm, and Oersted in 1843, G. Aimé adopted Poisson's calculation of the compressibility of glass: see his *Mémoire sur la compression des liquides. Annales de Chimie,* Tome 8, p. 258, 1843.

We shall again have to touch on this matter when we consider some experiments of Regnault.

[692.] G. H. Dufour. *Description du Pont suspendu en fil de fer construit à Genève.* Geneva, 1824. A second work of this author on suspension-bridges in general is said to have been published in 1831, but I have not been able to find a copy[1]. A memoir however on the subject of suspension-bridges by Dufour appeared in Tome 48, p. 254, of the *Bibliothèque universelle des sciences et arts,* published at Geneva. A *résumé* of his labours is given in the *Annales des ponts et chaussées,* 2$^e$ semestre, 1832, pp. 85—123. The Geneva bridge was the first bridge of any importance made of iron wire, and the various experiments made by Dufour on the iron wire used in its construction are all that concerns us here. They bring out various physical facts which are too often omitted in the theoretical consideration of traction problems, and for which no comprehensive mathematical theory has yet been propounded. We may note:

1°. Breaking load of annealed is roughly only little more than half that of unannealed wires: see Table in *Annales,* p. 87[2].

[1] Cf. Saint-Venant, *Historique Abrégé,* p. cclxxxvi, probably citing the *résumé* in the *Annales des ponts* (p. 85).

[2] We shall return to this point, when considering later experiments. The

2°. The breaking load of iron wire (unannealed) is almost one-third more than that of forged iron-bars of about the same diameter.

3°. The extension of a wire only becomes appreciable when it has received a load generally amounting to about $\frac{2}{3}$ of the breaking load. It is practically insensible to $\frac{1}{2}$ this load, and only what can be termed great when it amounts to $\frac{9}{10}$.

4°. A wire loaded nearly to rupture has only very feeble extension when loaded anew, and the breaking load does not differ perceptibly from what it would amount to on the first occasion.

5°. Annealed wires extend very considerably with smaller loads.

6°. The wires "thin down" at the point of rupture, but this *striction* (*étranglement*) is formed only at the instant of rupture (*l'étranglement s'est toujours formé instantanément*). Probably Dufour's testing apparatus was not delicate enough to distinguish that the stricture really precedes the rupture.

7°. Some experiments on impulse,—very inconclusive.

8°. Some experiments on the effect of temperature on absolute strength, which are not very conclusive. We cite the short account of these experiments given in the *Annales*, pp. 91—92:

Il a d'abord fait passer dans un manchon qui contenait de la glace à la température de $-22\frac{1}{2}°$ centigrades, un fil Laferrière n°. 4; il s'est cassé deux fois sous une charge de 46 kilogrammes et une fois sous une charge de 47, mais toujours hors du manchon. Il a ensuite introduit dans ce manchon de l'eau chaude à $92\frac{1}{2}°$: le fil s'est rompu une fois hors du manchon sous un poids de 45·5 kilogrammes et une seconde fois dans le manchon sous le poids de 46·5 kilogrammes. Enfin il fait passer le fil de fer dans deux manchons distans de 0m. 60: dans l'un se trouvait de l'eau à la température de $92\frac{1}{2}°$ et dans l'autre de la glace à $-22\frac{1}{2}°$. La rupture s'est fait entre les deux manchons sous le poids de 45·5 kilog. La force absolue du fil n°. 4 n'est que de 48 kilog., ainsi on ne peut pas conclure de ces expériences que le refroidissement diminue la tenacité.

ratio varies with the tenacity of the wire, and in the case of steel with the amount of carbon: see Art. 830.

We have in the above results evidence of the existence of the *yield-point*, the influence of the worked state on the strain, and of the temperature element in the breaking load. All these are matters which were gradually impressing themselves on the practical men, and will have to be considered in the mathematical theory of cohesion, if it is to keep pace with physical experience.

[693.] Ignaz, Edler von Mitis. Several papers containing experimental results are due to this physicist. We may note the following:

(I.) *Versuch über die absolute Festigkeit einiger österreichischen Stahlegattungen, und Vorschlag, dieses Material statt des Eisens zu Kettenbrücken und Ankertauen zu verwenden. Baumgartner's Zeitschrift für Physik und Mathematik.* Bd. III. 1827, pp. 1—17.

(II.) *Versuche über die Stärke und Elasticität des Eisens und Stahles, mit Rücksicht auf die Verwendung dieser Materialien zu Ketten und Balken. Ibid.* Bd. IV. 1828, pp. 129—171.

(III.) *Beiträge zur Kenntniss der Eigenschaften des Guss- und Stab-Eisens und des Stahls. Ibid.* Bd. VI. 1829, pp. 43—88.

(I.) Contains nothing novel, except perhaps the proposal with which the title closes; this is based upon the greater strength of steel as evidenced in the experiments. The writer remarks that the decrease in diameter of steel bars, even at rupture, is remarkably small as compared with that of all the kinds of iron which he has tested, even long before the iron bars have received their maximum load (p. 16)[1]

(II.) Contains a further consideration of the relative strength of steel and iron. At first the Edler von Mitis treats of breaking loads, and the results of his previous paper are confirmed. He then passes to the limit of elasticity, and finds that steel does not receive set till under a much greater load than wrought iron. He again notices how little its diameter decreases even at the instant of rupture (p. 155). As mean results stretches of 1/919 and 1/609 are given for the elastic limit of iron and steel respectively. The author argues that this 'greater elasticity' of steel ought to

[1] The steel here considered is steel in the old sense of the word with 7 or more p. c. of carbon.

enable it the better to withstand variations of temperature. At the same time expanding more than iron for the same load the oscillations (in the case of a suspension bridge etc.) would be greater (see p. 168).

The results for steel bars are compared with those obtained by Tredgold and Duleau, and on the whole found to agree.

(III.) Contains a wider range of experiments on a greater variety of iron and steel bars and wires, but offers nothing to concern us here.

[694.] 1828. *Annales des Mines.* T. III. *Deuxième série* pp. 510—516. A short article without the author's name, entitled: *Note sur la maniere de calculer les epaisseurs des chaudières en rôle des machines à vapeur.* The boiler consists of a right cylindrical body with hemispherical ends; the latter it is said must be one-half as thick as the cylindrical part for the thickness of this part a formula with factor of safety is given which presents nothing new. For the tenacity of materials for boilers of various metals, experiments of MM. Tremery and Poirier Saint-Brice, Cagniard Latour and George Rennie are cited.

[695.] *Gehler's Physikalisches Wörterbuch.* 1826—1845. (*Neubearbeitet von Brandes, Gmelin* etc.) The articles on *Anziehung* (I. p. 324) and *Cohäsion* (II. p. 113) contain nothing original, but may be consulted for the range of historical references given, particularly the latter: see pp. 149—150.

[696.] J. Rondelet. *Traité théorique et pratique de l'Art de Bâtir.* Paris. A first edition of this work appeared in 1812, and a sixth edition considerably altered in 1830—1832. To this latter edition a *Supplément* was published in 1847—1848. The major portion of the work is of a character which does not concern us here. At the same time it contains a vast amount of experimental detail on the strength of materials, and is continually referred to by writers of this period. Our references are to the sixth edition.

Tome I. *Deuxième section. Résultat d'expériences faites pour déterminer la force des matériaux* (pp. 203—321), is the portion which chiefly concerns us. It is included under the general heading: *Connaissance des matériaux.* We may draw attention to pp. 218—

225, where some of the first experiments on adherence are recorded. At the same time they are confined to mortar and plaster. The experiments were made in 1787. There does not seem to be any theoretical conception of the general nature of strain as existing in all elastic solids. So far as theory is concerned what little occurs in the work is of the old Bernoulli-Eulerian type. As Bevan with glue (see Art. 374) so Rondelet found with mortar, that its adherence was greater than its cohesion:

> C'est a dire que si on soumet à un effort de traction deux pierres unies par du mortier, la séparation s'opérera au milieu de l'épaisseur du joint, et non suivant les surfaces, et que le contraire a lieu pour le plâtre.

Adherence with Rondelet means the force with which one material adheres to another when they are acted upon by a traction normal to the surfaces in contact. This is not, it will be observed, adherence in the sense of Morin: see Art. 905.

[697.] Liebherr. *Beschreibung einer Maschine zum Zerreissen, Zerdrücken, Verdrehen, und Biegen, oder überhaupt zu Versuchen über die absolute und relative Festigkeit der Metalle.* This paper is to be found in the *Kunst und Gewerbe-Blatt des polytechnischen Vereins für das Königreich Bayern.* Munich, 1830 columns 233—237. Except for the variety of purposes to which it can be turned, the machine appears inferior to the hydraulic apparatus used by Lagerhjelm, and referred to in Art. 364.

[698.] While on the subject of apparatus, it may be as well to mention an instrument invented by Brewster, in which the fringes obtained on passing a ray of polarized light through a bit of strained glass are used to investigate the nature of a strain. He called his instrument a *Chromatic Teinometer* and it occurs in two forms. In the first form a plate or bar of glass is subjected to flexure, and the fringes observed by examining polarised light passed through it at different points determine the position of the neutral line. Their frequency is also a measure of the degree of flexure. In the second form a standard glass plate is placed between two metal plates, so arranged that they tend to give it flexure in opposite directions. The degree of curvature of the glass, if not perceptible to the eye, is revealed by the polariscope, and the maximum tint[1] gives a measure of the difference of the elasticities of the two plates. A full description of the Teinometer will be found in Brewster's edition of *Ferguson's Lectures on Select Subjects*, 3rd Ed. 1823, Vol. II. pp. 232—234. See also *Edin. Royal Soc. Trans.* Vol. III. p. 369.

[699] W. Weber. *Bemerkung über ein von Hrn. Poisson für die Extension elastischer Drähte aufgestelltes Theorem. Poggendorff's Annalen*, Bd. XIV 1828. pp. 174—176.

This is an experimental confirmation of Poisson's formula

$$n' = (2{\cdot}05610)\,\frac{ne}{l}$$

given in our Art. 429.

Poisson's theory gave 337·9 vibrations in a second; Weber found experimentally 334·7 vibrations.

[700.] W. Weber. *Theorie der Zungenpfeifen. Poggendorff's Annalen*, Bd. XVII. 1829, pp. 192—246.

This paper belongs to the theory of sound and contains a theoretical investigation of the notes of reed-pipes. The comparison of theory with experiment is based on results published by Weber in various volumes of the *Annalen* for 1828 and 1829 (XIV. p. 397 and XVI. pp. 193 and 415). I am not aware whether these papers have been considered by recent writers on sound, but attention might very well be drawn to them. I only intend to quote here a passage from that at the head of this article, describing how Weber ensured constancy of stress in the cords of a peculiar monochord invented by him (described *Annalen*, XV. 1). He refers to this method of treating iron wires in the memoir considered in Art. 702.

Die feine Eisensaite, welche in diesem Monochorde gebraucht werden sollte, war vorher eine Zeit lang der grössten Spannung unter worfen gewesen, die sie, ohne zu reissen, vertrug, und ich hatte diese Saite darauf einer doppelten Prüfung unterworfen. Bei einer Saite nämlich, welche nicht dieser grössten Spannung unterworfen gewesen war, hatte ich gefunden, dass, nachdem sie sich bei zunehmender

[1] By 'maximum tint' we are to understand the highest in Newton's scale of colours: see our footnote to F. E. Neumann's memoir of 1841 in Chapter VIII.

Spannung verlängert habe, sie bei abnehmender Spannung sich nicht wieder bis zu demselben Punkte zusammenziehe. Umgekehrt hatte ich gefunden, dass, nachdem eine Saite jenem Maximo der Spannung unterworfen gewesen war, ihre Länge eine Function bloss von der gegenwärtigen Spannung derselben sey (p. 226).

A table of experimental results amply confirms this (p. 227). It will be noted that Weber is here clearly conscious of the importance of reducing his material to a state of ease limited by a higher stress than he intends to apply.

[701.] W Weber. *Ueber die specifische Wärme fester Körper insbesondere der Metalle. Poggendorff's Annalen*, Bd. xx. 1830, pp 177—213. This is, so far as I am aware, the first memoir that drew attention to the importance of considering the influence of temperature on stress[1]. It seems to have escaped Saint-Venant, who in the *Historique Abrége*, p. clxviii, attributes to Duhamel the first consideration of the mutual relations of stress, strain and temperature. Some introductory remarks of Duhamel (see Art. 880) suggest to me that he had read Weber's paper Duhamel's memoir is the mathematical correlative to Weber's physical conceptions.

The phenomenon to which Weber draws attention is of the following nature. An iron wire is suddenly stretched to a

[1] Perhaps the earliest investigations of the relation of heat to elasticity are due to John Gough, who was the first to point out the somewhat singular thermo-elastic properties of caoutchouc. His paper is entitled:

*A Description of a Property of Caoutchouc or Indian-rubber; with some Reflections on the Cause of the Elasticity of this Substance, in a Letter to Dr Holme.* It is given in the *Manchester Philosophical Memoirs*, Second Series, Vol. I. 1805, pp. 288—295.

Gough's experiments go to prove:

(1) That the temperature of Indian-rubber increases when it is stretched, but decreases almost at once on allowing it to resume its natural size.

(2) That if a weight be suspended by a slip of Indian-rubber the slip will become 'shorter with heat and longer with cold.'

(3) 'If a thong of caoutchouc be stretched in water warmer than itself it retains its elasticity unimpaired; on the contrary, if the experiment be made in water colder than itself it loses part of its retractile power, being unable to recover its former figure; but let the thong be placed in hot water, while it remains extended for want of spring, and the heat will immediately make it contract briskly.'

(4) The specific gravity of caoutchouc is diminished by an increase of temperature and is increased by extension. Gough's theory of the phenomena based upon the mutual attraction of caoutchouc and caloric' need not be considered here.

definite length by a certain traction. The wire being now maintained at the *same length,* it is noticed that the traction diminishes during the first six seconds. This diminution of traction Weber attributes to a rise of temperature. As there is no external cause for this rise, he attributes it to an internal cause, i.e supposes that there is an inner source of heat producing this increase of temperature. We must hold, he writes:

dass diese innere Warmequelle daher entspringe, dass eine und die selbe Warmemenge, die in dem Drahte vorhanden ist, bei verschiedenen Spannungen des Drahts verschiedene Temperaturen erzeugen konne. Man sagt aber von einem Korper, in welchem durch gleiche Warmemengen unter verschiedenen Umstanden *verschiedene Temperaturen* entstehen, dass seine *specifische Warme* von diesen Umstanden abhänge und *nicht immer gleich gross sey* (p. 180).

[702.] Weber's apparatus and method of experimenting were as novel and ingenious even as those by which he investigated the elastic after-strain. In order to produce a sudden traction he had a very simple means of connecting the terminal of the iron-wire with that of another already subjected to the required traction (p. 183). After the wire had then been clamped at the length due to this traction, its new traction was calculated by the number of its vibrations in a second The apparatus was so arranged as to allow a double observation in each experiment, one on the primary, and one on the supplementary wire; for the particulars of this we must refer the reader to the original memoir. The experiments were made on iron, copper, silver and platinum wires. Care was taken (1) to eliminate the possible influence of compressional waves along the two wires (p. 193), and (2) to separate the effects considered from a possible after-strain (p. 195). This latter result Weber believed he had achieved by first treating his wires to a process which he had communicated in *Poggendorff's Annalen*, Bd. XVII.: see Art. 700.

[703.] He draws the following conclusions from his experiments: When the wire is sensibly $\left\{\begin{matrix}\text{extended}\\ \text{decreased}\end{matrix}\right\}$ in length and retained in this condition the traction $\left\{\begin{matrix}\text{diminishes}\\ \text{increases}\end{matrix}\right\}$ This result is due to $\left\{\begin{matrix}\text{an increase}\\ \text{a decrease}\end{matrix}\right\}$ in the temperature of the wire. This $\left\{\begin{matrix}\text{increase}\\ \text{decrease}\end{matrix}\right\}$ of temperature follows because the temperature of the wire has been $\left\{\begin{matrix}\text{reduced below}\\ \text{raised above}\end{matrix}\right\}$ that of the surrounding air by the $\left\{\begin{matrix}\text{extension}\\ \text{decrease}\end{matrix}\right\}$ in length (pp. 192—198).

[704.] In order to calculate the number of degrees which the temperature will sink or rise owing to a change of traction $p$ Weber proceeds as follows:

Let $n$ be the number of vibrations per second when the traction is $T$, and $n-\nu$ the number when the traction is reduced to $T-p$; $w$ = weight of unit length of wire, supposed originally of length $l$, $\omega$ the area of the section, and $g$ the acceleration of gravity[1].

Then
$$n=\frac{1}{l}\sqrt{\frac{g}{w}T},$$
$$n-\nu=\frac{1}{l}\sqrt{\frac{g}{w}(T-p)}.$$

Hence
$$p=\frac{n^2l^2w}{g}-\frac{(n-\nu)^2l^2w}{g}=\frac{2n\nu l^2w}{g},$$
since $\nu/n$ is very small.

Let $\epsilon$ be the extension of the wire by unit increase of traction, then by Art. 465, on the uni-constant theory:
$$\epsilon=\frac{2l}{5\mu\omega}.$$

The change of length due to $p$ is therefore
$$=p\epsilon=\frac{2l}{5\mu\omega}\,\frac{2n\nu l^2w}{g}.$$

But if $k'$ be the extension-coefficient of the material for one degree of temperature this change of length $=lk't$
$$\therefore t=\frac{1}{k'}\cdot\frac{4n\nu l^2w}{5g\mu\omega}\ \text{(p. 202)}.$$

[1] Weber takes $g$ equal to the distance fallen from rest in the first second. I have replaced his symbol by the current one, which is just twice Weber's in magnitude.

[705.] On p. 208 Weber states definitely a fact on which Duhamel afterwards insisted, namely:

The specific heat of metals by constant volume is different from their specific heat under constant stress (pressure).

Accepting the results of Dulong for the specific heat $\beta$ at constant pressure Weber obtains from his own experiments the specific heat $\beta'$ at constant volume (p. 211). Thus:

| Water $\beta=1$ | $\beta$ Dulong | $\beta'$ Weber |
|---|---|---|
| Iron | 0·1100 | 0·1026 |
| Copper | 0·0949 | 0·0872 |
| Silver | 0·0557 | 0·0525 |
| Platinum | 0·0314 | 0·0259 |

[706.] Weber calculates these specific heats from the formula $\beta' = \beta - 3k'/r \cdot a$, which he deduces in the following fashion (p. 208):

Angenommen nun, dass der Körper seine *Temperatur* beibehielte, so heisse $a$ die Wärme, welche der Körper bei $r$-maliger Vergrösserung oder Verkleinerung seines Volumens in sich aufnehmen oder herausgeben muss. Ferner heisse $\beta$ die *Wärme*, welche ein Körper, um eine $1^{0}$ höhere Temperatur zu erhalten, in sich aufnehmen muss. Nun ist mit der Erhöhung der Temperatur um $1^{0}$ zugleich eine $3k$ malige Vergrösserung des Volumens verbunden, wenn $k'$ die Längenausdehnung ist. Wenn wir die der $3k$-maligen Vergrösserung des Volumens entsprechende Wärme, $= 3k'/r \cdot a$, von $\beta$ abziehen, so erhalten wir die Wärme $\beta - 3k'/r \cdot a$ welche der Körper, um eine $1^{0}$ höhere Temperatur zu erhalten, in sich aufnehmen muss, wenn *sein Volumen constant ist.*

The value of $a$ must of course be deduced from that for $t$ given in Art. 704 above. Weber assumes here as elsewhere Poisson's relation between longitudinal and lateral stretch.

Seebeck and Clausius (*Poggendorff's Annalen*, Bd. LXXVI. S. 61) consider that some part of the difference in traction observed by Weber is due to elastic after-strain This seems to me improbable. Weber does not get impossible values for $\beta/\beta'$ like Wertheim in his memoir of 1842: see also Saint-Venant in his edition of Navier's *Leçons*, p. 745.

[707.] W. Weber. *De fili Bombycini vi elastica*, Göttingen, 1841. This is an offprint of a paper communicated to the Göttingen *Königliche Gesellschaft der Wissenschaften* in 1835. It is printed in

Vol. 8 (pp. 45—80) of the *Commentationes Recentiores*, which bears the date 1841. My references are to the offprint. An abstract of the paper appeared in *Poggendorff's Annalen*: see our Art. 719; the abstract has been usually cited, the original being perhaps hard to procure.

[708.] The paper is historically of very great importance, for in it attention is first drawn to the remarkable phenomenon now termed in Germany the *elastische Nachwirkung*, but which is here called by the discoverer *prolongatio vel contractio secundaria*. As the subject has hitherto been principally considered both experimentally and theoretically in Germany, no definite English equivalent has yet been generally adopted; we shall refer in future to the phenomenon indifferently as the *Weber effect* or the *elastic after-strain*[1]. The elastic after-strain is dependent on the time element; it differs from elastic fore-strain in that it requires a certain duration (as well as magnitude) of load; it differs from set in that if the load be removed for a certain period the after-strain disappears. This seems to be in keeping with Weber's own view of the matter; at the same time it might be well to recognise that a prolonged load *might* produce after-strain, which would still consist of two parts, *elastic after-strain* and *after-set*; the former would then correspond to the Weber effect or *elastische Nachwirkung*. The *after-set* appears by a previous higher loading to have been eliminated from these experiments.

[709.] Weber had been led to consider the 'elastic force' of silk threads, owing to the important part which it plays in magnetic and other physical apparatus.

Weber's testing machine, the construction of which had been suggested by Gauss, is ingenious, and deserves to be noted by physicists engaged in like investigations. The body to be tested is

[1] Sir William Thomson has investigated certain time effects in the torsional oscillations of wires, which he classes under 'viscosity of solids.' He remarks of one of his observations that 'it was in fact as it would be if the result were wholly or partially due to imperfect elasticity or "elastische Nachwirkung"—elastic after working.' The term viscosity as well as the identification of imperfect elasticity with the Weber effect seems to me open to objection. Art. *Elasticity*, *Encycl. Brit* §§ 29—36. The only English theoretical writers on this phenomenon are, I think, Clerk Maxwell (Art. *Constitution of Bodies*, *Encycl. Brit.*) and J. G. Butcher (*Proc. Lond. Math. Soc.* VIII. p. 103).

attached at one end by a horizontal string to some point in another vertical string which supports a weight, and at the other end to a micrometer screw the body itself being horizontal. The load is obtained by turning the screw till the vertical string is pulled out of the horizontal. The advantages are a continuously increasing or decreasing load, and that extensions of the body are always accompanied by a decrease of load. The apparatus is of course subject to the disadvantage that where a load is producing extension, it requires continual turning of the screw to maintain anything like constancy. (Cf. pp. 3, 38 and the plate.)

[710.] Weber's experiments on absolute strength and set need not detain us, as they offer nothing novel. We may remark that in the experiments on after-strain set had been removed by several times applying a greater load than that which was to produce the *elastic* after-strain.

[711.] Weber having questioned whether the ordinary law of elasticity explains all phenomena then proceeds as follows (p. 7):

Lex elasticitatis notissima ad rationem eam refertur, quae in statu aequilibrii intercedit inter *prolongationem* et *tensionem* fili, quae ratio ex illa lege in eodem *filo semper sibi constat*, h. e illa ratio neque a *magnitudine* tensionis, neque a *tempore*, quo tensio initium cepit, pendet. *Independentia* illius rationis cum a magnitudine tensionis tum a tempòre, quo tensio initium ceperit, in lege illa elasticitatis notissima proposita rerum natura *non confirmatur*. Imo experimentis demonstrari potest, post factam tensionem, quacum magna fili prolongatio conjuncta fuit, per temporis lapsum novam aliquam prolongationem paulatim subsequi, ita ut, quoad fili longitudinem, *duplex tensionis effectus* discerni possit, alter *primarius* seu *momentaneus* ac *subitus*, alter *secundarius seu subsequens* et *continuatus*......

[712.] It is then noted that this *prolongatio continuata* is not permanent. On the hypothesis of an elastic after-strain the author attempts to explain (i) why the oscillations of a body suspended by a thread in a vacuum have notable decrease, (ii) why the same body (a lead cylinder) supported, first by a metal wire and then by a horse-hair adjusted to have the same torsional resistance, gave the same periods of oscillation, but very different rates of diminution in the amplitude (pp. 8—9).

[713.] The author propounds the following theory to explain the phenomenon of after-strain. He supposes the ultimate particles of the body to have three axes, and the angles made by these axes with the central distances of adjacent particles to be capable of variation. The complete equilibrium for any applied load denotes a certain relative position of these axes for neighbouring particles, but this position can only be attained after a long interval of time. Let $x$ be the elongation of the thread at time $t$, then $dx\ dt$ will be a function of the difference in position of the axes of any particle from their position in perfect equilibrium, but this difference itself must be a function of $x$ accordingly Weber writes

$$\frac{dx}{dt} = f(x) \qquad \text{(pp. 9 and 26).}$$

[714.] He then makes various suppositions as to the form of $f(x)$. The one which agrees best with his experimental results is $f(x) = -bx^m$. This leads to a formula of the form

$$e = a \mp (\overline{m-1}\, b)^{\frac{1}{1-m}} (t+c)^{\frac{1}{1-m}} \quad \text{...............(i),}$$

where $e$ is the elongation and $m, a, b, c$ are constants. $b$ and $m$ are taken to be constants of the material, while $a$ and $c$ depend upon the load and length of the thread. For silk Weber finds (p. 32)

$$\left.\begin{aligned} 1/(1-m) &= -0{\cdot}17192, \\ (\overline{m-1}\, b)^{\frac{1}{1-m}} &= 137{\cdot}97 \end{aligned}\right\} \text{...............(ii).}$$

[715.] This theory of Weber's has been commended by Clausius (*Poggendorff's Annalen*, Bd. LXXVI. S. 65—66), but even in the amended form proposed by Kohlrausch can hardly be considered as in complete accordance with experimental fact.

[716.] A remark of Weber's on p. 36 suggests that no *qualitative* distinction can be made between elastic fore-strain and elastic after-strain:

Admodum probabile est, has primarias illasque secundarias contractiones vel prolongationes, quas observationis causa discrevimus, revera non ita esse sejunctas, ut certum quoddam temporis momentum definiri possit, quo illa desinat, haec incipiat. Contractio vel prolongatio, quam primariam appellavi, re vera magna quidem celeritate efficitur, non autem uno temporis momento, nec differt a secundaria s. subsequente nisi majore celeritate. Majorem autem hanc contractionis vel prolongationis celeritatem in illam minorem sensim ita commutari, ut celeritas omnes gradus intermedios accipiat, rerum natura poscit, quae saltum in phaenomenis non admittit.

[717] This remark does not seem to be true in the light of F. Braun's experiments, which have shewn that elastic after-strain differs from primary elastic strain in the non-applicability of the principle of superposition. (*Poggendorff's Annalen,* Bd. CLIX. S. 390.) This supposed relation between fore- and after-strain leads Weber to remark that the conception of the modulus of elasticity is necessarily vague and uncertain unless it be reckoned for complete equilibrium, namely that elongation which includes elastic after-strain; otherwise it would be necessary to determine the exact instant at which elastic after-strain commences.

[718.] A second paper by Weber occurs in the same volume of the Göttingen *Commentationes* and is entitled *De tribus novis librarum construendarum methodis.*

Section III. of this paper is entitled: *De librâ compensatoriâ laminis elasticis suspensâ.* The compensating principle depends on the flexure of an elastic lamina, and the theory adopted for the bending of this lamina under a *longitudinal* load is the approximate, if not insufficient, Bernoulli-Eulerian hypothesis (pp. 12—21).

[719.] W. Weber. *Ueber die Elasticität der Seidenfäden. Poggendorff's Annalen,* Bd. XXXIV. 1835, pp. 247—257 (and *Göttingen Gelehrte Anzeige,* 1835).

*Ueber die Elasticität fester Körper. Ibid.* Bd. LIV. 1841, pp. 1—18. These papers amount practically to a German translation of the Latin memoir presented to the Göttingen Royal Society in 1835 and reviewed in our Art. 707. The first considers the less exact formula, which would be obtained by putting $m = 2$ in equation (i) of Art. 714, and the second, based on a wider range of experiments, the formula given by the value of $m$ in equation (ii). The only point to be noticed is that Weber here terms the elastic after-strain *elastische Nachwirkung.*

[720.] G. Coriolis. *Expériences sur la résistance du plomb à l'écrasement, et sur l'influence qu'a sur sa dureté une quantité inappréciable d'oxide. Annales de chimie et de physique,* T. 44, pp. 103—111, Paris 1830.

These experiments are of importance as referring to the time-element and to the skin conditions in affecting strain. The existence of after-strain in lead cylinders subject to terminal tractive load is very conclusively proved:

On voit qu'après une heure le plomb est loin d'être arrivé à un état stable; il continue de s'écraser bien au delà de ce temps (p. 111).

With regard to the experiments on skin-influence Coriolis writes:

Tout incomplètes qu'elles sont, elles ont néanmoins l'avantage de montrer qu'en fondant du plomb pendant le peu de temps qui suffit à la fusion, même en employant des désoxidations la quantité inappréciable d'oxide qui se forme a la surface change sensiblement la dureté de la masse; et que, pour obtenir du plomb dont la ductilité ne soit pas altérée, il faut le fondre à couvert, en le tirant de fond sans qu'il cesse d'être à l'abri du contact de l'air. (*Ibid.*)

[721.] 1830. Vicat. *Description du pont suspendu construit sur la Dordogne à Argentat.* Paris 1830. This tract is only of interest in so much as it involves Vicat's first indictment of the mathematicians. This indictment had so far a basis that in the ordinary theory then current in practical books, the consideration of slides was entirely neglected. The introduction of the slide-modulus into *practical* elasticity must be attributed to Saint-Venant. It will be of interest to reproduce a portion of Vicat's charge here, because it has considerable similarity with that which practical men of the present day occasionally raise against the mathematicians, and which the latter would do well to recognize.

Les questions que l'on peut se proposer sur les ponts suspendus sont en effet de deux sortes: les unes, qui dépendent presque exclusivement de la statique rationnelle, ont été à peu près épuisées dans le savant mémoire que M. Navier a publié sur cette matière[1] mais les autres, qui ont pour objet certains calculs d'équilibre étroitement liés à notions

[1] See our Article 272.

de résistance, d élasticité, de dilatation, de frottement, etc., ne se résolvent d'une manière complete qu à l'aide de coefficients donnés par l'expérience , encore les solutions ainsi obtenues ne sont-elles pas toujours certaines, parce que les conceptions mathématiques dont elles dérivaient s'appuient elles-mêmes sur les hypotheses touchant la structure intime des corps, leur mode d'agrégation, de rupture, etc., et que ces hypothèses sont quelquefois très-éloignées de la vérité (p. 2).

Vicat quotes as example of this the formulae given by various authors for pulley axles and rivets, which give resistances infinitely great when the force acts in the *plan d'encastrement*—an evidently false result. He concludes therefore that:

La loi de continuité et le raisonnement suffiraient seuls pour in firmer toutes ces théories que l'on attribue à Galilée a Mariotte ou à Leibniz, si l'expérience n'en démontrait d'ailleurs l'insuffisance (p. 3).

[722.] Saint-Venant refers to this charge of Vicat's in the *Historique Abrégé* (p. ccxcvi) and points out that before Vicat drew attention to the omission of shear and slide in the ordinary theory, various experimenters had made them the subject of their investigations. These investigations, however, seem to have been confined to stone and mortar, in which materials the phenomena were treated of under the heading of *adherence*; see Art 696, and also Coulomb's erroneous theory cited in Art. 120.

[723.] Vicat. *Ponts suspendus en fil de fer sur le Rhône. Annales des ponts et chaussées*, 1831, 1[er] semestre, pp 93—144.

This belongs to the long series of memoirs on suspension-bridges. Experimental results are given on the last three pages, but no new physical fact is clearly brought out either by the text or by the tables, so we need not concern ourselves with this paper here. It has bearing only on the practical question, alluded to in Arts. 692 and 817, as to whether wrought iron bar or iron wire is the better material out of which to form the chains of suspension bridges.

[724.] Vicat. *Recherches expérimentales sur les phénomènes physiques qui précèdent et accompagnent la rupture ou l'affaissement d'une certaine classe de solides.* This memoir was presented by Vicat to the Academy of Sciences, and a report upon it was drawn up by de Prony and Girard. These elasticians reported unfavourably. Some account of the results seems to have been presented at an earlier date to the Academy and then reported on not unfavourably by de Prony, Dupin and Girard. This latter report will be found in the *Annales de Chimie*, T. 36, 1827, pp. 96—100. To the report of de Prony and Girard Vicat gave a rejoinder These papers and reports will be found at length in the following parts of the *Annales des ponts et chaussées*:

(*a*) Vicat's Memoir. 1833, 2$^{e}$ semestre, pp. 201—268.

(*b*) de Prony and Girard's *Rapport*, 1834, 1$^{er}$ semestre, pp. 293—304.

(*c*) Vicat's Rejoinder (*Observations adressées à l'Académie des sciences, sur le rapport. par M. Girard*). *Ibid.* pp. 305—314.

[725.] Vicat's memoir is a continuation of the indictment started in 1830 and referred to in our Art. 721. Viewing the whole controversy from the standpoint of our modern theoretical knowledge, we must confess that Vicat had strong grounds for attacking the then current theories; that, although his charges occasionally depended upon a misapplication of the theories, yet the report of de Prony and Girard does not clear the theoretical elasticians from the blame cast upon them. Saint Venant considers Vicat's indictments in the *Historique Abrégé* (pp. cclxxxviii and ccxcvi). He praises highly the experimental results, but remarks:

> Mais le but en étant surtout polémique, il convient d'étudier comme réponse le *Rapport* de MM. de Prony et Girard qui suffit, même après la réplique de M. Vicat, pour venger la théorie de ses attaques, portant d'ailleurs sur d'autres points que ceux où elle pose des affirmations, puisqu'elle ne prétend pas que ses formules s'appliquent jusqu'à l'instant où il y a rupture.

With all due deference to the opinion of such an authority we must venture to differ, and hope to shew cause for doing so in the following remarks.

[726.] The chief merit of Vicat's memoir is its insistancy on the importance of taking account of *shearing force*, and on the distinction which must be drawn between *instantaneous and permanent loads.*

Shear he terms *force transverse* and defines as resistance to

l effort qui tend à diviser un corps, en faisant glisser, pour ainsi dire, une de ses parties sur l'autre, sans exercer ni pression ni tirage hors de la face de rupture (p. 201).

He terms two other forms of resistance by new names to which the *Rapporteur* strongly objects: see (*b*) p. 302. These are resistance to extension = *force tirante* (*résistance absolue* of other French writers), and resistance to compression = *force portante* (*résistance à l'écrasement* of other French writers).

Of these he remarks that:

Ces trois forces ou résistances sont *permanentes* ou instantanées si, par exemple, un cube de pierre d'un centimetre de côté s'écrase quelques minutes, ou même quelques heures, apres avoir porté un poids de 100 kilog., ces 100 kilog. ne sont que l'expression d'une force *instantanée*, et conséquemment relative. Mais si le même cube peut, au contraire, porter indéfiniment sans se briser 30 kilog et pas plus de 30 kilogs., ce chiffre mesure sa véritable force portante absolue ou *permanente* ((*a*) p. 201).

[727.] The permanent forces are those which in the case of structures it is important to know Of course a distiuction might have been drawn between what is needed for a temporary scaffold, for a bridge on which there is a frequently repeated but not a persistent load, and for a permanent structure with persistent load. The *Rapporteur* has here no criticism to offer. With these preliminary remarks Vicat states the object of his memoir, namely

D etudier plus particulièrement qu'on ne l'a fait jusqu'à ce jour, les phénomènes physiques qui se manifestent dans les principaux cas de rupture des corps solides, pour déduire de cet examen, si la chose est possible, les causes de l'imperfection des théories connues, et prémunir ainsi contre les dangers de ces théories les constructeurs qui, n'ayant pas eu l'occasion de les vérifier, seraient portés à leur accorder une certaine confiance ((*a*) p. 202).

[728.] The *Rapporteur* observes that Vicat's results *confirm* the theories he continually describes as inexact. Also that he has experimented on material (*barreaux de plâtre gaché et de brique crue* etc. etc.) which do not conform to the ordinary theory: see (*b*) pp. 302, 303.

Vicat rejoins that he has drawn arguments also from experiments on wood and iron and that these by no means confirm the current theories: see (*c*) pp. 310—313.

[729.] Vicat considers the *forces portantes instantanées*. He cites Coulomb's theory (see Art. 120) which leads to the formula

$$P = 2\gamma ab$$

for the force of rupture $P$, $\gamma$ being the shear-strength per unit area of the material (*la force transverse*), $a$ the depth and $b$ the breadth of the right prism on rectangular base experimented on. He also quotes a formula due to Navier which includes friction. A table of results shew that Navier's and Coulomb's formulae give far too great results, especially the former. Further he remarks:

> La division en deux parties à biseaux, ainsi que l'a entendue Coulomb pour les solides terminés par des faces verticales, ne s'est pas présentée une seule fois dans le cours de nos nombreuses expériences ((*a*) pp. 203 *et seq.*).

In fact looking at Vicat's figures (Plate LXIX. figs. 6—10) we observe that, allowing for variations in material and possibly in uniformity of load, they entirely confirm the modern theory that rupture by pressure is produced by lateral extension. They entirely refute the now rejected hypothesis of Coulomb. Notwithstanding this the *Rapporteur* speaks of the *satisfactory* explanation of Coulomb, considers that theory beyond question and even confirmed by Vicat's experiments on the compression of 'rollers' ((*b*) pp. 294, 300, 301.) The rejoinder on this point is completely convincing ((*c*) pp. 307—309) and for this reason alone we cannot agree with Saint-Venant, who, strange to say, has elsewhere been among the first to repudiate Coulomb's theory: see his edition of Navier's *Leçons*, p. 7.

[730.] Some interesting experiments on spheres and rollers used as 'buffers'—namely, compressed between parallel tangent planes, will be found in (*a*) pp. 213—215.

For the rollers the *forces portantes instantanées* were found proportional to the product of the axes by the diameters, and for the spheres to the squares of the diameters. The surfaces of rupture in these two cases are given on Plate LXIX (figs. 13 and 16) and as the strain admits in both cases of mathematical calculation, it would not be unprofitable to compare experiment with theory. According to the *Rapporteur* these are the first experiments of this kind: see (*b*) p. 303.

There are a few, but somewhat insufficient remarks on the *forces portantes permanentes* for very ductile metals on p. 218 (only *lead* is taken).

[731.] The next results which we need notice are on the force of torsion: see (*a*) pp. 227—236. There is here a discrepancy between theory and practice which it is hard to account for, because it has not occurred with other experimentalists. The *Rapporteur* suggests briefly that it may be due to the want of homogeneity and the 'hygrometric properties of the materials used: see (*b*) p. 296. Vicat rejoins that this cannot account for the divergence: see (*c*) p. 309. As he is only experimenting on bodies for which *so far as he uses them* the results of the old theory are true, it is hard to explain the divergency unless we attribute it to the fact that his measurements were made at or near the point of rupture, i.e. after the beginning of set.

[732.] We next come to a series of experiments on *Résistances relatives instantanées*: see (*a*) pp 236—249. Vicat's results are obtained from comparatively short beams of a non-fibrous material built-in at one or both ends and subjected to transverse load. They do not agree with the then-current theory and he considers that this is due to the neglect of the shear (p. 249). He points out that the results of the ordinary theory become less and less true as the piece becomes shorter and shorter and the shear replaces the traction as producing the strain. The *Rapporteur* merely remarks that the current theory required a longer piece and a different material to that adopted by Vicat in order to be applicable ((*b*) p. 298). On this Vicat's rejoinder ((*c*) p. 310—312) may be consulted.

[733.] Saint-Venant draws attention to Vicat's experimental results on perfect and imperfect building-in as among the *choses precieuses* of this memoir. These results will be found (*a*) pp. 241—243. They shew that when the rupture takes place at a built-in end the surface is always curved. That if the horizontal faces only are fixed then the surface of rupture is cylindrical, but if all four faces are built-in the surface of rupture belongs to the class of spheroids. These surfaces are always *très prononcés* for a short beam, but cease to be appreciable when the length becomes considerable. Here again the influence of shear makes itself felt. Interesting figures of the various surfaces of rupture at a built-in end will be found on Plate LXX. (figs. 10—21 and 25—27).

[734.] Finally we may mention a series of experiments on what Vicat terms *forces instantanées d'arrachement*. They are thus defined: *Supposons une tige mi-plantée ou retenue dans un milieu solide par l'effet d'une tête ou d'un scellement quelconque, nous appelons force ou résistance d'arrachement celle que le milieu solide oppose a la sortie ou à l'extraction de la tige* see (*a*) p. 250.

A conception may be formed of the nature of the experiments made by Vicat in the following fashion: conceive a plate of definite thickness supported at its edges; in one face of this a right cylindrical hole, and in this hole another body also of right cylindrical shape supposed to represent the 'head' of a fastening of some sort, which it is required to pull or push through the plate. The cylinder being circular Vicat found that it tore out a piece of the plate nearly in the form of a hyperboloid of revolution of one sheet truncated above its median plane. Experiments of this kind appear to have been quite novel: see (*b*) p. 303. The memoir concludes with numerous tabulated results of the various series of experiments.

[735.] Vicat's work seems to me of great importance; it was the final blow of practice to the old theory. It drew attention to the questions of shear and of the time-element in language so strong that the theorists were compelled to take them into consideration. Within five years after its publication Saint-Venant gave the elements of a truer theory of flexure, and both mathematicians and practical men began to be more careful in their language as to the various kinds of resistance. It is characteristic that Girard, who fitly closed our first chapter devoted to the old theory of beams, should be found as *Rapporteur* thirty years later upon a memoir of this kind. We wonder that his report was not more hostile than it was.

[736.] 1834. Vicat. *Note sur l'allongement progressif du fil de fer soumis à diverses tensions. Annales des ponts et chaussées*, 1er semestre, 1834, pp. 40—44.

This is an extremely interesting paper for it contains some of the first well-considered experiments on after-strain Vicat took four pieces of the same unannealed wire and submitted them for a period of 33 months to different tractions. His results given on pp 42—43 were as follows:

1° Le fil de fer non recuit, tendu au $\frac{1}{4}$ de sa force tirante, telle qu'on la mesure ordinairement, et soustrait à tout mouvement trépidatoire, reçoit une première extension, mais ne s'allonge pas sensiblement ensuite.

2°. Le même fil tendu dans les mêmes circonstances au $\frac{1}{3}$ de sa force tirante, s'est allongé de $2{\cdot}75^{mm}$ par metre, en 33 mois, non compris l'allongement instantané dû au premier effet de la charge.

3° Le même fil, tendu au $\frac{1}{2}$ de sa force, s'est dans le même temps et les mêmes circonstances, allongé de $4{\cdot}09^{mm}$.

4°. Le même fil enfin, tendu aux $\frac{3}{4}$ de sa force, s'est allongé toujours dans le même temps et les memes circonstances de $6{\cdot}13^{mm}$

5°. A partir du moment où l'effet instantané de la charge est termine, les vitesses des allongemens subséquens restent à très peu près proportionnelles aux temps.

6° Les quantités d'allongement pour les brins chargés au delà du $\frac{1}{4}$ de leur force sont, après des temps égaux, sensiblement proportionnelles aux torsions (? tensions).

It should be observed that the fourth thread subjected to $\frac{3}{4}$ of its breaking load broke at its point of attachment on April 15, 1833, after hanging 33 months (from July 12, 1830). Vicat attributes the breaking to the fixing, as the wire nowhere else showed signs of rupture. With regard to the set of experiments Vicat concludes that:

La mesure de la résistance des matériaux, telle qu'on l'obtient dans les experiences ordinaires qui ne durent que quelques minutes ou quelques heures, est donc, comme on l'a déjà dit dans un mémoire présenté à l'examen de l'Académie des sciences, *tout-à-fait relative à la durée de ces expériences.*

La mesure des résistances absolues qu'il importerait de connaître, exigerait que les matériaux fussent soumis à des épreuves de plusieurs mois, et qu'on observat, avec des instrumens tres-précis, si pendant ce temps-là ils obéissent à l'action des forces qui les sollicitent (p. 43).

The memoir referred to is probably that mentioned in our Art. 725. With regard to Vicat's experiments we may remark on their practical importance (e.g. in iron wire suspension bridges), but physically they neither determine the exact nature of the after-strain (elastic or set?), nor enable us to say how far this after-strain is a result of the *worked condition* of the wire, or how far it is accompanied by a reduction to the *raw stage.* See Arts. 831 and 858 and the Appendix to this volume.

[737.] William Whewell. Certain chapters on beam problems were introduced by this writer into his *Elementary Treatise on Mechanics.* We may refer the reader to the *Supplement to the Fourth Edition: Analytical Statics,* Cambridge, 1832, for what Whewell has written concerning our subject. Chapters VI. and VII. (pp. 108—152) are respectively entitled: *The Equilibrium of an Elastic Body* and *The Strength of Materials.* There is nothing original in these chapters, their matter being drawn from Hodgkinson, Barlow, Tredgold and of course Poisson's *Traité.* We may however note that Whewell follows Tredgold (see Art. 198) in giving the true position to the neutral line of a beam subject to a load partly longitudinal. If $2a$ be the depth of a beam of rectangular section, $h$ the distance of the load-component parallel to the beam axis from the centre of gravity of any section, then the distance $n$ of the neutral line from the centre of gravity is for that section $a^2/3h$: see Art. 198, and compare Whewell pages 117—120.

If $\rho$ be the radius of curvature of the axis of the beam, Whewell falls into the grave error (p. 119) of calling $\rho + n$ the radius of curvature of the neutral line. He in fact supposes the neutral line parallel to the central axis, which in almost all cases of a partially longitudinal load, it is not. Further he states (p. 128) that we may neglect $n$ when $a$ or the dimension of the beam in the plane of flexure is small; as $h$ may and does in many such cases vanish, this seems to me an erroneous assumption.

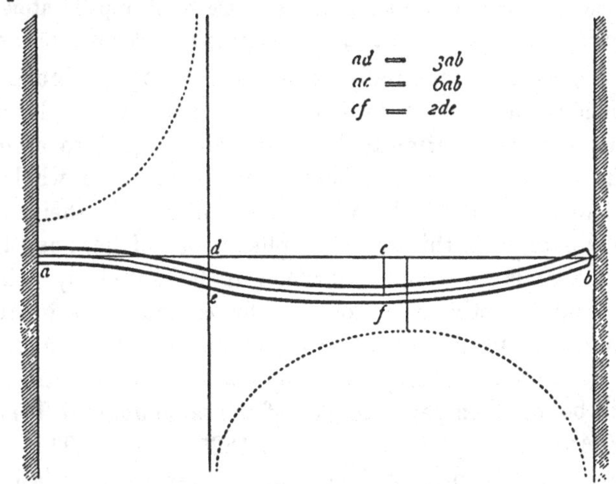


In the accompanying sketch the dotted line represents the neutral line for a beam built-in at one end and pivoted at the other; $n$ is in places infinite however small $a$ may be. The position of the neutral axis in the case of longitudinal load has been frequently misconceived by English writers since Whewell and Tredgold.

[738.] Ampère. *Idées de M. Ampère sur la chaleur et sur la lumière. Bibliothèque Universelle de Genève. Sciences et Arts*, Tom. 49, pages 225—235, 1832. This is a *résumé* of ideas expressed by Ampère, some as early as 1814. It is interesting as containing some conceptions which are not always associated with Ampère's name. He seems to have been the first to clearly distinguish between particles, molecules and atoms. A particle is an infinitely small portion of a body, which still retains the nature of the body. Further:

Les *particules* sont composées de molécules, tenues à distance 1° par ce qui reste, à cette distance, des forces attractives et répulsives propres aux atomes, 2° par la répulsion qu'établit entr'elles le mouvement vibratoire de l'éther interposé, 3° par l'attraction en raison directe des masses et inverse du carré des distances. Il appelle *molécules* un assemblage d'atomes tenus à distance par les forces attractives et répulsives propres à chaque atome, forces qu'il admet être tellement supérieures aux précédentes que celles-ci peuvent être considérées relativement comme tout à fait insensible . Ce qu'il appelle atomes ce sont les points matériels d'ou émanent ces forces attractives et répulsives.

The molecule is essentially solid, and cannot be broken up by mechanical force, only by chemical forces. The only state ment that can with certainty be made with regard to atoms is that they are absolutely indivisible. Molecules as a whole can vibrate, from this arise the phenomena of sound; their atoms can also vibrate, from this arise the phenomena of heat and light. Hence, if heat is an atomic vibration it is caused by existing repulsive and attractive intermolecular forces, thus it is irrational to attribute the repulsive force between atoms to heat (see Arts. 543, 597, 701 footnote). A further note by Ampère to much the same effect will be found on pages 26—37 of the same Journal, Tom. 59, 1835, or *Annales de Chimie*, Tom. 58, 1835, p. 432. This slight notice of Ampère's ideas will be found of service in relation to Mossotti's paper (see Art. 840), and to our discussion of uni constant isotropy in the articles devoted to Green's memoirs.

[739.] Ostrogradsky *Sur l'intégration des équations à différences partielles relatives aux petites vibrations d'un milieu élastique. Mémoires de l'Académie...des Sciences de St Pétersbourg* Sixième serie. Tom. I. 1831, pp. 455—461. This memoir was read on the 10th of June, 1829. It is that referred to by Poisson: see Art. 564.

The object of the article is to express by definite integrals the solution of the body shift equations for an infinite elastic solid, supposing the shifts and their first time-fluxions to be given at any epoch, say that denoted by $t=0$. The equations to be solved are

$$\left.\begin{array}{lll} u=f(x, y, z), & v=F(x, y, z), & w=\mathrm{f}(x, y, z) \\ \dfrac{du}{dt}=f_1(x, y, z) & \dfrac{dv}{dt}=F_1(x, y, z) & \dfrac{dw}{dt}=\mathrm{f}_1(x, y, z), \end{array}\right\} \quad \ldots\ldots(a),$$

when $t=0$.

$$\left.\begin{aligned}\frac{d^2u}{dt^2} &= k^2\left\{\nabla^2 u + 2\frac{d}{dx}\left(\frac{du}{dx}+\frac{dv}{dy}+\frac{dw}{dz}\right)\right\}\\ \frac{d^2v}{dt^2} &= k^2\left\{\nabla^2 v + 2\frac{d}{dy}\left(\frac{du}{dx}+\frac{dv}{dy}+\frac{dw}{dz}\right)\right\}\\ \frac{d^2w}{dt^2} &= k^2\left\{\nabla^2 w + 2\frac{d}{dz}\left(\frac{du}{dx}+\frac{dv}{dy}+\frac{dw}{dz}\right)\right\}\end{aligned}\right\} \quad \ldots\ldots(b).$$

The solutions—obtained by a fairly easy and brief process—are of the type

$$\begin{aligned}4\pi u = \frac{d}{dt}\int_0^\pi\int_0^{2\pi} f\,(x + kt\cos p\sin q,\ y + kt\sin p\sin q,&\\ z + kt\cos q)\ t\sin q\, dp\,dq&\\ + \int_0^\pi\int_0^{2\pi} f_1\,(x + kt\cos p\sin q,\ y + kt\sin p\sin q,&\\ z + kt\cos q)\ t\sin q\, dp\,dq&\\ + k^2\left(\frac{dY}{dx} + \int_0^t \frac{dU}{dx}\,dt\right),&\end{aligned}$$

where

$$\begin{aligned}Y = \int_t^{t\sqrt{3}}\Bigg[\frac{d}{dx}\int_0^\pi\int_0^{2\pi} f\,(x + k\tau\cos p\sin q,\ y + k\tau\sin p\sin q,&\\ z + k\tau\cos q)\sin q\, dp\,dq,&\\ + \frac{d}{dy}\int_0^\pi\int_0^{2\pi} F\,(x + k\tau\cos p\sin q,\ y + k\tau\sin p\sin q,&\\ z + k\tau\cos q)\sin q\, dp\,dq,&\\ + \frac{d}{dz}\int_0^\pi\int_0^{2\pi} \mathrm{f}\,(x + k\tau\cos p\sin q,\ y + k\tau\sin p\sin q,&\\ z + k\tau\cos q)\sin q\, dp\,dq\Bigg]\,\tau d\tau;&\end{aligned}$$

and

$$\begin{aligned}U = \int_t^{t\sqrt{3}}\Bigg[\frac{d}{dx}\int_0^\pi\int_0^{2\pi} f_1\,(x + k\tau\cos p\sin q,\ y + k\tau\sin p\sin q,&\\ z + k\tau\cos q)\sin q\, dp\,dq,&\\ + \frac{d}{dy}\int_0^\pi\int_0^{2\pi} F_1\,(x + k\tau\cos p\sin q,\ y + k\tau\sin p\sin q,&\\ z + k\tau\cos q)\sin q\, dp\,dq,&\\ + \frac{d}{dz}\int_0^\pi\int_0^{2\pi} \mathrm{f}_1\,(x + k\tau\cos p\sin q,\ y + k\tau\sin p\sin q,&\\ z + k\tau\cos q)\sin q\, dp\,dq\Bigg]\,\tau d\tau.&\end{aligned}$$

[740.] Ostrogradsky concludes with the following remarks on the case when only a certain finite portion of the medium is disturbed.

Il est intéressant de connaître l'instant où le mouvement commence, et celui où le mouvement finit, dans un point donné de l'espace. Pour déterminer ces instants considérons une des fonctions $f(x, y, z) \ldots f_1(x, y, z)$, par exemple la premiere. Pour savoir si

$$f(x + r\cos p \sin q,\ y + r\sin p \sin q,\ z + r\cos q)$$

est sensible ou non, il n'y a qu'à décrire du point $(x, y, z)$ comme centre et avec le rayon $r$ une surface sphérique, la fonction

$$f(x + r\cos p \sin q,\ y + r\sin p \sin q,\ z + r\cos q)$$

sera différente de zéro pour toute la partie de la surface sphérique qui sera comprise dans le volume primitivement dérangé donc cette fonction commence a avoir des valeurs sensibles quand $r$ sera égal à la plus petite distance du point $(x, y, z)$ au volume dérangé, et se réduira de nouveau à zéro quand $r$ deviendra égal à la plus grande distance du même point au même volume. Il en est de même pour les autres fonctions $F, f \ldots f_1$.

Cela posé, il est évident que les quantités $Y$ et $U$ deviendront sensibles quand $t = R_0/ . k\sqrt{3}$ et cesseront de l'être quand $t = R_1/k$, $R_0$ et $R_1$ étant la plus grande et la plus petite distance du point $(x, y, z)$ au volume primitivement mis en mouvement. Les parties de $u, v, w$ indépendentes de $Y$ et $U$ deviendront sensibles plus tard, savoir quand $t = R_0/k$ et elles disparaîtront en même temps que $Y$ et $U$ Donc le mouvement au point quelconque commence quand $t = R_0/ . k\sqrt{3}$, il finit quand $t = R_1/k$, et par conséquent le mouvement dure pendant le temps $= (R_1\sqrt{3} - R_0)/ k\sqrt{3}$. En sorte que la durée du mouvement est en raison inverse de l'élasticité $k$ et ne dépend point du dérangement primitif. (pp. 460—461.)

741. Ostrogradsky: *Mémoire sur l'intégration des équations à différences partielles relatives aux petites vibrations des corps élastiques;* par M. Ostrogradsky. This memoir occupies pages 339—371 of the *Mémoires de l'Académie...de St Pétersbourg*, Vol. 2, 1833: it is said to have been read to the Academy on the 27th of June, 1382, the year being an obvious misprint for 1832.

The object of the memoir is to integrate the equations of motion for an elastic body in the form which involves one constant of elasticity. The first of these equations is thus expressed by Ostrogradsky,

$$\frac{d^2u}{dt^2} = k^2 \left\{\frac{d^2u}{dx^2} + \frac{d^2u}{dy^2} + \frac{d^2u}{dz^2} + \frac{d\theta}{dx}\right\},$$

and the others similarly: here $\theta$ is the dilatation.

The integrals obtained as the author states on his page 360, correspond with those given by Poisson in his memoir of October 11, 1830. The process is by no means simple, and involves the use of various important formulae belonging to the higher parts of the Integral Calculus,—such as those of Arts. 280, 285, 331 of my treatise on that subject.

742. There is nothing novel except a few pages at the beginning which amount to an extension of *Legendre's Coefficients.* I will indicate the nature of this. Let

$$r^2 = x^2 + y^2 + z^2, \text{ and } \rho^2 = a^2 + b^2 + c^2;$$

put $Q$ for $\quad \{1 - 2\beta(ax + by + cz) + \beta^2 r^2 \rho^2\}^{-\frac{1}{2}};$

the quantities $a, b, c, x, y, z, \beta$ being all independent of each other.

It is easy to see that $Q$ satisfies the relation

$$\frac{d^2Q}{da^2} + \frac{d^2Q}{db^2} + \frac{d^2Q}{dc^2} = 0 \text{.....................}(1).$$

Suppose we expand $Q$ in powers of $\beta$, say

$$Q = P_0 + P_1\beta + P_2\beta^2 + P_3\beta^3 + \dots;$$

then all the functions $P$ will satisfy (1). Now proceeding as in my *Laplace's Functions,* Art. 7, we shall find that

$$P_n = \frac{1.3.5\ldots(2n-1)}{\underline{|n}}(ax + by + cz)^n$$

$$- \frac{1.3.5\ldots(2n-3)}{2\,\underline{|n-2}}(ax + by + cz)^{n-2} r^2 \rho^2$$

$$+ \frac{1.3.5\ldots(2n-5)}{2.4\,\underline{|n-4}}(ax + by + cz)^{n-4} r^4 \rho^4$$

.......

Now reciprocally $(ax+by+cz)^n$ might be expressed in terms of the functions $P$; for if we change in the preceding equation $n$ successively into $n-2$, $n-4$, $n-6$..., we shall obtain as many equations as we require for eliminating all the powers of $ax+by+cz$ except the $n$th. But we may proceed more simply in another way. Suppose we restrict ourselves to an *even* power; change $n$ into $2n$, and put

$$(ax+by+cz)^{2n} = A_{n,0}P_0 r^{2n}\rho^{2n} + A_{n,1}P_2 r^{2n-2}\rho^{2n-2},$$
$$+\ldots+ A_{n,i}P_{2i}r^{2n-2i}\rho^{2n-2i}+\ldots+A_{n,n}P_{2n} \quad \ldots\ldots\ldots\ldots(2).$$

It is evident from the expression for $P_n$ that

$$A_{n,n} = \frac{\lfloor 2n}{1.3.5..(4n-1)} \quad \ldots\ldots\ldots\ldots\ldots\ldots (3).$$

Perform on (2) the operation $\frac{d^2}{da^2}+\frac{d^2}{db^2}+\frac{d^2}{dc^2}$; we shall obtain

$$n(2n-1)(ax+by+cz)^{2n-2} = n(2n+1)A_{n,0}P_0 r^{2n-2}\rho^{2n-2}$$
$$+(n-1)(2n+3)A_{n,1}P_2 r^{2n-4}\rho^{2n-4}+\ldots$$
$$+(n-i)(2n+2i+1)A_{n,i}P_{2i}r^{2n-2i-2}\rho^{2n-2i-2}$$
$$+\ldots+(4n-1)A_{n-1,n-1}P_{2n-2}.$$

In order to obtain this we must remember that the functions $P$ satisfy the condition (1); and that

$$a\frac{dP_m}{da}+b\frac{dP_m}{db}+c\frac{dP_m}{dc} = mP_m$$

But we have also

$$(ax+by+cz)^{2n-2} = A_{n-1,0}P_0 r^{2n-2}\rho^{2n-2} + A_{n-1,1}P_2 r^{2n-4}\rho^{2n-4}$$
$$+\ldots+A_{n-1,i}P_{2i}r^{2n-2i-2}\rho^{2n-2i-2}+\ldots+A_{n-1,n-1}P_{2n-2}.$$

Hence, comparing the two formulae, we get

$$n(2n-1)A_{n-1,i} = (n-i)(2n+2i+1)A_{n,i};$$

this gives

$$A_{n,i} = 2^i\frac{1.3.5\ldots(4i+1)}{\lfloor 2i}\,\frac{n(n-1)(n-2)\ldots(n-i+1)}{(2n+1)(2n+3)\ldots(2n+2i+1)}A$$

But the value of $A_{i,i}$ is found from (3) by putting $i$ for $n$. Thus finally

$$A_{n,i} = 2^{2i}(4i+1)\,\frac{(n+i)(n+i-1)(n+i-2)\ldots(n-i+1)}{(2n+1)(2n+2)\ldots(2n+2i+1)}$$

Put $\gamma$ for $r^2\rho^2$; then we have[1]

$$(ax+by+cz)^{2n} = P_0 \frac{\gamma^n}{2n+1} + 2^2 . 5 P_2 \frac{(n+1) n\gamma^{n-1}}{(2n+1)(2n+2)(2n+3)}$$

$$+ 2^4 . 9 P_4 \frac{(n+2)(n+1) n (n-1) \gamma^{n-2}}{(2n+1)(2n+2)(2n+3)(2n+4)(2n+5)}$$

$$+ \ldots + 2^{2i}(4i+1) P_{2i} \frac{(n+i)(n+i-1)(n+i-2)\ldots(n-i+1)\gamma^{n-i}}{(2n+1)(2n+2)\ldots(2n+2i+1)}$$

Ostrogradsky gives also the formula for an *odd* power of $ax+by+cz$; and formulae for

$$\cos(ax+by+cz) \text{ and } \sin(ax+by+cz).$$

743. After having arrived at the required integrals on his page 360, Ostrogradsky puts them in a different form: see his pages 360—364. He simplifies these forms in the case in which the following expressions are perfect differentials when $t=0$:

$$udx + vdy + wdz,$$

$$\frac{du}{dt} dx + \frac{dv}{dt} dy + \frac{dw}{dt} dz.$$

See his pages 365 and 366.

Finally he verifies that the general integrals which he has obtained satisfy the differential equations and the initial conditions: see his pages 367—371. He concludes thus:

Nous nous proposons de revenir sur l'intégration des équations à différences partielles et de faire voir comment les formules de l'article 1^er^ généralisées peuvent servir à trouver les intégrales d'équations plus composées que celles que nous avons traitées dans ce mémoire.

This intention does not seem to have been realised[2]

[1] [There does not seem to me any novelty in this result; it is a very simple corollary from the expansion of $\cos^n\theta$ in the Legendre's coefficients $P_0$ (cos $\theta$), $P_1$ (cos $\theta$)......, cos $\theta$ being taken equal to the angle between lines whose direction-cosines are proportional to ($a$, $b$, $c$) and ($x$, $y$, $z$) respectively. Ostrogradsky's $P_i$ is obviously equal to $r^i\rho^i P_i(\cos\theta)$. Ed.]

[2] The following errata may be noted:

p. 370, supply $d/dt$ in the value of $d''$. For $d^2M/dt^2$ read $d^2M/dr^2$

p. 371, for $d^2M/dt^2$ read $d^2M/dx^2$.

[744.] Ostrogradsky. *Note sur l'équilibre d un fil élastique. Bulletin Scientifique*, No. 4, May, 1832, St Petersbourg. (Usually bound with the *Mémoires*, Sixième serie, Tome II. 1833.)

This refers to the supposed error of Lagrange pointed out by Schultén and alluded to in our Arts. 536—539. Ostrogradsky says that Schultén had not explained the inexactitude because he had not said why we cannot, like Lagrange, put $\delta ds = 0$.

Ostrogradsky points out why in varying

$$e = \frac{\sqrt{[(d^2x)^2 + (d^2y)^2 + (d^2z)^2 - (d^2s)^2]}}{ds}$$

we must vary the denominator as well as the numerator, writing

$$\delta e = \frac{d^2x\,\delta d^2x + d^2y\,\delta d^2y + d^2z\,\delta d^2z - d^2s\,\delta d^2s}{ds\sqrt{(d^2x)^2 + (d^2y)^2 + (d^2z)^2 - (d^2s)^2}} - e\frac{\delta ds}{ds},$$

but Lagrange has omitted the terms $\dfrac{e\,\delta ds}{ds}$ and $d^2s\,\delta d^2s$. Now I do not understand this, for Lagrange has distinctly included these terms in his Art. 52 (p. 151 of Bertrand's edition; they also occur in the earlier editions). There seems no occasion for the remarks of either Schultén or Ostrogradsky.

[745.] We may notice two interesting papers by two well-known English physicists carrying on the labours of Chladni, Savart and Strehlke: see Arts. 329, 352 to 360.

The first is entitled: *On a peculiar class of Acoustical Figures; and on certain Forms assumed by groups of particles upon vibrating elastic surfaces.* By M. Faraday. *Phil. Trans.*, 1831, Part II., pp. 299—340. This paper is a criticism of Savart's of 1827 (*Annales de Chimie* XXXVI.: see our Art. 329), and shows that the secondary mode of motion which is there discussed (as pointed out by figures delineated by lycopodium or other *light* powder) is really due to the nature of the medium in which the vibrating plate and powder are placed, i.e. to the currents established in it by the motion of the plate. The paper does not really concern our subject.

[746.] The second is entitled: *On the Figures obtained by strewing sand on vibrating surfaces, commonly called Acoustic*

*Figures* By Charles Wheatstone. *Phil. Trans.*, 1833, Part II. pp. 593—633.

The paper was communicated by Faraday. Wheatstone commences with an historical notice, in which he remarks that Galilei had noticed that small pieces of bristle laid on the sounding-board of a musical instrument were violently agitated on some parts of the surface, while on others they did not move. He gives however no reference: see Art. 358 Dr Hooke had also proposed to observe the vibrations of a bell by strewing flour upon it. But the sole merit of discovering the symmetrical figures is due to Chladni[1]. Then follows a long consideration of Chladni's results considered from their theoretical aspect, or on a 'principle of superposition'. Euler's results are alluded to, those of Strehlke (see Arts. 354—355) discarded as untenable, while James Bernoulli's are described as entirely unsuccessful: see Art. 121. As for the authors considered in our Chapters III.—V. Wheatstone writes:

> The various mathematicians who have more recently undertaken to investigate the laws of vibrating surfaces, as Poisson, Cauchy, Mademoiselle Germain, etc., do not appear to have taken into consideration anything resembling the theory of superposition. (p 607 )

He attributes this principle to the brothers Weber in their work, the *Wellenlehre*, published in 1825. The memoir concludes with some just praise of Savart's researches on wooden plates (see Art. 339) and a considerable number of plates of calculated figures. I cannot find more in Wheatstone's discovery of this theory of superposition than the fact, well known to the mathematicians above mentioned, that the differential equation for the vibrations of a plate is *linear*. The modes of vibration adopted must then be such that they individually satisfy the contour conditions. I cannot see that Wheatstone's assumed modes of vibration are really possible for an elastic plate.

[747.] Parrot. *Expériences de forte compression sur divers corps. Mémoires de l'Académie de St Pétersbourg*, Sixième série. Tome II. 1833, pp. 595—630.

[1] Chladni's three works are: *Entdeckungen über die Theorie des Klanges*, 1787, *Die Akustik*, 1802, and *Neue Beyträge zur Akustik*, 1817. We have omitted all consideration of them as belonging properly to the Theory of Sound.

Parrot seems to have been assisted by E Lenz in these experiments, the latter however in an appended note disclaims all responsibility for opinions expressed. The results do not seem of any great value except in drawing attention to the compressibility of the bulbs of glass thermometers, and hence (p 629) the remark that the measurements of the temperature of the sea at great depths have all been erroneous and far too great. Parrot seems to believe that he was the first to prove the compressibility of glass; it might, he fancies, have had all its known elastic properties and yet been only extensible. He finds the compressibility of glass proportional to the pressure up to 100 atmospheres. There is nothing of much apparent value in the memoir: see our Arts. 686—690.

[748.] Karl Karmarsch. *Versuche und Bemerkungen über das Drahtziehen. Jahrbücher des k. k. polytechnischen Institutes in Wien* Bd. 17, 1832, pp. 320—336.

*Versuche über die absolute Festigkeit der zu Draht gezogenen Metalle Ibid.* Bd. 18, 1834, pp. 54—115

These papers contain a series of interesting experiments on a great variety of metal wires. I must observe however that Karmarsch apparently considered that to draw wires of *different material through the same hole ensured their having the same diameter* [*sämmtlich durch Ein Ziehloch gezogen, um ihrer gleichen Dicke vom Neuen versichert zu seyn* (p. 321).] This is not generally true, see Art. 830. Valuable as these experiments may be for practical purposes they do not seem to contain the statement of any physical fact not previously noted.

[749.] A further paper by this author entitled: *Ueber die Festigkeit und Elastizitat der Darmsaiten,* will be found on columns 245—250 of the *Mittheilungen des Gewerbe-Vereins für das Konigreich Hannover, Jahrgang* 1840—1841. Hannover, 1841.

The results obtained may be thus summed up:

(1) The limit of perfect elasticity is not very far from the breaking load (e.g. Expt. No. 3 set began between 128 and 143 pds., and the breaking load was 146 pds.). The loads were however instantaneous, not permanent.

Karmarsch takes $\frac{5}{6}$ of the breaking load as within the limit of perfect elasticity.

(2) The gut can be expanded between 9 and 10 *per cent.* of its natural length before set begins.

There is no question raised of whether the gut may not in its preparation have been reduced to a state of ease up to $\frac{5}{6}$ of the breaking load, nor is the possibility of elastic after-strain considered.

[750.] Héricart de Thury. It appears that the firm of ironmasters, Gandillot and Roy of Paris and Besançon, (following probably the process of casting previously suggested by an Englishman of the name of Thompson) had endeavoured to introduce hollow iron cylinders to be used as beams or struts into the trade. They entrusted a number of hollow cylinders of their manufacture to de Thury to be experimented on, and the results of his experiments were given in the *Bulletin de la Société d'Encouragement*, Feb. 1832, p. 41. A translation of this paper will be found in the *Polytechnisches Journal*, edited by Dingler, Bd. 44, 1832, pp. 273—285. De Thury's results shew the superiority in strength of the hollow over the solid cylinder, and the resulting gain in material; a fact easily ascertained by theory. He seems however to consider that the hollow cylinder would suffer more than the solid from the effect of oxidation, if at all exposed to the weather. For the interior of buildings he believes that the hollow form will generally be adopted.

[751.] Another series of papers due to Bevan will be found in the *Philosophical Magazine*. We may note:

(*a*) The *Philosophical Magazine*, Vol. XI. (new and united series), 1832, p. 241. Here are some observations on experiments of Barlow on wood reported on pp. 179—183 of the same volume. Barlow's results have no theoretical importance, nor Bevan's either, —except in the one point that he insists on the importance of considering the *time-element*, especially when the load as in Barlow's experiments amounts to nearly $\frac{2}{3}$ of the breaking load.

(*b*) The *London and Edinburgh Philosophical Journal*, Vol. I. 1832, p. 53. Some remarks on a paper by John White which appeared on pp. 333—339 of the volume cited in (*a*), being entitled: *On Calcareous Cements.* Neither in the original paper nor in the remarks is there anything of physical importance.

(*c*) In the same volume as I have cited in (*b*) are, p. 17, *Additional Experiments* by Barlow, and p. 116, additional remarks by Bevan, of no value for our present purpose.

(*d*) In the same *Journal*, Vol. II. p. 445 and Vol. III. p. 20, are two short letters by Bevan giving the results of his experiments on the elasticity of gold. They are for the modulus of:

Pure gold—11,690,000 pounds p. sq. inch, or
1,390,000 feet. (i.e. the height modulus: see Art. 137.)

Standard quality used in the British coinage—
12,226,000 pounds...in one direction and
11,955,000 „ ...in the other.

Mean modulus = 12,110,500 pounds or 1,480,000 feet.

Bevan remarks of the last result that 'it agrees very nearly with the calculated modulus as deduced from the proportioned modulus of gold and its alloy. This result suggests an important inquiry on the properties of alloys in general, and is deserving of the attention of the experimentalist.'[1]

He also gives the modulus of copper = 4,380,000 feet, and corrects the Supplement to the *Encyclopaedia Britannica* where it is given as 5,700,000 feet without any authority.

A statement made by Bevan in his first letter, I do not understand, because even supposing *uni-constant* isotropy, the note would also depend on the density and possibly on the nodal system set up. The statement runs:

Those who are in the daily habit of taking gold coin soon acquire a knowledge of the proper sound or note given upon striking a piece of money upon a table or hard substance: this well-known though undefined note or sound depends upon the modulus of elasticity of the metal, as well as upon the diameter and thickness. A piece of coin, of the same dimensions, both as to diameter and thickness, of silver, will

[1] Such attention was given by Wertheim in a memoir of 1844 to be considered in Chap. VIII.

give a note about a major fifth higher than one of gold, when a similar coin of copper will give a note an octave above that of gold, but if made of steel would give a note a minor third above that of copper.

[752.] 1832. Giuseppe Belli. *Riflessioni sulla legge dell' attrazione molecolare.* In 1832 was published at Milan the first volume of a work entitled: *Opuscoli matematici e fisici di diversi autori.* It contains the above memoir by Belli and that by Piola referred to in Art. 759. The second volume of this work, published in 1834, contains chiefly analytical dissertations, and has nothing bearing on our present subject. The memoir we are about to consider is divided into four parts, thus distributed through the volume: I., pp. 25—50; II., pp. 50—68; III., pp. 128—168 and 237—261; IV., pp. 297—326. An abstract will be found in the *Annali delle scienze del regno Lombardo-Veneto*... Vol. II. Padova, 1832, pp. 289—297 and 313—325.

[753.] The memoir is of the same nature as that we have already noticed in Art. 163, and so does not very closely affect our subject. The first paragraph states the relation of the present to the earlier memoir:

Io aveva procurato di dimostrare in una memoria inserita già nel *Giornale di Fisica di Pavia*, che l' attrazione alle minime distanze, detta molecolare, non segue la medesima legge della universale secondo che opinava Buffon e più ricentemente Laplace, ma bensı, come credette il medesimo Newton scopritore di questa forza, e poscia sostenne il Clairaut, decresce all' aumentarsi delle distanze con una legge di gran lunga più rapida, cioè ch' ella segue, secondo che mi parve poter dedurre da diversi fenomeni, una legge più rapida di quella delle quarte potenze reciproche delle distanze, o anche delle quinte. Avendo però fatto uso di calcoli semplicamente approssimativi, la cui legittimità non bene potevasi da tutti sentire, n' è venuto che parecchi Fisici, sebbene avessero avuto sott' occhio quel mio lavoro, continuarano ad attenersi alle idee di Buffon, o a quelle di Laplace, parendo loro più consentanee alla semplicità delle operazioni della natura. Per la qual cosa avendo io sempre tenuto presente al pensiero questo punto controverso della fisica speculativa, e parendomi di aver trovato delle dimostrazioni rigorose in appoggio dell' opinione da me abbracciata, mi sono creduto in debito verso il pubblico di darla alla luce, affine di rischiarare e forse terminare interamente una tale questione. (p. 25.)

[754.] The first article of the paper is entitled: *Insufficienza dell' attrazione astronomica per produrre la coesione e l' adesione dei corpi, nell' ipotesi della continuità della materia.*

The insufficiency is proved in the following fashion. First the attractive force between two equal cubes placed in contact is calculated by a long and complex process of integration, followed by considerable numerical calculations (pp. 31—46); the unit of astronomical attraction is calculated from the labours of Cavendish and Maskelyne, and finally the following result for the attractive force between the two cubes is obtained:

$$F = \cdot 000340753 \frac{\Delta \delta h^4}{\delta' g},$$

where $\Delta$, $\delta$ express the density of the two cubes; $\delta'$ is the mean density of the earth; $h$ the side of the cube and $g$ the velocity acquired in a second by a body falling under gravity. The system being metric, $F$ will be given as a *number* of kilogrammes (p. 46).

Several calculations are made to show the insufficiency of this result; the most striking is based upon an experiment of Rumford's, who showed that a bar of good wrought iron could sustain a load of 4470 kilogrammes when its section was a square centimetre. Calling this load $L$, Belli shows that

$$L = 1059{,}570000{,}000000\, F,$$

and remarks:

> Donde apparisce chiarissimamente quanto poca parte avrebbe l' attrazione universale nella coesione de' corpi, se questi potessero riguardarsi come formati di materia continua. (p. 48.)

[755.] The second article is entitled: *Estensione delle precedenti consequenze ad altre ipotesi sulla costituzione dei corpi, e insufficienza dell' ipotesi imaginata da Laplace.* Belli notes that the hypothesis of the continuity of matter adopted in his first article is not generally accepted by physicists. There are three current hypotheses:

(i) Bodies are formed of minute extended particles separated by distances not much greater than their diameters (Newton: see our Art. 26).

(ii) Molecular distances are incomparably greater than molecular diameters (Laplace, *Systeme du Monde,* Livre IV., chap. XV., édit. de l'an IV—i.e 1796).

(iii) Bodies do not consist of discontinuous molecules in a vacuum, but of discontinuous vacua in a continuous matter. (See our Arts. 35 and 94.)

The author, ingeniously adapting the formula of his first article to these various cases, proceeds to dismiss them as improbable.

On the first hypothesis he deduces that in the case of an iron bar there would be 23,000,000 times more vacuum than matter, which is hardly consonant with the hypothesis itself (p. 58) This result does not contradict Laplace's hypothesis, but Belli shews first that, if the ultimate particles are spheres, this hypothesis gives a far too weak force of cohesion (p. 63), and secondly extends this result to a molecule of any shape (p. 65) by means of a proposition of which he gives no proof and which does not seem to me obvious. Taking a right prism of his material on a square base, of height twice the side of the base, he divides it by a plane parallel to the base into two equal parts $A$ and $B$. He then continues:

> Egli è certo che in ciascuna molecola appartenente alla parte $A$ esiste un punto, dove se tutta la massa di essa molecola si venisse a concentrare, non si alterebbe la sua attrazione verso $B$, considerando qui pure forza nella sola direzione perpendicolare alle basi del prisma.

On a second objection to Laplace's hypothesis, Belli quotes the work of Nobili referred to in our Art. 211 (or the *History of the Theories of Attraction,* Art. 1615).

The third hypothesis is briefly shewn to involve the same contradictions as the first (p. 67).

[756.] The third article, entitled: *Di alcune ipotesi le quali considerate dal lato della Meccanica potrebbero essere atte a conciliare le due attrazioni,* is occupied with the consideration whether any mechanical arrangements of the atoms would enable us to attribute cohesion to ordinary gravitation. It involves a long approximation to the attracting force between two equal right prisms on square bases and of any height, when placed base to base. It is shewn that the force between them can be made as great as we please by sufficiently increasing the ratio of the height to a side of the base (p. 146). This would explain cohesion in a *fibre.* A fibrous structure is then assumed for the body or, as the author terms it, *un tessuto fibroso o reticolare.* It is then proved that the density of these thin prismatic threads would have to be simply enormous (p. 148). It is not even then explained how *isotropic* bodies could possess cohesion. After various other arrangements have been suggested and considered Belli draws the conclusion that in every disposition of matter it would be necessary to suppose bodies to have

> una enorme rarita di tessuto e una enorme densità della materia, se vi vuole che la coesione possa dipendere dalla gravitazione (cf. pp. 152 and 168).

On the whole his investigation in these first three articles, if not always entirely convincing, yet affords very strong arguments against the efficacy of universal attraction to throw light on the phenomenon of cohesion (cf. p. 260).

[757.] The fourth and final article is entitled: *Delle leggi di attrazione a cui e d uopo ricorrere per conservare le piu ricevute nozioni sulla costituzione de' corpi* The law of universal gravitation having been found wanting, unless extravagant hypotheses are made as to the structure of bodies, Belli holds it best to seek for another force of attraction which shall coincide with the law of gravity at distances sensibly greater than those which separate molecules. He notes that many hypothetical laws of this kind might be invented, but conceives it the best plan to allow the atomic arrangement of bodies to be discovered by the researches of chemists and crystallographers; for the phenomenon of cohesion however to assume a force following a law of more rapid variation than that of the inverse square:

> rimettendo però la precisa determinazione di questa sua legge a quel tempo nel quale ci venga ciò permesso da una più chiara cognizione de' fenomeni. (pp. 297—298.)

On this point Belli refers to Poisson's memoir of October, 1829.

He gives as hypothetical examples of possible attractive force, on p. 299, the functions

$$\left(\frac{a}{r}\right)^2 e^{\left(\frac{a}{r}\right)^2}, \text{ and } \log_e \left\{\frac{1+\left(\frac{a}{r}\right)^2}{1-\left(\frac{a}{r}\right)^2}\right\},$$

where $a$ is a quantity little different from the molecular diameter, and $r$ the distance between the points containing attracting matter. We must note however that these laws of force, like that of Poisson (see Art. 439), do not give the repulsive force, which we must suppose to exist between molecules when brought very closely together.

[758.] In conclusion the author considers the question as to whether adding this new attractive force does not complicate Nature's proceedings. He remarks that we ought to admire simplicity in Nature, but not gratuitously presuppose it. This simplicity can exist in a greater or less degree; and where the Creator has not found one means sufficient for his purpose He sometimes employs two.

Se talora non trovò sufficiente un solo mezzo, Egli ne avra impiegato due, tre, secondo che avrà stimato più conveniente. (p. 300.)

A long *Nota* appended to the memoir (pp. 303—326) contains various propositions in attractions used in the body of the work, but which have no connection with our present subject.

Probably no physicist now-a-days attributes cohesion to gravitating force; how far Belli's memoir may have assisted in forming a general opinion of this kind, we are unable to judge. The memoir seems to have escaped notice in the *History of the Theories of Attraction.*

759. 1833. Piola. *La Meccanica de' corpi naturalmente estesi trattata col calcolo delle variazioni di Gabrio Piola. Memoria prima estratta dal fascicolo terzo degli Opuscoli matematici e fisici.* This memoir consists of 36 quarto pages besides the title page. In the *Opuscoli* it occupies pages 201—236 of the first volume.

760. In the first paragraph Piola alludes to the recent researches of Poisson and Cauchy respecting the theory of the elasticity of solid bodies; the former contained in Vol. VIII. of the Paris Memoirs, and the latter in the *Exercices de mathématiques*; he refers especially to a passage on page 561 of Poisson's memoir, which expresses the desire to reconstruct the science of mechanics on the new principles. Now Piola had been led in his youth to a close study of the *Mécanique Analytique*, and, as we have seen in Art. 362, he had published a prize essay on Lagrange's methods. He had formed a very high idea of the power and the generality of those methods, which, however, seemed to him to be almost neglected by writers on mechanics. Accordingly his object is substantially to invite the attention of mathematicians to the study and application of the principles of the *Mécanique Analytique;* and with this view he proposes to discuss, after the manner of Lagrange, what we call the theory of elasticity.

761. The present memoir is devoted to the establishment of the equations which hold for equilibrium or motion at every point of the interior of a solid body; that is, of the body stress-equations. The process is a good specimen of Lagrange's methods; the analysis is fully developed, and is easily intelligible. Some remarks will now be made as to details.

762. In Article 8 of the memoir Piola notices with great earnestness a point as to which he differs from Lagrange; he says

> Agevolmente potra il lettore persuadersi che avendo contraria l' autorità di Lagrange io mi sono posto e riposto molte volte a meditare un tal punto nella piena disposizione d' animo de trovar vera la sua asserzione e falsa la mia.

The matter in question is contained in Art. 16 of the fourth section of the first part of the *Mécanique Analytique;* here Lagrange states that there cannot be more than *three* equations of condition holding; Bertrand in a brief note in his edition says that there cannot really be so many as three. Piola holds that there may be more than three in the case he contemplates; but it seems to me that this case differs entirely from that which Lagrange has in view, and so there is no real contradiction.

763. A reference in a note on page 220 to two volumes of Cauchy's *Exercices de mathematiques* is incorrect: it should be to Vol. II., page 111, and Vol. III., page 166. This I judge from a correction in the copy I have of Piola's memoir and also from page 61 of the next memoir by Piola.

764. A very important difference between Lagrange's method, as used by Piola, and that of Poisson and Cauchy, must be carefully noticed. According to the latter writers the symbols $\widehat{xx}$, $\widehat{yy}$, $\widehat{zz}$, $\widehat{yz}$, $\widehat{zx}$, $\widehat{xy}$, which occur in the body stress equations denote certain *stresses*, which have been defined and explained beforehand. In the corresponding formulae obtained by Piola these symbols enter into the investigation merely as *indeterminate multipliers*; they may be said to denote *stresses* or *forces*, but this is only by a figure of speech common with Lagrange, to which Bertrand in his notes justly calls attention more than once Now Piola seems to think that this is a great merit in Lagrange's method, inasmuch as we are not compelled at the beginning of our investigations to consider and define the forces acting; it seems to me, on the contrary the great defect of Lagrange's methods. In fact the Calculus of Variations is very prominent but the mechanical principles are left in obscurity. There can be no doubt I think, that both Poisson and Cauchy took a view quite contrary to that of Piola as to the merit of this peculiarity of Lagrange's method.

765. On pages 228—230 there is a note respecting the well known six equations connecting the nine cosines which occur in the formulae for passing from one set of rectangular axes to another. There are two forms of these equations, and the object of the note is to deduce one form from the other by a purely algebraical process. The investigation is good, but not so simple as one which will be found on page 88 of Griffin s *Treatise on the Motion of a Rigid Body*, a work very useful in its day at Cambridge.

766. This memoir is called *memoria prima*, and the intention is expressed of following it up by other memoirs. This intention does not seem to have been carried into effect, though another memoir by Piola will now come under our notice.

767. 1836. *Nuova Analisi per tutte le questioni della meccanica molecolare del Signor Dottore Don Gabrio Piola.* This memoir is published in Vol. XXI. of the *Memoirs of the Italian Society of Sciences at Modena,* 1836. It occupies pages 155 321 of the volume. The copy I have used is dated 1835, and is paged from 3 to 171 both inclusive, except that there are no pages numbered 115 and 116: it has a title leaf. The great length of the memoir is due to the very diffuse manner in which it is written and printed.

768. We have already seen that Piola published a memoir recommending the use of Lagrange's method for the investigation of the theory of elasticity, and promising to continue to exemplify the method in future memoirs: see Art. 760. The present memoir however does not fulfil this engagement, but approaches nearer to the method employed by Poisson. After a few introductory sentences the memoir passes to the first of the seven sections of which it consists.

769. The first section is entitled: *Principio generale per l' applicazione del calcolo alle questioni relative al moto ed all' equilibrio de' corpi* it occupies pages 4—18. The principle may be said to amount to this: an expression may be found which for *n* different values of a variable shall be equal to *n* assigned numerical quantities respectively. We have in fact considerations of the same kind as are involved in the well-known process of Lagrange which serves as a foundation for the expansion of a function in a series of sines or cosines of multiple angles: see my *Integral Calculus* Art. 306. The process given by Piola is tedious; and I think it might with advantage have been replaced by a brief reference to that of Lagrange.

770. The second section is entitled: *Nuova Analisi del moto e dell' equilibrio de' corpi omogenei considerati come ammassi di molecole;* it occupies pages 19—51.

The commencement of this section shews, as I have stated, that the author recedes from the method of Lagrange and approaches that of Poisson.

Non ammetto in questa analisi alcuna equazione di condizione cui debbano soddisfare le coordinate dei diversi punti del corpo. Questa maniera con cui Lagrange cerco di esprimere i legami fisici e reciproci delle diverse particelle de' corpi, parve al Sig. Poisson troppo astratta: egli vorrebbe ridurre tutto alle sole azioni molecolari. Io mi conformo a questo voto non ammettendo appunto oltre le forze esterne, che un' azione reciproca di attrazione o repulsione fra le diverse molecole espressa per una funzione incognita della distanza. Non e gia che io creda da abbandonarsı l' altra maniera usata da Lagrange, chè anzi io sono d' avviso che eziandio con essa si possano vantaggiosamente trattare molte moderne questioni, ed ho gia pubblicato un saggio di un mio lavoro che puo in parte provare questa mia asserzione.

There is a note to the end of the extract which refers to the memoir of 1833 already noticed in Art. 759. There is a note to the name Poisson which refers to page 361 of his memoir of April 1828 in the former memoir Piola gave 561 as the page.

The section has for its object to find expressions for what we call the stresses. Like Poisson the author adopts *finite summation* and not *integration;* but he makes much use of the Calculus of Finite Differences, and this forms the speciality of the section.

771. A general theorem in Finite Differences may be noticed as presenting itself to a careful reader of the section; Piola himself however does not give the general theorem, but merely investigates the first three cases of it which correspond to $n = 1$, $n = 2$, and $n = 3$ respectively. The following is the theorem: let $\phi(x)$ and $u$ denote any functions of $x$; let $h$ denote the increment of the independent variable $x$; then will

$$\phi(x)\,\Delta^n u = \Delta^n\{u\phi(x-nh)\} - n\Delta^{n-1}\{u\Delta\phi(x-nh)\} + \frac{n(n-1)}{1.2}\Delta^{n-2}\{u\Delta^2\phi(x-nh)\}\ldots\ldots(1).$$

This may be verified when $n = 1$ or 2, and then the general truth may be shewn by an inductive process. For, assume that (1) is true when $n$ has an assigned value, whatever $u$ may be then changing $u$ into $\Delta u$ it will still be true: thus we have

$$\phi(x)\,\Delta^{n+1} u = \Delta^n\{\Delta u\phi(x-nh)\} - n\,\Delta^{n-1}\{\Delta u\,\Delta\phi(n-nh)\} + \frac{n(n-1)}{1\cdot 2}\Delta^{n-2}\{\Delta u\,\Delta^2\phi(x-nh)\}\ \ldots\ldots\ldots\ldots(2).$$

Now by an elementary theorem in Finite Differences we have

$$\Delta\{v\chi(x)\} = \Delta v\,\chi(x+h) + v\Delta\chi(x)$$

where $v$ and $\chi(x)$ are any functions of $x$. Thus

$$\Delta v\,\chi(x+h) = \Delta\{v\chi(x)\} - v\Delta\chi(x)$$

Now of the last formulae the following are examples

$$\Delta u\,\phi(x-nh) = \Delta\{u\phi(x-nh-h)\} - u\Delta\phi(x-nh-h),$$

$$\Delta u\,\Delta\phi(x-nh) = \Delta\{u\Delta\phi(x-nh-h)\} - u\Delta^2\phi(x-nh-h),$$

$$\Delta u\,\Delta^2\phi(x-nh) = \Delta\{u\Delta^2\phi(x-nh-h)\} - u\Delta^3\phi(x-nh-h)$$

......................

In this way we see that each term on the right-hand side of (2) can be separated into two; and then by putting together like terms we obtain

$$\phi(x)\,\Delta^{n+1}u = \Delta^{n+1}\{u\phi(x-nh-h)\} - (n+1)\,\Delta^n\{u\Delta\phi(x-nh-h)\} + \frac{(n+1)n}{1.2}\,\Delta^{n-1}\{u\Delta^2\phi(x-nh-h)\}.$$

This is the same as (1) with $n$ changed into $n+1$; and hence the theorem is universally true.

The theorem may also be expressed with some change of notation We know that in Finite Differences a symbol $E$ is used in the sense defined by $\phi(x+h) = E\phi(x)$. Using this symbol we may on the right-hand side of (1) put $E^{-n}\phi(x)$ instead of $\phi(x-nh)$ Or suppose we put $\psi(x)$ instead of $\phi(x-nh-h)$ on the right-hand side of (1), then instead of $\phi(x)$ on the left-hand side of (1) we must put $E^{n+1}\psi(x)$.

The formula in the Differential Calculus which corresponds to (1) is well known: see my *Differential Calculus*, Art. 83.

772. Page 32 of the memoir contains a number of formulae which involve serious errors. We have, for example, by the theorem just noticed

$$\phi(x)\Delta^2u = \Delta^2\{u\phi(x-2h)\} - 2\Delta\{u\Delta\phi(x-2h)\} + u\Delta^2\phi(x-2h).$$

Piola then practically takes

$$\Delta\{u\Delta\phi(x-2h)\} = \Delta^2\{u\phi(x-2h)\},$$

so as to incorporate the first and second of the three terms which represent $\phi(x)\Delta u$: but this cannot be done. However, in what follows Piola does not use those formulae of page 32 which involve the errors: see his page 103.

773. The third section is entitled; *Principio generale per passare alle espressioni di uso: sue prime applicazioni*: it occupies pages 50—64. The *general principle* consists of the following theorem: Suppose $x, y, z$ to be each a function of the three variables $a, b, c$; then by inversion we may express $a, b, c$ each as a function of the three variables $x, y, z$. Let $L, M, N$ be each a function of $a, b, c$; and suppose $K_1, K_2, K_3$ determined by the equations

$$\left.\begin{aligned} K_1 &= \frac{1}{H}\left(L\frac{dx}{da}+M\frac{dx}{db}+N\frac{dx}{dc}\right)\\ K_2 &= \frac{1}{H}\left(L\frac{dy}{da}+M\frac{dy}{db}+N\frac{dy}{dc}\right)\\ K_3 &= \frac{1}{H}\left(L\frac{dz}{da}+M\frac{dz}{db}+N\frac{dz}{dc}\right)\end{aligned}\right\} \quad\text{............(1)},$$

where $H$ is what we now call the determinant

$$\begin{vmatrix} \dfrac{dx}{da}, & \dfrac{dx}{db}, & \dfrac{dx}{dc} \\ \dfrac{dy}{da}, & \dfrac{dy}{db}, & \dfrac{dy}{dc} \\ \dfrac{dz}{da}, & \dfrac{dz}{db}, & \dfrac{dz}{dc} \end{vmatrix} \quad\text{........................(2)}.$$

Then will

$$\frac{dL}{da}+\frac{dM}{db}+\frac{dN}{dc} = H\left(\frac{dK_1}{dx}+\frac{dK_2}{dy}+\frac{dK_3}{dz}\right) \quad\text{.........(3)},$$

where on the right side we suppose $K_1, K_2, K_3$ expressed as functions of $x, y, z$ by substituting for $a, b, c$ their values in terms of $x, y$, and $z$.

We give the demonstration as a specimen of the author's method.

From (1) by solution we obtain

$$\left.\begin{aligned} L &= \alpha K_1+\alpha' K_2+\alpha'' K_3,\\ M &= \beta K_1+\beta' K_2+\beta'' K_3,\\ N &= \gamma K_1+\gamma' K_2+\gamma'' K_3\end{aligned}\right\} \quad\text{..................(4)},$$

whence

$$\left.\begin{array}{lll} \alpha=\frac{dy}{db}\frac{dz}{dc}-\frac{dz}{db}\frac{dy}{dc}, & \alpha'=\frac{dz}{db}\frac{dx}{dc}-\frac{dx}{db}\frac{dz}{dc}, & \alpha''=\frac{dx}{db}\frac{dy}{dc}-\frac{dy}{db}\frac{dx}{dc} \\ \beta=\frac{dy}{dc}\frac{dz}{da}-\frac{dz}{dc}\frac{dy}{da}, & \beta'=\frac{dz}{dc}\frac{dx}{da}-\frac{dx}{dc}\frac{dz}{da}, & \beta''=\frac{dx}{dc}\frac{dy}{da}-\frac{dy}{dc}\frac{dx}{da} \\ \gamma=\frac{dy}{da}\frac{dz}{db}-\frac{dz}{da}\frac{dy}{db}, & \gamma'=\frac{dz}{da}\frac{dx}{db}-\frac{dx}{da}\frac{dz}{db}, & \gamma''=\frac{dx}{da}\frac{dy}{db}-\frac{dy}{da}\frac{dx}{db} \end{array}\right\}\ldots(5).$$

Multiply the value of $\alpha$ by $\frac{dx}{da}$, that of $\beta$ by $\frac{dx}{db}$, and that of $\gamma$ by $\frac{dx}{dc}$; in this way we obtain the first of the following equations, and all the rest follow in a similar way:

$$\left.\begin{array}{ll} \alpha\frac{dx}{da}+\beta\frac{dx}{db}+\gamma\frac{dx}{dc}=H, & \alpha'\frac{dx}{da}+\beta'\frac{dx}{db}+\gamma'\frac{dx}{dc}=0, \\ & \alpha''\frac{dx}{da}+\beta''\frac{dx}{db}+\gamma''\frac{dx}{dc}=0, \\ \alpha\frac{dy}{da}+\beta\frac{dy}{db}+\gamma\frac{dy}{dc}=0, & \alpha'\frac{dy}{da}+\beta'\frac{dy}{db}+\gamma'\frac{dy}{dc}=H, \\ & \alpha''\frac{dy}{da}+\beta''\frac{dy}{db}+\gamma''\frac{dy}{dc}=0, \\ \alpha\frac{dz}{da}+\beta\frac{dz}{db}+\gamma\frac{dz}{dc}=0, & \alpha'\frac{dz}{da}+\beta'\frac{dz}{db}+\gamma'\frac{dz}{dc}=0, \\ & \alpha''\frac{dz}{da}+\beta''\frac{dz}{db}+\gamma''\frac{dz}{dc}=H \end{array}\right\}\ldots\ldots(6).$$

From (5) we can deduce the following:

$$\left.\begin{array}{l} \frac{d\alpha}{da}+\frac{d\beta}{db}+\frac{d\gamma}{dc}=0 \\ \frac{d\alpha'}{da}+\frac{d\beta'}{db}+\frac{d\gamma'}{dc}=0 \\ \frac{d\alpha''}{da}+\frac{d\beta''}{db}+\frac{d\gamma''}{dc}=0 \end{array}\right\}\ldots\ldots\ldots\ldots\ldots\ldots\ldots(7).$$

Now we have

$$\frac{dK_1}{da}=\frac{dK_1}{dx}\frac{dx}{da}+\frac{dK_1}{dy}\frac{dy}{da}+\frac{dK_1}{dz}\frac{dz}{da},$$
$$\frac{dK_1}{db}=\frac{dK_1}{dx}\frac{dx}{db}+\frac{dK_1}{dy}\frac{dy}{db}+\frac{dK_1}{dz}\frac{dz}{db},$$
$$\frac{dK_1}{dc}=\frac{dK_1}{dx}\frac{dx}{dc}+\frac{dK_1}{dy}\frac{dy}{dc}+\frac{dK_1}{dz}\frac{dz}{dc},$$

where, as the equations imply, $K_1$ is supposed to be expressed as a function of $a, b, c$ on the left-hand side, and as a function of $x, y, z$ on the right-hand side.

Multiply these equations by $\alpha, \beta, \gamma$ respectively, and add; then by virtue of three of equations (6) we obtain the first of the following three, and the others come similarly:

$$\left.\begin{aligned}\alpha \frac{dK_1}{da} + \beta \frac{dK_1}{db} + \gamma \frac{dK_1}{dc} = H \frac{dK_1}{dx}\\ \alpha' \frac{dK_2}{da} + \beta' \frac{dK_2}{db} + \gamma' \frac{dK_2}{dc} = H \frac{dK_2}{dy}\\ \alpha'' \frac{dK_3}{da} + \beta'' \frac{dK_3}{db} + \gamma'' \frac{dK_3}{dc} = H \frac{dK_3}{dz}\end{aligned}\right\} \text{............(8)}.$$

By combining the first (7) and (8) we obtain

$$\frac{d\alpha K_1}{da} + \frac{d\beta K_1}{db} + \frac{d\gamma K_1}{dc} = H \frac{dK_1}{dx}.$$

In like manner, by combining the other equations of (7) and (8), we obtain

$$\frac{d\alpha' K_2}{da} + \frac{d\beta' K_2}{db} + \frac{d\gamma' K_2}{dc} = H \frac{dK_2}{dy},$$

and
$$\frac{d\alpha'' K_3}{da} + \frac{d\beta'' K_3}{db} + \frac{d\gamma'' K_3}{dc} = H \frac{dK_3}{dz}.$$

Add and re-arrange the terms on the left-hand side; thus

$$\frac{d}{da}(\alpha K_1 + \alpha' K_2 + \alpha'' K_3) + \frac{d}{db}(\beta K_1 + \beta' K_2 + \beta'' K_3) \\ + \frac{d}{dc}(\gamma K_1 + \gamma' K_2 + \gamma'' K_3) = H\left(\frac{dK_1}{dx} + \frac{dK_2}{dy} + \frac{dK_3}{dz}\right),$$

that is, by (4),

$$\frac{dL}{da} + \frac{dM}{db} + \frac{dN}{dc} = H\left(\frac{dK_1}{dx} + \frac{dK_2}{dy} + \frac{dK_3}{dz}\right).$$

774. In this section the author arrives at the body stress-equations of elasticity, supposing however that there are *six* shears instead of three; that is, at present, he has not obtained the equations, equivalent to Cauchy's theorem, by which the six shears are reduced to three: see Art. 610 (ii).

775. Piola makes an allusion on his page 62 to results which

he had obtained in the manner of his former memoir; I do not suppose that he had published these. He says:

Il bisogno di considerare le coordinate $p$, $q$, $r$ dell' equilibrio siccome funzioni di altre tre variabili, mi si fece noto fin da quando trattai le questioni di Meccanica alla maniera de' Geometri nostri maestri: infatti ottenni allora con questo mezzo alcuni notabili resultati che non so se poteansi egualmente avere senza tale considerazione.

776. The fourth section is entitled: *Massa; densità; equazione detta della continuità; teorica delle condensazioni.* It occupies pages 66—93. .

The main results obtained in this section are three in number. One is the *equation of continuity* in the motion of fluids. Another is the relation between the density $\rho$ after the deformation of a body, and the density $\sigma$ before the deformation, which is now expressed by the equation $\rho = \sigma\left(1 - \frac{du}{dx} - \frac{dv}{dy} - \frac{dw}{dz}\right)$. The third result is that a small spherical portion of a body becomes by such a deformation as we contemplate converted into an ellipsoid: see Art. 617, page 332. On page 78 Piola refers to his former memoir which we have considered in Art. 759; but this is a mistake, the reference should be to his prize essay of 1825, which we have noticed in Art. 362.

777 The fifth section is entitled: *Riduzione delle equazioni generali dietro le proprietà fisiche dell' azione molecolare;* it occupies pages 93—123.

Piola considers on his page 101 that he *demonstrates* what Poisson assumed as an hypothesis, namely that the sphere of molecular activity, though of insensible extent, must be held to comprise within it an excessively great number of particles; but the proposition, from its very nature, is not susceptible of mathematical demonstration, unless we assume quite as much as we propose to establish.

In this section the three equations are established which reduce the six shears to three: see page 104 of the memoir.

On page 104 of the memoir Piola commences a useful summary of the results already obtained. These are equivalent practically to the body stress-equations for equilibrium or motion; the stresses are represented by some complicated expressions which involve triple summation.

On page 110 these expressions involving summation are converted by a rude approximation into triple integrals, the limits of the integrations being $+\infty$. However no inferences of importance are drawn from the new form there given to the expressions.

778. Some formulae at the end of the section which relate to the motion of fluids may be noticed. Let $x, y, z$ be the coordinates at the time $t$ of a particle of fluid in motion. Each of the three $x, y, z$ may be considered as a function of the time $t$ and of the original coordinates $a, b, c$ of the particle considered. Now we know that a certain equation, involving the density $\rho$, exists called the *equation of continuity*, namely

$$\frac{d\rho}{dt}+\frac{d\rho u}{dx}+\frac{d\rho v}{dy}+\frac{d\rho w}{dz}=0 \quad\text{.................(9).}$$

779. Piola then shews, by a process similar to that which he employs in demonstrating the equation (9), that he can obtain the following:

$$\left.\begin{aligned}
\frac{d}{dx}\left(\rho\frac{dx}{da}\right)+\frac{d}{dy}\left(\rho\frac{dy}{da}\right)+\frac{d}{dz}\left(\rho\frac{dz}{da}\right)&=0\\
\frac{d}{dx}\left(\rho\frac{dx}{db}\right)+\frac{d}{dy}\left(\rho\frac{dy}{db}\right)+\frac{d}{dz}\left(\rho\frac{dz}{db}\right)&=0\\
\frac{d}{dx}\left(\rho\frac{dx}{dc}\right)+\frac{d}{dy}\left(\rho\frac{dy}{dc}\right)+\frac{d}{dz}\left(\rho\frac{dz}{dc}\right)&=0
\end{aligned}\right\}\text{......(10).}$$

The meaning of the differential coefficients of $x$, $y$, and $z$ must be carefully observed. We consider for instance $x$ as a function of $a, b, c, t$; from this we get $\frac{dx}{da}$ as a function of the same variables, and then we transform this into a function of $x, y, z, t$, and we put for $\frac{dx}{da}$ in (10) the expression so transformed. Similar meanings must be attached to the other differential coefficients. We will shew briefly how Piola obtains the equations (10), taking the first of them only, as the others can be obtained in a similar way.

Let $H$ have the same meaning as in Art. 773, and also $\alpha, \beta, \gamma$. Then we obtain by differentiation

$$\left.\begin{aligned}\frac{dH}{da} &= \alpha\frac{d^2x}{da^2} + \beta\frac{d^2x}{da\,db} + \gamma\frac{d^2x}{da\,dc}\\ &+ \alpha'\frac{d^2y}{da^2} + \beta'\frac{d^2y}{da\,db} + \gamma'\frac{d^2y}{da\,dc}\\ &+ \alpha''\frac{d^2z}{da^2} + \beta''\frac{d^2z}{da\,db} + \gamma''\frac{d^2z}{da\,dc}\end{aligned}\right\}\text{............(11).}$$

Now let $\epsilon$ stand for $\frac{dx}{da}$ when transformed into a function of $x, y, z, t$; similarly let $\theta$ stand for $\frac{dy}{da}$ and $\tau$ for $\frac{dz}{da}$ Then

$$\frac{d^2x}{da^2} = \frac{d\epsilon}{dx}\frac{dx}{da} + \frac{d\epsilon}{dy}\frac{dy}{da} + \frac{d\epsilon}{dz}\frac{dz}{da},$$

$$\frac{d^2x}{da\,db} = \frac{d\epsilon}{dx}\frac{dx}{db} + \frac{d\epsilon}{dy}\frac{dy}{db} + \frac{d\epsilon}{dz}\frac{dz}{db},$$

and so on. By the aid of nine equations which we can thus form we have from (11),

$$\begin{aligned}\frac{dH}{da} = \frac{d\epsilon}{dx}\left(\alpha\frac{dx}{da} + \beta\frac{dx}{db} + \gamma\frac{dx}{dc}\right) + \frac{d\epsilon}{dy}\left(\alpha\frac{dy}{da} + \beta\frac{dy}{db} + \gamma\frac{dy}{dc}\right)&\\ + \frac{d\epsilon}{dz}\left(\alpha\frac{dz}{da} + \beta\frac{dz}{db} + \gamma\frac{dz}{dc}\right)&\\ + \frac{d\theta}{dx}\left(\alpha'\frac{dx}{da} + \beta'\frac{dx}{db} + \gamma'\frac{dx}{dc}\right) + \frac{d\theta}{dy}\left(\alpha'\frac{dy}{da} + \beta'\frac{dy}{db} + \gamma'\frac{dy}{dc}\right)&\\ + \frac{d\theta}{dz}\left(\alpha'\frac{dz}{da} + \beta'\frac{dz}{db} + \gamma'\frac{dz}{dc}\right)&\\ + \frac{d\tau}{dx}\left(\alpha''\frac{dx}{da} + \beta''\frac{dx}{db} + \gamma''\frac{dx}{dc}\right) + \frac{d\tau}{dy}\left(\alpha''\frac{dy}{da} + \beta''\frac{dy}{db} + \gamma''\frac{dy}{dc}\right)&\\ + \frac{d\tau}{dz}\left(\alpha''\frac{dz}{da} + \beta''\frac{dz}{db} + \gamma''\frac{dz}{dc}\right)&\end{aligned}$$

By aid of equations (6) of Art. 773 this reduces to

$$\frac{dH}{da} = H\left(\frac{d\epsilon}{dx} + \frac{d\theta}{dy} + \frac{d\tau}{dz}\right)$$

Now it is known from works on Hydrodynamics that $\rho H$ is a constant, so that $\frac{1}{H}\frac{dH}{da} = -\frac{1}{\rho}\frac{d\rho}{da}$: see Kirchhoff's *Vorlesungen... Mechanik*, page 162. Thus

$$\frac{d\rho}{da} = -\rho\left(\frac{d\epsilon}{dx} + \frac{d\theta}{dy} + \frac{d\tau}{dz}\right).$$

But we have also

$$\frac{d\rho}{da} = \frac{d\rho}{dx}\epsilon + \frac{d\rho}{dy}\theta + \frac{d\rho}{dz}\tau,$$

therefore

$$\frac{d\rho\epsilon}{dx} + \frac{d\rho\theta}{dy} + \frac{d\rho\tau}{dz} = 0;$$

which was to be proved.

However, Piola says that equation (9) is really an unimportant identity, and he does not suppose that (10) will be of any more value, though he had been much gratified when he first discovered them.

780 The sixth section is entitled: *Equazioni ai limiti;* it occupies pages 124—135. The object of this section is to investi gate the equations which must hold at the surface of a body At the end of the section Piola says he might introduce the theory given in Cauchy's *Exercices* relative to the pressure in the interior of bodies; but this he omits partly because it is not of importance in his method, and partly because it would occupy too much space. I do not know what is the precise theory of Cauchy's to which Piola alludes.

781. The seventh section is entitled: *Teorica dei fluidi* it occupies pages 136—171. This does not strictly concern us, but as it is of a very peculiar kind I will notice a few points.

782. Piola indicates at the outset that with respect to the motion of fluids he holds views which differ from those of other mathematicians. He says:

Comincio dalla applicazione al moto de' fluidi e per questa memoria mi limito ad essa, assicurando però d' avere già in pronto altre formole spettanti al moto di altri corpi e principalmente quelle relative ai moti oscillatorj e vibratorj che diversificano dalle trovate dai moderni Geo metri. La sola teorica del moto de' fluidi involge tante novità, che io non amo moltiplicare le applicazioni prima di sentire il voto de' Geometri su di questa. Vedremo che la teorica finora ammessa pel moto de' fluidi è ben lontana dall' essere perfetta.

783. Piola is not satisfied with the definitions hitherto given of a fluid; he criticises those of Lagrange, Laplace, and Poisson, and offers his own thus on page 138:

Io chiamerò *fluido quel corpo le cui molecole vicine si tengono in ogni movimento a tali reciproche distanze che non differiscono fra loro se non per quantità di second' ordine.*

784. Piola gives a very singular process on his pages 137—143 for arriving at equations which shall be distinctively characteristic of a fluid. I do not understand it. He seems for instance to assert that such an equation as

$$t_1 \xi^2 + t_2 \eta^2 + t_3 \zeta^2 + 2t_4 \xi\eta + 2t_5 \xi\zeta + 2t_6 \eta\zeta = C$$

represents an ellipsoid which is not referred to its centre as origin; and that, if the centre is the origin, we must have

$$t_4 = 0, \quad t_5 = 0, \quad t_6 = 0;$$

this is quite wrong. He obtains as a result that in fluid motion the following equations must hold:

$$\frac{dx}{da}\frac{dx}{db} + \frac{dy}{da}\frac{dy}{db} + \frac{dz}{da}\frac{dz}{db} = 0,$$

$$\frac{dx}{db}\frac{dx}{dc} + \frac{dy}{db}\frac{dy}{dc} + \frac{dz}{db}\frac{dz}{dc} = 0,$$

$$\frac{dx}{dc}\frac{dx}{da} + \frac{dy}{dc}\frac{dy}{da} + \frac{dz}{dc}\frac{dz}{da} = 0,$$

where the differential coefficients have the meaning assigned in Art. 779. But these equations are quite arbitrary and inadmissible

785. Piola refers to a point we have noticed in Art. 765 of the account of his former memoir. He quotes some equations due to Monge, which are demonstrated by Lacroix in his *Traité du Calcul*, Vol. I., p. 533; these Piola says furnish a simpler solution of the problem than that which he had himself given in his former memoir

786. By a process based on the unsound foundation of Art. 784 Piola arrives at three equations connecting the differential coefficients of the $u$, $v$, and $w$ which occur in investigations respecting fluid motion; these equations however are quite unknown to writers on Hydrodynamics and cannot be accepted as true: see his pages 161 and 166.

787. Piola objects to the hypothesis often used with respect to fluid motion, that $udx + vdy + wdz$ is a perfect differential; and he suggests instead that $\frac{du}{dt}dx + \frac{dv}{dt}dy + \frac{dw}{dt}dz$ should be considered a perfect differential, where the differential coefficients are *total*: see his page 163.

[788.] F. E. Neumann. Two articles by this writer bearing on our subject will be found in *Poggendorff's Annalen.* They are entitled:

(I) *Die thermischen, optischen und krystallographischen Axen des Krystallsystems des Gypses.* Bd. XXVII., 1833, pp. 240—278.

(II) *Ueber das Elasticitätsmaass krystallinischer Substanzen der homoëdrischen Abtheilung.* Bd. XXXI., 1834, pp. 177—192.

[789.] The object of (I) is to prove by a comparison of various measurements on gypsum that the elastic, thermal, optic and crystallographic axes of all crystalline forms symmetrical with regard to three planes at right angles coincide. The reasoning used is rather of a general physical than of a mathematical kind, and is not in itself quite conclusive.

[790.] Neumann begins by remarking that the 'elastic axes' of Fresnel are based upon an examination of the varying velocity of light in different directions in a crystalline structure. They are what we term 'optic axes,' and have not been deduced by Fresnel from what Neumann terms the 'axes of cohesion' or what we term 'elastic axes.' Neumann, referring to a paper of his own on double refraction (in the *Annalen*, Bd. XXV.), points out that the assumption of the symmetrical division of the medium at any point by three rectangular planes is the most reasonable for all crystalline media having three rectangular optical axes. This division leads to the existence of three 'axes of cohesion' (elastic axes) which must coincide with the optic axes.

Further, whatever be the mechanical construction of a homogeneous medium, if its surface be submitted to a uniform tractive load, a sphere will be distorted into an ellipsoid. The axes of this ellipsoid, Neumann terms the *Hauptdruckaxen*—directions of principal tractions,—and he argues that since they depend on the elastic structure of the body it is the simplest and most reasonable hypothesis that they coincide with the elastic axes. He quotes experimental result in favour of this:

Wenn man annimmt, und es ist die einfachste Annahme, die sich darbietet, und eine durch die bekannten Lichtphänomene des comprimirten Glases, und durch die Untersuchungen von Savart über die Schwingungen krystallinischer Scheiben sehr unterstützte Annahme, dass das System von Elasticitätskräften, welches in den Schallschwingungen wirksam ist, aus einem, dem Medium inhärirenden Cohäsionssystem hervorgeht, welches dasselbe Gesetz der Verschiedenheit nach den verschiedenen Richtungen befolgt, als dasjenige, aus welchem das System von Elasticitätskräften hervorgeht, welches in den Lichtschwingungen die Bewegungen fortpflanzt, so ergiebt sich strenge, dass in allen den krystallinischen Medien, welche drei auf einander rechtwinklige optische Axen haben die Hauptdruckaxen zusammenfallen mit diesen optischen Axen (pp. 243—244).

[791.] The next step taken by Neumann is ingenious if not very rigid. He notes the physical fact that the effect of heat on a body in the case of increased temperature is exactly like the effect of a uniform tractive load. Owing to change of temperature the relative position of the elements of a body are altered in the same way as they would be by such a load. Points which were originally on a sphere after change of temperature are to be found on an ellipsoid; the axes of this ellipsoid are the *thermal axes*, and the above mechanical conception of the effect of change of temperature as the same as a uniform tractive load leads obviously to the identity of the thermal axes with the directions of principal traction or with the optic axes also (p. 245).

[792.] Then follows the identification (for the case of symmetry about three rectanglar planes) of the crystalline axes with the other sets of axes. Neumann's proof is of this kind. All other lines except the thermo-elastic axes have a relative position with regard to these lines which depends on the particular conditions of pressure (e. g. atmospheric) and of temperature and is changeable with these. Hence, if crystallography defines the axes of a crystal as having unchangeable position they must coincide with the thermo-elastic axes. This argument leads to the hypothesis that such three rectangular axes (crystalline axes) exist even in those forms of crystals where they are not directly determinable by the symmetry (p. 246). Neumann sums up his results as follows:

Es giebt also in allen krystallinischen Formen ein rechtwinkliges krystallographisches Axensystem, und diess ist dasselbe, als das thermische und das der Hauptdruckaxen, von denen vorher die Identität mit dem optischen Axensysteme und dem Axensysteme der Cohäsionskräfte unter einer sehr einfachen und wahrscheinlichen Voraussetzung nachgewiesen ist (p. 247).

[793.] The rest of the paper is occupied with an investigation as to whether these results are true for gypsum. Neumann makes use of Mitscherlich's experiments on the position of the thermal axes, of Biot's experiments on the position of the optic axes, and of a revision of some of Phillip's on the crystalline axes, to show that these results are true for gypsum within the limits of observational error.

[794.] The second paper begins by a statement that for crystalline bodies symmetrical about three rectangular planes or those of the homoëdric type there are *six* elastic constants (Poisson's theory), but that hitherto these do not appear to have been determined for any crystal. The importance of such determination, Neumann holds, would lie to a great extent in the possible discovery of relations between these six constants, and thus their reduction to a lesser number. He proposes in this paper to give the chief elastic phenomena presented by such crystals in order that those desiring to experiment may have a means of determining these constants.

[795.] Neumann supposes $a$, $b$, $c$ to represent lengths taken parallel to the crystalline or elastic axes, and the body to receive a uniform tractive load which produces stretches in the direction of these axes represented by $M$, $N$, $P$.

Considering more closely a right six-face whose edges are parallel to the axes and equal to $a$, $b$, $c$, he supposes it first to receive a traction $T$ parallel to $a$. This produces stretches which he denotes by $M_a$, $N_a$, $P_a$.

Similarly tractions $T$ parallel to $b$ and $c$ give the systems

$$M_b,\ N_b,\ P_b,$$

and

$$M_c,\ N_c,\ P_c.$$

We have then $$M = M_a + M_b + M_c,$$

and similar relations for $N$ and $P$.

Among the nine stretches we have the following relations:

$$M_b = N_a,\ M_c = P_a,\ N_c = P_b.$$

If the traction $T$ be of unit magnitude, it will be seen that the system

| | | |
|---|---|---|
| $M_a$ | $M_b$ | $M_c$ |
| $N_a$ | $N_b$ | $N_c$ |
| $P_a$ | $P_b$ | $P_c$ |

represents the inverses of a system of *stretch-moduli* for direct and transverse stretch. It corresponds exactly in Saint-Venant's notation to:

| | | |
|---|---|---|
| $\frac{1}{E_a}$ | $\frac{1}{F_c}$ | $\frac{1}{F_b}$ |
| $\frac{1}{F_c}$ | $\frac{1}{E_b}$ | $\frac{1}{F_a}$ |
| $\frac{1}{F_b}$ | $\frac{1}{F_a}$ | $\frac{1}{E_c}$ |

where $F_a = E_b/\eta_{bc} = E_c/\eta_{cb}$ and similar relations hold for $F_b$ and $F_c$, $\eta_{bc}$ being the ratio of the stretch in direction $c$ to the stretch in direction $b$ produced by a traction in the direction $b$. (See Saint-Venant, Navier's *Leçons*, p. 809, or his *Clebsch*, p. 80.)

[796.] If $A$, $A_{,}$, $A_{,,}$, $B$, $C$, $D$, be the six elastic constants, Neumann then shews (p. 183) that:

$$M_a = -\frac{BC - A_{,}^2}{T}, \quad M_b = N_a = -\frac{AA_{,} - BA_{,,}}{T},$$

$$N_b = -\frac{DB - A^2}{T}, \quad M_c = P_a = -\frac{A_{,}A_{,,} - CA}{T},$$

$$P_c = -\frac{CD - A_{,,}^2}{T}, \quad N_c = P_b = -\frac{AA_{,,} - DA_{,}}{T},$$

where $\quad T = BCD + 2AA_{,}A_{,,} - BA_{,,}^2 - CA^2 - DA_{,}^2,$

results agreeing with those obtained later by Saint-Venant.

[797.] From these results, by putting

$$A = A_{,} = A_{,,} = \tfrac{1}{3}B = \tfrac{1}{3}C = \tfrac{1}{3}D = L,$$

Neumann shews that the result proved by Poisson for a wire holds for every right prismatic body: see Art. 469.

[798.] On p. 184 we have again the statement that for a uniform tractive load the directions of principal traction coincide with the elastic axes. If $\mu$, $\nu$, $\omega$ are the stretches of the axes of principal traction (here apparently *assumed* to be those of principal stretch), the stretch $\Delta\rho/\rho$ in direction ($\alpha$, $\beta$, $\gamma$) is given by

$$\frac{\Delta\rho}{\rho} = \mu\alpha^2 + \nu\beta^2 + \omega\gamma^2.$$

There are always two planes in which the stretch is the same in all directions: namely those of the circular sections of the ellipsoid

$$1 = \mu x^2 + \nu y^2 + \omega z^2.$$

The position of an element of the body being determined by reference to rectangular coordinates $x$, $y$, $z$, and the body being subjected to any load not producing flexure (*Biegung*), the stretch is given by

$$\frac{\Delta\rho}{\rho} = M\alpha^2 + N\beta^2 + P\gamma^2 + p\alpha\beta + n\alpha\gamma + m\beta\gamma, \quad \text{......(i)},$$

where $M$, $N$, $P$ are the stretches in the direction of $x$, $y$, $z$ and

$$p = \cos(xy), \quad n = \cos(xz), \quad m = \cos(yz)$$

are really the slides. Here again there exist three principal tractions determined by a cubic equation (p. 186).

[799.] Neumann now supposes a right prism cut from a crystal such that the axis of the prism $r$ makes angles with $a, b, c$ whose cosines are $C_a$, $C_b$, $C_c$. This prism is subjected to a tractive load $D$ on the bases. Then we find:

$$M = D\{M_a C_a^2 + M_b C_b^2 + M_c C_c^2\}, \text{ ...........(ii)},$$

and two similar relations for $N$ and $P$; also,

$$m = D\frac{C_b C_c}{A_{\prime}}, \quad n = D\frac{C_a C_c}{A}, \quad p = D\frac{C_b C_a}{A_{\prime\prime}}, \text{ .........(iii)},$$

where $M, N, P$ are the stretches in directions of the elastic axes and $m, n, p$ the angles between those axes in their distorted positions.

Let $E_r$ be the ratio of the tractive load $D$ to the stretch in direction of the prismatic axis ($C_a$, $C_b$, $C_c$), or the stretch-modulus,

$$\therefore \frac{1}{E_r} = \frac{\Delta\rho}{\rho}\frac{1}{D}.$$

Applying (i), (ii) and (iii), we obtain (p. 189),

$$\frac{1}{E_2} = M_a C_a^4 + N_b C_b^4 + P_c C_c^4 + 2\left(N_a + \frac{1}{2A_{\prime\prime}}\right) C_a^2 C_b^2$$
$$+ 2\left(M_c + \frac{1}{2A}\right) C_a^2 C_c^2 + 2\left(P_b + \frac{1}{2A_{\prime}}\right) C_b^2 C_c^2.$$

It will be seen that this ***stretch-modulus quartic*** is a particular form of that considered by Cauchy (see our Art. 658) and later by Rankine and Saint-Venant.

[800.] In order to know the value of $E_r$ for every direction it will be sufficient to determine it for six directions. Neumann suggests this should be done by experimenting on the flexure of thin rods. Neumann then gives (p. 190) a formula for the deflection of such a prism, deduced from that usual for a beam of which the longitudinal stretch-modulus is $E_r$; but when the crystalline axes are in any direction with regard to the axis of the rod, it does not seem to me obvious that this formula can be applied. For example, let one axis of the rod coincide with a crystalline axis, let the rod be of rectangular section as Neumann supposes, and the other elastic axes *not parallel* to its sides: I do not think the ordinary formula would hold, for the rod would owing to unsymmmetrical slide undergo torsional strain and so not necessarily receive *uniplanar* flexure.

[801.] Neumann suggests as six good directions the axes of the crystal and those which bisect the angles between each pair of them. If $E_a$, $E_b$, $E_c$, $E_{(ab)}$, $E_{(ac)}$, $E_{(bc)}$ represent the corresponding moduli, we find (p. 191):

$$1/E_a = M_a, \qquad 1/E_{(ab)} = M_a + N_b + 2N_a + 1/A_{\prime\prime},$$
$$1/E_b = N_b, \qquad 1/E_{(ac)} = M_a + P_c + 2P_a + 1/A,$$
$$1/E_c = P_c, \qquad 1/E_{(bc)} = N_b + P_c + 2P_c + 1/A_{\prime}.$$

By cutting the sides of the rectangular prism in the first three cases parallel to the other two axes of the crystal, and in the last case the faces parallel to the load perpendicular to the third axis, it might be possible to calculate these moduli from experiment on flexure; but I do not think without these restrictions, which Neumann does not make, that his flexure experiments would lead to any accurate result; it is not enough merely to apply the ordinary formula for deflexion.

A second method for finding the constants of the equation for $E_r$ is by measuring the change in angles between the sides and again between the sides and base of a rectangular prism when it is subject to load in the direction of its axis. Such experiments however only give the difference of the constants in the equation for $E_r$, and one observation by the flexure method would still have to be made (pp. 191—192).

The memoir, notwithstanding some obscurities, is suggestive, and contains I believe for the first time the quartic for the stretch-modulus in any direction: see however Art. 658.

[802.] 1833. Cagniard-Latour. *Journal de chimie médicale*, Paris, 1833, T. IX., p. 309. See also *Poggendorff's Annalen*, T. XXVIII., 1833, p. 239. We have here the results of certain experiments due to this well-known physicist. They have some bearing on the elastic properties of bodies.

1°. The note produced by the longitudinal vibrations of a metal wire is not altered by hard hammering (*écrouissement*).

2°. A tempered (*trempé*) steel-wire gives in vibrating longitudinally a deeper note than an untempered wire. The same, only in a lesser degree, holds for iron.

3°. A hard-hammered (*écroui*) iron tuning-fork gives in vibrating transversally a deeper note than one which has not been thus treated.

These statements are not in accordance with the results of Lagerhjelm's experiments cited in Art. 366, nor with those of Wertheim to be considered in Chapter VIII.

[803.] 1833. Franz Joseph, Ritter von Gerstner. *Handbuch der Mechanik. Erster Band. Mechanik fester Körper.* Prag, 1833—34. I have only examined the second edition of this work, which is however a mere reproduction of the first (1831). The third chapter, entitled: *Festigkeit der Körper*, treats of our subject. It occupies pp. 241—384. The book is edited by the son, Franz Anton.

The following parts seem original:

[804.] A set of experiments on piano-forte wires, which were procured of peculiarly uniform iron; individual cases which presented any irregularity in the tone were rejected (p. 259).

A relation between longitudinal stretch and load of the form

$$L = As + Bs^2 + Cs^3 + Ds^4 + \dots$$

was empirically assumed. Here $L$ is the load and $s$ the total stretch, whether elastic or set. Gerstner's experiments led him to the conclusion that $C$, $D$, etc. are all zero and $B$ is negative. Let $L$ be the maximum load corresponding to a maximum extension $s'$, then

$$\frac{dL}{ds} = 0 \text{ gives } A + 2Bs' = 0,\ L' = As' + Bs'^2.$$

Whence we deduce

$$\frac{L}{L'} = \frac{s}{s'}\left(2 - \frac{s}{s'}\right) \quad \text{......................(i)}$$

for the relation between load and stretch.

It must however be remarked that the maximum load does not necessarily correspond to the maximum stretch. If stricture appears the load begins to decrease after a certain stretch;—at least, if the load be calculated on the basis of the original sectional area, which it certainly must be in practice till we are better acquainted with the phenomenon of stricture.

[805.] Gerstner remarks that when $s/s'$ is small the load is proportional to the stretch. This condition he terms 'perfect elasticity' (p. 265). For any load the stretch is given by

$$\frac{s}{s'} = 1 - \sqrt{1 - \frac{L}{L'}} \quad \text{......................(ii).}$$

[806.] We now come to the statement of those facts with regard to set to which continental writers have given the name of *Gerstner's Law*[1]. Gerstner finds, namely, that the elastic strain even after the beginning of set remains proportional to the load.

Let $s''$ be the total stretch due to the load $L''$. Then, by the formula just found and the fact just stated, we have:

$$\text{Total stretch } s'' = s'\left\{1 - \sqrt{1 - \frac{L''}{L'}}\right\},$$

$$\text{Elastic stretch} = \frac{s'L''}{2L'},$$

$$\text{Set} = \frac{s'}{2}\left\{1 - \sqrt{1 - \frac{L''}{L'}}\right\}^2 \quad \text{.........(iii).}$$

Further, Gerstner finds that no set takes place when such a wire is reloaded, till the load becomes greater than $L''$. Up to that the stretch obeys the law of proportionality.

From the formula (iii), by putting $L'' = L'$, the author concludes that the set at time of greatest extension $= s'/2$ or half the greatest stretch. This however depends upon Gerstner's assumption that

[1] This 'law' has sometimes been attributed to Leslie, and a vague reference given to his *Elements of Natural Philosophy*, Edinburgh, 1823 (e.g. Saint-Venant *Historique Abrégé*, p. cclxxxix.). I can find nothing in the poor section of that work (Vol. I. pp. 243—275, 2nd ed.) devoted to our subject which would justify this. He merely states (p. 245) that after a certain load, stretch increases more rapidly than traction. Saint-Venant also remarks that it might with justice be ascribed to Coulomb, who in his memoir on torsion (see Art. 119) made some experiments on set, and found that the elastic strain remained nearly the same after set had begun. Gerstner's results however bring out two points: the permanency of the elasticity and the law of the set. This is the law confirmed ten years after by Hodgkinson: see Art. 969. The Coulomb-Gerstner Law is a law of *elasticity;* the Gerstner-Hodgkinson law, one of *cohesion.* We may term the first after Coulomb and the second after Gerstner. An English account of Coulomb's experiments and his views on elasticity and cohesion will be found in Brewster's edition of *Ferguson's Lectures on Select Subjects*, 3rd ed. 1823, Vol. II. pp. 238—241. See also Note A in the Appendix to this volume.

the maxima of stretch and load coincide, which is possibly not true when stricture intervenes before rupture (pp. 272—276).

Gerstner remarks on the importance of loading all material used in suspension bridges to as great a load as it is intended to stand before using it in the construction, but on the other hand practical reasons might suggest a factor of safety in the possibility of set. It must also be noted that Gerstner's results are entirely based on experiments made on a highly-worked material.

[807.] On pp. 327—364 we have an account of a set of experiments by Gerstner himself on the flexure of wooden beams. This is an attempt to establish a formula for the set which accompanies considerable flexure. He takes an empirical formula of the form

$$L = u(A - Bu) \text{ ........................(iv)},$$

where $L$ is the load which placed in the middle of the beam produces a depression $u$, $A$ and $B$ are constants. He inverts this into

$$u = \frac{A}{2B}\left\{1 - \sqrt{1 - \frac{4BL}{A^2}}\right\} \text{ .................(v)}.$$

He finds the formula (v) exhibits greater differences between experimental and calculated results than was the case with (iii). These differences would, he implies, be increased if after-strain were taken into account, i.e. if the loads were allowed to remain several days or weeks (p. 332).

[808.] A remark on p. 334 may be quoted as it contains a truth not always recognised:

Da überhaupt das Holz, Eisen und alle gebogenen Körper nur durch ihre zusammenhängende Kraft dem Bruche widerstehen, so folgt von selbst, dass diejenige Biegung, welche ein Körper durch früher aufgelegte Gewichte bereits angenommen hat, auch nicht als eine Wirkung der neu aufgelegten Last betrachtet werden kann; es muss daher in jedem Falle das Tragungsvermögen nur von derjenigen Kraft, welche widersteht, d. h. von dem wirklich vorhandenen oder sich äussernden Elastizitätsvermögen bestimmt werden, woraus folgt, dass überhaupt ein Stab, der keine Elasticität hätte, nach einander nachgeben und gar keine Last zu tragen fähig seyn würde. Man sieht hieraus, dass die Meinung derjenigen sehr gegründet ist, welche behaupten, dass die Festigkeit der Körper nur von der elastischen Kraft derselben bestimmt werde, und dass man überhaupt einem Körper keine Last anvertrauen könne, die seinen elastischen Widerstand übersteigt.

The last passage is somewhat obscure, but it would seem to denote that the load must be such that it falls within the limits which we call the state of ease. It must be noted that any load outside these limits would not necessarily produce rupture, but possibly only a certain amount of set and extension of the state of ease; the load itself would thus always remain practically the limit to the state of ease—which of course would hardly be a desirable state of things in a structure. An opinion expressed by Gerstner on the next page, that a load which exceeds 'what the elasticity of a beam can bear' will if it be applied long enough produce rupture, seems to be connected with the above. It is not however quite clear what the expression 'elasticity' here means; if it means the range in which permanent set does not appear, it does not seem to me necessarily true.

[809.] Other experiments on cast and wrought iron follow for which the constants of the formula (v) are calculated. They offer nothing of note.

[810.] Certain experiments on the torsion of wooden cylinders will be found on pp. 377—381. Before remarking upon them we may note that Gerstner as all writers before Cauchy (see Art. 661) supposes square prisms to follow the same law of torsion as circular prisms (p. 377).

By adopting a formula like (iii) and (v) the calculated results are found *after a certain load* to be in far closer accordance with experiment than when the ordinary formula is used. That is to say by a proper choice of the two constants set was to a great extent allowed for. The divergence however of calculation from experiment shewed even then a *steady*, if small, increase. Gerstner makes the same remark for torsional load as he has done for flexional: see Art. 808.

[811.] 1834. N. Persy. *Cours de stabilité des constructions à l'usage des élèves de l'École d'application de l'artillerie et du génie.* Metz, 1834. This is a *Cours lithographié*, and apparently reached at least four editions[1]. With one exception it seems to contain nothing of original value, reproducing the old Bernoulli-Eulerian theory with Lagrange's results, Coulomb's theory of *écrasement*, etc. The one exception is a remark on p. 24, namely, that it is a fundamental condition for the uniplanar flexure of a beam that the neutral surface contain not only the centroid of the section but one of the principal axes at that point.

La coexistence des équations

$$\int y d\omega = 0, \quad \int xy d\omega = 0,$$

selon la théorie des moments d'inertia signifie que la ligne *ax* est l'un des deux axes principaux d'inertia répondant au centre de gravité de la section.

*ax* is the neutral axis and $d\omega$ an element of sectional area. There is no attempt to investigate what would happen if the second condition were not fulfilled or the flexure ceased to be uniplanar.

[812.] Two volumes of another *Cours lithographié* entitled: *Principes de Mécanique,* and containing lectures delivered at the *École des Mineurs de St Étienne* (*Loire*), have reached me from a Berlin bookseller. There is no statement of the lecturer's name or date, but he may possibly have been M. S. Pitiot, whose name occurs at the bottom of the first page (or is this the lithographer?). The date 1831 is placed beneath the name of one of the designers who prepared the plates at the end of the second volume.

The second volume contains a *Résumé des leçons sur les constructions,* in which there is a good deal of experimental fact (partly drawn from Barlow) and a few pages of theory (e.g. pp. 31—36). It needs the perusal of such a *Cours* as this, or that of Art. 811, for anyone to comprehend the advance made by Saint-Venant in his lectures of 1837—38, faulty as he himself admits the latter to be: see our Chapter IX.

[813.] As we are treating of lithographed courses of lectures, we may here dismiss a third example which is in our possession,

[1] I owe to the kindness of M. de Saint-Venant the opportunity of examining this work.

although it belongs to a somewhat later date. The lectures were delivered at the *École des Ponts et Chaussées* in 1842—1843, and the work is entitled: *Notes sur la mécanique appliquée aux principes de la stabilité des constructions et à la théorie dynamique des machines.* No name appears in the work, but Duhamel is written, I do not know on what authority, on the fly-leaf.

The work commences with certain material which will be found in ordinary dynamical or statical text-books of the old type. Chapter 3 (pp. 26—59) is entitled: *De la résistance des matériaux employés dans les constructions.* In treating of the stretch produced by pure traction, the author considers the case of a vertical bar supporting a weight which has a certain velocity. In this manner it is shewn that the kinetic as distinguished from the statical stretch of the bar may exceed the elastic limit: see the results of Poncelet and of Sonnet, Arts. 988 and 938, (5).

Les considérations qui précèdent font comprendre qu'une charge insuffisante pour dépasser la limite d'élasticité d'une pièce peut produire cet effet si une cause accidentelle lui imprime une vitesse, et expliquent ce qu'on appelle improprement *l'influence ou l'action du temps.* Aussi une tige qui ne perd son élasticité que sous une charge de 12 à 15 kil. par millim. carré n'est-elle considérée que comme pouvant porter une charge permanente de 6 à 8 kil. par millim. carré (p. 29).

The remark is not without value, although it cannot be said to explain the phenomenon of time-effect, i.e. after-strain.

[814.] The consideration of flexure and torsion is entirely on the old lines and by no means up to date; thus, slide (see our Chapter IX.) is entirely neglected as well as Persy's remark as to the position of the principal axes in uniplanar flexure: see Art. 811.

[815.] There is an Appendix entitled: *Résumé de quelques leçons sur la flexion et la résistance des pièces courbes.* It occupies 43 pages.

The first part of this Appendix is entitled: *Introduction. Étude de la répartition d'une force sur la section droite d'un prisme* (pp. 1—15). It may be said to contain the whole theory of *cores* as applied to the beam-problem, and possibly for the first time. The core (*noyau central* or *Kern*) of a section is here termed *partie centrale de la section.*

A right prism being subjected to a terminal tractive load $P\omega$, the traction at the point $x, y$ of the section $\omega$ is given by

$$t = \frac{P}{\omega}\left(1 + \frac{ax}{\kappa'^2} + \frac{by}{\kappa^2}\right).$$

Here the principal axes at the centre of gravity of a section are taken as axis of coordinates. $\kappa$ and $\kappa'$ are the radii of gyration about the axes of $x$ and $y$ respectively, while $a$, $b$ are the coordinates of the point in which the direction of $P$ cuts the section (p. 2). This result is thrown into various geometrical forms and the chief properties of the neutral axis for any values of $a$, $b$ investigated. Finally the cores of various sections (rectangle, circular and elliptic annuli, etc.) are determined (pp. 7—9). We then have the usual application of the theory of cores to the problem of cohesion and a series of examples (pp. 9—15).

[816] The next two chapters are devoted to the flexure of curved beams of uniform section, and the treatment is such as will be found in Navier's *Leçons* (see Art. 279). Chapters 3 and 4 contain a discussion of the uniplanar flexure of curved pieces, which are originally circular. Various problems differing slightly from those solved by Navier (see Art. 278) and Saint-Venant (see Chapter IX.) are considered, but there is no attempt, as in the latter writer's work, to include the effect of slide (pp. 27—43).

[817.] 1834. E. Martin. *Emploi du fer dans les ponts suspendus, Annales des ponts et chaussées.* 1834, 2[e] *semestre*, pp. 157—168, with observations by Vicat, pp. 169—172.

We are not concerned here with the point in controversy between Martin and Vicat, but we may refer to some experimental results of Barbé and Bornet on iron such as is employed in the cables of ships. These experiments shew clearly, although on a limited scale that:

1°. in bars of iron subjected to traction, the load after a certain value ceases to be proportional to the strain, but that in this case the strain is nearly all set: see Note A, (2) (ii) in the appendix to this volume.

2°. the existence of a time-influence as exhibited in after-strain.

There does not appear however to have been any consideration of whether the after-strain was after-set or not. The experiments were made in 1829. See Tables III. and IV., pp. 167 and 168. Saint-Venant refers to these experiments in the *Historique Abrégé*, p. cclxxxix.

818. De Schultén. *Déduction des équations de l'équilibre des fils élastiques au moyen d'une méthode nouvelle;* par M. de Schultén, Professeur de Math. à l'Univ. imp. d'Alexandre en Finlande. This is published in the *Mémoires présentés à l'Académie de St Pétersbourg par divers Savans*, Vol. 2, 1835; it occupies pages 49—73. The memoir was read on the 26th of November, 1828.

It relates to the ordinary statical problem of the equilibrium of a flexible cord, and does not bear on our subject. The solution of the problem is quite different from that which is now usually given in works on statics, and does not seem to possess any special advantages. On page 71 the author refers to a criticism which he had already offered on Lagrange's treatment of the problem: see our Arts. 536—539, 744.

[819.] 1835. C. D. Arosenius. *De soliditate columnarum disquisitio.* This is a Swedish degree dissertation published in two parts, 24 pages in all, at Upsala.

It presents us with one of those instances, by no means infrequent, wherein a mathematician treating of a purely ideal body apparently views with complacency results having absolutely no physical value. The writer starts from Euler's paper of 1757 (see Art. 65), and, apparently ignorant of Euler's later memoirs (see Arts. 74—85) and those of Lagrange (see Arts. 97—113), obtains approximations to the load and deflexion of an impossible column which are by no means so close as theirs. In his last section he considers the problem of a 'column' in the form of a truncated cone, and seems to be quite unaware that Lagrange had already treated this case: see Art. 112.

The value of the dissertation may be judged from the fact that the moment of the load about any point on the axis of the column is taken equal to the curvature at the point multiplied by a constant $k$, *quae e dimensionibus et materia laminae pendens, his constantibus, ipsa constans permanet.* The author also asserts that: *Experientia tamen docet......ut fibrae illae, quae nec dilatantur nec comprimuntur, locum vere medium teneant.* Considering that these fibres have no existence in the cases treated by the author, and that the neutral line which corresponds to them is a curve having one or more points at infinity,—considering also that Robison had fully exposed the futility of Euler's theory of columns (see Arts. 145 and 146), we may, I think, dismiss this writer to the obscurity from which we have only drawn him in order that he might serve as a warning to one or two modern English writers, who still blindly follow Euler.

[820.] J. M. M. Peyré. *Notes sur le mouvement vibratoire longitudinal de quelques corps solides. Bibliothèque Universelle de Genève: Sciences et Arts.* Tome 60, 1835, pp. 161—196. This consists of extracts from memoirs presented to the *Société des Sciences naturelles de Seine-et-Oise.* I am unaware whether the original memoirs were ever printed. Peyré seems to have followed up Savart's investigations of 1822 on the nodal surfaces of bodies vibrating longitudinally (see Art. 327). The bodies he experimented upon were cylindrical tubes and prisms, and some figures are given of the nodal surfaces. I do not think the mathematical theory of such vibrating bodies has ever been worked out; as a general rule bodies of the kind experimented upon could hardly be treated as isotropic. One conclusion (p. 196) obtained may be cited:

Les surfaces de rupture des tubes de verre qui vibrent ou qui ont vibré paraissent être les mêmes que les surfaces nodales.

[821.] *Die Lehre von der Cohäsion, umfassend die Elasticität der Gase, die Elasticität und Cohärenz der flüssigen und festen Körper und die Krystallkunde.* Breslau, 1835.

This is a very considerable work of upwards of 500 pages by M. L. Frankenheim, a Breslau professor. It is written mainly from the physical and experimental side. The object of the writer, expressed in the preface, is to present a systematic text-book on the whole subject of cohesion. First giving fact and experimental statistic and then the law and theory so far as any have been suggested. With regard to his own hypotheses Frankenheim remarks:

Kein Physiker kann in Beziehung auf seine Hypothesen skeptischer sein als ich, und mit gleichem Vergnügen werde ich sie widerlegt oder bestätigt sehen, wenn dadurch die Wissenschaft gewinnt.

A healthy doctrine not always observed by modern physicists.

[822.] The first two parts of this book, entitled: *Cohäsion der Gase* and *Cohäsion der Flüssigen*, contain much information which is of first-class interest to the historian of mathematics (note especially the completeness of bibliographical detail), but they do not, except in the minor matter of intermolecular force, concern our present purpose.

[823.] The third section of the book—entitled: *Cohäsion der Festen*, and divided into (i) *Elasticität*, (ii) *Gestalt oder Krystallkunde*, (iii) *Cohärenz*—is a physical treatment of our subject in the light of the then existing knowledge. It deserves some consideration at our hands.

[824.] The chapter on elasticity, pp. 238—277, contains a fair amount of reference, a mass of experimental details (wherein however Frankenheim has not always modified the experimenter's own results on quite clear principles), and finally the mere statement of a modicum of the theoretical results of Poisson, Cauchy and others. The consideration of the stretch-modulus and 'specific stress' (*specifische Spannkraft*) on p. 243 do not seem to me peculiarly fortunate. Thus, probably misled by Bevan's and Cagniard-Latour's results[1], Frankenheim remarks that elasticity does not seem to be a characteristic distinction between solid and fluid bodies; for the 'specific stress' of the same body in the two states is not very different. The great difference is in the matter of cohesion, and this is attributed to the phenomenon of crystallisation (p. 277).

[825.] The treatment of crystallography (pp. 277—362) does

[1] See our Art. 372, and *Poggendorff s Annalen*, T. XXVIII. 1833, page 239, where the velocity of sound in water and ice is said to be the same.

not afford much material for our subject. Pp. 277—309 are occupied with formal as distinguished from molecular crystallography. In 'crystallophysics' Frankenheim reproduces some of Savart's results (see our Arts. 335—346), but his statements as to the elastic properties of crystals are not suggestive. On the subject of 'hardness' he refers frequently to an earlier work of his own: *De crystallorum cohaesione*, 1829; the experimental results quoted seem to have been principally obtained by scratching with a sharp point in different directions the several faces of various crystals.

[826.] The portion of the work (pp. 363—502) devoted to the *Cohärenz der Festen*, opens with the consideration of the 'structure' and 'texture' of solid bodies upon which their ultimate cohesion depends. The remarks upon abnormal structures or bodies in a state of internal stress (as many kinds of glass) are not without historical interest and give a wide field of reference. Then follows a section (pp. 408—431) entitled: *Elementare Bewegung durch äussere Kräfte entstanden*, or as we may translate, *change of molecular structure induced by load.* This part of the work is of value as shewing the course of discoveries relating to *after-strain*, in fact to the influence of the time-element in the phenomena of cohesion, for the set here considered is that produced by long-continued load, and except in the reference to W. Weber's experiments (p. 419 and our Art. 710) there is no evidence that these early experimenters investigated whether the time-strains were really set or elastic strain. Here again it would seem important to distinguish clearly in *after-strain* between *after-set* and *e.astic after-strain:* see Art. 708.

[827.] We give a brief account of the course of discovery noted by Frankenheim.

After-strain seems to have been first noted in the bulbs of glass thermometers. The change in the magnitude of these bulbs due to continued atmospheric pressure was first remarked by Bellani. The best discussion of the subject before 1830 is due to Egen: see *Poggendorff's Annalen*, XI. 1827, p. 347. How far the after-strain in this case consists of after-set does not seem to have been investigated. Coriolis next experimented on lead (see

Art. 720), and his results seem to me to point to elastic after-strain. Then follow the interesting experiments of Vicat on iron-wire (see Art. 736), here the extension per year was after a certain load nearly uniform and on the whole suggests an after-set. Some experiments of Eytelwein on wood are not quantitatively very valuable, they merely prove the existence of after-strain. Frankenheim notes Weber's discovery of the elastic after-strain and attempts by its means to explain not only the decrease in amplitude of oscillating bodies, but also the small intensity of the note given by materials such as lead. We may quote here Frankenheim's view of the relation of the time-element to the breaking load:

Es ist möglich, dass das kleinste der die Brechung hervorbringenden Gewichte erst in sehr langer, ja unendlicher Zeit wirke. Diese Gewichte sind also Functionen der Zeit, je grösser diese ist, desto kleiner können jene sein, und wenn es möglich wäre ein Gewicht nur eine unendlich kleine Zeit wirken zu lassen, so würde es, so gross es auch sein mag, keine bleibende Wirkung zurücklassen. Man darf sich jedoch durch diese Betrachtungen nicht verleiten lassen, jedes Gewicht, wenn es nur hinlänglich lange wirkt für hinreichend zum Zerbrechen zu halten; denn obgleich die bleibenden Veränderungen eine unendliche Zeit fortwachsen, so ist es dennoch sehr wahrscheinlich, dass die Summe dieser Aenderungen eine endliche Grösse ist, und nur da Brechen eintritt, wo diese Grösse den Körper bis zu irgend einem von seiner Beschaffenheit abhängigen Puncte, die Grenze seiner Elasticität, gebogen hat. (p. 417.)

[828] The author concludes this portion of his work by some remarks on the influence of temperature on the elastic condition (pp. 421—424), and by a general summary in which he attributes these molecular changes to *Krystallisationskraft*, a force arising as he believes from the absolute position of the axes of crystallisation in the elementary parts of the material: see pp. 425 and 357.

[829.] Pages 431—502 of the book are occupied with the consideration of absolute strength (*Festigkeit*). The results cited are selected from the innumerable experimental investigations of the previous fifty years and offer nothing of consequence. The treatment of impact and adhesion, which also falls into this section, may also be passed over, and we may conclude our review of the work by merely noting the last paragraph, where the author suggests a possible relation between magnetism and cohesion[1]. The correspondence pointed out between the difference of cohesive force in the different directions of a crystal and magnetic polarity does not seem peculiarly lucid (p. 502).

[830.] A. Baudrimont. *Recherches sur la Ductilité et la Malléabilité de quelques Métaux, et sur les Variations que leurs Densités éprouvent dans un grand nombre de circonstances. Annales de chimie et de physique.* T. 60, pp. 78—102. Paris, 1835. This paper is very suggestive in that it states facts with regard to metal wires which might naturally be expected, but which seem to have been much neglected by later physicists, who have thought to deduce general laws from these highly 'worked' forms of material. The author shews how even the process of annealing in various gases affects the breaking load (pp. 96—97). Annealing also affects, apparently in the most irregular fashion, the diameters, length and tenacity of wires (pp. 92, 94, 99): see our Art. 692, 1°. Elastic after-strain is noted in the following paragraph:

Il est simple de penser que, lorsqu'un fil passe forcément dans l'ouverture d'une filière, en vertu de l'élasticité qui lui est propre, ses molécules reviennent en partie sur elles-mêmes, et que son diamètre s'accroît au delà de l'ouverture de cette filière; mais j'ai trouvé un fait auquel je m'attendais peu, et sur lequel je ne conserve pas le moindre doute, c'est que le diamètre d'un fil s'accroît lentement, et qu'il est sensiblement plus grand au bout d'un mois, qu'il ne l'était quelques heures après avoir été étiré. J'ai encore remarqué que des fils écrouis

[1] The first suggestion of a possible relation between these two phenomena appears due to Kirwan in a memoir entitled: *Thoughts on Magnetism*, published in the *Transactions of the Royal Irish Academy*, Vol. VI., Dublin, 1797, pp. 177—191. Kirwan attributes crystallisation and magnetism to the same physical cause, the difference being one of degree not of kind. He writes:

"By crystallisation I understand that power by which the integrant particles of any solid possessing sufficient liberty of motion unite to each other, not indiscriminately and confusedly but according to a peculiar uniform arrangement, so as to exhibit in its last and most perfect stage regular and determinate forms. This power is now known to be possessed by all solid mineral substances." (p. 179.)

The way in which the face of one crystal attracts a first face and repels a second face of another crystal is compared with magnetic action. I suspect Frankenheim drew his ideas from Kirwan: see his p. 502.

qui avaient été pliés en plusieurs sens, et que j'avais redressés avec beaucoup de soin, perdaient leur rectitude en un jour ou deux. (p. 95.)

Reference is then made to Savart's memoir in the *Annales* of 1829: see Art. 332. We may also refer to the remarks of Brix: see Art. 858. (vii).

[831.] The general results of the memoir are given on pp. 101—102 and we reproduce them here, drawing especial attention to the first three. The author concludes:

1°. Qu'un fil métallique est généralement très irrégulier dans son étendue.

2°. Qu'il est impossible, en usant de précautions semblables, d'obtenir des fils de différentes natures et de mêmes diamètres en les étirant dans une seule ouverture de filière.

3°. Que, lorsqu'on les étire, ils subissent un allongement aux dépens de leur diamètre, et, quelquefois, par l'augmentation de la distance qui sépare les molécules qui les constituent.

4°. Que la densité des fils qui ont moins que $0^{mm}$, 5 de diamètre est très considérable, si on la compare à celles des autres préparations des mêmes métaux.

5°. Que les fils de $0^{mm}$, 5 de diamètre, et au-dessus, sont moins denses que les lames qui sont préparées en laminant ces fils, soit après, soit avant le recuit.

6°. Que l'écrouissement augmente la tenacité des métaux d'une manière considérable.

We have here clearly pointed out the very peculiar elastic character of wires. Physicists were growing conscious that material in its 'worked state' does not present uniformity in its elastic constants. For example, in the case of a wire or in the case of a bar of cast-iron it is certain that at the surface of the material the stretch-modulus has a value quite different from that which it has at some distance inwards. According to Saint-Venant, Ardant in a work entitled: *Études théoriques et expérimentales sur l'établissement des charpentes à grande portée*, Metz, 1840, points out particularly the irregularities in iron-wire. I have not been able to find a copy of this work: see however Art. 937.

[832.] *Transactions of the Institution of Civil Engineers*, Vol. I. London, 1836.

On p. 175 we have: *An Elementary Illustration of the Principles of Tension and of the Resistance of Bodies to being torn asunder in the Direction of their Length;* by the late J. Tredgold.

The object of this paper is very simple, and singularly important in its practical results. Tredgold states that the resistance of a bar subjected to terminal tractive load is usually taken as proportional to the cross section. This however supposes the load to be uniformly distributed or else axial; if the load be applied at a point not quite in the axis, the bar will not be uniformly strained and the resistance is no longer proportional to the cross section. The practical difficulty of making the load exactly axial is obvious. Tredgold's mathematical investigation (pp. 177—178) is, I think, unsatisfactory; like Dr Whewell he seems to imagine that the neutral axis throughout the length of the bar will be at a constant distance from the central line. He assumes the $y$ of his notation to be a constant, but the bar will receive flexure and $y$ will then vary from point to point. The equation to the central axis then takes an exponential form, and the distance of the neutral axis from the central line being equal to $\kappa^2$ divided by the deflection will also be of an exponential form. Here $\kappa$ represents the radius of gyration of the section about a line through its centre perpendicular to the plane of flexure.

Let the bar be pivoted at the fixed end, and let $P$ be the load supposed to be applied in a principal axis of the terminal section, $c$ the semi-diameter of the section in the plane of flexure, $\omega$ the area of the section, $b$ the distance from the centre at which $P$ acts, then by Art. 815 the greatest stress will be

$$T' = \frac{P}{\omega}\left(1 + \frac{cb}{\kappa^2}\right),$$

and will occur in the outermost fibres of the section upon which $P$ acts. We should thus expect a rod subject to a non-central terminal traction to break at its terminal.

Further, if $b$ were zero, we should have

$$T = \frac{P}{\omega}$$

Hence $$T' = T\left(1 + \frac{cb}{\kappa^2}\right).$$

Let $T_0$ be the greatest traction any fibre will sustain without set, then the greatest central tractive load will be $P_0 = \omega T_0$ while the greatest eccentric tractive load $P_b$ at distance $b$ is given by

$$P_b = \frac{\omega T_0}{1 + \dfrac{cb}{\kappa^2}};$$

hence $$P_b = P_0/\left(1 + \frac{cb}{\kappa^2}\right).$$

For a rectangular section $\kappa^2 = \dfrac{c^2}{3}$, or

$$P_b = P_0/\left(1 + \frac{3b}{c}\right).$$

For a circular section $\kappa^2 = \dfrac{c^2}{4}$, or

$$P_b = P_0/\left(1 + \frac{4b}{c}\right).$$

We see from this how much of the strength may be lost by eccentric loading.

[833.] Tredgold comes to the result that $\frac{1}{2}$ the strength of the bar may be lost, if the force be $\frac{1}{8}$ of the semi-diameter from the centre. This result agrees with my calculation in the preceding article if the bar be of rectangular section. He insists on the importance of the longitudinal load being exactly central in all sectional bars subject to traction.

In making a joint to resist traction, the surfaces in contact should be so formed as to render it certain that the direction of tractive load may be exactly, or at least very nearly, in the sectional centre of the bars which have to resist it.

Tredgold proceeds with practical remarks and then quotes a passage from Dr Robison (see Art. 145) which criticised Euler's theory of columns and affirmed that very little was known about resistance; he concludes thus:

Such was Dr Robison's view of the subject, but the question did not long remain in that state. Our celebrated countryman Dr Thomas Young, soon discovered the proper mode of investigation, which was published in 1807, and yet, strange as it may seem, the popular writers on mechanics in this country, as well as on the continent, either have not seen, or do not comprehend, the brief but elegant demonstration Dr Young has given. We can attribute it only to the difficulty of following the inquiries of that able philosopher without a most extensive knowledge of mathematics and of nature. (See Young's *Miscellaneous Works*, Vol. II. p. 129.)

[834.] Tredgold is undoubtedly right in asserting that Euler's theory fails because the neutral line is not necessarily in the material, but he himself falls into error in calculating its position.

[835.] On pp. 231—235 we have some experiments made under the direction of Messrs Bramah and Sons on the force requisite to fracture and crush stones. They have no bearing on theory.

We pass to Vol. II. of the same *Transactions*.

[836.] Pp. 15—32. *A series of Experiments on different kinds of American Timber*, by W. Denison. These have no theoretical value.

[837.] Pp. 113—135. *A series of Experiments on the Strength of Cast Iron*, by Francis Bramah. These experiments do not possess much novelty either in the sectional form of the beams or in the method of testing. A good many are on 'open beams' and merely shew their weakness. The results are not comparable with Hodgkinson's, and a good criticism of the latter's upon them will be found on p. 461 of his addition to Tredgold's *Cast Iron:* see Art. 970.

[838.] Lastly in Vol. III. of the same *Transactions* we have on pp. 201—218 a memoir entitled: *On the Expansion of Arches*, by George Rennie. The writer gives various experiments on the expansion of iron and stone bridges by heat, and the effect of this in straining bridges, but there is no attempt to investigate any thermo-mechanical laws, nor has the memoir any theoretical value.

[839.] With this volume the *Transactions* come to an end. Various brief notices, papers and discussions on the strength and elasticity of materials, will be found in the *Minutes of the Proceedings* of the same Institution from 1837—1841, all included in Vol. I. The majority are not of a valuable character and possess no theoretical interest.

[840.] O. F. Mossotti. *On the Forces which regulate the Internal Constitution of Bodies.* This memoir appears to have been printed at Turin in 1836 under the title: *Sur les forces qui régissent la constitution intérieure des corps, aperçu pour servir à la détermination de la cause et des lois de l'action moléculaire.* Of this work itself I have not seen a copy, but the English translation will be found in *Taylor's Scientific Memoirs,* Vol. I. pp. 448—469, 1837. It is there merely described as: 'From a Memoir addressed to M. Plana, published separately.' It is communicated by Faraday.

[841.] The memoir presents some interesting mathematical analysis, though its physical results would probably not at the present time be thought of a very valuable order. Starting from the hypothesis of Franklin—who explained statical electricity by supposing that the molecules of bodies are surrounded by a quantity of fluid or aether, the atoms of which, while they repel each other, are attracted by the molecules—Mossotti proposes the following mathematical problem:

If several material molecules, which mutually repel each other, are plunged into an elastic fluid, the atoms of which also mutually repel each other, but are at the same time attracted by the material molecules, and if these attractive and repulsive forces are all directly as the masses and inversely as the square of the distance, it is proposed to determine whether the actions resulting from these forces are sufficient to bring the molecules into equilibrium and keep them fixed in that state. (p. 452.)

[842.] The question is whether the molecular action which arises will explain cohesion. The analysis is interesting and involves some general applications of spherical-harmonic analysis. Mossotti afterwards proceeds to narrow his results by supposing: (1) his molecules are uniform and spherical, (2) of small volume and at mutual distances considerable as compared with their dimensions. Finally, the force between two molecules is found to be in the line joining their centres and to be represented by the function:

$$gv(\omega+q)\,v_1(\omega_1+q_1)\frac{(1+\alpha r)\,e^{-\alpha r}}{r^2}-(g-\gamma)\,\frac{v\omega\,.\,v_1\omega_1}{r^2},$$

where

$r$ is the molecular distance.

$g$, the accelerative force of attraction between atoms of aether and the matter of the molecules at unit distance.

$\gamma$, the accelerative force of repulsion between the matter of the molecules at unit distance.

$\omega$, $\omega_1$, the two molecular densities.

$v$, $v_1$, the two molecular volumes.

$q$, $q_1$, the densities of the aether at or in the immediate neighbourhood of the molecules.

$\alpha$, a constant depending on the nature of the atoms of aether and probably very great.

Mossotti remarks on this result that $g-\gamma$ is to be supposed small compared with $g$. Hence when $r$ is small the first term is all-important, but when it is great the second or gravitating term. There will be a position of equilibrium obtained by equating this force to zero, a position which will be found to be stable.

These molecules present a picture in which the *hooked atoms* of Epicurus are as it were generated by the love and hatred of the two different matters of Empedocles (p. 467).

With regard to Mossotti's repulsive force we must remark on the similarity in its kind to that suggested by Poisson: see Art. 439. But his theory leaves cohesion to be explained by an attractive force which is even *less* than the ordinary force of gravitation, and therefore still more subject to the objections of Belli: see Arts. 752—758.

[843.] The memoir is followed by an editorial note citing a passage in Roget's *Treatise on Electricity* noticed by Faraday in a Royal Institution lecture with reference to Mossotti's views. Although the view here propounded would hardly find acceptance today, it may be interesting to reproduce it for historical purposes:

It is a great though a common error to imagine, that the condition assumed by Aepinus, namely that the particles of matter when devoid of electricity repel one another, is in opposition to the law of universal gravitation established by the researches of Newton; for this law applies, in every instance to which inquiry has extended, to matter in its ordinary state, that is combined with a certain proportion of electric fluid. By supposing indeed, that the mutual repulsive action between the particles of matter is, by a very small quantity, less than that between particles of the electric fluid, a small balance would be left in favour of the attraction of neutral bodies for one another which might constitute the very force which operates under the name of gravitation; and thus both classes of phenomena may be included in the same law.

[844.] Amedeo Avogadro. *Fisica de' corpi ponderabili ossia Trattato della costituzione generale de' corpi.* Turin, 1837. Vol. I. Pp. 1—331 are more or less concerned with our subject. There is not much original work in these pages, which present however a very fair *resumé* of the labours of Poisson, Biot, Cauchy, Savart, Weber, etc. In the first chapter we find a general discussion of molecular force. The function suggested by Poisson is mentioned (see our Art. 439) and as usual the repulsive force attributed to the existence of the fluid *caloric*: see our Arts. 543, 597, 701 (footnote). At the same time the then novel view of Ampère[1], that the quantity of caloric might be replaced by the *vis-viva* of the molecular vibrations, is considered (p. 15). The second chapter, which concludes the first book, is occupied with densities and is based chiefly upon Biot.

[845.] The second book is divided into two sections, the first treating of molecular forces and the second of crystallisation. In the physical part of the first section we have a consideration of the important experimental results of Baudrimont, Dufour, Gerstner, Tredgold, Vicat, Weber, Lagerhjelm, etc., all of which will be found referred to in our present work. The theoretical part chiefly follows Poisson, giving however Coulomb's theory of torsion with the experimental work of Bevan, Savart and Ritchie (see Arts. 378, 333, 408). We may note however that this is the *first text-book* which contains the consideration of the general

[1] See our Art. 738.

equations for the equilibrium of an elastic solid (pp. 159—182). Poisson's singular result of 1829 (see Art. 556) and Mossotti's of 1836 (see Art. 842) are also reproduced. It may not be without interest to quote Avogadro's opinion of Mossotti's results:

> In generale le diverse supposizioni su cui il sig. Mossotti ha fondata la sua analisi, dedotte da considerazioni relative alle attrazioni e ripulsioni dei corpi elettrizzati, nell' ipotesi di un solo fluido elettrico, condensato o rarefatto alla loro superficie, possono incontrare difficoltà ad essere ammesse, come realizzate nelle forze molecolari; ma il lavoro del sig. Mossotti ci presenta un primo saggio di applicazioni del calcolo a questioni di questo genere, che potrebbe servir d' esempio a calcoli fondati sopra altre ipotesi che si credessero più conformi alla natura di queste forze (p. 203).

[846.] In the part that follows, on the vibrations of bodies, we have the results of Chladni, Weber, Savart, Strehlke, Wheatstone, Plana, Fourier and Poisson all analysed. Chapter IV. of this section (pp. 304—331) is occupied with some discussion of the relations between the distance and size of the molecules of solid bodies, and with their density. It is chemical rather than physical in character. Avogadro here draws attention to his own researches. In the second part of this work, on crystallisation, the only portion which really concerns our subject is the article entitled: *Diversità di coesione e di elasticità de' cristalli nelle diverse direzioni* (pp. 745—793). This contains a complete reproduction of F. Neumann's papers and an analysis of Savart's researches on aeolotropic plates: see our Arts. 788—801, 335—345.

[847.] As a model of what a text-book should be it is difficult to conceive anything better than Avogadro's. It represents a complete picture of the state of mathematical and physical knowledge of our subject in 1837. No trace of that divorce between physics and mathematics which is more or less to be found in Poisson (see Art. 568) and Biot (see Art. 181) and in innumerable text-books of to-day can be discovered. It may be read even at the present time with great pleasure, and in my opinion marks an epoch in physico-mathematical text-book writing.

[848.] A. F. W. Brix. *Abhandlung über die Cohäsions- und Elasticitäts-Verhältnisse einiger nach ihren Dimensionen beim Bau der Hängebrücken in Anwendung kommenden Eisendräthe des In- und Auslandes. Berlin*, 1837, pp. 1—36 text, 39—118 experimental results.

[849.] This work on the elasticity and strength of iron wires describes one out of the many series of experiments made in the first half of this century to test the material used in suspension bridges: see Arts. 692, 721, 723, 817, 936. As Lagerhjelm had compared Swedish and English iron so Brix, at the instance of the *königlich technischen Gewerbe-Deputation*, instituted a comparison of French and German iron. His statement of general results has considerable interest as confirming previous discoveries or suggesting new facts. The testing machine employed was a non-hydraulic lever machine, a modified form of one used by Lagerhjelm; it is described on pp. 1—6 and very fully figured on two plates at the end of the work. The experiments seem to have been made with considerable care, and a time-element amounting in some cases to twelve hours was allowed for the loads to produce a full effect.

[850.] We may note the following results:

(i) In all but one experiment (No. 49 in which set began with the least load) the iron wires seem to have been in a state of ease, i.e. in a condition of perfect elasticity, at the commencement of the experiment. Till set began Hooke's Law was found to hold and there is very striking agreement obtained for the moduli of various sorts of iron wire, even after the process of annealing. This agrees with the experiments of Coulomb, Tredgold and Lagerhjelm. The method of calculating the modulus (p. 22) is perhaps not altogether satisfactory.

[851.] A singular historical mistake occurs on p. 17, where the discovery of the proportionality of stress and strain is attributed to Mariotte and Leibniz, while the experimental confirmation alone of this relation is ascribed to Hooke.

[852.] (ii) The strain after the beginning of set is found to consist of two parts: (*a*) a purely elastic part which follows Hooke's Law,—Brix thus confirms the important discovery of Coulomb and Gerstner: see our footnote p. 441; (*b*) a set for which the author can discover no law (p. 19). He finds his experiments confirm no known theory, nor do they satisfy an empirical formula which expresses the load in ascending powers of the stretch (p. 29). Still less do they confirm Lagerhjelm's empirical relation

$$C\sqrt{\Delta} = \text{a constant}.$$

See our Art. 365.

[853.] (iii) As a result of (ii) Brix objects to the beginning of set being termed the 'limit of elasticity.' He is doubtful from his own and Lagerhjelm's experiments whether there is any limit of elasticity short of the limit of cohesion. He accordingly terms the usual limit of elasticity the beginning of set (*Anfangsgrenze der Verschiebbarkeit*) (pp. 19—20).

[854.] (iv) If a wire has once received a load which produces a certain amount of set, then that load being removed and the wire gradually reloaded no fresh set is produced till the first load is again reached (p. 20). Brix opposes the view originally propounded by Young and Coulomb, and afterward made universal by the sanction of Tredgold, that a load which produces set is generally capable of bringing about rupture if only maintained long enough, or, in Tredgold's words, that the limit of absolute strength is the beginning of set. Brix holds that his experiments confirm Lagerhjelm's and prove that a given load will only produce a definite amount of set however long maintained. At the same time he admits that Tredgold's rule gives the best limit for practical purposes (p. 34).

[855.] (v) The *yield-point* seems clearly marked in the tabulated experimental results, and yet the writer has evidently not discovered its existence. Thus he remarks that it is incontrovertible that the set must be a definite function of the load, the length and the sectional area of the wire, but his experiments have not led him to ascertain its form' (p. 29). He does not then recognise that the set may at all events at first, depending largely on the working to which the wire has been subjected, be very irregular and only become such a function as he imagines after the yield-point has been reached.

[856.] On the other hand he notes (p. 35) how, especially in annealed wires, the set received suddenly a great increase before it attained that maximum which could balance the applied load:

Hier nahm die Verlängerung in der Regel so bedeutend zu, dass man häufig jeden Augenblick ein Zerreissen des Drathes erwartete; allein plötzlich hörte diese Zunahme auf, das Gleichgewicht zwischen Kraft und Widerstand war eingetreten und der Draht erlitt, obgleich er manchmal zehn bis funfzehn Stunden lang unter derselben Spannung erhalten wurde, fast gar keine nachträgliche Verlängerung mehr.

There can be little doubt that the phenomenon described is that of the yield-point.

[857.] (vi) The 'beginning of set' is, after Tredgold, determined by Brix as the ratio of the first load which produces set to the breaking load. It may be doubted whether this measure is quite satisfactory. Brix's results on this point are in accordance with those of Telford, Brown, de Traitteur and Barbé (pp. 26—28).

[858.] (vii) We may finally note some interesting points with regard to annealed iron wire (pp. 32—34). Namely that the elastic and cohesive properties of such wire are identical with those of bar iron (*Stabeisen*). Brix remarks that the mechanical working of rolled iron produces a change in the elastic conditions of its *surface*. The effect of the hammer in wrought iron in producing a superficial elastic change was first noted by Rondelet[1], confirmed by Lagerhjelm, and theoretically allowed for by Bresse and Saint-Venant[2]. This change extends to a very small depth in wrought iron and probably disappears at the yield-point. Brix notes that a similar result, extending however to a much greater depth, is produced by the process of wire-drawing. A sort of epidermis is formed thicker and firmer than the rest of the wire:

Diese Epidermis trägt, nach der, von den Mitgliedern der französischen Akademie der Wissenschaften, de Prony, Fresnel, Molard und Girard, in ihrem Bericht über die Seguinischen Versuche aufgestellten Ansicht, hauptsächlich mit zu der grossen Festigkeit bei, welche die Eisendräthe im Vergleich mit dem Stabeisen besitzen und da sie bei den dünnern Dräthen verhältnissmässig einen grösseren Theil des

[1] *L'Art de Bâtir*, Paris, 1827, T. I. p. 281.

[2] See his edition of Navier's *Leçons*, p. 20.

Querschnittes einnimmt, als bei den stärkeren, so liegt darin der Grund der bekannten Erfahrung, dass ein Drath desto mehr an Festigkeit gewinnt, je feiner das Kaliber ist, bis zu welchem er gezogen wird. (p. 33.)

Brix points out how his results confirm this statement. The report of the Academicians is attached to Seguin's book, *Des ponts en fil en fer*, Paris, 1824: see our Art. 984. This condition of the epidermis of iron and other wires deserves fuller consideration from those English physicists who investigate the molecular and elastic properties of materials by means of experiment on those materials when in the peculiar 'worked' condition of wire.

[859.] An important memoir by P. H. Blanchet, presented to the French Academy in 1838 and published in Liouville's *Journal*, Vol. v., 1840, will be best considered in conjunction with other memoirs by the same scientist in our Chapter VIII.

860. Mainardi. *Sulle vibrazioni di una sfera elastica; Esercizio del prof. suppl. Gasparo Mainardi.* This memoir is published in the *Annali delle Scienze del Regno Lombardo-Veneto*, Vol. 8, Padova, 1838: it occupies pages 122—133.

The memoir begins by referring to the writings of Poisson and Cauchy on the theory of elastic bodies. As the subject acquires every day fresh interest Mainardi thinks it important to fix the attention on any principle which may require other proofs, or may present exceptions, and for the subject of the present memoir he has selected a special problem already discussed by Poisson: see Art. 449 in the account of Poisson.

861. From the beginning of the memoir to the end of page 127 Mainardi is occupied in establishing a differential equation which coincides with that of my Art. 450; the process is laborious and uninviting, and it involves that mode of expression by integrals which Poisson condemns: see Art. 436.

862. On pages 128—130 we have what may be considered the essence of the memoir. Mainardi in obtaining the general differential equation just noticed confined himself to an internal particle; he now considers the case of a particle *on the surface* of the sphere. We know that the theory of elasticity includes the investigation of the conditions which must hold at the *surface* of a

body; for instance in the problem under consideration we obtain equations like those of Arts. 451, 461; but Mainardi attempts to investigate the matter from the beginning. His process seems to me quite unsatisfactory, and moreover his equations (9) and (13) are not accurate.

863. On his page 130 Mainardi alludes to the memoirs of Belli in 1812 and 1832, which we have noticed in our Arts. 163 and 752. Mainardi seems to desire to shew that the molecular force must vary inversely as the fourth power of the distance. He uses Newton's *Principia*, Lib. I. Prop. 87, Cor. 1; and he combines this with the following statement: Suppose two equal homogeneous cubes placed with a pair of faces in contact, then if the matter always remains the same the force of adherence varies as the square of the length of an edge. For this statement he refers in a note to Rondelet, *Art de bâtir*, T. 1, page 44.

864. Pages 131—133 of the memoir are devoted to the integration of the differential equation obtained on page 127. One example is worked out fully; in this the condition which he obtains with respect to the surface agrees with an equation of our Art. 451, provided we suppose $\lambda = 0$. There is nothing in the integration requiring notice, except the statement on the third line of page 133; this is quite untrue, the asserted identities do not exist.

On the whole the memoir seems to me to add nothing to the discussion which Poisson had already given of the problem to which it refers[1].

[1] We may note the following misprints and errors:

123. last line, for $r$ read $r^2$.

124. line 2, for $\frac{r}{R}$ read $\frac{y}{R}$.

126. last line, for $2\cos\phi$ read $\cos\phi$: this error is repeated.

127. line 2, before $r^2$ supply $\cos\phi$.
Equation (10), read $\sin^4\phi\cos\phi - \sin^2\phi\cos\phi - \frac{2}{3}\cos^3\phi$.

128. Equation (13), for $2\cos\phi\left\{1 + \frac{2y}{R} + y'\right.$ read $\cos\phi\left\{\frac{2y}{R} + y'\right.$.
Equation (16), for $-\left(2 + \frac{y}{R} + y'\right)$ read $-\frac{1}{2}\left(1 + \frac{y}{R} + y'\right)$.

129. Equation (17), for $\pi + d\int$ read $-\pi\delta\int$.

132. line 10, for (22) read (23).

865. 1837. Möbius. A work in two volumes octavo was published at Leipsic in 1837, entitled *Lehrbuch der Statik.* Pages 246—313 of the second volume form a chapter entitled *Vom Gleichgewichte an elastischen Fäden;* there is however nothing here which strictly belongs to our subject. After a few pages relating to the case of a flexible extensible thread Möbius proceeds to that of an elastic rod or curve. But there is no intelligible account of the elastic forces; the equations are deduced by a vague process of analogy from the investigation previously given in the volume respecting the equilibrium of a flexible thread. The main results obtained by Möbius are the equations on his page 281, of which a particular case is given on his page 269; these agree practically with those in Poisson's *Traité de Mécanique,* Vol. II. pages 626 and 630; but Poisson's method is far superior to that of Möbius.

[866.] Adam Burg. *Ueber die Stärke und Festigkeit der Materialien: Jahrbücher des k. k. polytechnischen Institutes in Wien,* Bd. 19, Wien, 1837, pp. 41—93. This is merely a *résumé* of other people's work, notably Lagerhjelm's. It does not appear to be of original value even in its criticisms.

A continuation of it appears on pp. 183—300 of Bd. 20 of the same periodical. It contains a certain amount of historical information and analytical calculation of the formulae for flexure. Much of the matter here considered is reproduced in the *Compendium* of the same writer, which although of a later date we may refer to in the next Article.

Earlier papers by Burg merely giving a German record of English experimental investigations will be found in Bd. 5 (pp. (215—329) and Bd. 17 (pp. 45—111).

[867.] *Compendium der populären Mechanik und Maschinenlehre,* Wien, 1849 (2nd edition). This book contains a chapter entitled: *Von der Festigkeit der Materialien,* Vol. I. pp. 215—244. It is of no value and hardly level with the knowledge of its day: see for example the section on torsion, page 238.

A *Supplement-Band zum Compendium* appeared in 1850. It contains a section on elasticity, pp. 97—139; this is devoted to the mathematical deduction of the formulae cited in the *Compendium.* It does not appear to present any novelty.

868. *Mémoire sur le calcul des actions moléculairès développées par les changements de température dans les corps solides;* par J. M. C. Duhamel. This memoir is published in the *Mémoires...par divers savans,* Vol. v., 1838: it occupies pages 440—498.

The object of the memoir is very simple. Duhamel assumes the theory of elasticity as given in Poisson's memoir of April, 1828, and investigates the modifications which the formulae must undergo when we allow for change of temperature.

869. The function denoted by $f(r)$ in our Arts. 296 and 305 will now involve the temperature; thus if at the point considered the temperature is increased by $q$ the original value of $f(r)$ receives an increment equal to the product of $q$ into the differential coefficient of $f(r)$ with respect to the temperature. Thus when the summations are effected as in the Articles cited we obtain a term, which we may denote by $-\beta q$, to be added to the expression for the stress across a given plane, or on a particle in a given direction. Thus we must add to the three tractions $\widehat{xx}$, $\widehat{yy}$, $\widehat{zz}$ a term $-\beta q$. This is a brief sketch of what Duhamel effects in the section entitled: *Recherche des équations générales,* which occupies his pages 445—457.

Duhamel refers to Poisson's memoir for what he requires from the ordinary theory of elasticity, and would not be easily intelligible alone; he does not follow Poisson quite closely, but ultimately the two agree. The pages 448—450 of Duhamel refer to pages 372—376 of Poisson. On Duhamel's page 450 he quotes a formula from Poisson with $1/30$ instead of $2\pi/15$ for coefficient; I do not understand this. The $K$ of Duhamel is the $k$ of Poisson.

870. Duhamel's formulae involve two constants, namely one coefficient of elasticity, and the quantity $\beta$; he shews on his page 462 how these are to be determined by experiment. On his page 463 he remarks:

> ...M. Poisson a prouvé, par un calcul très-simple, que, si l'on considère un cylindre à base quelconque, dont les bases soient soumises à une augmentation donnée de tension, tandis que la surface reste soumise à la pression primitive, la dilatation dans le sens de la longueur est double de ce qu'elle serait si la même tension était appliquée à la surface entière.

I do not know where Poisson distinctly states this, though it may be deduced from pages 402 and 451 of his memoir. We may illustrate this from Lamé's *Leçons*, p. 114, where we find for the dilatation in a stretched prism $\delta' = \frac{(\lambda+\mu)}{\mu(3\lambda+2\mu)}$; but if we wish to have a uniform tension over the whole surface we have $\delta = \frac{1}{3\lambda+2\mu}$: thus, taking $\lambda = \mu$, the former value $\delta'$ is double the latter value $\delta$.

871. Duhamel discusses on his pages 469—476 the following example: To investigate the equilibrium of a hollow sphere when the temperature is expressed by a given function of the distance from the centre. The formulae obtained include as a particular case those given by Poisson for the case when no regard is paid to the temperature: see my account of Poisson, 1828. Next Duhamel on his pages 476—479 supposes the hollow sphere to be composed of two different substances. Finally on his pages 479—485 he considers the equilibrium of an indefinite cylindrical tube. In all these cases the introduction of the term depending on the temperature does not greatly complicate the problem.

872. On his pages 486—498 Duhamel discusses the problem of the vibrations of a sphere the temperature of which is variable. The temperature is supposed to be initially an arbitrary function of the distance from the centre, and to be determined at any subsequent epoch by the formulae given in the theory of the conduction of heat; the problem thus becomes much more elaborate than those which have been previously discussed in the memoir. It is worked out completely and forms a very interesting piece of analysis.

873. On his page 493 Duhamel arrives at the equation

$$\tan pR = \frac{4pR}{4 - 3p^2 R^2},$$

and he adds:

Cette équation s'était déjà présentée à M. Poisson, qui en a déterminé par approximation les premières racines dans un de ses mémoires sur les corps élastiques.

This is a very vague manner of giving the reference; Duhamel should say precisely: See pages 409 and 420 of the memoir of April, 1828[1].

[874.] The difference between Duhamel's first and second memoirs (see Art. 877) may be briefly described as follows. In the first memoir he obtains (pp. 455—456) the body and surface shift-equations (2) and (3) of our Art. 883, but he does not seem to have recognised the necessity of the equation (1) connecting the flow of heat with the dilatation. That is to say, he practically assumes the equality of the two specific heats, and supposes the flow of heat given by the ordinary Fourier form. Thus, when treating of the vibrations of a solid sphere, he assumes for the temperature $q$ the form (p. 487):

$$q = \Sigma A e^{-Kn^2t/(c_v\rho)} \cdot \frac{\sin nr}{r}.$$

He does not seem clearly to recognise the assumption he is really making as to the equality of the specific heats.

[875.] A point made by Duhamel on pp. 467—469 deserves to be noted in addition to what has been said above. Examining equations (2) of Art. 883, he notes that the effect of change of temperature in the interior of a body is the same as if a body-force with components

$$-\frac{\beta}{\rho}\frac{dq}{dx}, \quad -\frac{\beta}{\rho}\frac{dq}{dy}, \quad -\frac{\beta}{\rho}\frac{dq}{dz}$$

were added to the force-system, or, as he remarks, a force at each

[1] We may note the following misprints, etc.

447. There is no reason for putting $-c_1$, instead of $c_1$, in the three formulae near the bottom of the page.

470. Last line, for $\frac{dv}{\rho}$ put $\frac{dv}{d\rho}$.

478. If $R=0$ it is best to work out independently from the beginning; we find $\phi=$constant. Then $\phi$ and the pressure must be equal at the common surface; and the pressure must have a given value at the outside surface of the sphere.

480. Last line, for $9\rho$ read $3\rho$.

481. The value of $C$ is incorrect; we must put $R'^2$ to $(p-p')$, and multiply by $R^2$ see the next page.

487. Line 9, for $\frac{\tan nR}{nR}$ read $\frac{nR}{\tan nR}$.

494. Line 2 from bottom, for 9 read 0.

point having the same direction as, and proportional to the maximum flow of heat at the point.

Further, regarding the surface shift-equations (3) of Art. 883, Duhamel adds, that the effect of temperature is the same as if a tractive load equal to $\beta q$ or proportional to the temperature were added to the load at each point of the surface.

[876.] Duhamel practically *assumes* that the constant $\beta$ of Art. 869 will not be a function of the body-force or the load, and that a body will return to its primitive shape when stress and temperature have again their primitive values.

A very instructive paper by Pictet of a much earlier date, which is primarily of thermal importance, deserves at least a brief notice from this stand-point. It is entitled: *Sur les variations que peut éprouver dans sa longueur une barre de fer soumise à l'action de diverses forces* (*Bibliothèque universelle de Genève, Sciences et Arts,* Tom. I., 1816, pp. 171—200); and is an extract from a memoir read by Pictet before the *Société de physique et d'histoire naturelle de Genève* in 1806. Pictet had been induced to publish it, by an account of some experiments due to Biot on the extension of metals by heat. Biot had come to the conclusion that a body heated from the freezing to the boiling point and then again cooled to the freezing point takes its primitive dimensions (see p. 100 of the same volume of the *Bibliothèque*). Pictet however by experimenting on a bar of iron had long previously come to different conclusions. We cite his results, which shew how near he was to the discovery of elastic after-strain (p. 199).

(1) Que le fer exposé à des changemens brusques dans sa température, ne reprend pas sa dimension exacte, même au bout de plusieurs jours, par le retour lent à la température primitive.

(2) Qu'une pression extérieure modifie sensiblement les effets, soit de la force dilatante du feu, soit de la cohésion des molécules du métal, lorsque cette pression conspire avec l'une de ces deux forces, ou lorsqu'elle lui est opposée.

(3) Qu'abstraction faite de tout changement dans la température, la pression produit sur une barre de fer, dans le sens de sa longueur, un refoulement, dont une partie disparaît quand la pression cesse, et

dont une autre partie demeure permanente; au moins pendant un certain temps.

The second result seems to have considerable bearing on Duhamel's thermo-elastic theory, for it would make the coefficient $\beta$ a function of the load: see Art. 869.

[877.] J. M. C. Duhamel. *Second mémoire sur les phénomènes Thermo-mécaniques: Journal de l'École Polytechnique,* 25[e] *Cahier* (Tome XV.), 1837; pages 1—57. This memoir was read to the Academy of Sciences on February 23, 1835. It appears to be *second* to the memoir published in 1838 and considered in our Art. 868. The memoir is an important one.

[878.] Duhamel begins with the remark that Poisson had recognised the existence of thermal variations in the elastic constants, but thought that, except for great differences of temperature and where extreme accuracy is necessary, it was better to neglect them: an additional argument with Poisson being the want of sufficient experiments to determine the laws of variation. Duhamel then continues:

Je pense aussi, comme tous les géomètres qui ont traité le même sujet, que l'on peut, dans des limites assez étendues, considérer tous les coefficiens spécifiques comme constans; mais je me propose ici d'avoir égard à une circonstance que l'on a négligée, et dont l'influence est certainement beaucoup plus sensible. On admet généralement que tous les corps dégagent de la chaleur quand on les comprime, et en absorbent quand on les dilate; d'où il résulte qu'il y a une différence sensible entre les chaleurs spécifiques à volume constant et à pression constante. C'est ce principe qui sert de base à ma théorie; et j'admets que la quantité de chaleur dégagée est proportionnelle à l'accroissement qu'a subi la densité, pourvu que cet accroissement soit très-petit (pp. 1—2).

[879.] Duhamel seems to be ignorant of Weber's results (see Art. 705), and remarks with regard to the ratio of the specific heats:

Aucune tentative n'a encore été faite pour déterminer sa valeur dans les substances solides; mais on verra que ma théorie donne plusieurs moyens d'y parvenir (p. 2).

[880.] The introductory pages of the memoir are interesting historically and physically, and we quote pages 2—4.

Examinons d'abord si la théorie de la propagation de la chaleur dans les solides est complétement en harmonie avec ce fait incontestable d'une différence entre les deux chaleurs spécifiques d'une même substance. Fourier et tous les géomètres qui se sont occupés après lui de la théorie mathématique de la chaleur, ont supposé dans leur calcul que les molécules conservaient les mêmes distances respectives, malgré les variations de la température. Or, en regardant cette hypothèse comme admissible, il aurait fallu considérer la chaleur spécifique à volume constant, et non celle que les expériences des physiciens avaient fait connaître, et qui se rapportait à une pression constante. Mais il est facile de voir qu'après avoir fait cette première correction, il en reste à faire une seconde relative au changement de densité.

En effet, on peut supposer d'abord que l'on calcule l accroissement de température que subirait un élément infiniment petit du solide, dans un temps très-court, pendant lequel on maintiendrait tous les points dans les mêmes positions; mais ensuite il faut laisser prendre à ces points les positions relatives au nouvel équilibre mécanique et ajouter à la température déjà calculée celle qui résulte du changement de densité qu'a subi l'élément que l'on considère. Peut-être pensait-on qu'en laissant s'opérer les dilatations ou contractions intérieures, il était exact de prendre la chaleur spécifique à pression constante: mais cette opinion ne saurait être admise aujourd'hui. La théorie que j'ai donnée des effets mécaniques de la chaleur dans les corps solides, montre que pendant les changemens de température, la pression change à chaque instant au même point; que par conséquent la dilatation d'une particule du corps ne s'exécute pas comme si elle était isolée du reste et qu'il peut même arriver qu'il y ait contraction en même temps qu'élévation de température. Il était donc nécessaire de faire subir une modification à l'équation de la propagation de la chaleur; et cette modification consiste à substituer d'abord la chaleur spécifique à volume constant à celle relative à la pression constante; puis à ajouter l'effet produit par le changement de la densité. Quant à l'équation relative à la surface, elle ne subit aucun changement.

La théorie de la propagation de la chaleur se trouve ainsi dépendante de la théorie mécanique qui détermine les changemens de position qu'entraîne l'équilibre intérieur d'un corps inégalement échauffé; et réciproquement la seconde théorie dépend de la première, de sorte qu'aucune d'elles ne peut être traitée séparément. J'ai fait connaître dans un autre mémoire les équations générales qui déterminent l'équilibre ou le mouvement des différens points d'un corps d'après son état thermométrique (see our Art. 868). Ces équations ne subiront aucune modification; mais l'état thermométrique ne pourra plus être déterminé séparément; à moins que ce ne soit dans une première approximation, dont l'exactitude ne saurait encore être bien appréciée.

Ainsi se trouvent liées intimement les deux grandes théories physiques qui depuis quelques années occupent le plus l'attention des géomètres. Les changemens de température pourront encore dans bien des cas se déterminer par les équations de Fourier; par exemple, toutes les fois que les températures, quoique inégales, seront invariables. De même aussi l'équilibre et le mouvement des corps élastiques seront déterminés par les équations de M. Navier, toutes les fois que la température sera invariable dans toute l'étendue de ces corps. Dans tout autre cas ce sont nos équations qu'il faudra employer.

Et même, lorsque l'on considère le mouvement vibratoire des molécules d'un corps dont la température est partout la même, les équations de M. Navier exigent une modification que ce savant géomètre a bien prévue, et dont il a parlé dans son rapport sur la théorie que j'ai donnée des effets mécaniques de la chaleur dans les corps solides. Cette modification tient à la chaleur développée ou absorbée dans les changemens de densité qui peuvent accompagner les vibrations; elle ne pouvait être calculée que par la théorie que je viens de rappeler. Je traite cette question dans ce mémoire et je fais voir comment la vitesse de propagation du son se trouve altérée par ce développement de chaleur. Mais il y a ici une observation à faire. M. Poisson dans un de ses derniers mémoires, a démontré l'existence de deux espèces d'ondes sphériques dans les milieux solides dont l'élasticité est la même dans tous les sens. Dans l'une, la densité est la même que dans l'état primitif; dans l'autre elle est différente (see our Arts. 526 and 565). Or, il est évident que dans le premier cas les équations de M. Navier subsistent, puisqu'il ne peut y avoir changement de température: c'est donc aux ondes de la seconde espèce que se rapportera la modification dont il est question.

(This refers to Duhamel's hypothesis stated above: see our Art. 874.)

La formule que j'ai fait connaître à cet effet, est très différente de celle que Laplace a donnée pour les gaz. En la comparant à la vitesse de propagation déduite de l'expérience, on déterminera le rapport des deux chaleurs spécifiques pour tous les solides; et comme on connaît déjà la chaleur spécifique à pression constante, on en conclura la seconde, et l'on possédera toutes les données nécessaires au calcul des phénomènes que la chaleur et l'attraction moléculaire peuvent produire dans les solides homogènes (p. 4).

[881.] Besides the problems on vibrating bodies, Duhamel in this memoir treats other questions relative to the heat liberated when solids are suddenly compressed. With respect to these cases he makes the following hypothesis: *that each infinitely small particle of the solid takes instantly all the heat which the compression which it undergoes can give it.* Duhamel remarks that this hypothesis has been admitted by all physicists who have treated of the like case in a gas, and it ought to be recognised as having even a better foundation in the case of solids, owing to the slowness with which they conduct heat (p. 5).

[882.] With regard to the experimental verification of his results Duhamel remarks:

J'avais tenté quelques applications numériques de mes formules, en partant des expériences de Chladni sur la vitesse de propagation du son dans les solides, et j'en avais déduit le rapport des deux chaleurs spécifiques de diverses substances. Mais les résultats que j'ai obtenus dans certains cas, m'ont porté à croire que la compression des solides n'avait pas été déterminée par les physiciens avec toute l'exactitude qu'exige ce genre de recherches, et je me suis abstenu de les mentionner dans ce mémoire (p. 6).

[883.] We will now give some account of the analysis and results of the memoir.

Duhamel first determines the general body and surface thermo-elastic equations (pp. 8—12). We reproduce these equations in somewhat different notation.

Let $u$, $v$, $w$ and $\theta$ be the shifts and the dilatation at the point $x$, $y$, $z$ at time $t$; $q$ the temperature at the same point; $k$ the thermal conductivity of the body; $\rho$ its density; $c_p$ and $c_v$ the specific heats for constant pressure and constant volume respectively, and $\gamma = c_p/c_v$; $\delta$ the dilatation (cubical) produced by an elevation of temperature equal to unity, $\delta'$ the dilatation produced by unit tractive load exercised over the entire surface of the body, and $\beta = \delta/\delta'$.

Then, since on Poisson's theory of uni-constant isotropy $\mu = 3/(5\delta')$, we have for the thermo-elastic body shift-equations of an isotropic solid the following:

$$\frac{dq}{dt} = \frac{k}{c_v \rho} \nabla^2 q - \frac{\gamma - 1}{\delta} \frac{d\theta}{dt} \quad \text{.................(1)},$$

$$\left.\begin{aligned} \rho\left(\frac{d^2u}{dt^2} - X\right) &= \mu\nabla^2 u + 2\mu\frac{d\theta}{dx} - \beta\frac{dq}{dx} \\ \rho\left(\frac{d^2v}{dt^2} - Y\right) &= \mu\nabla^2 v + 2\mu\frac{d\theta}{dy} - \beta\frac{dq}{dy} \\ \rho\left(\frac{d^2w}{dt^2} - Z\right) &= \mu\nabla^2 w + 2\mu\frac{d\theta}{dz} - \beta\frac{dq}{dz} \end{aligned}\right\} \quad \text{.......... (2)}.$$

And the following thermo-elastic surface shift-equations:

$$\left.\begin{aligned} X' &= \left(2\mu\frac{du}{dx} + \mu\theta\right)\cos l + \mu\left(\frac{du}{dy} + \frac{dv}{dx}\right)\cos m \\ &\qquad + \mu\left(\frac{du}{dz} + \frac{dw}{dx}\right)\cos n - \beta q \cos l \\ Y' &= \mu\left(\frac{dv}{dx} + \frac{du}{dy}\right)\cos l + \left(2\mu\frac{dv}{dy} + \mu\theta\right)\cos m \\ &\qquad + \mu\left(\frac{dv}{dz} + \frac{dw}{dy}\right)\cos n - \beta q \cos m \\ Z' &= \mu\left(\frac{du}{dz} + \frac{dw}{dx}\right)\cos l + \mu\left(\frac{dw}{dy} + \frac{dv}{dz}\right)\cos m \\ &\qquad + \left(2\mu\frac{dw}{dz} + \mu\theta\right)\cos n - \beta q \cos n \end{aligned}\right\} \quad \text{...(3)}.$$

Here $X$, $Y$, $Z$ are body-forces at $x, y, z$, and $X'$, $Y'$, $Z'$ the load-components at the element of surface whose direction is given by the angles $l$, $m$, $n$.

[884.] Of these equations (2) and (3) are given by Duhamel in his earlier memoir and there obtained in the manner we have described in Art. 869. It is obvious that if we write $\theta' = \theta - \beta/2\mu \,.\, q$, we shall obtain equations of the same form as in the ordinary theory of elasticity; $\theta'$ will however no longer have the value

$$\frac{du}{dx} + \frac{dv}{dy} + \frac{dw}{dz}.$$

[885.] Equation (1) is deduced by two separate methods on pp. 9—10; we reproduce one method of obtaining it.

If the body did not expand owing to the change of temperature the increase of temperature at $x, y, z$ in time $dt$ would by Fourier's method be equal to

$$\frac{k}{c_v \rho} \nabla^2 q \, dt.$$

But owing to the increase in volume the temperature has not so great an increase. Let $\chi$ be the increase of temperature which would in general result from a small increase of the density due to the diminution $\epsilon$ in unit volume. Now if this unit volume cooled till it became $1-\epsilon$ under a constant pressure, the temperature would decrease by $\epsilon/\delta$ and the quantity of heat given off would be $c_p \epsilon/\delta$. But this last quantity is exactly that, which restored to the constant volume $1-\epsilon$, would raise it to the primitive temperature increased by $\chi$, or actually raise its temperature by $\epsilon/\delta + \chi$; therefore this amount of heat can also be represented by $c_v (\epsilon/\delta + \chi)$.

Hence $$c_p \epsilon/\delta = c_v (\epsilon/\delta + \chi),$$

or, $$\chi = \frac{c_p - c_v}{c_v} \epsilon/\delta = (\gamma - 1) \epsilon/\delta.$$

Now in our case the density *diminishes* or $\chi$ is negative for

$$\epsilon = -\frac{d\theta}{dt} dt.$$

Thus the total gain in temperature is

$$\frac{dq}{dt} dt = \frac{k}{c_v \rho} \nabla^2 q \, dt - \frac{\gamma - 1}{\delta} \frac{d\theta}{dt} dt,$$

or, $$\frac{dq}{dt} = \frac{k}{c_v \rho} \nabla^2 q - \frac{\gamma - 1}{\delta} \frac{d\theta}{dt},$$

which is the required equation (1).

[886.] The first problem considered by Duhamel is entitled: *Propagation du mouvement dans un solide indéfini* (pp. 12—19).

He begins with the following remarks:

La superposition des effets, qui s'observe dans cette théorie, permet de supposer le solide à une température uniforme, et libre de toutes forces étrangères. Les résultats de ce calcul devraient être augmentes de ceux que ces forces et le changement des températures primitives auraient produits; mais lorsqu'il s'agira d'un temps très court, comme dans le cas où l'on considère la propagation du mouvement, il sera inutile de faire cette correction.

Cela posé, nous concevons un solide homogène indéfini, dont tous les points ont la même température et ne sont soumis qu'à leur action mutuelle; une petite partie de ce solide est dérangée de l'état d'équilibre; les déplacemens et les vitesses de ses points sont donnés; il s'agit de déterminer la vitesse avec laquelle le mouvement se propagera. La chaleur développée par la compression ne se répandra dans le solide qu'avec une grande lenteur, et par conséquent, n'influera pas par sa diffusion sur la vitesse avec laquelle le mouvement gagnera les points primitivement en repos: elle n'influera que par les forces qu'elle produira dans les parties où il y aura changement de densité, et lorsque l'on considère le phénomène pendant un temps quelconque, c'est encore là la seule influence de la chaleur dégagée, lorsqu'au même point les condensations et dilatations successives sont sensiblement égales comme dans les vibrations sonores. Dans ces divers cas, les molécules conservent toujours la même quantité de chaleur sans pour cela conserver la même température; et les choses se passent comme si la conductibilité de la substance était nulle.

This last hypothesis amounts to putting $k=0$, and we then find:

$$\frac{dq}{dt} = -\frac{\gamma-1}{\delta}\frac{d\theta}{dt},$$

or, taking the initial temperature as zero, we have

$$q = -\frac{\gamma-1}{\delta}\theta.$$

Using this, equations (2) give us by differentiating respectively with regard to $x, y, z$ and adding

$$\frac{d^2q}{dt^2} = \frac{4+5\gamma}{3\rho}\mu\nabla^2 q.$$

[887.] Duhamel then solves this equation by Poisson's method. He remarks that the velocity of the wave motion

$$=\sqrt{\frac{(4+5\gamma)\mu}{3\rho}}$$

This value is very different from $\sqrt{\mu/\rho}$, the velocity of that form of sound wave where there is no dilatation. It also differs from Poisson's form (see Art. 526), as he supposes $\gamma = 1$, and obtains $\sqrt{3\mu/\rho}$. The above form is obtained from Poisson's value by multiplying by $\frac{1}{3}\sqrt{4+5\gamma}$, and not by $\sqrt{\gamma}$ as in the case of gases or liquids. This suggests a means of calculating from sound experiments the ratio $c_p/c_v$; a means adopted with a curious error by Wertheim in a memoir of 1844. The error was first pointed out by Clausius: see our Chapter VIII.

[888.] The next problem considered by Duhamel is entitled: *Équilibre d'un corps de figure quelconque, subitement comprimé* (pp. 19—20).

As before, $q = -\frac{\gamma - 1}{\delta}\theta$, and this value must be substituted in equations (2) and (3), the left-hand side of those equations being suppressed. They will then be satisfied if we take

$$u = Ax, \qquad v = By, \qquad w = Cz,$$

and we easily deduce, $p$ being the uniform tractive load,

$$A = B = C = -\frac{p}{5\mu\gamma},$$

or,
$$u = -\frac{px}{5\mu\gamma}, \qquad v = -\frac{py}{5\mu\gamma}, \qquad w = -\frac{pz}{5\mu\gamma},$$

$$q = p\frac{(\gamma - 1)}{\gamma\beta}$$

Hence we conclude:

Whatever may be the form of a body whose surface is suddenly subjected to a uniform tractive load, the stretch will be the same in all directions, and the value of this stretch will be obtained by dividing by the ratio of the two specific heats that value which it would have had if no heat had been liberated.

[889.] The next case treated is an easy and important one, namely that of a wire or bar whose terminal sections are subjected to a sudden uniform tractive load (pp. 20—23).

The body-equations,—here the same as those of the preceding

question,—and the surface-equations, if the traction for unit area of base be $T$, can be satisfied by

$$u = Ax, \qquad v = By, \qquad w = Cz,$$

where

$$A = \frac{T}{15\mu}\left(5 + \frac{1}{\gamma}\right), \qquad B = C = \frac{T}{30\mu}\left(\frac{2}{\gamma} - 5\right),$$

where $x$ is the direction of the axis of bar or wire.

Finally $$q = -\frac{\gamma - 1}{3\gamma\beta} T.$$

Now the stretch of the wire after the equilibrium of temperature is equal to $\frac{2T}{5\mu}$ on uni-constant theory.

Hence the difference between the immediate and final stretches is

$$\frac{T}{15\mu}\left(1 - \frac{1}{\gamma}\right),$$

or, the ratio of this difference to the final stretch is equal to

$$1 - \frac{1}{\gamma} : 6.$$

[890.] This offers another means of determining $\gamma$, from the measurement of the *immediate* and *final* extensions of a bar or wire[1]. This distinction, we must remark, is quite apart from elastic after-strain and does not seem to have been sufficiently regarded in experiments to determine the elastic modulus from the extension of bars, especially in cases where the stress-strain curve is plotted out for a continuous and rapid increase of the terminal tractive load.

[891.] On p. 23 we have the problem of the equilibrium of a spherical shell or hollow sphere subjected to sudden uniform tractive loads $p$ and $p'$ on the interior and exterior surfaces respectively. The analysis is here very simple. It is shewn that the increase of temperature and density is the same at all points, but that the

[1] We may note that using Weber's results in Art. 705 the ratio of the difference between immediate and final elongation of an iron bar to its final elongation amounts to something more than 1/100.

stretch is not the same at all points or in all directions. Duhamel remarks with regard to these results:

Il serait nécessaire d'avoir égard à ces circonstances dans les expériences sur la compressibilité des liquides, si l'on ne s'assurait pas que l'enveloppe est rigoureusement parvenue à la température du liquide environnant (p. 25).

See our Arts. 686—691.

Duhamel on page 25 refers to his former solution of the more general equations for the thermo-elastic condition of a solid sphere compressed and then left to itself.

[892.] We then have a problem which is thus stated:

Considérons un corps de forme quelconque qui soit d'abord à la température zéro, et supposons qu'on y introduise une quantité de chaleur telle que la température de chaque point s'élèverait d'une quantité exprimée par $F(x, y, z)$, si la densité restait partout la même qu'auparavant; admettons de plus que l'équilibre s'établisse instantanément entre les forces moléculaires développées par la chaleur, et proposons-nous de déterminer ce premier état d'équilibre qui sera bientôt altéré sensiblement par la propagation de la chaleur (p. 26).

In this case

$$q = F(x, y, z) - \frac{\gamma - 1}{\delta}\theta,$$

so that the body shift-equations become of the type:

$$\mu\nabla^2 u + \mu\frac{(1 + 5\gamma)}{3}\frac{d\theta}{dx} - \beta\frac{dF}{dx} = 0,$$

and the surface shift-equations of the type:

$$\left(2\mu\frac{du}{dx} + \mu\frac{5\gamma - 2}{3}\theta\right)\cos l + \mu\left(\frac{du}{dy} + \frac{dv}{dx}\right)\cos m + \mu\left(\frac{du}{dz} + \frac{dw}{dx}\right)\cos n - \beta F . \cos l = 0.$$

[893.] A simple case considered by Duhamel is when $F(x, y, z)$ has the constant value $\chi$, we then have the solutions

$$u = v = w = \frac{\beta}{5\mu\gamma}\chi,$$

$$\text{and } q = \frac{\chi}{\gamma}.$$

This result might easily have been foreseen; for, the temperatures $q$ and $\chi$ being produced by the same quantity of heat, but the first supposing the pressure constant, the second the volume constant, ought to be reciprocally proportional to the specific heats; while the linear dilatation ought to have for value $\frac{1}{3}q\delta$, or

$$\frac{1}{3}\frac{\chi\delta}{\gamma}=\frac{\beta}{5\mu\gamma}\chi.$$

[894.] We have next the application of these results to a hollow sphere, the temperature to which each point is raised being a function of the radius only. This follows very simply. Duhamel arrives at the following somewhat remarkable result:

Quel que soit le rapport des deux chaleurs spécifiques de la substance qui compose une sphère creuse; si l'on y introduit une quantité quelconque de chaleur dont la loi de distribution soit représentée par une fonction arbitraire de la distance au centre, l'équilibre s'établira entre les forces développées par cette chaleur, de telle sorte que les deux surfaces extrêmes auront subi la même dilatation que si la chaleur avait été distribuée uniformément. La dilatation ou contraction des autres couches dépendra au contraire de la loi de cette distribution et du rapport des deux chaleurs spécifiques, ou de la chaleur développée par la contraction (p. 30).

[895.] The succeeding problem is more complex and is treated by Duhamel in two modes of which the second seems somewhat the better. The problem is: To determine the rate of cooling of a free sphere taking account of the heat liberated by the contraction. We have to find the temperatures, shifts and stresses after any time has elapsed. Duhamel limits the problem by supposing the initial temperature a function only of the distance from the centre and that the surface is maintained at uniform zero-temperature. The investigation occupies pages 31—48 and has some interesting analysis of the kind which occurs in the treatment of the distorted sphere by Poisson: see Arts. 449—463.

[896.] Pages 49—57 contain a discussion of the same problem with a different surface-condition,—the surface being in this case exposed to the action of a medium of which the temperature is invariable. The analysis used in the solution is like that of the previous case.

Throughout this memoir Duhamel makes frequent references to his earlier memoir, especially to his consideration of the sphere and cylinder.

[897.] A number of further memoirs by Duhamel on subjects more related to the theory of sound than that of elasticity will be found analysed in the volumes of the *Comptes Rendus.* They embrace numerous problems on the motion of strings to which weights are attached, various considerations on the general types of vibratory motion, and a long discussion on the *vibrations produced by the bow of a fiddle*—a point which Duhamel appears to have been the first to treat of[1].

[898.] A last memoir on our subject due to this writer may fitly be considered here, although it belongs to a much later date. It is entitled:

*Mémoire sur le mouvement des différents points d'une barre cylindrique dont la température varie.* It occupies pp. 1—33 of the 36$^e$ *Cahier* (*Tome* XXI.) of the *Journal de l'École Polytechnique,* Paris, 1856.

The object of this memoir is to treat a special case of the earlier memoirs in a simpler manner; it is expressed in the following words:

Il est peu de phénomènes où les corps solides n'éprouvent des changements de température, et ne développent, par suite, des forces dont il est souvent indispensable de tenir compte. J'ai fait connaître, il y a longtemps, les équations générales au moyen desquelles ces effets peuvent être calculés dans les corps élastiques homogènes; et j'en ai fait alors différentes applications. Je me propose de montrer comment, dans certains cas simples, on peut se dispenser de recourir à ces équations compliquées, qui, en donnant à la solution une plus grande exactitude théorique, ne la rendraient pas réellement plus propre à la mesure des effets observables.

[1] An interesting memoir, which deserves at least to be noted if only for a peculiarity in the form of a solution of a differential equation, is that of 1865. It is entitled: *Mouvement d'un fil élastique soumis à l'action d'un courant de fluide animé d'une vitesse constante.* *Comptes Rendus*, Vol. LVI. pp. 277—288.

[899.] The method adopted here is similar to Fourier's treatment of the motion of heat in a bar, where the temperature is assumed to be uniform across a section and thus reduced to a function of one instead of three co-ordinates. Duhamel makes a like supposition for the thermo-elastic state of a cooling bar. He announces his first problem thus:

On donne une barre cylindrique ayant pour section orthogonale une figure quelconque. Ses deux bases sont soumises à des tractions égales, constantes ou variables avec le temps, suivant une loi donnée. Dans l'état initial, les températures varient arbitrairement d'un point à un autre, ainsi que les déplacements et les vitesses. Cette barre est placée dans une enceinte dont la température est invariable et l'on demande le mouvement de chacun de ses points pendant la durée indéfinie du refroidissement, ainsi que l'état final vers lequel il converge (p. 2).

The general equations are:

$$\left.\begin{aligned} \frac{dq}{dt} &= \frac{k}{\rho c}\frac{d^2q}{dx^2} - \frac{pk'}{\omega\rho c}q \\ \rho\frac{d^2u}{dt^2} &= E\frac{d^2u}{dx^2} - \eta\frac{dq}{dx} \end{aligned}\right\} \dots\dots\dots\dots\dots\dots\dots(1),$$

where $q$ is the difference of temperature at any point between the rod and the surrounding medium; $u$ is the displacement at time $t$ of a point which is distant $x$ from the central cross-section of the bar when $q=0$ throughout; $\rho$ is the density; $\omega$ the sectional area and $p$ its perimeter; $m$ is the interior-conductivity and $k'$ the exterior-conductivity (radiation-coefficient); $E$ is the stretch-modulus, and $\eta$ the ratio of the stretch produced by unit elevation of temperature to the stretch produced by unit traction (i.e. to $1/E$).

With regard to $c$, Duhamel merely terms it *the specific heat.* He appears accordingly in this memoir to suppose *the two specific heats equal,* which causes a term involving $u$ in the first of the above equations to disappear. He does not justify this in his memoir, or even note the assumption.

For brevity we write:

$$\frac{k}{\rho c} = a, \quad \frac{pk'}{\omega\rho c} = b, \quad \frac{E}{\rho} = \mu^2, \quad \frac{\eta}{\rho} = \nu.$$

[900.] The first special case considered by Duhamel is that of a bar (length $2l$ at zero temperature) raised to uniform temperature $q_0$ and allowed to cool under the action of no load, initial shifts and velocities being given. Here obviously:

$$q = q_0 e^{-bt}$$

and

$$\frac{d^2u}{dt^2} = \mu^2 \frac{d^2u}{dx^2};$$

while we must satisfy the special relations:

$$\left.\begin{aligned} &u = 0 \text{ for } x = 0, &\quad u = f(x) \\ &E\frac{du}{dx} = \eta q \text{ for } x = l, &\quad \frac{du}{dt} = \phi(x) \end{aligned}\right\} \text{for } t = 0.$$

Assuming $u = Ue^{-bt} + w$, where $U$ and $w$ are periodic functions only, we easily deduce

$$U = q_0 \frac{\mu\eta}{Eb} \cdot \frac{\sinh bx/\mu}{\cosh bl/\mu}.$$

$$w = \sum_0^\infty \left[A_n \sin \frac{(2n+1)\pi\mu t}{2l} + B_n \cos \frac{(2n+1)\pi\mu t}{2l}\right] \sin \frac{(2n+1)\pi x}{2l},$$

and find to determine $A_n$ and $B_n$ the Fourier's series,

$$\sum_0^\infty B_n \sin \frac{(2n+1)\pi x}{2l} = f(x) - U,$$

$$\frac{\pi\mu}{2l} \sum_0^\infty (2n+1) A_n \sin (2n+1) \frac{\pi x}{2l} = \phi(x) + bU.$$

We see that $w$ represents the final condition of the bar, which is obviously the same as if we supposed the initial condition to be one of uniform temperature the same as the surrounding medium but the initial shifts to be given by $f(x) - U$ and the initial shift-velocities by $\phi(x) + bU$.

We may note the case where the final state is one of repose or where the shifts and their velocities are given respectively by $U$ and $-bU$.

Duhamel calculates (p. 10) the values of $A_n$ and $B_n$ when the initial velocities are zero, and the shifts are those due to a uniform permanent temperature $q_0 \left(\text{i.e. } f(x) = \frac{q_0 \eta x}{E}\right)$.

[901.] The next problem is a more general case of the first; namely, the initial temperature $q_0$ is not supposed uniform but equal to $F(x)$; the terminal sections are also subjected to a uniform tractive load $T$ (pp. 11—17). The solution is slightly more complex, although of the same character as in the first problem. The difference lies in the fact that we have now to solve equations (i) with the terminal conditions

$$\left.\begin{aligned} k\frac{dq}{dx} + k'q = 0 \\ E\frac{du}{dx} - \eta q = T \end{aligned}\right\} \text{for } x = l,$$

and a more complex initial state.

The solution of a third problem, that in which the tractive load is a function of the time is indicated, but not effected (pp. 17—18).

[902.] Duhamel then passes to the second section of his paper which is occupied with the cooling of bars whose terminals are fixed to moveable masses (*obstacles mobiles*). In order to simplify the problem, the centre of the bar is supposed to be either fixed or a centre of symmetry, and the resistance of the masses attached to the terminals is supposed to vary in a *given* manner with the terminal displacements; further the temperature of the bar (as in the first problem of the first section) is supposed the same throughout its length.

With regard to the general problem Duhamel remarks:

Le refroidissement des barres métalliques a été employé à produire des efforts considérables, et à opérer des rapprochements entre des corps qui y opposaient de grandes résistances.

Il peut être intéressant, au point de vue mécanique, de calculer la quantité de travail produite ainsi dans des circonstances données. Mais c'est plutôt au point de vue de la théorie que de l'utilité pratique que nous l'envisagerons. Nous poserons d'abord le principe d'après lequel la question mécanique doit être traitée, et vu la nouveauté du sujet, nous en suivrons le développement mathématique et physique au delà de ce qui semble intéresser la simple pratique. Et ce ne sera pas seulement dans le but de faire des applications curieuses d'analyse à des questions nouvelles, c'est parce qu'il est toujours important de bien connaître toutes les conséquences des principes physiques d'où l'on part, et qu'on parvient ainsi à se rendre compte, d'une manière générale, d'effets qu'il eût été quelquefois difficile de prévoir, et qui, étant nécessaires à l'intelligence complète des phénomènes, peuvent par suite se trouver utiles à la pratique elle-même (pp. 18—19).

With the limitations imposed by Duhamel we evidently have

$$q = q_0 e^{-\delta t} \text{ .............................(i).}$$

Further if $U$ be the terminal shift at time $t$, and $\zeta$ the stretch produced by unit increase of temperature, the stretch produced by the tractive load at the terminal $= (U - lq\zeta)/l$ and therefore the tractive load $= E(U - lq\zeta)/l = EU/l - \eta q$. But this tractive load is a function of the shift $U$, or of the form $\psi(U)$. Thus we have

$$\psi(U) - EU/l + \eta q = 0 \text{ .................... (ii).}$$

Equations (i) and (ii) theoretically give $U$ for all values of $t$, and may generally be solved by approximation (p. 22). Duhamel then proceeds to investigate the work accomplished by the bar in cooling and shows that this work does not depend only on the bar itself, but also on the nature of the resistance overcome, i.e. on the function $\psi(U)$. He points out however that the work is always less than

$$\frac{l\eta^2 q_0^2}{2E},$$

whatever the resisting mass may be (p. 23). The general expression for the work is easily seen to be

$$\int_{U_1}^{lq_0\delta} \psi(U)\, dU \text{ ......................(iii).}$$

where $U_1$ is given by

$$\psi(U_1) = EU_1/l.$$

[903.] This is applied (pp. 25—26) to the simple case in which the tractive load may be supposed produced by a spring the force of which is proportional to the displacement or

$$\psi(U) = A(lq_0\zeta - U),$$

where $A$ is a given constant.

Duhamel deduces for the work the expression

$$\frac{Al^2\eta^2 q_0^2}{2(E + Al)^2},$$

which has a maximum value when $A = E/l$, a result easily interpreted physically.

The displacement at any point of the bar at time $t$ is calculated from the following equations:

$$\frac{d^2u}{dt^2} = \mu^2 \frac{d^2u}{dx^2},$$

$u = 0$, when $x = 0$,

$u = \phi(t) = U$ obtained from equations (ii), when $x = l$,

$u = xq_0\zeta$, $du/dt = 0$, when $t = 0$.

The value given is

$$u = \frac{x}{l}\phi(t) + \frac{2}{\pi}\sum_1^\infty \frac{\cos n\pi}{n} \sin\frac{n\pi x}{l}\int_0^t \phi'(\epsilon)\cos\frac{n\mu\pi(t-\epsilon)}{l}\,d\epsilon.$$

For this value, Duhamel refers to his memoir: *Sur les vibrations d'un système quelconque de points matériels.* This will be found in the *Journal d'École polytechnique, Cahier* 23, Paris, 1834, pp. 1—36: see p. 36. He applies it to the simple case stated at the beginning of this article (pp. 25 and 27).

[904.] The memoir concludes with a section entitled: *Refroidissement d'une barre dont l'extrémité est liée à une verge élastique à une température constante* (pp. 28—33). This is the interesting case where a cooling bar is employed to stretch a second bar, the temperature of which is that of the surrounding medium. The analysis presents no greater difficulties than such as arise in considering the longitudinal vibrations of a stretched string composed of two segments of different materials. Duhamel brings his memoir to a conclusion with the words:

Nous avons voulu seulement appeler l'attention sur un nouveau genre de questions, et ouvrir la voie à ceux qu'elles pourraient interesser.

[905.] Arthur Morin. *Nouvelles Expériences sur l'adhérence des pierres et des briques posées en bain de mortier ou scellées en plâtre.* Paris, 1838.

These experiments were made in 1834, and an account of them presented to the Academy but, owing to the delay caused by the death of Navier who had been appointed *Rapporteur*, Morin decided to publish his memoir independently in 1838[1]. We have

[1] Saint Venant in the *Historique Abrégé*, p. ccxcvii. gives 1828 as the date of this memoir: it is important to note that it follows Vicat's memoir of 1833: see our Art. 729.

seen that Coulomb had given a false theory of slide and that Navier had tried to make his theory more in accord with experience by introducing the element of friction into the formula deduced from Coulomb's hypothesis: see Art. 729. Morin in his experiments on friction had been led to inquire how far friction plays a part in the resistance to slide of two materials united by mortar. Some earlier experiments of his proved that after eight days the resistance of cohesion began to be much stronger than the frictional resistance. The traction was applied in the plane of the joint and the resistance to shear is termed by Morin *adhérence;* the distinction between the forces of friction and adherence is determined by the variation or non-variation of the resistance of the joints to slide with (1) the extent of the surfaces in contact, and (2) the amount of normal pressure.

For the first few days the resistance of the joint was found to vary with (2) and not with (1), but after a certain period to vary with (1) and not with (2). This period marks the transition from friction to adherence or to cohesion. Morin believes that these two resistances to slide are successive and not simultaneous, hence we may say Coulomb's formula as extended by Navier has no practical value (see Art. 729). These experiments seem to require some confirmation, especially having regard to Morin's other, and erroneous results on the nature of friction. It would seem probable that (1) and (2) have simultaneous effects in an early stage, but that the ratio of these effects varies. The author confirms Rondelet's result that for stones which take mortar well (*qui prennent bien le mortier*) we have to do with the cohesion of the mortar, but that with plaster (*plâtre*) we have to consider its adherence to the face of the material: see Art. 696.

[906.] J. P. G. v. Heim. *Ueber Gleichgewicht und Bewegung gespannter elastischer fester Körper. Mit einem Anhange über die Berechnung des Widerstandes und die vortheilhafteste Gestalt der Eisenbahn-Schienen.* Stuttgart und Tübingen, 1838.

As we advance in our historical investigations the increasing number of text-books on the subject of elasticity become a marked feature, and the difficulty of discovering what amount of novelty they contain becomes more and more considerable. The present work consists of upwards of 650 pages and the author claims that it contains the solutions of certain problems which presented themselves to him in his practical work, problems which he apparently holds were at that time unsolved. The appendix on *Eisenbahn-Schienen* is characteristic of the period upon which we are entering; the construction of bridges had for a long time dictated the direction of experiment and theory in the matter of elasticity, railway needs now begin to take their part in this field of investigation.

[907.] The work itself belongs essentially to the old school; that is, the general equations of elasticity are not considered, but by means of special hypotheses problems on beams, struts or rods (bodies having a central axis—*Centrallinie*) are solved. In this respect the work is of the same character as the treatises of Girard and Eytelwein, or the *Leçons* of Navier. The author adopts the Bernoulli-Eulerian hypothesis that the normal-sections of a beam remain plane and normal to the central axis after flexure; he also expands stress in terms of strain by Taylor's theorem without the least qualifying remark. The shear and slide are in fact completely neglected. At the same time while following in the footsteps of Euler and Lagrange he has solved the equation for flexure of a beam subjected to a terminal load more generally and to a higher degree of approximation than any previous and possibly than any subsequent writer. He puts (p. 53) this equation in a slightly different form than that usually adopted,—the flexure being supposed to take place in one plane that of the central axis,—namely:

$$\frac{\dfrac{d^2y}{dx^2}\left(1-\dfrac{P}{E\omega}\dfrac{dy}{ds}\right)}{\left\{1+\left(\dfrac{dy}{dx}\right)^2\right\}^{\frac{3}{2}}}=\frac{P}{E\kappa^2\omega}x \qquad \text{(i)}.$$

Here $P$ is the terminal load, $E$ the stretch-modulus, $\omega$ the sectional area, and $\kappa$ the radius of gyration of the section $\omega$ about a line through the central axis perpendicular to the plane of flexure. The direction of $P$ is taken as axis of $y$, and $x$ is perpendicular to it and thus not necessarily coincident with the unstrained position of the central axis; $s$ is as usual the length of this axis measured from a fixed point (p. 53).

[908.] Of this equation a first integral is easily obtained, namely, writing $\frac{dy}{dx} = \tan\phi$ and $C$ for the constant,

$$\sin\phi + P/(2E\omega) \,.\, \cos^2\phi = Px^2/(2E\kappa^2\omega) + C \,\ldots\ldots\ldots\text{(ii)}.$$

For the solution of this equation (ii) various methods of approximation in series involving the use of Maclaurin's and Ohm's Theorems are given (see pp. 55—178). The approximation to be adopted for any particular case depends largely on the character of the constant $C$, and the equation to the central axis is given not only in algebraic, but in trigonometrical and logarithmic forms. The approximations involve very long analytical work and seem unnecessarily fine for any practical application. In fact the only case in which we can suppose the Bernoulli-Eulerian hypothesis to approach the actual state of affairs is that wherein the length of the beam is very great as compared with the linear dimensions of its section; in this case the majority of terms involved in these approximations are insensible.

[909.] It is somewhat startling to find a practical man like Heim (he was colonel of Artillery) even asserting that the equation (i) is perfectly valid when the normal section varies from point to point of the central line (p. 179) and applying it to various surfaces of revolution, for example (p. 181) the frustrum of *any* right circular cone. This is much on a par with the formula he gives (p. 28) for the extension by a longitudinal load of any beam of varying section, namely:

$$\text{extension} = P\int\frac{dx}{E\omega},$$

where $\omega$ is of course a function of $x$.

[910.] On p. 190 Heim gives an equation for the flexure when not only a terminal load, but in addition any distribution of load in the plane of flexure is added from point to point of the beam. He applies this to Euler's problem of columns bent by their own weight and considers also the case of the maximum height for a conical column (pp. 191—221). Heim's numerical result for a cylindrical column on p. 209 does not agree with Euler's (see our

Art. 85), but he is treating a somewhat different problem. Heim's result in Euler's notation is that

$$h > \sqrt[3]{7{\cdot}837325\, m},$$

or, roughly speaking, only *one-third* of Euler's value. Greenhill finds $h > \sqrt[3]{7{\cdot}9524\, m}$: see his paper, *On height consistent with stability. Camb. Phil. Proc.* 1882, p. 67. The difference considering the character of the physical hypothesis on which the solution rests, is scarcely worthy of further investigation.

[911.] We may here make a remark which is not without its importance in relation to later investigations. Is the load to be given that value which it has in the strained condition or in the unstrained condition of the body? At first sight the question might seem immaterial, as the shifts are in every case supposed to be very small quantities. But in the problem of a rod under flexure by a terminal force, the consideration becomes important. Take, for example, Clebsch's treatment of Saint-Venant's problem (*Theorie der Elasticität f. Körper*, S. 95). Here the load applied at the free end of the rod consists of forces whose sums parallel to three chosen axes through the fixed end of the rod are $A$, $B$, $C$ and whose moments about these axes are $A'$, $B'$, $C'$; in the case of a negative tractive load all these vanish except $C$, and according to the formulae on p. 102 there would be no flexure. Any the least force $A$ will produce flexure; now suppose a small force $A$ applied and a considerable force $C$; obviously $C$ will now not only produce compression, but also (if greater than a certain magnitude) flexure. Hence if we hold that *theoretically* $C$ alone would only produce compression and $A$ alone flexure, it is evident that we cannot here apply the usual theory of elasticity as the superposition of strains is no longer legitimate. In this case we must either assert that the mathematical theory does not hold for a normal load of such magnitude as to be capable of producing flexure, or else we must calculate the load for the *strained* position of the body. This is possible in the example we have considered, but it is conceivable that the strained position could not in some cases be determined till it was obtained from differential equations themselves involving the load. We may remark in addition that although $C$ might not in itself be sufficient to maintain flexure, it might considerably modify an existing flexure, and thus our first alternative of excluding certain loads from consideration would be useless. Further if $C$ and $A$ be tractive and shearing loads at the free end of a beam, the bending moment of $A$ will practically be unaffected by the shift produced by $C$, on the other hand the shift produced by $A$ may be the cause of the bending moment of $C$. We are thus compelled to consider $C$ as applied to the body in its strained state.

These remarks are called forth by the very copious treatment which under the *old* theory of flexure a strut subjected to longitudinal load received,—Heim following in the footsteps of Euler and Lagrange devotes a great amount of space to the matter. This case although practically very important is not embraced by Saint-Venant's flexure researches. If indeed we suppose $C$ to produce flexure and calculate $B'$ for the strained position of the rod, we have $B' = aC$, where $a$ would denote the total deflection of the free end, and we obtain from Clebsch's equations (86) p. 102 and (72) p. 85 the purely *algebraical* form for the deflection of a point on the central axis; $u = z\,(d\Omega/dx)_0 - a_1 z^2/2$, but there can be small doubt that the *transcendental* form of the old theory: $u = a\,(1 - \cos \beta z)$ is at least an approximation to the real state of affairs. The former value applies in fact only in the immediate neighbourhood of the fixed end, giving indeed the curvature at that end when $(d\Omega/dx)_0$ is zero. The inapplicability to this case of Saint-Venant's solution arises from the usual theory of elasticity assuming the superposition of strains.

[912.] Chapter IV. of Heim's book (pp. 234—266) is devoted to the consideration of cases wherein the central axis becomes a curve of double curvature. The treatment is not only exceedingly cumbrous but incorrect and leads to equations which the author does not integrate.

[913.] Chapter V. is entitled: *Vom Gleichgewichte elastischer fester Körper mit ursprünglich krummer Centrallinie.* When the load is applied in the plane of the unstrained central axis the equation obtained is

$$\left(\frac{1}{r-k} - \frac{1}{r_1 - k}\right)\left(1 - \frac{P}{E\omega}\frac{dy}{ds}\right) = \frac{Px}{EN} \quad \text{........(iii).}$$

Here $r_1$ and $r$ are the radii of curvature of the unstrained and strained central axis at $x$, $y$, $k$ is the distance from the central axis of the "neutral line," and $N$ is a complex integral which reduces to $\omega k^2$ when $r_1$ is large as compared with $k$ and the dimensions of the section; the other quantities having the same values as in equation (i). Heim makes no attempt to integrate this equation generally: see p. 279.

It is not in accordance with the result given by Grashof (*Theorie der Elasticität und Festigkeit*, 1878, p. 254), nor with Navier's equation: see Art. 278, but reduces to the latter if we neglect $k/r$ and $dy/ds$ as small, the first of which we certainly cannot generally neglect. Here as in obtaining equation (i) Heim does not point out that the plane of flexure must cut all the normal sections in one of their principal axes. The analysis is extremely clumsy, not to say obscure, so that being unable to follow several steps, it is impossible for me to say where the error really arises. There is an addition to this chapter in the form of a sixth chapter, the only part that need be noticed here is contained on pp. 334—336, where the writer is evidently conscious that all is not quite physically accurate in the Bernoulli-Eulerian theory.

[914.] Chapter VII. occupies itself with the equilibrium of elastic beams supported or built-in at both ends and loaded between. The methods of this chapter are similar to those given in Navier's *Leçons* or Poisson's *Mécanique* and still to be found in the text-books. The results, obtained in some cases by very long analysis, are tabulated on pp. 409—411 and might be useful for comparison and reference.

[915.] Chapter VIII. is concerned with the old problem of Galilei—the solid of equal resistance. As the effect of shearing stress is here neglected we have the anomalous solids presented by many early writers recurring, *i.e.* beams which at their points of support or at their loaded terminals have a vanishing section: See p. 425, § 190, for example.

[916.] Chapter IX. Here the vibrations of various elastic beams are discussed at great length of analysis. Neither the methods employed, nor the results are of a kind that need detain us. The chapter is inordinately long (pp. 430—577). The book concludes with the appendix on the best form of rails which we have previously referred to. It is really an addition to Chapter VIII. The kind of stress to which a portion of this appendix is devoted arises when the ends of a loaded beam are supported or built-in; upon this the writer remarks:

Zu den Korpern, welche dieser Art von Spannung unterliegen, gehören namentlich die Schienen, welche die Geleise der unter dem Namen: *Eisenbahnen*, bekannten neuen Gattung von Kunststrassen bilden (p. 578).

The analysis with which Heim attacks the problem before him is simply astounding in its magnitude, and I have not felt it necessary to enter into an examination of equations which occasionally fill entire pages. The final form of his rails has for under-contour two lines inclined to the horizontal and a portion of a hyperbola. There can be no doubt that a line made with such unpractical rails as these would be indeed a *Kunststrasse*.

917. We must now notice two memoirs by George Green relating to the theory of Light published in the *Transactions of the Cambridge Philosophical Society* Vol. VII. The first memoir is entitled: *On the Laws of the Reflexion and Refraction of Light at the common surface of two non-crystallized media;* this occupies pages 1—24 of the volume; it was read to the Society on December 11, 1837, and published in 1839. A *Supplement* to this memoir occupies pages 113—120 of the volume; it was read to the Society on May 6, 1839, and published in 1839. The second memoir is entitled: *On the propagation of light in crystallized media;* this occupies pages 121—140 of the volume; it was read to the Society on May 20, 1839, and published in 1842. The papers are contained in the octavo volume published in 1871, entitled: *Mathematical Papers of the late George Green:* see pp. 243—269, 281—290, 291—311.

918. An account of these memoirs would find a more appropriate place in a history of the wave theory of light than in our work; but their importance will justify some notice of them here. What they offer to our attention is essentially a demonstration

of the body shift-equations for free vibrations of an elastic solid; this demonstration occupies about eight pages. There is no mention of stress; the equations are given as necessarily holding for the motion of a set of particles under certain suppositions. One supposition is that the whole mass is composed of "two indefinitely extended media, the surface of junction when in equilibrium being a plane of infinite extent." The other supposition is that the medium is uncrystallised or isotropic. With respect to the mechanical principle used the author says in the early part of the memoir:

The principle selected as the basis of the reasoning contained in the following paper is this: In whatever way the elements of any material system may act upon each other, if all the internal forces exerted be multiplied by the elements of their respective directions, the total sum for any assigned portion of the mass will always be the exact differential of some function. But this function being known, we can immediately apply the general method given in the *Mécanique Analytique*, and which appears to be more especially applicable to problems that relate to the motions of systems composed of an immense number of particles mutually acting upon each other. One of the advantages of this method, of great importance, is that we are necessarily led by the mere process of the calculation, and with little care on our part, to all the equations and conditions which are *requisite* and *sufficient* for the complete solutions of any problem to which it may be applied.

The mechanical principle which Green uses, we see, is that which was formerly called briefly that of *Vis Viva*; more fully stated it is: that the vis viva of a system can be expressed in terms of the co-ordinates of its parts for such forces as occur in nature. It is now sometimes called the principle of the *Conservation of Work*.

919. Green shews that the function of which he speaks, when he confines himself to a certain approximation, involves in general 21 coefficients; these reduce to 9 when the mass is symmetrical with respect to three planes at right angles to each other, to 5 when moreover the mass is symmetrical about an axis, and further to 2 when the mass is symmetrical about two other axes at right angles to each other and to the first, i.e. where there is isotropy.

920. An abstract of a memoir by Professor MacCullagh is given in the *Philosophical Magazine* for March, 1840, pp. 229—232; from this it appears that he did not entirely agree with Green. We read:

It will be perceived that this theory employs the general processes of analytical mechanics, as delivered by Lagrange[1]. The first attempt to treat the subject of reflection and refraction in this manner was made by Mr Green, in a very remarkable paper, printed in the *Cambridge Transactions*, Vol. VII., Part I.......

[This is Green's *first* memoir.]

Such is certainly the great advantage of starting with that general principle; but the chief difficulty attending it, namely, the determination of the function $V$, on which the success of the investigation essentially depends, has not been surmounted by Mr Green, who has consequently been led to very erroneous results, even in the simple case of *uncrystallized* media to which his researches are exclusively confined. In this case Mr MacCullagh's theory confirms the well-known formulae of Fresnel, one of which Mr Green conceives to be inaccurate, and proposes to replace by a result of his own, which however will not bear to be tested numerically, p. 232.

[921.] We cannot enter here into the discussion between Green and MacCullagh as to the point connected with the wave theory of light, but MacCullagh's objection to Green's method of obtaining the equations of elasticity deserves some attention. We must in fact point out the leading features of the elastic-constant controversy, which plays such a large part in the future of our subject[2]. The questions are these: Is elastic isotropy to be marked by one or two constants? Is elastic aeolotropy to be marked by fifteen or twenty-one constants? For brevity we may speak of the first alternative as the *rari-constant* theory, and the second alternative as the *multi-constant* theory.

[922.] The rari-constant theory is based upon the assumption that a body consists of molecules and that the action between

[1] [Lagrange's principles, as we have seen, were used by Navier in the first instance and afterwards by Piola: see our Arts. 268, 761. ED.]

[2] This is the discussion to which I have referred on p. 224.

two molecules or ultimate parts of a body is in the line joining them. Navier and Poisson, who first obtained the general equations of elasticity on this hypothesis made additional assumptions: (1) both assumed that the action was a function of the molecular distance; (2) Poisson assumed that 'the irregular action of the molecules in the immediate neighbourhood of the one considered may be neglected in comparison with the total action of those more remote, which is regular.' These additional assumptions are both unnecessary, the rari-constant equations have been obtained by Weyrauch supposing the action to be central but not necessarily a function of the intermolecular distance only (see his *Theorie elastischer Körper*, S. 132—149); they have been obtained by Cauchy without neglecting the irregular action: see our Art. 616. Hence an objection raised by Stokes in 1845, to the rari-constant theory falls to the ground; he does not appear at that time to have seen Cauchy's memoir of 1828: see *Mathematical and Physical Papers*, Vol. I. p. 120. It is true that Cauchy also makes an assumption: namely, that the molecules are symmetrically distributed, but as Clausius has shewn in a memoir of 1849 (see our Chapter VIII.) this is perfectly legitimate as we require the value of the constants not at one individual point, but their mean or average value for an indefinite number of points. Cauchy's arrangement then gives the mean or normal value of the elastic constants.

[923.] It would seem then that if intermolecular action is in the line joining reacting molecules, the rari-constant equations must hold. How far is this assumption necessary or true? There seems no reason why intermolecular action should not be of a polar character, indeed many phenomena of crystallisation favour this assumption: see Arts. 828—829. But in treating isotropic bodies we are compelled to consider their crystallisation as 'confused' (see Art. 332), and it is difficult to see how the mean results would be affected by the polar property. This is in fact the position taken up by Clausius in his very able article on this matter, to which we shall return in its proper place. We may further remark that many proofs of the conservation of energy are closely associated with the hypothesis that natural (or intermolecular) forces are functions of the distance and act in that distance[1]. Among rari-constant elasticians we must include Navier, Poisson, Cauchy (see however Art. 927), Clausius, F. E. Neumann, Haughton, Castigliano, Baer, Grashof, and last, but by no means least, Saint-Venant. It is quite true that these various physicists limit very considerably the area of uni-constant isotropy, and the majority rightly remark that substances like cork and other vegetable matter with cells and tissues, or like india-rubber and jellies containing fluid, cannot be treated as elastic bodies, and that arguments deduced from them against uni-constant isotropy are invalid. The exact views of these writers I shall consider in detail when I treat of their memoirs or treatises on our subject, but I may at present refer the reader to Saint-Venant's thorough discussion of the whole subject in the *Appendice* v. to his edition of Navier's *Leçons* (pp. 645—762).

[924.] It would seem that this weight of authority at least demands that we should examine carefully into the basis of the rari-constant theory, and we cannot hold it entirely satisfactory to dismiss that theory in the brief manner adopted by some English writers:

The only condition that can be theoretically imposed upon these coefficients (the 21 of Green) is that they must not permit $w$ (the work) to become negative for any values, positive or negative, of the strain-components....Under Properties of Matter we shall see that an untenable theory (Boscovich's), falsely worked out by mathematicians[2], has led to relations among the coefficients of elasticity which experiment has proved to be false. (Thomson and Tait's *Treatise on Natural Philosophy*, 2nd edn., Part II., p. 214.)

[1] We may refer the reader to Schell, *Theorie der Bewegung und der Kräfte*, Vol. II. pp. 494, 542; Riemann, *Partielle Differentialgleichungen*, p. 216 (who appeals to this very principle in establishing the body shift-equations by Green's method!); C. Neumann, *Die mechanische Theorie der Wärme*, p. 9; Clausius, *Die mechanische Wärmetheorie*, 2[te] Ausg. Bd. I., S. 14—17 (especially the concluding remarks on p. 17), etc., etc.

[2] The reader of our chapters on Poisson and Cauchy will observe that the assumption made is *not necessarily* that of Boscovich, and further that if the assumption made be untenable, there has been no evidence produced that it is falsely worked out. We have been unable to find any flaw in Cauchy's reasoning in Art. 616.

Elsewhere these two distinguished physicists refer to Stokes' memoir of 1845, state that 'clear elastic jellies and india-rubber present familiar specimens of isotropic homogeneous solids' (p. 221), appeal to experiments on *wires*, and to the properties of cork (p. 222) to demonstrate the 'utter worthlessness' of the rari-constant theory. We are no supporters of the rari-constant theory, but it cannot be met by experiments upon bodies, which as we shall see in the sequel, its upholders either exclude from their theory, or for what appear very valid reasons, refuse to treat as isotropic. The latter case applies peculiarly to experiments on *wires* and *plates*. The metals are substances for which rari-constant elasticians assert their theory holds, but metals in the form of wires and thin plates are, as is well known to all practical engineers (see our Arts. 332, 830, 858), the last form in which metal can be considered isotropic. These bodies possess owing to their method of manufacture not only a cylindrical or planar system of elasticity, but extremely often an initial state of stress, both conditions which lead rari-constant theorists to bi-constant formulae. On the other hand it is found that large bars of metal *when they have been reduced to a state of ease*, and when the experiments are conducted with the greatest care (see, for example, the caution necessary in the simple case of tension, Arts. 832 and 940) offer a close approach to uni-constant isotropy. We shall have occasion later to give details of experiments of this kind, we may however remark that Clapeyron, as a result of his practical experience held that for the metals used in construction it was safest to adopt uni-constant isotropy as the nearest approach to truth: see also Art. 973.

[925.] We shall return again to Stokes' memoir of 1845, but it is needful to consider here the arguments he brings against the rari-constant theory. Our references will be to the pages of the first volume of his collected papers. The argument of p. 120 is perfectly valid against Poisson's treatment, but viewed in the light of Cauchy's memoir of 1828 is no objection to rari-constancy. On p. 122 we have some experiments of Lamé's appealed to. These again were made on *wires* and considered even by Lamé himself as unsatisfactory: see our Art. 1034. On p. 123 there is a reference to Oersted's experiments (see our Art. 690). Now we have pointed out that Oersted's theory of compression was entirely erroneous, and that he did not allow, or allow properly, for the compressibility of the material of his piezometer. Further we may remark that glass and metal bottles, especially the former, are owing to their method of preparation anything but isotropic. If further evidence be wanted of the unsatisfactory nature of Oersted's results, it may be found in their total disagreement with those of Wertheim and Regnault in memoirs which we shall consider in Chapter VIII.

[926.] The strongest argument however for multi-constancy in Stokes' memoir is undoubtedly to be found in the transition he would make from viscous fluids to elastic solids. He in fact draws no line between a plastic solid and a viscous fluid. The formulae for the equilibrium of an isotropic plastic solid would thus be bi-constant. Now the strain of a body due to a given system of load consists of two parts, elastic strain and set; these two parts follow quite *different* laws and we can find materials which for any definite load have very varying amounts of elastic strain and set. When for a given load there is no set, we say the body is an elastic solid in its state of ease; when the strain is all set, we say the body is a plastic solid, or a viscous fluid. Because the limit of elasticity can be made to vary with the load, and so a solid pass imperceptibly into a viscous fluid, it does not follow that elastic strain and set have the same number of independent physical constants (see my remarks on the memoirs of Maxwell and Clausius in Chapter VIII.). There is in fact a very distinct difference in the physical characteristics of a metal in its primary elastic and its plastic stages. The 'tendency to a rearrangement of molecules' which is essentially the characteristic of the latter is wanting in the former, or only appears in the neighbourhood of the elastic limit. So soon as it does appear stress ceases in many cases to be proportional to strain and we have bi-constant formulae of which we may take the Gerstner-Hodgkinson law as an example: see Arts. 803—807, and 969. The constants we add are not due to bi-constant isotropy, but to the fact that the square of the strain is, in the case of set, no longer negligible.

[927.] So much having been said for the supporters of rari-constant elasticity we may next turn to those of multi-constancy. We have seen that bi-constant isotropy was first propounded by Cauchy in a memoir of 1828 (see Art. 614) and treated in a manner very similar to that afterwards adopted by Stokes and Maxwell. We have also pointed out how in later memoirs, proceeding from the equilibrium of a single molecule without regard to initial stress, Cauchy arrived at bi-constant isotropy, but that proceeding by a definition of stress and calculation of the stresses he arrives at uni-constant isotropy in the case of no initial stress (Arts. 616—617). Hence it arises that as Cauchy uses bi-constant formulae throughout the greater part of the third volume of the *Exercices*, we must look upon him as essentially the first to introduce multi-constancy. He was followed by Poisson, who, as will be seen from our Art. 553, adopted a method similar to that of Stokes to deduce the relation between stress and strain, which in the case of isotropy leads to two constants. We have spoken of both Poisson and Cauchy as supporters of rari-constancy because the great memoirs of both lead up to this result, but we must remark that Cauchy afterwards conceived it quite possible that a relation might hold between the two isotropic constants different from that to which his earlier memoirs had led him: see our Arts. 682 and 683.

[928.] This being the state of affairs Green comes upon the scene and propounds a new method of obtaining the body shift-equations, a method which has practically been followed by the majority of the upholders of multi-constancy ever since; for example, by Thomson and Kirchhoff. This consists in forming an expression for the work and expanding it in powers of the strain-components. The primary assumption of this method is the manner in which the work is expanded in *integer powers* of the strain-components. Green gives no physical or theoretical reason for this assumption on his part; Thomson follows Green and expands by Maclaurin's theorem without comment: see *Mathematical and Physical Papers*, Vol. I. page 301 *et seq.* Maxwell lays down in two axioms a generalised form of Hooke's law not very different from Cauchy's hypothesis in Art. 614; he states the experiments which he holds confirm these axioms in a note appended to his memoir; see our Chapter VIII. A similar method, without however experimental data, is followed by Clebsch. Stokes with keen physical insight grasped even before Maxwell that the theory of elastic bodies must be based on a physical axiom the result of experiment, and not upon a theorem in pure mathematics:

The capability which solids possess of being put into a state of isochronous vibration shews that the pressures called into action by small displacements depend on homogeneous functions of those displacements of one dimension. I shall suppose, moreover, according to the general principle of the superposition of small quantities, that the pressures due to different displacements are superimposed, and consequently that the pressures are linear functions of the displacements. (*Ibid.* p. 114.)

The fact stated in these lines may be taken as the physical basis for the generalised Hooke's law, and till they were written Green's method was only a chain of arbitrary assumptions. If we assume however intermolecular action to be central, Stokes' axiom is seen to be a result of our molecular hypothesis, and we should expect the linearity of the relation between small stress and strain. Because we accept Stokes' axiom and adopt Green's method it does not follow that the multi-constant theory is true, there may be other physical axioms (*e.g.* that of central molecular action) which we have not considered.

[929.] Although we should be inclined ourselves to accept Stokes' axiom as proving the linearity of the stress-strain relation, we may yet remark that the following criticism of Saint-Venant is not without weight and may tend to throw light on the peculiarity of the stress-strain relation for certain materials.

Nous reconnaissons, dans ce fait de "l'isochronisme des vibrations" attesté par les phénomènes du son, etc., une excellente preuve que les forces développées par les déplacements relatifs, c'est-à-dire par les changements très-petits de distances mutuelles des points, sont proportionnelles à ces changements. Mais quelles sont ces forces qui se trouvent réellement en jeu une fois que le corps, légèrement déformé par une application momentanée de quelque force extérieure, est abandonné à lui-même? Nous ne voyons pas que ce soient les *pressions*, qui ne sont apres tout que des forces fictives comme toutes les résultantes ou les sommes de composantes. Ce qui agit pour produire ou pour continuer le mouvement vibratoire, ce sont les actions moléculaires individuelles, dont la considération, loin de pouvoir être éludée, a paru nécessaire, *même pour définer les pressions*, aux savants qui ont voulu en donner la définition d'une manière rationnelle; et le fait des petites vibrations isochrones des solides nous semble la plus forte preuve de l'existence de ces actions mutuelles à distance. (*Appendice* v. to Navier's *Leçons*, p. 720.)

We must further remark that there exist certain materials for which even in a state of ease the stress-strain relation is not linear; that is to say the stress-strain curve of a bar of that material is not a straight line even for very small elastic strains: see Note D. upon this point and the plotted curves at the end of the present volume. Green's assumption that the work can be expanded in integer powers of the strain components and the second powers only retained is thus not valid for these materials.

It will be seen, then, that Green's method of arriving at multi-constant equations is by no means opposed to the possibility of rari-constancy, it is only more agnostic, more empirical. So far as experiments go they have been so repeatedly made on bodies like *wires*, that it is not possible to say that they have absolutely settled the controversy in favour of multi-constancy.

[930.] The bi-constant argument drawn from the nature of the ether, which nature, indeed, gave rise to Green's method, seems to me to be a very fallacious one. It begins by *assuming* that the ether is an elastic solid and then argues that an elastic solid must on that account be multi-constant; we cannot find that there is really sufficient evidence for this so-called 'jelly theory of the ether.' That the equations for its motion bear some similarity in form to those of an elastic solid, is very probable, but this does not allow us to make the great jump to the 'jelly theory.' Stokes, in the well-known memoir: *On the constitution of the luminiferous ether*[1], gives an illustration of the possible nature of the ether, which makes its constitution approach that of a viscous fluid; that a substance like

[1] *Cambridge Philosophical Transactions*, Vol. xxxii. p. 343, 1848, or the reprint in the *Papers*, Vol. ii. pp. 8—13

a thin jelly may have two independent constants is a statement which I believe the supporters[1] of rari-constancy would not deny, for such a jelly is one of the materials they exclude from their consideration: see Art. 923, p. 498.

[931.] The strongest argument in favour of multi-constancy seems to me to arise when we admit the action between molecules to be in the normal case central, but argue that the action between two molecules $A$ and $B$ may depend on the relative position, internal vibration and possibly oscillation of third molecules such as $C$—that is, the action of $A$ on $B$ depends directly on the state of $A$ and $B$, but indirectly on the action of $C$, on $A$ and $B$. This would be the case for instance if $A$, $B$ and $C$ were pulsating spheres in a fluid medium. On this supposition of indirect action Jellett has shewn that multi-constancy follows. At the same time we must note that it is hard to conceive any dynamical system in which $A$'s action on $B$, due to the indirect action of $C$, would not be of a much higher order than $A$'s direct action on $B$, and thus at least to a first approximation negligible. The *sum* of the indirect actions of all such molecules as $C$ might however be commensurable with the direct action of $A$ on $B$. In this case, as Clausius has suggested from a different standpoint, we might possibly have rari-constant equations for vibrational movement of elastic solids, but multi-constant equations for their equilibrium. This would help to explain the support sound experiments give to rari-constancy and the divergence in the value of the elastic constants as obtained by vibrational and statical methods: see my remarks on Clausius's and Jellett's memoirs in Chapter VIII.

[932.] It might be thought that the adoption of multi-con-

[1] In a letter to the editor dated September, 1885, Saint-Venant thus replied to the question of whether he continued to support the rari-constant hypothesis:

Je réponds *oui* pour les vrais solides (supposés isotropes) commes sont les métaux ordinairement, ainsi que le marbre, le verre; mais *non* si l'on veut absolument par un motif quelconque que je ne conçois guère, appliquer les formules de l'élasticité au caoutchouc, aux gommes molles, aux gelées, et aux autres corps mous et élastiques, car ces corps-là ne sont que les mélanges de tissus cellulaires, de membranes élastiques, et de fluides visqueux que leurs cellules contiennent.

stant equations would always be on the side of safety, as experiment would take account of any relation between the constants; but apart from the multiplication of the processes of testing involved in determining these constants, there is a more important point still which must not be left out of consideration. The use of two constants may enable us in an individual case to fit our theory to experimental result, and may thus disguise the fact that the material we are treating is *aeolotropic;* a fact which may lead us to quite erroneous results if we apply the constants calculated upon our theory to a different case. It is a very practical question whether a metal wire or plate shall be treated as isotropic with two constants, or as aeolotropic on the rari-constant theory. There cannot be the slightest doubt that to give many bodies cylindrical or other symmetrical distribution of elasticity on the rari-constant theory will lead to truer results than the supposition of a bi-constant isotropy. It is possibly the assumption of isotropy, not that of rari-constancy, which has led to contradictory results, such as we shall have to lay before the reader in remarking on many later experimental memoirs.

[933.] Green, as we have seen, divides his work-function into homogeneous functions of the six strains $s_x, s_y, s_z, \sigma_{yz}, \sigma_{zx}, \sigma_{xy}$ which he represents by $\phi_0, \phi_1, \phi_2, \phi_3 \ldots$ and says are of the degrees 0, 1, 2, 3 etc., '*each of which is very great compared with the next following one*' (*Collected Papers* p. 249). He himself only treats of $\phi_2$. In an Appendix (p. 332) his editor calculates the values of $\phi_1, \phi_2, \phi_3 \ldots$ in terms of $s_x, s_y, s_z, \sigma_{yz}, \sigma_{zx}, \sigma_{xy}$. It would appear from the reasoning in the Appendix, that the stretches and slides are given values of the form:

$$s_x = \frac{du}{dx}, \quad \sigma_{yz} = \frac{dw}{dy} + \frac{dv}{dz}.$$

Yet, if as in the case of Hodgkinson's law for cast iron (see Chap. VIII.), we need to retain terms of the order $\phi_3, \phi_4 \ldots$ *etc.*, we must not use the values of the strains given in this Appendix, but their complete values, *i.e.* those containing the squares of the shift-fluxions. It then follows that $\phi_1\ \phi_2\ \phi_3 \ldots$ etc. will be of ascending order in terms of the complete strains $s_x, \ldots, \sigma_{yz}, \ldots$ but not in terms of the shift-fluxions; they cease to be homogeneous functions of the shift-fluxions, and therefore successive $\phi$'s may contain parts of the same order[1]. Sometimes perhaps the easiest method of expanding the work-function will be to introduce the twists

$$\tau_{yz} = \tfrac{1}{2}\left(\frac{dw}{dy} - \frac{dv}{dz}\right), \tau_{zx} = \tfrac{1}{2}\left(\frac{du}{dz} - \frac{dw}{dx}\right), \tau_{xy} = \tfrac{1}{2}\left(\frac{dv}{dx} - \frac{du}{dy}\right)$$

If we do this it is not necessary to change the values of $s_x$, ..., $\sigma_{yz}$, ..., but we may assume the work-function

$$= \chi(s_x, s_y, s_z, \sigma_{yz}, \sigma_{zx}, \sigma_{xy}, \tau_{yz}, \tau_{zx}, \tau_{xy}) = \chi_1 + \chi_2 + \chi_3 + \chi_4 + \ldots\ldots$$

where the $\chi$'s are homogeneous not only in $s_x$, $s_y$, $s_z$, $\sigma_{yz}$, $\sigma_{zx}$, $\sigma_{xy}$, $\tau_{yz}$, $\tau_{zx}$, $\tau_{xy}$, but also in the first shift-fluxions, or each $\chi$ is very great compared with the next following one. Obviously $\chi_1 = \phi_1$, and $\chi_2 = \phi_2$.

934. Green's memoirs are frequently cited or alluded to by Saint-Venant in Moigno's *Statique*: see pages 640, 660—666, 709, 711, 719—722. On his page 640 Saint-Venant seems to say that the quantities which he calls *dilatations* and *glissements* had received names from Green: but this is not the case; Green uses symbols for these quantities but not names. On page 720 Saint-Venant gives in French, with marks of quotation, a passage as if from Green. The original will be found on page 293 of the Reprint. "Our problem thus becomes...*accurately* in the front of the wave." Saint-Venant thus sums up his opinion of Green's method:

Cette manière de l'illustre physicien anglais est large et simple. Mais elle s'appuie sur une suite d'hypothèses singulières, et en tous cas bien moins justifiées que n'est la loi physique des actions entre molécules suivant leurs lignes de jonction....

935. Bellavitis. *Considerazioni di Giusto Bellavilis* (sic !) *sulle formule per l'equilibrio di una verga elastica che si leggono nella 2*[da] *edizione* (1833) *della Meccanica del Poisson.* This is published in the *Annali delle Scienze del regno Lombardo-Veneto*, Vol. IX. 1839, Venezia: it occupies pages 202—207.

At the beginning of the memoir Bellavitis gives a reference to a memoir (Vol. VIII., p. 117 of the *Annali*) in which he had already alluded to the matter which he now proposes to treat more in detail;

[1] Thus in many cases the strains will not all be of the same order; for example in the motion of a plane wave to the third order, the stretch varies as the square of the slide. Hence parts of $\phi_4$ cannot be neglected if $\phi_2$ is retained in full.

the matter is that which I notice in Art. 571 of my account of Poisson's *Mécanique.* Bellavitis objects to the result contained in Arts. 308 and 316 of Poisson's work, namely that the tension is *constant* if there is no force applied along the tangent; and also to the result contained in Art. 318, namely that the *moment* of torsion is constant. Bellavitis shews how the errors of Poisson arise. Suppose a normal section at any point of the rod; the action of one part of the rod on the other may be supposed to amount to forces in three directions at right angles to each other, and couples round these directions as axes; take for the three directions the tangent at the point, the principal normal at the point, and the normal to the osculating plane at the point: then Poisson *assumes* that the force along the principal normal is zero, and also the couple which has this straight line for axis[1].

[The reader should compare the results obtained by Binet, Wantzel and Saint-Venant: see our Chapters VIII. and IX. Bellavitis gives the true expressions for the moments of the elastic forces and shares with Saint-Venant the honour of discovering the 'third moment'. Saint-Venant, however, was the first to put the whole matter in a clearer light with regard to the 'inertial isotropy or aeolotropy' of the cross section: see his memoir of 1843 discussed in our Chapter VIII. ED.]

[936.] 1839. Leblanc. *Pont de la Roche-Bernard sur la Velaine. Expériences sur la résistance du fil de fer et la fabrication des câbles. Annales des ponts et chaussées* 1839, 2e *semestre,* pp. 300—334. This memoir was followed by a book on the same bridge in 1841 entitled: *Description du pont suspendu de la Roche-Bernard.* I have examined the memoir but not the book. There is little to note in it.

1°. Long wires have a less breaking load than short ones of the same diameter because there is a greater possibility of small flaws (p. 303). They generally break where 'built-in' or bent (p. 304 *et seq.*).

2°. Wires stretched nearly to rupture (presumably not to the state of perfect plasticity?) by a load applied during a very short time do not thereby lose any of their primitive strength (have the same breaking load?) (p. 334 *et seq.*).

[1] Page 206, for 3 in many places read *r*.

3°. A wire can support during a space of three months a load equal to 9/10 of that which would break it without losing its primitive strength (*la force primitive*) (pp. 319—323).

It is necessary to remark that the *Pont de la Roche-Bernard* built by Leblanc collapsed owing to the oscillations produced by a storm in October, 1852. An account of this accident will be found in the *Annales des ponts et chaussées*, T. XXVI., 1859, pp. 249 *et seq.*

It is interesting to note how the practical needs of engineers in the construction of suspension-bridges led to the discovery of many of the physical properties of elastic materials, thus to the influence of time and vibrations in producing strain.

[937.] Ardant. *Études théoriques et expérimentales sur l'établissement des charpentes à grandes portées.* Metz, 1840.

I have not been able to consult the original work, but a report by de Prony, Arago, Coriolis, Rogniat and Poncelet upon it, when it was presented as a memoir to the Academy of Sciences will be found in the *Comptes rendus*, t. IX. pp. 200—210, 1839. According to the report the experiments were made on beams and circular arches of as much as 12 metre span and on frame works (*assemblages de charpentes*) made of arcs or straight pieces. The author also wished to test the results of Navier on the laws of flexural resistances and of resistance to rupture.

Que les théories établies à ce sujet, par M. Navier, et les expériences citées de M. Reibell[1], ne paraissent point avoir amené une conviction entière dans l'esprit des constructeurs (p. 201).

The main part of the experimental results seem to have referred to the strength of frameworks made up of circular arcs; and of the theory of such frameworks Ardant appears in his book to have treated in notes.

Les formules rapportées dans le texte du Mémoire de M. Ardant, se trouvent justifiées dans les Notes fort étendues qui l'accompagnent. La marche analytique suivie par l'auteur est analogue à celle qui a été

[1] Some experiments of Reibell's are to be found in the *Annales maritimes et coloniales*, 22e année, 2e série, Tome XI., but I have not had access to this publication.

exposée par M. Navier, dans l'ouvrage déjà cité; mais elle se trouve appliquée à des cas que ce savant ingénieur n'avait point considérés, et qui concernent les assemblages des formes droites avec les cintres circulaires continus. Les questions de cette espèce sont très délicates, et conduisent à des expressions analytiques fort compliquées; elles ne peuvent être abordées d'une manière un peu simple, qu'à l'aide de suppositions plus ou moins arbitraires sur le mode d'action des forces qui sollicitent les différentes parties des assemblages; on ne peut s'attendre à des résultats qui offrent tous les caractères d'une exactitude mathématique. Ceux auxquels M. Ardant est parvenu dans ses Notes, sont subordonnés aux hypothèses admises; ils nous paraissent suffisamment appropriés à la nature particulière de la question, quoique le rôle des résistances y ait peut-être été un peu exagéré (p. 204).

The work of Navier referred to is of course the *Leçons* (see Art. 279). The paragraph may give some idea of Ardant's method and its results. The report speaks very highly of his work (p. 210).

[938.] H. Sonnet. *Sur les vibrations longitudinales des verges élastiques*, Paris, 1840. This is a *Thèse de Mécanique* presented to the Paris Faculty of Science. It contains probably the most complete treatment of the stretch vibrations of thin rods at that time published. Although a good deal of the matter considered is now common property of the elementary text-books, yet some of Sonnet's problems do not seem to be mentioned by later writers.

The case considered by the author is that of a *vertical* rod built in at one end, the other end being either free, loaded or subject to impact; various forms of initial conditions are treated. We may note the following points:

(1) An investigation of the possibility of nodes when the initial shifts and velocity are not represented by simple harmonic terms of the same argument (pp. 11—12).

(2) In the further consideration of (1) the author proves that a wave reflected from a fixed end is of the same nature as the incident wave, but that reflected at a free end it changes its nature, i.e. a wave of extension becomes one of compression and *vice versa* (pp. 17—18).

(3) By means of the interference of two sets of waves a system of agitation is determined which produces nodes. This is used to explain how a note higher than the fundamental one may be obtained from a violin string:

C'est à ce mode d'ébranlement que semble devoir se rapporter l'action de l'archet lorsqu'on le fait glisser perpendiculairement à l'éxtremité d'une verge élastique ($l$). Ses aspérités exercent sur la verge, indépendamment d'une flexion transversale dont nous faisons abstraction ici, des pressions longitudinales; ces pressions se succèdent à des intervalles très rapprochés, eu égard à nos moyens de mesurer le temps, mais qui peuvent ne pas l'être par rapport à la vitesse $a$ de la propagation des ondes. Si ces ébranlements sont sensiblement identiques, se reproduisent à des intervalles sensiblement égaux à leur propre durée, et qu'enfin chacun de ces intervalles de temps soit un sous-multiple impair de $2l/a$, il y aura formation de nœuds et par suite, comme nous l'avons vu, production d'un son plus élevé que le son fondamental (p. 20).

(4) The author now treats the rod as *heavy*. He shews that, in the case where the rod has no initial shift or velocity, the greatest extension produced would require a tension equal to half the weight of the rod (p. 26). The maximum vis viva is equal to 2/3 of the work necessary to raise the rod a height equal to this maximum extension (p. 27).

(5) The free extremity of the rod is loaded with a weight $P$ to which a velocity $V_0$ is given. It is proved that if the weight of the bar is small compared with $P$, the bar executes sensibly isochronous oscillations of period $2\pi l\sqrt{P}/(a\sqrt{p})$, where $p$ is the weight of the rod (p. 31). When the velocity $V_0$ is zero it is shewn that the maximum extension is the sum of the statical extension which would be produced in the bar acted upon only by its own weight together with twice the extension which would be statically produced by the load $P$ (p. 32). See Poncelet's results, Art. 988.

(6) The paper concludes with the consideration of the vibrations which would be produced in a bar loaded with a weight $P$ as in (5), if this load received a blow from a second weight $P'$ so that two united by the blow acted afterwards as a combined load (pp. 32—35).

[939.] Eaton Hodgkinson. *On the Effect of Impact on Beams*, and *On the direct tensile strength of Cast Iron*. Report of the Third Meeting of the British Association (Cambridge, 1833), 1834 (pp. 421—424).

The first paper contains some interesting experiments on *lateral* impact. A cast iron ball of 44 lbs. weight is suspended by a cord of 16 feet and allowed to fall through various arcs in such manner as to strike normally the side of a horizontal bar of 4 feet supported at its ends. The ball just touches the bar when at rest. Deflections were measured by the depth which a long peg, touching the back of the bar, had been driven by the blow into a mass of clay placed there.

The following results were obtained:

(1) The deflections were nearly as the chords of the arcs through which the weight was drawn, that is as the velocities of impact.

(2) The same impact was required to break the beam, whether it was struck in the middle or at quàrter span.

This result was generalised by Hodgkinson in a later memoir: see our Art. 942, (v).

(3) When the impacts in the middle and at quarter span were the same, the deflection at the latter place was to that at the former nearly as 3 : 4.

These deductions Hodgkinson found to be in agreement with a theory based on the following suppositions, which however do not seem to me of an entirely satisfactory character: (*a*) that the form of a beam bent by small impacts was the same as if it had been bent by pressure through equal spaces. (*b*) That the ball and beam when struck proceeded together after impact as one mass. Further from this theory the following results were deduced:

(4) The power of a heavy beam to resist impact is to the power of a very light one, as the sum of the inertias of the striking body and of the beam is to the inertia of the striking body.

(5) The time required to produce a deflection, and consequently the time of an impact, between the same bodies, is always the

same, whether the impact be great or small. The time, moreover, is inversely as the square root of the stiffness of the beam. (Is this stiffness in Euler's, Young's or Poncelet's sense? See our Arts. 65, 138 and 979.)

(6) The result of calculations, comparing pressure with impact, gave deflections agreeing with the observed ones, within an error of about 1/8 or 1/9 of the results.

We have drawn attention to these experiments, because it is the problem of resilience which will occupy much of our attention in the next period.

[940.] The second paper has for object the determination of the disputed magnitude of the tensile strength of cast-iron: see Art. 377. According to Hodgkinson its very diverse values as given by different experimenters are due to the difficulty of applying the tractive load exactly at the centre of the terminal section of a bar[1].

[941.] Eaton Hodgkinson. *On the Collision of imperfectly Elastic Bodies. Report of the Fourth (or Edinburgh) Meeting of the British Association in* 1834, London, 1835 (pp. 534—543).

This is an investigation of the dynamic coefficient of elasticity in Newton's sense, that is, the ratio of the relative velocity of recoil to the relative velocity of impact in the case of the direct collision of spherical bodies. Hodgkinson in fact repeats the experiments given by Newton in the *Principia* (Scholium to Corollary VI.). The object of the paper is to connect the coefficients of elasticity in Newton's sense with the stretch-moduli. If $e$ and $e'$ be the impact elasticities for, say, glass against glass and brass against brass, and $E$, $E'$ the corresponding stretch-moduli, Hodgkinson gives for the impact elasticity of glass against brass the expression

$$\frac{Ee' + E'e}{E + E'}$$

[1] Professor A. B. W. Kennedy tells me, however, that different qualities of cast-iron, especially if melted a different number of times (which is always involved in the use of 'scrap' cast-iron), have very varying tenacities even when tested in precisely the same manner.

His experiments in no case shew a difference greater than 1/9 between the calculated and observed values of the impact coefficient.

[942.] Eaton Hodgkinson. *Impact upon Beams. Report of the Fifth (or Dublin) Meeting of the British Association in* 1835. London, 1836, pp. 93—116.

This paper is a continuation of the paper considered in our Art. 939, and its object is 'an inquiry into some of the effects of impact upon beams when struck by bodies of different weight, hardness and elastic force.'

The following conclusions are first arrived at from experimental data:

(i) If different bodies of equal weight, but differing considerably in hardness and elastic force, be made to strike horizontally with the same velocity against the middle of a heavy beam supported at its ends, all the bodies will recoil with velocities equal to one another.

(ii) If, as before, a beam supported at its ends be struck horizontally by bodies of the same weight, but different hardness and elastic force, the deflection of the beam will be the same whatever body be used.

(iii) The quantity of recoil in a body, after striking against a beam as above, is nearly equal to (though somewhat below) what would arise from the full varying pressure of a perfectly elastic beam as it recovered its form after deflection.

(iv) The effect of bodies of different natures striking against a hard flexible beam seems to be independent of the elasticities of the bodies, and may be calculated, with trifling error, on the supposition that they are inelastic.

Hodgkinson here notes that Young in his *Natural Philosophy*, and Tredgold in his *Treatise on the Strength of Cast Iron*, make this assumption without, apparently, noticing that it is one: see our Art. 999 and Note A. (3) of the Appendix. The explanation given by the author is hardly satisfactory.

(v) The power of a uniform beam to resist a blow given horizontally is the same in whatever part it is struck: see our Art. 939, (2).

(vi) The power of a heavy uniform beam to resist a horizontal impact is to the power of a very light one as half the weight of the beam, added to the weight of the striking body, is to the weight of the striking body alone.

(vii) The power of a uniform beam to resist fracture from a light body falling upon it (the strength and flexibility being the same) is greater as the weight of the beam increases, and greatest when the weight of half the beam, added to that of the striking body, is nearly equal to one-third of the weight which would break the beam by pressure.

The use of the word 'power' in V.—VII. is extremely vague, taken to mean that it requires the same blow to break the beam, whatever the point struck, it is only a rough approximation to fact: see Note E. in the Appendix.

[943.] Hodgkinson then proceeds to unfold a theory which shall be in harmony with all these conclusions. In the light of more recent work this theory cannot be considered as more than a first approximation. There is a difference occasionally of 1/6 between the calculated and observed values of both recoil and deflection. A certain constant $r$, termed the inertia of the beam' and deduced from experiment to be one-half the weight (p. 101), occurs also by Tredgold, but does not seem to me clearly defined[1]. Thus if $w$ be the weight of ball giving the impact $w+r$ is used as 'the mass moved,' or apparently, it should be the weight of the mass moved. Hodgkinson obtains the following results:

Greatest velocity of recoil of the ball striking horizontally

$$= b\sqrt{\frac{gp}{e\,(w+r)}}.$$

Deflection due to horizontal impact

$$= b = w\sqrt{\frac{2he}{p\,(w+r)}}.$$

[1] A more accurate value for $r$ than $\frac{1}{2}$ the weight has been deduced by Homersham Cox in a memoir which we shall consider in Chap. VIII. See also Saint-Venant in the *Historique Abrégé*, p. ccxxii. Both these writers find $r = 17/35$ of the weight. Hodgkinson's experimental results seem fairly in accordance with the more complete theory of Cox and Saint-Venant. Saint-Venant's theory will be found fully discussed in his edition of *Clebsch*, pp. 490—627 We shall of course consider the original memoirs in our second volume.

Here $p$ is the pressure which if applied to the middle of a beam would produce a deflection $e$; $h$ is the height fallen through by the ball, and $g$ is the measure of gravity.

Somewhat more complex formulae are obtained for a vertical impact, but it does not seem necessary to consider them here. The paper concludes with some experiments on the effect which loading a wire has in increasing its resistance to a blow. The blow appears to spend itself on the inertia of the load. Hodgkinson argues that this shews clearly the benefit of giving considerable weight to elastic structures subject to impact and vibration (p. 116).

[944.] *British Association Reports.* There are several reports in the *Transactions of the British Association* for the years 1833—1837 which concern our subjects. We will briefly note them.

[945.] *Report on the Present State of our Knowledge respecting the Strength of Materials. Report of Third (or Cambridge) Meeting of* 1833—1834, pp. 93—103. This report is by Peter Barlow, and possesses no present value; indeed it can hardly be considered a contribution to the history of the subject. Barlow exhibits here as elsewhere a want of theoretical grasp. It may be noted that he acknowledges Eaton Hodgkinson's correction of his error in the earlier editions of his *Essay on the Strength of Timber* (see our Arts. 192, 233 and p. 96 of the Report).

[946.] *On the difference between the Composition of Cast Iron produced by the Cold and Hot Blast.* By Thomas Thomson. *Report of Seventh (or Liverpool) Meeting of* 1837—1838, pp. 117—126. This is a purely chemical investigation of the composition of various kinds of iron and is not very closely related to the two following reports.

[947.] *On the relative Strength and other mechanical Properties of Cast Iron obtained by Hot and Cold Blast.* By Eaton Hodgkinson, *Ibid.* pp. 337—375. This report, which possesses very considerable experimental value and corresponding theoretical interest, is to a great extent reproduced in the *Experimental Researches on the Strength and other Properties of Cast Iron*, London, 1846, which forms a second part to *Tredgold's Practical*

*Essay on the Strength of Cast Iron*, 4th edition, by Hodgkinson in 1842: see our Arts. 966—973.

[948.] The chief point to be noted in the paper is the results obtained from short prisms of various shapes broken by a crushing force. We have already noted Coulomb's theory of this crushing force (Art. 120), as well as Rennie's and Vicat's experiments (Arts. 186, 729). The mode of fracture is extremely interesting, and the forms on the plate facing p. 346 should be noted and compared with Vicat's results (see Art. 729). Hitherto a mathematical theory does not seem to have been given for the extremely regular shapes of fracture which constantly recur; we except of course Coulomb's erroneous hypothesis. Yet, as Hodgkinson remarks:

When a rigid (*sic*) body is broken by a crushing force which is prevented from acting after it has effected a rupture, it will be found not to be crumbled or reduced to a shapeless mass, but to be divided according to mathematical laws, and sometimes into very interesting forms of fracture. The accompanying plate will shew how the fracture was effected in a variety of cases, and that they were all subject to one pervading law.

[949.] The 'pervading law,' according to Hodgkinson, seems to be the formation of a cone or wedge which slides off at a nearly constant angle. In each prism two cones or wedges will be formed which do not meet directly but have sharp points and slip past each other to effect the destruction of the piece of which they are formed (pp. 348—349). Hodgkinson measured the angle of the wedge or cone and found with certain variations the mean angle varied from 53°. 30′ to 56°. 43′. He remarks (p. 350):

From the preceding examination of the angles obtained from specimens of different forms and lengths, it appears that amidst great anomalies, there is, taking the *mean* results, a considerable approach to equality, as is more particularly shewn from the angles of the cylinders and rectangular prisms; and this approach would doubtless have been greater and the anomalies less if the specimens had always been longer than the wedge. The defect in the angle from this cause is evident in the shorter rectangular prisms and has been alluded to before.

We may assume therefore, without assignable error, that in the crushing of short cast-iron prisms of various forms longer than the wedge, the angle of fracture will be the same. This simple assumption, if admitted, would prove at once, not only in this material but in others which break in the same manner, the proportionality of the crushing force in different forms to the area, since the area of fracture would always be equal to the direct transverse area multiplied by a constant quantity dependent on the material.

[950.] Experiments on bone, marble and timber confirmed these results (p. 352), which are in tolerable accordance with those of Vicat: see Art. 729.

[951.] For the developement of a mathematical theory experiments are required to shew how far the surface of fracture depends (i) on the direction of casting, (ii) on a change of elastic constants towards the surface, but especially, (iii) *on the fact that the ends of the prism are practically incapable of expanding owing to the friction of the compressing surfaces.*

[952.] Following the experiments on crushing are some on transverse force. These confirm the results of the *Manchester Memoirs* (see Arts. 230, 237, 244) that the strengths of the bottom and top ribs of a cast-iron beam should be as 6 or $6\frac{1}{2}$ to 1; a ratio found to be nearly that of the tensile to the compressive strength of the iron (pp. 361 and 364). Hodgkinson also found:

(i) That with cast-iron beams subject to transverse load there is no load however small which will not produce some set. It does not appear however from his mode of experimenting whether any part of this was really elastic after-strain. He concludes, I think somewhat too generally:

> It is evident that the maxim of loading bodies within the elastic limit has no foundation in nature; but it will be considered as a compensating fact, that materials will bear for an indefinite time a much greater load than has hitherto been conceived.

There is here no evidence that the writer had any conception of the 'state of ease,' or had attempted to reduce his bars to that state. He found little difference in the quantity of the set, whether it arises from tension or compression; the latter being however somewhat less (p. 363).

(ii) That Emerson's paradox (see our Art. 187) is no paradox, or the frustrum is weaker than the triangle (p. 364). Here it seems to me that theoretically the triangle would certainly shew rupture at a less load than the frustrum, but it does not follow because the vertex of the triangular prism was ruptured that the triangle itself would break at a *less* load than the frustrum. I do not think Hodgkinson's experiments on this point are at all conclusive: see Saint-Venant's edition of Navier's *Leçons*, pp. 94—102.

(iii) That the assertion of practical men, that if the hard skin at the outside of a cast-iron bar be removed, its strength, comparatively with its dimensions, will be much reduced, is not true (p. 364). Hodgkinson's experiments seem hardly conclusive; even if the strength be not reduced (which is in itself questionable) there can be little doubt that the change in the elasticity towards the surface would effect the strain (e.g. deflections) within the limit of elasticity.

The general conclusions of the paper as to the effect of hot and cold blast will be found on p. 375, but they do not concern our present purpose.

[953.] *On the Strength and other Properties of Cast Iron obtained from the Hot and Cold Blast.* By W. Fairbairn. *Ibid.* pp. 377—415. This contains the results of experiments with similar aims to those of Hodgkinson. Some points which bear upon theory may be noted here.

(i) *Time Effect.* The experiments were on deflection by transverse load and the weights were such as nearly to produce fracture, the sets increased gradually with the time, but there is no sufficient evidence to prove how much may have been really elastic after-strain. The limit of the load for which set ceased after a time to increase does not seem to have been ascertained from the experiments. With regard to the general result Fairbairn remarks:

There cannot be a doubt that the phenomenon of cohesive force is strongly developed in the preceding Tables; the minute crystalline particles of the bars are acted upon by loads, which, in the heavier weights, are almost sufficient to produce fracture: yet fracture is not (except in one instance) produced, and to what extent the power of resistance may yet be carried is left for time to determine. It nevertheless appears from the present state of the bars (which indicate a slow but progressive increase in the deflections) that we must at some period arrive at a point beyond their bearing powers; or otherwise to that position which indicates a correct adjustment of the particles in equilibrium with the load. Which of the two points we have in this instance attained is difficult to determine: sufficient data are however adduced to shew that the weights are considerably beyond the elastic limit, and that cast-iron will support loads to a much greater extent than has usually been considered safe, or beyond that point when a permanent set takes place (pp. 204—405).

Obviously little is determined by Fairbairn's experiments. He suggests that Hodgkinson or Barlow should investigate these matters mathematically, thus shewing how unconscious he was of the infinite difficulties which beset the subject of after-strain.

(ii) *Temperature Effect.* The breaking load was found to decrease considerably with increase of temperature, but I can find in the tables no obvious and general law for the stretch-modulus. It appears in fact with cold-blast iron to increase from 26° to 190° Fahrenheit, but with hot-blast iron to decrease between the same temperatures.

Some further experimental details of Fairbairn's researches will be found in the *Manchester Memoirs,* Vol. VI., 1842, pp. 171—273 and 524—560.

[954.] Eaton Hodgkinson. *Experimental Researches on the Strength of Pillars of Cast Iron and other materials. Philosophical Transactions,* 1840, Part II., pp. 385—456, with three plates. The object of this very important memoir is to discover the laws governing the strength of pillars. Hodgkinson was induced to undertake the investigation by Robison's remarks on Euler's theory (see Art. 145) and Barlow's strongly expressed opinion as to the want of satisfactory rules for practice. After a description of apparatus, Hodgkinson begins the account of his experiments with the following words:

In order to ascertain the laws connecting the strength of cast-iron pillars with their dimensions, they were broken of various lengths, from five feet to one inch; and the diameters varied from half an inch to two inches, in solid pillars; and in hollow ones, the length was increased to seven feet six inches, and the diameter to three inches and a half. My first object was to supply the deficiencies of EULER's theory of the strength of pillars, if it should appear capable of being rendered practically useful; and if not, to endeavour to adapt the experiments so as to lead to useful results. (p. 386.)

[955.] The cast-iron pillars were all of one material, and according to the writer the method of casting makes little or no difference in the strength, a result we should hardly have expected. The first series of experiments are a comparison of the strength of pillars with rounded ends and with flat ends. Here the results are extremely interesting. So long as the length was about 80 times or more the diameter the relative strength of pillars with flat and pillars with rounded ends was in the mean as 3·167 : 1. This ratio increased slowly when the length became a greater multiple of the diameter. It decreased rapidly on the other hand when the length was less than 30 times the diameter; in these cases however the breaking was by crushing rather than by flexure, or at all events the load produced by its crushing force a *set* before flexure began (p. 388). This was confirmed by experiments on wrought iron and wood. Hodgkinson terms the 'crushing load' that which would crush a short prism perfectly built in at the ends, and concludes from his experiments that:

About one-fourth of the crushing weight is the greatest load which a cast-iron pillar, flat at the ends, will bear without producing a crushing or derangement of the materials, which would lessen its breaking weight; and that the length of such a pillar should be thirty times the diameter or upwards. Pillars whose length is less than in this proportion, give the ratio of the strengths of those with rounded and with flat ends, from 1 : 3 down to 1 : 1½ or less, according as we reduce the number of times which the length exceeds the diameter, as will be seen by the abstract (see pp. 387 and 389).

[956.] A point mentioned by Hodgkinson on p. 389 should also benoted with regard to the rounded ends:

It became necessary to render those which were rounded at the ends more flat there than if the ends had been hemispheres; whilst in the experiments upon pillars, whose length was greater with respect to the diameter than these, the ends were more prominent than in the hemispherical form. This change became necessary on account of the splitting of the ends of the short pillars; it having been found that the pillars whose diameter was 1/13th of the length or upwards with rounded ends failed in many instances by the ends becoming split. In these cases a portion of the rounded end of the pillar formed the base of a cone, whose vertex was in or near the axis of the pillar. This cone acting as a wedge whose sides were in the angle of least resistance, and having its vertex sharp, split and cut up the sides of the pillar of which it formed a part.

[957.] I do not follow the reasoning of the last remarks, but I draw attention to the paragraph as shewing that Hodgkinson had not made his ends true pivots, so that all the force on the pillar was necessarily in the axis. Euler's theory in its improved form tells us that if the crushing force be really in the axis a doubly pivoted strut will break in the centre only (supposing a considerable ratio of length to diameter and the pivots of sufficient strength). On the other hand I find that a doubly built-in strut would break in that theory not only in the middle but at the two ends at the same time[1]. A strut with *flat* ends differs from a doubly built-in strut in that it may before or just at breaking swing round on the edges of its end sections. If it breaks before swinging round it will break as a doubly built-in strut, if after swinging round as a doubly-pivoted strut. I cannot find that Hodgkinson has noted these important distinctions, but they seem very essential in judging of the theory. I find that the strut would swing round on its edges when the deflection was equal to the diameter. A statement also of the writer's on p. 390, that till recently in all inquiries respecting the strength of materials, bodies have been assumed to be incompressible, is by no means historically accurate. The next step in Hodgkinson's experiments is of peculiar value, he casts large discs at the ends of his pillars and then finds their breaking loads; here he is practically treating doubly built-in struts. The result he arrives at for *long* cast-iron struts is: *that a long uniform cast-iron pillar, with its ends firmly fixed, has the same power to resist breaking as a pillar of the same diameter and half the length with the ends rounded or turned so*

[1] Such a strut broken at the ends and the centre is figured on our frontispiece. It is one out of several recently tested in the Engineering Laboratory of University College.

*that the force would pass through the axis.* If as in Euler's theory the breaking force is inversely as the length squared, we find that a doubly-pivoted strut has *one-fourth* the breaking load of a doubly built-in strut. Examining Hodgkinson's figures we find it always a *little* less. Hodgkinson next considers struts with one end pivoted and the other flat or built-in. The general conclusion he arrives at is: *that the strength of a pillar with one end pivoted and the other built-in is always an arithmetical mean between the strengths of pillars of the same dimensions with both ends pivoted and both built-in* (p. 393). We note that in Hodgkinson's results this mean is generally somewhat *greater*. The breaking point of a pivoted built-in beam was a little more than 1/3 from the pivoted end.

[958.] We have now before us Hodgkinson's first general results. How far are they in accordance with theory? This is how Hodgkinson himself judges the matter:

The theory of the strength of pillars, as given by Euler and Lagrange and afterwards pursued by Poisson and others, furnishes us with little information upon these subjects....

The strength is much influenced, as has been previously observed, by the quantity of compression which the pillar sustains; and consequently, by the position of the neutral line when the pillar is bent The strengths, too, are different in their definitions in the two cases. In the theory of Euler, the strength is estimated by the greatest weight which a pillar would bear without flexure; whilst in the present case, the estimate is formed upon the weight which would break the pillar by flexure. I have sought, on many occasions, but without success, to determine experimentally some fixed point, according to the definition of the continental theory. So far as I can see, flexure usually commences at very small weights, such as could be of little use to load pillars with in practice. It seems to be produced by weights much smaller than are sufficient to render it capable of being measured. I am, therefore, doubtful whether such a fixed point will ever be obtained, if indeed it exists. With respect to the conclusions of some writers, that flexure does not take place with less than about half the breaking weight, this as is evident from my experiments, taken in general, could only mean a large and palpable flexure; and it is not improbable that the writers were in some degree deceived from their having generally used specimens thicker, compared with their lengths, than have usually been employed in this paper. Some results of the continental theory we shall, however, find of great service further on.

[959.] So far Hodgkinson, but I venture to think Euler's theory in a modified form may help us even in the results cited above. If we *assume* that Euler's bending load is proportional to the final breaking load for struts fixed in various fashions, we shall obtain very similar results to Hodgkinson. The reasons for this assumption will more clearly appear from the consideration of Lamarle's memoir in our Chapter VIII.

I have calculated the following results by modifying Euler's theory so as to allow for the compression and for the varying position of the neutral axis.

Force which will just bend a doubly built-in strut

$$= E\omega \frac{\frac{\pi^2\kappa^2}{l^2}}{1 + \frac{\pi^2\kappa^2}{l^2}},$$

Force which will just bend a built-in pivoted strut

$$= E\omega \frac{\frac{\pi^2\kappa^2}{4l^2}\,2{\cdot}04}{1 + \frac{\pi^2\kappa^2}{4l^2}\,2{\cdot}04},$$

Force which will just bend a doubly pivoted strut

$$= E\omega \frac{\frac{\pi^2\kappa^2}{4l^2}}{1 + \frac{\pi^2\kappa^2}{4l^2}}.$$

*Absolute* strength as deduced from these expressions seems often to have *very* erroneous values, but comparative results, as those numbered (1)—(3) below, appear to be experimentally verified.

Point of a pivoted built-in strut where the traction in the fibres is greatest = nearly $\cdot 35l$ from the pivoted end.

Here $\omega$ is the area of the section, $\kappa$ its radius of gyration about an axis through its centre of gravity perpendicular to the plane of flexure, $l$ the length of the strut, and $E$ the stretch-modulus. From these results we deduce on the assumption made above that:

(1) The breaking load of a doubly built-in strut is very little *less* than four times as great as that of doubly-pivoted beams, provided the ratio $\kappa/l$ is small, but decreases as this ratio increases.

(2) The breaking load of a pivoted built-in strut is approximately the arithmetic mean of the breaking loads of doubly-pivoted and doubly-built in struts of the same size. The mean is however somewhat greater.

(3) A pivoted built-in strut will break at a little more than a third from the pivoted end. Hodgkinson gives as an example one case (p. 395) in which it broke at ·34$l$. Our theory gives ·35$l$.

These remarks seem to me sufficient to point out that there is a value in the Eulerian hypothesis, if it be properly *modified* so as to include the effect of compression. I conceive Hodgkinson's results would be even more in accordance with this theory had his pivoted ends been more satisfactory, and had notice been taken of a possible tendency of the flat-ended struts to swing round on their edges. It is easy to calculate the theoretical load at which this happens.

[960.] The next point to which Hodgkinson turns is the *strength of long pillars as dependent upon their dimensions* (p. 395). Here he remarks that Euler's computations, although referring to *incipient flexure*, do not seem very widely different from those which apply to the breaking point. Euler's theory makes the strength proportional to $\frac{\omega\kappa^2}{l^2}$, the modification of it to $\frac{\omega\kappa^2}{l^2}\left(1-\frac{\pi^2\kappa^2}{\gamma l^2}\right)$ nearly, if $\frac{\kappa}{l}$ be small, where $\gamma$ is a constant depending on the terminal conditions. Euler thus makes the strength vary as the 4th power of the diameter; the modification introduces another term varying as the 6th power, and also the terminal conditions affect the result. Hodgkinson makes it vary as the *nth* power of the diameter, where the mean value of $n = 3·7$, about, for both circular and square sections. If it be noted that the mathematical theory is not properly applicable to fracture, and that Hodgkinson gets considerable variations in his $n$ (3·4 to 3·9) depending on the pillars from which it is calculated, it does not seem unreasonable to suppose that the modified theory would give as accurate results as the empirical assumption $n = 3·7$.

The result $n = 3·7$ is obtained for pivoted ends; for flat ends Hodgkinson gives $n = 3·5$ (or 3·6), and for built-in ends 3·7 (see his pages 396—397). It will be observed that the modified theory also distinguishes between these differently treated struts.

[961.] We then have an investigation of what the inverse power $m$ of the length may be. Hodgkinson finds:

$$\left.\begin{array}{ll}\text{For pivoted struts} & m = 1·799 \text{ to } 1·583 \\ \text{... flat-ended ...} & m = 1·625 \text{ to } 1·587\end{array}\right\} \text{mean values.}$$

The highest value obtained is 1·914 and the lowest is 1·537. He assumes 1·7 as a working value, but remarks that $x$ increases regularly in value as $d/l$ diminishes or the breaking load becomes less, $d$ being the diameter. Its limit is probably 2. Here again it does not seem unreasonable to suppose the modified theory might be as satisfactory as this empirical value.

Finally then Hodgkinson takes $d^{3·76}/l^{1·7}$ as a comparative measure of the breaking weight for pivoted struts and $d^{3·55}/l^{1·7}$ for flat-ended struts (p. 400 *et seq.*). The first he approves only for pillars where $d/l < 1/15$, the greatest error in his experiments is about 1/8 the breaking load; the second only for pillars where $d/l < 1/30$, and the greatest error is about 1/9.

[962.] The further remarks on breaking strength of short and of similar struts are interesting, but must be passed over here (pp. 403—408) as having no great bearing on theory. Hodgkinson gives an empirical formula usually attributed to Gordon or Rankine.

[963.] Pages 409—417 are principally occupied with experiments on hollow cylindrical struts, and others of varying section; an empirical formula for hollow cylinders is obtained on p. 414, similar to those given above. The general results of these experiments seem to me fairly in accordance with theory, at least in its modified form; for example, the conclusion that local reductions of the thickness of a strut can be made without reducing the breaking load (p. 412). This, as well as the modified theory, completely refutes Lagrange's statement as to a strut of uniform circular section being that of the strongest shape: see Art. 113.

[964.] We have the time-element taken into consideration in four experiments on p. 418. The deflections were measured for more than a year in the case of four struts loaded with 448, 784, 1120, 1456 lbs. respectively, the struts were practically equal and had an immediate breaking load of 1500 lbs. The first two do not seem to have experienced any continuously increasing deflection; in the case of the third a slight increase is noticeable (·215 inches to ·38); the fourth had a deflection which increased rapidly from ·25 to ·955, and it broke just before the six months were completed. The experiments are not however sufficiently exhaustive for us to draw any important inference from them.

[965.] Hodgkinson then proceeds to calculate various physical constants of the cast-iron used by him (pp. 418—420); to give the results of some experiments on wrought-iron columns, for which he gives $n$ the same value as for cast-iron, but puts $m = 2$ (pp. 420—424); and finally to consider formulae for timber columns based partly on results of his own, partly on those of other experimenters. Here for long columns he adopts Euler's theory, for short, that propounded by himself (see Art. 962). He cites also the *British Association Report* for 1839: see our Art. 950 (pp. 424—430). The remainder of the paper (pp. 431—456) is occupied with the tables of experimental results. The theoretically interesting part of these tables, as of the plates, is the confirmation they give to theory in the variable position of the neutral surface, especially as evidenced in some cases by a wedge being broken out in the middle, i.e. theory shews us that the neutral surface first enters the beam at the centre, and in the case of doubly built-in struts simultaneously at the two ends: see the footnote to Art. 75.

[966.] *Experimental Researches on the Strength and other Properties of Cast Iron.* This forms a second volume by Eaton Hodgkinson to the fourth edition of Tredgold's *Strength of Cast Iron,* and was published in 1846. It is so closely connected with the contents of the memoirs which we have been considering in this chapter, that it seems best to notice it here[1]. It contains practically an abstract of the results obtained by Hodgkinson and Fairbairn, together with others given by Rennie, Navier and Bramah: see our Arts. 185, 279, 837.

[967]. The *Manchester Memoirs*, the British Association papers, and that of the *Philosophical Transactions* are here substantially reproduced. On p. 375 some later experiments of Fairbairn on the effect of long continued load are given. The duration of the load amounted in some cases to five years. It would appear from these experiments that the deflections of the beams increased considerably for the first twelve or fifteen months, after which time there was a smaller increase in their deflections although from four to five years elapsed:

> The beam in Experiment 8, which was loaded nearest to its breaking weight, and which would have been broken by a few additional pounds laid on at first, had not, perhaps up to the time of its fracture, a greater deflection than it had three or four years before; and the change in deflection in Experiment 1, where the load is less than 2/3 of the breaking weight, seems to have been almost as great as in any other; rendering it not improbable that the deflection will, in each beam, go on increasing till it becomes a certain quantity beyond which, as in that of Experiment 8, it will increase no longer, but remain stationary.

These results are fairly in accordance with Vicat's: see Art. 736.

[968.] Some further experiments of Fairbairn on the effect of temperature are given. They do not bear out those of Art. 953, (ii). For with some anomalies, it would seem that cast-iron has practically a not widely different breaking load for all temperatures from 16° to 600°. A fuller description of these experiments will be found in the *British Association Report* for 1842, pp. 88—92.

[969.] On pp. 407—411 we have a section entitled: *Defect of Elasticity*. This is an abstract of a paper laid before the British Association at its Cork meeting in 1843, a fuller account of

[1] A French translation, in a somewhat abridged form by Pirel, will be found in the *Annales des ponts et chaussées*, 1855, 1[er] *semestre*, p. 1.

which will be found in the *Report* for that year, Part II. pp. 23—25. In the *Report* for 1844 will also be found an account of some similar experiments (Part II. pp. 25—27) undertaken with more accurate apparatus, in order to meet some objection which had been raised against the previous series. In these experiments the set in cast-iron and soft stone was found to vary nearly as the square of the load; it followed a different law, which Hodgkinson had not fully determined, in wrought-iron and steel. He expresses his opinion that no body is perfectly elastic, but that the slightest change of form produces some, however small, set.

The main point to be noticed is, that Hodgkinson found the set of cast-iron to be proportional to the square of the load:

> If, therefore, $x$ represents the quantity of extension or compression which a body has sustained, and $ax$ the force producing that extension or compression on the supposition that the body was perfectly elastic; then, the real force $f$, necessary to produce the extension or compression $x$ will be smaller, than on the supposition of perfect elasticity, by a quantity $bx^2$; and we shall have $f = ax - bx^2$.

It does not appear from Hodgkinson's experiments that he investigated whether any of his set was of the nature of elastic after-strain. This law of Hodgkinson's will be found reproduced in many text-books, but it will be seen that it is nothing more than what had been given by Gerstner ten years previously: see Art. 806. It should be noticed that by *defect of elasticity* here Hodgkinson means something quite different to what he understands by that term in the *Report* of the Iron Commissioners: see our Chapter VIII.

[970.] Pp. 464—474 contain a discussion of some experiments of Bramah's (see Art. 837) and others of Cubitt's published in 1844. The latter would appear at first sight to condemn Hodgkinson's beam of strongest section (see Art. 244), but Hodgkinson's criticism of them is perfectly just, Cubitt had made a beam of stronger section than Hodgkinson by *increasing its depth, one of the constants of Hodgkinson's problem.* The experiments therefore are comparatively worthless and we shall not refer to them again.

[971.] There is a theoretical investigation of the position of the neutral line (pp. 483—494) on the assumption that the positive traction at a distance $x$ from the neutral line is proportional to

$$x - \frac{x^v}{na},$$

and the negative traction at a distance $x$ from the neutral line to

$$x - \frac{x^{v'}}{n'a'},$$

where $v$, $v'$, $n$, $n'$ are constants depending on the material and $a$, $a'$ the distances of the neutral line from the top and bottom of the beam respectively. The analysis is very straightforward, but it may be doubted if it be very valuable. A modified form of Hodgkinson's hypothesis is given by Saint-Venant, Navier's *Leçons*, pp. 176—184.

[972.] The volume concludes with a chapter on the Resistance to Torsion (pp. 495—504), in which one or two remarks deserve notice.

We have here (p. 496) the first appearance of a new formula for the torsion of square and rectangular beams. Namely the result obtained by Cauchy and mentioned in Art. 661. This result has been proved to be inexact by the later researches of Saint-Venant.

If $T/\mu$ be the limit of safe slide, $\mu$ being the slide-modulus, and $R/E$ be the limit of safe stretch, $E$ being the stretch-modulus, then for a perfectly isotropic body on the uni-constant hypothesis

$$\mu = 2E/5, \qquad T = 4R/5.$$

Further, we have for the couple $M$, producing the torsion $\tau$ of a prism of unit length, and for its limit $M_1$, the prism being upon a square base of side $d$:

On the old theory

$$M = \frac{1}{6}\mu d^4\tau, \qquad M_1 = \frac{\sqrt{2}}{6} d^3T;$$

On Saint-Venant's theory

$$M = \cdot 14058\, \mu d^4\tau, \qquad M_1 = \cdot 20817\, d^3 T.$$

Hodgkinson takes the first of these results and considers them in relation to experiments of Rennie, Bramah, Tredgold and Bevan: see Art. 378.

[973.] With regard to Bevan, Hodgkinson commends his results as carefully obtained and deduces theoretically one of his experimental conclusions (p. 500). If the reader will refer to Art. 378, he will see that Bevan defines his modulus of torsion, which he terms $T$ (not the same $T$ as that used above), by

$$\frac{M}{d^4T}=\frac{\delta}{r},$$

where we have supposed the prism of unit length; $\delta$ being the deflection and $r$ the arm of the distorting force. Now Hodgkinson remarks that 'as $\frac{\delta}{r}$ is the deflection at unit distance and very small it may be taken for the arc $\tau$.' Hence he finds:

$$T=\frac{M}{d^4\tau}$$

$$=\tfrac{1}{6}\mu \text{ on the old theory,}$$

or

$$=\cdot 14058\,\mu \text{ on Saint-Venant's theory.}$$

But on Poisson's supposition

$$\mu=2E/5,$$

$$\therefore\quad T=\tfrac{1}{15}E=\cdot 0\dot{6}\,E.$$

Bevan found that for metals $T=\frac{1}{16}E=\cdot 0625\,E.$
If we take the more correct results of Saint-Venant,

$$T=\cdot 056232\,E.$$

Thus the old hypothesis gives the theoretical value of $T/E$ as differing from Bevan's experimental value by ·0042 and the more correct theory by ·0063. As Bevan's 1/16 is only a round number, we may assume that for metals we can practically regard the relation

$$\mu/E=2/5$$

as very approximate.

The volume concludes with an abstract of Savart's memoir on torsion: see Art. 333, and reference to a paper of Rennie's on the effect of thermal expansion and contraction on iron structures: see Art. 838.

[974.] 1839. Poncelet. This distinguished mathematician gave courses of lectures at Metz in the years 1827—1829, wherein were first treated several points relating to our subject. These lectures were lithographed under the title: *Cours de Mécanique industrielle, fait aux artistes et ouvriers messins.* They appeared in parts. Part I., 1827—1828. Part II., 1828—1829 (in a second edition, 1831). Part III., 1831. The first part was also printed at Metz in 1829. A volume entitled: *Introduction à la Mécanique industrielle, physique et expérimentale,* was published at both Metz and Paris in 1839. It bears on the title page the words: *Deuxième Édition entièrement corrigée et contenant un grand nombre de considérations nouvelles.* The Metz book of 1829 is, I suppose, the first edition, but as that seems to contain only 248 pages and this 719 pages, the additions must be very considerable. A third edition edited by M. X. Kretz and forming a part of the proposed complete edition of Poncelet's works appeared in 1870, at Paris. It is almost identical with the second, the paragraphs but not the pages are the same. As the paragraphs are long, my references are to the pages of the second and third (K) editions; the former is that which Saint-Venant cites simply as Poncelet's *Mécanique industrielle*[1]. Various books published at Liege and Brussels entitled: *Mécanique industrielle* and bearing Poncelet's name were unauthorised compilations based on the lithographed editions. Poncelet refers to these compilations in the preface to his second edition. In the same place (p. vi) he also remarks that various causes have delayed the publication, so that the printing,

> parvenue à la page 224 dès l'année 1830, fut seulement reprise et continuée jusqu'à la page 273 en 1835, et jusqu'à la page 520 en 1838.

It is useful to note this fact as bearing on Poncelet's relation to contemporary work. We proceed to note the salient features of the *Mécanique industrielle.*

[1] I have endeavoured in vain to find a copy of the lithographed course. I owe to the kindness of M. Gustav Plarr an opportunity of examining the second printed edition. A bibliography of Poncelet is given after his life by Didion (*Notice sur la vie et les ouvrages du Général J. V. Poncelet,* Paris, 1869), but it is imperfect and inaccurate. I may take this opportunity of acknowledging the courtesy of M. le Bibliothécaire, of the Académie de Besançon, who kindly forwarded me a *Cours lithographié* of Poncelet belonging to the Bibliothèque universitaire. It proved to be his treatise on the mechanics of machinery which unfortunately contains nothing on our subject.

[975.] In the *Principes fondamentaux*, after some consideration of the molecular structure of bodies and their physical states and properties, we have a section entitled: *Élasticité des corps* (p. 12, K. 20). Poncelet remarks after his definition of elasticity that two kinds are to be distinguished, 'l'élasticité de *forme* et l'élasticité de *volume*.' This is very like the language to be found in current English text-books, and suggests bi-constant isotropy. On p. 14 (K. 22) the limit of elasticity is defined, but it is made to depend not on a limiting stretch, but on a limiting traction.

[976.] The next part of the work which engages our attention is entitled: *Des résistances que les corps opposent à l'action directe des forces* (p. 250, K. 269). This part treats in the first place of what we are to understand by the forces of affinity, adherence and cohesion, and involves some discussion of molecular forces, crystallisation and the influence of the now discarded caloric: see our Art. 844. On p. 260 (K. 280) we begin with the application of these results to the phenomena of elasticity, and the following pages 260—465 (K. 280—503) are occupied with our subject.

[977.] On pp. 260—272 (K. 280—293) will be found a discussion by graphical methods of the law of molecular force. Poncelet draws curves to represent respectively the repulsive and attractive elements of the force; such for example as might be obtained by tracing the relation between force and distance given in the footnote on our page 222. He deduces from his curves explanations of cohesion, perfect and imperfect elasticity etc. He even appears to arrive on p. 264 (K. 284) at something like the generalised Hooke's Law. His reasoning however seems to me invalid as it really refers to intermolecular force and not to stress. The use of the term *la force élastique* apparently for intermolecular force and afterwards for stress is very misleading.

[978.] On pp. 276—277 (K. 296—298) we have some remarks on differences of elasticity in the same body. The want of isotropy in wrought-iron bars and in wires is especially noted. A result of Savart's is quoted, which I have not come across elsewhere:

Un fait, d'ailleurs très-digne de remarque, c'est que, dans les corps cristallisàbles obtenus par la fusion, dans le plomb notamment, l'état d'agrégation et par conséquent d'élasticité, peut se modifier d'une manière extrêmement lente avec le temps, et sans qu'il s'en manifeste extérieurement aucune trace appréciable par les moyens ordinaires d'observation.

[979.] The next section is entitled *Notions et principes concernant la résistance des prismes aux allongemens, à la compression et à la rupture.* We may note (p. 279, K. 299) that Poncelet terms the *force ou résistance élastique* the *raideur* or stiffness per unit length; these terms are confusing A notable feature however is the representation of the stress-strain (in this case a traction-stretch) relation graphically by a curve. This method, perhaps introduced by J Bernoulli in 1694, had apparently been forgotten till Poncelet revived it see our Appendix, Note A. (1). Some of these curves have been plotted out by Poncelet (figs. 47—48), the abscissae and ordinates representing respectively the stretches and the corresponding tractions. The following remarks seem to indicate that Poncelet was acquainted with the *yield-point* as well as the *stricture* or local thinning down before rupture:

Ces courbes montrent, par l'inclinaison de leurs tangentes sur l'axe horizontal des abscisses, que la résistance élastique, qui d'abord reste sensiblement constante, diminue souvent d'une manière très-rapide à partir d'un certain terme, sans néanmoins devenir rigoureusement nulle, même pour des allongements très voisins de la rupture. Or cette derniere circonstance tient, sans aucun doute, à la difficulté qu'on éprouve à observer les états d'équilibre instables; à la rapidité avec laquelle la résistance du prisme décroît dans les instants où s'opère la séparation complète des parties; enfin à ce que, vers ces instants, les allongements cessent de s'opérer uniformément sur l'étendue entière de la barre, et n'ont plus lieu sensiblement que sur la portion, souvent très-courte pour les corps raides, où se fait la séparation définitive des molécules, portion dont l'altération élastique est masquée par la force de ressort que conservent encore les autres parties, et qui se manifeste clairement après la rupture complète. (p. 283, K. 304.)

The *états d'équilibre instables* are states suggested to Poncelet by the possible existence of points of intersection of the curves which represent the repulsive and attractive elements of the force between *two* molecules other than the initial state of rest I do not see that, even if they exist for these curves, it follows that there will be like unstable stages in the relation of stress to strain.

[980.] Pages 284—288 (K. 305—309) are devoted to a consideration of lateral stretch and cubical dilatation; they reproduce Poisson's results based on uni-constant isotropy. Reference is made also to Cagniard-Latour and to the memoir of Colladon and Sturm see our Arts. 368, 688. These pages are followed by a discussion of the condition of rupture, and Poncelet notes the uncertainty which the formula for the rupture of a prism under terminal tractive load, namely

$$\text{Load} = \text{area} \times \text{limiting traction},$$

involves; for, owing to the rapid change of diameter about the period of rupture, it is doubtful what area is to be taken (p. 290, K. 310).

[981.] § 247, entitled: *Notions sur la résistance vive des prismes,* introduces Poncelet's characteristic conception of work. His terminology is somewhat different from that now in use, but as we owe to him the introduction into practical mechanics of the principle of energy, it may not be out of place to give his definitions:

Nous nommons plus spécialement *résistance vive d'elasticité,* le travail dynamique qui répond à l'intervalle où, l'élasticité étant parfaite, les allongements demeurent sensiblement proportionnels aux efforts de réaction correspondans, et *résistance vive de rupture,* celle qui a été développée, par ces efforts, au moment où ils ont atteint leur plus grande valeur et où le prisme se trouve entièrement rompu (p. 291, K. 313).

It is shewn how the work may be obtained from the area of the stress-strain curve (p. 292, K. 313).

If $E$ be the stretch-modulus, $s$ the stretch produced by terminal tractive load in a prism of length $l$ and section $\omega$, Poncelet shews that the work for a purely *elastic* strain $= \frac{1}{2} Es^2 . \omega l$.

[982.] In considering the work required to rupture a prism, Poncelet supposes that if $W_0$ be the work required to rupture a prism of unit length and unit section, then $W = W_0\ \omega l$ will be the work requisite to rupture one of length $l$ and section $\omega$. The *résistance vive*, or resilience of a beam subjected to impact, is thus proportional to its volume, a theorem first proved for transverse impact by Young and for longitudinal impact by Tredgold: see our Appendix, Note A. (3) and our Art. 999. At the same time Poncelet remarks that this assumes the stretch to be uniform throughout the prism, which

> n'est nullement admissible pour les instants qui précèdent immédiatement la rupture, et réclamerait des expériences spéciales relatives à l'influence de la longueur des prismes (p. 293, K. 315).

On the following pages we have a slight sketch of the method in which this conception of work can be applied to the problem of resilience, an application first made, as Poncelet states, by the English writers Young and Tredgold. Poncelet however adds that this principle of energy is not wholly satisfactory as we do not know *what part is played by heat* when elastic materials are subjected to impact (p. 294, K. 316).

[983.] Pages 295—305 (K. 317—327) are principally occupied with the consideration of the time-effect and with methods of experiment. On the former head experiments of Savart, Vicat and Ardant are cited: see our Arts. 332, 735. Ardant seems to have made experiments at Poncelet's request on iron wires and to have concluded that:

> des fils de fer chargés de poids capables d'altérer, d'une manière notable, leur élasticité, non-seulement ne s'allongeaient pas indéfiniment, mais encore reprenaient, sous la charge et un repos suffisamment prolongé, un degré d'élasticité ou de raideur plus grand que celui qu'ils montraient à l'instant où l'allongement apparent avait cessé (p. 297, K. 320).

[984.] The next section of Poncelet's work is entitled: *Résultats de l'experience concernant la résistance directe des solides.* It occupies pp. 305—385 (K. 328—418), and is a *résumé* of experimental papers by various authors; to the majority of these we have referred in the course of our work. Here and there Poncelet criticises these experimental results. Thus he remarks that the results of Gerstner on the rate of extension after the traction has ceased to be proportional to stretch do not seem generally true: see our Art. 806. He cites in support of his opinion the experiments of Seguin[1], Bornet[2], and Ardant.

L'ensemble des résultats de ces expériences montre seulement qu'en deçà d'une certaine limite, les allongements sont, en effet, sensiblement comme les charges, et qu'au delà ils croissent dans une progression d'autant plus rapide que le métal, soumis à l epreuve de la tension, est plus doux, plus ductile de sorte que, jusqu a présent du moins, il n'est pas permis de dire que la loi de cette progression soit la même dans tous les cas, ni aussi simple que tendraient à le faire croire les expériences déjà citées de MM. Leslie et Gerstner (p. 344, K. 375).

We have the same vague reference to Leslie's *Elements of Natural Philosophy*, which has been followed by later writers probably copying out of Poncelet: see Art. 806.

[985.] Poncelet represents by curves the results of the experiments cited by him (figs. 47 and 48). I believe this is the first occasion on which we have anything like an accurate graphical representation of the traction-stretch relation for bars subjected to terminal tractive load. These curves are principally based on some careful experiments of Ardant, communicated by him to Poncelet, who here discusses at considerable length their singularities and suggests physical explanations (pp. 347—351, K. 378—382).

Wire of hard brass (*fils de laiton durs*) has a traction-stretch curve which can be closely represented by the equation,

$$s = T(a + b \cdot c^T),$$

where $s$ is the stretch, $T$ the traction, and $a$, $b$, $c$ constants depending on the material (footnote, p. 348, K. 379: see our Appendix, Note D).

[986.] On p. 353 (K. 384) Poncelet quotes results confirming Lagerhjelm's statement as to the moduli of steel and various kinds of wrought iron: see our Art. 366. He shews however that Ardant's

[1] *Des ponts en fil de fer*, 2e édit., Paris, 1826: see also the *Annales de Chimie*, T. 25, 1825, p. 190 (experimental details on strength of iron wires).

[2] *Du fer dans les ponts suspendus*, par MM. Martin et Fourchambault. This work and that referred to in the previous note were not accessible to me: see however Martin's memoir referred to in our Art. 817.

experiments do not accord with the formula given by Lagerhjelm as connecting the limits of elasticity and cohesion: see our Arts. 365, 852. He writes:

On ne doit se servir qu'avec beaucoup de réserve, de semblables relations, établies sur un trop petit nombre de faits, pour être considérées comme suffisamment exactes (p. 355, K. 386).

[987.] Pages 365—385 (K. 397—418). This is a section entitled: *Questions particulières relatives à la résistance des matériaux*. It contains the discussion of various problems, with numerical examples. We may note the consideration of the best form for a column. Poncelet does not refer to the labours of Euler and Lagrange: (Arts. 65, 106,) nor consider the possibility of flexure. He advocates, on grounds which are not very clear, the conchoid form of the Greeks: see p. 374 (K. 406). A simple case of thermal effect is treated on pp. 376—380 (K. 409—412). It is among the earliest examples on this subject.

[988.] We now reach what is Poncelet's most substantial contribution to the theory of elasticity namely the section entitled: *Examen des principales circonstances du mouvement oscillatoire des prismes sous l'influence de charges constantes et de chocs vifs*, pp. 385—465 (K. 418—503). Poncelet first treats the problem of a bar, which is supposed without inertia, to the end of which a weight is attached and allowed to fall without initial velocity. He proceeds from the principle of energy and easily deduces the now well-known proposition that the weight will produce a maximum stretch in the bar *double* of that which it statically could produce. He calculates the velocity of the weight, the traction and stretch of the bar at any time, and represents the whole (simple harmonic) motion by an elegant geometrical construction (pp. 385—396, K. 418—432). The discussion is followed by a numerical example[1].

[989.] The next problem considered is that of the load having an initial velocity. Suppose $L$ the load, $s$ the maximum stretch,

[1] The same problem is really discussed by Poisson in the *Journal de l'École polytechnique*, 18e Cahier, pp. 476—489. See also Sonnet, Art. 938, (5).

$v_0$ the initial velocity, then we have for a bar of length $l$ and section $\omega$,

$$\frac{L}{g}\frac{v_0^2}{2} + Lls = \tfrac{1}{2}\omega l\ \ Es^2,$$

or
$$s = \frac{L}{\omega E} + \sqrt{\frac{L^2}{\omega^2 E^2} + \frac{Lv_0^2}{\omega Egl}}.$$

This stretch in structures liable to be acted upon in this fashion must not exceed the limit of elasticity. Poncelet discusses the entire motion (its amplitude, period, etc.), but the reader will have no difficulty in reproducing for himself the description of this simple motion, in which the bar is supposed to have *no inertia or to stretch instantaneously as a whole* (pp. 398—410, K. 436—448).

[990.] In a footnote on p. 410 (K. 445) the more general case is treated in which the bar is supposed to have mass and vibrate. If $D$ be the specific gravity, $z$ the displacement in direction of the bar of a section originally at distance $x$ from the fixed end, the equations to be solved are obviously

$$\frac{d^2z}{dt^2} = g + \frac{gE}{D}\frac{d^2z}{dx^2},$$

$$\frac{L}{g}\frac{d^2z}{dt^2} = L - \omega E\frac{dz}{dx}, \text{ when } x = l$$

The solution is easily found to be

$$z = \frac{l\omega D + L}{\omega E}x - \frac{D}{2E}x^2 - \frac{L}{\omega E}\Sigma A_m \frac{\sin mx}{m}$$
$$\times\left(\cos\sqrt{\frac{gE}{D}}mt - m\sqrt{\frac{E}{gD}}v_0 \sin\sqrt{\frac{gE}{D}}mt\right),$$

where
$$A_m = \frac{4\sin ml}{2ml + \sin 2ml},$$

and the $\Sigma$ denotes a summation with regard to all values of $m$ which are roots of

$$ml \tan ml = \frac{l\omega D}{L}.$$

If the load $L$ is large compared with the weight of the bar, $m$ has the values approximately

$$\sqrt{\frac{\omega D}{Ll}}, \quad \frac{n\pi}{l} + \frac{\omega D}{n\pi L},$$

where $n$ is a positive integer.

Somewhat similar formulae were first given by Navier in the book referred to in Art. 272.

[991.] The next problem treated is that of a bar to the permanent load of which a blow is given (pp. 410—424, K. 448—461). Poncelet supposes the bar without inertia, so that every stretch is instantaneously uniformly distributed. The analysis is again of an easy kind. The bar being at rest with a load $L$, the weight $P$ is allowed to fall upon $L$ and thus give it an initial velocity. This initial velocity may be calculated on the principles of impact; the case practically considered by Poncelet is that in which $L$ and $P$ have no 'elasticity.' The maximum extension is easily obtained from the principle of energy, and a geometrical representation is given. In a footnote (p. 424, K. 460) Poncelet gives the solution for the case when the inertia of the bar is taken into account. If $V$ be the velocity of the weight $P$ at the moment of impact and $V_1 = VP/(L+P)$, then with the same notation and with the same equation for $A_m$ as in Art. 990:

$$z = \frac{l\omega D + P + L}{\omega E} x - \frac{D}{2E} x^2 - \Sigma A_m \frac{\sin mx}{m}$$

$$\times \left[\frac{P}{\omega E} \cos \sqrt{\frac{gE}{D}}\, mt - \frac{P+L}{\omega E}\, m V_1 \sqrt{\frac{E}{gD}} \sin \sqrt{\frac{gE}{D}}\, mt\right],$$

where $ml \tan ml = l\omega D/(L+P)$.

[992.] The last section of Poncelet's work which concerns us now is entitled: *Conséquences et applications diverses concernant les effets des mouvements imprimés aux prismes* (pp. 425—465, K. 461—503). It is occupied with various numerical examples and extensions of the theory of resilience previously developed, with special reference to suspension-bridges[1], and to the comparison of experimental results with theory.

[993.] We see that the problem of resilience in the case of a bar subjected to longitudinal impact has been fairly worked out by Poncelet; the problem of course had been previously considered by Navier. To Young and Tredgold's contributions we have before

[1] Besides the work of Navier on this subject referred to in Art. 272 the reader may consult a memoir by Vicat, *Annales de Chimie*, 2e Série, T. 27. 1826, p. 70.

referred and shall return when we consider later memoirs which treat more completely the case of bars subjected to transverse impact: see also Appendix, Note A. (3).

[994.] Poncelet gave a course of lectures on the Resistance of Solids to the Faculty of Sciences of Paris in 1839. These lectures have never been lithographed or printed, but exist only in manuscript. A few extracts from them, involving a simplification in the method of obtaining the ordinary flexure formulae, are printed by Saint-Venant in Appendice I., pp. 512—519, of his edition of Navier's *Lecons*. Poncelet's *feuilles inédites* are referred to on several occasions in the same work (e.g. pp. 374, 381, etc.) A further extract will be found in Schnuse's German translation of Poncelet's *Mécanique appliquée aux machines* (Bd. II. Note 1 to § 220). It is reproduced by Rühlmann as a footnote on p. 398 of his excellent work: *Vorträge über Geschichte der technischen Mechanik*. The extract is of considerable historical interest. According to Saint Venant Poncelet's methods have been adopted by Général Morin in his *Lecons sur la résistance des matériaux*[1].

[995.] We may note that the true condition of rupture, the maximum stretch as opposed to the maximum stress, was distinctly laid down by Poncelet in this course. Thus the condition of rupture by compression must be sought for in the lateral stretch. Generally the stretch-quadric and not the stress-quadric gives the directions of danger. Mariotte seems to have propounded the stretch condition of rupture, but its practical adoption is due to Poncelet and Saint-Venant. We may remark that Lamé, Clebsch and innumerable English writers have fallen into the error of taking the maximum-stress: see our Chapters VIII. and IX., also Saint-Venant's *Historique Abrégé*, pp. cxcix—ccv.

Poncelet in this course also adopted Saint-Venant's formula for the condition of rupture in a bar subjected to flexure, i.e. that which takes account of the slides: see our Chapter IX.

[996.] *Examen critique et historique des principales théories ou*

[1] Au reste les diverses méthodes de Poncelet ont été reproduites dans les Leçons bien connues de Morin. Extract from letter to Editor.

*solutions concernant l'équilibre des voûtes. Comptes rendus*, T. XXXV 1852, pp. 494—502, 531—540 and 577—587. This is a most interesting historical paper on the various theories of the arch. It only so far concerns our present purpose as Poncelet recognises in his last two pages a connection between the theory of arches and that of elasticity. In the latter theory he commends the then recent work of Saint-Venant, Lamé and Wertheim, and trusts it may react upon the former theory.

[997.] It will be seen that Poncelet's services to the subject of elasticity are considerable, although his influence was as much oral as by his writings.

[998.] Joseph Colthurst. *Experiments for determining the position of the neutral axis of rectangular beams of cast and wrought iron and wood, and also for ascertaining the relative amount of compression and extension at their upper and under surfaces, when subjected to transverse strain.* This is a memoir presented to the Institution of Civil Engineers, April 20th, 1841. It is reproduced, with the discussion upon it, on pp. 359—365 of the *London Journal of Arts and Sciences*, Vol. XIX. (*Conjoined Series*), 1842. The experiments were undertaken to determine the mooted point of the position of the neutral axis in beams subject to transverse load. The results obtained are briefly as follows:

(1) Position of neutral line in materials of the form stated in the title does not differ perceptibly from the central line.

(2) Amounts of extension and compression in the case of *wrought*-iron continue to be equal up to the complete destruction of the elasticity (beginning of set ?).

(3) They are only equal in the case of *cast*-iron up to about 2/3 of breaking load, after this load 'extension yielded in a higher ratio than compression.'

(4) With fir battens, extension and compression were equal up to 3/4 of the breaking load, but after this compression yielded in a much higher ratio than extension.'

(5) Amounts of extension and compression are in direct

proportion to the strain (i.e. stress[1]), within the limits of elasticity, and even after those limits are greatly exceeded, and up to 3/4 of the strength of a beam they do not sensibly differ.

In the course of the discussion reference was made by J. Horne to some experiments of his own laid before the Institution in 1837 shewing like results as to the position of the neutral line.

[999.] Thomas Tredgold. *A Practical Essay on the Strength of Cast Iron and other Metals.* We have already referred to the first edition of this work: see Art. 197. A second edition appeared in 1823; a third in 1831, after Tredgold's death, superintended through the press by Barlow; and finally a fourth, edited but not re-written by Eaton Hodgkinson in 1842: see Art. 966.

The last edition of Tredgold's book labours under the same defects as the first. The beginning of set is taken as the proper limit of strength, regardless of the fact that this epoch is most variable and in the case of cast-iron can hardly be said to exist (pp. 3—5): see our Arts. 853, 952, 969. Erroneous conclusions as to the absolute strength of cast-iron are obtained by flexure experiments, because set, and a consequent shifting of the neutral line to the compressed side of the beam, have been neglected. Tredgold thus obtains a tensile strength nearly three times too great. Further he supposes a triangular prism to be equally strong whether the base or vertex of the section be compressed. Finally we have Tredgold's theory of resilience with its somewhat gratuitous assumptions[2], and a characteristic sneer at the method of fluxions (p. 255); the *modulus of resilience* is here defined as 'the product of the force which produces permanent alteration and the corresponding extension' (p. 260).

On the whole, notwithstanding Hodgkinson's occasional notes of warning, Tredgold's *Essay* is as poor a book in the fourth as in the first edition.

[1000.] *Summary.* We have endeavoured to bring together

[1] Strain was used in the sense of stress in all engineering memoirs up to a very recent date.

[2] Ce n'était de la part du célèbre ingénieur qu'un tâtonnement.' Saint-Venant in his edition of *Clebsch.* Footnote, p. 547.

in this chapter a great body of mechanical and physical facts which have formed the basis of many of the later developments of the mathematical theory; we note also that many of these facts have not even yet been subjected to mathematical treatment. The characteristic mark of the period is that the practical needs of engineers and physicists are discovering a wider range of phenomena and demanding a more comprehensive theory. The uni-constant isotropy of Navier and Poisson after being questioned by Cauchy, receives at the hands of Green and Stokes a severe criticism. The general theory of thermo-elasticity is given by Duhamel, and new conceptions of the stretch-modulus in aeolotropic bodies are developed by Neumann. Poncelet advances the theories of resilience and of cohesion. Molecular theories of cohesion receive attention from Ampere, Belli and Mossotti, but the results of the two latter physicists are of a negative character. The influence of time on strain is considered, and Weber discovers the phenomenon of elastic after-strain. Set is investigated by Gerstner and Hodgkinson, while the latter physicist in England and Vicat in France contribute to forward our science by long series of valuable experiments on absolute strength. There is probably no period in the history of our subject which is richer in physical results than this, and the influence of these results will be strongly felt in the theory of the next decade.

# CHAPTER VII.

## LAME AND CLAPEYRON. LAMÉ.

[1001.] THE first paper we have to notice is entitled: *Sur les ponts de chaînes (de Russie) et sur les résistances des fers employés dans leur construction.*

This is an extract from a letter written to M. Baillet by Lamé, who was at that time a 'Major du Génie au service de Russie.' It is dated St Petersburg, Oct. 12/24, 1824, and is printed on pp. 311—330 of the *Annales des Mines*, T. x. Paris, 1825.

It is a contribution of the well-known elastician to the subject of the most suitable iron for suspension bridges; a subject which drew forth innumerable memoirs, many of considerable physical value, in the first half of the present century: see Arts. 692, 721, 723, 817, 848, 936.

Clapeyron and Lamé seem to have been engaged at this time in making researches on the strength of the iron used for the construction of such bridges in Russia, and this note consists of a theoretical investigation of the traction in the elements of such a bridge and of a description of a machine invented by Lamé to test pieces of iron.

The theoretical investigation is limited to a very special case, the general problem as stated by Lamé[1] not being solved. It involves nothing of importance, being merely a problem in elementary statics. The machine invented by Lamé was constructed in

[1] L'ouverture du pont étant donnée, on propose de déterminer la flèche à donner aux chaînes pour que la dépense totale du pont soit la moins forte possible (p. 322).

St Petersburg from the plans of General Bétancourt. It was a variation of the hydraulic testing-machine which had become usual since the experiments of Lagerhjelm : see Art. 364. Lagerhjelm's machine was invented by Hammarschioeld, but is only a modification of a machine which had been some time in use for testing chains.

[1002.] *Mémoire sur l'emploi du fer dans les ponts suspendus. Journal des Voies de Communication.* No. 5, pp. 19—43, and No 9, pp. 29—55. St Petersburg, 1826—1827.

This memoir announces the results of a series of experiments made by a commission to investigate the advantages of the various kinds of iron used for suspension bridges. It is drawn up by Lieutenant-Colonel Henry, but it would appear that Colonel Traitteur, Lieutenant-Colonels Lamé and Clapeyron, and General Bazaine (Pierre Dominique) were members of the commission; there is no list of the members in the memoir, but references to the memoir by Lamé and others lead me to this conclusion.

This memoir is worthy of note as emphasising not only the want of homogeneity in iron bars, but also a change in the elastic nature as we pass from the surface to the interior:

> Or, ces essais indiquant que la résistance du fer est plus grande à la paroi qu'à l'intérieur, on doit en conclure que les fils de fer résisteront plus que les fers en barre, puisqu ils ont plus de surface à proportion de leur volume; ce que l'expérience confirme. (No. 5, p. 26.)

Thus the want of isotropy in iron bars and wires was noticed as early as 1826, although it is continually forgotten to-day. The peeling off of the oxide of iron (first noticed by Robert Stevenson, and which may be used to mark the position of the neutral axis) and the stricture in the case of traction after the elastic limit is passed were also remarked (No. 5, p. 28). The increase of temperature during the plastic stage was noted (p. 29).

[1003.] Certain irregularities in the conduct of bars under traction which were noticed in the experiments were clearly due to the bar not having been reduced to the state of ease, and to the peculiar conditions of the yield-point. The following results state concisely the properties of the state of ease:

(1°) Que, quand il y a eu allongement et reprise, ou que l'élasticité s'est manifestée, elle subsistera dans toutes les épreuves subséquentes.

(2°) Que, quand la reprise est totale, elle se maintiendra dans les essais suivants.

(3°) Que, lorsque l'allongement est plus grand que la reprise de l'expérience précédente, la reprise ne sera que partielle. (No. 5, p. 35.)

These laws held for certain classes of iron, and we might well say that the discovery of the state of ease is due to this little band of French engineers in the Russian service.

A series of general results will be found drawn up on pp. 47—50 of No. 9. The authors again insist on the non-homogeneous character of iron, and point out that bars of square and rectangular section break more suddenly than those of circular section, which usually exhibit tendency to rupture by stricture considerably before actual rupture. This difference between bars of square and circular section is, if true, interesting, and does not appear to have been theoretically investigated.

[1004.] There is an article by Lamé in the *Journal du Génie Civil*, tom. I, p. 245, Paris, 1828, but I am unable to give any account of its contents, as no library was accessible to me in which such an early series of this *Journal* was to be found.

1005. *Mémoire sur l'équilibre intérieur des corps solides homogènes:* par MM. Lamé et Clapeyron, Ingénieurs des mines. This memoir occupies pp. 465—562 of the *Mémoires présentés par divers Savans*, Vol. IV, 1833. The date at which the memoir was presented to the French Academy is not recorded here, but the *Report* on it by Poinsot and Navier, to whom it was referred, was made on September, 29th, 1828: see Crelle's *Journal*, Vol. VII, p. 145. The memoir may be described in general terms as an elementary treatise on the subject of elasticity with numerous important applications; there is no reference to any preceding writer on the theory of elasticity. The memoir consists of an introduction and four sections.

1006. The *Introduction* occupies pp. 465—469. It says that writers on Statics had hitherto confined themselves to an investigation of the relations which must hold among the forces applied to a body in order that the body may remain in equilibrium; these relations are independent of the internal constitution of the body But it is the object of the present memoir to investigate the way in which the interior of a body is affected by the transmission through it of the action of forces. On p 468 the memoir speaks of a solid body as consisting of an infinite number of particles much in the same way as Lamé writes on p. 5 of his *Lecons,* only that in the memoir the particles are said to be *equidistant,* a condition which is omitted in the *Lecons.*

1007. The first section is entitled: *Équations différentielles;* it occupies pages 470—486. The general equations of equilibrium which must hold at every point of the interior of an elastic body are investigated. The same assumption with respect to the nature of the molecular force is made as by Navier; and the equations obtained involve *one* constant which represents a certain integral: see Art. 266 of my account of Navier. [The method by which the body shift-equations are established is very like that of Navier and involves the same errors; namely the assumption that intermolecular force may be represented by a function of the form $m_1 m_2 f(r)(r' - r)$ and the replacement of summations by integrations: see our Arts. 266, 443, 531, 615. The method adopted by Lamé and Clapeyron is in fact Navier's first investigation wherein the equations are obtained from calculation of the stresses and not by the Calculus of Variations; there is, I think, some cause for Navier's expression of feeling on the subject: see our Art. 277. We may also remark that the definition of stress across a plane in terms of intermolecular action is that often used by Cauchy but objected to by Saint-Venant in his *Cours lithographié* and elsewhere: see our Chapter IX.] Nothing is explicitly said with respect to the equations which must hold at the *surface* of the body; they are however implicitly involved in § 24 and the matter is treated properly, where necessary, in the applications of the theory which form the third and fourth sections.

1008. The second section is entitled: *Theoremes sur les pressions;* it occupies pages 486—508. This goes over much the same ground as the fourth and fifth of Lamé's *Lecons;* and we may mention in particular various results obtained. The second of Cauchy's theorems is obtained in its most general form: see my Art. 610 (ii), and Saint-Venant's *Torsion,* page 250 The equation to the *ellipsoid of elasticity* or *Lamé's stress-ellipsoid* is obtained: see Art. 1059. The existence of *principal tractions* is established: see Art. 603. The property with respect to conjugate diameters is established: see the *Leçons,* p. 56. The equations to the *stress director-quadric* and to the *shear-cone* are obtained: see Art. 1059. The special cases of the principal tractions are also discussed. This is the first appearance of Lamé's stress-ellipsoid (unfortunately named the ellipsoid of elasticity), the stress director-quadric and the shear-cone; there is little doubt that the other properties were rediscovered by the authors for themselves.

1009. A simple result is given on p. 490 which is perhaps not formally reproduced in elementary works. Take the three equations (i) of my Art. 659, square and add; thus we obtain the value of $p_1^2$. Similarly let $p_2^2$, which corresponds to another position of the line *OL* of that Article, be obtained; and $p_3^2$, which corresponds to a third position. Then if the three normals are mutually at right angles we get $p_1^2 + p_2^2 + p_3^2 = (\widehat{xx})^2 + (\widehat{yy})^2 + (\widehat{zz})^2$. Thus the sum of the squares of the stresses across three plane elements which are mutually at right angles is constant for the same point of a solid body, whatever may be the position of the system of the three elements[1]

1010. The third section is entitled: *Cas simples;* it occupies pp. 508—541. This section consists of various simple applications of the general equations; they appear here for the first time[2] but have all since been reproduced in the elementary

[1] [This result appears due to Cauchy, for in the *Exercices de mathématiques,* Seconde Année, 1827, p. 53, we find:

Théorème. Si par un point donné d'un corps solide on fait passer trois plans rectangulaires entre eux, la somme des carrés des pressions ou tensions supportées par ces mêmes plans sera une quantité constante, égale à la somme des carrés des pressions ou tensions principales.

Cauchy's result was thus *printed* in 1827. Ed.]

[2] [Not all of them, as may be seen by an examination of Poisson's great memoir of April, 1828. Ed.]

works, and treated with greater generality by introducing *two* constants of elasticity instead of *one*. A student concerned with practical details would however find some useful and interesting remarks in the memoir which are not in the more recent discussions. We will now briefly notice the various cases considered in the memoir.

1011. The first case is entitled: *Cas d'un prisme indéfini*; it occupies pp. 508—515: see Lamé's *Leçons*, p. 74. In the memoir some numerical values are given for the constant which corresponds to the $\lambda$ or $\mu$ of the *Lecons*, the two being here assumed to be equal. Thus for forged iron the value expressed in kilogrammes per square millimetre is about 8000, and for brass 2510. The experimental data of Duleau and Tredgold are taken as a basis. On p. 514 of the memoir we have some notice of an experiment made by Perkins to determine the compressibility of water. The water was put in a vessel over mercury, and subjected to a pressure of 2000 atmospheres. The vessel was assumed to have remained unchanged in volume, and by the observed ascent of the mercury in the vessel it was inferred that the volume of the water was diminished by $\frac{1}{12}$. Lamé and Clapeyron observe that this result requires correction on account of the compressibility of the vessel which contained the water. Let $\delta$ denote the coefficient of compressibility of the vessel, $\delta'$ the coefficient of compressibility of the water found on the supposition that the volume of the vessel remained unchanged; then they say that the true coefficient of compressibility of the water is $(1 - \delta)(\delta + \delta')$, that is approximately $\delta + \delta$ It is not obvious how the expression is obtained which is here given as exact. Let $x$ denote the true coefficient; then a volume originally $V$ of water becomes under compression $V(1 - x)$. The internal volume of the vessel, supposed to be originally $V$, becomes under compression $V(1 - \delta)$; of this the water is observed under compression to occupy the fraction $1 - \delta'$; therefore $V(1 - x) = V(1 - \delta)(1 - \delta')$. Thus the *exact* expression seems to be $x = \delta + \delta' - \delta\delta'$, which differs slightly from that of the memoir, though leading to the same approximate result: see our Arts. 686—690.

1012. The second case is entitled: *Cas d'un cylindre creux indéfini*; it occupies pp. 516—525: see Lamé's *Leçons*, p. 188. The cylinder in the present case, and the prism in the first case, are supposed to be *indefinite*, apparently, in order to avoid any consideration of the complex conditions which must hold at the *ends*. The writers suppose in fact the ends of the cylinder to produce a uniform longitudinal traction throughout the material of the cylinder measured by $(P_0 r_0^2 - P_1 r_1^2)/(r_1^2 - r_0^2)$. If we suppose the cylindrical shell cut by a plane parallel to the axis we obtain two rectangles. The traction at any point $(r)$ of the section is given by the expression

$$\frac{r_0^2 P_0 - r_1^2 P_1}{r_1^2 - r_0^2} + \frac{r_0^2 r_1^2 (P_0 - P_1)}{r^2 (r_1^2 - r_0^2)},$$

where $P_1$, $P_0$ are respectively the outside and inside tractive loads and $r_1$, $r_0$ the corresponding radii.

Multiply this by $dr$, integrate from $r = r_0$ to $r = r_1$; and double the result; thus we obtain the whole force tending to burst the cylindrical shell along a meridian plane, estimated on a portion of the height equal to a unit in length. The result is $2(P_0 r_0 - P_1 r_1)$, as found on p. 519 of the memoir; the authors add the remark:

Ce résultat est égal à celui que l'on obtiendrait, en cherchant directement, d'après des principes connus d'hydrostatique, la force qui tend à briser un tube soumis, intérieurement et extérieurement, à des pressions différentes.

1013. Lamé deduces from his formulae the limits of the pressures which can be safely applied to the interior of metal vessels[1]. For different metals these limits range between 400 and 1400 atmospheres. Lamé and Clapeyron add on p. 522:

[1] [The calculation of these limits is based upon the maximum traction not exceeding what is termed the elastic limit, they ought to have been obtained from a consideration of the maximum stretch not exceeding $T_0/E$, where $T_0$ is the limit of safe tractive load to be ascertained by pure traction experiments, and $E$ is the stretch-modulus. Proceeding from the stretch-quadric, I find that the necessary condition is that

$$2\mu\,[P_0(r_1^2 + r_0^2) - 2r_1^2 P_1] + 3\lambda r_1^2 (P_0 - P_1) < T_0\, 2(\lambda + \mu)(r_1^2 - r_0^2)$$

Hence,

$$\left(\frac{r_1}{r_0}\right)^2 > \frac{2\mu P_0 + 2T_0(\lambda + \mu)}{2T_0(\lambda + \mu) - (3\lambda + 2\mu)(P_0 - P_1) + 2\mu P_1}$$

M. Perkins, dans les expériences, dont les journaux ont rendu compte, sur la compressibilité des liquides, a poussé la pression dans l'intérieur d'un cylindre métallique jusqu'a 2000 atmosphères. Ce fait n'est pas une objection au résultat que nous venons d'annoncer; il est possible que si l'expérience eût été prolongée pendant un intervalle de temps suffisant, le cylindre se fût brisé, ou qu'il fût arrive quelque chose d'analogue à ce qui se passe quand on suspend à une tige de fer un poids qui dépasse 14 kilogrammes par millimètre.

1014. The third case is entitled: *Cas d'un cylindre soumis a une torsion;* it occupies pp. 526—532: see Lamé's *Leçons*, p. 186. A formula is obtained in the memoir equivalent to the $\alpha = \frac{2M}{\pi\mu R^4 l}$ of the *Leçons*, p. 188. Thus if we determine by experiment the amount of torsion produced by assigned forces we can calculate the value of $\mu$. The authors of the memoir thus find that for forged iron it is 7493 and for brass it is 2248; these values agree closely with those assigned in Art. 1011, and they say on their p. 529,

> ...nous croyons voir dans la coincidence approchée de deux résultats déduits d'expériences d'un genre si different, l'une des preuves les plus frappantes de l'exactitude de notre théorie

1015. The fourth case is entitled: *Cas d'une sphère dont toutes les parties gravitent les unes sur les autres;* it occupies pp. 532—536. This is reproduced by Lamé in his *Leçons*, p. 213, and is given substantially on the eighth page of his memoir in Liouville's

In the case where $\lambda=\mu$ if $P_0>P_1$ I find that no cylindrical shell however thick can sustain an internal load

$$P_0 > \frac{4T_0+7P_1}{5}.$$

In the case of $P_1=0$, $P_0$ must be $<4T_0/5$; these results do not agree with those of Lamé and Clapeyron on pp. 521—525 of their memoir or with those of Lamé's *Leçons*, p. 191. I consider their results to be vitated by taking traction rather than stretch as determining the limit of elasticity. We see also that the cylinder will first receive set *internally*. These results are extremely interesting in relation to the experiments of Messrs Amos and Easton considered in our Chapter VIII. ED.]

[1] [The reader will note that these results suppose the uni-constant isotropy of metals; the writers' results are based on the experimental data of Biot and Coulomb. ED.]

*Journal*, 1854, and in his *Coordonnées curvilignes*, p. 351: see Art. 1118 of my account of the memoir.

1016. The fifth case is entitled: *Cas d'une sphère creuse;* it occupies pp. 536—541: see Lamé's *Leçons*, p. 211, and my account of the *Leçons* in Arts. 1093—1094. On p. 537 of the memoir an expression is obtained for the resultant stress normal to a diametral plane. This is $\iint \psi_3 \, r dr d\psi$ with the notation of the *Leçons*, p. 212; the integration is to extend over the area of the circular section; $\psi_3$ is the $\widehat{\phi\phi}$ of Art. 1113 of my discussion of *Lamé's Problem.* The result is $\pi(P_0 r_0^2 - P_1 r_1^2)$; the authors add the remark:

Qui est effectivement égale à celle que l'on obtiendrait directement en se fondant sur des principes connus d'hydrostatique

1017. The fourth section of the memoir is entitled: *Cas généraux;* it occupies pp. 541—562: these are more complex than the cases of the third section, and are not reproduced in the *Leçons*, but references to them occur on pp. 164 and 191 of that work. We will now notice these cases, which are three in number.

1018. The first case is entitled: *Cas d'un plan indéfini;* it occupies pp. 541—548: a body is supposed to be bounded by an

[1] [On p. 540 Lamé and Clapeyron obtain an inequality similar to that of Art. 1013, namely $P_0 < 2T_0 + 3P_1$, a relation quite independent of the radii and of the elastic constant. This result is again obtained by treating the maximum traction and not the maximum *stretch* as giving the proper condition of cohesion. Working from the stretch-quadric, I find:

$$4\mu(r_0^3 P_0 - r_1^3 P_1) + (3\lambda + 2\mu) r_1^3 (P_0 - P_1) < 4(\lambda + \mu) T_0 (r_1^3 - r_0^3).$$

Hence $$\left(\frac{r_1}{r_0}\right)^3 > \frac{4\mu P_0 + 4(\lambda+\mu) T_0}{4(\lambda+\mu) T_0 - (3\lambda + 2\mu)(P_0 - P_1) + 4\mu P_1}$$

A result not agreeing with that given by Lamé and Clapeyron, or by Lamé in the *Leçons*, p. 213. It leads at once to the condition that no spherical shell however thick can resist an internal load

$$P_0 > \frac{4(\lambda+\mu) T_0 + 3(\lambda + 2\mu) P_1}{3\lambda + 2\mu}.$$

Or, in the case of $\lambda = \mu$, we must have

$$P_0 < \tfrac{8}{5} T_0 + \tfrac{9}{5} P_1.$$

In both the cases of the spherical and the cylindrical shells the stretch condition, it should be noticed, leads to a less limit for $P_0$ than that adopted by Lamé, whose condition is thus not even on the side of safety. Ed.]

infinite plane on which is a given distribution of tractive load. The analysis depends mainly on the use of Fourier's theorem, by the aid of which an assigned function of two variables is expressed as a quadruple integral. The only point in the process to which I need draw attention is on p. 544; the authors obtain twelve equations, noted as (7), and these they seem to say lead to four more noted as (8); it seems to me that (8) must hold, but that they are not deducible from (7) But from (7) four equations can be deduced which are something like the equations (8), but not identical with them. Thus instead of the first of (8), which is $-l_1q+h_2p-f_3r=0$, we get from (7) this equation

$$-l_1q+h_2p-2f_3pq/r=0.$$

But this does not invalidate any of the final results of the memoir; for we have $f_3=0$.

[1019.] These final results are the following: the plane face being taken for that of $xy$, $F(xy)$ giving the law of tractive load and $\lambda$ being the coefficient of uni-constant isotropy.

$$u=\frac{1}{4\pi^2\lambda}\iiiint e^{-\sqrt{p^2+q^2}z}\left(\frac{p}{4(p^2+q^2)}-\frac{pz}{2\sqrt{p^2+q^2}}\right)\times$$
$$F(\mu,\nu)\sin p(x-\mu)\cos q(y-\nu)\,dp\,d\mu\,dq\,d\nu,$$

$$v=\frac{1}{4\pi^2\lambda}\iiiint e^{-\sqrt{p^2+q^2}z}\left(\frac{q}{4(p^2+q^2)}-\frac{qz}{2\sqrt{p^2+q^2}}\right)\times$$
$$F(\mu,\nu)\cos p(x-\mu)\sin q(y-\nu)\,dp\,d\mu\,dq\,d\nu,$$

$$w=\frac{1}{4\pi^2\lambda}\iiiint e^{-\sqrt{p^2+q^2}z}\left(-\frac{3}{4\sqrt{p^2+q^2}}-\frac{z}{2}\right)\times$$
$$F(\mu,\nu)\cos p(x-\mu)\cos q(y-\nu)\,dp\,d\mu\,dq\,d\nu,$$

$$\theta=\frac{1}{4\pi^2\lambda}\iiiint e^{-\sqrt{p^2+q^2}z}\,\tfrac{1}{2}F(\mu,\nu)\cos p(x-\mu)\cos q(y-\nu)\,dp\,d\mu\,dq\,d\nu.$$

The quadruple integrations are all from $-\infty$ to $+\infty$: see p. 546 of the memoir. A curious result is obtained on page 546. Suppose the infinite plane to be that denoted by $z=0$; and let the assigned normal force at the point $(x,y)$ be denoted by $F(x,y)$; then the value of the dilatation at that point of the plane is $F(x,y)/2\lambda$, where $\lambda$ is the dilatation-coefficient, which is taken throughout the memoir to be the same as the slide-modulus $\mu$.

1020. The second case is entitled: *Cas de deux plans parallèles*; it occupies pp. 548—552: a body is supposed to be bounded by two infinite parallel planes on which are given distributions of tractive load. This case depends almost entirely on the first.

[1021.] The results in this case are of a more complex form, but can be all represented by a single function $\phi$.

$$u = -\frac{d\phi}{dx},\ v = -\frac{d\phi}{dy},\ w = 3\int \theta dz - \frac{d\phi}{dz},$$

$$\theta = \tfrac{1}{2}\nabla^2\phi,$$

where

$$\phi = \frac{1}{4\pi^2\lambda}\iiiint\left[\left(\frac{\frac{1}{2\sqrt{p^2+q^2}}B - Aa}{AB - a\sqrt{p^2+q^2}}\cdot\frac{F-f}{4} + \frac{Az}{AB+a\sqrt{p^2+q^2}}\cdot\frac{F+f}{4}\right)\zeta\right.$$
$$\left.+\left(\frac{\frac{1}{2\sqrt{p^2+q^2}}A - Ba}{AB+a\sqrt{p^2+q^2}}\cdot\frac{F+f}{4} + \frac{Bz}{AB-a\sqrt{p^2+q^2}}\cdot\frac{F-f}{4}\right)\xi\right]$$
$$\frac{\cos p\,(x-\mu)\cos q\,(y-\nu)}{\sqrt{p^2+q^2}}\,dp\,d\mu\,dq\,d\nu.$$

Here $F$ and $f$ stands for $F(\mu, \nu)$ and $f(\mu, \nu)$, and $F(x, y)$ and $f(x, y)$ give the distribution of tractive load over the parallel faces of the plate supposed equally distant from the plane of $xy$. Further its breadth $= 2a$; $\zeta = \sinh\sqrt{p^2+q^2}\,z$, $\xi = \cosh\sqrt{p^2+q^2}\,z$; $A = \sinh\sqrt{p^2+q^2}\,a$ and $B = \cosh\sqrt{p^2+q^2}\,a$.

The last paragraph is this:

Les formules précédentes, pour être obtenues en séries numériques et immédiatement applicables, exigent la connaissance des valeurs d'un genre particulier d'intégrales définies, dont il ne nous paraît pas que les géomètres se soient encore occupés. Nous avons fait à ce sujet un travail que nous nous proposons de publier incessamment.

It does not seem however that this proposed memoir was ever published.

1022. The third case is entitled: *Cas général du cylindre indéfini*; it occupies pp. 552—562. A body is supposed to be in the form of an infinite cylinder, solid or hollow, and having its curved surface, or surfaces, acted on by given distributions of load. In all the other cases discussed in the memoir the equations of equilibrium of an elastic body referred to *rectangular* coordinates were sufficient, but for the present problem what the authors call semi polar coordinates are more suitable; they give without demonstration the required formulae which are worked out in the fourteenth of the *Leçons*. The investigations in the memoir are carried on so far as to indicate how the problem is to be solved; but they are worked out fully only for a comparatively simple case of a shearing load varying along a generator, but constant round a circular section. [The whole process is interesting, and forms practically a solution in Bessel's functions. It is the first attempt to apply cylindrical coordinates to the shift equations. At the same time we may remark that the reader who attempts to apply Lamé and Clapeyron's formulae to such a simple case as a cylindrical column of which the ends are *built-in* and subject to uniform pressure will soon find himself involved in analytical difficulties which he will not easily master.]

1023. The whole memoir is reproduced in Crelle's *Journal für die...Mathematik*, Vol. VII., 1831, pp. 150—169, 237—252, 381—413: there are many misprints in the original, and nearly all of them are reproduced. On pp. 145—149 we have the *Rapport* on the memoir by Poinsot and Navier made to the French Academy on the 29th of September, 1828; I presume that this was originally published in France, but I have not seen it in any place except this volume of Crelle's *Journal*. The whole *Rapport* is very interesting, and contains allusions to the researches on the subject which Navier had already made; but I must confine myself to a few extracts from it. Lamé and Clapeyron are styled *Colonels du Génie au service de Russie*.

1024. On p. 146 of the *Rapport* we have the following paragraph:

Quant à l'établissement des équations différentielles destinées à représenter les conditions générales de l'équilibre des parties des solides élastiques, MM Lamé et Clapeyron ont admis le même principe, et procédé de la même manière que l'un de nous l'avait fait dans un mémoire qui a été présenté à l'Académie en 1821, publié par extrait dans le *Bulletin des sciences de la société philomatique* en 1822, et imprimé en entier l'année dernière dans le tome VII. de nos mémoires. Ils parviennent à des équations semblables à celles qui avaient été données dans cet écrit. On doit seulement remarquer qu'ils ont déduit directement du principe dont il s'agit les équations déterminées relatives aux points de la surface du corps, et qui font connaître aussi les pressions ou tensions intérieures; tandis que, dans le mémoire de 1821, ces équations avaient été obtenues par les méthodes de la *mécanique analytique.* Nous sommes obligés de rappeler ici que ce dernier mémoire contient la première idée de la question, et les fondements des recherches relatives aux corps élastiques. MM. Lamé et Clapeyron n'en faisant aucune mention, on doit penser qu'ils n'en avaient pas connaissance, et qu'ils sont parvenus de leur côté aux mêmes résultats Ils ont d'ailleurs poursuivi avec succès le genre de recherches qui était l'objet de cet écrit.

The preceding paragraph says that the memoir contains the equations which must hold at any point of the surface; but this is wrong, for the memoir does not explicitly notice this matter.

On p. 147 we read:

La partie des *Exercices de mathématiques* de M. Cauchy qui a été publiée l'année derniere contient plusieurs propositions relatives aux pressions intérieures qui ont lieu dans un corps solide, analogues aux précédentes, et dont quelques unes avaient été données antérieurement par M. Fresnel. MM. Lamé et Clapeyron remarquent que la théorie exposée dans leur ouvrage diffère essentiellement de celle qu'avait adoptée M. Cauchy. Nous nous abstenons de parler des recherches qui ont été publiées après la présentation du mémoire qui est l'objet de ce rapport.

There is apparently some mistake here, for Lamé and Clapey ron in the course of their memoir make no allusion whatever to Cauchy.

On p 148 there is a good summary of the most curious results obtained in the third section of the memoir.

...Ces solutions conduisent à diverses conséquences remarquables. On trouve, par exemple, que lorsque la pression intérieure qui a lieu dans un cylindre creux (qui est supposé d'une longueur infinie) dépasse une certaine limite, il n'est pas possible, quelle que soit l'épaisseur donnée à ce cylindre, de le rendre capable de résister à la rupture Dans le cas d'une sphère pleine dont les points s'attirent en raison inverse du carré des distances, on trouve qu'au centre la pression intérieure est égale au poids qu'aurait, à la surface de la sphere, une colonne de la matiere dont cette sphère est formée, la longueur de cette colonne étant les 11/30 du rayon. A une petite profondeur au-dessous de la surface, la pression dans le sens du plan tangent est égale au poids d'une semblable colonne dont la hauteur serait les 2/15 du rayon de la sphère. Les parties d'un semblable globe voisines de la surface sont donc comprimées latéralement avec une force extrêmement grande comparativement au poids de la colonne de matiere dont elles sont chargées dans le sens du rayon[1]

[1] The following are some of the principal misprints of the memoir. Those marked with an asterisk are corrected in Crelle's *Journal*.

*518. For $U$ read $V$.
In the last line but one for $R - R'^2$ read $R'^2 - R^2$.
In the last line for $R'^2R^2$ read $R'^2 - R^2$.

*520. Line 7, for + read −.

*526. Line 4 from foot, for $w$ read $\omega$.

529. Line 2 from foot, for 32 read 33.

533. Line 5, for $\frac{dU}{dr} + \frac{2}{r} U$ read $\frac{1}{3}\left(\frac{dU}{dr} + \frac{2}{r} U\right)$
Line 13, for $C$ read $c$. Line 4 from foot, for $\frac{11c}{3A}$ read $\frac{11c}{30A}$.

537. Line 9, for $\frac{1}{r^3}$ read $\frac{1}{2r^3}$. Line 2 from foot, for $3y^2 - z^2$ read $3y^2 - r^2$.

541. Line 2, for $R'^3$ read $R'^2$.

545. Line 4, for $\frac{1}{2\pi}$ read $\frac{1}{(2\pi)^2}$.

546. First line, for $k$ read $k_1$.

549. In equations (18) and (20) for $g_1 r$ read $g_1 p$.

550. After $F(\mu, \nu)$ insert "and $f(\mu, \nu)$ respectively." Then $A$ should not be used, for it is already appropriated.

*551. Line 6, in the numerator for $A$ read $B$.
Last line, for $s_y$ read $c_y$.

552. In equations (28) read $w = 3\int \theta dz - \frac{d\phi}{dz}$.

*556. In equation (7) (that for $\zeta$) for $\frac{dU}{dr}$ read $\frac{dU}{dz}$.

* In the third of equations (8), for $\frac{d\Omega}{dz}$ read $\frac{d\Omega}{dy}$, and for $\frac{1}{r^2}\frac{d\zeta r}{dr}$ read $\frac{1}{r}\frac{d\zeta r}{dr}$.

557. In equations (11) for $\frac{dW}{dz}$ read $\frac{dW}{dr}$

1025. *Mémoire sur les lois de l'équilibre du fluide éthéré.* This is published in the *Journal de l'École...Polytechnique,* 23rd Cahier, 1834; it occupies pp. 191—288 This is one of the numerous attempts to bring the phenomena of light under the dominion of the theory of elasticity, and may deserve attention in a history of Physical Optics, but has little connection with our subject. I have not examined it closely, and shall give only a brief notice of it. The memoir consists of five parts.

1026. The first part is entitled: *Equations différentielles de la lumière;* it occupies pp. 191—214. The process bears some resemblance to Cauchy's molecular methods; equations are obtained which involve *one* constant of elasticity, but which differ from the usual equations for uni constant isotropy, by involving differential coefficients of the density $\rho$, which is not taken to be uniform. Lamé follows the example of Poisson in supposing that two of the summations may be converted into integrals: see Art. 441 of my account of Poisson's memoir of April, 1828. In this respect, and in the use of only one constant of elasticity Lame s later judgment would not have approved of the process of his present memoir. One of the most important results of the first part is that the density $\rho$ must be such that $\nabla^2 \log \rho = 0$.

1027. The second part of the memoir is entitled: *Surfaces orthogonales conjuguées;* it occupies pp 213—246. This is a chapter in pure mathematics, reproduced substantially by Lamé himself in later works, and also discussed by other writers: see Art 265 of my *Treatise on Laplace's Functions.*

1028. The third part of the memoir is entitled: *Sur les équations différentielles de la lumière rapportées aux couches de l'éther;* it occupies pp. 247—261. This is analogous to the process, soon to be noticed, by which Lamé transformed the equations of elasticity to curvilinear coordinates.

559. Line 9, for $p, \mu, p, \nu$ read $p, \mu, q, \nu$.
Line 3 from foot, for $a$ read $A$.
561. Line 9, for $\zeta$ read $\rho$.
562. Line 4 from foot, for $T$ read $\tau$.

1029. The fourth part of the memoir is entitled: *Sur l'intégration des équations différentielles de la lumière;* it occupies pp. 262—281. Lamé obtains on his p. 269 the following result:

Ainsi, la densité moyenne de l'éther dans les corps diaphanes est moindre que celle du même fluide dans le vide; et l'action que la matière pondérable exerce sur l'éther est répulsive.

Lamé considers in detail the case of a single particle of ponderable matter, spherical and homogeneous; and finishes the fourth part of the memoir thus:

J'ai aussi considéré le cas d'une particule hétérogène, ou de forme non sphérique, distribuant l'éther autour d'elle en couches ellipsoidales; les résultats que j'ai obtenus reproduisent en quelque sorte les phénomènes connus de la polarisation et de la double réfraction, dans les cristaux à un et à deux axes, ils en fournissent même l'explication la plus complète et la plus naturelle. Mais ces recherches analytiques feront l'objet d'un autre travail dont je m'occupe à present, et dans lequel j'essaierai de prouver, en multipliant les vérifications, la réalité de la théorie nouvelle que j expose dans ce Mémoire.

1030. The fifth part of the memoir is entitled: *Sur les lois générales de l'équilibre de l'éther;* it occupies pp. 282—288. This seems to be an addition made after the memoir had been sent to press. Lamé now recognises that a result obtained in his first part is not absolutely correct; instead of having $\nabla^2 \log \rho$ zero all the conditions are satisfied if $\nabla^2 \log \rho$ is a function of $\rho$. Accordingly Lamé says, on his p. 282:

Il importe d'interpréter cette nouvelle loi, et de définir les changemens qu'elle peut apporter aux conclusions de ce Mémoire.

1031. Lamé himself seems to imply that he was not satisfied with this memoir as a contribution to the theory of light, for he never refers to it in the course of the long discussion on this subject which forms, as we shall see, a part of his Lectures on the theory of elasticity. I do not remember to have seen any reference to the memoir by other writers, except one by Thomson and Tait on p. 250 of their *Natural Philosophy*, Part II.

[1032.] *Mémoire sur les lois de l'équilibre de l'éther dans les corps diaphanes. Annales de Chimie*, T. LV. 1833, pp. 322—335.

This is an account of a paper read before the Academy of Sciences on April 28, 1834. The paper is that of our Art. 1025, and the results are here stated generally without the use of mathematical symbols or language.

[1033.] *Mémoire sur les vibrations lumineuses des milieux diaphanes. Annales de Chimie*, T. LVII. 1834, pp. 211—219.

This is an account of a paper read to the Academy of Sciences on September 22, 1834. I can find no other account of this paper; it is possible that we have here the paper published in full. It involves no mathematical symbols or language. Lamé refers to the results obtained in the previous memoir on equilibrium namely that the 'elasticity of the ether' is proportional to its density, and that:

Les particules pondérables agissent sur la portion d'éther située aux lieux où les vibrations lumineuses peuvent se propager par une force répulsive dont l'intensité varie en raison inverse du carré de la distance.

With these conclusions, based upon the assumptions that: (1) the ether exists in transparent bodies, (2) that the molecules of ether in luminous vibrations oscillate in the surface of the wave (see Art. 1029) Lamé seems still perfectly content. He goes on in this memoir to consider the wave vibrations which may produce light, and looks upon his work, not as a new theory of light, but as a supplement to that of Fresnel. The memoir which belongs properly to the History of Light does not seem to have been noted by the historians of that subject

[1034.] Lamé. *Cours de Physique*. Paris, 1836. The *Septième Leçon* (T. I. pp. 139—164) of this work is concerned with the elastic and cohesive properties of bodies. It is merely descriptive and of no present value. Historically we may note that Lamé in 1836 accepted the Navier Poisson theory of uni constant isotropy (p. 144). Some experimental results on the value of the elastic constant $A$[1] for cast-iron, gun-metal, brass, tin and lead are given on p. 149, but Lamé remarks with regard to them:

[1] $A$ is the slide-modulus or the constant Lamé afterwards denoted by $\mu$; $\frac{5}{2}A$ is thus the stretch-modulus on the uni-constant theory.

Mais ces expériences doivent être répétées avec tout le soin nécessaire pour donner les valeurs exactes de $A$.

Lamé evidently puts no confidence in their exactitude. On the same page Lamé deduces from the above results that the compressibilities—measured by $3/(5A)$—of different materials placed in order of increasing magnitude are iron, cast-iron, glass gun-metal, brass, tin, lead.

These experiments of Lamé are cited by Stokes in his memoir of April, 1845: see our Art. 925, and Chapter VIII.

[1035.] *Sur les causes des explosions des chaudières dans les machines à vapeur.* Extrait d'une lettre de M. Lamé, professeur à l'Ecole polytechnique à Paris, à M. Kupffer. This extract was read to the Academy of Sciences in St Petersburg on November 1, 1839. It is published in the *Bulletin Scientifique publié par l'Académie des Sciences,* T VI. 1840, cols. 380—382. Lamé as member of a commission on steam engines had been called upon to investigate a memoir of M. Jaquemet of Bordeaux. This memoir he conceived to be of such importance that he wrote an account of its contents to M. Kupffer. The memoir considers only the action of the water and steam at the instant of explosion and does not treat of the elastic properties of the material out of which the boiler is constructed; it has therefore no relation to the memoir we shall consider in Art. 1038, as the name might possibly induce some to imagine.

1036. In the *Comptes Rendus,* Vol. VII., 1838, we have the following heading on p. 778: Mémoires présentés. Mécanique. *Mémoire sur les surfaces isostatiques dans les corps solides en équilibre d'élasticité,* par M. G. Lamé. (Commissaires, MM. Poisson, Cauchy.) Then on pp. 778—780 we have an abstract of the memoir by Lamé himself. This abstract closely resembles the introductory remarks in the memoir published in Liouville's *Journal de mathématiques,* 1841; thus we may conclude that the memoir presented to the Academy is the same as that published in the Journal; to this we now proceed.

1037. *Mémoire sur les surfaces isostatiques dans les corps solides homogènes en équilibre d'élasticité*, par G. Lamé. *Journal de Liouville*, Vol. VI., 1841. This is substantially reproduced in an improved form by Lamé in his *Lecons sur les coordonnées curvilignes*, 1859, pp. 257—298, with the use of *two* constants of elasticity instead of *one;* we shall notice the *Leçons* hereafter. At present we will give a brief account of a matter which occurs on pp. 54—57 of the memoir and is not reproduced in the volume.

Suppose that the shifts are assigned by $u = -ax$, $v = -ay$, $w = -az$, which correspond to a uniform stretch (here a negative stretch or *squeeze*). Then we know that $\theta = -3a$; that the shears vanish, and that the tractions are all equal. Lamé indicates the steps by which the same results can be obtained from the formulae referred to curvilinear coordinates.

Suppose that one set of orthogonal surfaces, such as Lamé treats of, are *closed* surfaces; consider one of them as $\rho_1$, and suppose the origin of coordinates to be within it. There are two ways in which the decrease of the volume included by $\rho_1$ can be expressed. One way is putting it equal to $-\iint v_1 ds_2 ds_3$, where $v_1$ is the normal shift and $ds_2$, $ds_3$ elements of the lines of curvature on $\rho_1$, the integral being extended to all the values of the parameters $\rho_2$ and $\rho_3$. The other way is putting it equal to the product of $-\theta$ into the whole volume within the closed surface. Thus we have a theorem which may be expressed thus:

$$-\iint v_1 ds_2 ds_3 = -\theta \iiint ds_1 ds_2 ds_3$$

By the aid of equations like those marked (3) on p. 226 of the *Leçons sur les coordonnées curvilignes* we get:

$$v_1 = \frac{1}{h_1}\left(u\frac{d\rho_1}{dx} + v\frac{d\rho_1}{dy} + w\frac{d\rho_1}{dz}\right)$$

$$= -\frac{a}{h_1}\left(x\frac{d\rho_1}{dx} + y\frac{d\rho_1}{dy} + z\frac{d\rho_1}{dz}\right).$$

Also $\theta = -3a$, so that the theorem becomes

$$\iint\left(x\frac{d\rho_1}{dx} + y\frac{d\rho_1}{dy} + z\frac{d\rho_1}{dz}\right)\frac{1}{h_1}\,ds_2 ds_3 = 3\iiint ds_1 ds_2 ds_3$$

By the aid of equations like (7) on p 11 of the same work this may be written thus:

$$\iint \frac{1}{h_1 h_2 h_3}\left(x\frac{d\rho_1}{dx} + y\frac{d\rho_1}{dy} + z\frac{d\rho_1}{dz}\right) d\rho_2 d\rho_3 = 3\iiint \frac{\rho_1 d\rho_2 d\rho_3}{h_1 h_2 h_3}.$$

For an application of this theorem by himself Lamé refers to Liouville's *Journal* for November, 1838.

[1038.] *Note sur les épaisseurs et les courbures des appareils à vapeur. Comptes Rendus*, T. XXX., pp. 157—161. Paris, 1850. This note was read before the French Academy, February 18, 1850.

Lamé commences this paper by remarking on the few complete solutions hitherto obtained in the consideration of elastic bodies. Yet if it is not possible to treat completely cases which arise in practice, simple forms approaching to the actual can be considered and consequences, exact within certain limits, deduced. He considers a boiler as approximately a right cylindrical shell with spherical ends, and questions what are the proper thicknesses to be given to the cylindrical and spherical parts, as well as what curvature the ends ought to have in order that they may not be distorted. He remarks:

> Ces questions peuvent être résolues par la théorie mathématique de l'élasticité, qui conduit à des règles extrêmement simples. Ces règles n'ont rien d'empirique; elles supposent, il est vrai, l'homogénéité parfaite de l'enveloppe solide; mais on corrige aisément cet excès de perfection, en réduisant, dans les applications, le nombre qui exprime la force ou la résistance du métal employé

Lamé then reproduces for the thicknesses the rules Clapeyron and he had already obtained in their memoir; see our Arts. 1013 and 1016. These results, as we have seen, are quite erroneous, depending on maximum-stress and not on maximum stretch. Results similar to those of our footnotes to the same articles ought to replace Lamé's first and second rules for practical purposes; or, still better, like results might easily be calculated on the supposition that the envelope is endowed with a cylindrical and spherical distribution of homogeneity.

Lamé's third rule is a safeguard against the spherical ends tending to curve the cylindrical or *vice versa*. This is analytically expressed by making the displacements of both at the surface of junction the same. It is found approximately, Lamé states, that

Le triple du rayon de courbure du fond, divisé par l'épaisseur de ce fond, doit égaler sept fois le rayon du cylindre, divisé par l'épaisseur de ce cylindre. (p. 161.)

This result may be compared with that mentioned in our Art. 694[1]. It cannot be said that Lamé has placed the problem on a satisfactory footing, though the full treatment of the subject, at least on Lamé's hypothesis, may be easily worked out by the reader for himself, if he adopts maximum-stretch not maximum-stress as giving the elastic limit.

[1039.] On p. 187 of the same volume of the *Comptes Rendus*, we have a note by Lamé saying that it had been pointed out to him by M. Piobert that his first two results as to the curvature of boilers might be deduced on hydrostatical principles, and had been verified by numerous experiments. Lamé then lays stress upon the third result which he considers as by far the most important. The others, he regards, as at least a verification of the mathematical theory of elasticity.

1040. In the *Comptes Rendus*, Vol. XXXV. 1852, we have a *Note sur la théorie de l'élasticité des corps solides;* par M. Lamé; it occurs on pp. 459—464. Lamé presented to the Academy a copy of his *Lecons sur la théorie mathématique de l'élasticité des corps solides,* and gave an account of the object and distinctive characteristics of the work. The whole statement is very interesting, but I must restrict myself to a few extracts.

...C'est, en grande partie, une œuvre de coordination; car les éléments de la science dont il s'agit se trouvent disséminés dans les travaux des géomètres de notre époque, même dans ceux d'analyse pure, qui, souvent, ont plus aidé aux progrès de la physique mathématique que les Mémoires spéciaux.

Les premiers pas de cette science, toute nouvelle, ont été incertains. Des discussions se sont élevées entre d'illustres géomètres de cette Académie, sur les principes posés, sur la nature des actions moléculaires, et sur les fonctions qui peuvent les représenter. Les objections et les

[1] With bi-constant isotropy I find that instead of *le triple du* we must read '$(\lambda+2\mu)$ *fois le,*' and instead of *sept fois* '$(3\lambda+4\mu)$ *fois.*'

réponses, également obscures et incomplètes, ont inspiré des doutes sur la réalité de la nouvelle théorie; doutes que sont venus confirmer plusieurs épreuves expérimentales, constatant l'inexactitude de certains nombres déduits de cette théorie. Aujourd'hui, toutes ces discussions sont sans objet, ces doutes ne peuvent plus exister, et les épreuves expérimentales serviront à faire connaître des coefficients spécifiques, que la théorie seule ne saurait déterminer.

Lamé gives his reasons for thinking that the constants of elasticity cannot safely be reduced to *one*, and says

Dans les Leçons que je publie, j'admets deux coefficients distincts au lieu d'un seul, ce qui n'amène d'ailleurs aucune complication notable dans les formules.

1041. Lamé alludes to Clapeyron's theorem, but does not give any reference to the work in which it had been published

M. Clapeyron a découvert un théorème général, qui mérite le nom de *principe du travail des forces élastiques.*

Perhaps the theorem appeared for the first time in Lamé's work.

1042. At the end of Lamé's statement we have these words:

M. Cauchy demande en quoi les résultats que M. Lamé a indiqués, et auxquels il est parvenu, en appliquant la théorie des corps élastiques aux vibrations lumineuses, diffèrent des résultats obtenus par M. Cauchy lui-même, en 1830.

M. Lamé répond que, si aucune différence essentielle n'existe dans les résultats, il était néanmoins utile de chercher, le plus possible, à présenter cette application d'une manière élémentaire.

We now proceed to give an account of the work to which this *Note* relates.

1043. In the year 1852 a work was published entitled: *Leçons sur la théorie mathématique de l'élasticité des corps solides,* par M. G. Lamé. This is an octavo volume of 335 pages besides xvi pages containing the title and prefatory matter; there is one plate of diagrams; a second edition appeared in 1866. There are some remarks on the subject of the present volume in Lamé's *Leçons sur les coordonnées curvilignes,* 1859. The second edition of the work on elasticity is practically a reprint of the first.

1044. The treatise is divided into 24 Lectures; and this division is most convenient for a student, as each Lecture discusses some well defined portion of the subject; the whole forms an admirable specimen of a clear and attractive didactic work. It may be said roughly to consist of four parts. The first part comprises the first six Lectures, in which the general equations of elasticity referred to rectangular coordinates are established; Lamé himself says on his p. 76:

Ici se termine ce que nous avions à dire pour démontrer les équations générales de l'élasticité, et particulierement celles qui appartiennent aux solides homogènes et d'élasticité constante.

The second part comprises the next seven Lectures and contains various applications of the equations already established. The third part comprises the fourteenth, fifteenth, and sixteenth Lectures; here we have the general equations of elasticity referred to cylindrical coordinates, and to spherical coordinates, together with applications. The fourth part comprises the last eight Lectures, and consists of an application of the theory of elasticity to the double refraction of light.

1045. Lamé gives on his p. 8 his definition of *elastic force* (i.e. stress across a plane at a point); it resembles that of Cauchy to which Saint-Venant has taken objection: see our account of his *Cours lithographié* in Chapter IX. It introduces the 'little cylinder' but omits the clause, *the directions of which cross this face*: see our Art. 678. Lamé then refers on his p. 10 to a second definition which he says is apparently more simple, but he does not think that it gives a very clear idea of the stress across a plane; finally, however, having explained it suitably, he adopts it on his p. 12; it is the definition given by Cauchy: see Art. 616.

1046. Lamé in his second Lecture establishes the stress-equations of equilibrium of elasticity. He takes an infinitesimal rectangular parallelepiped, and starts with *nine* stress-symbols. To this elementary solid he applies the six equations of statical equilibrium. The first three give the body stress equations; the last three, namely those respecting moments, reduce the *nine* stresses to *six* Lamé afterwards establishes the conditions which must hold at the surface by the usual method of the tetrahedron. He shews that when the equations which he obtains are all satisfied then any *finite* portion of the body will satisfy the ordinary six equations of statical equilibrium.

1047. A few interesting points may be noticed in the second Lecture. Lamé adopts the term *forces of inertia* for the quantities denoted by $-\frac{d^2x}{dt^2}, -\frac{d^2y}{dt^2}, -\frac{d^2z}{dt}$ which occur in the use of D'Alembert's theorem: see his p. 13. Then he adverts to the two modes of arranging a course of study on Mechanics, the old mode which began with Statics, and a modern mode which begins with Dynamics He thinks that for a person who can go thoroughly into the subject the order is not of importance, but with respect to those who can acquire only partial knowledge he says,

Reste à savoir si, pour les étudiants qui sont forcés de s'arrêter en route, il est préférable d'avoir des idées saines en Dynamique, et de très obscures en Statique, ou, au contraire, de connaître à fond les lois de l'équilibre, et fort peu celles du mouvement. L'expérience répondra.

1048. In the process of finding the body stress-equations the symbol $\omega\ \widehat{xx}$ denotes a certain force acting on an element of area $\omega$. Lamé says that by an extension, of which many examples are to be found in the applications of Mechanics, especially in the theory of fluids, we may speak of $\widehat{xx}$ as a force; it is only necessary to suppose that on a unit of area forces are exerted of the same relative intensity as over the area $\omega$: see p. 17 of the volume.

1049. Towards the end of the Lecture Lamé gives his reasons for not citing the names of the inventors of the various ideas which constitute the theory; one reason is that most of these ideas present themselves so naturally as really to belong to all. He concludes thus:

L'unique but de notre travail est de mettre hors de doute, et l'utilité de la théorie mathématique de l'élasticité et la nécessite de l'introduire dans les sciences d'application. Quand ce but important sera atteint, fasse qui voudra le partage des inventions, et, quelque peu qu'on nous en attribue, nous ne réclamerons pas.

1050. The third Lecture first establishes a formula for the stretch $s_r$ in direction $r$ equivalent to that in our Art 663; and then formulae for the stresses like those of Art. 553; the method resembles that of Poisson in Art. 553 rather than that of Poisson in Arts. 589—591. Lamé objects to the process by which some of the earlier writers attempted to put definite integrals in the expressions instead of signs of summation: see his p. 38; the method which he now condemns was followed by Lame and Clapeyron in the memoir of which we have given an account: see our Art. 1007.

[1051.] Lamé while he does not actually calculate the stress across a plane on the basis of his definition (see Art. 1045) yet appeals, p. 33, to the conception of intermolecular force as central and as a function of the distance. This would undoubtedly have led him to rari constant equations. He seems to think on p. 77 that it is the use of definite integrals which alone leads to such equations, and talks somewhat disparagingly of the *ancienne méthode*. His own process on p 33, so far from having the merits he claims (pp. 76—78) for it, seems to me extremely obscure, being based on no clearly stated physical axiom.

1052. The expressions for the stresses noticed in the preceding Article involve *thirty-six* coefficients in all it is the object of the fourth Lecture to shew that in the case of a homogeneous body of constant elasticity in all directions they reduce to two distinct coefficients. A large part of the Lecture consists in investigations of the formulae for various transformations of the stresses; these transformations are necessary in the process by which Lamé effects the reduction of the number of coefficients to two. The process adopted by Lamé is not quite the same, perhaps not quite so good, as that of Saint Venant in his memoir on *Torsion*, p. 253 etc.; but one has no great advantage over the other. Lamé himself says on his p. 43

...les formules que nous allons établir sont longues à écrire, plutôt qu a démontrer....

1053. We shall exemplify the way in which Lamé conducts the application of the transformations; but we will first shew, in a manner substantially agreeing with that of Lamé, how from two examples of given displacements he draws inferences as to the values of some of the stresses. We may however observe that he speaks unfavourably of these and proposes to make changes with respect to them: see pp. 365 and 366 of his *Coordonnées curvilignes.*

1054. The first example begins on p. 39. Suppose that the shifts of the particles of a solid body are assigned by

$$u = 0,\ v = 0,\ w = cz\ ;$$

then in the notation of our p. 322, out of the three stresses $\widehat{xx}$, $\widehat{xy}$, $\widehat{zx}$, the latter two must vanish.

For take an element parallel to the plane of $yz$. Let $P$ denote the centre of the element; let $A$, $B$, $C$, $D$ be four equal particles on the same side of the element forming a rectangle parallel to the element, so that $P$ is equally distant from the four particles; and let $AB$ and $CD$ be parallel to the axis of $z$: all this refers to the original state of the body before any displacement. Now suppose displacements assigned as above, which correspond to a uniform stretch parallel to the axis of $z$. By virtue of the *relative* shifts the distances $PA$, $PB$, $PC$, $PD$ will all be equally and symmetrically changed; and so by symmetry the resultant stress at $P$ arising from the relative shifts will act at *right angles* to the element. In this way, considering the aggregate of all the particles on the same side of the element, we see that the final resultant at $P$ must be at right angles to the element. This shews that $\widehat{xy}$ and $\widehat{zx}$ must vanish so that only $\widehat{xx}$ remains.

1055. Next suppose that the shifts of the particles of a solid body are assigned by

$$u = -\,\omega yz,\ v = \omega xz,\ w = 0,$$

where $\omega$ is a very small constant; then all the three $\widehat{xx}$, $\widehat{xy}$, $\widehat{zx}$ will vanish for a point $P$ on the axis of $x$.

For take $A$, $B$, $C$, $D$ as before; and let the plane $zx$ lie midway between $AB$ and $CD$. Consider first an element at $P$ perpendicular to $x$. The shift in this case is equivalent to a rotation round the axis of $z$ of each slice which is perpendicular to the axis of $z$, the amount of rotation for each slice being proportional to the ordinate $z$; it corresponds to a torsion round the axis of $z$. Suppose the direction of rotation to be such as we may describe by saying, that it is from the axis of $y$ towards that of $x$ Then owing to the *relative* shift the distances of $A$ and of $C$ from $P$ are equally diminished, while those of $B$ and $D$ from $P$ are just as much increased: hence the molecular action arising from $A$ and $B$ is exactly balanced by that arising from $C$ and $D$ In this way, considering the aggregate of all the particles on the same side of the element, we see that the final resultant at $P$ is zero; that is $\widehat{xx}$, $\widehat{xy}$ and $\widehat{zx}$ all vanish.

Moreover assuming the same shifts we shall also have $\widehat{zz} = 0$ For take an element at $P$ parallel to the plane of $xy$. Then in the manner we have just exemplified, using the same position for $A$ and $D$ as before, we see that the molecular forces from $A$ and $D$ will be equal but of opposite signs; their resultant will be a force at $P$ *parallel to the axis* of $y$; and this holds for every such pair of particles like $A$ and $D$ on the same side of the element. Thus the resultant force reduces to the single stress $\widehat{yz}$; thus we have $\widehat{zz} = 0$, as well as $\widehat{zx} = 0$, which we knew already[1].

1056. We have now arrived at this stage: we have by Art. 1052 three formulae of the type

$$\widehat{xx} = A\frac{du}{dx} + B\frac{dv}{dy} + C\frac{dw}{dz} + D_{\iota}\left(\frac{dv}{dz} + \frac{dw}{dy}\right) + E_{\iota}\left(\frac{dw}{dx} + \frac{du}{dz}\right) + F_{\iota}\left(\frac{du}{dy} + \frac{dv}{dx}\right),$$

and three of the type

$$\widehat{yz} = G_{\iota}\frac{du}{dx} + H_{\iota}\frac{dv}{dy} + J_{\iota}\frac{dw}{dz} + K_{\iota}\left(\frac{dv}{dz} + \frac{dw}{dy}\right) + L_{\iota}\left(\frac{dw}{dx} + \frac{du}{dz}\right) + M_{\iota}\left(\frac{du}{dy} + \frac{dv}{dx}\right);$$

in each type $\iota$ will have a new value when we interchange $x, y, z$ in the stress on the left. We have then to establish relations between the coefficients, so as to reduce the number of them.

Suppose we are concerned with $\widehat{xx}$. In regard to this the shift $u$ occupies a distinct position, while the shifts $v$ and $w$ occupy coordinate positions; so we infer that $B_1$ and $C_1$ will be

[1] [I consider the method of this and the previous article extremely unsatisfactory. Lamé appeals to regularity of molecular distribution and centrality of intermolecular force, which if carried to their legitimate outcome would have led him to rari-constant equations: see our Arts. 923 and 1051. ED.]

identical, while $A_1$ will be different. Similarly we may infer that $E_1$ and $F_1$ will be the same but different from $D_1$. Then as a change of coordinate axes transforms $\widehat{xx}$ into $\widehat{yy}$ without any change in the circumstances the same coefficients must recur; we must have in $\widehat{yy}$ the same coefficients for $\frac{dv}{dy}$ as we have for $\frac{du}{dx}$ in $\widehat{xx}$, and the same for $\frac{du}{dx}$ and $\frac{dw}{dz}$ as we have for $\frac{dv}{dy}$ and $\frac{dw}{dz}$ in $\widehat{xx}$. Proceeding in this way we see that $\widehat{xx}$, $\widehat{yy}$, $\widehat{zz}$ can involve only *four* different coefficients, and we may express these tractions thus:

$$\widehat{xx} = A\frac{du}{dx} + B\frac{dv}{dy} + B\frac{dw}{dz} + D\left(\frac{dv}{dz} + \frac{dw}{dy}\right) + E\left(\frac{dw}{dx} + \frac{du}{dz}\right) + E\left(\frac{du}{dy} + \frac{dv}{dx}\right),$$

$$\widehat{yy} = B\frac{du}{dx} + A\frac{dv}{dy} + B\frac{dw}{dz} + E\left(\frac{dv}{dz} + \frac{dw}{dy}\right) + D\left(\frac{dw}{dx} + \frac{du}{dz}\right) + E\left(\frac{du}{dy} + \frac{dv}{dx}\right),$$

$$\widehat{zz} = B\frac{du}{dx} + B\frac{dv}{dy} + A\frac{dw}{dz} + E\left(\frac{dv}{dz} + \frac{dw}{dy}\right) + E\left(\frac{dw}{dx} + \frac{du}{dz}\right) + D\left(\frac{du}{dy} + \frac{dv}{dx}\right).$$

By similar considerations of symmetry it is shewn that $\widehat{yz}$, $\widehat{zx}$, $\widehat{xy}$ can involve only four different coefficients, and we may express these shears thus:

$$\widehat{yz} = G\frac{du}{dx} + H\frac{dv}{dy} + H\frac{dw}{dz} + K\left(\frac{dv}{dz} + \frac{dw}{dy}\right) + L\left(\frac{dw}{dx} + \frac{du}{dz}\right) + L\left(\frac{du}{dy} + \frac{dv}{dx}\right)$$

$$\widehat{zx} = H\frac{du}{dx} + G\frac{dv}{dy} + H\frac{dw}{dz} + L\left(\frac{dv}{dz} + \frac{dw}{dy}\right) + K\left(\frac{dw}{dx} + \frac{du}{dz}\right) + L\left(\frac{du}{dy} + \frac{dv}{dx}\right)$$

$$\widehat{xy} = H\frac{du}{dx} + H\frac{dv}{dy} + G\frac{dw}{dz} + L\left(\frac{dv}{dz} + \frac{dw}{dy}\right) + L\left(\frac{dw}{dx} + \frac{du}{dz}\right) + K\left(\frac{du}{dy} + \frac{dv}{dx}\right)$$

We have thus reduced our expressions so as to involve only *eight* different coefficients; we proceed further in the next Article.

1057. Suppose now that the shifts are those assigned in Art. 1054; we have shewn that then

$$\widehat{xy} = 0, \text{ and } \widehat{zx} = 0.$$

But by Art. 1056, we have then

$$\widehat{xy} = G, \text{ and } \widehat{zx} = H.$$

Hence we must have $G = 0$, and $H = 0$.

Next suppose that the shifts are those assigned in Art. 1055 we have shewn that for a point on the axis of $x$,

$$\widehat{xx} = 0, \quad \widehat{xy} = 0, \quad \widehat{zx} = 0, \quad \widehat{zz} = 0,$$

But by Art. 1056, we have then

$$\widehat{xx} = D\omega x, \quad \widehat{xy} = \widehat{zx} = L\omega x, \quad \widehat{zz} = E\omega x.$$

Hence we must have

$$D = 0, \quad L = 0, \quad E = 0.$$

Put $\qquad B = \lambda, \quad A = \lambda + 2\mu, \quad \theta = \dfrac{du}{dx} + \dfrac{dv}{dy} + \dfrac{dw}{dz},$

thus our formulae become

$$\widehat{xx} = \lambda\theta + 2\mu \frac{du}{dx}, \quad \widehat{yy} = \lambda\theta + 2\mu \frac{dv}{dy}, \quad \widehat{zz} = \lambda\theta + 2\mu \frac{dw}{dz},$$

$$\widehat{yz} = K\left(\frac{dv}{dz} + \frac{dw}{dy}\right), \quad \widehat{zx} = K\left(\frac{dw}{dx} + \frac{du}{dz}\right), \quad \widehat{xy} = K\left(\frac{du}{dy} + \frac{dv}{dx}\right).$$

We have thus reduced our expressions so as to involve only *three* different coefficients.

1058. We shall now shew that we must have $K = \mu$. Since we have assumed that the body is homogeneous and of constant elasticity, if we change the axes of coordinates we ought to arrive at precisely the same forms for the expressions relative to the new axes. Now by Art. 659 we have

$$\widehat{rr} = \widehat{xx} \cos^2(rx) + \widehat{yy} \cos^2(ry) + \widehat{zz} \cos^2(rz)$$
$$+ 2\widehat{yz} \cos(ry) \cos(rz) + 2\widehat{zx} \cos(rz) \cos(rx) + 2\widehat{xy} \cos(rx) \cos(ry).$$

Substitute from the formulae at the end of Art. 1057 thus

$$\widehat{rr} = \lambda\theta + 2\mu \left\{\frac{du}{dx}\cos^2(rx) + \frac{dv}{dy}\cos^2(ry) + \frac{dw}{dz}\cos^2(rz)\right\} + 2K\phi,$$

where $\phi$ stands for

$$\cos(ry)\cos(rz)\left(\frac{dv}{dz} + \frac{dw}{dy}\right) + \cos(rz)\cos(rx)\left(\frac{dw}{dx} + \frac{du}{dz}\right) + \cos(rx)\cos(ry)\left(\frac{du}{dy} + \frac{dv}{dx}\right).$$

In Art. 547 we have the following result for the stretch $s_r$ in direction $r$, although it is there expressed in a different notation,

$$s_r = \frac{du}{dx}\cos^2(rx) + \frac{dv}{dy}\cos^2(ry) + \frac{dw}{dz}\cos^2(rz) + \phi.$$

Hence we deduce

$$\widehat{rr} = \lambda\theta + 2\mu s_r + 2(K - \mu)\phi.$$

To make this correspond exactly with the value of $\widehat{xx}$ given in Art. 1057, we must have $K = \mu$.

Hence in the results at the end of Art. 1057, we must put $\mu$ for $K$; and thus we obtain the well-known forms with two constants for an isotropic medium.

1059. Lamé in his fifth Lecture introduces the *ellipsoid of elasticity* or, what we term, *Lamé's stress-ellipsoid* which gives the *magnitude* of the stress in any direction round a point, and also the surface of the second degree which gives the direction of the stress across any element; we may term it the *stress director-quadric*, its asymptotic cone, if one exists, is the *shear-cone*: see Art. 1008.

Lamé seems to view the introduction of these surfaces with great satisfaction. He says on his p. 64:

En un mot, les surfaces et les courbes du second ordre, pourvues de centre, viennent remplir, dans la théorie de l'élasticité, un rôle aussi important que les sections coniques en Mécanique céleste; elles lui appartiennent aux mêmes titres, elles en traduisent les lois avec autant de clarté, et même plus rigoureusement, car les lois des forces élastiques autour d'un point ne subissent aucune perturbation. Si, dans l'avenir, la Mécanique rationelle, courant plus rapidement sur les problèmes, aujourd'hui complétement résolus, du monde planétaire, se transforme pour s'occuper avec plus d'étendue de physique terrestre, la théorie que nous avons exposée dans cette leçon formera l un de ses premiers chapitres, et des plus importants, comme la suite du Cours le dé montrera.

1060. Lamé in his sixth Lecture gives the general equations for the equilibrium of an elastic solid which he proposes to use in the applications discussed in his work; he restricts himself to the case of homogeneous isotropic elastic solids: see his p. 65.

1061. Let us denote the body forces by $X_0$, $Y_0$, $Z_0$ respectively; and let us suppose them to be such that

$$X_0 = \frac{dF}{dx}, \quad Y_0 = \frac{dF}{dy}, \quad Z_0 = \frac{dF}{dz},$$

where $F$ is some function of $x$, $y$, $z$; moreover let us assume that $F$ satisfies the equation

$$\frac{d^2F}{dx^2} + \frac{d^2F}{dy^2} + \frac{d^2F}{dz^2} = 0,$$

which for brevity we will express as

$$\nabla^2 F = 0 \quad\ldots\ldots(1).$$

The equations of equilibrium of an elastic body are

$$\left.\begin{aligned}(\lambda+\mu)\frac{d\theta}{dx} + \mu\nabla^2 u + \rho\frac{dF}{dx} = 0\\ (\lambda+\mu)\frac{d\theta}{dy} + \mu\nabla^2 v + \rho\frac{dF}{dy} = 0\\ (\lambda+\mu)\frac{d\theta}{dz} + \mu\nabla^2 w + \rho\frac{dF}{dz} = 0\end{aligned}\right\} \quad\ldots\ldots(2).$$

Differentiate the first of these with respect to $x$, the second with respect to $y$, and the third with respect to $z$, and add; thus

$$(\lambda+\mu)\nabla^2\theta + \mu\nabla^2\left(\frac{du}{dx} + \frac{dv}{dy} + \frac{dw}{dz}\right) = 0,$$

that is $$(\lambda + 2\mu)\nabla^2\theta = 0 \quad\ldots\ldots(3).$$

Hence $\theta$ must be such a function as satisfies the condition which we have assumed in (1) that $F$ satisfies.

Now perform the operation denoted by $\nabla^2$ on the first of (2); thus

$$(\lambda+\mu)\frac{d}{dx}\nabla^2\theta+\nabla^2(\nabla^2 u)+\frac{d}{dx}\nabla^2 F=0;$$

and since both $\nabla^2\theta$ and $\nabla^2 F$ vanish we get

$$\nabla^2(\nabla^2 u)=0.$$

In the same manner from the second and third of (2) we deduce

$$\nabla^2(\nabla^2 v)=0, \text{ and } \nabla^2(\nabla^2 w)=0.$$

1062. It is known that if $R=\sqrt{\{(x-\alpha)^2+(y-\beta)^2+(z-\gamma)^2\}}$, and

$$F=\iiint\frac{f(\alpha,\beta,\gamma)}{R}\,d\alpha\,d\beta\,d\gamma \;\ldots\ldots\ldots\ldots\ldots\ldots\; (4),$$

where the triple integral is extended throughout any volume which does not contain the point $(x, y, z)$, then $\nabla^2 F=0$. Now let

$$\phi=\iiint f(\alpha,\beta,\gamma)\,R\,d\alpha\,d\beta\,d\gamma \;\ldots\ldots\ldots\ldots\ldots\; (5);$$

then $\phi$ will satisfy the equation $\nabla^2(\nabla^2\phi)=0$.

For

$$\frac{d\phi}{dx}=\iiint f(\alpha,\beta,\gamma)\frac{dR}{dx}\,d\alpha\,d\beta\,d\gamma=\iiint f(\alpha,\beta,\gamma)\frac{x-\alpha}{R}\,d\alpha\,d\beta\,d\gamma,$$

therefore $$\frac{d^2\phi}{dx^2}=\iiint f(\alpha,\beta,\gamma)\left\{\frac{1}{R}-\frac{(x-\alpha)^2}{R^3}\right\}d\alpha\,d\beta\,d\gamma,$$

Similar expressions hold for $\frac{d^2\phi}{dy^2}$ and $\frac{d^2\phi}{dz^2}$; and thus

$$\nabla^2\phi=2\iiint\frac{f(\alpha,\beta,\gamma)}{R}\,d\alpha\,d\beta\,d\gamma;$$

therefore by what was stated above we have

$$\nabla^2(\nabla^2\phi)=2\nabla^2\iiint\frac{f(\alpha,\beta,\gamma)}{R}\,d\alpha\,d\beta\,d\gamma=0.$$

Lamé thinks that it would be convenient to call a function $F$, determined by (4), an *inverse potential*, and a function $\phi$, determined by (5), a *direct potential*; or to call $F$ a potential of the *first kind* and $\phi$ a potential of the *second kind*.

1063. The equations (2) are linear and with constant coefficients; but besides the second shift fluxions they contain the given terms $\frac{dF}{dx}, \frac{dF}{dy}, \frac{dF}{dz}$. It is convenient and, in general, easy, to simplify the equations by getting rid of these terms in the following way.

Assume $u = u' + u_0$, $v = v' + v_0$, $w = w' + w_0$, and $\theta = \theta' + \theta_0$; then the first of (2) may be written

$$(\lambda + \mu)\frac{d\theta'}{dx} + (\lambda + \mu)\frac{d\theta_0}{dx} + \mu\nabla^2 u' + \mu\nabla^2 u_0 + \rho\frac{dF}{dx} = 0.$$

Then find $u_0$ such that

$$(\lambda + \mu)\frac{d\theta_0}{dx} + \mu\nabla^2 u_0 + \rho\frac{dF}{dx} = 0 \quad\text{.........}\quad (6),$$

and we are left with

$$(\lambda + \mu)\frac{d\theta'}{dx} + \mu\nabla^2 u' = 0 \quad\text{..................}\quad (7).$$

Now we need not seek for the most general solution of (6); any particular solution will suffice. Then we can seek for the most general solution of (7), which is simpler in form than the first of (2). In a similar manner we may proceed with the other equations of (2).

1064. As an easy example suppose that $\frac{dF}{dx}, \frac{dF}{dy}, \frac{dF}{dz}$ are constant; denote them by $a, b, c$ respectively. Then (6) and the two analogous equations will be satisfied by

$$u_0 = -\frac{\rho a x^2}{2(\lambda + 2\mu)}, \quad v_0 = -\frac{\rho b y^2}{2(\lambda + 2\mu)}, \quad w_0 = -\frac{\rho c z^2}{2(\lambda + 2\mu)}$$

Again suppose that $F$ is determined by (4); take

$$\phi = -\tfrac{1}{2}\iiint f(\alpha, \beta, \gamma)\, R\, d\alpha\, d\beta\, d\gamma,$$

and examine if (6) and the two analogous equations can be satisfied by

$$u_0 = k\frac{d\phi}{dx}, \quad v_0 = k\frac{d\phi}{dy}, \quad w_0 = k\frac{d\phi}{dz},$$

where $k$ is some constant. The left hand member of equation (6) becomes

$$k(\lambda+\mu)\frac{d}{dx}\nabla^2\phi + k\mu\frac{d}{dx}\nabla^2\phi + \rho\frac{dF}{dx} = 0,$$

that is
$$k(\lambda+2\mu)\frac{d}{dx}\nabla^2\phi + \rho\frac{dF}{dx} = 0,$$

or
$$-k(\lambda+2\mu)\frac{dF}{dx} + \rho\frac{dF}{dx} = 0;$$

and this vanishes if $k = \dfrac{\rho}{\lambda+2\mu}$

Similarly this supposition satisfies the two equations analogous to (6)

1065. Lamé makes a few remarks as to the relation between the coefficients $\lambda$ and $\mu$. Suppose that an elastic body is subjected to a uniform tractive load $P$ over all the surface; it is natural to assume that the body changes its size only so as to remain similar to itself; thus we may put

$$u = ax, \quad v = ay, \quad w = az.$$

These will lead in the manner of Art. 463, to

$$a = \frac{P}{3\lambda+2\mu}, \quad \theta = 3a = \frac{3P}{3\lambda+2\mu}$$

Thus if we denote the dilatability of the unit of volume under the unit of tractive load by $\beta$, we have

$$\beta = \frac{3}{3\lambda+2\mu} \quad\text{.......................} \quad (1).$$

Again suppose a body in the form of a bar or prism to be under the influence of a uniform terminal tractive load $F$; this gives rise to a stretch in the direction of the axis, and to a squeeze in directions at right angles to the axis; thus we may put

$$u = -ax, \quad v = -ay, \quad w = cz.$$

This leads, as may be easily shewn, to

$$c = \frac{\lambda+\mu}{\mu(3\lambda+2\mu)}F, \qquad a = \frac{\lambda}{2\mu(3\lambda+2\mu)}F.$$

Thus, if we denote by $\gamma$ the stretch of a prism, whose section is the unit of area, when under the unit of load, we have

$$\gamma = \frac{\lambda + \mu}{\mu\,(3\lambda + 2\mu)} \quad \text{.....................(2).}$$

If by experiment we can find the values of $\beta$ and $\gamma$, then from (1) and (2) we shall be able to determine $\lambda$ and $\mu$. Some writers hold that $\lambda = \mu$; among these we have Navier, also Poisson, and Lamé and Clapeyron in their earlier researches: see Arts. 266, 438, 1007 Wertheim from some experiments inferred that $\lambda = 2\mu$. Lamé holds that $\lambda$ is neither always equal to $\mu$ nor to $2\mu$, but that the ratio of $\lambda$ to $\mu$ may vary with the nature of the body: see his p. 76[1]

1066. The expression $\gamma$ is called by Lamé and several other French and Russian writers the *coefficient of elasticity;* but it is more usual now to term the reciprocal of this expression, that is, $\mu\,(3\lambda + 2\mu)/(\lambda + \mu)$,—Young's modulus or the modulus of elasticity; in this work it is termed the *stretch-modulus.* It is equal to $F/c$, and is therefore a quantity of the same nature and dimensions as $F$.

1067. Lamé's seventh Lecture is devoted mainly to *Clapeyron's Theorem.* Lamé establishes the equation which I express thus:

$$\Sigma\,(X_0 d_0 + Y_0 v_0 + Z_0 w_0)\,\omega = \iiint \left(\frac{I_1^2}{E} - \frac{I_2}{\mu}\right) dx\,dy\,dz,$$

where $X_0$, $Y_0$, $Z_0$ are the load-components producing shifts $u_0$, $v_0$, $w_0$ at their point of application, and $\Sigma$ is a surface summation for elements $\omega$; $E$ is the stretch-modulus and $\mu$ the slide-modulus; $I_1 = \widehat{xx} + \widehat{yy} + \widehat{zz}$ and $I_2 = \widehat{yy}\,\widehat{zz} - \widehat{yz}^2 + \widehat{zz}\,\widehat{xx} - \widehat{zx}^2 + \widehat{xx}\,\widehat{yy} - \widehat{xy}^2$.

Lamé says that it is this equation which constitutes Clapeyron's Theorem[2]. I shall make a few remarks on the Lecture.

1068. Lamé states that the left-hand side represents double of the external work (p. 79), but he does not prove this except for the simple case of a wire under terminal traction. Further he states, as it seems to me, abruptly, on p. 82 that $(I_1^2/E - I_2/\mu)\,dx\,dy\,dz$ is

[1] [See however the paragraph I have inserted as Art. 1051. Ed.]

[2] [$I_1$ and $I_2$ are of course the stress-invariants of the first and second order. Ed.]

the *double of the internal work of the element dxdydz*: see Moigno's *Statique*, 662, 716.

1069. As an example of Clapeyron's Theorem Lamé proposes to find the most favourable proportions to be given to the parts of an isosceles triangular frame which is to stand in a vertical plane and to support a weight at the highest point. After discussing this example Lamé adds on his p. 91:

> Il est facile de traiter de la meme manière des assemblages plus complexes, formés de pièces de bois, de fer ou de fonte, destinées à s'opposer à des efforts d'autre nature. Dans tous ces cas divers, on déduit du théorème de M. Clapeyron, que l'on peut appeler *principe du travail des forces élastiques*, les dispositions les plus avantageuses de la construction qu'on étudie. Jamais, je crois, on ne s'était approché aussi près de la solution générale du fameux problème des *solides d'égale résistance*, qui préoccupait tant Girard, et dont la nature a donné des exemples si remarquables. Nous aurons l'occasion d'appliquer le principe du travail des forces élastiques à divers cas d'equilibre d'élasticité.

I do not know to what applications the last sentence refers; Lamé alludes to Clapeyron's Theorem on pp. 99 and 187 of his volume, but with respect to matters of no great importance.

1070. In Clapeyron's Theorem it is supposed that there are no body forces; Lamé on his p. 84 briefly alludes to the modification which the Theorem undergoes when we take account of these. We must introduce on the left-hand side of the equation in Art. 1067 the term $\Sigma(Xu + Yv + Zw)\rho\varpi$, where $X$, $Y$, $Z$ are the components of body-force on the element of volume $\varpi$. As an example he notices the case in which a heavy elastic wire is hung up by one end and a weight is fastened at the other. He obtains a result which seems to be equivalent to one well known in the ordinary theory of Statics: see my *Analytical Statics*, Art. 199. This he says may be obtained by the aid of formulae which he has given and which we have reproduced in Art. 1064. Lamé has not worked this out.

[I do not know by what process Lamé reaches his results, but the following may perhaps replace it. Consider a solid vertical cylinder of material; let its section supposed of any shape have an area $\omega$, its length be $l$, its weight $W$, and let tractive loads $F/\omega$, $F'/\omega$ where $F' = F + W$ be applied to its terminal sections. Let the axis of $z$ coincide with the axis of the cylinder, and the origin be taken in the top terminal section. Then by an application of Saint-Venant's *méthode semi inverse* I find for a possible system of shifts:

$$\left.\begin{matrix} u \\ v \end{matrix}\right\} = -\frac{\lambda}{2\mu(2\mu+3\lambda)}\left(\frac{F'}{\omega} - \frac{Wz}{\omega l}\right)\left\{\begin{matrix} x \\ y \end{matrix}\right\}$$

$$w = \frac{F'z}{\omega E} - \frac{W}{\omega E}\frac{z^2}{2l} - \frac{W}{\omega}\frac{\lambda}{2\mu(2\mu+3\lambda)}\frac{x^2+y^2}{2l}.$$

Hence we see that all the cross-sections become paraboloids of revolution and the curved surface of the cylinder a cone A variety of interesting cases can be obtained by slight modifications (e g. $F$, or $F' = 0$), and applying the stretch condition for set or rupture; thus we can find the form taken by a standing or pendant column of pitch or other material for which the ratio $W/(\omega E)$ is not very small.

In the case of a wire we may neglect the third term of $w$, and we then find by Clapeyron's Theorem:

$$\Sigma Zw \quad \varpi + \Sigma Z_0 w_0 \omega = \int (zz^2/E)\, \omega dz,$$

$$\frac{W}{l}\int_0^l w dz + F\left(F + \frac{W}{2}\right)\frac{l}{\omega E} = \frac{\omega}{E}\int_0^l \left(\frac{F'}{\omega} - \frac{Wz}{\omega l}\right)^2 dz,$$

an equation easily shewn to be identically satisfied. If $w_0$ be the whole extension, the work equals half the left side

$$= \frac{Ww_0}{4} + \frac{Fw_0}{2} + \frac{W^2 l}{24\omega E}, \text{ since } w_0 = \left(F + \frac{W}{2}\right)\frac{l}{\omega E},$$

$$= \frac{Ww_0}{4} + \frac{Fw_0}{2} + \frac{Ww_0'}{12},$$

if $w_0'$ be that part of the whole extension due to the weight of the string. Lamé in his result omits the last term. ED.]

1071. Lamé in his eighth Lecture discusses the equilibrium and motion of an elastic string. The equations relating to this subject were obtained by the older writers on special principles but Lamé here deduces them from his general theory. He says on his p. 93,

Il importe de faire voir aujourd'hui que la mise en équation de ces anciens problèmes rentre dans la théorie générale. C'est ce qu'a pensé Poisson et ce que nous essayerons après lui, d'une manière plus rapide et peut être plus simple.

The allusion to Poisson applies I believe to pages 422—442 of his memoir of April, 1828. The whole of Lamé's Art. 39 which constitutes the introduction to this Lecture is very good.

1072. Lamé in his ninth Lecture discusses the equilibrium and motion of an elastic lamina. He arrives with ease at the general equations. Lamé says that he has sought in vain for a motive which could have induced Poisson to adopt the very long and complicated course he did: but see Art. 473 of my account of Poisson. He goes on to make the following remark on his p. 111,

> Lorsqu'on parvient à un résultat simple par des calculs compliqués, il doit exister une manière beaucoup plus directe d'arriver au même résultat; toute simplification qui s'opère, tout facteur qui disparaît dans le cours du calcul primitif, est l'indice certain d'une méthode à chercher où cette simplification serait toute faite, où ce facteur n'apparaîtrait pas.

Such a remark as this the late Professor De Morgan made with great emphasis to his students many years before the publication of Lame's work. The motion of an elastic membrane is illustrated by taking the case in which the membrane is originally in the form of a plane rectangle in the plane of $xy$, with two edges coinciding with the axes of $x$ and $y$ respectively, and is then displaced into the vaulted form determined by $z = kx(d-x)y(b-y)$ where $k$ is small.

[1073.] On p. 114, Lamé finds the ratio of the stretches in the three cases: of an isotropic solid subjected to uniform tractive load ($s'$); of an isotropic membrane subjected to uniform contour tractive load ($s''$); of an isotropic thread or wire subjected to uniform terminal tractive load ($s'''$). He finds

$$s' : s'' : s''' :: 2\mu : 2\mu + \lambda : 2\mu + 2\lambda.$$

Lamé adds:

> Suivant Poisson, qui admet la relation fausse $\lambda = \mu$, on aurait
>
> $$s' : s'' : s''' :: 2 : 3 : 4.$$

Suivant Wertheim, qui admet la relation douteuse $\lambda = 2\mu$, on aurait

$$s' : s'' : s''' :: 1 : 2 : 3 ;$$

c'est à-dire que la dilatation serait en raison inverse du nombre des dimensions qui la subissent.

See our Arts. 922 *et seq.*

It must be noted that if Lamé had followed his own assumptions to their legitimate conclusions, he would also have found $\lambda = \mu$: see Art. 1051.

1074. Lamé in his tenth Lecture examines the nature of the terms of the series which he had obtained in his preceding Lecture for the vibrations of a plane membrane; he takes in succession a square membrane, a rectangular membrane, and a membrane in the form of an equilateral triangle. It is known that we have for the shift at right angles to the membrane, supposing the initial velocity zero, a series of terms of the type $G \cos \gamma t \sin \frac{i\pi x}{l} \sin \frac{j\pi y}{b}$, where $i$ and $j$ are integers connected by the relation $\gamma^2 = c^2\pi^2 \left(\frac{i}{l^2} + \frac{j^2}{b^2}\right)$, which when $l = b$ becomes $\gamma^2 = \frac{c^2\pi^2}{b^2}(i^2 + j^2)$. In virtue of the last equation the *Theory of Numbers* is serviceable, as pointing out what values of $i$ and $j$ will occur with an assigned value of $\gamma$. The *Theory of Numbers* fascinated Lamé with the charms which have been felt by so many of the greatest mathematicians, and he obviously was much gratified with this application of a subject which he had himself cultivated with singular zeal.

1075. Lamé gives interesting discussions relative to the *nodal lines* of a membrane such as will be found now in most text-books on Sound. If however $i$ and $j$ are very large numbers the theoretical development of all the forms of the nodal lines would become very laborious; and Lamé says on his p 129,

Le très-petit nombre de systèmes nodaux de la membrane carrée que nous avons décrits, comparé au nombre infini de tous ceux dont l'analyse indique l'existence, laisse un champ vaste, et l'occasion d'une sorte de triomphe, aux expérimentateurs qui ont pris pour sujet de leurs recherches les figures si variées que le sable dessine sur les surfaces vibrantes; il faut nous résigner à cette défaite.

1076. In the case of the membrane in the form of an equilateral triangle Lamé employs coordinates of the same kind as those we now usually term *trilinear coordinates.* We denote these frequently at present by the letters $\alpha$, $\beta$, $\gamma$; there is a linear relation between them which for the equilateral triangle becomes $\alpha+\beta+\gamma=\lambda$, where $\lambda$ is the height of the triangle. Lamé takes in fact for his variables $\alpha-\frac{\lambda}{3}$, $\beta-\frac{\lambda}{3}$, $\gamma-\frac{\lambda}{3}$ instead of $\alpha$, $\beta$, $\gamma$ respectively; so that the sum of his variables is zero. Lamé presents his results briefly, referring for more detail to a memoir which he had formerly published on the equilibrium and motion of heat within a regular triangular prism; there is however no difficulty in verifying all that he gives respecting the vibrations of the triangular membrane without consulting this memoir.

1077. Lamé says on his p. 136, very naturally

L'objet de cette Leçon paraîtra sans doute fort peu important aux ingénieurs qui s'intéressent spécialement à l'équilibre d'élasticité.

He proceeds in his usual interesting manner to justify his course; he suggests that what we consider to be a body at rest may really involve very rapid vibrations of the molecules which never cease; and he concludes thus:

Ces considérations nous semblent mettre hors de doute l'utilité de l'étude des vibrations; et répétons-le, cette étude, reconnue nécessaire, serait superficielle et incomplète, si l'on n'avait pas recours aux propriétés des formes quadratiques des nombres entiers, à cette *théorie des nombres*, si souvent anathématisée par les détracteurs de la science pure, par les praticiens exclusifs.

1078. In his eleventh Lecture Lamé treats of *plane waves.* Two classes of these are obtained; in one class the vibrations are at right angles to the front of the wave, and in the other they are in the front of the wave. Each class of wave is propagated with its own special constant velocity, that of the former being the greater. The substance of the whole investigation is due to Poisson (see

Art. 526), but is here treated with bi-constant notation and in much fuller detail.

1079. In his twelfth Lecture Lamé treats of the elastic equilibrium of a right six-face. The complete solution of this problem, when given forces act on the faces, has not yet been obtained. Lamé considers the problem, although in a form not quite general, and proceeds to a certain extent with the solution. He proposes the problem in this form: we have to find $u$, $v$, $w$ so as to satisfy the equations

$$\frac{d\theta}{dx} + \epsilon\nabla^2 u = 0, \quad \frac{d\theta}{dy} + \epsilon\nabla^2 v = 0, \quad \frac{d\theta}{dz} + \epsilon\nabla^2 w = 0 \ldots\ldots(1),$$

where $$\theta = \frac{du}{dx} + \frac{dv}{dy} + \frac{dw}{dz}, \text{ and } \epsilon = \frac{\mu}{\lambda + \mu}.$$

Then the six stresses are to be found by the usual equations; and they are to satisfy the surface conditions

$$\left.\begin{array}{llll} \widehat{xx} = \phi_1, & \widehat{xy} = 0, & \widehat{zx} = 0 & \text{for } x = \pm a \\ \widehat{xy} = 0, & \widehat{yy} = \phi_2, & \widehat{yz} = 0 & \text{for } y = \pm b \\ \widehat{zx} = 0, & \widehat{yz} = 0, & \widehat{zz} = \phi_3 & \text{for } z = \pm c \end{array}\right\} \ldots\ldots\ldots (2),$$

where $\phi_1$, $\phi_2$, $\phi_3$ are given functions of the variables which they respectively involve, namely the first of $y$ and $z$, the second of $z$ and $x$, and the third of $x$ and $y$. There is of course considerable limitation of the problem in thus making the forces which act on one face precisely correspond point for point with those which act on the opposite face. Of course the *aggregate* of the forces on one face must be equal to the *aggregate* of the forces on the opposite face, by reason of one of the six equations of equilibrium of ordinary Statics; but the present limitation is much more stringent than this. Lamé also assumes that the functions $\phi_1$, $\phi_2$, $\phi_3$ are *even* functions of the variables which they respectively involve.

1080. Now Lamé finds expressions which satisfy the differential equations (1); these expressions involve sines and cosines of the variables, and also what are called the *hyperbolic* sines and cosines,—the hyperbolic sine of $t$ being defined to be $\frac{1}{2}(e^t - e^{-t})$, and the hyperbolic cosine to be $\frac{1}{2}(e^t + e^{-t})$ Lamé adjusts his expressions so as to satisfy out of (2) the conditions relating to the vanishing of the shears at the faces; but there is no known method for determining the coefficients of the expressions so as to satisfy the remaining equations of (2).

1081. Lamé refers at the end of the Lecture to the memoir in which Clapeyron and he had treated two general problems respecting the equilibrium of an elastic body: see our Arts. 1018 and 1020, in the account of the memoir. He says however:

Nous nous dispensons de reproduire ici ces solutions analytiques, malgré quelques conséquences dont l'énoncé est simple et qui pourraient être utilisées. Ce ne sont là que des essais entrepris dans le but de chercher une solution générale, et qui ne paraissent pas placés sur la route qui doit y conduire.

1082. Lamé in his thirteenth Lecture discusses the vibrations of a rectangular prism. He begins by remarking that although the question of the equilibrium of elasticity of a rectangular prism is very difficult, yet it is quite otherwise with respect to the question of the vibrations. He says:

En général, sauf quelques cas simples, les problèmes relatifs à l'équilibre d'élasticite sont incomparablement plus difficiles à traiter par l'analyse mathématique que les problèmes relatifs aux vibrations; aussi les travaux des géomètres sont-ils fort nombreux sur les seconds, très-rares sur les premiers.

But the greater comparative difficulty of the problem relating to equilibrium arises apparently from the fact that the surface conditions for a chosen type of vibration are more easily satisfied than those which arise from any but the simplest distributions of load.

1083. Lamé's results in this Lecture are peculiarly interesting in their reference to Savart's experiments: see Arts. 323—330. He shews that in the longitudinal vibration of a rectangular prism the section cannot remain unstrained unless a tractive load varying with the time be applied to those faces of the prism to which the section is perpendicular (p. 167). This could not hold for a rectangular prism vibrating in the air.

Further (p. 168) he considers transverse vibrations and shews that, even if a rectangular prism has only two parallel faces free, it cannot vibrate transversely like a membrane. The transverse vibrations must be accompanied by longitudinal vibrations. As type of such vibrations he takes the shifts given by

$$u = m\epsilon \cos mx \sin mz \cos mt\omega\sqrt{2}, \quad v = 0,$$
$$w = -m\epsilon \sin mx \cos mz \cos mt\omega\sqrt{2},$$

where $a$ is the length of the prism; $\omega = \sqrt{\mu/\rho}$; and $m = i\pi/a$, $i$ being any integer. In this case $w = 0$ for $x = 0$ and $x = a$, or we may suppose the prism supported at its ends.

1084. Lamé then proceeds to remark that if we give $i$ only odd integral values, this solution will be suitable for a prism of length $a/2$ which is *encastrée* at the origin and free at the other end. I do not see how the condition denoted by *encastrée* is fulfilled; I should have thought $u$ and $w$ would both be required to vanish over the section. Again take the lamina originally of the length $a$; then it would seem that we might cut off a piece of any length according to the reason which he gives, namely that the stress across such a section will be *normal*, and therefore can be furnished by the air. But how can we get from the air *positive*, even if we can get *negative* tractions? I do not understand this at all. I suppose *encastrée* may mean put in an iron collar, so that $u$ may change but $w$ be zero; this is then apparently equivalent to *pincée*, as used on page 172. [Lamé's treatment of vibrations seems to me vitiated by his statement in § 64 (p. 148) that the only condition at a surface exposed to the air is the vanishing of the shearing load. *La composante normale*, he writes, *doit rester variable et non-déterminée*. He supposes any traction can be supplied by the air, and repeatedly uses this principle. Thus it is assumed in Arts. 1083 and 1090.]

[1085.] On p. 172 we have a consideration of Savart's *vibration tournante*; see Art. 326. A vibration is represented by the forms $v = m\epsilon \sin my \sin mz \cos mt\omega\sqrt{2}$, $w = m\epsilon \cos my \cos mz \cos mt\omega\sqrt{2}$, where $m = (2i+1)\pi/b$, $b$ being the breadth of the prism. It is obvious that these vibrations are in the section of the prism perpendicular to its length $a$. Lamé says that this is Savart's vibration when $i = 1$; I do not understand why it *must* be.

On p. 173 Lamé refers to another result of Savart's and endeavours to interpret it analytically. He remarks:

On sait qu'une lame de verre, pincée en son milieu, et que l'on frappe à l'une de ses extrémités, de manière à la faire vibrer longitudinalement exécute en même temps des vibrations transversales, dont l'existence se manifeste par des lignes nodales que le sable dessine sur les faces latérales.

The transverse vibrations are then taken to be represented by the forms of the preceding Article, while the longitudinal are those he has considered on p. 167: see Art. 1083. He makes the periodic times of both coincide. This would seem to involve the relation

$$(2j+1)\frac{\pi\Omega}{a} = (2i+1)\frac{\pi\omega\sqrt{2}}{b},$$

where
$$\Omega = \sqrt{\frac{\lambda+2\mu}{\rho}},$$

or
$$2i+1 = (2j+1)\frac{b}{a}\sqrt{\frac{\lambda+2\mu}{2\mu}}.$$

Lamé however writes down the result $m = (2j+1)\frac{\pi}{a}\sqrt{\frac{\lambda+2\mu}{2\mu}}$ without apparently recognising that the value of $m$ is already settled, and that we are led to the condition that $\frac{b}{a}\sqrt{\frac{\lambda+2\mu}{2\mu}}$ must be a proper fraction of the type $\frac{2i+1}{2j+1}$. Further how can such results as those of p. 167 be applied to bodies vibrating in the air?

The whole treatment seems remarkably obscure, not to say erroneous, and still leaves a wide field for analytical investigation in Savart's results.

1086. Lamé gives in the latter part of his Lecture a classification of the vibrations of rectangular prisms depending on whether they produce or do not produce dilatations. He says that the cases of vibration which have been studied by experiment, such as those relating to the figures formed by sand on a vibrating plate of glass, are small in number compared with those which must exist and form in the interior nodal surfaces that give no evidence by which the experimenter can become aware of their existence. After discussing the examples he concludes thus with respect to the two kinds of motion.

Ne sont-ils qu'une preuve de plus de la fécondité de l'analyse mathématique? ou existent-ils réellement dans la nature? A cette question, nous nous dispensons de répondre.

1087. Lamé devotes his fourteenth Lecture to the transformations which are necessary when instead of the ordinary system of rectangular coordinates we substitute what may be called, from its most important feature, a cylindrical system. Instead of the ordinary $x$, $y$, $z$ we now use $r$, $\phi$, $z$; that is we change $x$ into $r \cos \phi$, we change $y$ into $r \sin \phi$, and we leave $z$ unchanged. Lamé works out the transformations fully, though the new formulae were not absolutely necessary for the few and simple examples which he discusses. With respect to this point he says on his page 186

Mais nous avons pensé qu'il était utile d'établir ces nouvelles équations dans toute leur généralité, pour faciliter des recherches plus difficiles, où leur emploi serait indispensable. Quelquefois, dans les travaux de Physique mathématique, on abandonne une idée accessoire et qui mériterait d'être poursuivie, parce que l'on n'a pas à sa disposition les relations analytiques nécessaires, et que leur recherche, exigeant trop de temps, ferait perdre de vue l'idée principale. C'est alors que l'impatience peut conduire à l'erreur....

1088. Lamé refers to the memoir already cited in Art. 1022 as containing a general solution of the problem of the equilibrium of elasticity of an indefinite cylindrical shell, when the surfaces are under the action of given forces; he adds on his page 192:

Nous ne reproduisons pas ici cette solution, encore trop compliquée. D'ailleurs nous avons pour but, non de donner un traité complet, mais de montrer, par des exemples simples et variés, l'utilité et l'importance de la théorie mathématique de l'élasticité.

1089. Lamé gives at the end of his Lecture a few words to the vibrations of solids of cylindrical form. He supposes the body-forces to vanish, and puts as before $\Omega^2$ for $(\lambda + 2\mu)/\rho$

and $\omega^2$ for $\mu/\rho$. Lamé after this proceeds, as elsewhere, in what seems to me too positive a manner; he says that certain things *must be* when it would be safer to say that they *may be* in particular cases: we will however follow his process. According to the theory expounded in Art. 526, there are two distinct classes of vibrations.

For the vibrations of the first class the terms in $\omega^2$ must disappear from the equations; hence we have $U$, $V$, $W$ being the shifts corresponding to the variable $r$, $\phi$, $z$:

$$\frac{dVr}{dz}=\frac{dW}{d\phi}, \qquad \frac{dW}{dr}=\frac{dU}{dz}, \qquad \frac{dU}{d\phi}=\frac{dVr}{dr} \quad \ldots\ldots(1).$$

From these we infer that

$$U=\frac{dF}{dr}, \qquad Vr=\frac{dF}{d\phi}, \qquad W=\frac{dF}{dz} \quad \ldots\ldots\ldots(2),$$

where $F$ is some function of $r$, $\phi$, $z$. These lead to

$$\theta=\nabla^2 F \ldots\ldots\ldots\ldots\ldots\ldots\ldots\ldots\ldots(3).$$

Then we easily find:

$$\frac{d^2F}{dt^2}=\Omega^2\nabla^2 F \quad \ldots\ldots\ldots\ldots\ldots\ldots\ldots\ldots(4).$$

For the vibrations of the second class the terms in $\Omega^2$ must disappear from the fundamental equations, and we have

$$\theta=\frac{1}{r}\frac{dUr}{dr}+\frac{1}{r}\frac{dV}{d\phi}+\frac{dW}{dz}=0.$$

1090. Lamé says that the longitudinal vibrations which are produced in a cylindrical rod (*tige*) by clasping it at the middle and rubbing it parallel to its length are of the first class; but I do not see why. They are *longitudinal*, but in a sense different from that in which the vibrations of the first class are longitudinal[1].

According to Lamé the shift being parallel to the axis we have $U=0$, $V=0$, so that $W$ alone exists; hence the $F$ of the preceding Article becomes independent of $r$ and $\phi$, and we see from (4) that $W$ must satisfy the equation

$$\frac{d^2W}{dt^2}=\Omega^2\frac{d^2W}{dz^2}$$

[1] [Lamé's discussion of vibrations seems to me here as elsewhere unsatisfactory; I think I am justified in saying *erroneous*. Ed.]

The tangential components of the stress are zero everywhere. If $a$ is the length of the rod, and $n$ any integer, a suitable solution of the last equation, when the ends are free, is given by

$$W = \cos\frac{(2n+1)\pi z}{a}\cos\frac{(2n+1)\pi\Omega t}{a}.$$

1091. An example of a vibration of the second class is furnished when $V$ alone exists, and does not vary with $\phi$. Thus

$$U=0, \qquad W=0, \qquad \frac{dV}{d\phi}=0;$$

these lead to $\theta=0$.

Then the shear $\widehat{rz}$ vanishes[1], while the shear $\widehat{r\phi}$ reduces to $\mu\left(\frac{dV}{dr}-\frac{V}{r}\right)$, and this will also vanish if we take $V=rf$, where $f$ contains only $z$ and $t$. If $f$ satisfies the equation

$$\frac{d^2f}{dt^2}=\omega^2\frac{d^2f}{dz^2},$$

then the body-shift equations are satisfied. The shear $\widehat{z\phi}$ reduces to $\mu\frac{dV}{dz}$; this must vanish at the ends if they are free. This, and all the other conditions will be satisfied if we take

$$V = r\cos\frac{n\pi z}{a}\cos\frac{n\pi\omega t}{a},$$

where $n$ is any integer.

The tractions $\widehat{rr}$, $\widehat{\phi\phi}$, $\widehat{zz}$ are zero. Thus this kind of motion does not involve any elastic action at the surface, either traction or shear; the molecules on the surface vibrate without quitting it, and the cylinder undergoes no periodic deformation. Lamé concludes the Lecture thus:

Il serait difficile d'imaginer un état vibratoire plus silencieux et plus imperceptible. Les expériences sur le pendule, faites au Panthéon par M. Foucault, ont constaté un mouvement de cette nature, par les oscillations tournantes de la boule sphérique attachée au long fil pendulaire.

[1] [The system of stresses for cylindrical coordinates (tractions $\widehat{rr}$, $\widehat{\phi\phi}$, $\widehat{zz}$ and shears $\widehat{\phi z}$, $\widehat{zr}$, $\widehat{r\phi}$) readily explains itself. ED.]

It seems to me that Lamé falls into an anti-climax when he excites our interest in this mysterious motion, described as far beyond the cognizance of our senses, and then says that the experiments of Foucault have established the existence of such a phenomenon.

1092. Lamé devotes his fifteenth Lecture to the transformations which are necessary when instead of the ordinary system of rectangular coordinates we substitute the ordinary system of polar coordinates, which may be called a spherical system.

1093. In this Lecture formulae are obtained which are suitable for problems relative to the elastic equilibrium, or to the vibrations, of a sphere or of a body bounded by concentric spherical surfaces. The formulae are not fully worked out, but the steps to be followed are sufficiently indicated for the guidance of a resolute student. A case of vibrations of a spherical shell is worked out at the end of the Lecture. In considering the amount of originality in this and the following Lectures, as well as in Lamé's memoir of 1854 (see Art. 1111), the work of Poisson and Mainardi ought to be duly weighed: see Arts. 449 and 860. Lamé was the first to apply Laplace's coefficients to the problem.

1094. In his sixteenth Lecture Lamé applies the formulae of the fifteenth to the elastic equilibrium of a spherical shell. There are two cases; in the first case no internal applied force is considered; in the second case the force arising from the attraction of the particles themselves is considered. Thus the second case includes the first: both are parts of the general problem which I have called *Lamé's Problem*, and shall discuss in Art. 1111.

1095. Lamé makes an application of his formulae to the case of the earth, and then on his page 221 proceeds thus:

Ce résultat paraît s'accorder avec l'idée de M. Élie de Beaumont, sur la formation des chaînes de montagnes, à la suite d'un affaissement genéral, dû au refroidissement: car il semble résulter de cette idée que la pression intérieure doit aller en diminuant, d'un cataclysme au suivant, du dernier à celui vers lequel nous avançons.

Cette application de la théorie de l'élasticité à l'équilibre intérieur de l'écorce terrestre pourra paraître trop hasardée; mais s'il en est ainsi, les applications de la théorie analytique de la chaleur au refroidissement du globe, faites par Laplace, Fourier, Poisson, doivent inspirer les mêmes scrupules. Et nous ne nous défendrons pas d'avoir imité ces illustres géomètres, lors même qu'ils se seraient trompés, en appliquant aux sublimes questions de la Mécanique céleste de simples formules de Physique mathématique. Au reste, la question que nous avons abordée peut se traiter d'une manière plus complète ou plus voisine de la réalité: nous avons constaté, par des recherches analytiques, qu'en considérant la terre comme un sphéroïde peu différent de la sphère, qu'en prenant pour les deux parois de l'enveloppe solide des ellipsoïdes homofocaux, ce qui donne une épaisseur variable; enfin, qu'en ayant égard aux variations de la pesanteur dues à la force centrifuge, on arrivait dans tous les cas, aux mêmes conclusions.

I suppose that these investigations were never published, for in Lamé's memoir on the subject which appeared in Liouville's *Journal* in 1854 the earth is treated as originally a sphere.

1096. On his page 222 Lamé gives a brief account of the memoir in which he applied his *curvilinear coordinates* to the theory of elasticity; the memoir was published in Liouville's *Journal de mathématiques* for 1841: see Art. 1037. The memoir is substantially reproduced by Lamé in his *Leçons sur les Coordonnées curvilignes*, where however he uses *two* constants of elasticity instead of one as in the memoir.

1097. Hitherto Lamé has confined himself to the case of a medium which is both homogeneous and isotropic; but he now gives up this restriction. The remainder of his volume occupying pp. 225—335 is devoted to the subject of light, and forms a theory of the double refraction of light based on the principles of elasticity, the medium being supposed homogeneous but aeolotropic. The name of Lamé does not occur in the paper by Stokes entitled: *Report on Double Refraction*, published in the *British Association Reports* for 1862.

1098. Lamé in his seventeenth Lecture investigates the relations which must hold among the thirty-six coefficients of elasticity in order that double refraction may be possible. These coefficients are reduced to twelve in number; then six are suppressed because they occur multiplied by the dilatation which is here assumed to vanish; of the remaining six it is shewn that three will disappear by a suitable choice of axes of co-ordinates, leaving finally three. The axes of co-ordinates which must be used to reduce the coefficients to three are called *axes of elasticity*. This Lecture carries the investigation to the point which presents the quadratic equation giving the two velocities for a plane wave propagated in a certain direction through a crystal.

1099. Lamé begins as usual with some very interesting general remarks; they occur on his pp. 225 and 226:

Jusqu'ici nous avons traité la théorie de l'élasticité comme une science rationnelle, donnant l'explication complète et les lois exactes de faits qui ne peuvent pas évidemment avoir une autre origine. Nous allons maintenant la présenter comme un instrument de recherches, ou comme un moyen de reconnaître si telle idée préconçue, sur la cause d'une certaine classe de phénomènes, est vraie ou fausse. C'est sous ce dernier point de vue que Fresnel l'avait considérée, lors de ses belles découvertes sur la double réfraction, et ses commentateurs auraient dû suivre plus scrupuleusement son exemple. La théorie physique des ondes lumineuses porte certainement en elle l'explication future de tous les phénomènes de l'optique; mais cette explication complète ne peut être atteinte par le seul secours de l'analyse mathématique, il faudra revenir, et souvent, aux phénomènes, à l'expérience. Ce serait une grave erreur que de vouloir créer, dès aujourd'hui, une théorie mathématique de la lumière; cette tentative, inévitablement infructueuse, jetant des doutes sur le pouvoir de l'analyse, retarderait les vrais progrès de la science.

Il nous paraît donc utile de bien préciser le rôle que doit remplir l'analyse mathématique dans les questions de physique, et, pour cela, nous ne saurions choisir un meilleur exemple que celui du travail de Fresnel; mais nous présenterons ce travail comme il aurait été fait, si la théorie de l'élasticité des milieux solides avait été aussi bien établie qu'elle l'est aujourd'hui......

1100. I think with Lamé that his method is a decided improvement on that originally given by Fresnel; it seems to me to be more intelligible with respect to mechanical principles and to involve fewer assumptions. I am not sufficiently acquainted with the history of the subject to say how much of this part of Lamé's work is due to himself: see, however, Art. 1042 and various memoirs in our Chapter VIII.

1101. Lamé's eighteenth Lecture consists of mathematical investigations of the consequences which follow from the fact of the double velocity of propagation of plane waves. The equation to the wave surface is obtained; and the existence of *conjugate points* is established: these are such that each is the *pole* relative to a certain ellipsoid of the tangent plane to the wave surface at the other. The Lecture begins thus:

Dans cette Leçon et celles qui la suivent, nous laissons de côté la théorie de l'élasticité, pour rechercher toutes les conséquences qui résultent de la double vitesse de propagation des ondes planes, telle que l'analyse nous l'a donnée; pour définir les propriétés optiques que ces conséquences assignent aux milieux biréfringents; enfin, pour exposer les règles capables de déterminer à priori la marche de la lumière dans ces corps diaphanes. Quand cette théorie analytique sera aussi complète que possible, nous la rapprocherons de la théorie physique des faits connus, et scrutant avec soin leurs concordances, leurs désaccords, nous essayerons d'en déduire des réponses aux diverses questions posées, savoir: si ce sont les molécules pondérables d'un cristal qui exécutent et communiquent les vibrations lumineuses; si les coefficients des $\widehat{xx}$, $\widehat{yz}$ sont constants ou variables; et, en outre, si l'approximation qui limite l'influence des déplacements à leurs premières dérivées est réellement suffisante.

1102. Lamé's nineteenth Lecture is devoted to the mathematical discussion of the form of the wave surface; this includes an investigation of the singularities which it exhibits connected with the intersection of the two sheets which compose it.

1103. Lamé's twentieth Lecture gives the explanation of the reflection and refraction of light on the wave theory; substantially this coincides with what we find in various elementary works, though Lamé invests the most familiar subjects with a special charm. After speaking of the construction given by Huyghens for double refraction, and its accordance with facts Lamé observes on his pp. 277 and 279:

...Mais cette conception hardie, si bien justifiée par les faits, laisse en dehors la cause même de la double réfraction et de la polarisation qui accompagne ce phénomène; aussi la construction d'Huyghens n'a-t-elle été regardée, pendant longtemps, que comme une règle empirique, due à un heureux hasard. C'était méconnaître un trait de génie, et Fresnel ne s'y est pas trompé. Le fait de la double réfraction du verre comprimé lui fit penser que la bifurcation de la lumière réfractée et sa polarisation dépendaient d'une différence d'élasticité dans les directions diverses. Et c'est en poursuivant cette idée, en l'étudiant avec le concours de l'analyse, que Fresnel a été conduit à sa principale découverte....

Fresnel n'avait pour but que d'expliquer, par la théorie physique des ondes lumineuses, la construction d'Huyghens, qui coordonnait les faits optiques des cristaux biréfringents à un axe, seuls connus à cette époque. Il s'attendait donc à trouver, pour les deux nappes de l'onde multiple déduite de la théorie mathématique, une sphère et un ellipsoïde de révolution ayant l'axe commun de double réfraction. Il trouva une surface du quatrième degré qui ne se décomposait, de manière à donner la sphère et l'ellipsoïde, que dans des cas particuliers; il conclut de là que le fait général de la double réfraction n'était encore qu'imparfaitement connu, et qu'il devait exister des milieux cristallisés où l'onde multiple, serait indécomposable, comme dans sa formule. L'expérience est venue justifier cette prévision hardie: les phénomènes optiques de la topaze, et d'autres cristaux biréfringents, dits à deux axes, découverts par Fresnel, ont donné à sa théorie de la double réfraction une réalité incontestable, que sont venues confirmer avec éclat les découvertes faites par Hamilton, et vérifiées par Lloyd, des réfractions coniques et cylindriques dont nous parlerons dans la Leçon suivante.

1104. Lamé makes some comparison between two kinds of theories; one which includes with accuracy a certain assemblage of facts, but leads to no new discovery; and another founded on some novel idea of the cause of a class of phenomena, which indicates the existence of other phenomena of a similar kind, afterwards confirmed by experiment: see his p. 279. He obviously estimates the second kind far more highly than the first: thus with respect to the wave theory of light, notwithstanding the difficulties which it has not yet conquered, the predictions which it has uttered, and has seen fulfilled give it the greatest claims to attention.

1105. In his twenty-first Lecture Lamé explains the *conical* refraction deduced by Hamilton from theory and verified experimentally by Lloyd. With respect to these phenomena Lamé says on his p. 285:

La vérification complète de ces conséquences extrêmes donne à la réalité de la théorie de Fresnel, une certitude qu'aucune théorie mathématique de phénomènes naturels n'a certainement point dépassée.

Lamé gives mathematical formulæ for determining the refracted rays which correspond to a given incident ray. He also shews that the difference of the squares of the velocities of the two waves which have a common direction of propagation varies as the product of the sines of the angles which that direction makes with the optic axes. Lamé's investigations seem somewhat indirect; the main fact is that if a *ray* be incident normally on a crystal the refracted *wave* in the crystal is parallel to the surface of the crystal, that is, has the direction of the incident *ray* for the normal to its plane.

1106. The twenty-second Lecture is devoted to an examination of the conditions necessary for the propagation in an elastic medium of a wave of Fresnel's form; it begins thus:

L'explication des phénomènes optiques des cristaux biréfringents repose sur ce principe, qu'une molécule de la surface, atteinte par la lumière, devient le centre d'un système d'ondes à deux nappes. Il est donc nécessaire, pour la vérité de cette explication, qu'un pareil système puisse exister. Nous allons chercher les conditions que la théorie de l'élasticité impose à ce système isolé.

The discussion is given partly in this Lecture, and partly in the next.

1107. The twenty-third Lecture continues the mathematical discussion of the twenty-second; the main result is the expression obtained for the amplitude of vibration.

1108. The twenty-fourth Lecture contains some general remarks on the theory of light. The value found for the amplitude becomes infinite at the origin; and some observations are made with the object of explaining this difficulty. Attention is drawn to the difference which exists between Fresnel's own theory of light, and that developed by Lamé; according to Fresnel the direction of a vibration at any point is the same as that of the projection of the radius vector on the tangent plane to the wave surface at that point; according to Lamé the direction is at right angles to this projection; that is to say parallel to the plane of polarisation. Lamé names MacCullagh and Neumann as coming to the same conclusion as himself, while Cauchy on the other hand agrees with Fresnel. [Lamé's work is in some respects similar to Green's; he supposes the direction of transverse vibrations to be accurately in the front of the wave, but he does not introduce the 'extraneous pressures' (better 'initial stresses') of Green, and hence cannot arrive at Fresnel's result that the direction of vibration is *perpendicular* to the plane of polarisation. His discussion is much fuller, and more suggestive than Green's, yet of course he had the advantage of working after Cauchy, Green, MacCullagh and Neumann. The objections which have been raised to Green's second memoir, notably by Saint-Venant (see Navier's *Leçons* Appendice v. pp. 721—731), would I suppose be equally valid against Lamé's treatment. The discussion of this matter would however carry us beyond our limits. Lamé's relation to Cauchy is noted in the *Comptes Rendus,* Vol. XXXV.; according to Lamé himself he simplified Cauchy's methods: see our Art. 1042.]

Lamé insists strongly that the existence of the imponderable medium which we call the ether is fully established. He closes his work thus:

Quoi qu'il en soit, l'existence du fluide éthéré est incontestablement démontrée par la propagation de la lumière dans les espaces planétaires, par l'explication si simple, si complète des phénomènes de la diffraction dans la théorie des ondes; et comme nous l'avons vu, les lois de la double réfraction prouvent avec non moins de certitude que l'éther existe dans tous les milieux diaphanes. Ainsi la matière pondérable n'est pas seule dans l'univers, ses particules nagent en quelque sorte au milieu d'un fluide. Si ce fluide n'est pas la cause unique de tous les faits observables, il doit au moins les modifier, les propager, compliquer leurs lois. Il n'est donc plus possible d'arriver à une explication rationnelle et complète des phénomènes de la nature physique, sans faire intervenir cet agent, dont la présence est inévitable. On n'en saurait douter, cette intervention, sagement conduite, trouvera le secret, ou la véritable cause des effets qu'on attribue au calorique, à l'électricité, au magnétisme, à l'attraction universelle, à la cohésion, aux affinités chimiques; car tous ces êtres mystérieux et incompréhensibles ne sont, au fond, que des hypothèses de coordination, utiles sans doute à notre ignorance actuelle, mais que les progrès de la véritable science finiront par détrôner.

1109. The work of Lamé cannot be too highly commended; it affords an example which occurs but rarely of a philosopher of the highest renown condescending to employ his ability in the construction of an elementary treatise on a subject of which he is an eminent cultivator[1]. The mathematical investigations are clear and convincing, while the general reflections which are given so liberally at the beginning and the end of the Lectures are conspicuous for their elegance of language and depth of thought. The work is eminently worthy of a writer whom Gauss is reported to have placed at the head of French mathematicians, and whom Jacobi described as *un des mathématiciens les plus pénétrants. Comptes Rendus*, XV. p. 907.

1110. In the *Comptes Rendus*, Vol. XXXVII., 1853, we have on pp. 145—149 an account by Lamé of his *Mémoire sur l'équilibre d'élasticité des enveloppes sphériques.* The memoir itself is published in Liouville's *Journal de mathématiques*, Vol. XIX., 1854; and the present account forms substantially the introductory part of the memoir. The last page of the account is not reproduced in the *Journal* and we will give it here.

...Je terminerai cette Note par quelques indications sur les recherches qu'il faudrait faire pour hâter les progrès de la théorie de l'élasticité, et multiplier ses applications. Dans le Mémoire actuel, j'ai

[1] [These remarks are perfectly just when we regard Lamé as a mathematician, but it seems to me that a certain want of 'physical touch' somewhat reduces the value of his contributions to the science of elasticity. ED.]

considéré l'enveloppe sphérique *complète*, d'où résulte que les séries ne doivent pas contenir les termes qui deviendraient infinis pour certaines valeurs particulières de la latitude ou de la longitude.

Mais si l'on voulait considérer le cas d'une sorte de dôme, découpé dans l'enveloppe sphérique par un cône d'égale latitude, ou, plus généralement, celui d'un voussoir compris entre deux sphères concentriques, deux cônes d'égale latitude et deux plans méridiens, les séries ou les intégrales générales admettraient les nouveaux termes. Elles contiendraient alors d'autres suites de constantes arbitraires, que devraient déterminer les nouvelles forces appliquées sur les faces coniques et méridiennes.

Quand ces nouveaux cas seront traités généralement, et complétement résolus, on pourra sans doute transformer leurs formules, de manière à déduire le cas du cylindre droit et celui du parallélipipède, que l'on ne sait pas encore traiter directement. Mais, avant d'entreprendre ces nouvelles recherches, il faut étudier avec soin les propriétés des termes additionnels, qui n'étant pas utilisés dans la *Mécanique céleste*, sont généralement peu connus.

Cette étude indispensable est commencée : elle fait l'objet principal de Mémoires récemment publiés. D'ailleurs, l'histoire de la science manifeste comme une loi du progrès, que les géomètres habituellement occupés d'analyse pure, soit par prévision, soit par une sorte de logique instinctive, s'exercent précisément sur les sujets qui, dans une époque prochaine, seront réclamés comme instruments par les sciences d'application. Il est donc permis de l'espérer, le travail préliminaire que j'ai défini sera bientôt achevé. Alors, la théorie de l'équilibre intérieur des corps solides élastiques, s'appuyant sur ces découvertes de l'analyse, pourra devenir la branche la plus féconde de la physique mathématique.

1111. The memoir to which the preceding Article relates is published in Liouville's *Journal de mathématiques*, Vol. XIX., 1854; it occupies pp. 51—87. It is reproduced almost entirely in the *Leçons sur les coordonnées curvilignes* pp. 299—336 : see our Arts. 1154—1163.

It was read before the Paris Academy on August 1, 1853, and contains the solution of a very interesting problem; namely the investigation of the conditions for the equilibrium of a spherical elastic envelope or shell subjected to given distribution of load on the bounding spherical surfaces, and the determination of the resulting shifts. The problem is the only completely general one in the theory of elasticity which can be said to be completely solved; see Thomson and Tait's *Natural Philosophy*, Art. 696. It may be justly called by Lamé's name, although as we shall see in the sequel it has also engaged the attention of Resal, Thomson, Chevilliet, and others.

1112. Lamé, after giving a general description of his method, starts (p. 55) with the following equations established on p. 200 of his *Leçons sur l'Élasticité*:

$$\left.\begin{aligned}(\lambda+2\mu)\,r^2\cos\phi\frac{d\theta}{dr}+\mu\left(\frac{d\gamma}{d\phi}-\frac{d\beta}{d\psi}\right)+\rho r^2\cos\phi R_0=0\\ (\lambda+2\mu)\cos\phi\frac{d\theta}{d\phi}+\mu\left(\frac{d\alpha}{d\psi}-\frac{d\gamma}{dr}\right)+\rho r\cos\phi\Phi_0=0\\ (\lambda+2\mu)\frac{d\theta}{d\psi}+\mu\cos\phi\left(\frac{d\beta}{dr}-\frac{d\alpha}{d\phi}\right)+\rho r\cos\phi\Psi_0=0\end{aligned}\right\}\ldots\ldots(1),$$

where
$$\left.\begin{aligned}\alpha&=\frac{1}{r\cos\phi}\left(\frac{dv}{d\psi}-\frac{dw\cos\phi}{d\phi}\right)\\ \beta&=\frac{1}{\cos\phi}\left(\frac{drw\cos\phi}{dr}-\frac{du}{d\psi}\right)\\ \gamma&=\cos\phi\left(\frac{du}{d\phi}-\frac{drv}{dr}\right)\end{aligned}\right\}\ldots\ldots\ldots\ldots\ldots\ldots\ldots(2).$$

Here $r$ is the radial distance of any point from the centre of the shell, $\phi$ is the latitude and $\psi$ is the longitude of the point; we shall use $s$ for $\sin\phi$ and $c$ for $\cos\phi$ when convenient; $u, v, w$ are the shifts of the particle of the shell which was originally at the point $r, \phi, \psi$, namely $u$ along the radius, $v$ along the tangent to the meridian, and $w$ along the tangent to the circle of latitude; these displacements are considered positive when they increase the coordinates $r, \phi, \psi$ respectively. $R_0, \Phi_0, \Psi_0$ are the components, referred to the unit of mass, of the body-forces in the same directions as $u, v, w$ are estimated respectively; $\rho$ is the density of the body, $\lambda$ and $\mu$ are the two constants which measure its elasticity, and $\theta$ is the dilatation. This notation differs slightly from that of Lamé; his does not seem very inviting, and has I think not been adopted by others.

1113. Lamé also quotes from his *Lecons* (p. 199) the following expression for the dilatation:

$$\theta = \frac{1}{r^2}\frac{dr^2u}{dr} + \frac{1}{rc}\frac{dcv}{d\phi} + \frac{1}{rc}\frac{dw}{d\psi} \quad \text{.....................(3)},$$

and the following expressions for the various stresses:

$$\left.\begin{aligned}
\widehat{rr} &= \lambda\theta + 2\mu\frac{du}{dr} \\
\widehat{\phi\phi} &= \lambda\theta + 2\mu\left(\frac{u}{r} + \frac{1}{r}\frac{dv}{d\phi}\right) \\
\widehat{\psi\psi} &= \lambda\theta + 2\mu\left(\frac{u}{r} - \frac{s}{c}\frac{v}{r} + \frac{1}{rc}\frac{dw}{d\psi}\right) \\
\widehat{\phi\psi} &= \mu\left(\frac{1}{rc}\frac{dv}{d\psi} + \frac{1}{r}\frac{dw}{d\phi} + \frac{s}{c}\frac{w}{r}\right) \\
\widehat{\psi r} &= \mu\left(\frac{dw}{dr} - \frac{w}{r} + \frac{1}{rc}\frac{du}{d\psi}\right) \\
\widehat{r\phi} &= \mu\left(\frac{1}{r}\frac{du}{d\phi} + \frac{dv}{dr} - \frac{v}{r}\right)
\end{aligned}\right\} \quad \text{............(4)}.$$

The stress notation explains itself.

1114. Besides the body-forces which act throughout the shell we suppose given loads to act on the bounding surfaces. Thus for the outer surface let the load at any point be resolved in the same directions as $u$, $v$, $w$ are estimated respectively; then $\widehat{rr}$, $\widehat{r\phi}$, $\widehat{r\psi}$ must be equal respectively to these components. Similarly $-\widehat{rr}$, $-\widehat{r\phi}$, $-\widehat{r\psi}$ must have given values over the inner surface. Thus finally the problem to be solved is this: values of $u$, $v$, $w$ must be found which satisfy (1), and then each of the expressions $\widehat{rr}$, $\widehat{r\phi}$, $\widehat{r\psi}$ must be equal to a given function of $\phi$ and $\psi$ over the outer surface and also over the inner surface. That is we have six conditions to enable us to determine the arbitrary quantities which enter into the general solution of (1): see the memoir, p. 57.

1115. In treating the problem it is convenient to find first some particular solution of (1) as in our Art. 1063; let $u_0$, $v_0$, $w_0$ denote such a set of values of the shifts. Then put $u = u_0 + u'$, $v = v_0 + v'$, $w = w_0 + w'$, and substitute in (1); we thus obtain for $u'$, $v'$, $w'$ equations precisely like (1) except that the terms in $R_0$, $\Phi_0$, $\Psi_0$ *will not occur*, so that our equations will be to that extent simpler than (1). Lamé gives three examples of this preliminary process of simplification.

1116. Suppose the spherical shell is acted on by a *constant force;* denote it by $q$, and suppose its direction to be parallel to the polar axis. We have then $R_0 = -qs$, $\Phi_0 = -qc$, $\Psi_0 = 0$. Now as the action of the constant force will not tend to displace any particle out of its original meridian plane we have $w = 0$, and $u$ and $v$ independent of $\psi$. Thus the equations (1) become

$$\left.\begin{aligned}(\lambda+2\mu)\,r^2\frac{d\theta}{dr}+\frac{\mu}{c}\frac{d\gamma}{d\phi}&=qr^2\rho s\\(\lambda+2\mu)\frac{d\theta}{d\phi}-\frac{\mu}{c}\frac{d\gamma}{dr}&=q\rho rc\end{aligned}\right\}$$

where $$\theta=\frac{1}{r^2}\frac{dr^2u_0}{dr}+\frac{1}{rc}\frac{dcv_0}{d\phi},\quad \gamma=c\left(\frac{du_0}{d\phi}-\frac{dv_0r}{dr}\right)$$

These equations are satisfied by

$$\theta=\frac{q\rho}{\lambda}rs,\qquad \gamma=\frac{q\rho}{\lambda}r^2c^2;$$

and these lead to the particular solutions

$$u_0=\frac{q\rho r^2s}{10\lambda},\qquad v_0=-\frac{3q\rho r^2c}{10\lambda},\qquad w_0=0.$$

1117. Again, suppose that the shell is in relative equilibrium when rotating with spin $\omega$ round a fixed axis; take this for the polar axis; then from the so-called centrifugal force, we have

$$R_0=\omega^2rc^2,\qquad \Phi_0=-\omega^2rcs,\qquad \Psi_0=0.$$

In the same way as in the preceding example we obtain

$$\theta=-\tfrac{7}{2}Qr^2c^2,\quad \gamma=0,\quad \text{where } Q=\frac{\rho\omega^2}{7(\lambda+2\mu)};$$

and these lead to the particular solutions (p. 58)

$$u_0=(\tfrac{1}{5}-c^2)\,Qr^3,\;\; v_0=\tfrac{1}{2}Qr^3cs,\;\; w_0=0.$$

1118. Finally, Lamé supposes that the shell is acted on by a force varying as the distance, directed towards the centre; denote this by $-qr/r_1$, so that its numerical value is $q$ at the outer surface, the radius of which we denote by $r_1$. Thus (1) reduces to

$$(\lambda+2\mu)\frac{d}{dr}\frac{1}{r^2}\left(\frac{dr^2u_0}{dr}\right)=\frac{qr\rho}{r_1};$$

and this leads to the particular solutions (p. 59)

$$u_0=\frac{q\rho r^3}{10(\lambda+2\mu)r_1},\quad v_0=0,\quad w_0=0.$$

1119. Lamé having thus exemplified how it is possible to free his equations from body-forces throws them (p. 59) into the form:

$$\left.\begin{aligned}\frac{d\beta}{d\psi}-\frac{d\gamma}{d\phi}&=\frac{\lambda+2\mu}{\mu}r^2c\frac{d\theta}{dr}\\ \frac{d\gamma}{dr}-\frac{d\alpha}{d\psi}&=\frac{\lambda+2\mu}{\mu}c\frac{d\theta}{d\phi}\\ \frac{d\alpha}{d\phi}-\frac{d\beta}{dr}&=\frac{\lambda+2\mu}{\mu}\frac{1}{c}\frac{d\theta}{d\psi}\end{aligned}\right\}\quad\ldots\ldots\ldots\ldots(5).$$

To these Lamé adds the following equations (p. 60), which the reader will easily deduce from (5) and (2):

$$\frac{d}{dr}\left(r^2\frac{d\theta}{dr}\right)+\frac{1}{c}\frac{d}{d\phi}\left(c\frac{d\theta}{d\phi}\right)+\frac{1}{c^2}\frac{d^2\theta}{d\psi^2}=0\ldots\ldots\ldots\ldots(6),$$

$$\frac{dr^2\alpha}{dr}+\frac{1}{c}\frac{dc\beta}{d\phi}+\frac{1}{c^2}\frac{d\gamma}{d\psi}=0\quad\ldots\ldots\ldots\ldots\ldots\ldots(7).$$

The following is the process of solution which these equations suggest: find $\theta$ from (6); then $\alpha$, $\beta$, $\gamma$ from (5) and (7); then $u$, $v$, $w$ from (2) and (3).

1120. Put $\theta=\frac{\mu}{\lambda+2\mu}F$; then (6) may be written

$$\frac{d}{dr}\left(r^2\frac{dF}{dr}\right)+\frac{d}{ds}\left\{(1-s^2)\frac{dF}{ds}\right\}+\frac{1}{1-s^2}\frac{d^2F}{d\psi^2}=0\ \ldots\ldots(8).$$

This, as Lamé remarks (p. 61), is a well-known equation frequently employed in the *Mécanique céleste* and in the *Théorie analytique de la chaleur*. As an integral sufficiently general for our purpose we may take

$$F=\Sigma\Sigma\left\{(Ar^n+Br^{-n-1})\cos l\psi+(Cr^n+Dr^{-n-1})\sin l\psi\right\}P_n^l,$$

where $P_n^l$ in Heine's terminology is an associated function of the first kind satisfying the differential equation

$$\frac{d}{ds}\left\{(1-s^2)\frac{dP_n^l}{ds}\right\}-\frac{l^2P_n^l}{1-s^2}+n(n+1)P_n^l=0 \text{ ......(9)}.$$

$A$, $B$, $C$, $D$ are constants with respect to $r$, $\phi$, $\psi$ but are functions of $l$ and $n$; the two letters $\Sigma$ denote summation with respect to integral values of $l$ from 0 to $n$, and with respect to integral values of $n$ from 0 to $\infty$.

1121. Lamé remarks (p. 63) that we may arrange according to a different order the operations denoted by the two letters $\Sigma$ of Art. 1120. We may collect in a group all the terms which involve the same value of $l$, and in which $n$ takes all integral values from $n=l$ to $n=\infty$; and afterwards take all the groups which exist from $l=0$ to $l=\infty$. We may adopt either order of operation which is convenient in any case, and we will use the single letter $S$ to denote the double summation. Thus we have the following symbolical relation

$$S=\sum_{n=0}^{n=\infty}\sum_{l=0}^{l=n}=\sum_{l=0}^{l=\infty}\sum_{n=l}^{n=\infty}.$$

1122. Let $\xi$ and $\zeta$ be two functions of the variable $\psi$ such that

$$\xi=A\cos l\psi+C\sin l\psi,\quad \zeta=B\cos l\psi+D\sin l\psi \text{ ....(10)}.$$

Then the expression for $F$ in Art. 1120 may be written

$$F=S(\xi r^n+\zeta r^{-n-1})P_n^l \text{ ........ .........(11)}.$$

We shall have occasion for other functions verifying, like $F$, the equation (8); and we shall denote such functions by $F'$, $F''$,.... We shall put in like manner accents on the constants of the equations like (10).

1123. Lamé then proceeds to equations (5) and (7) and in the former substitutes $\dfrac{\mu F}{\lambda+2\mu}$ for $\theta$. This gives him four equations for $\alpha$, $\beta$, $\gamma$ of which three only are independent.

1124. In treating equations (5) and (7) Lamé adopts the method already introduced in Art. 1115. He puts $\alpha = \alpha_0 + \alpha'$, $\beta = \beta_0 + \beta'$, $\gamma = \gamma_0 + \gamma'$, where $\alpha_0$, $\beta_0$, $\gamma_0$ form a particular solution, and $\alpha'$, $\beta'$, $\gamma'$ form a solution of the equations supposing the right-hand members to be zero. It is easy to see that for $\alpha'$, $\beta'$, $\gamma'$ we may take the differential coefficients with respect to $r$, $\phi$, $\psi$ respectively of any function $F'$ which verifies equation (8); for with these values the first members of equation (5) vanish. Thus we have

$$\alpha' = \frac{dF'}{dr}, \qquad \beta' = \frac{dF'}{d\phi}, \qquad \gamma' = \frac{dF'}{d\psi},$$

where $F' = S(\xi' r^n + \zeta' r^{-n-1}) P_n^l$; and $\xi'$ and $\zeta'$ are formed as in (10) when the constants $A$, $B$, $C$, $D$ are affected with accents. It remains then to find $\alpha_0$, $\beta_0$, $\gamma_0$.

1125. Lamé states on p. 65 that he has found particular values for these quantities by a method of integration which he says is too long to develop but of which he gives some hint in the latter part of his memoir. These values are

$$\alpha_0 = 0, \qquad \beta_0 = S\left(-\frac{d\xi}{d\psi}\frac{r^{n+1}}{n+1} + \frac{d\zeta}{d\psi}\frac{r^{-n}}{n}\right)\frac{P_n^l}{c},$$

$$\gamma_0 = S\left(\frac{\xi r^{n+1}}{n+1} - \frac{\zeta r^{-n}}{n}\right)c^2\frac{dP_n^l}{ds}$$

It is easy to verify by substitution that these give a particular solution.

1126. Thus, collecting the two parts of which $\alpha$, $\beta$, $\gamma$ respectively consist according to the assumptions of Art. 1124, and effecting the differentiations of $F'$ which are indicated, we obtain for the general integrals of (5)

$$\left.\begin{aligned}
\alpha &= S\{n\xi' r^{n-1} - (n+1)\zeta' r^{-n-2}\} P_n^l, \\
\beta &= S\left\{-\frac{d\xi}{d\psi}\frac{r^{n+1}}{n+1} + \frac{d\zeta}{d\psi}\frac{r^{-n}}{n}\right\}\frac{P_n^l}{c} + S\{\xi' r^n + \zeta' r^{-n-1}\}\, c\,\frac{dP_n^l}{ds}, \\
\gamma &= S\left\{\frac{\xi r^{n+1}}{n+1} - \frac{\zeta r^{-n}}{n}\right\}c^2\frac{dP_n^l}{ds} + S\left\{\frac{d\xi'}{d\psi} r^n + \frac{d\zeta'}{d\psi} r^{-n-1}\right\}P_n^l.
\end{aligned}\right\} \quad (12).$$

1127. The next step, as stated in Art. 1119, is to find $u$, $v$, $w$ from equations (2) and (3), which may now be written

$$\left.\begin{aligned} \frac{drv}{d\psi} - \frac{drcw}{d\phi} &= r^2 c\,(\alpha_0 + \alpha') \\ \frac{drcw}{dr} - \frac{du}{d\psi} &= c\,(\beta_0 + \beta') \\ \frac{du}{d\phi} - \frac{drv}{dr} &= \frac{1}{c}(\gamma_0 + \gamma') \end{aligned}\right\} \text{..............(13)},$$

$$\frac{dr^2 u}{dr} + \frac{1}{c}\frac{dcrv}{d\phi} + \frac{1}{c^2}\frac{dcrw}{d\psi} = \frac{\mu}{\lambda + 2\mu} r^2 F \text{ ........ (14)},$$

where the functions on the right-hand sides of the equations are to have the values which have been already found. The equations (13) amount to only *two* distinct equations, for if two of them are satisfied the third is necessarily satisfied.

Lamé (p. 67) separates the expressions which solve (13) and (14) into three parts by putting

$$u = u_0 + u' + u'', \quad v = v_0 + v' + v'', \quad w = w_0 + w' + w'' \text{ ...(15)}.$$

Here we shall take $u_0$, $v_0$, $w_0$ to form a particular solution when the right-hand members of (13) involve only $\alpha_0$. $\beta_0$, $\gamma_0$; we shall take $u'$, $v'$, $w'$ to form a particular solution when the right-hand members of (13) involve only $\alpha'$, $\beta'$, $\gamma'$; and we shall take $u''$, $v''$, $w''$ to form the general solution when the right-hand members are zero.

1128. The values of $u''$, $v''$, $w''$ can be assigned in precisely the same manner as $\alpha'$, $\beta'$, $\gamma'$ were obtained in Art. 1124. Let $F''$ denote another function which verifies equation (8); then we may put

$$u'' = \frac{dF''}{dr}, \quad v'' = \frac{1}{r}\frac{dF''}{d\phi}, \quad w'' = \frac{1}{rc}\frac{dF''}{d\psi},$$

where $F'' = S\,(\xi'' r^n + \zeta'' r^{-n-1}) P_n^i$; and $\xi''$ and $\zeta''$ are formed as in (10) when the constants $A$, $B$, $C$, $D$ are affected with double accents.

1129. Next for the values of $u'$, $v'$, $w'$ Since the quantities $\alpha'$, $\beta'$, $\gamma'$ are respectively partial differential coefficients of $F'$, it is obvious that the particular solutions $u'$, $rv'$, $rcw'$ may be found from $F'$ just as $\alpha_0$, $\beta_0$, $\gamma_0$ were found from $F$. Thus by Art. 1125 we obtain

$$u'=0,\qquad v'=S\left(-\frac{d\xi'}{d\psi}\frac{r^n}{n+1}+\frac{d\zeta'}{d\psi}\frac{r^{-n-1}}{n}\right)\frac{P_n^i}{c},$$

$$w'=S\left(\frac{\xi' r^n}{n+1}-\frac{\zeta' r^{-n-1}}{n}\right)c\frac{dP_n^i}{ds}.$$

1130. Finally for the values of $u_0$, $v_0$, $w_0$. Since $\alpha_0$ is zero the first of equations (13) shews that $rv$ may be the partial differential coefficient of some function with respect to $\phi$, and $rcw$ the partial differential coefficient of the *same* function with respect to $\psi$. The following forms may then be assumed on trial as a solution; when $h$, $h'$, $k$, $k'$ denote constants as yet undetermined,

$$u_0=S(-h\xi r^{n+1}-h'\zeta r^{-n})P_n^i,$$

$$v_0=S(-k\xi r^{n+1}+k'\zeta r^{-n})\frac{dP_n^i}{d\phi},$$

$$w_0=S\left(-k\frac{d\xi}{d\psi}r^{n+1}+k'\frac{d\zeta}{d\psi}r^{-n}\right)\frac{P_n^i}{c}.$$

Substitute these values in equations (13) and (14), the right-hand members being reduced to contain only the terms in $\alpha_0$, $\beta_0$, $\gamma_0$; then by comparing the parts which involve the same $\xi$ we obtain two equations for finding $h$ and $k$, and by comparing the parts which involve the same $\zeta$ we obtain two equations for finding $h'$ and $k'$. These equations give, putting for simplicity $\lambda+\mu=e$, and $\lambda+2\mu=a$:

$$h=\frac{(n+2)e-2a}{2(2n+3)a}\qquad k=\frac{(n+1)e+2a}{(2n+2)(2n+3)a},$$

$$h'=\frac{(n-1)e+2a}{2(2n-1)a},\qquad k'=\frac{ne-2a}{2n(2n-1)a}.$$

The value of $h'$ can be obtained from that of $h$ by changing $n$ into $-n-1$; and the value of $-k'$ can be obtained from that of $k$ by the same change (p. 68).

1131. Substitute in (15) the values found in the preceding three Articles; effect the differentiations of $F''$ which are indicated; introduce the constants according to (10) and the analogous equations with accents, and employ the following abbreviations:

$$nA''r^{n-1} - hAr^{n+1} - (n+1)B''r^{-n-2} - h'Br^{-n} = G,$$

$$nC''r^{n-1} - hCr^{n+1} - (n+1)D''r^{-n-2} - h'Dr^{-n} = G',$$

$$A''r^{n-1} - kAr^{n+1} + B''r^{-n-2} + k'Br^{-n} = H,$$

$$C''r^{n-1} - kCr^{n+1} + D''r^{-n-2} + k'Dr^{-n} = H',$$

$$\frac{A'r^n}{n+1} - \frac{B'r^{-n-1}}{n} = K,$$

$$\frac{C'r^n}{n+1} - \frac{D'r^{-n-1}}{n} = K';$$

then we obtain

$$u = S(G\cos l\psi + G'\sin l\psi)P_n^l,$$

$$v = S(H\cos l\psi + H'\sin l\psi)c\frac{dP_n^l}{ds} - S(K'\cos l\psi - K\sin l\psi\frac{lP_n}{c},$$

$$w = S(H'\cos l\psi - H\sin l\psi)\frac{lP_n^l}{c} + S(K\cos l\psi + K'\sin l\psi)c\frac{dP_n^l}{ds}.$$

These values furnish the complete integrals of equations (1) when we omit $R_0$, $\Phi_0$, and $\Psi_0$ (p. 69).

1132. The values of $u$, $v$, $w$ just given involve three quadruple sets of arbitrary constants, introduced by the series for $F$, $F''$ and $F'''$. These constants must be determined by the aid of the three conditions which hold at each bounding surface as found in Art. 1114. To the abbreviations already noticed in Art. 1131 let us add these,

$$Ar^n + Br^{-n-1} = J, \qquad Cr^n + Dr^{-n-1} = J';$$

then, since $\theta = \frac{\mu}{\lambda+2\mu}F = \frac{\mu}{a}F$, we have by (11),

$$\theta = S\left(\frac{\mu}{a}J\cos l\psi + \frac{\mu}{a}J'\sin l\psi\right)P_n^l.$$

1133. Lamé now substitutes the values of $u$, $v$, $w$, $\theta$ as found in Arts. 1131 and 1132 in the expressions for the stresses which are mentioned in Art. 1113. We shall use $b$ for $3\lambda + 2\mu$, so that

$a+b=4e$. We may also use for abbreviation six new symbols as follows:

$$2n(n-1)\,\mu A''r^{n-2}-\frac{\mu}{a}\frac{n(n-1)e-b}{2n+3}Ar^n$$
$$+2(n+1)(n+2)\,\mu B''r^{-n-3}+\frac{\mu}{a}\frac{(n+1)(n+2)e-b}{2n-1}Br^{-n-1}=L,$$

a similar expression with $C''$, $C$, $D''$, $D$ instead of $A''$, $A$, $B''$, $B$ respectively will be called $L'$;

$$2(n-1)\,\mu A''r^{n-2}-\frac{\mu}{a}\frac{(n+1)^2e-a}{(n+1)(2n+3)}Ar^n-2(n+2)\,\mu B''r^{-n-3}$$
$$-\frac{\mu}{a}\frac{n^2e-a}{n(2n-1)}r^{-n-1}B=M,$$

a similar expression with the changes just mentioned will be called $M'$;

$$\frac{n-1}{n+1}\mu A'r^{n-1}+\frac{n+2}{n}\mu B'r^{-n-2}=N,$$

$$\frac{n-1}{n+1}\mu C'r^{n-1}+\frac{n+2}{n}\mu D'r^{-n-2}=N'.$$

Then we find that

$$\widehat{rr}=S(L\cos l\psi+L'\sin l\psi)P_n^l,$$

$$\widehat{r\phi}=S(M\cos l\psi+M'\sin l\psi)\,c\frac{dP_n^l}{ds}-S(N'\cos l\psi-N\sin l\psi)\frac{lP_n^l}{c},$$

$$\widehat{r\psi}=S(M'\cos l\psi-M\sin l\psi)\frac{lP_n^l}{c}+S(N\cos l\psi+N'\sin l\psi)\,c\frac{dP_n^l}{ds}.$$

Now each of the three expressions which occur on the right-hand side must have a given value over the whole outer surface, and also a given value over the whole inner surface. Call these given values $X_0$, $Y_0$, $Z_0$ respectively for the inner surface, and $X_1$, $Y_1$, $Z_1$ respectively for the outer surface; and also use subscripts in a similar manner with respect to the six quantities $L$, $L'$, $M$, $M'$, $N$, $N'$. Then we must have at every point of the inner surface

$$S(L_0\cos l\psi+L_0'\sin l\psi)P_n^l=X_0,$$

and at every point of the outer surface

$$S(L_1\cos l\psi+L_1'\sin l\psi)P_n^l=X_1.$$

We shall sometimes find it convenient to use the general equation

$$S(L\cos l\psi + L'\sin l\psi)P_n^l = X \dots\dots\dots\dots\dots (16)$$

as an abbreviation for the two.

Similar remarks hold with respect to $Y$ and $Z$.

1134. Thus, corresponding to any given pair of values of $l$ and $n$, we have in the expressions of the preceding Article twelve quantities, namely $L_0$, $M_0$, $N_0$, $L_0'$, $M_0'$, $N_0'$, $L_1$, $M_1$, $N_1$, $L_1'$, $M_1'$, $N_1'$; and when these are known we have twelve equations in the same Article for finding the twelve constants $A, A', A'', B, B', B'', C, C', C''$, $D, D', D''$: for each of the equations by which we define $L, L'$, $M, M', N, N'$ is to hold both at the inner surface where $r$ becomes $r_0$, and at the outer surface where $r$ becomes $r_1$. Lamé then proceeds to shew how the twelve quantities $L_0$, $M_0$,...... are obtained.

1135. The method of obtaining $L_0$, $L_1$, $L_0'$ $L_1'$ involves only well known applications of the Integral Calculus. Take equation (16), multiply both sides first by $\cos l\psi$, and next by $\sin l\psi$, and integrate for $\psi$ between the limits 0 and $2\pi$. Thus we get

$$\Sigma LP_n^l = \frac{1}{\varpi}\int_0^{2\pi} X\cos l\psi\, d\psi,$$

$$\Sigma L'P_n^l = \frac{1}{\varpi}\int_0^{2\pi} X\sin l\psi\, d\psi;$$

where $\Sigma$ denotes a summation with respect to $n$ from $n = l$ to $n = \infty$; also $\varpi$ stands for $\pi$ if $l$ is not zero, and for $2\pi$ if $l$ is zero. Moreover each of these equations represents two, for we may apply the suffix 0 or the suffix 1.

1136. It will be unnecessary henceforth to repeat a remark like that which has just been made respecting the double forms of the equations; we may proceed as if we were concerned with only one bounding surface instead of two.

1137. We have thus determined $\Sigma LP_n^l$ and $\Sigma L'P_n^l$; we wish however to find the values of the separate terms denoted in general by $L$ and $L'$. We proceed thus; it will be sufficient to consider one of the two, say $L$.

We know then that

$$\Sigma L P_n^l = \frac{1}{\varpi}\int_0^{2\pi} X \cos l\psi\, d\psi;$$

the right-hand member may be considered known, and we will denote it by $X'$. Hence a fundamental property of $P_n^l$ leads us to the result $L_{l,n}\int_{-1}^{+1} (P_n^l)^2\, ds = \int_{-1}^{+1} P_n^l X' ds$, which determines $L$ for any given values of $l$ and $n$. Lamé arrives at a result similar to this on p. 74.

1138. We now proceed to obtain the terms denoted by $M, M'$, $N, N'$. Take the equations

$$S(M\cos l\psi + M'\sin l\psi)\, c\frac{dP_n^l}{ds} - S(N'\cos l\psi - N\sin l\psi)\frac{lP_n^l}{c} = Y,$$

$$S(M'\cos l\psi - M\sin l\psi)\frac{lP_n^l}{c} + S(N\cos l\psi + N'\sin l\psi)\, c\frac{dP_n^l}{ds} = Z;$$

and in them let $l$ have any assigned integral value different from zero. Four new equations are obtained by multiplying each of these first by $\cos l\psi$, and next by $\sin l\psi$, and integrating for $\psi$ between the limits 0 and $2\pi$. Of these four equations we shall use only two which will find $M$ and $N'$; the other two may be applied in the same way to find $N$ and $M'$. Thus we have

$$\left.\begin{aligned} \Sigma\left\{M_{l,n}\, c\frac{dP_n^l}{ds} - N'_{l,n}\frac{lP_n^l}{c}\right\} &= \frac{1}{\varpi}\int_0^{2\pi} Y\cos l\psi\, d\psi \\ \Sigma\left\{-M_{l,n}\frac{lP_n^l}{c} + N'_{l,n}\, c\frac{dP_n^l}{ds}\right\} &= \frac{1}{\varpi}\int_0^{2\pi} Z\sin l\psi\, d\psi \end{aligned}\right\}\ldots(17),$$

where $\Sigma$ denotes summation with respect to $n$ from $n = l$ to $n = \infty$, and we have individualised the constants by means of the subscripts $l$, $n$. The right hand expressions may be regarded as known quantities, say $Y'$ and $Z'$ respectively.

Multiply the former by $c\frac{dP_m^l}{ds}$, and the latter by $\frac{lP_m^l}{c}$, and integrate with respect to $s$ between the limits $-1$ and $1$; then subtract. Thus

$$\Sigma M_{l,n} \int_{-1}^{1} \left\{ c \frac{dP^l_n}{ds} c \frac{dP^l_m}{ds} + \frac{l}{c} P^l_n \frac{l}{c} P^l_m \right\} ds$$
$$- \Sigma N'_{l,n} l \int_{-1}^{1} \left\{ P^l_n \frac{dP^l_m}{ds} + P^l_m \frac{dP^l_n}{ds} \right\} ds$$
$$= \int_{-1}^{1} \left\{ Y' c \frac{dP^l_m}{ds} - Z' \frac{lP^l_m}{c} \right\} ds.$$

The right-hand member may be considered known; the left-hand member is easily shewn to reduce to

$$m(m+1) M_{l,m} \int_{-1}^{1} \{P^l_m\}^2 ds,$$

so that $M_{l,m}$ becomes known, namely from

$$M_{l,m} = \frac{\int_{-1}^{+1} \left\{ Y' c \frac{dP_m}{ds} - Z' \frac{lP^l_m}{c} \right\} ds}{m(m+1) \int_{-1}^{+1} \{P^l_m\}^2 ds}$$

The value of any $N'$ comes from that of the corresponding $M$ by interchanging $Y'$ and $Z'$: see Lamé, p. 76.

1139. We have excluded the case of $l = 0$, so that it now remains for consideration. The equations (17) are then not two simultaneous equations with two unknown quantities, but reduce to

$$\Sigma M_{0,n} c \frac{dP^0_n}{ds} = Y', \qquad \Sigma N'_{0,n} c \frac{dP^0_n}{ds} = Z' \quad \text{........(18).}$$

Also, by supposing $l = 0$, we easily find

$$\int_{-1}^{+1} c^2 \frac{dP^0_n}{ds} \frac{dP^0_m}{ds} ds = 0 \text{ if } m \text{ and } n \text{ are unequal,}$$

and
$$= m(m+1) \int_{-1}^{1} (P^0_m)^2 ds \text{ if } m \text{ and } n \text{ are equal.}$$

Multiply the first of equations (18) by $c \frac{dP^0_m}{ds}$ and integrate between the limits $-1$ and $1$ for $s$. Thus we get

$$M_{0,m} = \frac{\int_{-1}^{+1} Y' c \frac{dP^0_m}{ds} ds}{m(m+1) \int_{-1}^{+1} \{P^0_m\}^2 ds}$$

Similarly from the second of equations (18) we can determine $N'_{0m}$, namely zero, for $Z'$ is zero when $l = 0$. We see by comparing these special results with the general formulae obtained in Art. 1138 that these formulae hold when $l = 0$ (see p. 77)[1].

1140. Thus Lamé has, as he proposed, determined all the twelve quantities $L_0$, $M_0$, .. corresponding to any assigned pair of values of $l$ and $n$; at least he has expressed these quantities as definite integrals.

On pp. 78—80 Lamé evaluates certain integrals of the associated functions $P_n^l$ which have occurred in the earlier parts of his memoir. The results obtained are now well known and need not be considered here.

1141. We have now, according to Art. 1134, to determine the values of the twelve constants, $A$, $A'$, $A''$,... corresponding to a given pair of values of $l$ and $n$. The equations are given towards the beginning of Art. 1133. They resolve themselves into four groups.

The first group finds the values of $A''$, $A$, $B''$, $B$ in terms of $L_0$, $L_1$, $M_0$, $M_1$. The second group finds the values of $C''$, $C$, $D''$, $D$ in terms of $L_0'$, $L_1'$, $M_0'$, and $M_1'$. The third group finds the values of $A'$ and $B'$ in terms of $N_0$ and $N_1$. The fourth group finds the values of $C'$ and $D'$ in terms of $N_0$ and $N_1'$

So long as $n$ is greater than unity there is no difficulty; but when $n = 1$ some of the terms vanish from the left-hand sides of the equations in Art. 1133, and thus we shall find that some of the constants remain undetermined; also when $n = 0$ there are peculiarities to notice. These points we now proceed to examine.

1142. Put $n = 1$ in the first four equations of Art. 1133; thus

$$\frac{\mu b}{5a} Ar + \frac{12\mu}{r^4} B'' + \frac{\mu(6e - b)}{ar^2} B = L,$$

$$\frac{\mu b}{5a} Cr + \frac{12\mu}{r^4} D'' + \frac{\mu(6e - b)}{ar^2} D = L',$$

$$-\frac{\mu b}{10a} Ar - \frac{6\mu}{r^4} B'' - \frac{\mu(e - a)}{ar^2} B = M,$$

$$-\frac{\mu b}{10a} Cr - \frac{6\mu}{r^4} D'' - \frac{\mu(e - a)}{ar^2} D = M'.$$

[1] In this Article $P_m^0$ is of course the $m$th Legendre's coefficient, usually denoted simply by $P_m$. Lamé uses somewhat different factors for the associated functions and for the Legendre's coefficients to those adopted by Heine.

The constants $A''$ and $C''$ have disappeared, so that we cannot determine them; and, as in each of these equations we are to put $r_1$ and $r_0$ in succession for $r$, we have *eight* equations to determine the *six* constants $A, B'', B, C, D'', D$. Thus two conditions must hold among the quantities supposed known. To the first of these equations add twice the third, and put $r_1$ and $r_0$ in succession for $r$; thus

$$\frac{3\mu}{r_1^2} B = L_1 + 2M_1, \qquad \frac{3\mu}{r_0^2} B = L_0 + 2M_0.$$

So in like manner from the second and fourth equations

$$\frac{3\mu}{r_1^2} D = L_1' + 2M_1', \qquad \frac{3\mu}{r_0^2} D = L_0' + 2M_0'.$$

Thus when $n = 1$ we must have these two conditions holding,

$$\left.\begin{aligned} r_1^2 (L_1 + 2M_1) &= r_0^2 (L_0 + 2M_0) \\ r_1^2 (L_1' + 2M_1') &= r_0^2 (L_0' + 2M_0') \end{aligned}\right\} \ldots\ldots\ldots\ldots(19).$$

Next put $n = 1$ in the fifth and sixth equations of Art. 1133; thus

$$\frac{3\mu}{r^3} B' = N, \qquad \frac{3\mu}{r^3} D' = N'.$$

The constants $A'$ and $C'$ have disappeared, so that we cannot determine them; and, as in each of these equations we are to put $r_1$ and $r_0$ in succession for $r$, we have *four* equations to determine the *two* constants $B'$ and $D'$. Thus two conditions must hold among the quantities supposed known; these conditions obviously are

$$N_1 r_1^3 = N_0 r_0^3, \qquad N_1' r_1^3 = N_0' r_0^3 \ldots\ldots\ldots\ldots(20).$$

1143. When $n = 1$ we may have $l = 0$ or $l = 1$; we proceed to take these cases separately.

Suppose $n = 1$, and $l = 0$. Then

$$P_1^0 = s, \qquad c\frac{dP_1^0}{ds} = c, \qquad \int_{-1}^{+1} (P_1^0)^2\, ds = \tfrac{2}{3}, \qquad \varpi = 2\pi.$$

By Art. 1135

$$L_{0,1} = \frac{3}{4\pi} \int_{-1}^{+1} \int_0^{2\pi} Xs\, ds\, d\psi.$$

Here $L_{0,1}$ is the $L$ of Equation (16); it becomes $(L_{0,1})_1$ or shortly $L_1$, when we put $X_1$ for $X$, and $(L_{0,1})_0$, or shortly $L_0$, when we put $X_0$ for $X$.

Again, by Art. 1139,

$$M_{0,1}=\frac{3}{8\pi}\int_{-1}^{1}\int_{0}^{2\pi}Ycdsd\psi.$$

Here $M_{0,1}$ is the $M$ of equation (17); it becomes $(M_{0,1})_1$, or shortly $M_1$, when we put $Y_1$ for $Y$, and $(M_{0,1})_0$, or shortly $M_0$, when we put $Y_0$ for $Y$.

Thus the first of equations (19) gives us

$$r_1^2\iint(X_1s+Y_1c)\,dsd\psi-r_0^2\iint(X_0s+Y_0c)\,dsd\psi=0\ldots(21),$$

where the limits of integration are 0 and $2\pi$ for $\psi$, and $-1$ and $+1$ for $s$.

1144. This condition can now be easily interpreted. It asserts that the forces applied to the surface of the shell must satisfy one of the ordinary six equations of statical equilibrium, namely that which expresses that the applied forces resolved parallel to the polar diameter must vanish.

It is natural to conjecture that (19) and (20) will on the whole amount to just the ordinary six equations of statical equilibrium among the applied forces; and this will be found to be the case as we proceed.

The second of equations (19) gives no result when $l=0$, for then $L'$ and $M'$ both vanish: see Arts. 1135, 1139.

Let us then take equations (20); the second gives no result when $l=0$, for then $N'$ vanishes. It will be found that

$$N=\frac{3}{8\pi}\int_{-1}^{+1}\int_{0}^{2\pi}Zcdsd\psi,$$

so that the condition becomes

$$r_1^3\iint Z_1cdsd\psi-r_0^3\iint Z_0cdsd\psi=0,$$

the limits of integration being the same as before. This is the equation of the moment of the applied forces round the polar diameter.

1145. We now suppose $n=1$ and $l=1$; and then from (19) and (20) we shall obtain the other four equations of statical equilibrium. The following will be the expressions which have to be used with appropriate suffixes in (19) and (20). When $n=1$ and $l=1$, we have $P_1^1 = c$, $c\frac{dP_1^1}{ds} = -s$, $\int_{-1}^{+1}(P_1^1)^2\,ds = \frac{4}{3}$,

$$L = \frac{3}{4\pi}\iint Xc\cos\psi\,ds\,d\psi,$$

$$L' = \frac{3}{4\pi}\iint Xc\sin\psi\,ds\,d\psi,$$

$$M = -\frac{3}{8\pi}\iint (Ys\cos\psi + Z\sin\psi)\,ds\,d\psi,$$

$$N' = -\frac{3}{8\pi}\iint (Y\cos\psi + Zs\sin\psi)\,ds\,d\psi,$$

$$M' = \frac{3}{8\pi}\iint (-Ys\sin\psi + Z\cos\psi)\,ds\,d\psi,$$

$$N = \frac{3}{8\pi}\iint (Y\sin\psi - Zs\cos\psi)\,ds\,d\psi;$$

all the integrals being taken between limits as before.

Then using relations (19) and (20) Lamé finds:

$$r_1^2\iint(X_1c\cos\psi - Y_1s\cos\psi - Z_1\sin\psi)\,ds\,d\psi$$
$$- r_0^2\iint(X_0c\cos\psi - Y_0s\cos\psi - Z_0\sin\psi)\,ds\,d\psi = 0,$$

$$r_1^2\iint(X_1c\sin\psi - Y_1s\sin\psi + Z_1\cos\psi)\,ds\,d\psi$$
$$- r_0^2\iint(X_0c\sin\psi - Y_0s\sin\psi + Z_0\cos\psi)\,ds\,d\psi = 0,$$

$$r_1^3\iint(Y_1\sin\psi - Z_1s\cos\psi)\,ds\,d\psi$$
$$- r_0^3\iint(Y_0\sin\psi - Z_0s\cos\psi)\,ds\,d\psi = 0,$$

$$r_1^3\iint(Y_1\cos\psi + Z_1s\sin\psi)\,ds\,d\psi$$
$$- r_0^3\iint(Y_0\cos\psi + Z_0s\sin\psi)\,ds\,d\psi = 0.$$

Of these four conditions the first asserts that the loads vanish when resolved along a diameter in the equator in the plane of the first meridian, and a second asserts that these loads vanish when resolved along a second diameter in the equator at right angles to the former. The other two conditions are the equations of moments of the loads round these two axes (see pp. 81—85).

1146. We must now advert to the constants which, as we said in Art. 1142, disappear from the equations and so cannot be determined. These are denoted by $A''$, $C''$, $A'$ and $C'$.

Let us write down the values of $u, v, w$ from Art. 1131, so far as they depend on these constants. We have

$$u = [A''] s \ + A''c \cos\psi + C''c \sin\psi,$$
$$v = [A''] c \ - A''s \cos\psi - C''s \sin\psi + \tfrac{1}{2}A'r \sin\psi - \tfrac{1}{2} C'r \cos\psi,$$
$$w = \tfrac{1}{2} [A'] rc - A'' \sin\psi + C'' \cos\psi - \tfrac{1}{2} A' rs \cos\psi - \tfrac{1}{2} C'rs \sin\psi;$$

where $[A'']$ and $[A']$ correspond to $n = 1$ and $l = 0$, while the other constants correspond to $n = 1$ and $l = 1$.

These formulae explain why the constants which occur remain undetermined; each constant corresponds to a certain shift of the body *as a whole*, which leaves the *relative* positions of the particles unchanged, and so calls no stress into operation. Thus $[A'']$ corresponds to a translation parallel to the polar diameter, $A''$ to a translation parallel to the first equatorial diameter, $C''$ to a translation parallel to the second equatorial diameter, $[A']$ to a rotation round the polar diameter, $A'$ to a rotation round the first equatorial diameter, and $C'$ to a rotation round the second equatorial diameter.

1147. We have now to attend to the peculiarities which occur, as we stated in Art. 1141, when $n = 0$, in the equations of Art. 1133.

The constants $A''$ and $C''$ disappear; while the coefficients of $B, D, B', D'$ become infinite. But when $n = 0$ we have $l = 0$, and $P_0^0 = 1$, and $\frac{dP_0^0}{ds} = 0$. Hence the series for $\widehat{r\phi}$ and $\widehat{r\psi}$ in Art. 1133 have no terms which correspond to $n = 0$; and the same remark holds for the equations obtained by putting these expressions equal to $Y$ and $Z$ respectively, and also for the values of $v$ and $w$ in Art. 1131. Hence we see that when $n = 0$ we are concerned with only the first equation in Art. 1133, which then becomes

$$\frac{\mu b}{3a} A + \frac{4\mu}{r^3} B'' = L.$$

For $B$ must necessarily be zero; for if $B$ be not zero we get infinity occurring.

By Art. 1137 we find that when $n = 0$ and $l = 0$, we get

$$L = \frac{1}{4\pi} \int_{-1}^{+1} \int_0^{2\pi} X \, ds \, d\psi.$$

Thus we get by putting in succession $r_1$ and $r_0$ for $r$,

$$\frac{\mu b}{3a} A + \frac{4\mu}{r_1^3} B'' = \frac{1}{4\pi} \int_{-1}^{+1} \int_0^{2\pi} X_1 \, ds \, d\psi,$$

$$\frac{\mu b}{3a} A + \frac{4\mu}{r_0^3} B'' = \frac{1}{4\pi} \int_{-1}^{+1} \int_0^{2\pi} X_0 \, ds \, d\psi.$$

Thus $A$ and $B''$ can be found.

The constant $A''$ will not occur, as it will disappear from the expression for $G$ in Art. 1131, and so will not appear in the expression for $u$ in that Article.

1148. Thus the solution is completed for the case in which we suppose no forces to be applied within the body. When such forces do occur we must get a particular solution corresponding to them; and then to the values $u$, $v$, $w$ of Art. 1131 we must add terms expressing the values of the shifts for the particular solution. The consequent modification of the process for determining the constants of which the types are $A$, $A'$, $A''$, $B$,... can be easily traced.

If instead of a spherical *shell* we have a complete sphere, then all the coefficients of the *negative* powers of $r$ must vanish to avoid infinite expressions at the centre; and this consideration must be used instead of the conditions at the inner surface.

1149. We now come to the work by Lamé published in 1859, entitled: *Leçons sur les coordonnées curvilignes et leurs diverses applications.* This is an octavo volume of 368 pages, besides xxvii pages of introductory matter; it gives the theory of curvilinear coordinates and their application to Mechanics, to Heat, and to the theory of Elasticity. The part of the volume of which we have to give an account occupies pages 257—368, and consists of the Lectures from the fifteenth to the twentieth both inclusive. Lamé himself on page xviii of his Preliminary Discourse gives the following brief sketch of the contents of this part of his volume:

La dernière partie du Cours sera consacrée à la théorie mathématique de l'élasticité. Elle comprendra: 1° la transformation, en coordonnées curvilignes, des équations de cette théorie, la loi des surfaces isostatiques, et son application à la résistance des parois sphériques, cylindriques, ou planes (XV^e et XVI^e leçons); 2° la solution complète du problème de l'équilibre d'élasticité des enveloppes sphériques, comme exemple de la marche à suivre, dans l'intégration des équations générales (XVII^e, XVIII^e et XIX^e leçons); 3° enfin, l'examen des principes qui doivent servir de base à la théorie de l'élasticité (XX^e leçon).

[1150.] The fifteenth Lecture transforms in terms of curvilinear coordinates the three body stress-equations for the equilibrium or the motion of elastic bodies; thus results of the following kind are obtained:

If $\rho_1$, $\rho_2$, $\rho_3$ be three orthogonal curvilinear coordinates; $s_1$, $s_2$, $s_3$ the lines of intersection respectively of $\rho_2, \rho_3$; $\rho_3, \rho_1$; $\rho_1, \rho_2$, and thus $s_2$, $s_3$ the lines of curvature on $\rho_1$ and so forth; $v_1$, $v_2$, $v_3$ the three shifts of $(x, y, z)$ in the directions $s_1$, $s_2$, $s_3$; $r_2'$, $r_3'$ the radii of curvature of $\rho_1$ in the lines of curvature $s_2$, $s_3$ respectively, $r_3''$, $r_1''$ of $\rho_2$ in the lines $s_3$, $s_1$, and $r_1'''$, $r_2'''$ of $\rho_3$ in the lines $s_1$, $s_2$ respectively; $\Delta$ the density; $S_1$, $S_2$, $S_3$ the components of body-force in directions $s_1$, $s_2$, $s_3$ and the system of stresses be given by our usual notation, then Lamé finds (p. 272) equations of the type:

$$\frac{d\widehat{s_1s_1}}{ds_1} + \frac{d\widehat{s_1s_2}}{ds_2} + \frac{d\widehat{s_1s_3}}{ds_3} + \left(S_1 - \frac{d^2v_1}{dt^2}\right)\Delta$$
$$= \frac{\widehat{s_1s_1} - \widehat{s_2s_2}}{r_2'} + \frac{\widehat{s_1s_1} - \widehat{s_3s_3}}{r_3'} + \left(\frac{2}{r_1''} + \frac{1}{r_3''}\right)\widehat{s_1s_2} + \left(\frac{2}{r_1'''} + \frac{1}{r_2'''}\right)\widehat{s_1s_3}.$$

The transformation to curvilinear coordinates is necessarily long, but not difficult since Lamé has the formulae necessary for the purpose ready to his hand in the former part of the volume.

1151. Lamé gives at the beginning of his fifteenth Lecture two pages of formulae, quoted without demonstration from the ordinary theory of elasticity; he adds a short commentary on the formulae, marking with an asterisk topics which require development, and to these he returns in his twentieth. He notices on p. 263 that in the most general case there are 36 coefficients involved in the expressions for the stress at a point; but that on a certain hypothesis these reduce to 21, and on a more restricted hypothesis to 15. Then on his p. 264 he introduces the *two* coefficients $\lambda$ and $\mu$ which occur when the medium is homogeneous and isotropic; but he allows that these are necessarily equal if we adopt completely the restricted hypothesis to which we have just alluded.

1152. Lamé introduces the definition of what he calls an *isostatic surface*. This is a continuous surface having the property that the stress on the tangent plane at any point is normal to the plane at the point of contact; at least this seems to be his definition, but he is not very clear.

[Lamé appears to consider that as at any point there are three elementary planes upon which the stress is entirely normal, so these elementary planes will if there be no sudden or considerable change in their directions envelope three families of orthogonal surfaces which may be taken as our systems of curvilinear coordinates; these he terms *isostatic*. The equations therefore of the preceding Article reduce to the simple form

$$\frac{d\widehat{s_1s_1}}{ds_1} = \frac{\widehat{s_1s_1} - \widehat{s_2s_2}}{r_2'} + \frac{\widehat{s_1s_1} - \widehat{s_3s_3}}{r_3'}$$

when we study only the deformation produced by a load. These equations express a law, which is thus given by Lamé:

Dans tout système effectivement isostatique, chacune des trois forces élastiques principales éprouve, suivant sa direction même, une variation qui est égale à la somme de ses excès sur les deux autres, respectivement multipliés par les courbures correspondantes de la surface qu'elle sollicite. p. 274.]

The following remarks by Boussinesq are given in the *Comptes Rendus*, Vol. 74, 1872, p. 243:

Lamé a désigné par ce nom d'*isostatiques* les surfaces auxquelles ne sont appliquées que des actions normales: il croyait qu'un triple système orthogonal de surfaces pareilles existait toujours dans un corps; ce qui est une erreur, bien qu'il y ait en chaque point trois éléments plans rectangulaires sollicités par des forces normales, parce qu'il ne suffit pas, pour que ces éléments plans se raccordent de manière à former des surfaces, que leurs inclinaisons varient avec continuité d'un point aux points voisins. Toutefois, le beau théorème sur les surfaces isostatiques, qui se trouve démontré au § CXLIX des *Leçons sur les coordonnées curvilignes*, n'en subsiste pas moins pour les cas où ces surfaces existent: il serait assez facile de l'établir géométriquement....

[1153.] The sixteenth Lecture is restricted to the case of homogeneous isotropic bodies. Lamé here transforms into curvilinear coordinates the special equations which apply to this case, namely those which express the six stresses in terms of *two* constants and the shift-fluxions, and gives the value of the dilatation for this case.

These results are of the following forms:

$$\widehat{s_1 s_1} = \lambda\theta + 2\mu\left(\frac{dv_1}{ds_1} - \frac{v_2}{r_1''} - \frac{v_3}{r_1'''}\right),$$

$$\widehat{s_2 s_3} = \mu\left(\frac{dv_2}{ds_3} + \frac{dv_3}{ds_2} + \frac{v_2}{r_2'''} + \frac{v_3}{r_3''}\right),$$

$$\theta = \frac{dv_1}{ds_1} + \frac{dv_2}{ds_2} + \frac{dv_3}{ds_3} - v_1\left(\frac{1}{r_2'} + \frac{1}{r_3'}\right) - v_2\left(\frac{1}{r_1''} + \frac{1}{r_3''}\right) - v_3\left(\frac{1}{r_1'''} + \frac{1}{r_2'''}\right)$$

See p. 285. On pp. 289—292 Lamé obtains the body-shift equations. They may be found in a slightly different form by substituting the above values for the stresses in the equations of Art. 1150. The results are extremely complex.

Some applications to isostatic surfaces, in which approximate results are obtained, are given at the end of this Lecture, although they belong more properly to the preceding Lecture.

1154. The seventeenth, eighteenth, and nineteenth Lectures are devoted to the problem which I call Lamé's Problem; they amount substantially to a reprint of the memoir published by Lamé in Liouville's *Journal de mathématiques*, 1854, the difference between the two being very slight. I will make a few brief remarks on points where there is some variety in the republication.

1155. In the memoir Lamé starts with assuming the differential equations in terms of spherical coordinates which he had given in his treatise on elasticity: see Art. 1093 of my account of that work. In the present volume he deduces the equations for spherical coordinates from the general equations in curvilinear coordinates which he had previously investigated; these are given on his p. 300, and from this point the general curvilinear coordinates occur no more in the volume. It is curious to notice that although Lamé expresses in various places his opinion of the great advantages which follow from the use of his general curvilinear coordinates, yet he allows that the only problem with respect to an elastic body which has been thoroughly solved is this which involves only the ordinary spherical coordinates: see pp. 299 and 368 of the volume now under consideration[1].

1156. Lamé had remarked in his memoir that there are two modes of effecting a certain double summation, and that we may adopt the one which is most suitable in any particular case; see my Art. 1121. Here he adds on his p. 310:

> Le premier mode de groupement rappelle les fonctions $Y_n$ de Laplace, et est exclusivement employé dans la *Mécanique céleste.* En physique mathématique, et notamment dans la question qui nous occupe, il faut essentiellement adopter le second mode, quand il s'agit de déterminer isolément les constantes arbitraires.

1157. A little more development is given in the book of the matter contained in Art. 1130 of my account, that is to make out that the values there assigned for $u_0$, $v_0$, $w_0$ do really satisfy the equations which they ought to satisfy; but the steps are such as a reader may easily take for himself: see p. 315 of the volume.

1158. For the definite integral $\int_{-1}^{+1} (P_n^l)^2 ds$, which is of frequent occurrence, Lamé uses the symbol $\varpi$ with $l$ as a suffix and $(n)$ as an index: see his p. 329.

1159. Lamé gives on his p. 335 especially the values of certain coefficients when $l=0$, namely the coefficients which we

[1] [The case of the ellipsoidal shell has since been solved by M. Painvin; we shall consider his memoir in our second volume. Ed.]

denote by $L$, $L'$, $M$, $M'$, $N$, $N'$; see my Arts. 1137 and 1139; these special cases can be deduced immediately from the general forms.

1160. He considers that the success of the investigation depends mainly on the solution of the simultaneous equations discussed in my Art. 1138. He describes this as the *simultaneous* development of the two expressions $Y'$ and $Z'$ by the aid of the functions $P$: see his pp. 333 and 337. He says on the latter page:

Il y a tout lieu de penser, qu'on ne réussira, dans la même voie, avec un autre système orthogonal, qu'en lui découvrant, d'abord, la faculté analogue, de développer *simultanément* deux ou trois fonctions, de une ou de deux de ses coordonnées. Car la simultanéité dans les développements des fonctions données, paraît être nécessitée, et par la simultanéité des équations aux différences partielles à intégrer, et par la presence simultanée des fonctions intégrées dans les équations à la surface.

1161. On his pp. 354—357 Lamé adverts to two cases in which the solution becomes simplified. One is the case in which we have a full sphere instead of a spherical envelope; in this case, in order to avoid the occurrence of infinite quantities, all the terms must disappear which involve *negative* powers of the radius $r$, so that the constants which are the coefficients of such terms must be zero.

The other case is that in which we suppose no external boundary, that is we take a medium of indefinite extent in which a spherical cavity exists; in this case, in order to avoid the occurrence of infinite quantities, all the terms must disappear which involve *positive* powers of the radius vector $r$.

1162. The nineteenth Lecture finishes thus:

Pour compléter l'examen de la solution générale, il faudrait étudier successivement les termes les plus influents, ou ceux qui correspondent aux moindres valeurs des entiers ($l$, $n$), et faire ressortir les propriétés caractéristiques, et distinctes, de ces différents termes, desquels chacun pourrait exister seul, si les fonctions introduites ou les efforts extérieurs se prêtaient à cet isolement. On devrait, aussi, considérer particulièrement le cas des enveloppes sphériques minces, ou dont l'épaisseur $(r_1 - r_0)$ est une très-petite fraction du rayon $r_1$, ce qui permettrait de simplifier considérablement les séries finales. Enfin on pourrait citer un grand nombre d'applications spéciales et importantes. Mais nous passerons tout cela sous silence. Une digression trop étendue, sur une question particulière de la théorie mathématique de l'élasticité, pourrait donner quelque apparence de raison, à ceux qui ne veulent voir, dans la grande généralité de cette théorie, qu'une complication inextricable, et qui préfèrent et prônent des procédés hybrides, mi-analytiques et mi-empiriques, ne servant qu'à masquer les abords de la véritable science [1].

1163. The last Lecture is entitled: *Principes de la théorie de l'élasticité;* it occupies only ten pages, and the topics brought forward are not very fully discussed. The main idea seems to be this; we really know nothing about molecules or molecular action, and therefore we cannot fully rely on any theory which is based on hypotheses respecting molecular action. Accordingly we ought not to attach much importance to the process by which the thirty-six coefficients are reduced to fifteen: see Art. 594. Lamé seems to employ the phrase *l'ancien principe* to designate the Navier-Poisson hypothesis as to the nature of molecular action, and he presents his general conclusions thus on his p. 363:

On le voit, chaque partie du principe dont il s'agit, chaque mot de son énoncé donne lieu à un doute, déguise une hypothèse ou présuppose une loi. La théorie mathématique de l'élasticité ne peut donc faire usage de ce principe, sans cesser d'être rigoureuse et certaine. Pour être sûre de rester d'accord avec les faits, elle doit se restreindre: 1° aux équations générales déduites, avec Navier, des théorèmes fondamentaux de la mécanique rationnelle; 2° aux relations qui existent entre les forces élastiques autour d'un point, si bien définies par la loi de réciprocité, ou par l'ellipsoïde d'élasticité et qui résultent de l'équilibre du tétraèdre élémentaire, imaginé par Cauchy; 3° aux $\widehat{xx}$, $\widehat{xy}$ exprimés linéairement par les dérivées premières des déplacements, avec leurs coefficients indépendants, sous la forme essentielle établie par Poisson.

[1] [These last words read like a covert sneer at Saint-Venant's *méthode mixte ou semi-inverse*. No elastician now-a-days would hesitate to acknowledge the value of Saint-Venant's method, and the majority would probably endorse what he himself has written about Lamé's pursuit of that Will-o'-the-wisp—the solution for a perfectly general 'mathematical' distribution of load: see the *Historique Abrégé*, pp. clxxii—clxxiii. ED.]

Ainsi les belles recherches ultérieures de ces géomètres, partant de lois préconçues, sortent du champ des applications actuelles. Mais, elles ont admirablement préparé, et rendront faciles les applications futures, lorsque de nouveaux faits, et leur étude approfondie, auront conduit aux lois réelles des actions moléculaires.

The three points which Lamé here holds to be firmly established are apparently: (i) the body-stress equations; (ii) the theorems relative to the stresses discovered by Cauchy and deducible from an elementary tetrahedron, or as an equivalent the properties of Lamé's stress-ellipsoid: see Art. 1059; (iii) the expressions for the stresses involving thirty-six constants, as we have them in Art. 553 of my account of Poisson's memoir of October, 1829.

1164. Lamé then refers specially to the case of homogeneous isotropic bodies; he considers that the two lemmas which he gave in his *Leçons sur l'Élasticité* were not properly established, as they were based on the old ideas. These occur on pp. 39—42 of the work: see my Arts. 1054 and 1055. Accordingly he now substitutes new demonstrations for those formerly given of these lemmas, which he calls respectively the *lemma of simple traction* and the *lemma of simple torsion;* the new demonstrations do not appeal to the consideration of the action of a single molecule on another, but to some results which must obviously hold by symmetry with respect to the action of an aggregate of molecules.

When the new demonstrations of these lemmas are substituted for the old, Lamé says that the establishment of the formulae which relate to homogeneous isotropic bodies "est complètement dégagé de toute hypothèse, de toute idée préconçue": see his p. 367[1].

[1] [Lamé's statement here, together with that on p. 359, with regard to the easy establishment of the linearity of the stress-strain relation seem to me unsatisfactory. His lemmas do not *definitely* appeal to any physical axiom, and we have, precisely as in the case of Green, the apparent miracle of the theory of an important physical phenomenon springing created from the brain of the mathematician without any appeal to experience. The physical axiom or hypothesis of molecular force which Lamé uses in his *Leçons sur l'Élasticité* and which would undoubtedly have led him to rari-constancy if carried out (see our Art. 1051) is here dropped, and the only bridge over the void between the pure theory of quantity and the physical phenomenon is formed by these two lemmas, based upon considerations of symmetry, and a tacit assumption that the most sensible terms in stress are linear in strain. I have

1165. The work finishes thus:

Si quelque personne trouvait étrange et singulier, que l'on ait pu fonder un Cours de Mathématiques, sur la seule idée des systèmes de coordonnées, nous lui ferions remarquer que ce sont précisément ces systèmes qui caractérisent les phases ou les étapes de la science. Sans l'invention des coordonnées rectilignes, l'algèbre en serait peut-être encore au point où Diophante et ses commentateurs l'ont laissée, et nous n'aurions, ni le Calcul infinitésimal, ni la Mécanique analytique. Sans l'introduction des coordonnées sphériques, la Mécanique céleste était absolument impossible. Sans les coordonnées elliptiques, d'illustres géomètres n'auraient pu résoudre plusieurs questions importantes de cette théorie, qui restaient en suspens; et le règne de ce troisième genre de coordonnées spéciales ne fait que commencer. Mais quand il aura transformé et complété toutes les solutions de la Mécanique céleste, il faudra s'occuper sérieusement de la Physique mathématique, ou de la Mécanique terrestre. Alors viendra nécessairement le règne des coordonnées curvilignes quelconques, qui pourront seules aborder les nouvelles questions dans toute leur généralité. Oui, cette époque définitive arrivera, mais bien tard: ceux qui les premiers, ont signalé ces nouveaux instruments, n'existeront plus et seront completement oubliés; à moins que quelque géomètre archéologue ne ressuscite leurs noms. Eh! qu'importe, d'ailleurs, si la science a marché!

Lamé died on the 1st of May, 1870: see *Comptes Rendus*, Vol. 62, p. 961.

dwelt on this point, because we find even in mathematicians of the standing of Lamé not infrequently an omission to state clearly the physical principle upon which they base their calculations of a physical phenomenon. The history of mathematical elasticity gives many examples of this divorce between theory and physical fact; the mathematician has too often identified elasticity with the solution of certain differential equations, the constants of which are to be determined by a purely fanciful and often practically idle, if not impossible distribution of load. Ed.]

# CHAPTER VIII.

## INVESTIGATIONS OF THE DECADE 1840—1850[1]

### INCLUDING THOSE OF BLANCHET, STOKES, WERTHEIM, AND HAUGHTON.

[1166.] P. H. Blanchet: *Mémoire sur la propagation et la polarisation du mouvement dans un milieu élastique indéfini cristallisé d'une manière quelconque. Journal de mathématiques* (*Liouville*). Tome v. pp. 1—30. Paris, 1840. This memoir was presented to the Academy of Sciences on August 8, 1838, and a report on it by Poisson, Coriolis and Sturm appears in the *Comptes rendus*, Tome VII. p. 1143. The report speaks very favourably of the memoir; we quote the following remarks:

L'un de nous (Poisson, see our Art. 523) après avoir donné les équations différentielles de ce problème, les a intégrées complètement dans le cas d'un corps homogène non cristallisé, c'est-à-dire d'un corps dont la constitution et l'élasticité sont les mêmes en tous sens autour de chaque point. Il a conclu de ses formules que si l'ébranlement initial est circonscrit dans une petite portion du milieu, il donne naissance à deux ondes sphériques qui se propagent uniformément avec des vitesses différentes, et dont chacune a une constitution particulière. Dans le même Mémoire, on trouve aussi l'indication succincte de la méthode qu'il faudrait suivre pour traiter de la même manière le problème général qui a pour objet les lois du mouvement dans un milieu homogène élastique indéfini, cristallisé d'une manière quelconque, et qui a partout la même température. C'est ce problème général que

[1] The memoirs of Kirchhoff and W. Thomson due to this period will be found in the chapters of our second volume especially devoted to those writers. Some few memoirs of an earlier or later date are inserted for diverse reasons in this chapter.

M. Blanchet a résolu dans son premier Mémoire. Les équations différentielles auxquelles sont assujétis les déplacements d'un point quelconque du milieu écarté de sa position d'équilibre renferment 36 coefficients constants, qui dépendent de la nature du milieu, et qu'on ne pourrait réduire à un moindre nombre sans faire des hypothèses sur la disposition des molécules et sur les lois de leurs actions mutuelles.

Blanchet in fact assumes each stress as a linear function of the six strains, involving 6 independent constants (see our Art. 553). If these constants be reduced on the rari-constant theory to 15, or on the multi-constant theory to 21, his results are considerably simplified.

[1167.] After pointing out that in the solution of his equations Blanchet has used the methods of Poisson and Fourier, the *Rapporteur* continues:

Mais les moyens qu'il emploie dans tout le reste de son Mémoire pour réduire ultérieurement ses intégrales quadruples et pour en tirer les lois du mouvement vibratoire, lui appartiennent exclusivement et sont aussi simples qu'ingénieux......Le travail de M. Blanchet se recommande à l'attention des géomètres et des physiciens par l'importance et la difficulté du sujet et par le talent avec lequel l'auteur l'a traité. Les propositions qu'il a démontrées sur la propagation du mouvement ondulatoire dû à un ébranlement central et limité acquerront encore plus d'intérêt par l'application qu'on en pourra faire à la théorie des ondulations lumineuses.

[1168.] If $u$, $v$, $w$ be the shifts of the point $x$, $y$, $z$, Blanchet assumes them to be of the type:

$$u = U \,.\, expt.\, (l_1 x + m_1 y + n_1 z) \sqrt{-1},$$
$$v = V \,.\, expt.\, (l_2 x + m_2 y + n_2 z) \sqrt{-1},$$
$$w = W \,.\, expt.\, (l_3 x + m_3 y + n_3 z) \sqrt{-1},$$

where $U$, $V$, $W$ are functions only of the time and $l_1$, $l_2$, $l_3$, $m_1$, $m_2$, $m_3$, $n_1$, $n_2$, $n_3$ constants. He then proceeds by substitution to determine these constants, and obtains the value of the shifts in terms of sextuple integrals with limits $-\infty$ to $+\infty$: see his p. 7. The expressions for the shifts involve vibrations of three periods, or terms having for arguments $s't$, $s''t$ and $s'''t$, where $s'^2$, $s''^2$, $s'''^2$ are the roots of a certain cubic (p. 4, Equation 17).

On p. 10 Blanchet proceeds to the interpretation of his sextuple integrals, and by a change of variables and application of Fourier's theorem reduces them to quadruple integrals. A further reduction to double integrals is obtained by an ingenious change of variables which involves replacing one of them by the parameter of the cubic equation before referred to (pp. 13—16). This method is due entirely to Blanchet.

[1169.] After these preliminary reductions the following problem is attacked: to find the nature of the vibrations when the initial disturbance is limited to a certain portion of the space round the origin of coordinates, and when the time elapsed since the initial disturbance is very great (pp. 17—30).

The characteristics of the motion are very similar to those obtained by Poisson (see Arts. 565—567) and later by Stokes for an isotropic medium (see Art. 1268), as the following remarks will shew:

Il suit de là que pour connaître les points de l'espace en mouvement après le temps $t$, il faut déplacer la surface $\rho = Nt$ ($N$ is here a function of direction only) parallèlement à elle-même, de manière que l'origine du rayon vecteur $\rho$ se promène dans toute la portion de l'espace où l'ébranlement initial a eu lieu. Il y aura une surface enveloppe exterieure et une surface enveloppe intérieure à la surface $\rho = Nt$, dans ses diverses positions, et les points de l'espace non compris entre ces deux surfaces enveloppes seront en repos, si, d'ailleurs d'autres intégrales que l'intégrale $I$ (Blanchet is treating here only a part of the shift) ne donnent rien pour ces points.

La propagation se fera donc suivant une onde et les dimensions des surfaces limites de cette onde croîtront évidemment avec le temps.

La vitesse de propagation de l'onde sera constante dans chaque direction, du moins tant qu'on ne considérera que l'intégrale $I$. Elle changera, au contraire avec la direction, à cause de la variabilité de $N$ (p. 19).

It will be seen that Blanchet here uses a method which is now generally adopted for the motion of waves of any kind in space.

[1170.] In conclusion Blanchet sums up the results of his first memoir as follows:

1°. Dans un milieu élastique, homogène, indéfini, cristallisé d'une manière quelconque, le mouvement produit par un ébranlement central se propage par une onde plus ou moins compliquée dans sa forme.

2°. Pour chaque nappe de l'onde, la vitesse de propagation est constante dans une même direction, variable avec la direction suivant une loi qui dépend de la forme de l'onde.

3°. Pour une même direction, les vitesses de vibration sont constamment parallèles entre elles dans une même nappe de l'onde pendant la durée du mouvement, et parallèles à des droites différentes pour les différentes nappes, ce qui constitue une véritable polarisation du mouvement (p. 30).

[1171.] In a note appended to the memoir in Liouville Blanchet remarks that some parts of his analysis might be simplified. He states that he has applied his results in a second memoir to the case treated by Poisson (see Art. 565), and he finds that the same conclusions, as that 'great geometrician' has discovered, flow from them. Finally he promises to publish applications of his theory to various special cases, particularly for uniaxial crystals.

This second memoir does not seem to have been published.

[1172.] A third memoir entitled: *Mémoire sur la délimitation de l'onde dans la propagation générale des mouvements vibratoires,* appears to have been presented on June 14, 1841, to the Academy of Sciences. An extract by the author is given on pp. 1165—6 of the *Comptes rendus,* Tome XII. 1841. The *Commissaires* to whom it was referred were Cauchy, Liouville and Duhamel. The object of the memoir is expressed in the following words:

Dans le Tome X des *Mémoires de l'Académie des Sciences* M. Poisson a démontré les lois de la propagation sphérique des mouvements vibratoires. (See our Art. 564.) En se bornant à prendre les parties les plus considérables de ses intégrales, il a trouvé deux ondes sphériques. La forme des intégrales complètes montre qu'il peut y avoir entre les deux ondes des mouvements comparativement plus ou moins négligeables;

mais il n'y a rien en-deçà de la plus petite onde, rien au-delà de la plus grande. Les intégrales de M. Ostrogradski présentent aussi ce dernier caractère. L'Académie n'a pas oublié sans doute toute l'importance qu'y attachait le grand géomètre qu'elle a perdu.

Blanchet, at the invitation of Liouville, then states that he has endeavoured to obtain similar results for the case of a crystallised body. By means of Cauchy's *Calcul des résidus*, he has been able to obtain limits for the integrals of the first memoir and concludes that:

*Il n'y a, en général, ni déplacement ni vitesse au-delà de la plus grande nappe des ondes.*

The memoir itself is published in Liouville's *Journal des mathématiques*, Tome VII. 1842, pp. 13—22. It refers to Cauchy's discovery of a lower limit (p. 16) and then proceeds to the investigation of a superior limit. The last lines of the memoir point to a certain jealousy of Cauchy, and a claim to priority in results (p. 22).

[1173.] A fourth memoir on this subject was presented to the *Académie des Sciences* on July 5, 1841. It is entitled: *Mémoire sur une circonstance remarquable de la délimitation de l'onde.* A note upon it will be found in the *Comptes rendus*, Tome XIII. p. 18, but the memoir in full is given in Liouville's *Journal des mathématiques*, Tome VII. pp. 23—34.

[1174.] The principal object of the memoir is the consideration of the nature of the wave when two roots of the cubic obtained in the first memoir (see our Art. 1168) are equal. There are also various simplifications of the analysis of the earlier memoirs. The results appear to me of greater analytical than physical interest.

[1175.] In the same volume of the *Comptes rendus* (Tome XIII.) will be found other notes relating to this matter.

(*a*) On pp. 184—188 is a memoir by Cauchy on what he terms the *surface caractéristique* and the *surface des ondes.* He refers to Blanchet in a footnote, p. 185.

(*b*) On pp. 188—197 is a *Mémoire sur l'emploi des fonctions principales représentées par des intégrales définies doubles, dans la recherche de la forme des ondes sonores, lumineuses, etc.* by Cauchy, which has for its object the investigation of inferior and superior (interior and exterior) limits of the waves.

Ces conclusions s'accordent avec celles qu'a obtenues M. Blanchet, en appliquant le calcul des résidus à la détermination des intégrales triples (p. 197).

(*c*) On p. 339 is a note by Blanchet entitled: *Démonstration géométrique de l'identité de la limite extérieure de l'onde que M. Cauchy vient de donner avec celle que j'ai donnée précédemment.........*

Blanchet remarks:

Je connaissais depuis longtemps ces théorèmes que M. Cauchy vient d'imprimer le premier. Je suis bien aise de trouver une occasion d'en faire usage.

He then gives a geometrical proof of the limit obtained by Cauchy in (*b*), and remarks that it is a direct corollary from a result of his last memoir.

(*d*) On pp. 958—960 there is a note by Blanchet pointing out that Cauchy's results in a memoir to be found in the same volume (pp. 397—412) do not agree with his own.

(*e*) This is followed by an observation of Cauchy's (p. 960), that Blanchet has not shewn where his (Cauchy's) formulae are wrong. The question Cauchy holds to be a delicate one, and it appears to him necessary that all the calculations on the matter should be revised.

(*f*) Finally in this volume we have a letter of Blanchet on p. 1152. It is entitled: *Sur la propagation de l'onde.* Blanchet here states his general conclusions, and considers his methods to be of wider application than Cauchy's. He refers to a memoir he hopes soon to present to the Académie.

[1176.] The discussion is continued in the *Comptes rendus*, Tome XIV. We briefly note the various occasions.

(*g*) On pp. 8—13 is a memoir by Cauchy on the same subject which contains the following remarks on p. 13:

Les déplacements et par suite les vitesses des molécules s'évanouiront pour tous les points situés en dehors ou en dedans des deux ondes propagées. M. Blanchet a remarqué avec justesse qu'on ne pouvait, en général, en dire autant des points situés entre les deux ondes. Toutefois il est bon d'observer que, même en ces derniers points, les déplacements et les vitesses se réduisent à zéro quand on suppose nulle la dilatation du volume représentée par la lettre $v$, c'est-à-dire, en d'autres termes, quand les vibrations longitudinales disparaissent; et comme, dans la théorie de la lumière propagée à travers un milieu isotrope, on fait abstraction des vibrations longitudinales, en se bornant à tenir compte de celles qui ont lieu sans changement de densité, on pourra conclure des formules précédentes, appliquées à cette théorie, que les vibrations lumineuses subsistent seulement dans l'épaisseur de l'onde la plus lente.

(*h*) On pp. 389—403 is the report of Sturm, Liouville, Duhamel and Cauchy on Blanchet's memoirs on waves in crystalline media and on their delimitation. The report recommends their publication in the *Recueil des Savants étrangers.* To the report are added four notes by Cauchy principally relating to his own results.

[1177.] We have treated these notices shortly because their practical application belongs rather to the theory of light than to our present topic. We may however sum up the results in the following statement. Blanchet was the first to investigate fully the motion of a wave in a 36-constant elastic medium. He has demonstrated that, when such a medium is initially disturbed in any way about a point, the exterior limit of the vibrating portion is determined by the greatest, the interior limit by the least sheet of the wave-surface. He has in a later memoir (see our Art. 1174) extended his results to the case where the two sheets intersect. The important question then arises as to what the shifts and velocities between the two wave-sheets may be. Blanchet holds that they will exist, but that they will be very small relative to those which take place on the sheets of the wave-surface, provided the dimensions of the volume included by the wave-surface have become very large compared with the dimensions of the space initially disturbed. If the wave produces a 'sensible phenomenon,' that phenomenon may possibly cease to be sensible between the wave-sheets, when at a great distance from the origin of disturbance, but it will never absolutely disappear. Cauchy, considering the shifts between the wave-sheets to be infinitely small as compared with those on the sheets themselves, argues (*Comptes rendus*, XIII. pp. 397 and 960) that they may be considered as non-existent.

[1178.] A final memoir of Blanchet's must be noted. It is entitled: *Sur les ondes successives.* It was presented to the Académie on May 3rd, 1842. It was published in Liouville's *Journal des mathématiques,* Tome IX. pp. 73—96. Its object is to deduce the nature of the motion in a crystallised medium in which there is not an instantaneous but continuous central disturbance varying with the time. The author makes use of a principle due to Duhamel (*Journal de l'École polytechnique,* Cahier XXIII^e. p. 1) in order to apply the sextuple integrals of his first memoir to this case. He draws some general conclusions which are clearly stated in the extract by the author inserted in the *Comptes rendus* (Tome XIV. p. 634). We reproduce these partially:

1°. A une distance suffisamment grande, le mouvement, en chaque point, est la résultante statique des mouvements qu'y amènent trois systèmes partiels d'ondes successives *dont les vitesses de propagation sont différentes.*

2°. Dans chacun de ces systèmes les déplacements et les vitesses des molécules sont polarisés, suivant des directions variables avec celles des rayons vecteurs, menés d'un même point pris pour origine dans la partie de l'espace agitée par la force accélératrice.

3°. Chacune des propagations partielles se fait comme si le mouvement glissait en quelque sorte tout d'une pièce dans les différentes directions pendant que les déplacements, et les vitesses des molécules varieraient en raison inverse des distances à l'origine.

4°. Ce mouvement reste pour ainsi dire semblable à lui-même sur certaines surfaces concentriques et semblables entre elles, qui doivent être considérées comme les surfaces des ondes.

5°. La partie de l'espace agitée entre deux de ces surfaces très-voisines constitue l'onde élémentaire. L'épaisseur proprement dite des ondes ne peut être bien définie que dans le cas où la force accélératrice est périodique par rapport au temps.

Blanchet's other results appear to be only properties due to the superposition of small motions, and depend really on the linearity of the body shift-equations.

[1179.] J. Fr. L. Hausmann. *Ueber einige am Eisen, bei Versuchen über seine Elasticität, beobachtete Erscheinungen. Poggendorffs Annalen,* Bd. LI. 1840, pp. 441—443.

In 1834 a Commission was appointed to examine the elastic and cohesive properties of bar iron (*Stabeisen*) prepared in Hannover. A full account of these experiments will be found in the *Studien des göttingischen Vereins bergmännischer Freunde,* Bd. IV., Heft 3. The above communication of Hausmann has relation to a very remarkable phenomenon associated with that of stricture. After rupture had taken place at the section of stricture, the strictured ends of the bar were found to be strongly magnetised. This phenomenon did not appear to such a marked extent with '*Gussstahl von der schweissbaren Sorte,*' although the stricture in that case was still more marked. The magnetisation was of a very permanent kind, and was quite sensible after six years. The following are the results obtained:

1°. Nur das äusserste, bei dem Zerreissen verdunnte Ende der Stücke liess Magnetismus erkennen; weder am entgegengesetzten Ende der 3 bis 4 Zoll langen Stücke noch an anderen Stellen derselben zeigte sich eine Spur davon.

2°. An dem verdünnten Ende zeigten sich die Kanten und Ecken, so wie die hervorragenden Spitzen der Fadenbündel am Stärksten magnetisch.

3°. Die Stücke, an welchen sich unzweideutig Magnetismus wahrnehmen liess, gehörten der Mehrzahl nach zu den Stäben, welche bei dem Zerreissen im Verhältniss zur Ausdehnung, sich am Stärksten zusammengezogen hatten.

4°. Magnetismus zeigte sich vorzüglich an solchen Stücken, welche sich durch eine vollkommen fadige Textur auszeichneten.

The magnetisation was tested by the influence of the strictured section on iron filings.

[1180.] A number of papers on vibrating elastic bodies by A. Seebeck will be found in the volumes of *Poggendorffs Annalen* for the decade 1840—1850. They belong however to the history of the theory of sound. Some of these papers are reprints or abstracts of memoirs presented to the *Königliche sächsische Gesellschaft der Wissenschaften* (Leipzig), and will be found in the *Berichte* of that Society. We may note the paper in *Poggendorff*, Bd. LXXIII. 1848, pp. 442—448, or the *Berichte*, Bd. I. pp. 159 and 365. This paper contains a calculation of the nodal points and loops (*Knoten-* and *Wende-Punkte*) of vibrating rods. Seebeck notes that Duhamel's theory of N. Savart's result only applies to one special case, and not to a rod with both ends built-in: see our Art. 1228 and Lord Rayleigh's *Theory of Sound*, Vol. I. pp. 230—232.

[1181.] *Investigation of the Tendency of a Beam to break when loaded with weights. Cambridge Mathematical Journal*, Vol. I. 1840, pp. 276—278. This paper is initialled H. T., but in the table of contents is attributed to A. Smith. The writer remarks that at a point of transverse load the tendency to break changes discontinuously, and suggests a method of representing the tendency to break by one formula which contains discontinuous factors. The paper has no elastic importance.

In the same volume will be found another paper by A. Smith entitled: *The Propagation of a Wave in an Elastic Medium*, pp. 97—100. It belongs entirely to the theory of light and so does not concern us.

[1182.] *On the Form of a Bent Spring. Cambridge Mathematical Journal*, Vol. II. 1841, pp. 250—252. This paper is merely described as 'from a correspondent.' It notices the fact that in the ordinary Bernoulli-Eulerian theory of beams, no account is taken, when the load is not transverse, of the extending or compressing effect of the longitudinal component of the load. This effect had however been referred to by previous writers: see our Arts. 198 and 737. The paper is worthless, as may be shewn by the remark on the Bernoulli-Eulerian hypothesis (*i.e.* that the cross sections remain perpendicular to the longitudinal fibres):

Did this law not hold the laminae of the spring would have a sliding motion, and the form of the bent spring *could not* be made the subject of *mathematical* investigation.

As a matter of fact Saint-Venant had already commenced to investigate mathematically this slide: see our Chapter IX.

[1183.] Morin. *Note sur la résistance au roulement des corps les uns sur les autres, et sur la réaction élastique des corps qui se compriment réciproquement. Comptes rendus,* T. XIII. 1841, pp. 1022—1023.

I do not know whether the memoir of which this is an extract was ever published. Morin holds that the law that the resistance to rolling is proportional to the pressure is not general, but if there be no impulse this resistance will be sensibly independent of the velocity. He then cites general results which he says are partly due to experiment and partly to reasoning. These general results relate to the impact of elastic bodies which are endowed with different *vitesses de retour*. By this term Morin denotes the rate at which a body regains its primitive form. We quote the results here as bearing upon Hodgkinson's and Haughton's experiments: see our Arts. 939—943, and 1523.

1°. Que dans ce choc il y a toujours une perte de force vive ou de travail provenant de cette différence des vitesses de retour, abstraction faite de celle qui peut être due aux mouvements vibratoires;

2°. Que si des corps de même forme et de même poids parfaitement élastiques, mais doués de vitesses de retour différentes, choquent un même corps avec des vitesses égales, ils quitteront le corps choqué avec des vitesses différentes;

3°. Que si on laisse tomber de diverses hauteurs sur une surface plane horizontale des sphères de matière et de poids différents, le rapport de la hauteur de retour à la hauteur de chute est constant;

4°. Que quand le corps choqué est sensiblement plus compressible que le corps choquant, le rapport de la hauteur de retour à la hauteur de chute ne dépend que de la réaction élastique du corps choqué, et qu'il est, dans les limites des expériences, indépendant de l'élasticité, de la rigidité et de la masse du corps choquant;

5°. Qu'à l'inverse, quand c'est le corps choquant qui est le plus compressible et qui a la vitesse de retour la plus faible, le rapport de la hauteur de retour à la hauteur de chute est indépendant de la dureté et de l'élasticité du corps choqué.

[1184.] A. Masson. *Sur l'élasticité des corps solides. Annales de Chimie et Physique.* Tome III. 1841, pp. 451—462. This memoir was presented to the Académie des Sciences on November 15, 1841. A brief account of it will be found in the *Comptes rendus,* Tome XIII., 1841, pp. 961—963.

This memoir opens with the following words:

Malgré ses nombreuses expériences sur la physique moléculaire, Savart voyait avec regret qu'il ne pourrait jamais résoudre tous les problèmes qui sans cesse assaillaient son esprit. Livré depuis longtemps à l'étude des propriétés mécaniques des fluides, il voulait soumettre les solides à des études comparatives, afin d'établir les bases d'une mécanique générale des corps pondérables. Il m'avait chargé d'une partie de ce travail. Aidé de ses conseils, travaillant sous sa direction et dans ses cabinets, j'avais commencé des recherches qui ont été interrompues par le malheur qui a plongé tous ses amis et moi particulièrement dans un chagrin que rien ne saurait adoucir, sinon le souvenir de sa bienveillance et de son amitié.

The idea of Savart was to study the action of heat, electricity, etc. on the cohesive and elastic properties of bodies. It was never carried out, and the present memoir is only a very slight contribution to these great problems.

We may notice a few points.

1°. Masson experimented on the rods used by Savart in the memoir referred to in our Art. 347. The results diverge considerably from Savart's, and great irregularity was noticed in the extension.

On voit par ce tableau que les corps solides ne s'allongent pas d'une manière continue, mais par saut brusque (p. 454).

The reader will recognise that the rods had not been reduced to a state of ease, and also remark the influence of the yield-point.

2°. The stricture of bars subjected to longitudinal load is noted (les verges chargées de poids devaient *filer*, p. 454). There is especially early stricture in the case of zinc.

3°. In calculating the *coefficients d'élasticité* the time effect seems to have been taken into account (see p. 456), but there is no distinction made between elastic fore-strain and after-strain.

4°. The *coefficients d'élasticité* calculated from the velocity of sound were found to agree fairly with those calculated from the statical stretch. There is no consideration however of the distinction which ought perhaps to be made between the specific heats at constant volume and at constant stress (see our Art. 705).

5°. Between $-4^\circ$ and $+20^\circ$ C. no difference was found in the elasticity.

6°. Masson obtained the same result as Lagerhjelm that iron, tempered steel, and annealed steel present no notable difference in their elasticity: see our Art. 366.

7°. There is an attempt to set up a relation between the elasticity of a substance and its atomic weight. This is stated as follows:

En multipliant les coefficients d'élasticité des corps simples par un multiple ou sous-multiple de leur équivalent on obtient un nombre constant (p. 460).

This constant number has the mean value 2·45; the multiple is unity except in the case of iron when it is taken equal to 2. The experiments, made only on iron, copper, zinc, tin, and silver, cannot be considered very conclusive.

8°. Special experiments were made on the velocity of sound in lead. It was found to be 1443·48 metres and thus greater than that of water (1435 metres). This result is not in accordance with later experiments.

The *coefficient d'élasticité* is defined by Masson as the extension produced by unit weight in a rod of unit length and of unit section. This must be remembered in any comparison of his results with those of other investigators.

[1185.] 1841. F. E. Neumann. *Die Gesetze der Doppelbrechung des Lichts in comprimirten oder ungleichförmig erwärmten unkrystallinischen Körpern. Abhandlungen der k. Akademie der Wissenschaften zu Berlin. Aus dem Jahre* 1841. *Zweiter Theil.* Berlin, 1843, pp. 1—254, with plate. The last six pages are filled with a list of errata, remarkably long even for a German memoir of this period, but by no means exhaustive. This memoir is far more important for our subject than its name would seem to imply. Starting from Brewster's researches[1] Neumann developes a theory

[1] For the fuller understanding of Neumann's memoir as well as Clerk-Maxwell's memoir of 1850, I have placed in this footnote a short account of some of Brewster's results.

Brewster's researches on the polarising effect of strain on glass and other bodies will be found in the following memoirs:

(*a*) Strain produced by heat. *Phil. Trans.* 1814, p. 436, and *Phil. Trans.* 1815, p. 1.

(*b*) Strain produced by stress. *Phil. Trans.* 1815, p. 60; *Phil. Trans.* 1816, p. 156, and *Edinb. Roy. Soc. Trans.* Vol. VIII, 1818, p. 281, and p. 353.

The points which concern us in these memoirs are those connecting the stress at any point of a body with its polarising effect on light at that point. Brewster remarks in the *Phil. Trans.* for 1816 that his experiments furnish a method of rendering visible and even of measuring the mechanical changes which take place during the straining of a body (p. 160):

'The tints produced by polarised light are correct measures of the compressing and dilating forces, and by employing transparent gums, of different elasticities, we may ascertain the changes which take place in bodies before they are either broken or crushed.'

Brewster suggests that models of arches should be made of glass, and so the stress in different parts of the arch rendered visible by exposure to polarised light.

*Proposition* IV. on p. 161 of the same memoir is important. It runs:

'The tints polarised by plates of glass in a state of compression or dilatation, ascend in Newton's scale of colours as the forces are increased; and in the same plate, the tint polarised at any particular part is proportional to the compression or dilatation to which that part is exposed.'

By reference to Brewster's figure and text, it would appear that he denotes here by compression and dilatation, the stretch and squeeze in a bar subject to flexure, or as we may put it the tint ascends in the scale as the longitudinal stress in a fibre increases; thus the tints are not always the same throughout the length of a bar subject to flexure at the same distance from the neutral line.

*Proposition* VII. (p. 164) proves the superposition of small strains, by means of the superposition of optical effects.

The memoir suggests various instruments for measuring temperature and force by the polarising effect of glass submitted to strain.

In the *Edinb. Trans.* Vol. VIII. (pp. 362—364), will be found some account of the tints in the case of tubes and cylinders of glass. Pp. 369—371 give an account of the *Teinometer* to which we have referred in Art. 698, and explain how the Teinometer is to be made use of in practice. It is true that the maximum tint at an edge of the standard glass plate will measure its deflection, but Brewster does not enter into the necessary theoretical calculations which must be made before we can ascertain how this deflection may be used to compare the elasticities (i.e. stretch-moduli) of the materials of the two metal plates: see our Art. 698. The fuller theoretical consideration of his method was left for Neumann and Clerk-Maxwell.

for the analysis of strain by means of its double-refracting influence on light, the strain being due (1) to load, or (2) to unequal temperature, or (3) to set. The memoir also involves one of the first attempts to investigate the general equations of set. Throughout the memoir Neumann appears as a rari-constant elastician. In the Lectures of 1857—1860, recently published, Neumann seems to have maintained an agnostic attitude: see his *Vorlesungen über die Theorie der Elasticität der festen Körper und des Lichtäthers... Herausgegeben von Dr O. E. Meyer*, Leipzig, 1885, pp. 133—163.

[1186.] The memoir is divided into five parts. We have (1) the *Einleitung* (pp. 3—24) containing a general statement of method and results; (2) the discussion of the law of the double refraction of light in homogeneously strained bodies (pp. 25—61); (3) the consideration of the optical phenomena (colour-fringes) produced by passing polarised light through a heterogeneously strained body (pp. 61—85); (4) the consideration of the like phenomena produced by an unequal distribution of temperature (pp. 86—229); and (5) a theory of set (pp. 230—247). Neumann's chief object was to determine the fringes produced by a given strain; the elastician will regard his memoir as solving the converse problem: To analyse an unknown strain by means of the fringes it produces.

[1187.] Neumann commences his memoir by a consideration of the three possible modes in which the strain of a homogeneous isotropic body can affect its power of transmitting light. He holds that the most probable hypothesis for the double-refractive property of strained bodies, is the new arrangement of the particles of ether, produced by the shifts of the solid parts of those bodies. In bodies subject to homogeneous strain the re-arrangement of the ether-particles must be symmetrical about the three planes of strain-symmetry. Thus the 'optical axes of elasticity' will coincide in direction with the axes of principal stretch, and the lengths $v_1$, $v_2$, $v_3$, of these optical axes will be functions of the three principal stretches $s_1$, $s_2$, $s_3$. Since these stretches are supposed to be small we must have relations of the form

$$\left.\begin{aligned} v_1 &= V' + qs_1 + ps_2 + ps_3 \\ v_2 &= V' + ps_1 + qs_2 + ps_3 \\ v_3 &= V' + ps_1 + ps_2 + qs_3 \end{aligned}\right\} \quad \text{..................(i).}$$

Here $V'$ differs from $V$ – the velocity of light in the unstrained body—only by very small quantities which are functions of the square of the strain, and $p$, $q$ are two constants depending on the material of the body.

Of these equations Neumann remarks:

Die Grösse der Doppelbrechung hängt von der Differenz der optischen Elasticitätsaxen ab; sie hängt also nur von zwei Constanten $p$ und $q$ ab. Ob zwischen den Werthen von $p$ und $q$ noch ein konstantes Verhältniss stattfindet. oder ob auch ihr Verhältniss durch die individuelle Natur des comprimirten Körpers bedingt ist, lässt sich nicht weiter durch allgemeine Betrachtungen ermitteln, sondern muss der Entscheidung durch Beobachtungen überlassen bleiben. (p. 37.)

[1188.] Neumann then proceeds to consider the two surfaces, the radii-vectores of which are the reciprocals of $\rho'$, $v$, where:

$$\left.\begin{aligned} \rho'^2 &= a^2l^2 + b^2m^2 + c^2n^2 \\ v^2 &= v_1{}^2l^2 + v_2{}^2m^2 + v_3{}^2n^2 \end{aligned}\right\} \dots\dots\dots\dots\dots\dots\text{(ii)}.$$

In the first $\rho'$ represents the strained magnitude of the radius-vector $\rho$ whose direction cosines are $(l, m, n)$, and

$$a = \rho\,(1 + s_1), \quad b = \rho\,(1 + s_2), \quad c = \rho\,(1 + s_3);$$

thus the surface is what we term the strain-ellipsoid; Neumann terms it the *Elastizitätsfläche des Drucks*.

In the second $v$ represents the velocity of wave propagation for the direction of molecular displacement $(l, m, n)$. This surface is Fresnel's ellipsoid of elasticity. Neumann terms it the *optische Elastizitätsfläche* (p. 37).

It is then shewn that if the square of the strain be neglected, these two ellipsoids have the same directions of circular-section. Neumann for no very clear reason terms the normals to the circular-sections of the strain-ellipsoid *the neutral axes of pressure* (*neutrale Axen des Drucks*), and as Fresnel terms the corresponding normals for his ellipsoid of elasticity the optic axes, we have the proposition:

The neutral axes of pressure and the optic axes have the same directions. (p. 38.)

[1189.] The next step is to shew that in the sections by any plane of these two surfaces the directions of the principal axes coincide. Further if $O$, $E$ be the reciprocals of the principal axes of the section of Fresnel's surface of elasticity, and $\rho(1+s)$, $\rho(1+s')$, the reciprocals of the principal axes of the section of the strain-ellipsoid made by the wave front, an easy analysis leads us to the conclusion that

$$O - E = -(p-q)(s-s') \text{.................(iii).}$$

These results give us (1) the direction of the principal stretches in any planes as the directions of $O$ and $E$, (2) the magnitude of $s-s'$, the difference of the principal stretches or the maximum-slide (see our Art. 1368) for planes perpendicular to the wave-front: see our remarks on Maxwell's memoir, Art. 1544. Neither Neumann nor Maxwell seems to have remarked that the difference of the velocities of the ordinary and extraordinary rays depend solely on the maximum-slide of planes perpendicular to the wave front. Neumann expresses the result in a somewhat longer form on p. 40, and then states that:

The greatest radius-vector in a section of Fresnel's ellipsoid of elasticity coincides in direction with the smallest or greatest radius-vector of the same section of the strain-ellipsoid, according as $p-q$ is positive or negative.

[1190.] The remarks which follow on the same page, I shall quote in full; they suggest clearly the methods which the elastician must adopt in order to analyse a strain by means of the polariscope:

Aus diesen Sätzen folgen überraschende Analogien zwischen den lineären Dilatationen des gleichförmig comprimirten Körpers und den Fortpflanzungs-Geschwindigkeiten der Lichtwellen und ihren Polarisations-Richtungen. Eine Lichtwelle, welche senkrecht auf einer neutralen Axe des Drucks steht, hat nur einerlei Fortpflanzungs-Geschwindigkeit und die Richtung ihrer Polarisations-Ebene ist willkürlich; in allen Richtungen eines Schnittes aber, die senkrecht auf einer neutralen Axe stehen, haben auch die festen Theile des Körpers dieselben Dilatationen erlitten. In jedem andern Schnitt, welchen man durch den Körper macht, giebt es zwei auf einander rechtwinklige Richtungen, in welchen die Dilatation ein Maximum oder Minimum ist, eine Lichtwelle, welche sich parallel mit diesem Schnitt bewegt, ist entweder nach der einen oder der andern dieser beiden Richtungen polarisirt; die raschere Welle ist nach der Richtung der grössten Dilatation polarisirt, wenn $p-q$ einen positiven Werth hat, und nach der Richtung der kleinsten Dilatation, wenn $p-q$ einen negativen Werth hätte. Der Unterschied der grössten und kleinsten Dilatation in einem Schnitt ist proportional mit dem Unterschiede der beiderlei Geschwindigkeiten, mit welchen die mit dem Schnitt parallele Welle sich bewegen kann.

[1191.] Neumann then considers the case of a right six-face of sides $H, B, D$ subjected to a uniform tractive load (pressure) over two parallel faces perpendicular to $H$, when a ray of light parallel to $D$ and polarised in a plane making an angle of 45° with $H$ is passed through it. He easily deduces that:

$$2\delta = D\left(\frac{U}{O} - \frac{U}{E}\right) = \frac{p-q}{V'} \cdot \frac{U}{V'} s_3 (1+\eta) D \text{......(iv)},$$

where $\delta$ is the thickness of the air, which corresponds in the Newtonian scale to the colour produced by a given value of $D . s_3$; $s_3$ is the squeeze produced by the load, $\eta$ the ratio of lateral stretch to longitudinal squeeze, and $U$ the velocity of light in air. Neumann takes $U = 1$, and $\eta = 1/4$; thus he obtains:

$$\delta = \tfrac{5}{8} \frac{p-q}{V'^2} s_3 D = \tfrac{5}{8} \frac{p-q}{V^2} s_3 D, \text{ nearly} \quad \text{......(iv)}'.$$

As soon as the stretch-modulus for the material in question is known we have $s_3$ for a given load, and can thus determine $p-q$. Neumann does not directly find the value of $s_3$, but takes a case of non-homogeneous strain, namely that of a glass rod of rectangular section, subject to transverse load, and practically constructs the colour fringes for polarised light transmitted through it; these fringes are easily shewn to be of hyperbolic form. The stretch-modulus is then measured in terms of the deflection and thus the value of $\frac{p-q}{V^2}$ obtained. The two values of this expression which Neumann obtains from Brewster's and from his own experiments respectively differ somewhat. We must note that: (i) he has supposed his glass to possess uni-constant isotropy, (ii) he has made use of the old Bernoulli-Eulerian theory of beams to obtain his stretch-modulus, neither very satisfactory assumptions. He finds $(p-q)/V = {\cdot}082$: see his pp. 40—49.

[1192.] Equations (iv) and (iv)′ will obviously be of use in determining the elastic limit or the limit of cohesion in terms of $\delta$, that is in terms of a definite tint in Newton's scale. If set has begun the fringes will not disappear entirely on the removal of the load. On the other hand we can measure by means of the maximum tint reached before rupture the maximum stretch (limit of cohesion). This method was actually suggested by Brewster, and Neumann gives some numerical examples in a footnote: see his pp. 49—50 and our Art. 698.

[1193.] On pp. 50—58 Neumann describes an optical method of measuring the absolute values of $p$ and $q$; briefly it may be said to depend on the shifting of the diffraction fringes owing to the retardation of one of the interfering rays. Neumann's own calculations are again based on a measurement of the deflections of a glass bar supposed to obey the Bernoulli-Eulerian theory and to possess uni-constant isotropy (pp. 56—57). He finds $p/V = -0{\cdot}131$, and $q/V = -{\cdot}213$, or both negative. The equations (i) thus take the form

$$v_1 = V\{1 - 0{\cdot}213\, s_1 - 0{\cdot}131\, s_2 - 0{\cdot}131\, s_3\} \ldots\ldots\ldots \text{(v)}.$$

These equations should enable us, supposing the numerical factors to be correct, to analyse by the polariscope all forms of strain in glass: see Brewster's suggestions in our footnote, p. 640.

[1194.] We may note a point made on p. 59, which, however, belongs essentially to the history of optics. If a body be compressed by a uniform tractive load, $s_1 = s_2 = s_3 = \theta/3$;

$$\therefore v_1 = v_2 = v_3 = V(1 - {\cdot}158\, \theta),$$

thus the velocity of light in a medium is *increased* by *compressing* it, for $\theta$ is negative. If $\mu$ and $\mu'$ be the refractive indices before and after loading

$$\mu' = \mu\,(1 + {\cdot}158\, \theta)$$

or the refractive index *decreases* with increased density.

Neumann remarks that he has found, when the change in density is produced by a change in temperature instead of by a change in load that this law holds, although the coefficient of $\theta$ is in the case of temperature only about half as great as that given by the above theory for the mechanical load (p. 60). He gives a reference to some experiments of Fresnel's which do not at all agree with his results: see *Annales de Chimie*, T. XV, 1820, p. 385.

[1195.] The memoir now passes to the case where the strain is heterogeneous, or the direction and magnitude of the principal stretches change from point to point. Neumann remarks that this is the case when a body does not possess uniform temperature at all points, or when it possesses initial strain from rapid cooling. His investigation is principally of optical interest; as far as elasticity is concerned it involves the determination of the value of the quantity $s - s'$ in equation (iii) in terms of the first shift-fluxions. Neumann, however, does this only for a special choice of the coordinate axes. In fact, he has previously limited the problem by supposing the compressed body to be a plate bounded by parallel faces; the ray striking one of these obliquely is retarded and rotated as to its plane of polarisation by the various strata in different states of strain through which it passes; the ray is further supposed to remain very nearly straight. The grounds for this assumption are given on pp. 62—65. The formulæ finally obtained are somewhat lengthy, but the analysis by which they are deduced is fairly easy to follow. As usual there are numerous misprints not all enumerated in the *Errata*. Neumann considers one example only, that of a ray passing through the plane ends of a right circular prism, but not parallel to the axis, and works this case completely out only when the ray lies in the same plane as the axis. In this case there is no rotation of the plane of polarisation, and, as Neumann shews in a footnote, this special instance can be treated more easily by a direct investigation (pp. 85—88).

[1196.] The next section of the memoir is the investigation of the effect of a varied distribution of temperature in producing colour-fringes. Pp. 86—100 are occupied with the deduction of the thermo-elastic stress equations. These are practically identical with those of Duhamel, a term of the form $-\beta q$ being introduced into the three tractions: see our Arts. 869, 875.

Neumann in fact makes the following remark in his *Einleitung*, p. 9:

Uebrigens, obgleich ich seit vielen Jahren im Besitz dieser Gleichungen bin, hat Duhamel, der seinerseits zu denselben Gleichungen gekommen ist, die Priorität ihrer Publikation. Diese Gleichungen, welche, wie aus dem Folgenden erhellen wird, bei mir nur einen besondern Fall von viel allgemeineren Gleichungen bilden, können unmittelbar auf krystallinische Medien angewandt werden, nur müssen dann für die Molekular-Krafte die auf krystallinische Medien sich beziehenden Ausdrücke derselben gesetzt werden.

There is a sort of covert claim to priority here which is not very happy.

[1197] Pp. 100—110 treat the case of a sphere. We have first worked out as in Duhamel's memoir (see our Art. 871) the relation between stress and temperature when the latter is a function only of the central distance. Neumann then applies his earlier results (see our Art. 1195) to the discussion of the fringes produced when a polarised ray is passed through the sphere. The investigation has however only very remote bearing on our present topic.

[1198.] Neumann commences an important application of his previous work on p. 110; namely, the investigation of the stress and the corresponding colour-fringes for a thin plate, in which there is a non-uniform distribution of temperature, supposed however constant throughout the thickness of the plate at any point. The equations obtained are the same as those for an elastic membrane or plate stretched in its own plane, the contour-load and the body-forces being replaced by terms involving the temperature. Thus taking Cauchy's equations (69), and (71) of Art. 640, using Duhamel's results in Art. 875 and remembering that we are now dealing with the flow of heat in *two*, not three dimensions, we have

$$P' = P = -\beta q;$$

hence, the thermo-elastic traction at any point [Eqn. (71)]

$$= P/\kappa - P' = \beta q\,(1 - 1/\kappa),$$

and we must take

$$X_0 = -\beta \frac{dq}{dx}(1 - 1/\kappa), \qquad Y_0 = -\beta \frac{dq}{dy}(1 - 1/\kappa).$$

It follows that for the body shift-equations we have putting

$$\xi_0 = u, \ \eta_0 = v:$$

substituting for Cauchy's $\Omega$ and $\kappa$ their values in terms of the $\lambda$ and $\mu$ of our work, and reducing:

$$\left.\begin{aligned} 4(\lambda+\mu)\frac{d^2u}{dx^2}+(\lambda+2\mu)\frac{d^2u}{dy^2}+(3\lambda+2\mu)\frac{d^2v}{dxdy}&=2\beta\frac{dq}{dx}\\ 4(\lambda+\mu)\frac{d^2v}{dy^2}+(\lambda+2\mu)\frac{d^2v}{dx^2}+(3\lambda+2\mu)\frac{d^2u}{dxdy}&=2\beta\frac{dq}{dy}\end{aligned}\right\}\text{...(vi).}$$

And for the contour-equations:

$$\left.\begin{aligned} \left(4(\lambda+\mu)\frac{du}{dx}+2\lambda\frac{dv}{dy}\right)\cos\alpha+(\lambda+2\mu)\left(\frac{du}{dy}+\frac{dv}{dx}\right)\sin\alpha&\\ =2\beta q\cos\alpha&\\ \left(4(\lambda+\mu)\frac{dv}{dy}+2\lambda\frac{du}{dx}\right)\sin\alpha+(\lambda+2\mu)\left(\frac{dv}{dx}+\frac{du}{dy}\right)\cos\alpha&\\ =2\beta q\sin\alpha&\end{aligned}\right\}\text{...(vii).}$$

These agree with Neumann's equations (7) and (9) on pp. 113 and 114, if we suppose uni-constant isotropy, i.e. $\lambda=\mu$, $=k$ in Neumann's notation, and write for our $\beta$ its value in Neumann's notation $=p$. Neumann also writes $s$ for our $q$, and $\nu$ for Cauchy's $\alpha$; Cauchy's $\beta$ is of course $\pi/2-\alpha$.

[1199.] These equations Neumann transforms to polar coordinates (pp. 114—115) considering what the contour-conditions become in the special cases of a circular and of an elliptic contour (pp. 115—116). The reader will find no difficulty in deducing these equations which in the case of a circular contour take simple forms.

In the notation of this book, I find for the body shift-equations in polar coordinates,

$$\left.\begin{aligned} 2\beta\frac{dq}{dr}&=4(\lambda+\mu)\left(\frac{d^2u}{dr^2}+\frac{1}{r}\frac{du}{dr}-\frac{u}{r^2}\right)+(\lambda+2\mu)\frac{1}{r^2}\frac{d^2u}{d\phi^2}\\ &\quad+(3\lambda+2\mu)\frac{d^2v}{rdrd\phi}-(5\lambda+6\mu)\frac{1}{r^2}\frac{dv}{d\phi}\\ 2\beta\frac{dq}{rd\theta}&=(\lambda+2\mu)\left(\frac{d^2v}{dr^2}+\frac{1}{r}\frac{dv}{dr}-\frac{v}{r^2}\right)+4(\lambda+\mu)\frac{d^2v}{d\phi^2}\\ &\quad+(3\lambda+2\mu)\frac{1}{r}\frac{d^2u}{drd\phi}+(5\lambda+6\mu)\frac{1}{r^2}\frac{du}{d\phi}\end{aligned}\right\}\text{...(viii);}$$

and for the contour-equations in the case of a circle:

$$\left.\begin{aligned}\beta q &= 2(\lambda+\mu)\frac{du}{dr}+\lambda\left(\frac{u}{r}+\frac{1}{r}\frac{dv}{d\phi}\right)\\ \frac{1}{r}\frac{du}{d\phi}&+\frac{dv}{dr}-\frac{v}{r}=0\end{aligned}\right\}\ \ldots\ldots\ (\text{ix}),$$

where $r$, $\phi$ are the polar coordinates of the point, the shifts of which parallel and perpendicular to the radius-vector $r$ are $u$ and $v$ respectively. These equations reduce to Neumann's (p. 115), if we suppose $\lambda=\mu$.

[1200.] On the supposition that the plate is very thin we may suppose no sensible rotation of the plane of polarisation and shall then have from equation (iii) of Art. 1189 for the retardation:

$$=\int(1/O-1/E)\,d\tau=(1/O-1/E)\,\tau=\tau\,(p-q)\,(s-s')/V^2,$$

where $\tau$ is the thickness of the plate. It remains to find $s-s'$.

Here $s$ and $s'$ are obviously the principal stretches in the plane of the plate. Now obviously the invariants of the stretch-conic for that plane are

$$I_1=s_r+s_\theta=s+s'$$
$$I_2=\sigma_{r\theta}^2-4s_rs_\theta=-4ss'.$$

Where

$$\left.\begin{aligned}s_r=\frac{du}{dr},\qquad s_\theta&=\frac{u}{r}+\frac{1}{r}\frac{dv}{d\phi}\\ \sigma_{r\theta}=\frac{1}{r}\frac{du}{d\phi}+\frac{dv}{dr}&-\frac{v}{r}\end{aligned}\right\}\ \ldots\ldots\ldots\ldots\ldots(\text{x}).$$

Hence

$$s-s'=\sqrt{(s_r+s_\theta)^2+\sigma_{r\theta}^2-4s_rs_\theta}$$
$$=\sqrt{(s_r-s_\theta)^2+\sigma_{r\theta}^2},$$

which might have been written down at once since $s-s'$ is the maximum-slide.

Thus the retardation is measured by

$$\frac{p-q}{V^2}\,\tau\sqrt{(s_r-s_\theta)^2+\sigma_{r\theta}^2}\ \ldots\ldots\ldots\ldots\ldots\ldots\ (\text{xi}),$$

when we neglect the square of the strain. If we substitute from (x), this agrees with Neumann's Equation (I.) of p. 117, remembering that in his notation $O-E$ is the retardation of the extraordinary ray: see his (7) p. 79.

[1201.] Neumann then proceeds to apply these results to several fairly simple cases. Thus (1) to a circular plate with temperature symmetrical about the centre (pp. 117—125); (2) to a circular annulus (pp. 125—134). In this case a 'neutral zone' exists, the determination of which requires some rather complex analysis. Case (3) to which Neumann then turns is thus stated:

Ich werde in diesem §. die allgemeinen Gleichungen entwickeln für die Biegungen, welche ein sehr dünner und schmaler Kreisring oder ein Stück eines solchen erfährt, wenn die Temperaturvertheilung in ihm allein eine Funktion des Bogens ist. Diese Gleichungen enthalten sechs willkürliche Constanten, welche durch die verschiedenen Bedingungen, welche an den Enden des Ringbogens erfüllt werden müssen, ihre Bestimmungen erhalten. (p. 134.)

[1202.] The general analysis (pp. 134—145) is of a very interesting kind and involves one or two purely elastic theorems. Thus the expression under the radical of our equation (xi) is thrown into the form

$$\frac{1}{\mu^2}\left\{\left(\frac{\widehat{rr}-\widehat{\theta\theta}}{2}\right)^2+\widehat{r\theta}^2\right\},$$

which follows of course easily from our method of obtaining it: see Neumann's p. 144.

The stresses $\widehat{rr}$, $\widehat{\theta\theta}$, $\widehat{r\theta}$ are expanded in terms of ascending powers of the diameter of the section of the ring; the method is similar to that used by Poisson and Cauchy for the problem of the elastic plate: see our Arts. 479 and 632. The problem is really a thermo-elastic one although Neumann considers also the photo-elastic results. The solution involves six-constants, which Neumann determines in the following cases: (1) when both the terminals of the circular arc are fixed, (2) when one is fixed and the other loaded, (3) when the terminals are both attached to other bodies, i.e. as in the case of sextant, (4) when the circular arc is a complete ring and carried by $n$ spokes: see pp. 145—172. The latter case leads to a consideration of the stress in a 'spoke,' when the depth being constant the breadth, although small, is a function of its distance from the centre. The problem is here again solved by expanding the stresses in terms of ascending powers of the breadth of the spoke (pp. 156—167). I do not feel quite satisfied with the legitimacy of this expansion, nor see why the objections to Cauchy's method of treating the torsion problem (see our Art. 661), and p. 621, footnote, of Saint-Venant's edition of Navier's *Leçons*) do not also apply to this case. Neumann himself remarks:

Ich werde in einem spätern §. auf diesen Fall zurückkommen mit einer Analyse, welche die Entwickelbarkeit der Molekular-Componenten und der Temperatur nach den Potenzen von $y$ nicht voraussetzt. (p. 165.)

Neumann considers the special instances when the stress is produced (*a*) only by non-uniform temperature in the spoke; (*b*) only by a tractive load in the direction of the axis, the temperature being uniform; (*c*) only by a shearing load producing flexure at one terminal the temperature being uniform. He refers to Brewster's memoir of 1816 for a confirmation of his calculation of the fringes in the case (*a*).

[1203.] In § 18, we have a discussion of the case of two long thin rectangular plates of different substances cemented together along two edges at some definite temperature; at other temperatures the combination will be bent, and if transparent exhibit fringes. This combination is similar to that made use of in Breguet's metallic thermometer (pp. 172—185).

The investigation is of an extremely interesting kind, and we wish that our space would allow of its reproduction in full. Neumann determines the stresses and the shifts at all points of the two plates, on the assumptions that the plates possess uni-constant isotropy, and that this is really a case where it is possible to apply expressions for the stresses similar to those for a plate stretched in its own plane (see his pp. 173 and 113). He shews that the form taken by the common edge is that of a parabolic cylinder. It seems to me that it is the latus-rectum of this cylinder, rather than the diameter of a *grade Cylinderfläche*, which he gives at the bottom of p. 183. He also determines the deflection of the free end for a given temperature (p. 184). The two sets of fringes obtained in this case have each a neutral zone, and the fringes are all parallel to the curve formed by the common edge (p. 185).

[1204.] The following section is occupied with the consideration of a case which had played a considerable part in Brewster's memoir of 1816 (p. 114. Figs. 1—4 etc.). In this case a hot plate is placed with one edge upon a cold surface, or a cold plate with one edge upon a hot surface. The problem is to determine the resulting stresses at any point of the plate, and the corresponding system of fringes. Neumann remarks with regard to the problem:

Die Auflösung der Gleichungen von denen das Problem der innern Spannungen in einer solchen rechtwinklichen Platte bei ungleichförmiger Temperaturvertheilung abhängt, hat vollständig mir nicht gelingen wollen, indessen habe ich daraus in zwei Fällen, nämlich, wenn entweder die Breite der Platte in Beziehung auf ihre Höhe, oder umgekehrt ihre Höhe in Beziehung auf die Breite bedeutend ist, die innern Spannungen und die Gesetze der Farbenvertheilung im polarisirten Licht bis zur numerischen Berechnung abgeleitet. Die Schwierigkeit, welche ich nicht habe überwinden können, und welcher ich die Aufmerksamkeit eines Geometers zuwenden möchte, besteht in der Bestimmung der Coefficienten der Glieder einer Reihe, welche fortschreiten nach den Wurzeln einer transcendenten Gleichung. Solche Reihen sind häufig vorgekommen bei der Anwendung der Analysis auf physikalische Probleme, hier hat sich aber, zum ersten Male, wie ich glaube, der Fall dargeboten, wo sämmtliche Wurzeln der transcendenten Gleichung imaginär sind. Dieser Fall ist von allgemeinem Interesse, ein grosser Theil der Probleme in der Theorie der Elasticität, Akustik und Optik führt zu ähnlichen Reihen. (p. 185.)

[1205.] The investigation and the comparison of theory and experiment occupy pp. 186—229 of the memoir. The same somewhat doubtful process as we have referred to in Art. 1202, namely, of expanding some of the quantities which occur in the problem (in this case quantities akin to the stresses) in terms of integral powers of a variable coordinate, will be found to repeat itself on pp. 188 and 209.

A process of treating simultaneous differential equations of which the coefficients of the terms involving the differential coefficients are constant is given on pp. 189—193. The equations are of an infinitely high order and Neumann remarks:

Es ist mir nicht bekannt, dass das Verfahren ein solches System gewöhnlicher Differentialgleichungen mit constanten Coefficienten, wo auf der linken Seite eine beliebige Funktion der unabhängigen Variabeln sich befindet, irgendwo entwickelt ist. Ich werde daher ein allgemeines Verfahren hier auseinandersetzen. (p. 189.)

It is sufficient here to note the fact for the benefit of the pure mathematician.

For the comparison of his theory with experiment in the two cases, where he is able to approximate to the value of the stresses, Neumann cites results of Brewster, Fourier and Depretz: see pp. 202 and 223[1].

[1206.] The final pages of Neumann's memoir are entitled *Erläuterungen* (pp. 230—247). Their object is expressed in the following words:

Es hat mir zweckmässig geschienen, die in der Einleitung auseinandergesetzten Principien der Theorie der innern Spannungen, welche aus bleibenden Dilatationen in einem festen Körper entstehn, noch durch einige Formeln zu erläutern und einige ihrer einfachsten Anwendungen zu entwickeln. (p. 230.)

The part of the preface which describes briefly Neumann's theory is pp. 18—24.

[1207.] In the *Erläuterungen* the following process is suggested for the consideration of set. Let $S_r$ be the set part of the stretch in direction $r$ of which the direction-cosines are $\cos\alpha$, $\cos\beta$, $\cos\gamma$; let $\Sigma_{xy}$ be the set-slide parallel to $y$ of a face perpendicular to $x$, and $\Sigma_{yz}$, $\Sigma_{zx}$ have similar meanings. Then

$$S_r = S_x\cos^2\alpha + S_y\cos^2\beta + S_z\cos^2\gamma + \Sigma_{yz}\cos\beta\cos\gamma + \Sigma_{zx}\cos\alpha\cos\gamma + \Sigma_{xy}\cos\alpha\cos\beta.$$

The coefficients $S_x$, $S_y$, $S_z$, $\Sigma_{yz}$, $\Sigma_{yz}$, $\Sigma_{zx}$ are continuous or discontinuous functions of the coordinates. The value of these func-

[1] Neumann draws attention to a numerical error of Brewster's who has used the value $\frac{5}{16\cdot02} = \cdot312$ instead of its square $\frac{1}{10\cdot24}$. The slip appears to be on p. 355, line 8 from the bottom, where Brewster writes $D = \cdot312\,B^2$. His line above would give $D = \cdot312\,B$; possibly however this should be $D^2 = \cdot312\,B^2$. The square of $\frac{5}{16\cdot02}$ seems to be $\frac{1}{10\cdot26}$ and not $\frac{1}{10\cdot24}$ as Neumann has it: see *Edin. Trans.* Vol. VIII. p. 355 *et seq.*

tions must be deduced from the circumstances which produce the set. Neumann terms the three greatest values of $S_r$ the 'principal sets' (*die bleibenden Hauptdilatationen*).

[1208.] We next require a postulate as to the relation between set and elastic strain. Neumann proposes the following:

**If the set arises from elastic strain, owing to the elastic limit being exceeded, the principal sets can be taken, if this limit be not much exceeded, as linear functions of the principal elastic stretches. (p. 230.)**

Thus if the constants in these linear functions be determined by experiment, and the consideration that the principal sets and the principal elastic stretches take place in the same directions, then the above expression for $S_r$ is fully determined.

It must however be noted that Neumann's postulate for the set does not agree with Gerstner's Law, which makes the set vary as the square of the elastic strain, nor is it quite in accord with Hodgkinson's experimental value for the set given in the Iron Commissioner's *Report*: see our Arts. 806, 969 and 1411. It seems probable that for some materials set does not vary as the elastic strain even for small sets.

[1209.] If the elastic stretch be the same in all directions and equal to $s$, then $S_r$ will be a function of

$$s - \beta/5\mu \,.\, q \mp s_0,$$

where $q$ is the temperature, $\beta/5\mu$ the stretch produced by unit increment of temperature, and $s_0$ the elastic limit, the negative or positive sign being given to $s_0$ according as $s - \beta/5\mu \,.\, q$ is positive or negative. $s_0$ is a quantity to be determined by experiment, generally depending however on $s$, and perhaps on the sign which must be given to $s_0$ itself. Hence we may put

$$S_r = \nu\,(s - \beta/5\mu \,.\, q \mp s_0),$$

where $\nu$ is a constant which has only a value when $s - \beta/5\mu \,.\, q$ is greater than $s_0$.

[1210.] The next step is to express the stresses due to the set in terms of the components of set. These are the 'initial stresses' in the meaning of our Arts. 616 and 666. If we represent them by $\widehat{xx}_0$, $\widehat{yy}_0$, $\widehat{zz}_0$, $\widehat{yz}_0$, $\widehat{zx}_0$, $\widehat{xy}_0$, Neumann puts in our notation and on the theory of uni-constant isotropy,

$$\widehat{xx}_0 = \mu (3S_x + S_y + S_z),$$
$$\widehat{yz}_0 = \mu \Sigma_{yz},$$

with similar values for the other set-stresses[1]. This step seems to me of a somewhat doubtful character; for, even supposing uni-constant isotropy, I do not understand why the constant should be the same for both set and elastic strain. Neumann then gives body and surface stress-equations of the types:

$$o = X + \frac{d(\widehat{xx} + \widehat{xx}_0 - \beta q)}{dx} + \frac{d(\widehat{xy} + \widehat{xy}_0)}{dy} + \frac{d(\widehat{xz} + \widehat{xz}_0)}{dz},$$
$$X_0 = (\widehat{xx} + \widehat{xx}_0 - \beta q)\, l + (\widehat{xy} + \widehat{xy}_0)\, m + (\widehat{xz} + \widehat{xz}_0)\, n,$$

where $l, m, n$ are the direction-cosines of the normal to the element of surface and $X, Y, Z, X_0, Y_0, Z_0$ the components respectively of body-force and load.

The equations for the special case of a stretch-set $S$ uniform in all directions at a point are deduced from these. In this case

$$\widehat{xx}_0 = \widehat{yy}_0 = \widehat{zz}_0 = 5\mu S, \quad \widehat{zy}_0 = \widehat{xz}_0 = \widehat{yx}_0 = 0.$$

Hence

$$\beta \frac{dq}{dx} - X - 5\mu \frac{dS}{dx} = \frac{d\widehat{xx}}{dx} + \frac{d\widehat{xy}}{dy} + \frac{d\widehat{xz}}{dz},$$
$$X_0 + (\beta q - 5\mu S)\, l = \widehat{xx} l + \widehat{xy} m + \widehat{xz} n,$$

are the types of body and surface stress-equations (p. 232).

[1211.] If $u, v, w$ be the absolute shifts, Neumann terms $\frac{du}{dx}$, $\frac{dv}{dy}$, $\frac{dw}{dz}$, $\frac{du}{dy} + \frac{dv}{dx}$, etc., the *absolute* strains; the *relative* strains $\frac{dU}{dx}$, $\frac{dV}{dy}$, $\frac{dW}{dz}$, $\frac{dU}{dy} + \frac{dV}{dx}$, etc., are then given by equations of the types:

$$\frac{dU}{dx} = \frac{du}{dx} - S_x, \qquad \frac{dU}{dy} + \frac{dV}{dx} = \frac{du}{dy} + \frac{dv}{dx} - \Sigma_{xy}.$$

[1] Neumann takes throughout his memoir, *negative* tractions (pressures) with a *positive* sign.

These expressions for $dU/dx$, $dU/dy + dV/dx$, etc., are to be substituted in the formulae referred to in our Art. 1195 to get the effect of the strain on polarised light. Thus, if $S_x = S_y = S_z = S$ and the set-slides are all zero, the set disappears and the formulae remain unaltered.

[1212.] Neumann applies these formulae to calculate the set developed by the rapid cooling of a glass sphere (*Härtung einer Glaskugel*), pp. 233—240, and of a long right-circular cylinder of glass, pp. 240—247. The results are of a complex kind and not given in a form adapted to calculation. Their physical value seems to me somewhat doubtful, as I do not feel convinced of the correctness of Neumann's theory of set.

[1213.] The whole memoir deserves, however, very careful study; much of it might be expanded and rewritten in a somewhat more general form. It is one of the most important researches in our subject, since Poisson's great memoir of 1829, and indeed forms the chief contribution to both thermo-elasticity and photo-elasticity published before 1850.

1214. *Su le condizioni di equilibrio di una corda attorta e di una verga elastica sottile leggiermente piegata, memoria del Dottor Gaspare Mainardi.* This memoir is published in the *Memorie di matematica e di fisica della Società Italiana*...Modena, 1841; it occupies pp. 237—252. It was received on May 9, 1840. The memoir does not involve any of the modern theory of elasticity, so that a brief notice of it will suffice.

The first part of the memoir relates to what is called the *torsion balance;* this occupies pp. 237—246. The memoir begins by alluding to the important use made of the torsion balance by Coulomb, Cavendish and Gauss. The force of torsion was universally admitted to be proportional to the angle of torsion, and Gauss held it to be probable that the coefficient expressing the ratio would consist of two parts, one proportional to the weight stretching the cord, and the other to the number of threads forming the cord and to the tension which a thread could support. Mainardi proposes to investigate the admissibility of Gauss's conjecture. The process is one of approximation, is not of an inviting character, and is very badly printed; up to the end of p. 243 the mistakes may be corrected by a careful reader without much trouble, but after this they are so numerous as to render the investigation worthless.

1215. The second part of the memoir relates to the equilibrium of an elastic rod; it occupies pp. 246—252. The problem is treated on two special assumptions. One relates to the nature of the change of position of the particles of the rod produced by the action of the forces to which the rod is exposed. According to Mainardi this change amounts to supposing that a transverse section of the rod undergoes a translation and rotations about two axes, one in the transverse section and one at right angles to it: but his process seems really to use only the two rotations and not the translation. The second assumption is that the molecular force arising from the relative displacement of two particles acts along the line of this displacement and is proportional to the displacement. The whole process is obscured by mistakes or misprints, and seems to me of no value.

The following note is given on page 249:

Sequendo i principj della Meccanica molecolare, dietro le tracce dei chiar. sig. Poisson e Cauchy, facilmente si tratta il problema con maggiore generalità, mà non ho voluto recare qui un calcolo, la cui prolissità non è compensata dall' importanza dei risultati.

On the last page of the memoir Mainardi alludes to the unsatisfactory part of an investigation by Poisson which has already come under our notice: see our Arts. 571, 935 and 1601—1608. Part of Mainardi's process consists of a purely analytical proposition, which we will give.

Let $a$, $b$, $c$ be direction-cosines of one straight line, $a_1$, $b_1$, $c_1$ those of a second, and $a_2$, $b_2$, $c_2$ those of a third; suppose the original axes rectangular, and also the three straight lines mutually at right angles: then will

$$da_2 = a_1(a_1 da_2 + b_1 db_2 + c_1 dc_2) + a(a da_2 + b db_2 + c dc_2).$$

For the right-hand member

$= (a_1^2 + a^2)\, da_2 + (a_1 b_1 + ab)\, db_2 + (a_1 c_1 + ac)\, dc_2$

$$= (a_1^2 + a^2)\,da_2 + (a_2b_2 + a_1b_1 + ab)\,db_2 + (a_2c_2 + a_1c_1 + ac)\,dc_2 - a_2b_2db_2 - a_2c_2dc_2$$
$$= (a_1^2 + a^2)\,da_2 - a_2\,(b_2db_2 + c_2dc_2) = (a_1^2 + a^2 + a_2^2)\,da_2$$
$$= da_2.$$

Of course other similar formulae also exist.

Mainardi thus deduces fairly from his equations the following result in his notation:

$$\frac{dT}{ds} + \frac{a_2da_1 + b_2db_1 + c_2dc_1}{ds}\,U + \frac{a_2da + b_2db + c_2dc}{ds}\,V = 0.$$

Thus it would follow that we cannot have $\frac{dT}{ds} = 0$ unless a certain condition holds; but Mainardi, without any warrant, breaks up this condition into two, and says that we cannot have $\frac{dT}{ds} = 0$ unless either $U = 0$ and $V = 0$, or $\frac{a_2da_1 + b_2db_1 + c_2dc_1}{ds} = 0$ and $\frac{a_2da + b_2db + c_2dc}{ds} = 0$.

(The notation is not good. $A$ is used to denote a point, while on p. 248 it is the coefficient of $\tau$. $M$ is used for a point, while on p. 250 it represents a force. On p. 248 a certain length is denoted by $\tau$, while on p. 251 this letter is put for $A\theta$.)

[1216.] Ignace Giulio. *Expériences sur la résistance des fers forgés dont on fait le plus d'usage en Piémont. Memorie della reale Accademia delle Scienze di Torino,* Serie II. Tomo III. pp. 175—223. Turin, 1841. The paper was read July 5, 1840.

These experiments have principally a local and temporary value as a comparison in regard to elasticity and strength of the kinds of iron generally used at the date of the memoir in Italy. Of the general conclusions on p. 204 only the following three seem to me of general physical interest:

1°. Resistance to flexure is greater for circular than for bars of square section, and the mean value of the ratio of the resistance of these two forms is about 35 : 33.

2°. The resistance to rupture, on the other hand, is greater for bars of square than of circular section, the mean value of the ratio of the resistances differing little from 19 : 18.

The 'resistance to rupture' was calculated from flexure experiments, and is taken to be the average stress across unit area of section when rupture begins[1]. The resistance to flexure is said to be greater for circular bars, because the stretch-modulus as calculated from the flexure of such bars was found to be greater than the stretch-modulus obtained from the flexure of square bars.

3°. The elastic line obtained by experiments on cylindrical or prismatic bars with supported terminals, and with different positions of transverse load, was found to be in accordance with the Bernoulli-Eulerian theory that the moment of the elastic reaction is proportional to the curvature.

[1217.] *Expériences sur la force et sur l'élasticité des fils de fer. Ibid.* pp. 275—434. This paper was read December 20, 1840.

This paper has first like the last a local object, namely to collect statistics with regard to the iron-wire in local use, and secondly a wider aim, the consideration of Gerstner's Law. The latter part only has general physical interest. We note some of Guilio's conclusions:

1°. He holds that the duration of tractive load between fairly extended limits (2 or 3 minutes to 10 or 15 hours) makes little or no difference in the stretch given to an iron-wire.

Cette proposition, qui ne s'applique cependant qu'aux tensions qui ne sont pas de très-peu inférieures à celle qui produit la rupture, a déjà été démontrée par les expériences de M. le Colonel Dufour (see our Art. 692).

2°. In iron-wires set begins with almost the first tractive loads. Giulio thus confirms one of Gerstner's results: see our Art. 804.

[1] This method of measuring the resistance to rupture seems somewhat arbitrary. It neglects the change in sign of the stress. If $2d$ be the sectional diameter in the plane of flexure, $T_0$ the traction which will produce rupture in the bar subjected only to longitudinal traction, I find for $R$, Giulio's resistance to rupture, $R = T_0 \bar{x}/d$, $\bar{x}$ being the distance of the centroid of the half of the section above the neutral axis from the neutral axis. Hence the theoretical ratio is $3\pi : 8$, which differs from Giulio's 19 : 18.

3°. He does not find Gerstner's formula for the set verified (see our Art. 806). He remarks:

Je me crois donc autorisé à conclure, que l'équation proposée par M. de Gerstner pour exprimer la loi des allongements des fils élastiques sous des tensions croissantes, depuis zéro, jusqu'à la tension qui produit la rupture, donne des résultats, qui s'écartent sensiblement de ceux auxquels on parvient, en opérant sur des fils de fer *non choisis et tels qu'on les emploie dans les arts, et qu'on les trouve communément en commerce.* Je m'abstiendrai dans ce moment de chercher à expliquer ces différences, dont la cause ne paraît pouvoir être parfaitement éclaircie que par de nouvelles expériences, faites sur des fils de quelque matière beaucoup plus extensible que le fer (p. 431).

We must remark here the absence of any consideration of the disturbing factor due to variability of the yield-point; it is also probable that Gerstner's Law does not extend to the plastic stage, i.e. only from the yield-point to the point of stricture.

[1218.] *Sur la torsion des fils métalliques et sur l'élasticité des ressorts en hélices. Ibid.* Tomo IV. 1842, pp. 329—383.

The memoir opens with reference to the labours of Coulomb, Chladni, Savart, Poisson, and Duleau. Giulio's first experiments are upon torsion, and he makes use of the method of torsional vibrations. He assumes however the relation between the slide- and stretch-moduli which is based upon uni-constant isotropy, namely in our notation $E/\mu = 5/2$. This is hardly true for an iron-wire which has probably a cylindrical distribution of elasticity. Giulio finds for the value of $n/n'$ the quantity 1·5219, which is about the mean of those obtained by Chaldni and Savart and does not differ much from that obtained on the hypothesis of uni-constant isotropy (1·5811): see his p. 340, and our Art. 470.

[1219.] The second part of Giulio's memoir is devoted to springs in the form of helices (*ressorts à boudin*). Of this pp. 341—347 contain a theory of such springs, and the rest (pp. 347—383) is occupied with experimental detail. Giulio adopts the hypothesis of Mossotti that the points primitively on a generator remain on a generator (see our Art. 249). We reproduce the main points of his theory to serve as a comparison with those given by Mossotti, Saint-Venant, and later by Thomson and Tait.

Let $a_0$ be the radius of the cylinder on which the helix lies, $h_0$ the distance measured along a generator of this cylinder between two turns of the spiral, $l$ the length of a turn of the spiral, and $n$ the number of turns, $s$ the arc measured from the lower terminal $A$ to any point $P$ on the helix, $d\phi_0$ the angle between the osculating plane to the helix at $A$, and the consecutive osculating plane at $A'$, where $AA'$ equals $ds$, $R_0$ the radius of curvature; and $H$ the load parallel to the axis of the cylinder, supposed to be the only load on the upper terminal. Let the same letters with the subscripts removed denote the like quantities for the new helix which on Mossotti's supposition is the form taken by the old helix when strained.

Giulio proceeds as follows; he neglects the effect produced by direct traction in stretching the wire, and he does not note that unless the wire were of equi-momental section (possessed *inertial isotropy* in Saint-Venant's sense of the words: see our Art. 1602) he ought to introduce a term depending on the change in angle between the radius of curvature and a principal axis of the section at any point. With these limitations and one to be later noted his method seems to me legitimate and agrees as a particular case with Saint-Venant's work.

[1220.] Since $$R = \frac{l^2}{2\pi\sqrt{l^2-h^2}}, \qquad \frac{d\phi}{ds} = \frac{2\pi h}{l^2} \quad \text{............(i)},$$

$$\delta\frac{1}{R} = \frac{-2\pi h\,\delta h}{l^2\sqrt{l^2-h^2}}, \qquad \delta\frac{d\phi}{ds} = \frac{2\pi\,\delta h}{l^2} \quad \text{.........(ii)}.$$

Let $\epsilon$ be the *moment d'élasticité de flexion*, $\alpha$ the *moment de l'élasticité de torsion*, then Giulio obtains the equation of virtual moments:

$$Hn\delta h = \epsilon\int_0^{nl}\left(\frac{1}{R}-\frac{1}{R_0}\right)\delta\frac{1}{R}\,.\,ds + \alpha\int_0^{nl}\left(\frac{d\phi}{ds}-\frac{d\phi_0}{ds}\right)\delta\frac{d\phi}{ds}\,.\,ds.$$

Integrating, we easily obtain after dividing by $n\delta h$:

$$H = \epsilon\left(\frac{1}{R_0}-\frac{1}{R}\right)\frac{2\pi h}{l\sqrt{l^2-h^2}} + \alpha\left(\frac{d\phi}{ds}-\frac{d\phi_0}{ds}\right)\frac{2\pi}{l} \quad \text{......(iii)}.$$

By using (i):

$$\frac{l^3H}{4\pi^2} = \epsilon h \frac{\sqrt{l^2 - h_0^2} - \sqrt{l^2 - h^2}}{\sqrt{l^2 - h^2}} + \alpha (h - h_0) \text{.........(iv).}$$

If $h/l$, as will usually be the case, is so small that its fourth power may be neglected,

$$\frac{l^3H}{4\pi^2} = \epsilon h \frac{h^2 - h_0^2}{2l^2} + \alpha (h - h_0) \text{..................(v).}$$

If we may neglect the cubes of $h/l$,

$$\frac{l^3H}{4\pi^2} = \alpha (h - h_0) \text{........................(vi);}$$

a result which shews us that spiral springs act principally by torsion: see Arts. 175, 250 and 1382.

[1221.] Now Giulio supposes the section to be circular and of radius $r$, and thus writes:

$$\left.\begin{array}{l} \epsilon = \frac{1}{4} E\pi r^4 \\ \alpha = \frac{1}{5} E\pi r^4 \end{array}\right\} \text{........................(vii).}$$

This latter result is obtained on the assumption that $\mu/E = 2/5$. Equations (v) and (vi) thus become respectively

$$H = \frac{4\pi^3 E r^4}{5l^3} (h - h_0) \left\{1 + \frac{5}{8} \frac{h(h + h_0)}{l^2}\right\} \text{.........(v)',}$$

$$H = \frac{4\pi^3 E r^4}{5l^3} (h - h_0) \text{ ..................... .........(vi)'.}$$

Giulio throws this last equation into the form

$$E = \frac{5l^3}{4\pi^3 r^4} \cdot \frac{H' - H}{h' - h},$$

by supposing another load $H'$ and a corresponding turn-distance $h'$. Finally, since $h$ is small, we may put $l = 2\pi a$ nearly, or

$$E = \frac{10a^3}{r^4} \frac{H' - H}{h' - h} \text{.....................(viii).}$$

This is the formula which Giulio uses to find the stretch-modulus of a wire in the form of a helicoidal spring. We note that it depends on the assumptions of isotropy and of uni-constant isotropy. A better result would be

$$\mu = \frac{4a^3}{r^4}\frac{H' - H}{h' - h},$$

as an equation to find the slide-modulus.

See pp. 341—345 of the memoir.

[1222.] On pp. 345—347 Giulio considers the lower terminal fixed and a couple $C$ applied to the upper terminal in a plane perpendicular to the axis of the cylinder. In this case, if $n$ become $n + \delta n$, the virtual moment of the couple is $2\pi\delta nC$, and we have in place of (iii) the equation

$$2\pi\delta nC = nl\epsilon\left(\frac{1}{R} - \frac{1}{R_0}\right)\delta\frac{1}{R} + nl\alpha\left(\frac{d\phi}{ds} - \frac{d\phi_0}{ds}\right)\delta\frac{d\phi}{ds} \quad \text{...(ix).}$$

Now Giulio supposes that the length of the axis $b$ cannot vary and that the whole length of the spiral $nl = L$. Then

$$a = \frac{\sqrt{L^2 - b^2}}{2\pi n}, \qquad R = \frac{L^2}{2\pi n\sqrt{L^2 - b^2}}, \qquad \frac{d\phi}{ds} = \frac{2\pi nb}{L^2} \quad \text{...(x).}$$

Whence it easily follows,

$$\delta\frac{1}{R} = \frac{2\pi\sqrt{L^2 - b^2}}{L^2}\delta n, \qquad \delta\frac{d\phi}{ds} = \frac{2\pi b}{L^2}\delta n \quad \text{.........(xi).}$$

Applying (x) and (xi) to (ix) we find:

$$\frac{CL^3}{2\pi} = \{\epsilon(L^2 - b^2) + \alpha b^2\}(n - n_0) \quad \text{...........(xii).}$$

From the values (vii) by substitution

$$E = \frac{10L^3}{\pi^2 r^4(5L^2 - b^2)}\frac{C}{n - n_0}.$$

Now $b/L$ will usually be small, and if two couples $C'$ and $C$ are taken corresponding to $n'$ and $n$, we find

$$E = \frac{2L}{\pi^2 r^4}\frac{C' - C}{n' - n}.$$

See p. 347.

[1223.] Giulio modifies the result (xii) by giving $\epsilon$ and $\alpha$ their values for a rectangular section. In this case he adopts Cauchy's value for $\alpha$ (see our Art. 661), but he omits to notice the third term which must now be introduced: see Saint-Venant's consideration of the same problem in our Arts. 1599—1602. Formula (vi) coincides with Mossotti's assumption as to the proportionality of vertical load and vertical shift: see Art. 250.

[1224.] The remainder of the memoir is of considerable physical interest, being an experimental discussion of set and elastic after-strain in helicoidal springs. Giulio's results fully bear out the Coulomb-Gerstner Law, that the elasticity after set remains nearly the same as before[1]. The duration of the load does not affect the amount of the *immediate* loss of strain on removing the load, in other words the duration affects only the greatness of the set. There arises however a difficulty here, for what Giulio calls set (*allongement permanent*) appears to have been in great part elastic after-strain:

On voit encore que ce que j'ai nommé jusqu'ici *allongement permanent* est loin de mériter ce nom d'une manière absolue, puisqu'il disparait en grande partie après un temps suffisamment long, et se reproduit alors sous des tensions suffisantes et assez longtemps continuées.

See pp. 354, 356, 360.

We see again the great need for a careful experimental distinction between elastic after-strain and after-set. The important experiments of Gerstner and Hodgkinson, which tend to shew that set varies as the square of the stress, lose somewhat in value owing to this omission: see our Arts. 806, 969.

[1225.] We may cite Giulio's concluding remarks as especially instructive:

L'altération de forme produite par l'action d'une force extérieure sur un corps élastique se compose de deux parties: l'une indépendante

[1] Giulio remarks on Coulomb's claim to be considered as discoverer of this law: On pourrait conclure de quelques expressions de Coulomb, et surtout de quelques-unes des expériences qu'il rapporte, qu'il était parvenu à la même loi; je ne la trouve cependant nulle part explicitement énoncée dans son mémoire. Footnote, p. 354, to a statement of Gerstner's Law. See, however, Note A, (2) in the appendix to this volume.

de la durée de cette action et sensiblement proportionnelle à son intensité; l'autre, croissant plus rapidement que la force qui la produit, et suivant une fonction de la durée de son action. La force extérieure venant à cesser la première partie de l'altération qu'elle avait causée dans le corps disparaît instantanément: la seconde persiste, mais en diminuant continuellement avec le temps. Une nouvelle force plus ou moins intense que la première vient-elle à son tour agir sur le même corps? Les mêmes effets se reproduisent avec une intensité qui dépend de l'intervalle de temps qui s'est écoulé entre l'action des deux forces, de l'intensité de la seconde, et de la durée de son action. Quelle est la fonction de la tension et du temps suivant laquelle ces altérations se produisent et disparaissent? et cette fonction est-elle la même quelle que soit la matière des corps où elles se produisent? Voilà des questions auxquelles, malgré les travaux très-remarquables de plusieurs physiciens, il ne paraît pas que l'on soit en état de répondre encore complètement.

Giulio then refers to the labours of Leslie[1], Gerstner, and Weber: see our Arts. 806, 707. He remarks that more investigations are required and suggests his own method of experimenting on helicoidal springs because of the accuracy of which this method appears capable (p. 361).

[1226.] Giulio's conclusions are remarkable, if true, but we must to a certain extent qualify them by his not quite exact theory of the helicoidal spring and by various difficulties he met with in experiment. They would tend to shew that fore-set disappears with time on the removal of the load, but that a body having been reduced to a state of ease, there will be no fore-set, only after-set, which in its turn disappears with time on the removal of the load. In other words elastic after-strain only requires time to produce it when the body has been reduced to its state of ease; with or without the state of ease it requires time to disappear. This is not opposed to Weber's results, for he had reduced his threads to a state of ease: see Art. 708.

[1227.] V. Regnault: *Note sur la dilatation du verre. Annales de Chimie et de Physique*, T. IV. 1842, pp. 64—67.

[1] With the usual vague reference; in this case to Poncelet, *Mécanique Industrielle*, p. 343: see our Art. 984.

This paper gives some account of the dilatation of glass, which Regnault found necessary to determine for his experiments on the dilatation of water. He shews that, contrary to the experiments of Dulong and Petit, the dilatation varies very greatly with different kinds of glass. The following concluding remarks of Regnault are of extreme importance as shewing the errors introduced by supposing glass vessels to be isotropic: see our Arts. 686—691, 925, 1323 and 1358.

Les différences que l'on remarque dans les dilatations du même tube de verre, lorsqu'il est sous forme de tube ou bien quand il est soufflé en boules de diverses grosseurs, ne paraissent assujetties à aucune loi simple......

Le même verre soufflé en boule paraît avoir un coefficient de dilatation d'autant plus grand que son diamètre est plus considérable, ou peut-être que l'épaisseur de ses parois est plus petite.

En tous cas, l'on voit combien on s'expose à se tromper dans des expériences précises, en calculant la dilatation d'un appareil de verre, d'après le nombre obtenu dans une expérience directe faite sur un tube ou sur une boule soufflée avec la même matière, et, à plus forte raison, d'après la dilatation linéaire observée sur une tige du même verre, comme l'ont fait plusieurs physiciens distingués (p. 67).

[1228.] N. Savart: *Recherches expérimentales sur l'influence de l'élasticité dans les cordes vibrantes. Annales de Chimie et de Physique,* T. VI. 1842, pp. 1—19.

Duhamel: *Remarque à l'occasion du Mémoire de M. le Colonel Savart sur les cordes vibrantes. Ibid.* pp. 19—21.

N. Savart points out a relation between the numbers of vibrations in a second of a cord supposed non-elastic and subject to a given traction, and elastic but not subject to a given traction. The matter belongs properly to the theory of sound, but we may remark that Duhamel shews that the law deduced by N. Savart from a long series of experiments is in one case an easy deduction from the mathematical theory of vibrating cords: see our Art. 1180.

[1229.] G. H. L. Hagen: *Die Elasticität des Holzes. Bericht... der k. Preuss. Akademie der Wissenschaften,* Berlin, 1842, pp. 316—319. See also the *Annales de Chimie,* T. XI. 1844, pp. 112—115.

This physicist made a considerable number of experiments on the stretch-moduli of various kinds of wood in the direction of the fibres and perpendicular to the fibres. The moduli were deduced from flexure experiments on the supposition that the moduli for stretch and squeeze are equal. Very considerable differences indeed were found between the moduli in different directions. For a direction making an angle $\phi$ with the fibres Hagen gives the empirical formula

$$E_\phi = \frac{E_0 E_{\pi/2}}{E_0 \sin^3 \phi + E_{\pi/2} \cos^3 \phi},$$

where $E_\phi$ is the stretch-modulus for the direction $\phi$. He states that this formula has been confirmed by numerous experiments. Compare Saint-Venant's remarks in his edition of Navier's *Leçons*, pp. 817—825.

[1230.] Two papers on the law of molecular force will be found in Vol. VII. of the *Cambridge Philosophical Transactions*. They are entitled:

*On Molecular Equilibrium*, Part I., by P. Kelland, pp. 25—59.

*On the Nature of the Molecular Forces which regulate the Constitution of the Luminiferous Ether*, by S. Earnshaw, pp. 97—112.

The first of these papers is based upon Mossotti's hypothesis of two systems of particles repulsive towards atoms of their own kind, but each respectively attractive towards atoms of the other. Kelland calls one system *caloric* and the other *matter*. He concludes, after a lengthy analysis, that all the known laws of attraction and cohesion can be explained by the Newtonian hypothesis. *He does not, however, like Belli apply his results to any numerical calculations,* and therefore it is not obvious that the hypothesis does satisfy these laws. He criticises Mossotti on p. 28, I think justly: see our Art. 840.

Earnshaw, taking only a single system of particles, shews like Cauchy that the molecular force is *not* the Newtonian.

[1231.] A criticism of Mossotti's method by R. L. E. (? Robert Leslie Ellis) is printed on pp. 384—387 of the *Philosophical Magazine*, Vol. XIX. 1841. The writer points out that Mossotti's equations involve the conception of fluid pressure, which is itself but a mode of molecular action. He insists on the necessity of distinct ideas on the connection between the theory of molecular action and the ordinary principles of equilibrium. Kelland replied to this paper (*Phil. Mag.*, Vol. xx. pp. 8—10) and suggests that the required pressure might be produced by 'molecular contact.'

[1232.] This paper of Kelland's was immediately followed by a controversy between Earnshaw, O'Brien and Kelland on the law of molecular force. The various stages of this controversy may be followed in the volumes of the *Philosophical Magazine* for the years 1842 and 1843.

We have seen that Belli had denied the sufficiency of the law of gravitation to explain cohesion. This view was also that of Cauchy and Earnshaw, who arrived at it by consideration of the unstable equilibrium of a system of particles attracting according to that law. On the other hand Mossotti and Kelland, by introducing a second system of particles repelling each other and attracting the first system, endeavoured to prove the sufficiency of the law of the inverse square to explain cohesion and the phenomena of light. The controversy belongs to a great extent to the history of the undulatory theory, but it seems to me by no means favourable to those who assert, even with the assistance of *two* media, the sufficiency of the law of gravitating force to explain cohesive phenomena.

[1233.] H. Moseley. *The Mechanical Principles of Engineering and Architecture*, London, 1843.

This book contains a chapter (pp. 486—585) on the *Strength of Materials* and one on *Impact* involving some problems in resilience (pp. 586—603).

Moseley's chief merit is the introduction into England of the methods of the French elasticians, notably Poncelet's conception of work.

The first chapter referred to contains a number of beam problems, similar to those treated by Navier in his *Leçons* (see our Art. 279); Navier's results are generally obtained by Moseley in a shorter and simpler manner. He does not however take any account of slide, and on p. 550 introduces Navier's modification of Coulomb's erroneous theory of rupture by contractive load: see our Arts .120 and 729. Further we have (pp. 561—568) the old absurd forms of beams of greatest strength, and the erroneous torsion theory of Cauchy reproduced (pp. 583—585). An elementary treatment of the danger to a suspension bridge of isochronous vibrations (p. 494), and another of the position of the neutral line in beams subject to load not wholly transverse (p. 497), may be noted. The problems on resilience in the chapter on Impact are based upon Poncelet, but there are one or two interesting special applications, e.g. to the driving of piles (pp. 598—603). This latter investigation follows Whewell, but the method seems due to Airy.

The book contains a good deal which might be useful even to the practical student of to-day, for many of the problems considered are not to be found in the ordinary text-books.

[1234.] Another work by Moseley, entitled: *Illustrations of Mechanics*, reached a fourth edition in 1848. It also contains some references to our subject, but being merely a student's text-book has no claim to originality or to a place in the history of elasticity.

1235. 1843. F. Brünnow. *De Attractione Moleculari...Auctor Franciscus Bruennow.*

This is an academical dissertation published at Berlin in 1843; it is in quarto, and consists of 31 pages, besides the title and dedication which occupy 6 pages at the beginning, and a life of the author with a list of theses which occupy 2 pages at the end.

The subject of elasticity is not formally mentioned, but the dissertation may be noticed as relating to the nature of the molecular forces by which stress is produced.

1236. Brünnow considers that there are two erroneous opinions which have been maintained with respect to molecular force. According to the one opinion the force between two particles varies inversely as the square of the distance, so that the law coincides with the Newtonian law of attraction; according to the other opinion the force between two particles varies inversely as some power of the distance higher than the square, the third or the fourth for example. Brünnow briefly notices the views of the following writers: Newton, Keil, Freind, Maupertuis, Madame du Chatelet, Sigorgne, Le Sage, Clairaut, d'Alembert, Buffon, Laplace, Belli, Mollweide, Fries, Jurin, Munke, J. Schmidt, Emmet, J. T. Mayer, and Bessel. He makes a few observations which he considers refute the above two erroneous opinions, and he maintains that nothing more at present can be said about molecular force than was maintained by Laplace; namely, that the force is insensible at finite distances, and very powerful at infinitesimal distances.

1237. Brünnow investigates some ordinary problems of attraction on his pp. 23—30, namely the attraction of a straight line on a particle, of a straight line on a certain parallel straight line, of a rectangle on a certain parallel rectangle, and of a cylinder on a particle situated on the prolongation of the axis. His results enable him to correct erroneous statements made by J. Schmidt and J. T. Mayer, who both held that molecular force varied according to the Newtonian law.

1238. *Programme d'une Thèse de Mécanique: Sur la résistance des solides élastiques; présenté à la Faculté des Sciences de Montpellier.* Par J. P. Aimé Bergeron, Montpellier, 1844. This *Programme* consists of six quarto pages, besides the title. It contains general statements as to resistance and rupture, which I presume were developed either in writing, or *vivâ voce,* in some academic formality.

1239. *Die Fortschritte der Physik im Jahre* 1845. This was published in 1847; it is the first volume of an important series which has been continued to the present time. I shall occasionally have to refer to it for notices of memoirs on our subject; but I have always read such memoirs for myself, and formed my own judgment on them, before consulting the volumes of this series. I do not cite the notices in this series unless they seem to demand special attention.

1240. I now take three papers in conjunction which belong

to the same subject, namely the integration of the equations for the equilibrium of an elastic curve of double curvature.

(i) *Mémoire sur l'intégration des équations de la courbe élastique à double courbure;* par M. J. Binet (Extrait). *Comptes rendus*, Vol. XVIII., 1844, pp. 1115—1119.

(ii) *Note sur l'intégration des équations de la courbe élastique à double courbure;* par M. Wantzel. *Ibid.* pp. 1197—1201.

(iii) *Réflexions sur l'intégration des formules de la tige élastique à double courbure;* par M. J. Binet. *Comptes rendus*, Vol. XIX., 1844, pp. 1—3.

Lagrange stated that the integration of certain differential equations was probably impossible: see Art. 159 of my notice of the *Mécanique Analytique*. Now Binet succeeded in effecting the integration, and (i) is an abstract of his memoir, giving without demonstrations the results which he had obtained. In (ii) Wantzel simplifies the method of Binet, and actually demonstrates the process of integration; much of this Note is reproduced by Bertrand in his edition of the *Mécanique Analytique*, 1853: see Vol. I. pp. 401—405.

In (iii) Binet makes a few remarks on the subject, in the course of which he recognises the merit of Wantzel's process, and holds that some words of Lagrange are no longer applicable:

Jusqu'à présent il ne paraît pas qu'on ait été plus loin dans la solution générale du problème de la courbe élastique.

Binet in his articles refers to Poisson's researches on the problem, and finds no fault with them: see Art. 571. The memoir of which Binet's first article is an abstract seems not to have been published; it was, I suppose, superseded by the simpler process of Wantzel.

[It should be remarked that Binet only obtained the first integrals—i.e. the values of the direction-cosines of the tangent to the elastic line—in the form of elliptic functions, the actual coordinates he left to be found by the method of quadratures. Wantzel on the other hand by a choice of simpler axes obtained the coordinates themselves in the form of elliptic functions. Binet in the third paper cited above questions whether Wantzel's results are expressed entirely by integrals which can be really termed elliptic functions, but this does not affect the fact that Wantzel has obtained a complete solution of the equations. It is necessary to note that these equations are not the most general conceivable, for they suppose the shifts to be small, the central axis in the unstrained state to be a straight line and the section of the wire to possess inertial isotropy. The very needful introduction of the moment about the radius of curvature in the case where the section does not possess inertial isotropy is due to Bellavitis and Saint-Venant: see our Arts. 935 and 1599. Wantzel shews also in his memoir that a straight wire will take the form of a helix, if acted upon by terminal couples whose planes make a constant angle with its central axis. We must add however to his statement the condition of inertial isotropy in the section. ED.]

1241. 1845. *Mémoire sur la Théorie des Corps élastiques, présenté à l'Académie des sciences le* 18 *août*, 1845; par M. Ossian Bonnet. This memoir is published in the *Journal de l'École polytechnique*, 30th Cahier, 1845; it occupies pp. 171—191. The memoir is briefly noticed in the *Comptes rendus*, Vol. XXI. 1845: see pp. 434 and 1389.

The object of the memoir is to establish in a direct manner without any transformation of coordinates the formulae obtained by Lamé for expressing the equations of elasticity in terms of curvilinear coordinates. The memoir after a brief introduction is divided into three sections.

1242. The first section is entitled: *Pressions qui sollicitent des surfaces orthogonales tracées d'une manière quelconque dans le corps*: this occupies pp. 172—184. The final results obtained correspond to the values of the stresses given in my Art. 1153, supposing uni-constant isotropy.

1243. The second section is entitled: *Coefficient de dilatation cubique;* this occupies pp. 185—186. The result obtained is the value of $\theta$ in the form given in my Art. 1153.

1244. The third section is entitled: *Equations de l'équilibre et du mouvement des solides élastiques rapportées aux pressions;* this occupies pp. 185—191.

The results obtained are the three equations analogous to that of my Art. 1150.

1245. With respect to the mechanical principles adopted we have to observe that the stresses are calculated by means of the general formula

$$\widehat{ns} = \frac{\rho}{2} \Sigma m\, r f(r) \cos(r, n) \cos(r, s).$$

Here $n$ is the normal to an elementary plane $\omega$ at the point $M$ where the density is $\rho$; $\Sigma$ denotes a summation with regard to the distance $r$ from $M$ of all molecules ($m$), so that $\frac{1}{2}\Sigma$ will denote the sum with regard to molecules on *one* side only of $\omega$; $(r, n)$ denotes the angle between a distance $r$ and the normal $n$, and $(r, s)$ the angle between $r$ and a fixed straight line $s$.

For this Bonnet refers to Saint-Venant and Cauchy in the *Comptes rendus*, Vol. XXI.: see our Art. 679. Bonnet also admits the relation

$$|yyyy| = 3\,|yyzz|$$

see our Art. 615, the Appendix Note B, and Moigno's *Statique*, p. 703. Bonnet refers to Cauchy's *Exercices de mathématiques*, Vol. III., page 201. By virtue of this relation in the case of an isotropic body the constants $\lambda$ and $\mu$ of our notation are taken to be equal.

1246. The instrument of investigation mainly employed by Bonnet is infinitesimal geometry; and a reader will probably find it necessary to exert strict attention in order to feel confidence in the accuracy of the successive steps. On the whole the process is of interest as confirming the results obtained by Lamé, but could not with advantage be substituted for that of Lamé.

1247. G. B. Airy. *On the Flexure of a Uniform Bar supported by a number of equal Pressures applied at equidistant points, and on the Positions proper for the applications of these Pressures in order to prevent any sensible alteration of the length of the Bar by small Flexure.* This is published in the *Memoirs of the Royal Astronomical Society,* Vol. XV., 1846, pp. 157—163; it was read

January 10, 1845. There is an abstract of it in the *Monthly Notices of the Royal Astronomical Society*, Vol. VI., 1845, pp. 143—146.

The title sufficiently indicates the nature of this memoir: the subject is of great practical importance, because it is necessary to support bars, which serve as standards of length, in such a manner as to secure them from any appreciable change of length through the operation of their own weight. The memoir makes no use of the modern theory of elasticity, but rests on assumptions which are thus stated on p. 160:

The fundamental assumptions for this investigation are, that the flexure is so small that the mere curvature of a neutral line will not produce a sensible alteration in its length; that the extension of a surface is proportional to the momentum of the bending force; and that, when the momenta are equal, the extension produced by a bending force downwards, and the contraction produced by a bending force upwards, are equal.

The result obtained is simple in form, and was used by Baily in his contrivance for supporting the national standard of length.

[1248.] W. Sullivan. *On Currents of Electricity produced by the vibration of wires and metallic rods. Philosophical Magazine*, Vol. XXVII., pp. 261—264. London, 1845. This paper contains some experimental evidence of the vibrations of wires and rods producing electric currents in them. The current did not depend upon the heat produced, as it at once ceased with the vibration. Its existence if *satisfactorily* proved would point to a relation between the elastic and electric properties of a body[1].

[1249.] Brix. *Ueber die Dehnung und das Zerreissen prismatischer Körper wenn die spannende Kraft seitwärts der Schwerpunkts-Achse wirkt. Verhandlungen des Vereins zur Beförderung des Gewerbfleisses in Preussen.* Jahrgang 24. Berlin, 1845, pp. 185—192.

This is a paper bearing on the same important point as the memoir of Tredgold considered in our Art. 832. Brix refers to Tredgold's paper in a footnote on p. 186, as the only one which, so

[1] Sullivan failed to repeat his experiments in the presence of Beatson: see my footnote, p. 720.

far as he is aware, has treated of this point; namely, the non-axial application of a terminal tractive load.

Brix obtains an expression for the traction $T$ in a 'fibre' at distance $x$ from that diameter of the section which is perpendicular to the line joining the point where the direction of the load cuts the section to the centroid of the section (of area $\omega$). The total load $P$ being applied at a distance $b$ from the axis, and $\kappa$ being the radius of gyration of the section about the above diameter, he finds:

$$T = \frac{P}{\omega}\left(1 \pm \frac{bx}{\kappa^2}\right).$$

Compare Duhamel and Tregold's results in our Arts. 815 and 832.

Hence, if $\sigma$ be the greatest stretch which can be given to the body without rupture (better without exceeding the elastic limit), we have, $x_0$ being the greatest value of $x$,

$$P = E\omega\sigma/\left(1 + \frac{bx_0}{\kappa^2}\right).$$

This agrees with the result we found in Art. 832, and practically with Tredgold's conclusions.

Brix applies this to the case of the load $P$ being attached to the bar by a hook. If the diameter ($2r$) of the bar, supposed of circular section, be equal to the opening of the hook, we have

$$b = 2r, \quad x_0 = r, \quad \kappa^2 = \frac{r^2}{4},$$

$$\therefore \quad P = \frac{1}{9} E\omega\sigma.$$

Or, the strength of a hook is in this case only *one-ninth* the absolute strength (better elastic strength) of the bar out of which it is made. See Grashof's *Theorie der Elasticität und Festigkeit*, Berlin, 1878, p. 156.

[1250.] Luigi Pacinotti and Giuseppe Peri. *Esperienze sulla resistenza elastica dei legni. Il Cimento, Giornale di Fisica, Chimica e Storia Naturale.* Pisa, 1845, pp. 241—297.

This contains an important series of experiments on wood. The memoir opens with some remarks on Wertheim's experiments on metals. We may cite the following lines:

Alcune particolari osservazioni di confronto fra il coefficiente d' elasticità, e la densità stessa dei metalli, dal prelodato Wertheim sottoposti all' esperienza, ci hanno indotti a tentare l'elasticità di alcuni legni di maggior uso cementandoli in tre modi diversi: colla *Flessione*, cioè; coll' *Allungamento*, e colla *Torsione*. (p. 241.)

The object of the writers was to ascertain with what exactness the theoretical formulae for flexure give the stretch-modulus; to calculate this modulus for the woods experimented on, and to state a general rule for determining its true value; finally to compare the diverse methods among themselves, to explain why traction experiments give more consistent results than flexure and torsion experiments, and to shew how these latter experiments may be made available.

[1251.] The experimental details are given in *Sezione* I. (pp. 243—265). These are followed by a comparison with theory in *Sezione* II. (pp. 265—295). We cannot, however, give much weight to the torsional results, for Pacinotti and Peri experimented only on wooden bars of square section and used the old Coulomb theory to calculate the slide-modulus $\mu$. The values of the stretch-modulus $E$, obtained (i) from traction experiments, (ii) from flexure experiments, were then substituted in the ratio $E/\mu$, and the nearness of the resulting ratio to 5/2 was taken as a measure of the exactness attained in these different modes of experimenting. For the traction experiments the mean value of $2E/\mu$ was 4·792, for the flexure experiments 5·637. It was therefore held that the traction-experiments were more exact than the flexure. But it must be remembered that the relation $E/\mu = 5/2$ cannot be conceived as holding for a substance like wood which is certainly aeolotropic, and further that the values of $\mu$ require modifying by Saint-Venant's factor, though this would affect both sets of results. Of course if the theory used by Pacinotti and Peri were exact the values of $E/\mu$ should be, although not equal to 5/2, the same for both sets of experiments. We must, however, point out that there was no consideration of slide in the flexure calculations. Similar experiments by Duleau and Giulio (see our Arts. 226 and 1218) which led to like results are referred to.

Quindi deducesi che nella teoria della flessione oltre al coefficiente d' elasticita deve introdurci qualche altro elemento, o che il coefficiente d' elasticità varia nei medesimi corpi secondo l' operazione che vogliamo sopra il solido produrre. (p. 295.)

[1252.] The authors reach among other the following general conclusions:

1°. Not only elastic strain before set, but elastic strain after set has begun (obtained by deducting set from the whole strain) is proportional to stress.

2°. The stretch-modulus is proportional to the specific gravity, or Young's height-modulus is nearly a constant for all kinds of wood: see our Art. 137.

This is the result which Bevan had previously obtained: see Art. 379. It is denied by Wertheim, who observes that the velocity of sound in all kinds of wood would then be the same, which is certainly not the fact: see our Art. 1312. We may remark however that this supposes the equality of the two specific heats, which Weber and others have shewn is hardly probable: see Arts. 705 and 882. The authors also allow this result to be only an approximate one; the height-modulus being *about* 2000 for the C. G. S. system.

3°. The stretch-modulus obtained by flexure must be multiplied by 6/7 and that obtained by torsion by 2·396 in order to deduce the true modulus supposed to be given by traction experiments. (The stretch-modulus obtained from torsion pre-supposes uni-constant isotropy.)

According to a final remark of Pacinotti and Peri we must use a different stretch-modulus for each kind of strain which can be given to the material. It will be seen at once that this reduces the theory of elasticity to pure empiricism.

[1253.] 1845. E. Lamarle. *Mémoire sur la flexion du bois. Annales des Travaux publics de Belgique*, III. pp. 1—64 and IV. pp. 1—36. Brussels, 1845 and 1846. The first part treats of transverse, the second of longitudinal load.

The object of the first part is to justify the ordinary theory of flexure, which, within the limits of elasticity, assumes the stretch and squeeze-moduli to be the same. The author commences by considering—to a closer degree of approximation than is to be usually found in the text-books of that period—what he terms circular and parabolic flexure, i.e. flexure produced by a terminal couple and flexure produced by a terminal shearing load. The results which he finds are not novel and will be found in the works of Heim or Saint-Venant: see our Arts. 905 and 1615. He supposes the squeeze- and stretch-moduli to be different and calculates upon this assumption the deflection for certain cases of triangular prismatic beams. Experimentally, however, beams of wood give a deflection coinciding with theory only on the hypothesis of the equality of the moduli.

[1254.] Lamarle arrives at the following conclusions:

1°. That the hypotheses upon which the ordinary theory of the transverse flexure of wood rests, and the calculations based upon them, agree in a very remarkable manner with experimental facts.

2°. That the resistances to contraction and extension ought for equal changes of length, within the limits of elasticity, to be considered as identical, or at least as not differing by a sensible quantity.

3°. That transverse flexure offers a simple, convenient and very precise means of determining the stretch-modulus.

4°. The author holds that the mean value of the stretch-modulus for any kind of wood will not give results without large error for any specimen of that wood (p. 63).

The conclusion 3° can hardly be accepted; it is an admitted fact that the stretch-modulus as obtained by flexure experiments differs sensibly from that obtained by traction experiments. In 4° we have a statement of the well-known fact that the stretch-modulus of wood depends upon the part from which and the direction in which it has been cut out of the tree, and upon its dryness, etc.

[1255.] The second part of Lamarle's memoir seems to me of more interest. It is a consideration of the Euler-Lagrangian theory of struts: see our Arts. 65 and 106. There are several points in this memoir worthy of notice. If $Q$ be the contractive load, $\epsilon$ Euler's moment of stiffness, $E$ the stretch-modulus and $\omega$ the section, Lagrange arrives at an equation of the form:

$$\epsilon/\rho = Qy \text{ .............................. (i).}$$

Lamarle points out that the true equation on the Bernoulli-Eulerian hypothesis is

$$\left(1 - \frac{Q}{E\omega} \cdot \frac{dx}{ds}\right) \epsilon/\rho = Qy \text{ .................... (ii).}$$

He then solves this equation by a series and determines the least value of $Q$ for which flexure is possible, and the deflection for any given value of $Q$. The modification thus introduced into Lagrange's results gives, however, only a factor negligible in most practical cases (pp. 33—36): see our Art. 959. On the other hand Lamarle does not seem to have noticed that this more correct form of the equation solves Robison, Tredgold and others' difficulty as to the form and position of the neutral line in the case of struts, or rods subjected to contractive load.

[1256.] In the consideration of a doubly-pivoted strut (of length $l$), Lamarle follows Lagrange closely, proceeding in the same way to investigate equation (i) without neglecting $(dy/dx)^2$ as small. He shews that up to a load $Q = \pi^2\epsilon/l^2$ there will be no flexure, but that after this load the flexure changes continuously with the load (pp. 8—15): see our Art. 109.

[1257.] Then follows a remark of Lamarle's, on the convergency of the series used. We may note that the greatest negative stretch in any fibre would be given by $\frac{b}{\rho} + \frac{Q}{E\omega} - \frac{bQ}{\rho E\omega}$ and the greatest positive stretch by $\frac{b}{\rho} - \frac{Q}{E\omega} - \frac{bQ}{\rho E\omega}$, where $b$ is the distance of either extreme fibre of the beam from the central axis. Now we may neglect the term $\frac{b}{\rho}\frac{Q}{E\omega}$ in comparison with the others. Lamarle also neglects $Q/(E\omega)$ as compared with $b/\rho$. I think, however, that his set-conditions are wrong. Since the greatest stretch will be at the central cross-section of the beam, if $s_0$ be the greatest stretch which can be given without passing the elastic limit and $f$ be the central deflection I find by equation (i) that:

$$s_0 > \eta\left(\frac{Qbf}{\epsilon} + \frac{Q}{E\omega}\right),$$

and
$$> \left(\frac{Qbf}{\epsilon} - \frac{Q}{E\omega}\right),$$

where $\eta$ is the ratio of lateral stretch to longitudinal squeeze. Of these for a small deflection the former will be the greater. Hence the condition

$$\frac{Qfb}{\epsilon} < s_0/\eta$$

is really too favourable to $Q$.

Now the series in which Lagrange and Lamarle have solved their differential equation ascends in powers of $\frac{Qf^2}{4\epsilon}$ and therefore in powers of something less than $\left(\frac{s_0}{2\pi\eta}\frac{l}{b}\right)^2$ since $Q > \frac{\pi^2\epsilon}{l^2}$. Now in practice $l/b$ will be $< 100$ and $s_0$ at greatest equal to ·0006 for iron and wood. Hence the series ascends in powers of something not greater than $(\cdot 01/\eta)^2$, or $(\cdot 04)^2$ for uni-constant isotropy, and so is rapidly convergent.

[1258.] A result of Lamarle's on p. 20 is, I believe, entirely original and has not been sufficiently regarded. A contractive load $Q$ would produce a squeeze $Q/(E\omega)$. Lamarle states that this must be less than $s_0$. But this conclusion does not seem to me correct; if we suppose compressed bodies to reach set by lateral extension we ought to put $Q/E\omega < s_0/\eta$. Now the first load which produces flexure is equal to $\pi^2\epsilon/l^2$. Hence we must have for flexure which does not destroy the elasticity

$$\frac{\pi^2\epsilon}{l^2E\omega} < s_0/\eta.$$

Now $\epsilon = E\omega\kappa^2$ where $\kappa$ is the radius of gyration of the section about a line through its centre perpendicular to the plane of flexure. Hence it follows that

$$\frac{\kappa^2}{l^2} < \frac{s_0}{\eta\pi^2}.$$

Suppose $s_0 = \cdot 0006$. Then for a circular section of diameter $c$ we have

$$l/c > 32{\cdot}06 \times \sqrt{\eta}.$$

For a rectangular section of which the lesser side is $c$,

$$l/c > 37{\cdot}12 \times \sqrt{\eta}.$$

If we suppose uni-constant isotropy as holding for metal struts, $\eta = 1/4$, and the numbers on the right-hand side are respectively 16·03 and 18·51.

Thus we arrive at the important conclusion that:

There will be no flexure without set in the case of a strut of circular section pivoted at both ends unless its length is greater than 16 times its diameter; if it be built-in at both ends, it may be shewn that the length must be 32 times the diameter.

For a strut of rectangular section the length must be in the corresponding cases 18 and 37 times the least diameter. These numbers are those given by Lamarle, but I think his process is wrong.

[1259.] In the preceding article we have shewn when the limit of elasticity will be reached *before* flexure begins, but we can obtain even closer results. In Art. 1257 we have seen that provided $l/b < 100$ or $l/c < 50$, then

$$\sqrt{\frac{Q}{\epsilon}}\frac{f}{2} < {\cdot}01/\eta.$$

Now it may be shewn from our Art. 110 that

$$f = 4\sqrt{\frac{\epsilon}{Q}}\sqrt{\left(\frac{l}{\pi}\sqrt{\frac{Q}{\epsilon}} - 1\right)} \text{ nearly,}$$

hence we must have

$$\frac{l}{\pi}\sqrt{\frac{Q}{\epsilon}} < 1 + {\cdot}00002/\eta^2,$$

or for uni-constant isotropy, $Q < 1{\cdot}0006\,\frac{\pi^2\epsilon}{l^2}$.

But for flexure $$Q > \frac{\pi^2\epsilon}{l^2}.$$

Now these numbers are practically equal and we are able to draw the all-important conclusion:

That for values of $l/c$ up to 50, the load will not produce flexure without destroying the elasticity. This supposes the strut pivoted at both ends; if it were built-in at both ends we must read for values of $l/c$ up to 100.

Thus we see that up to a certain ratio of length to diameter the maximum load $Q_0$ ought to be calculated from the formula

$$Q_0 = E\omega s_0/\eta.$$

But beyond this ratio $Q_0$ ought to be calculated from

$$Q_0 = \frac{\pi^2 \epsilon}{l^2}.$$

A table for the strength of doubly-pivoted struts based on these principles is given by Lamarle on p. 24. It needs, however, modification for the corrections indicated in our Arts. 1257—1258.

On pp. 25—26 the exaggerated effect, which is produced if the load be not in the axis of the strut, is pointed out.

[1260.] Lamarle's theory is in accordance with the results of Duleau and other investigators, who have found in many cases that the breaking load is measured by Euler's formula for the deflecting load;—that is to say, if we suppose that the limit of elasticity and the rupture-stretch are nearly identical: see Arts. 228 and 959.

[1261.] The second part of this memoir was presented to the French Academy and reported on by Poncelet and Liouville, January 15, 1844. They speak of Lamarle's principles offering *dans leur ensemble, une solution satisfaisante de la question des pièces chargées debout.*

Without entirely agreeing with this statement we may yet admit that they introduce into the Euler-Lagrangian theory of struts an element which goes far to reconcile that theory with experience. It is noteworthy that Lamarle's results seem to have entirely escaped the consideration of later writers.

[1262.] They seem to be of such importance that we quote them in his own words (pp. 28—29):

1°. Que les charges que peuvent supporter, sans altération permanente, les pièces dont il agit, sont indépendantes de leur longueur et simplement proportionnelles à leur section, tant que le rapport entre la longueur et la plus petite dimension de l'équarrissage n'atteint pas une certaine limite;

2°. Qu'au delà de cette limite, et pour tous les cas d'application, la charge *maximum* peut atteindre et non dépasser l'effort correspondant à la flexion initiale;

(We may shew generally that $Q < \frac{\pi^2 \epsilon}{l^2} \left\{1 + 2\left(\frac{s_0}{2\pi\eta}\right)^2 \left(\frac{l}{c}\right)^2\right\}$, so that in our case for $l/c = 100$, $Q$ cannot exceed $\frac{\pi^2\epsilon}{l^2}$ (1 003) without set, and this surpasses the buckling load by ·003 only.)

3°. Que la théorie qui permet d'établir *à priori* ces deux principes essentiels et qui les rend applicables à l'aide de formules tres simples, se concilie d'ailleurs parfaitement avec les faits d'observation, lorsque l'on a pris les précautions convenables pour réaliser les hypothèses sur lesquelles elle repose.

1263. G. G. Stokes. We have now to consider the contributions to our subject of one of the foremost of English physicists. These, without being numerous, are important. So far as they bear upon the moot point of bi-constant isotropy we have considered them to some extent in Arts. 925 and 926 of Chapter VI.

1264. The first memoir we have to note is entitled: *On the Theories of the Internal Friction of Fluids in Motion, and of the Equilibrium and Motion of Elastic Solids* It was published in the *Transactions of the Cambridge Philosophical Society*, Vol. VIII. 1849 pp. 287—319 (*Mathematical Papers*, I. pp. 75—129), and was read on April 14, 1845. This important memoir is devoted mainly to the motion of fluids. The following sentences are from the introductory remarks.

In reflecting on the principles according to which the motion of a fluid ought to be calculated when account is taken of the tangential force, and consequently the pressure not supposed the same in all directions, I was led to construct the theory explained in the first section of this paper, or at least the main part of it, which consists of equations (13), and of the principles on which they are formed. I afterwards found that Poisson had written a memoir on the same subject, and on referring to it I found that he had arrived at the same equations. The method which he employed was however so different from mine that I feel justified in laying the latter before this Society....

Poisson, in the memoir to which I have referred, begins with establishing according to his theory the equations of equilibrium and motion of elastic solids, and makes the equations of motion of fluids depend on this theory. On reading his memoir I was led to apply to the theory of elastic solids principles precisely analogous to those which I have employed in the case of fluids....

The memoir by Poisson to which reference is here made is that of October, 1829; see our Art. 540.

1265. So far as relates to our subject Stokes' memoir offers the following matters: the equations for the equilibrium and motion of an elastic body are established involving *two* constants; the principles of Poisson's theory are criticised, and objections are urged especially against that hypothesis in virtue of which only *one* constant occurs in the equations; and finally the necessity for the introduction of two constants is urged on experimental grounds.

An allusion to this memoir as relating to the motion of fluids occurs in Moigno's *Statique*, p. 694.

[1266.] The parts which concern us are Sections III. and IV In the first of these sections the principle of the superposition of small quantities is made use of to deduce that stress is a linear function of strain. Thus the method adopted is equivalent to Cauchy's in our Art. 614. We have already had occasion to refer to the criticism of Poisson which occurs in Section IV., as well as to the experimental data cited from Lame and Oersted: see our Arts. 690 and 1034. Stokes supposes that the effect of the heat developed when the elastic solid is in a state of rapid vibration is the introduction of a term into the three tractions proportional to the temperature; the heat developed has accordingly no effect upon the three shears. Thus purely transverse vibrations would not be influenced by the thermal state. The equation for the conduction of heat is not introduced; the treatment of the thermo-elastic equations is thus not quite so general as that due to Duhamel: see our Art. 877. We have already referred to Saint Venant's remarks on Stokes' memoir: see Art. 929. The application of his equations, which Stokes has peculiarly in view, is to the luminiferous ether, which it is needful to consider as an elastic solid for one kind of motion and as a perfect fluid for another. This 'thin jelly theory of the ether is considered in a paper entitled: *On the Constitution of the Luminiferous Ether*, to be found in the *Philosophical Magazine*, Vol. XXXII. p. 343, 1848 (or *M. Papers*, II. p. 8). The subject belongs, however, rather to the theory of light than to the history of elasticity.

1267. *Remark on the theory of homogeneous elastic solids. Cambridge and Dublin Mathematical Journal,* Vol. 3, 1848, pp. 130—131. This is an historical note; and its nature will be obvious from the opening sentences:

In a paper on Elastic Solids published in the last number of this Journal, Professor Thomson speaks of my paper on the same subject as being the only work in which the equations of equilibrium or motion of a homogeneous elastic solid are given with the two arbitrary independent constants which they must contain. This is so far correct that there is no other work, of which I am aware, in which these equations are *insisted on* as being those which it is absolutely necessary to adopt; but the equations have been *obtained* by M. Cauchy, and a different method of arriving at them has been pointed out by Poisson.

[1268.] *On the Dynamical Theory of Diffraction. Camb. Phil. Trans.* Vol. IX. pp. 1—62 (or *Math. Papers,* II. pp. 243—328). This paper was read on November 26, 1849. The only part which concerns our present subject is Section II which is entitled: *Propagation of an Arbitrary Disturbance in an Elastic Medium.* (It occupies pp. 257—280 of the *Math. Papers,* II.) This contains an important solution of the equations of bi-constant isotropy for wave motion. The solution is somewhat similar to those obtained by Poisson, Ostrogradsky and Blanchet, but the method is far simpler: see Arts. 523—526, 740 and 1169

If $u, v, w$ be the shifts, $\tau_{yz}, \tau_{zx}, \tau_{xy}$ the twists defined by equations of the form

$$2\tau_{yz} = \frac{dw}{dy} - \frac{dv}{dx},$$

and $\theta$ the dilatation. Then, $a, b$ being constants, we have equations of the type

$$\frac{d^2u}{dt^2} = b^2\nabla^2 u + (a^2 - b^2)\frac{d\theta}{dx} \text{ .................. (i).}$$

From these we deduce for the dilatation

$$\frac{d^2\theta}{dt^2} = a^2\nabla^2\theta \text{ ..........................(ii),}$$

and for the twists three equations of the type

$$\frac{d^2\tau_{yz}}{dt^2} = b^2\nabla^2\tau_{yz} \text{ ........................ (iii).}$$

These equations have been previously solved by Stokes: see formulae (12) and (16) of his Section I. He there divides each shift $u$ into two parts $u_1$ and $u_2$; of these the first parts are such that $u_1dx + v_1dy + w_1dz$ is a perfect differential, or the corresponding twists are zero the second parts are due to twists $\tau_{yz}, \tau_{zx}, \tau_{xy}$ which have given finite values throughout a finite space and vanish elsewhere. $\theta$ will thus consist of two parts, the second however is zero. The solution is then given by equations of the type

$$u = \frac{1}{2\pi}\iiint (y\tau_{xy} - z\tau_{zx})\frac{dV}{r^3} - \frac{1}{4\pi}\iiint \frac{\theta_0}{r^2}\cos(rx)\, dV \ldots \text{(iv)},$$

where $r$ is the distance of the point whose shifts are $u, v, w$ from the point whose twist-components are $\tau_{yz}, \tau_{zx}, \tau_{xy}$ or dilatation $\theta_0$, and $x, y, z$ are the coordinates of the latter point *relative* to the former; the integration is to be taken all over the finite space $V$ wherein the twists have finite values.

[1269.] Stokes now proceeds to analyse the integrals of equations (ii) and (iii):

Let $O$ be the point of space at which it is required to determine the disturbance, $r$ the radius vector of any element drawn from $O$; and let the initial values of $\theta$, $d\theta/dt$ be represented by $f(r)$, $F(r)$ respectively[1], ......we have

$$\theta = \frac{t}{4\pi}\iint F(at)\, d\sigma + \frac{1}{4\pi}\frac{d}{dt}\, t\iint f(at)\, d\sigma \ldots\ldots\ldots\ldots\text{(v)}.$$

The double integrals in this expression vanish except when a spherical surface described round $O$ as centre, with a radius equal to $at$, cuts a portion of the space $T$ (that of the initial disturbance). Hence if $O$ be situated outside the space $T$, and if $r_1$, $r_2$ be respectively the least and greatest values of the radius vector of any element of that space, there will be no dilatation at $O$ until $at = r_1$. The dilatation will then commence, will last during an interval of time equal to $a^{-1}(r_2 - r_1)$, and will then cease for ever. The dilatation here spoken of is understood to be either positive or negative, a negative dilatation being the same thing as a condensation.

[1] This solution is due to Poisson: *Mémoires de l'Académie*, t. III. p. 130. The values of $\theta$ and $d\theta/dt$ at the epoch are supposed to be finite within a given finite space and to vanish outside; the double integration is over the boundary of this space. Further $\theta$ and $d\theta/dt$ may be initially functions of the direction as well as magnitude of $r$, but it is not necessary to express this analytically: see Stokes' Section I. § 4.

Hence a *wave of dilatation* will be propagated in all directions from the originally disturbed space $T$, with a velocity $a$. To find the portion of space occupied by the wave, we have evidently only got to conceive a spherical surface of radius $at$, described about each point of the space $T$ as centre. The space occupied by the assemblage of these surfaces is that in which the wave of dilatation is comprised. To find the limits of the wave, we need evidently only attend to those spheres which have their centres situated in the surface of the space $T$. When $t$ is small this system of spheres will have an exterior envelope of two sheets, the outer of these sheets being exterior and the inner interior to the shell formed by the assemblage of spheres. The outer sheet forms the outer limit to the portion of the medium in which the dilatation is different from zero As $t$ increases the inner sheet contracts, and at last its opposite sides cross, and it changes its character from being exterior, with reference to the spheres, to interior. It then expands, and forms the inner boundary of the shell in which the wave of condensation is comprised. It is easy to shew geometrically that each envelope is propagated with a velocity $a$ in a normal direction.

It appears in a similar manner from equations (iii) that there is a similar wave, propagated with a velocity $b$, to which are confined the rotations $\tau_{yz}$, $\tau_{zx}$, $\tau_{xy}$. This wave may be called for the sake of distinction, the *wave of distortion*, because in it the medium is not dilated or condensed, but only distorted in a manner consistent with the preservation of a constant density. (§§ 11, 12.)

We have here the clear expression of the manner in which the *wave of dilatation* and the *wave of twist* are propagated. The reader may compare our Arts. 740 and 1169.

[1270.] Paragraph 14 states a general dynamical theorem which Lord Rayleigh has termed *Stokes' Rule* in his *Theory of Sound* (Vol. I. p. 96). We give it almost in the words of the memoir:

When the expressions for the disturbance of a dynamical system at the end of time $t$ are linear functions of the initial displacements and the initial velocities, then the part of the disturbance which is due to the initial displacements may be obtained from the part due to the initial velocities by differentiating with respect to $t$, and replacing the arbitrary functions which represent the initial velocities by those which represent the initial displacements.

Stokes remarks that this result constantly presents itself in investigations like those which are the subject of his memoir, and that on considering its physical interpretation it will be found to be of extreme generality.

It presents itself in fact in all problems related to the vibrational motion of perfectly elastic solids.

[1271.] The application of the principle mentioned in the preceding article enables Stokes to treat only of initial velocities. He considers also separately the parts of the shifts due to dilatation and to twist.

Thus he finds for the shifts due to initial velocities of dilatation $\dot{u}_0, \dot{v}_0, \dot{w}_0$ equations of the type

$$u_1 = \frac{t}{4\pi}\iint l(\dot{q}_0)_{at}\, d\sigma + \frac{t}{4\pi}\iiint (\dot{u}_0 - 3l\dot{q}_0)\frac{dV}{r^3} \quad (r > at)\ldots(\text{vi}).$$

Here $\dot{q}_0 = l\dot{u}_0 + m\dot{v}_0 + n\dot{w}_0$ where $l$, $m$, $n$ are the direction-cosines of the line joining the point $O$ whose shifts are $u_1$, $v_1$, $w_1$, to the point whose initial velocity-components are $\dot{u}_0$, $\dot{v}_0$, $\dot{w}_0$. $(\dot{q}_0)_{at}$ means that we are to take $\dot{q}_0$ at the point distant $at$ from $O$; the triple integration is to be taken over all parts of $V$ for which $r > at$: see § 17

[1272.] In § 18 we have the consideration of the values of $u_2$, $v_2$, $w_2$ due to initial velocities of a twist character. Stokes easily finds that

$$u_2 = \frac{t}{4\pi}\iint (\dot{u}_0 - l\dot{q}_0)_{bt}\, d\sigma - \frac{t}{4\pi}\iiint (\dot{u}_0 - 3l\dot{q}_0)\frac{dV}{r^3} \quad (r > bt)\ldots(\text{vii}).$$

If the expressions for $u_1$ and $u_2$ be added we have finally for the shift $u$ due to initial velocity of any kind

$$u = \frac{t}{4\pi}\iint l(\dot{q}_0)_{at}\, d\sigma + \frac{t}{4\pi}\iint (\dot{u}_0 - l\dot{q}_0)_{bt}\, d\sigma$$

$$+ \frac{t}{4\pi}\iiint (3l\dot{q}_0 - \dot{u}_0)\frac{dV}{r^3} \quad (bt < r < at)\ldots\ldots\ldots\ldots(\text{viii}).$$

By application of the theorem stated in Art 1270 the shift due to initial displacement is found to be

$$u = \frac{1}{4\pi}\iint\left\{l\left(4\rho_0 + at\frac{d\rho_0}{dr}\right) - u_0\right\}_{at} d\sigma$$
$$+ \frac{1}{4\pi}\iint\left\{2u_0 + bt\frac{du_0}{dr} - l\left(4\rho_0 + bt\frac{d\rho_0}{dr}\right)\right\}_{bt} d\sigma$$
$$+ \frac{1}{4\pi}\iiint(3l\rho_0 - u_0)\frac{dV}{r^3} \quad (bt < r < at) \ldots\ldots\ldots\ldots\ldots (\text{ix})$$

Here $u_0$, $v_0$, $w_0$ are the components of the initial displacement, $\rho_0$ that displacement resolved in the direction $l$, $m$, $n$, and the subscripts $at$ or $bt$ denote that $r$ is supposed to be equal to $at$ or $bt$ after differentiation (§ 19).

[1273.] We have reproduced these results of Stokes' not only to indicate the method in which he has attacked the problem, but also for the sake of the following remarks, which have a bearing on the controversy between Blanchet and Cauchy considered in our Arts. 1169 *et seq.*

The first of the double integrals in equations (viii), (ix) vanishes outside the limits of the wave of dilatation, the second vanishes outside the limits of the wave of distortion. The triple integrals vanish outside the outer limit of the wave of dilatation, and inside the inner limit of the wave of distortion, but have finite values within the two waves and between them. Hence a particle of the medium situated outside the space $T$ (i.e. space of initial disturbance) does not begin to move till the wave of dilatation reaches it. Its motion then commences, and does not wholly cease till the wave of distortion has passed, after which the particle remains absolutely at rest (§ 20).

[1274.] In the following paragraphs Stokes treats special cases of these equations. § 21, no wave of distortion or twist; § 22, no wave of dilatation; §§ 23—26 consider the terms which are of importance at a great distance from the space originally disturbed. It is shewn that as the surface of the wave of twist grows larger, the inclination of the resultant shift of a particle to the wave's front decreases with great rapidity. §§ 28—29 contain the important bearing of these results on the undulatory theory of light, the ether being considered as an elastic solid, i.e. not strictly incompressible.

[1275.] We have omitted to mention § 27 because it contains a problem of great generality in its application to the motion of perfectly elastic solids and deserves a fuller consideration. The problem is the following one: to deduce the disturbance due to a given variable force acting in a given direction at a given point of an infinite elastic medium.

Let $O_1$ be the given point of the medium, we require to know the shifts $u$, $v$, $w$ at any point $O$ at time $t$. Suppose $f(t)$ the magnitude of the given force and its direction $l'$, $m'$, $n'$; let $(l, m, n)$ be the direction of $OO_1$ $(=r)$ and let

$$k = ll' + mm' + nn';$$

let $D$ be the density of the medium in equilibrium; then Stokes shews by means of the principle that the limit of a series of small impulses will be a continuous moving force, and by means of repeated application of equation (viii), that the shifts are given by three equations of the type

$$u = \frac{lk}{4\pi D a^2 r} f(t - r/a) + \frac{l' - lk}{4\pi D b^2 r} f(t - r/b) + \frac{3lk - l'}{4\pi D r^3} \int_{r/a}^{r/b} t' f(t - t')\, dt' \;\ldots\ldots(\text{x}).$$

An interesting discussion of this result follows.

[1276.] *Discussion of a Differential Equation relating to the breaking of Railway Bridges. Camb. Phil. Trans.* Vol. VIII. pp. 707—735 (or *Math. Papers*, Vol. II. pp. 178—220). The paper was read May 21, 1849.

It has reference to a very important practical problem in resilience; namely, the impulsive effect of a travelling load on a horizontal beam or girder. A Royal Commission had been appointed in 1847 "for the purpose of inquiring into the conditions to be observed by engineers in the application of iron in structures exposed to violent concussions and vibrations." Of this Commission the late Professor Willis was a member, and the following remarks with which Stokes opens his paper will explain the scope of Willis' investigations: see also our Arts. 1406 and 1417, wherein the experiments undertaken by the Commission and by Willis in particular are referred to.

The object of the experiments was to examine the effect of the velocity of a train in increasing or decreasing the tendency of a girder bridge over which the train is passing to break under its weight. In order to increase the observed effect, the bridge was purposely made as slight as possible: it consisted in fact merely of a pair of cast or wrought iron bars, nine feet long, over which a carriage, variously loaded in different sets of experiments, was made to pass with different velocities. The remarkable result was obtained that the deflection of the bridge increased with the velocity of the carriage, at least up to a certain point, and that it amounted in some cases to two or three times the central statical deflection, or that which would be produced by the carriage placed at rest in the middle of the bridge. It seemed highly desirable to investigate the motion mathematically, more especially as the maximum deflection of the bridge considered as depending on the velocity of the carriage, had not been reached in the experiments, in some cases because it corresponded to a velocity greater than any at command, in others because the bridge gave way by the fracture of the bars on increasing the velocity of the carriage. The exact calculation of the motion, or rather a calculation in which none but really insignificant quantities should be omitted, would however be extremely difficult, and would require the solution of a partial differential equation with an ordinary differential equation for one of the equations of condition by which the arbitrary functions would have to be determined In fact, the forces acting on the body and on any element of the bridge depend upon the positions and motions, or rather changes of motion, both of the body itself and of every other element of the bridge, so that the exact solution of the problem, even when the deflection is supposed to be small, as it is in fact, appears almost hopeless.

In order to render the problem more manageable Professor Willis neglected the inertia of the bridge, and at the same time regarded the moving body as a heavy particle. Of course the masses of bridges such as are actually used must be considerable; but the mass of the bars in the experiments was small compared with that of the carriage, and it was reasonable to expect a near accordance between the theory so simplified and experiment. This simplification of the problem reduces the calculation to an ordinary differential equation, which is that which has been already mentioned; and it is to the discussion of this equation that the present paper is mainly devoted (p. 707).

See our Art. 1419.

We thus see that Willis' equation solved by Stokes neglects the inertia of the bars, supposing them in fact to take at each instant the form due to the pressure exerted by the weight and centrifugal force of the moving body supposed to be collected into a particle.

[1277.] Stokes' theoretical results, while according fairly with experiments so long as the inertia of the bars was small, diverged considerably as it increased and began to equal that of the carriage. He accordingly in an *Addition* to his memoir states the general equations of the problem on two assumptions, and gives another solution for the case when the weight of the travelling load is extremely small as compared with that of the bars. Such remained the state of our knowledge on this important problem till the year 1855 when the subject was investigated by Phillips[1] He was followed in 1860 by Renaudot[2], and finally the matter has been very fully treated by Saint Venant[3]. The work of these elasticians in supplementing the investigations of Stokes will be considered in its proper place in our second volume; but it may be well to quote here Saint-Venant's concluding remarks on the problem:

Qu'il convient de ne plus dire, comme à une époque où la question n'avait pas encore été suffisamment étudiée, que les solutions telles que ($\omega$) et ($\psi$) ne conviennent qu'à deux cas extrêmes; celle-là pour $P/Q$ très petit, et celle-ci pour $P/Q$ très grand, ni que M. Stokes n'a eu d'autre intention, en 1849, que de traiter ce dernier cas. Le but primitif de l'éminent correspondant de l'Institut a été, comme l'exprime notre citation ci-dessus, de "tenir compte de l'inertie de la poutre"; ce but s'est trouvé atteint six ans après, en 1855, d'une autre manière, et sans changement hypothétique du mode de distribution des déplacements le long de la poutre, par des formules algébriques telles que ($\omega$); mais son beau et hardi travail a eu le précieux effet de fournir une évaluation approchée et suffisante de cette partie périodique ou oscillatoire des mouvements que la solution algébrique de 1855 était réduite à négliger; évaluation que nous croyons avoir montré être facile à étendre à tous les rapports habituels des poids $P$, $Q$ de la poutre et de la charge mobile qu'elle supporte pendant son trajet[4].

[1] *Annales des mines*, t. VII. pp. 467—506, 1855.

[2] *Annales des ponts et chaussées*, T. I. 4e serie, pp. 145—204, 1861.

[3] See his edition of Clebsch: *Théorie de l'élasticité des corps solides*, 1885, pp. 597—619.

[4] *Ibid.* p. 615, j.

We proceed to consider the various points of the memoir.

[1278.] The equation which Stokes set himself to investigate is that determined by Willis, namely:

$$\frac{d^2y}{dx^2} = \beta - \frac{\beta y}{(x - x^2)^2} \quad \text{.......................(i)},$$

where

$$\beta = \frac{gc^2}{4V^2S}, \qquad x' = 2cx, \qquad y' = 16\,Sy.$$

Here $2c$ is the length of the bar,

$V$ the velocity of the travelling load,

$S$ the central deflection when the travelling load is placed at rest on the centre of the bar,

$x'$ and $y'$ the coordinates of the travelling load, $x'$ being measured from the beginning of the bridge and $y'$ vertically downward,

$g$ the acceleration due to gravity.

See our Art. 1419.

[1279.] In §§ 3—6 Stokes solves this equation by means of a convergent series. In § 7 he finds a solution in definite integrals. This solution is:

$$y = -\frac{\beta}{m-n}\left\{x^m(1-x)^n\int_0^x x^n(1-x)^m dx - x^n(1-x)^m\int_0^x x^m(1-x)^n dx\right\} \text{...(ii)},$$

where $m$ and $n$ are roots of the equation

$$z^2 - z + \beta = 0.$$

The two arbitrary constants of the complete integral disappear by virtue of the terminal conditions:

$$y' = 0, \qquad \frac{dy'}{dx'} = 0, \qquad \text{when } x' = 0.$$

The following §§ 8—10 are occupied with the evaluation of the series and constants which express the deflection.

[1280.] In § 11 the vertical velocity of the body at the centre of the bridge is determined. The value of the vertical velocity at any time

$$= \frac{dy'}{dt} = \frac{d\,.\,16\,Sy}{dx}\,\frac{dx}{dt}$$

Now $\frac{dx'}{dt} = \frac{d \cdot 2cx}{dt} = V$ very nearly. Hence the vertical velocity

at the centre $$= \frac{8SV}{c} \cdot f'(\tfrac{1}{2}),$$

where $f'(\frac{1}{2})$ is the value of $dy/dx$ obtained from (ii) when $x = \frac{1}{2}$. It is shewn that

$$f'(\tfrac{1}{2}) = \frac{\pi\beta^2}{2\cos r\pi}, \text{ or } = \frac{\pi\beta^2}{e^{\rho\pi} + e^{-\rho\pi}},$$

where $r = \sqrt{\frac{1}{4} - \beta}$ and $\rho = \sqrt{\beta - \frac{1}{4}}$, according as $\beta <$ or $>$ ¼

The rest of the paragraph is occupied with an investigation of the maximum value of this central vertical velocity for varying values of $V$. It is found to take its maximum value $(= \cdot 6288 \sqrt{2gS})$ when $V = \cdot 4655c \sqrt{2g/S}$.

[1281.] In the following paragraph the tendency to rupture is expressed in terms of the variables. Stokes takes as the tendency to rupture at any point the moment of all the forces acting on the bridge on one side of the section at that point. It is greatest immediately under the travelling load, and is in the memoir measured in terms of its value at the centre of the bridge, when the load is placed there at rest.

Stokes, like later writers—Phillips, Renaudot and Saint-Venant—leaves shear entirely out of consideration and follows the Bernoulli-Eulerian hypothesis for the flexure of beams.

[1282.] §§ 13—17 are devoted to a table of the numerical values of the tendency to rupture and of the ratio $y'/S$ for the values of $\beta = 5/36,\ 1/4,\ 1/2,\ 5/4$; to the plotting of the corresponding trajectories, and a consideration of the general results. These have already been stated by Stokes on p. 708 of the memoir We reproduce them here because they shew clearly the points demanding further investigation:

It appears from the solution of the differential equation that the trajectory of the body is asymmetrical with respect to the centre of the bridge, the maximum depression of the body occurring beyond the centre. The character of the motion depends materially on the numerical value of $\beta$. When $\beta$ is not greater than the tangent to the

trajectory becomes more and more inclined to the horizontal beyond the maximum ordinate, till the body gets to the second extremity of the bridge, when the tangent becomes vertical. At the same time the expressions for the central deflection and for the tendency of the bridge to break become infinite. When $\beta$ is greater than $\frac{1}{4}$, the analytical expression for the ordinate of the body at last becomes negative, and afterwards changes an infinite number of times from negative to positive, and from positive to negative The expression for the reaction becomes negative at the same time with the ordinate, so that in fact the body leaps.

The occurrence of these infinite quantities indicates one of two things either the deflection really becomes very large, after which of course we are no longer at liberty to neglect its square; or else the effect of the inertia of the bridge is really important. Since the deflection does not really become very great, as appears from experiment, we are led to conclude that the effect of the inertia is not insignificant, and in fact I have shewn that the value of the expression for the *vis viva* neglected at last becomes infinite. Hence, however light be the bridge, the mode of approximation adopted ceases to be legitimate before the body reaches the second extremity of the bridge, although it may be sufficiently accurate for the greater part of the body's course (compare § 18 of the Memoir).

[1283.] In addition to the above remarks we may add the following from § 19:

There is one practical result which seems to follow from the very imperfect solution of the problem which is obtained when the inertia of the bridge is neglected. Since this inertia is the main cause which prevents the tendency to break from becoming enormously great, it would seem that of two bridges of equal length and equal strength, but unequal mass, the lighter would be the more liable to break under the action of a heavy body moving swiftly over it.

In § 21 the moment of rupture is calculated for a long bridge for which $\beta$ might be supposed great in practice; in this case the moment does not become infinite, but the motion is sensibly symmetrical about the centre of the bridge, and the moment has its greatest value at that point. Stokes holds that the inertia of the bridge would decrease rather than increase this maximum.

[1284.] The memoir concludes with an ingenious application

of the 'method of dimension-counting' to determine the conditions under which an *exact dynamical model* of a larger system may be obtained: see p. 721.

[1285.] In the *Addition* which follows (dated October 22 of the same year) Stokes calculates the values of the moment of rupture and the deflection for greater values of $\beta$, namely 3, 5, 8, 12 and 20. This was upon the suggestion of Willis, who remarked that such larger values were those occurring in practice. The corresponding trajectories were plotted by Willis and will be found on fig. (i) of the plate at the end of the memoir. For $\beta = 20$ the trajectory becomes sensibly symmetrical except in the close neighbourhood of the extremities, where however the depression itself is insensible.

[1286.] On p. 725 we have a criticism of Cox's paper on the basis of the remark we have cited above as to the *vis viva* becoming infinite: see our Arts. 1282 and 1433.

[1287.] Stokes also shews p. 724 that the ratio of the central dynamical deflection $D$ to the central statical deflection $S$ is given by

$$\frac{D}{S} = 1 + \frac{1}{\beta} = 1 + \frac{4V^2S^2}{gc^2},$$

to three places of decimals when $\beta$ is equal to or greater than 100.

[1288.] The main value of the *Addition* however depends upon the attempt made on pp. 727—733 to include the effect of the inertia of the bridge:

In consequence of some recent experiments of Professor Willis', from which it appeared that the deflection produced by a given weight travelling over the trial bar with a given velocity was in some cases increased by connecting a balanced lever with the centre of the bar, so as to increase its inertia without increasing its weight, while in other cases the deflection was diminished, I have been induced to attempt an approximate solution of the problem, taking into account the inertia of the bridge. I find that when we replace each force acting on the bridge by a uniformly distributed force of such an amount as to produce the same mean deflection as would be produced by the actual force taken alone, which evidently cannot occasion any very material error, and when we moreover neglect the difference between the pressure exerted by the travelling mass on the bridge and its weight, the equation admits of integration in finite terms (pp. 727—728).

[1289.] The result of Willis' experiments (see our Art. 1421) should be noted in connection with the statement made in Art. 1282 as to the deflection and moment of rupture becoming infinite. It does not seem so certain that these infinite values depend entirely on the neglect of the inertia, and that they will take *less* values if that inertia be included. It may possibly be related to the unsatisfactory nature of the Bernoulli-Eulerian hypothesis as to the elasticity of beams.

It will be seen that Stokes' solution of the problem including the inertia of the beam depends upon two assumptions. The second of these assumptions, namely that we may neglect the difference between the pressure on the bridge and the weight of the travelling load, is shewn in the course of the work to depend on the two conditions, (i) that $\beta$ be large, (ii) that the mass of the travelling load be small compared with that of the bridge: see p. 729.

[1290.] Stokes gives tabular and graphical results for the solution of the resulting equation on various numerical hypotheses. The variety in these solutions depends upon a certain quantity $q$. If $S_1$ be the central statical deflection that would be produced by a mass equal to that of the bridge,

$$q^2 = \frac{252gc^2}{31V^2S_1}.$$

Suppose the travelling load removed and the bridge depressed through a small space, then the period of a complete oscillation $P$ would be given by

$$P = 2\pi\sqrt{\frac{31S_1}{63g}}$$

Hence $$q = \frac{4\pi c}{PV} = 2\pi\frac{\tau}{P},$$

if $\tau$ be the time of the mass crossing the bridge.

The quantity $\tau/P$ will usually be very large, and Stokes then shews that the central deflection is liable to be alternately increased and decreased by the fraction $25/8q$ or by $112\sqrt{S_1}/\tau$ of the central statical deflection. Here the units of space and time are an inch and a second.

The value of the central deflection is calculated in the case of two examples drawn from the Ewell and Britannia Bridges.

[1291] We may conclude our examination of this important memoir by the following general conclusions of the *Addition*:

In this paper the problem has been worked out, or worked out approximately, only in the two extreme cases in which the mass of the travelling body is infinitely great and infinitely small respectively, compared with the mass of the bridge. The causes of the increase of deflection in these two extreme cases are quite distinct. In the former case, the increase of deflection depends entirely on the difference between the pressure on the bridge and the weight of the body, and may be regarded as depending on the centrifugal force. In the latter, the effect depends on the manner in which the force, regarded as a function of the time, is applied to the bridge. In practical cases the masses of the body and of the bridge are generally comparable with each other, and the two effects are mixed up in the actual result. Nevertheless, if we find that each effect, taken separately, is insensible, or so small as to be of no practical importance, we may conclude without much fear of error that the actual effect is insignificant. Now we have seen that if we take only the most important terms the increase of deflection is measured by the fractions $\beta^{-1}$ and $25/(8q)$ of $S$. It is only when these fractions are both small that we are at liberty to neglect all but the most important terms, but in practical cases they are actually small. The magnitude of these fractions will enable us to judge of the amount of the actual effect.

[1292.] G. Wertheim. We now reach a scientist whose labours in the field of physical elasticity are among the most important we have to deal with in this period. Wertheim's first memoir was presented to the Académie on July 18, 1842. It is entitled: *Recherches sur l'élasticité; Premier Mémoire*, and will be found on pp. 385—454 of the *Annales de Chimie*, T. XII. Paris, 1844.

This paper deals with the stretch-moduli and the elastic limits of the metals, as well as with the relations of these elastic to other physical quantities. It contains a very complete experimental consideration of the elasticity of lead, tin, cadmium, gold, silver, zinc, palladium, platinum, copper, iron and steel in a variety of states. The stretch-moduli obtained by Wertheim are those usually quoted in the text-books. At this time he accepted the uni-constant hypothesis, and treated wires as homogeneous and isotropic.

[1293.] The memoir opens with some historical notes and references to writers already considered in this book. There is also a footnote upon a sealed packet which the author had deposited with the Académie in 1841, containing a table of the elastic properties of certain metals, and a rather vague statement that: there appears to exist an intimate relation between the mechanical properties of bodies and their molecular distances We have then a reference to the various methods vibrational and statical for determining the stretch modulus. The three methods adopted by Wertheim are (i) from longitudinal vibrations, (ii) from lateral vibrations, (iii) from statical extension. In case (ii) no correction was made for the rotatory inertia (see Lord Rayleigh's *Theory of Sound*, I. § 186). The problems which the author set himself to answer were:

1° To determine the stretch-modulus, the velocity of sound, the limit of elasticity and the maximum-stretch for those metals for which they had not yet been determined.

We know that Lagerhjelm (Art. 366) held that the stretch-modulus was the same for iron (? wrought iron) and steel whether annealed or unannealed. Poncelet (*Mécanique industrielle*, p. 354, K. 384, and our Art 986) confirmed this. Hence the problem:

2° To determine whether the difference produced by annealing in the stretch-moduli of the metals is so small that it cannot be measured by the ordinary methods.

3° If the elasticity of a body changes with its density and its chemical nature, to determine the relation between these quantities.

4° To investigate if there be any relation between the limit of elasticity and the rupture-stretch similar to that propounded for iron by Lagerhjelm (see Art. 365).

5°. To investigate the temporary and permanent effects of heat on the elastic state of the metals.

See pp. 389—391.

[1294.] Wertheim after stating his problems proceeds to describe his apparatus and methods (pp. 391—406). Then follow long and careful experiments on each individual metal, with an account of its preparation and chemical constitution. We may note, however, that a distinction does not seem to have been made between the metal in the form of a rod and in the form of a wire (pp. 406—435).

[1295.] We have next experiments on the relation between elasticity and temperature, then those on the effect of annealing, and on the elastic limit and rupture-stretch. We may quote the following statements as physically interesting and important (p. 438):

En comparant la première colonne à la seconde, on voit que l'effet de l'étirage n'est pas le même sur tous les métaux; les uns se condensent, les autres se dilatent. En effet, l'étirage se compose de deux actions différentes: l'une qui tend à diminuer la densité par l'effort de traction, tandis que l'autre tend à l'augmenter par la compression latérale; ce dernier effort prédomine dans la plupart des métaux (see our Arts. 368—369 and the footnote to the latter).

Le recuit ramène les fils sensiblement à la même densité qu'ils avaient à l'état fondu; il s'opère donc une dilatation; le plomb, le zinc, le fer et l'acier en gros fils font exception, ils ont éprouvé une légère condensation (see our Arts. 692, 830, 858).

La troisième colonne montre que la traction sans compression latérale produit une diminution de densité sur les métaux étirés; mais cela n'a plus lieu pour les métaux recuits, qui se condensent, pour la plupart, par l'allongement.

This agrees with M'Farlane's experiments on annealed copper wire: see Thomson's Article Elasticity in the *Encycl. Brit.* § 3.

[1296.] Further, as to the elastic limit Wertheim writes:

J'ai pris pour limite d'élasticité, en suivant l'exemple de plusieurs auteurs, le poids qui produit un allongement permanent de 0·00005 par unité de longueur. Cette détermination est arbitraire, car on peut trouver des allongements permanents aussi petits que l'instrument peut les mesurer. J'ai opéré avec beaucoup de lenteur, surtout sur l'argent, l'or, le cuivre et le platine, et je n'ai augmenté les poids que de très-petites quantités à la fois. Je n'ai trouvé alors ni sauts ni saccades dans les allongements; ils croissent, au contraire, d'une manière continue, dès qu'ils sont devenus mesurables, ce dont on peut se convaincre en parcourant les colonnes (*a*) de mes expériences. Les allongements permanents sont, en outre, des fonctions inconnues du temps pendant lequel le poids a agi; tel poids qui ne produit pas d'allongement permanent mesurable, quand il n'agit que peu de temps, en produira un après une action assez prolongée. Il n'est pas probable que la même chose n'ait pas lieu pour les allongements plus petits que 0 00001 de l'unité de longueur, quoique nos instruments ne puissent pas les mesurer. On peut donc dire que les nombres exprimant les limites d'élasticité doivent diminuer à mesure que les instruments de mesure se perfectionneront, et qu'on laissera agir les poids pendant plus de temps.

Du reste, on voit que le recuit diminue très-considérablement les limites d'élasticité, sans qu'il y ait une relation constante entre la limite d'élasticité du métal recuit et celle du métal non recuit. La limite d'élasticité des métaux recuits ne change pas considérablement par l'élévation de température à 200 degrés (p. 439).

On voit que le recuit diminue très considérablement la resistance à la rupture, en même temps qu'il fait grandir les allongements maxima L'élévation de température jusqu'à 200 degrés ne change pas sen siblement ces quantités (p. 441).

See our Arts. 692 and 858.

It will be seen from the above remarks that Wertheim had not reached any conception of the principle which is involved in the 'state of ease. He believed set to begin with the most feeble stress. Further it is singular that he should not have discovered the phenomenon of the *yield point,* but he speaks too definitely of finding neither jumps nor jerks in the extension for us to believe he had any idea of it. He also seems to have been ignorant at this time of elastic after-strain, so that it is difficult to judge how far the *allongements permanents* which he describes as unknown functions of the time were genuine set and how far elastic after-strain. This point must be taken into consideration in our valuation of Wertheim's results

The elastic limit as measured by Wertheim is the stretch at which set commences. This is not necessarily the mathematical elastic limit, or stretch at which stress ceases to be proportional to strain: see our Appendix, Note C, (7).

[1297.] On pp. 441—445 we have the results of experiments on the velocity of sound in relation to the stretch-modulus. It is found that the velocities as obtained from the statically determined stretch modulus are always less than those obtained directly from the vibrations. *Iron alone appears to be an exception*[1] (p. 444) This difference is attributed to the heat developed in the vibrational method, which produces in the solids as in gases an acceleration of the velocity. The specific heats at constant stress and at constant volume are then referred to, and a formula of Duhamel's cited: see our Arts. 705 and 887 Wertheim seems to have been unacquainted with Weber's priority in this matter. If we compare Weber's results with Wertheim's we find that for the two metals they have both treated there is not much agreement.

Nor should we indeed expect much, for Wertheim is using for the velocity of sound in bars a formula of Duhamel's which applies only to the velocity of sound in an infinite solid: see our Art. 887. If, however, Wertheim's numbers are substituted in the proper formula, we arrive, as Clausius first pointed out, at still more inconsistent results; in fact we find in certain cases impossible values for the ratio of the two specific heats. Either, then, in this matter Wertheim's experimental method was defective, or the difference between the two specific heats is not sufficient to explain the difference in the values of the stretch-modulus. On p. 445 Wertheim himself remarks on the difficulty attaching to experiments of this kind see our Arts. 1350 and 1403.

[1298.] With regard to the influence of heat Wertheim s experiments go to shew that the stretch-modulus decreases continuously between $-15^\circ$ and $200^\circ$ Iron and steel however are exceptions their moduli increase from $-15^\circ$ to $100^\circ$, but at $200^\circ$

[1] The values of the stretch-modulus for two pieces of *steel* tested by A. B. W. Kennedy in 1880 were 30,950,000 and 31,100,000 lbs. per square inch respectively. Wüllner afterwards determined the stretch-modulus for the same pieces by the vibrational method, and he found values respectively 0·958 and 0·961 of Kennedy's. The difference is almost within the instrumental error, yet, if steel be included with iron, is, so far as it goes, opposed to Wertheim's result.

are not only less than at 100°, but sometimes less than at the ordinary temperature (p. 445)

Wertheim's investigations on iron and steel require careful repetition; it is not certain from his account of the experiments whether he had definitely ascertained that no *permanent* alteration of elasticity had been produced by the alteration in temperature. Such a permanent alteration is at least suggested by Cagniard Latour's results see our Art. 802.

[1299] An attempt is then made to find a relation between the stretch modulus and the molecular distance This is done by an appeal to a formula which occurs in the theory of uni-constant isotropy. If $f(r)$ express the law of molecular force, $a$ the molecular mean distance and $E$ the stretch-modulus, we find from Poisson's results cited in our Art. 442,

$$E=\frac{\pi}{3}\sum_{r=a}^{r=\infty}\left(\frac{r}{a}\right)^5\frac{d}{dr}\left(\frac{f(r)}{r}\right) \text{ .................. (i)}$$

As to $a$ Wertheim writes (p. 447):

Pour trouver la valeur de $a$ pour chaque métal, supposons que les poids relatifs des molécules des corps chimiquement simples soient exprimés par leurs poids atomiques. On sait combien cette hypothèse est devenue probable par la loi de Dulong et Petit sur les chaleurs spécifiques, et par les travaux de M. Regnault. MM. Avogadro et Baudrimont ont cherché à démontrer, depuis, que les équivalents déduits de la chaleur spécifique sont la vraie expression du poids des molécules. Soient $S$ le poids spécifique d'un corps simple, $A$ le poids de sa molécule, $a$ la distance moyenne relative de ses molécules, on aura

$$a=(S/A)^{-\frac{1}{3}} \text{ ..............................(ii).}$$

Wertheim finds that $Ea^7$ is approximately constant for the same metal, and its mean value sensibly the same for lead, cadmium, silver, gold, zinc, palladium, iron and steel; somewhat greater, however, for tin and platinum and less for copper. Its mean value for the former set of metals is 8·09907; for tin 8·26715; for platinum 8 31789; and for copper 7·91020.

If $Ea^7$ were accurately constant for all metals, we should have at once from equation (i) $f(r)$ varying as $r^{-5}$, or intermolecular force would vary as the inverse fifth power of intermolecular distance (p 449).

[1300.] Experiment does not, according to Wertheim, permit us to form this conclusion; for the quantity $E\alpha^7$ is found to vary somewhat with the temperature. But it does not seem impossible that $f(r)$ may vary not only as $r^{-5}$ but also as a function of the temperature[1] The nature of this function of the temperature is not determined; Wertheim only remarks, after tabulating the values of $E\alpha^7$ for several metals at 100° and 200°

En comparant ces nombres avec ceux du tableau précédent on verra que l'élasticité diminue par l'accroissement de température, dans un rapport plus grand que cela ne devrait être, en vertu de la seule dilatation. Mais il faudra des expériences faites sur une plus grande échelle pour exprimer l'élasticité des métaux en fonction de leur température (p. 450).

[1301.] Wertheim concludes his first memoir with a summary of results which we briefly reproduce as giving answers to some of the problems proposed in our Art. 1293.

1°. The stretch-modulus is not constant for the same metal, every circumstance which increases the density increases the modulus, and reciprocally.

2°. Lateral and longitudinal vibrations lead to the same stretch-modulus.

3°. But these vibrations give greater stretch-moduli than are obtained by the statical stretch.

4°. The difference noted in 3° can be used as a method of determining the ratio of the specific heat at constant stress to the specific heat at constant volume. This ratio is greater for annealed than for unannealed metals see however our Art. 366.

[1] I may perhaps be permitted to draw attention to the fact that if the ultimate *atoms* be considered as spheres pulsating in a fluid medium, then there will always be an attractive force between two *molecules* varying as the inverse fifth power of the distance. This force varies also as a function of the intensities of the atomic pulsations, upon which we may suppose temperature to be based. The part of the intermolecular force which depends on the inverse square may perhaps be considered as vanishing when substituted in expression (i) above for the stretch modulus, as it involves a cosine of the difference of phase in the atomic pulsations which can possibly take every conceivable value: see 'On a certain atomic hypothesis,' *Camb. Phil. Trans.* Vol. XIV. Part II. p. 106.

5°. The stretch-modulus decreases continuously as the temperature rises from $-15°$ to $200°$ centigrade. This is true for all metals but iron and steel, for which it varies in a remarkable manner: see, however, the qualifying remarks in our Art. 1298.

6°. The Coulomb-Gerstner Law as to the constancy of the stretch-modulus is confirmed for all metals: see Art. 806.

7°. Set does not proceed by jumps and jerks, but continuously. By modifying the load and its duration any set desired can be produced. (This seems doubtful: see our Art. 856 and Appendix, Note C (9).)

8°. There is no real elastic limit; what is termed this limit depends upon the exactness of the measuring instrument. (This seems doubtful: see Arts. 853—854, 1003.)

Wertheim thus rejects all relations similar to that stated by Lagerhjelm: see Art. 366.

9°. Resistance to rupture is considerably diminished by annealing, but elevation of temperature does not much diminish the cohesion of metals previously annealed: see our Art. 692.

10°. We may include under this heading the properties of $Ex^7$ mentioned in our Art. 1299, and their bearing on the elastic and molecular order of the metals. Namely, the metals follow the same order in relation to their stretch-moduli as in relation to their molecular distances. Only platinum is placed between copper and iron in relation to its stretch-modulus and between zinc and copper in relation to its intermolecular distance.

A further result of Wertheim's that magnetisation does not affect the stretch-modulus of iron is rendered doubtful by the author's own statement in a footnote.

[1302.] *Recherches sur l'élasticité: Deuxième Mémoire. Annales de Chimie,* T. XII. pp. 581—610. Paris, 1844. Wertheim's second memoir is devoted to the consideration of the elasticity and cohesion of alloys. He begins by remarking that the only experi mental results on the elasticity of alloys are those on brass due to Tredgold, Ardant and Savart, and on bell-metal (*métal de cloche*) due to Tredgold and Bevan. The cohesion of alloys has been treated by a number of writers, notably Musschenbroek and

Karmarsch. But these experiments have not led to the statement of any definite law connecting the mechanical properties of the alloy with its chemical constitution. The discovery of such law is the object of the present memoir. The experiments were conducted in much the same manner as in the first memoir, except that, longitudinal and lateral vibrations giving the same value for the stretch-modulus, the former only were made use of. Pp. 604—607 give results for a very great number of alloys.

[1303.] Wertheim does not succeed in drawing any very definite law from these results. He remarks indeed that the simple inspection of his tables shews that there is no regularity in the elastic limits or the rupture-stretches. This of course is only saying that the law, which must exist, is not easy to find. His general conclusions are the following:

1° If one supposes that all the molecules of an alloy are at the same distance from each other, then it is found that the smaller this distance the greater is the stretch-modulus.

(The quantity $E\alpha^7$ varies however more than in the case of the simple metals (p. 609): see our Art. 1299.)

2°. The stretch-modulus of an alloy is pretty nearly equal to the mean of the stretch-moduli of the constituent metals, so that the condensations or dilatations which take place during the formation of the alloy do not sensibly modify the modulus.

3° Neither the elastic limit, the rupture-stretch, nor the cohesion of an alloy can be determined *à priori* from the same quantities in the constituent metals (p. 610). (That is to say Wertheim has not determined the relation.)

[1304.] *Recherches sur l'élasticité: Troisième Mémoire.* This memoir occurs in the same volume of the *Annales*, pp. 610—624. Its sub-title is: *De l'influence du courant galvanique et de l'électromagnétisme sur l'élasticité des métaux.* It contains, I believe, the first attempt to find a relation between the elastic and electromagnetic properties of bodies.

Après avoir étudié dans les précédents Mémoires l'élasticité des métaux dans leur état naturel, puis à différents degrés de condensation et de dilatation, et enfin aux différentes températures, il nous reste à voir si les forces mécaniques et la chaleur sont seules susceptibles de modifier leur élasticité, ou bien si l'électricité et le magnétisme peuvent produire des effets analogues (p. 610).

[1305.] Wertheim commences by noticing the close connection between molecular and electric forces, so that it is only natural to suppose that an electric current would affect the elasticity. But the problem is a complex one, as an electric current generates heat and this of course affects the elasticity Knowing however, the influence of heat from the previous experiments, it will be possible to measure any direct modification of the molecular forces. Wertheim gives a fairly complete account of his apparatus and experimental methods, as well as the tabulated results of a considerable range of experiments on the metals. He does not hold his experiments, however, sufficiently exact to enable him to state the absolute relation of the stretch-modulus to the electric current (p. 618). He then proceeds to a series of experiments on the magnetisation of iron and steel wire.

[1306.] His general conclusions are the following :

1°. The electric current produces in the stretch modulus of metal wires, which it is made to traverse, a temporary diminution. This diminution is due to the direct action of the current and independent of the diminution produced by the rise in temperature. It disappears when the current ceases, whatever may have been the current's duration.

2°. The magnitude of this diminution depends on the strength of the current and probably also on the electric resistance of the metal.

3°. The cohesion of the wire is diminished by the current; yet the variable character of this property does not allow us to determine whether this diminution is due to the direct action of the current, or whether it may not rather be only a consequence of the rise in temperature.

4°. Magnetisation (*australe que boréale*), excited by the continuous passage of an electric current, produces a small diminution in the stretch-modulus of soft iron and steel; this diminution persists in part after the current has ceased.

Compare the conclusion at which Wertheim arrived in his first memoir: Art. 1334.

[1307.] G. Wertheim: *Note sur l'influence des basses températures sur l'elasticité des métaux. Annales de Chimie*, T. XV. pp. 114—120. Paris, 1845.

In his first memoir Wertheim had considered the elasticity of the metals at temperatures of 15°, 100° and 200°; he completes these researches in this memoir by further experiments at low temperatures. The temperatures considered vary from 10° to − 20°. It is not impossible that extreme cold makes several metals more brittle, but Wertheim confines his experiments to elasticity and does not enter on the question of cohesion (see our Art. 692, 8°). The experiments embrace gold, silver, palladium, platinum, copper, iron, steel and brass. As to lead, tin and zinc, we are told that they could not be considered,

> Parce qu'ils s'effilent trop vite pour qu'on puisse prendre des mesures exactes (p. 116)

I do not clearly understand what this statement means.

[1308.] The results are collected in a most valuable table on p. 119, from which it appears that from − 10° or − 17° up to 200° the stretch-modulus of all the metals mentioned, but iron and steel, continuously and sensibly diminishes. In iron and steel however it first begins to increase, reaches a maximum (at or before 100° in some varieties), and then decreases; for at 200° it is sensibly the same as at − 15°.

I must remark that these experiments were made on wires and it would be of value to have the results verified on larger and possibly more isotropic masses of metal.

[1309.] An interesting remark with which Wertheim closes his note ought not to escape our notice:

> Je ferai encore remarquer que l'action du froid ne paraît pas toujours être passagere; mais l'augmentation de densité, et par conséquent d'élasticité, semble persister encore en partie, après que le métal est revenu à sa température primitive. C'est, du moins, ce qui paraît résulter de quelques expériences que j'ai faites (p. 120).

(Here follows an account of experiments on copper wire)...

J'ai observé la même chose sur le platine; il paraît donc que les basses températures produisent un effet permanent analogue au recuit, mais en sens opposé pourtant, il faudrait opérer à des températures beaucoup plus basses avant de pouvoir énoncer ce fait d'une maniere positive et générale.

We have suggested in Art. 1298 that Wertheim's conclusions as to the unique effect of heat on the elastic character of iron and steel may be due to his not having examined whether a permanent change had arisen from the increase of temperature The remarks cited in this article shew, however, that he was fully alive to a permanent change possibly taking place.

[1310.] E. Chevandier and G. Wertheim. *Note sur l'élasticité et sur la cohésion des différentes espèces de verre. Annales de Chimie,* T. XIX. pp. 129—138. Paris, 1847. This note was presented to the Académie on June 2, 1845. A *Note supplémentaire* appears on p. 252 of the same volume of the *Annales.*

The authors mark the divergence in the values (i) of the stretch-modulus and (ii) of the velocity of sound in glass as determined by the experiments of Savart, Chladni, Colladon and Sturm They attribute these differences to variety in the kinds of glass used and to peculiarities due to shape and working. A remark on p. 136 as to Savart's theory of the coexistence of longitudinal and lateral vibrations might perhaps deserve attention from writers on the theory of sound: see our Art. 350.

[1311.] The general conclusions which the authors draw from their experiments are:

1°. The stretch-modulus and the density are increased at the same time by annealing. (Thus glass differs in this respect from several of the metals: see Art. 1295.)

2°. The stretch-modulus found from traction is less than that found by the vibrational method: see our Art. 1297.

3° The different kinds of glass have the same order when arranged according to their stretch-moduli or according to their absolute strengths. The greater stretch modulus corresponds to the greater cohesion.

4°. There is no difference as to elasticity or as to density between the same glass whether run or drawn (*coulé et étiré*), after it has been annealed.

The other conclusions, as well as the *Note supplémentaire*, have reference to the influence of the chemical constitution and of the presence of colouring matter on the cohesion and elasticity of glass.

[1312.] E. Chevandier and G. Wertheim. *Mémoire sur les propriétés mécaniques du bois.* Paris, 1848. This is an octavo pamphlet of 136 pages and two plates. The memoir was presented to the Académie on October 5, 1846[1]. An abstract will be found in the *Comptes rendus*, T. 23, 1846, pp. 663—674. The work contains an account of the most exhaustive experiments hitherto made on wood. It is divided into five parts: (i) History of the subject, (ii) Description of apparatus and detail of experiment, (iii) Calculation of experiments and discussion of the methods employed, (iv) Results obtained, (v) Conclusions from these results.

The historical part treats of experiments most of which we have referred to in our work, namely those of Parent, Art. 28, $\alpha$; Musschenbroek, Art. 28, $\delta$; Buffon, Art. 28, $\epsilon$; Duhamel du Monceau (*Traité de la conservation et de la force des bois*, 1780); Girard, Art. 131; Perronet, Art. 28, $\zeta$; Bélidor, Art. 28, $\beta$; Dupin, Art. 162; Bevan, Arts. 373, 378; Savart, Art. 339; Wheatstone, Art. 746; Poncelet, Art. 984 (*Méc. Ind.* K. 340—350); Ardant, Art. 937; Hodgkinson, Art. 235; Hagen, Art. 1229; and Pacinotti and Peri, Art. 1250. There is some criticism of these different physicists, and finally a long series of questions which Chevandier and Wertheim themselves propose to investigate in this memoir. These questions have relation in particular to the stretch-moduli of wood in various directions, in various parts of the tree, and for trees of the same kind in various soils and at different ages (pp. 13—17).

[1] The copy which I have examined is the one presented by the authors to Alexander von Humboldt, which had found its way in an uncut state into the British Museum. The memoir is apparently very scarce.

[1313.] The experiments made were on the velocity of sound, on the flexure, and on the cohesion (pp. 35—40) under varying conditions of age, humidity and so forth. On p. 47 an empirical formula is given for the stretch-modulus as varying with the humidity. It contains two arbitrary coefficients depending on the ratios of the changes of the density and of the sound-velocity to the change of humidity. The tabulated results of the experiments will be found on pp. 74—135. They shew great variety in the values of the stretch-modulus according to the position and to the direction which the block experimented on originally took in the substance of the tree. No formula like that of Hagen or Saint-Venant (see Art. 1229) is given, but general statements as to the change in the stretch-modulus will be found in the *Conclusions* 5°—7° on pages 70—71.

[1314.] We quote the following *Conclusions* as containing matter of physical as well as of practical value:

1°. The values of the stretch moduli deduced by the method of vibrations are greater than those obtained by means of static traction. The ratios of the numbers given by the two methods are sensibly the same for trees of the same kind, whatever may be their degree of humidity. These constant ratios enable us to find the static stretch modulus from the velocity of sound.

2°. The extensions produced by terminal tractive loads in the direction of the fibres are composed of an elastic part proportional very nearly to the load, and of a permanent part which is measurable even for relatively small loads, and of which the magnitude varies not only with the load, but with its duration. (Was this really set or possibly elastic after-strain ?)

3°. The result in 2° applies also to the flexure of even strong pieces, when supported at their terminals and loaded at their centres.

4°. The stretch modulus found by the flexure of a beam (*bille*) 2 metres long agrees with the mean value of those obtained by pure tractive experiments on a great number of rods (*tringles*) taken from this beam. The accordance does not hold for resinous trees, for which flexure always gives moduli much greater than those obtained from the traction of rods: see p. 69.

[1315.] G. Wertheim: *Mémoire sur l'élasticité et la cohésion des principaux tissus du corps humain. Annales de Chimie* T. XXI. pp. 385—414. Paris, 1847. This memoir was presented to the Académie on December 28, 1846. It contains a consideration of some of the purely mechanical properties of the constituent parts of the human body.

En effet, dans la chirurgie, dans l'orthopédie, et dans la médecine légale, il se présente beaucoup de cas où il serait important de pouvoir déterminer quelles forces extérieures on peut appliquer sans danger aux parties dures ou molles du corps; quelles sont les extensions ou les flexions qu'on peut faire subir à ces parties; si une force donnée a pu ou a dû produire une rupture; et, enfin, quelle peut être l'influence du sexe, de l'âge, etc. (p. 385).

[1316.] The above paragraph will serve to shew the value of Wertheim's experiments, which were made on all parts of the human frame and upon the bodies of persons of a great variety of ages. His tabulated results occupy pp. 397—414. On pp. 395—396 will be found his general conclusions on the elastic order of the human tissues, and the influence of age, etc., on cohesion and elasticity. He gives references to somewhat similar but less wide experiments by Musschenbroek, Clifton Wintringham, Hales[1] and Valentin[2].

[1317.] The physical interest of Wertheim's paper lies however in his discovery of after-strain in the human tissues. On p. 385 he refers to Weber's experiments on silk threads, and suggests that the proportionality of stress and strain may possibly not be true for other organic substances. After remarking that as a result of experiment the proportionality of stress to strain is found to hold for certain parts of the human body, he continues:

Il n'en est plus de même pour les parties molles du corps: ici nous devons remarquer avant tout que les contractions secondaires observées

[1] Alb ab Haller, *De partium corporis humani praecipuarum fabricâ et functionibus opus*, Bernae, 1778, T. I. pp. 142 and 244.

[2] *Lehrbuch der Physiologie des Menschen*, Braunschweig, 1844, B. I. p. 34.

**par M. Weber sur des fils de soie ont lieu également dans toutes les parties qui contiennent une grande quantité d'humidité, mais qu'elles deviennent moins sensibles à mesure que ces substances se dessèchent. Ces raccourcissements secondaires n'ont été, en général, que de quelques dixièmes de millimètre dans le premier quart d'heure après l'enlèvement de la charge (p. 389).**

[1318.] The curve by which Wertheim represents the stress-strain relation (*pour les parties molles du corps*) is hyperbolic and so neither of the kind adopted by Weber nor parabolic like the set curve of Gerstner and Hodgkinson: see our Arts. 714, 804, 969.

If $y$ be the stretch and $x$ the load, he puts:

$$y^2 = ax^2 + bx,$$

where $a$ and $b$ are constants of experiment (p. 389). He calculates their values from the least and greatest stretches and finds that they then give intermediate values very accurately (so close *qu'on puisse regarder ces formules comme la véritable expression de la marche de l'élasticité*). In the case of very great extensions, as for various vessels, they give results much too small. The constant $b$ appears to be always positive, and to diminish as the tissues lose their moisture, but its value is extremely variable for different tissues, sometimes being even greater than $a$: see Wertheim's Tables, pp. 393—394.

Since the stretch-modulus is the value of the traction when the stretch is unity, Wertheim obtains the equation

$$ax^2 + bx = 1$$

to determine this modulus. This equation has only one positive root, and he takes this to be the stretch-modulus. It must be borne in mind that the stretch here considered is wholly elastic, being either elastic fore-strain or elastic after strain. Set is excluded from the experiments (p. 391).

[1319.] G. Wertheim: *Mémoire sur l'équilibre des corps solides homogènes. Annales de Chimie*, T. XXIII. pp. 52—95. Paris, 1848. This memoir was presented to the Académie on February 10, 1848. While by no means the most valuable contribution this elastician has made to our subject, it yet has attracted considerable notice, because this is the first occasion on which he proposed that relation between the elastic moduli with which his name has been generally associated, namely $\mu/E = 3/8$, or $\lambda = 2\mu$, for uni-constant isotropy The memoir is very instructive as shewing the dangers into which a physicist may fall who has not thoroughly grasped the steps of a mathematical process.

[1320.] Wertheim commences with the statement that Cauchy has obtained the equations of equilibrium in a more general form than either Navier or Poisson. The memoir of Cauchy's to which he refers is probably that considered in our Art. 610, but we must note that the equations there obtained by Cauchy do *not depend at all on a consideration of intermolecular force*, but on an assumption as to the nature of the principal tractions. Wertheim remarks that the Navier-Poisson equations are a particular case of Cauchy's bi-constant forms,—

Mais pour pouvoir admettre les hypothèses fondamentales sur lesquelles sont basés tous ces calculs, il faudrait que toutes les conséquences et toutes les lois qui s'en déduisent fussent d'abord contrôlées par l'expérience (p. 52).

[1321.] The relation to which Wertheim specially draws the attention of the reader is that between the lateral and longitudinal stretches of a rod subjected only to uniform terminal tractive load. Their ratio ($\eta$) was first stated by Poisson to be 1/4 and this was experimentally verified by Cagniard Latour: see our Arts. 368 369. Wertheim criticises at considerable length, and not without reason, Cagniard Latour's experiment, and concludes that it is not sufficient to determine finally the point. Next he turns to the experiments of Regnault (see our Art. 1358) and points out how discordant are the results he has obtained by applying the uni-constant formulae of Poisson to spherical and cylindrical vessels. The stretch-moduli so determined are greater than those which Wertheim himself had obtained from the traction of rods of the same material.

M. Regnault ne s'est pas borné à signaler ce désaccord, ainsi que l'incertitude qui règne encore dans toute cette partie de la théorie de l'élasticité, mais, de plus, il a indiqué une méthode exacte pour déterminer les changements de volume des corps solides, et le rapport entre ces changements de volume et les allongements ou compressions correspondants. M. Regnault a bien voulu m'engager à m'occuper spécialement de cette question (p. 54).

[1322.] Before proceeding to make experiments by Regnault's method, Wertheim refers to some less accurate ones he has made on india-rubber (pp. 54—56). In these experiments ($\eta$) the ratio of the lateral to the longitudinal stretch is neither $\frac{1}{4}$ following Poisson nor $\frac{1}{3}$ as it should be if $\lambda = 2\mu$, though it approaches nearer to the latter than the former value.

[1323.] The next experiments are on hollow glass and metal tubes:

La méthode proposée par M. Regnault consiste dans l'emploi de cylindres creux que l'on soumet à des tractions longitudinales. On mesure à la fois l'allongement linéaire et le changement de volume intérieur. Ce dernier est donné avec beaucoup de précision par l'abaissement de la colonne liquide dans un tube capillaire qui communique avec la cavité du cylindre. La section intérieure de celui-ci étant relativement très grande, le moindre changement de volume produit un grand changement de niveau (p. 57).

The experimental details will be found on pp. 61—73. They verify neither Poisson's nor Wertheim's value for $\eta$ although they accord better with the latter. In fact all that this somewhat limited range of experiments can be said to accomplish is to shew the inapplicability of uni constant isotropy (i.e. any rigid relation between the stretch- and slide-moduli) to the materials experimented on. They force us to conclude either for the bi-constant isotropy, or for the aeolotropy of these materials: see our Art. 1358.

[1324.] Wertheim has however great faith in his arbitrary relation, and does not seem to perceive that it stands on a very different basis to Poisson's, which is not empirical but the result of a possible molecular theory.

Il est possible que cette égalité n'ait pas rigoureusement lieu pour tous les corps solides élastiques: il est même probable qu'elle n'existe pas dans des corps qui ont passé par la filière ou le laminoir, et qui ont, par conséquent, cessé d'être des corps vraiment homogènes. Toutefois, comme les expériences ne donnent que de très petits différences, et comme nous avons trouvé la même loi par les expériences sur le caoutchouc, nous pourrons, pour le moment, l'admettre comme rigoureusement exacte, et rechercher quelles modifications il faudra apporter à la théorie, afin de la mettre d'accord avec cette loi (p. 73)

[1325.] Wertheim proceeds then to examine what changes will arise in the elastic equations when the ratio $\eta$ is taken equal to 1/3. But we now find the most singular application of the equations of Cauchy given in our Art. 615. These equations, as we have pointed out on p. 330, involve a state of initial stress. This initial stress introduces a second constant $G \; (=\widehat{xx}_0)$, which does not appear in Navier's equations. Working upon Cauchy's equations for initial stress, Wertheim shews that on his hypothesis

$$R = -3G.$$

In other words, if his hypothesis were true, there would have to be an initial stress in all isotropic bodies in a constant ratio to their elasticity! Not perceiving this remarkable result Wertheim goes on to deduce that the intermolecular force must vary as the *inverse fourteenth power* of intermolecular distance (p. 79). The fact is, that he is, apparently quite unconsciously, applying Cauchy's results, based on the ordinary uni-constant theory, to his own hypothesis which in itself contradicts that theory It was perfectly legitimate for Wertheim to use the equations of Cauchy given in our Art. 614, but with his hypothesis he must reject entirely the equations given in our Arts. 615 and 616.

[1326.] On pp. 81—87 we have the ordinary formulae for rods, cylinders, spheres, etc., on the assumption that $\lambda = 2\mu$. It is then shewn that these formulae give results more in accordance with Regnault's experiments than those based on $\lambda = \mu$. But the accordance in itself is not so great as to carry conviction see Saint-Venant's criticism on pp. 665—681 of his edition of Navier's *Leçons.*

The memoir concludes with a correction of Oersted's statement in the *Annales de Chimie,* T. XXXVIII. (see our Art. 689), and a remark on the ratio of the velocities of sound (of the *dilatation* waves) in a solid mass and in a bar of the same material on the hypothesis $\lambda = 2\mu$ (i.e. the ratio $= \sqrt{3/2}$).

[1327.] Guillemin. *Observations relatives au changement qui se produit dans l'élasticité d'un barreau de fer doux sous l'influence de l'électricite. Comptes rendus*, T. 22, 1846, p. 264.

This is a short note to the effect that a horizontal iron bar surrounded by a coil has its elasticity augmented by magnetisation when a current is passed through the coil. This was proved by a horizontal bar, built in at one terminal, slightly raising on the passage of a current a weight suspended from its other terminal; that is, by a *flexure* experiment.

Cette action est peu énergique, mais elle est cependant assez sensible pour qu'on puisse la constater sans aucun appareil micrométrique, en se servant d'un seul élément de Bunsen et d un barreau de 1 centimètre de diamètre sur 20 ou 30 de long.

[1328.] This result was opposed to Wertheim's, namely, that the stretch-modulus was decreased by such a current: see our Art. 1306, 1°. The latter physicist accordingly criticised Guillemin's result in a *Note sur les vibrations qu'un courant galvanique fait naître dans le fer doux*, which will be found in the same volume of the *Comptes rendus*, pp. 336—339. He attributes the decreased flexure noted by Guillemin to the fact that his bar was not accurately in the axis of the coil and quotes some experiments of his own with a large coil (*une grande bobine*), from which

On voit donc qu'en rapprochant la barre des points correspondants de la circonférence de la bobine, on peut la faire fléchir horizontalement ou verticalement, ou dans une direction intermédiaire quelconque (p. 337).

Further, he seems to attribute the apparent decrease in the stretch-modulus which he had himself noted, to an extension of the bar produced by the magnetic forces between the bar and the coil. This conclusion was opposed to some results of de la Rive: see our Art. 1336.

[1329.] In a letter to Arago de la Rive replies to the remarks of Wertheim. An extract from this letter entitled: *Sur les vibrations qu'un courant électrique fait naître dans un barreau de fer doux*, is printed in the same volume of the *Comptes rendus*, pp. 428—432. De la Rive remarks:

M. Wertheim estime qu'il n'y a qu'une action mécanique dans le phénomène des vibrations qu'éprouve le fer doux par l'influence extérieure ou par la transmission intérieure d'un courant élastique, tandis que je vois dans ce phénomène une action moléculaire. Voilà en quoi gît la différence importante qui sépare la manière de voir de M. Wertheim de la mienne.

His note cites various experiments which he thinks conclusively prove a change in molecular condition.

[1330.] It is immediately followed by a letter of Guillemin's to Arago entitled: *Réponse aux remarques faites par M. Wertheim* ...pp. 432—433. Guillemin replies that his bar was axial, because he wound the coil upon it as core. He further remarks that the action of the elements of the coil on each other does not produce the change in flexure, for there is no change in flexure when the coil is placed on a bar of wood.

[1331.] On p. 544 of the same volume of *Comptes rendus* there is a short note to the effect that M. Wartmann, professor in Lausanne, has sent a letter saying that his experiments confirm de la Rive's results.

A paper of Wartmann's on the subject will be found in the *Philosophical Magazine*, Vol. XXVIII. pp. 544—546, 1846. He finds that the tenacity of wires is altered by a prolonged current of electricity, while their elasticity is altered by an intermittent current.

[1332.] Wartmann's remarks are immediately followed in the *Comptes rendus* by a note of Wertheim's entitled: *Réponse aux remarques de M. de la Rive sur une Note*... pp. 544—547. He points out that he never meant to deny the existence of a molecular action, and cites his memoir of July, 1844, to that effect: see our Art. 1304. But he believes that certain phenomena which are not explicable by the molecular can be explained by the mechanical action. He promises to investigate the whole matter in an early memoir. Finally he acknowledges that his criticism of Guillemin, owing to the latter's mode of experimenting, had no application. His own experiments had been made on wires and thin bands of soft iron. He intends to make investigations bearing on the experiment of Guillemin. The memoir to which Wertheim here refers is that considered in the following Article.

[1333.] The next memoir of Wertheim's that calls for our notice appears on pp. 302—327 of T. XXIII. of the *Annales*. It is entitled: *Mémoire sur les sons produits par le courant électrique*, and was presented to the Académie on May 1, 1848. It is his second memoir on the electro-elastic and magneto-elastic properties of bodies: see our Art. 1304.

In 1837 Page[1] had remarked that a bar of iron gives a sound the moment it is magnetised by the passage of an electric current. Marrian[2] had noticed that the influence of an 'external current' (traversing a helix or coil in the axis of which is placed the stretched iron bar or wire) produces a note identical with that obtained by striking the bar on one of its ends in the direction of its axis; further, that the same note is given by bars of the same dimensions of iron, tempered steel and steel previously magnetised —other metals give no sounds. Matteucci[3] determined the relation between the strength of the current and the intensity of the sound. De la Rive[4] and Beatson[5] discovered that a transmitted current produces also a sound. Guillemin[6] observed that an external current alters the transverse rigidity of a soft iron bar, thus extending and confirming Wertheim's results: see our Arts. 1306 and 1327. Wartmann[7] recognised that the sound does not depend on the electrical resistance of the bar, or that heat plays only a very insignificant part in the phenomenon. De la Rive[8] communicated to the Royal Society the discovery that all conductors when placed under the influence of a strong electro-magnet give a very pronounced sound on the passage of a current. Joule[9] made more careful

[1] *American Journal of Science*, Vol. XXXII. 1837, p. 396 (*Galvanic Music*), and XXXIII. 1838, p. 118.

[2] *L'Institut*, No. 576, p. 20. 1845. *Electrical Magazine*, Vol. I. p. 527.

[3] *L'Institut*, No. 609, p. 315, 1845.

[4] *Comptes rendus*, T. XX., 1845, pp. 1287—1291. Translation in the *Electrical Magazine*, Vol. II. pp. 28—33.

[5] *Electrical Magazine*, April, 1845. Vol. I. p. 557.

[6] *Comptes rendus*, T. XXII., p. 264. 1846.

[7] *Comptes rendus*, T. XXII., p. 544. 1846.

[8] *Phil. Trans.*, Part I. p. 31, 1847; *Annales de Chimie*, T. XIX. p. 378.

[9] *Philosophical Magazine*, 1846, Vol. XXX. pp. 76—87, and pp. 225—241. Joule refers to an experiment made by him in 1841. The paper contains a mass of experimental statistics of the influence of magnetism on the length and bulk of wires and bars.

investigation of the action of magnetisation on the volume of iron and steel bars. He concluded that magnetisation extends iron and steel bars; that this strain is partly temporary and partly permanent; that each of these parts is proportional to the square of the magnetic force of the bar; that after reaching a certain charge however the bar begins to contract; that this contraction is proportional to the strength of the current and to the magnetic intensity of the bar; and finally that the extension is probably due to the molecular forces of the bar, while the contraction arises from the attraction of the coil (the external current) on the magnetised molecules. Beatson[1] obtained results similar to those obtained by Wertheim for an external current (see our Art. 1328) for means of a transmitted current; the extension was distinct from that produced by the heating of the bar, but was sensible only for iron.

Such, taken in conjunction with Wertheim's previous results (see our Arts. 1304—1306), is a brief historical description of the state of knowledge with regard to the electro-elastic and the magneto-elastic properties of bodies, when Wertheim returned to the subject in 1848: see his account, pp. 302—310 of the memoir.

[1334.] He made some careful experiments on the influence of external currents (when the bar was or was not in the axis of the coil— central or eccentric'), transmitted currents and combinations of the two, and draws the following conclusions:

1°. A current traversing a coil exercises upon a mass of iron placed within it a mechanical attraction identical with that which, according to the discovery of Arago, a wire conductor exercises upon iron-filings.

2°. The stress can be considered as composed of two forces, one longitudinal and the other transverse.

3°. It is proportional to the intensity of the current and to the mass of the iron.

4°. The longitudinal component can tend, according to the position of the coil, to extend or contract the bar of iron.

[1] *Electrical Magazine*, April, 1846, Vol. II. pp. 296—300. We have already referred to Sullivan's memoir of 1845: see our Art. 1248. C. V. Walker writing in the *Electrical Magazine*, Vol. I. 1844, p. 528 states that Marrian had suggested that a mechanical vibration or note would produce electricity. Beatson's experiments, however, do not confirm Sullivan's results: see p. 298 of the above memoir of 1846.

5°. The transverse components, the mechanical equivalent of which can be easily expressed in pounds when the iron is in an eccentric position, are null when the iron is central to the coil.

6° A transmitted current produces an impulse (*un choc brusque*) on traversing an iron conductor.

7°. There is a complete analogy between the action of a current and that of a purely mechanical force acting in the same direction.

8°. The above propositions explain all the sounds which can be produced by an external or by a transmitted current in bars, wires, or plates of iron and steel[1] (pp. 326—327).

Wertheim propounds the following questions as deserving investigation:

(*a*) Does stretch take place in a mass of iron owing to its magnetisation independently of the mechanical action of the coil?

(*b*) Magnetised iron appears to be no longer mechanically isotropic. What are the position and ratio of its axes of elasticity?

(*c*) How can a current transmitted along a wire produce a 'mechanical impulse'? Does this take place by the mutual action of molecules magnetised perpendicularly to the current?

(*d*) The last question is as to the nature of a *bruit de ferraille* which is sometimes produced and does not seem (see p. 321) due to the longitudinal or transverse vibrations but appears to 'run along the wire'

[1335.] Besides the points we have noted there are (pp. 314 and 316) some valuable experiments on the influence of magnetisation on the stretch in a bar due to a terminal tractive load and on the flexure due to a terminal shearing load.

The memoir is among the more important of the earlier physical papers which treat of the magneto-elastic properties of bodies.

[1336.] A paper by de la Rive entitled: *Nouvelles recherches sur les mouvements vibratoires qu'éprouvent les corps magnétiques et*

[1] Have we not here the beginnings of telephonetic discovery?

*les corps non magnétiques sous l'influence des courants électriques extérieurs et transmis,* will be found in the *Annales de Chimie,* T. 26, pp. 158—174, 1849. It contains some reference to Wertheim's memoir of 1848 and to the controversy in the *Comptes rendus:* see our Arts. 1327—1333. The writer accepts Wertheim's compromise of a direct molecular as well as a mechanical action on the bar due to an electric current.

A translation of this paper will be found in the *Phil. Mag* Vol. XXXV. 1849, pp. 422—434. The volumes of the same periodical for this decade contain various other memoirs, as those by Wartmann, which treat of this subject, but their bearing on elasticity is only indirect.

[1337.] G. Wertheim: *Mémoire sur la vitesse du son dans les liquides. Annales de Chimie,* T. XXIII. pp. 434—475. Paris, 1848. This memoir concerns us only very slightly, but we must draw attention to one or two statements made in it.

[1338.] The argument of this memoir is of a rather singular kind. The velocities of sound in large masses of liquid have been calculated *directly* only for water. Wertheim notices that experiments on *columns* of water give a much less result. The question which he now sets himself to solve is the following: What is the relation of the velocity of sound in a large mass of liquid to its velocity in a column of the same liquid? In other words, How can the first velocity be deduced from the second, and so the second compared with the result obtained by calculation from the compressibility of the given liquid? Now Wertheim finds that for water the ratio of these two velocities is very nearly equal to $\sqrt{3/2}$. But this is what, *upon Wertheim's own hypothesis* (see our Art. 1326), the ratio ought to be in the case of an unlimited elastic mass and a column of the same material. He then argues as follows:

La coïncidence de ces deux nombres prouve que la loi s'applique réellement aux liquides, que par conséquent l'égalité de pression en tout sens n a pas lieu pendant leurs vibrations sonores, et qu'une colonne liquide vibrant longitudinalement donne le même son que rendrait une barre solide dont la matière aurait la même compressibilité cubique que le liquide (p. 466).

Lastly we may cite the following paragraph:

Il s'ensuit également que les lois de l'équilibre des corps solides s'appliquent aux liquides pendant un très-court intervalle de temps après l'application des forces extérieures. Ainsi donc, si on pouvait suspendre librement une colonne liquide, si on pouvait appliquer à ses deux extrémités une traction instantanée, et si on pouvait dans ce moment mesurer sa longueur et son volume, l'augmentation de volume serait égale à un tiers de l'allongement, et on pourrait calculer l'une et l'autre d'après la compressibilité cubique. Enfin, la loi de l'attraction moléculaire doit être la même pour les liquides que pour les solides (p. 467).

Wertheim's theory is thus based upon two hypotheses; one is apparently suggested by the single experiment on water, the other is the peculiarly doubtful $\lambda = 2\mu$.

[1339.] G. Wertheim: *Note sur la torsion des verges homogènes. Annales de Chimie,* T. XXV. pp. 209—215. Paris, 1849. The object of this memoir is to shew that all torsion experiments confirm the supposition made by Wertheim in his memoir of February, 1848, considered in our Art. 1319; namely, that with our notation $\lambda = 2\mu$, or the stretch-modulus bears to the slide-modulus the ratio of 8 : 3.

[1340.] Wertheim begins his memoir by the remark that the stretch-moduli as calculated from the torsion experiments of Coulomb, Duleau, Savart, Bevan and Giulio (see our Arts. 119, 229, 333, 378, 1218) do not agree within the limits of experimental error with those obtained from traction experiments. Accordingly Wertheim takes Poisson's result for a cylindrical rod and modifies it on the above supposition. Referring to Saint-Venant's paper in the *Comptes rendus* (T. XXIV. p. 486)—to be considered in our next chapter—he remarks on the correction of Cauchy's formula for rectangular rods by the introduction of a numerical constant independent of the elasticity. For comparison with experiment he then adopts Saint-Venant's formula with his own value of the ratio between the stretch- and slide-moduli.

[1341.] The larger portion of the *Note* is an endeavour to shew that the formulae thus obtained agree better with experiment than those deduced from the usual uni constant theory $\lambda = \mu$. The experiments Wertheim makes use of are those of Coulomb, Duleau, Savart, Giulio, and finally in the matter of torsional vibrations some fresh ones of his own. He replaces Poisson's relation $n/n' = \sqrt{5/2}$ (see our Art. 470) by $n/n' = \sqrt{8/3} = 1{\cdot}6330$. This value he holds agrees better with Savart's result 1·6668 and the mean of his own, 1·6309.

[1342.] The *Note* concludes with a statement that the author had intended to verify his relation, $\lambda = 2\mu$, by applying it to elastic plates:

Mais M. Kirchhoff ayant annoncé à l Académie qu'il s'occupait de ce sujet, je crois devoir attendre que cet habile géomètre ait publié les résultats de ses recherches (p. 215).

Kirchhoff's memoir was published in January, 1850, and will be considered in the chapter devoted to that physicist. It may suffice here to remark that the pitch of the notes given by elastic plates agree better with Poisson's than with Wertheim's hypothesis, though differing considerably from both; the radii of the nodal circles agree better with Wertheim's than Poisson's. But Kirchhoff himself remarks that the difference is so small as to be no argument against Poisson's assumption. Possibly it is an argument against uni-constant isotropy or rather for the aeolotropy of the plates.

[1343.] This *Note* of Wertheim's led to a polemic with Saint-Venant. In a later and more considerable work on Torsion, the memoir of 1855, Wertheim returns to the same subject, and even supports Cauchy's erroneous theory of the torsion of rect angular prisms. It must be noted that in this controversy there are three points to be considered

1°. Were the prisms experimented on really isotropic?

2°. If they were isotropic ought we to put $\lambda = \mu$ with Poisson or $\lambda = 2\mu$ with Wertheim, or to accept neither relation?

3°. Is Saint-Venant's numerical factor' a true correction of the ordinary theory?

Saint-Venant in the fourth *Appendice* to his edition of Navier's *Leçons* conclusively answers the question 3°. We have seen in considering the memoir of 1848 that it is extremely improbable *à priori*, and certainly contrary to much experiment *à posteriori*, that $\lambda$ should equal $2\mu$. Either the difference between theory and experiment must be accounted for by bi-constant isotropy, or perhaps more satisfactorily by supposing the wires experimented on to be non-isotropic see our Arts. 831, 858, and Saint-Venant's fifth *Appendice* to Navier's *Leçons* referred to in our Art. 923.

[1344.] 1849. G. Wertheim. *Mémoire sur les vibrations des plaques circulaires. Annales de Chimie*, Tom. 31, 1851, pp. 1—19. This memoir was presented on October 1, 1849. It is occupied with a further consideration of Wertheim's theory of the relation which should hold between the two constants of elasticity in the case of isotropy.

[1345.] The memoir begins with a reference to the torsional experiments of Kupffer: see our Art. 1389. These experiments had given values of the stretch-moduli considerably smaller than those obtained by a terminal tractive load. Wertheim endeavours to shew that with his relation between the constants there is identity in the results obtained by the two methods. The divergence however is most probably due to the non-isotropic character of the wire employed.

[1346.] Wertheim then proceeds to the main question of his memoir, whether on his hypothesis ($\lambda = 2\mu$) the calculated and observed values of the radii of the nodal circles will not be in greater agreement than on Poisson's theory. It will be seen on referring to our Arts. 519 and 520, that Savart's value was too small for the second nodal radius, and too great for the first and third. Wertheim proceeds to calculate the value of these radii on his theory, and, as Kirchhoff has noted, the values do not differ very much from those obtained on Poisson's supposition:

Toutes les différences, à l'exception de celle qui existe sur la première valeur de $\mu^2$ [$=\lambda_1^2$, of our Art. 518], sont tellement petites, qu'elles tombent nécessairement entre les limites des erreurs d'expérience; c'est donc le son fondamental surtout qu'il faudra chercher à déterminer avec exactitude.

On voit, du reste, qu'indépendamment du changement de formule, il a suffi de calculer, avec plus d'exactitude, les valeurs de $y$ pour faire presque disparaître les différences constantes qui existaient encore entre le calcul de Poisson et les expériences de Savart (p. 11).

If in the equation of Art. 518 we put $\frac{1}{3}$ instead of 3/8 as the coefficient of the long bracket, we have Wertheim's equation; his $\mu^2$ being then Poisson's $\lambda_1^2 (= 4x_1)$. His $y$ is the same as the $y_1$ of Art. 519. We have already noted the want of accuracy in Poisson's calculation: see our footnote, p. 266.

[1347.] Wertheim apparently holds that his experimental results confirm his theory, but this is hardly the fact. The discordance between theory and observation is most probably due either (1) to isotropy being bi-constant, or (2) to the plates experimented upon being really aeolotropic. This latter view receives some confirmation from Wertheim's remarks on p. 13. Referring to some tabulated results he writes:

Nous avons marqué par des astérisques les cercles dans lesquels on commençait à apercevoir une légère ellipticité......En général, l'excentricité de l'ellipse qui remplace le premier cercle, augmente à mesure que le son s'élève dans la série des harmoniques de la plaque, et son grand axe ne se place que suivant deux diamètres déterminés, et perpendiculaires l'un sur l'autre; pour obtenir les sons aigus avec facilité, il faut soutenir la plaque par les deux points dans lesquels un cercle nodal rencontre l'un de ces deux diamètres. *Dans les plaques en laiton, ces deux diamètres, et par conséquent les axes de l'ellipse, font un angle de* 45° *avec la direction du laminage.* Ces faits sont analogues à ceux que Savart a observés en produisant l'ébranlement par un point de la circonférence; nous aurons à les étudier, lorsque nous nous occuperons des plaques dont l'élasticité n'est pas la même dans tous les sens.

[1348.] Wertheim refers to Kirchhoff's memoir, which had only been published in part at the time his own paper was presented. In the interval which elapsed before the publication of the latter in the *Annales*, Kirchhoff's memoir was published in Crelle's *Journal* (B. XL.), and on p. 7 Wertheim refers to it in a footnote. The reader of Kirchhoff's memoir, having regard especially to the last few pages, will find Wertheim's footnote not wholly satisfactory. It is as follows:

La comparaison des résultats de son calcul avec les mesures données par M. Strehlke a conduit M. Kirchhoff à divers résultats analogues à ceux que j'ai obtenus moi même.

Kirchhoff, as we shall see when considering his memoir in our second volume, does not by any means accept Wertheim's hypothesis ($\lambda = 2\mu$).

On the whole Wertheim's memoir has greater value for its experimental results, than for the support it gives to his hypothesis.

[1349.] G. Wertheim. *Mémoire sur la propagation du mouvement dans les corps solides et dans des liquides. Annales de Chimie,* T. 31, 1851, pp. 19—36. This memoir was presented to the Académie on December 10, 1849.

Wertheim, referring to the memoirs of Poisson, Cauchy and Blanchet on the nature of waves in an infinite isotropic medium, shews that, on his hypothesis of $\lambda = 2\mu$, the velocity of propagation of the longitudinal wave would be *double* that of the transverse wave. He strives to find evidence in favour of this from the sounds produced by bars which are vibrating longitudinally and transversely, and argues from experiment that the same ratio of velocities will hold for a bar as for an infinite solid. His reasoning does not seem to me very clear. He again introduces (as in the memoir referred to in our Art. 1337) liquids as behaving absolutely like solids in relation to sound vibrations (p. 23), and he states that the same differential equations apply to both cases and the same ratio of the velocities for the two waves. I do not understand this.

Some remarks on the two waves which ought to be expected in the case of earthquakes are of interest (pp. 22—23).

[1350.] G. Wertheim. *Note sur la vitesse du son dans les verges. Annales de Chimie,* T. XXXI. 1851, pp. 36—39.

Wertheim in his memoir of 1842 had found that the stretch-modulus when obtained by vibration experiments had always a greater value than when obtained from traction experiments[1]. To

[1] See, however, my footnote, p. 702.

explain this difference he supposed the sound to be accelerated by the heat given off, and made use of a formula stated by Duhamel (see our Arts. 887 and 1297) to obtain the ratio of the two specific heats by means of the ratio between the two velocities of sound. Clausius published a memoir in which he shewed that the formula used by Wertheim applied to the propagation of spherical waves in the interior of an elastic solid, and not to wave motion along a rod: see our Art. 1403. Clausius also pointed out that the true formula when applied to Wertheim's experiments led to impossible values for the ratio of the specific heats: see the Article referred to above.

In the present note Wertheim gives rather a lame excuse for this slip on his part; namely, that the distinction between the propagation of sound in a rod and in an unlimited mass was not then admitted by physicists. An excuse which will hardly be accepted by those acquainted with Poisson's and Cauchy's re searches.

Wertheim now apparently rejects the explanation which might be derived from the specific heats, and seeks to explain the difference between theory and experiment by considering the difference in the velocity of sound in a bar and in a plate. Thus he writes

> Mais maintenant qu'il ne reste plus de doute sur la distinction à établir entre les deux vitesses de propagation, l'hypothèse de l'accélération du son dans les corps solides, par suite de la chaleur dégagée, me semble d'autant moins soutenable, que cette accélération n'a positivement pas lieu dans les liquides, quoique ces derniers se comportent, par rapport aux vibrations, absolument comme les corps solides; il faudra donc chercher à expliquer autrement la différence entre la vitesse théorique et expérimentale (p. 37).

[1351.] Wertheim then quotes from the third volume of Cauchy's *Exercices*, which be it noted was published in 1828. Taking our Articles 649 and 654, we have what Wertheim reproduces, with however the alteration introduced by his own hypothesis ($\lambda = 2\mu$). Thus he makes the ratio of the velocity of sound in a plate to that in a rod $= \sqrt{\frac{9}{8}}$, while Cauchy and Poisson give it as $\sqrt{\frac{16}{15}}$. These ratios refer to a plate of *indefinite* extent and to dilatation vibrations. The concluding remarks of Wertheim I do not understand, nor do his experimental results seem to be at all in harmony with his theory (pp. 38—39). They point either to bi-constant isotropy or to aeolotropy in the material experimented on.

A further series of memoirs due to Wertheim will be considered in our second volume.

[1352.] Oliver Byrne. *A new Theory of the Strength and Stress of Materials. The Civil Engineer and Architect's Journal*, Vol. IX. London, 1846, pp. 163—167, and pp. 231—232. A criticism of this paper, presumably by the editor of the *Journal*, appears on pp. 204—205, and a letter of the author as rejoinder on p. 257. The 'new theory' seems to be of a rather confused nature and its propounder a somewhat self-confident character. Two extracts will suffice:

He must be a very clever man indeed who determined the modulus of elasticity of pipe-clay. Mere book-makers like Hall and Moseley, of King's College, cannot be offended; but men like Barlow and Hodgkinson, who have lost their time experimenting to find them, may be a little indignant to find their favourite numbers spoken so lightly of.

In the next number will be pointed out the erroneous principle upon which Hooke's law is founded.

I do not know whether Professor Byrne ever carried out this latter intention, as his contributions to this *Journal* seem to have ceased.

[1353.] W. R. Johnson. *Effect of Heat on the Tenacity of Iron The American Journal of Science and Arts*, edited by Silliman and Dana. Second series, Vol. I. pp. 299—300. 1846.

A committee seems to have been appointed by the Franklin Institute to draw up a report on the strength of materials for steam-boilers. This report was published in 1837, but I have not consulted it. A short notice, with a table of some of the experiments embodied in the report, is given in the above *Journal*. The table exhibits the effect of heat on thirty-two varieties of malleable iron The experiments go to shew that a great traction applied to an iron-bar at a high temperature increases its absolute strength when cold again; the process is here termed 'treatment with thermo-tension.' The average gain by the treatment was 17·85 p.c. (from 8·2 to 28·2 per cent.), the average temperature at which the effect was produced being 573°·7 Fahr. Further, the absolute strength was found to be greater at high temperatures than at low: see our Art. 1524[1].

The experiments also confirmed the fact that the total elongation of a bar of iron, broken in its original cold state, is from two to three times as great as the *same force* would produce upon it if applied at a temperature of 573°, which force will moreover not break the bar at that temperature. An average difference of 5·9 per cent. was found in the absolute strength of the bar when hot and cold.

[1354.] In the *Artizan* for 1846, p. 127, there will be found some remarks of Fairbairn and Hodgkinson on the *Strength of Wrought Iron Pillars.* The only copy accessible to me was that in the British Museum; pp. 86—207 are, however, missing.

1355. In 1846 the Paris Academy of Sciences proposed a question in the theory of elasticity as the subject for the great prize of mathematics to be awarded in 1848: see the *Comptes Rendus,* Vol. XXII. p. 768. The problem is thus enunciated:

Trouver les intégrales des équations de l'équilibre intérieur d'un corps solide élastique et homogène dont toutes les dimensions sont finies, par exemple d'un parallélépipède ou d'un cylindre droit, en supposant connues les pressions ou tractions inégales exercées aux différents points de sa surface.

I presume that no satisfactory essay was contributed in competition for the prize, as the problem is held to be unsolved in Lamé's *Leçons sur la théorie...de l'élasticité...* 1852 see p 162 of the work.

We have in the *Comptes Rendus,* Vol. 38, 1854, p. 223, a notice respecting this matter. The subject was proposed for 1848, and again for 1853. One memoir was sent in but no prize awarded.

[1356.] Ludwig Wilhelmy. *Die Wärme als Maass der Cohäsion.—Inaugural-Dissertation.* Heidelberg, 1846. This is an octavo pamphlet of 27 pages divided into two parts, respectively

[1] Some unpublished experiments of A. B. W. Kennedy's shew that the absolute strength of *steel* at 450° is 37/30 of the absolute strength at 70°, but at 600° only 33/30; in other words, it reaches a maximum between these temperatures.

entitled: *Resultate der Berechnung* (pp. 1—18) and *Theoretische Betrachtungen zur Erläuterung der Rechnungs-Resultate* (pp. 19—27). To the work are attached two tables of experimental data.

The object of this dissertation is expressed in the following words:

Der bekannte Zusammenhang der Cohäsion und der Wärme, wonach die Cohäsion in ihren Wirkungen,—in denen sie die Erscheinungen der Capillarität und Adhäsion, die Festigkeit und Dichtigkeit der Korper zur Folge hat,—geschwächt wird durch eine Steigerung der Wärme, veranlasste mich den Versuch zu machen, beide durch ein gemeinschaftliches Maass auszudrücken (p. 1)....Vermehrung des Drucks wirkt wie Verminderung der Wärme, und umgekehrt. Beide—Streben des Zusammenhangs und der Ausdehnung—sind sich gerade entgegengesetzt, man kann sie daher auch durch ein gemeinschaftliches Maass und durch dieselbe Einheit ausdrücken, also von Cohäsions- wie von Wärme-Graden sprechen, so dass durch Hinzufügen eines Warmegrades ein Cohäsionsgrad aufgewogen wird und umgekehrt. Die Dichtigkeit eines Körpers wird bei gleicher Temperatur direckt proportional sein der Anzahl seiner Cohäsionsgrade, die Ausdehnung, das Volum, bei gleichem Werthe der Cohäsion, der Anzahl der Wärmegrade. Da sich bei ungeänderten Massen die Dichtigkeiten umgekehrt verhalten müssen, wie die Volume, so folgt daraus der Zusammenhang, in welchem anderseits Wärme und Dichtigkeit, Cohäsion und Volum mit einander stehen (pp. 19—20).

From these principles Wilhelmy attempts to deduce the laws which connect for gases, fluids and solids, the volume, density and temperature. These deductions do not seem to me satisfactory, because I am unable to follow the course of the argument on p. 21, wherein a certain quantity $R$ is used as a common measure of atomic density, of the sphere of atomic influence and finally of heat and cohesion. I suppose that, if the results on p. 21 were even admitted, they must be considered as flowing from a somewhat limited atomic hypothesis, which, however, does not seem to be anywhere clearly stated.

[1357.] V. Regnault. *Septième Mémoire. De la compressibilité des liquides, et en particulier de celle du mercure. Mémoires de l'Académie* Tome XXI. 1847, pp. 429—464.

This memoir has interest for us as involving the question of the relation between the dilatation- and the stretch-moduli. The reader may compare Art. 1227.

Regnault begins by referring to the labours of Oersted, Colladon and Sturm, and G. Aimé. He refers to Poisson's mathematical investigation and states that physicists have in general accepted his results: see our Arts. 686—691. Next Oersted's results are quoted as disagreeing with Poisson's theory, and Regnault remarks:

On ne peut pas se dissimuler que les formules mathématiques, du genre de celles dont nous nous occupons, ne présentent de grandes incertitudes, par suite des hypothèses que l'on est obligé de faire sur les forces moléculaires, pour établir les équations différentielles du problème. Ces hypothèses s'éloignent probablement beaucoup de la réalité. Ainsi les géomètres admettent que les molécules d'un corps solide se meuvent avec une égale facilité dans tous les sens, et qu'un déplacement égal, suivant une direction quelconque, développe toujours une force de réaction égale. Cette proposition est certainement inexacte, même dans les corps à crystallisation confuse. Il est très-probable qu'une molécule d'un corps solide éprouve des résistances très-inégales dans ses déplacements en différents sens. On peut, jusqu'à un certain point, se représenter ces molécules, comme formant des espèces de systèmes articulés, chaque système prenant le mouvement qui lui est le plus facile, lorsqu'une pression s'exercant à la surface extérieure du corps détruit l'équilibre moléculaire (p. 432).

[1358.] Regnault then proceeds to direct experiments on the compressibility of solids. On pp. 438—442 will be found a note by Lamé on elastic formulae for spherical, cylindrical, and hemispherically terminated cylindrical shells on the supposition of isotropy. These formulae are applied to the very careful experimental data obtained by Regnault. He finds that the 'cubical compressibility' of certain metals as obtained by direct experiment is less than that which would be obtained on Poisson's uniconstant hypothesis from Wertheim's values of the stretch-moduli for the same metals. This result has been used as an argument by some writers against uni-constant isotropy. It appears to be rather an argument against the isotropy of the vessels used by Regnault and of the bars and wires experimented on by Wertheim. Indeed Regnault's remarks quoted above seem rather to point to his attributing the divergence to aeolotropy. At the same time the divergence is by no means great, and he concludes:

Les expériences qui précèdent ne peuvent donc pas être considérées comme établissant l'exactitude des formules mathématiques; mais on ne peut pas non plus les regarder comme condamnant ces formules, parce que l'on peut attribuer les divergences à ce que les piézomètres que nous construisons, s'éloignent trop des conditions géométriques et physiques qui ont été admises dans l'établissement des formules (p. 456).

A very careful consideration of Regnault's experiments is given by Saint-Venant in his edition of Navier's *Leçons,* pp. 650, 665—676. He shews that they agree quite as well with Poisson's as with Wertheim's relation, and attributes the divergence which exists to the aeolotropic character of the vessels employed.

[1359.] We may mention in connection with Regnault's memoir, a paper by Grassi which will be found in the *Annales de Chimie,* T. XXXI. 1851, pp. 437—476. It is entitled: *Recherches sur la compressibilité des liquides.* The writer applies to the theoretical results, with which Lamé provided Regnault, the hypothesis of Wertheim, $\lambda = 2\mu$: see our Arts. 1326 and 1358. This is the basis upon which he calculates the compressibility of his piezometer. He remarks:

M. Wertheim a fait voir l'inexactitude de la loi de Poisson, car les résultats que l'on en déduit ne s'accordent pas avec ceux que donnent les expériences directes faites sur le changement de volume des corps soumis à différentes pressions ou tractions. Ces expériences prouvent, en effet, que la compressibilité ou la dilatabilité cubique est égale à la compressibilité ou dilatabilité linéaire (p. 440).

The last sentence expresses a far greater confidence in the results of Wertheim's experiments than seems to me justifiable. I may note that Grassi found very little change in the compressibility of glass due to temperature. In one of his piezometers there was a slight increase of the compressibility with the temperature (p. 453).

[1360.] 1847. M. O'Brien. *On the Symbolical Equation of*

*Vibratory Motion of an Elastic Medium, whether crystallized or uncrystallized.* *Camb. Phil. Trans.*, Vol. VIII. pp. 508—523. The paper was read March 5, 1847.

The object of the following Paper is twofold; *first*, to shew that the equations of vibratory motion of a crystallised or uncrystallised medium may be obtained in their most general form, and very simply, without making any assumption as to the nature of the molecular forces; and, *secondly*, to exemplify the use of the symbolical method and notation explained in two Papers read before the Society during the present academical year.

The symbolical method and notation referred to is practically that of the Quarternion Calculus.

[1361.] If $\alpha, \beta, \gamma$ be direction units, or three lines each of unit length drawn parallel to the axes, $V = \alpha u + \beta v + \gamma w$, where $u, v, w$, are the shifts, and $\mathfrak{D}$ be taken for the operation:

$$\alpha \frac{d}{dx} + \beta \frac{d}{dy} + \gamma \frac{d}{dz};$$

further, if

$$\sigma = \alpha x + \beta y + \gamma z,$$
$$\sigma' = \alpha x' + \beta y' + \gamma z',$$

the symbol $\Delta\sigma'\sigma$ standing for $xx' + yy' + zz'$; then, the general equation for vibratory motion of an isotropic medium is

$$\frac{d^2V}{dt^2} = B(\Delta\mathfrak{D} \,.\, \mathfrak{D}) \,.\, V + (A - B)\,\mathfrak{D}\Delta\mathfrak{D} \,.\, V,$$

where $A$ and $B$ are elastic constants of the medium (p. 515).

[1362.] O'Brien shews that the symbol $\mathfrak{D}$ written before any quantity $v$ which is a function of $x, y, z$ has a remarkable signification: 'the *direction unit* of the symbol $\mathfrak{D}v$ is that direction *perpendicular* to which there is no variation of $v$ at the point $x, y, z$, and the *numerical magnitude* of $\mathfrak{D}v$ is the *rate of variation* of $v$, when we pass from point to point *in that direction.*'

He also shews that $\Delta\mathfrak{D} \,.\, V$ is a numerical quantity representing the *degree of expansion*, or what is called the *rarefaction* of the medium at the point $x, y, z$': see p. 510.

[1363.] In investigating the motion of a crystalline medium, O'Brien follows closely the relations between the constants suggested by Fresnel's theory of transverse vibrations; thus the equations at which he arrives belong rather to the theory of light than to that of elasticity proper. He starts with 15 constants; by assuming six relations essential to Fresnel's theory he gets rid of 6 of these; 3 others do not appear in the equations for transverse vibrations, so that he is left with 6. Finally these 6 are proved to be pair and pair equal and so reduced to three in the case of biaxial crystals (p. 522). It is shewn how Fresnel's and MacCullagh's equations are contained in the symbolical forms.

[1364.] Further, O'Brien points out that if the above six relations necessary to Fresnel's theory be introduced into the equations for a vibrating elastic medium based upon the theory of intermolecular force being central, those equations reduce to the equations for an isotropic medium. 'From this it follows that M. Cauchy's hypothesis cannot be applied to any but uncrystallised media. In fact, it may easily be proved that, if the equations derived from M. Cauchy's hypothesis be true, a crystallised medium is incapable of propagating transverse vibrations' (p. 510).

The reader should, however, on this point consult Saint-Venant's paper: *Sur les diverses manières de présenter la théorie des ondes lumineuses; Annales de Chimie*, T. XXV., pp. 335—385, 1872; also the same writer's *Appendice* V. to Navier's *Leçons* pp. 729—732, footnote; and Glazebrook's *Report on Optical Theories, British Association Transactions*, 1885, p. 164 *et seq.* London, 1886. These writers do not, indeed, directly refer to O'Brien. The matter is beyond our present field of investigation.

[1365.] For the transverse vibrations of a crystalline medium the equation found is:

$$\frac{d^2V}{dt^2} = D\mathfrak{D} \, . \left\{ \left( B_2 \frac{dv}{dz} - B_3' \frac{dw}{dy} \right) \alpha + \left( B_3 \frac{dw}{dx} - B_1' \frac{du}{dz} \right) \beta + \left( B_1 \frac{du}{dy} - B_2' \frac{dv}{dx} \right) \gamma \right\},$$

where, $B_1$, $B_2$, $B_3$, $B_1'$, $B_2'$, $B_3'$ are constants such that for biaxial crystals $B = B'$, and $D\sigma' \, . \, \sigma$ denotes the operation

$$(zy' - z'y)\,\alpha + (xz' - x'z)\,\beta + (yx' - y'x)\,\gamma.$$

It is thus the symbol of a line perpendicular to $\sigma'$ and $\sigma$; hence the above equation indicates: 'that the force $d^2V/dt^2$ is perpendicular to the direction of $\mathfrak{D}$, and that direction, as we have seen, is the direction of propagation.' The equation being based on the six relations between the constants before referred to, it follows that: 'the forces brought into play by transverse vibrations are always perpendicular to the direction of propagation' (p. 521).

1366. *On the Internal Pressure to which Rock Masses may be subjected, and its possible Influence in the Production of the Laminated Structure.* By W. Hopkins, Esq., M.A., F.R.S., &c. This is published in the *Camb. Phil. Trans.* Vol. VIII. 1849, pp. 456—470. It was read on May 3, 1847.

We are concerned with the first ten pages only of this memoir, namely, the introductory remarks and the Section I. which is entitled: *Relative positions of the lines of maximum and minimum tension, and planes of maximum tangential force in the interior of a continuous mass.* Two propositions relative to our subject are established. The first is the existence of axes of principal traction: see Art. 603. The second is an investigation of the numerically greatest shear; the investigation is rather complex. The first proposition was already well known; the second proposition appears here, I think, for the first time: the only previous work cited on the subject is Cauchy's *Exercices de mathématiques*, Vol. II.

[1367.] I quote the following remark, as it bears upon a point not always sufficiently regarded. Taking an elementary plane $s$ at a point $P$, Hopkins writes:

If this plane assume different positions by moving about $P$ as a fixed point, the normal and tangential forces acting on it will have different values, assuming maxima or minima values for certain determinate positions of $s$, and it is on these particular positions of $s$ that the distortion of a small portion of the mass about $P$, and that of any organic form contained in it will depend. Generally, the linear dimensions of the element will be altered by extension or compression, and it will also be *twisted*, so that if it were originally a rectangular parallelopiped it will become an oblique-angled one, and these changes of form will be indicated by the corresponding distortions of the organic remains. Now, if the directions of the cleavage planes were originally determined by the state of internal tension and pressure of the mass, it would seem probable that they would be perpendicular to the directions of greatest, or to those of least normal pressure, or that they would coincide with the planes of greatest tangential action (p. 456).

It is to the last statement I wish to draw attention, as I think it may contain a fallacy; if the planes of cleavage are planes of *set* they would be surfaces orthogonal to the original directions of greatest stretch,—to the greatest *positive* as distinguished from *negative* stretch (or squeeze). The author nowhere shews that these directions of greatest stretch coincide with the directions of maximum stress. In fact they need not necessarily do so, and he thus may have fallen into the same error as Coulomb did in discussing the problem of cohesion: see our Arts. 120 and 729.

[1368.] Hopkins' results as to maximum shear may be thus briefly expressed. Let $x, y, z$ be any three rectangular directions at a point, and $x', y', z'$ a second set of rectangular directions; then it is well known that:

$$\begin{aligned}\widehat{x'y'} = \widehat{xx}\cos(x'x)\cos(y'x) + \widehat{yy}\cos(x'y)\cos(y'y) + \widehat{zz}\cos(x'z)\cos(y'z)\\ + \widehat{yz}\,\{\cos(x'y)\cos(y'z) + \cos(x'z)\cos(y'y)\}\\ + \widehat{zx}\,\{\cos(x'z)\cos(y'x) + \cos(x'x)\cos(y'z)\}\\ + \widehat{xy}\,\{\cos(x'x)\cos(y'y) + \cos(x'y)\cos(y'x)\}.\end{aligned}$$

Now taking $z'$ to coincide with $z$ and $x'$ to bisect the angle between $x$ and $y$, we find:

$$\widehat{x'y'} = \tfrac{1}{2}(\widehat{yy} - \widehat{xx}).$$

Hence it follows that $\widehat{x'y'}$ will be greatest when $\widehat{yy} - \widehat{xx}$ is greatest, or when $\widehat{yy}$, $\widehat{xx}$ are principal tractions. Or, the greatest shear will be in a plane which contains two principal tractions and across a face which bisects the angle between them.

If the three principal tractions are all of the same sign the greatest shear lies in the plane of the greatest and the least and is equal to half their difference; if the principal tractions are not all of the same sign, the greatest shear is the arithmetical sum of the two which would be called the greatest and the least algebraically and lies in their plane.

1369. Section II. is an application of the results obtained to a point in geology. Organic remains, such as shells, are found in distorted forms, and the point for consideration is how these forms "may indicate the directions which must have been those of maximum and minimum tension or pressure, and the position of the planes of maximum tangential action at some former epoch, posterior to the elevation which raised the general mass into anticlinal ridges."

[1370.] C. G. Page. *Singular Property of Caoutchouc, illustrating the value of Latent Heat in giving Elasticity to solid bodies, and the distinct functions in this respect of latent and free or sensible heat. Silliman's American Journal of Science,* Vol. IV. 1847 pp. 341—342.

This paper notes the rediscovery by Page of a fact remarked by Gough in 1805, that a strip of caoutchouc, stretched and quickly cooled, loses its elasticity; 'it resembles a piece of frozen rubber in some respects, although not quite so rigid': see the footnote to our p. 386, (3). A further experiment on the same point as to the result of compressing portions of such a strip is worth noting:

If successive portions of the inelastic strip be pinched between the thumb and finger, it contracts powerfully in these parts *b*, leaving the others *a* unaffected, and presenting the appearance of a string of knots or beads, which may be preserved in this state for any length of time, if not handled, and kept at a moderate temperature. Upon examination by a sensitive thermometer the portions *a* and *b* are found to be of the same temperature. As regards the amount of heat contained, the portions *a* and *b* differ considerably, and in respect to latent heat, *a* may be said to be positive and *b* negative. The function of the two portions continues abrupt and well defined, showing that there is no tendency to distribution or equilibrium of latent heat between the two portions.

[1371.] 1848. Séguin. There are three memoirs of this period by the above-named physicist on the nature and law of molecular force. They will be found in the *Comptes rendus*, T. XXVII. 1848, pp. 314—318, T. XXVIII. 1849, pp. 97—101, and T. XXIX. 1849, pp. 425—430. I shall refer to these memoirs when considering later papers of the same writer in the second volume of this work, as they will be best treated together.

[1372.] Andrew Bell. *On the Determination of the Modulus of Elasticity of a rod of any material, by means of its musical note. Cambridge and Dublin Mathematical Journal,* Vol. III. 1848, pp. 63—67.

It is proposed in this paper to determine the modulus of elasticity of any material, by means of the musical note obtained from a rod of the material. The modulus being determined, it will of course thence be possible to ascertain the weight a column of the material can support before beginning to bend, and other elements dependent on the modulus.

The writer seems ignorant that the method he proposes had been very generally used during the past twenty years from Lagerhjelm to Wertheim : see Arts. 370 and 1293. Also he appears still to have faith in the Eulerian theory of columns: see Arts. 65, 146 and 954. On p. 67 he notes that some acceleration of the propagation of disturbance in a rod might be expected from the 'disengagement of free caloric.' To this there is a footnote by Sir William Thomson:

Our ignorance of the amount of this effect, and our consequent inability to make the necessary correction for it, are such that the practical application suggested by this paper, cannot, in the present state of science, be considered as likely to lead to very accurate results.

On this point we must refer to the corrections made by Weber and Duhamel and to the practical applications made by various experimentalists, notably Wertheim : see our Arts. 701, 885—890 and 1297.

[1373.] 1848. Max Becker. *Die gusseisernen Brücken der badischen Eisenbahn,* 1848. The original work was not accessible to me, and probably contains little of importance for our present purpose. An account of it will be found on pp. 441—464 of *Der Ingenieur,* Bd. I., Freiberg, 1848. We have noted the work here only because it contains the results of experiments on the deflections of 11 cast-iron bridges subjected to a load travelling with various velocities. These experiments are similar to those made somewhat later by the English Iron Commissioners, and are interesting in the light of Stokes' researches: see our Arts. 1287 and 1290.

[1374.] The deflections were measured by an exceedingly ingenious arrangement, in which a plunger in a reservoir of mercury forced the fluid along a capillary tube and so magnified the deflection twenty-fold. It seems to me that it would be easy to construct by a like arrangement and the use of photography an apparatus for tracing automatically the stress-strain curve for very small elastic strains. An automatic apparatus of this kind is needed for the investigation of the elastic constants for very small stresses.

[1375.] The results obtained from the 11 bridges, all of different construction, were as follows:

(1) The deflection is smallest when the locomotive is at rest, it becomes larger when the locomotive moves and increases with its velocity.

(2) The deflection increases with the magnitude of the adhesion between the driving wheels and the rails (*die Grösse der zwischen den Triebrädern und den Bahnschienen stattfindenden Adhäsion*).

(3) The deflections which the greatest velocities (60 ft. per second) of a locomotive produce do not very largely exceed the statical deflection of the same locomotive. In all 11 bridges they did not exceed the latter by $\frac{5}{12}$ inch. (In the only numerical example given the excess was a little less than $\frac{1}{2}$ the statical deflection.)

(4) The deflections due to the greatest velocities were within the elastic limits.

(5) An impact of any kind produced by unevenness in the rails increases the deflection. (*Findet über der Mitte des Schienenträgers ein Schienenstoss statt, so wird die Zunahme der Senkung bei dem Darüberrollen der Locomotive vergrössert.*)

[1376.] Becker gives the following empirical formula to calculate the deflection (see our Appendix, Note A):

$$f = \alpha n L v + \beta L v^2,$$

where $f$ is the kinetic deflection, $L$ the weight of the locomotive, $v$ its velocity, $n$ the 'coefficient of adhesion,' and $\alpha$, $\beta$ two constants to be determined for each individual bridge. To $f$ we must add, in order to obtain the whole deflection, the statical deflection due to the load $L$ at the centre. Becker takes $n = 1/10$ (p. 462).

[1377.] 1848. J. Weisbach. *Die Theorie der zusammengesetzten Festigkeit.* *Der Ingenieur*, Bd. I., Freiberg, 1848, pp. 252—265. The limits of cohesion having been found for simple strains, Weisbach considers how these limits must be compounded for complex strains, as when longitudinal traction and flexure or either of these and torsion are combined. He remarks on p. 253:

Man hat seither auf die zusammengesetzten Festigkeiten fast gar nicht Rücksicht genommen, sondern in den Fällen, wo zwei Festigkeiten zugleich in Anspruch genommen werden, jede einzeln betrachtet, als wenn die andere nicht da wäre, und bei Berechnung der Dimensionen der Körper von den erhaltenen Doppelwerthen alle Mal den grössten ausgewählt......Dass dies nicht richtig ist, und dass man hiernach zu kleine Werthe für die Dimensionen der Körper erhält, ist leicht zu ermessen......Welche Dimensionen aber in solchen Fällen die angemessenen sind, wird in folgendem kurzen Aufsatze gezeigt werden.

This statement is a little too sweeping in the light of Saint-Venant's researches (see our Chapter IX.), still it draws attention to a very important practical point; the superposition of strain naturally demands that the resultant strain as a whole shall be less than the cohesive (or elastic) limit.

[1378.] Weisbach proceeds to apply this to various cases of compound strain. (i) Traction and flexure (p. 254). (ii) Contraction and flexure (p. 255). Here I consider his results erroneous, because it is not the compression (lateral stretch) in the undermost fibres of the beam which in this case *must* produce rupture; the place of rupture depends upon the ratio of the contractive to the flexural load; we must here as elsewhere take the maximum positive stretch as our limit[1]: see our Art. 1567. (iii) Non-central

[1] I find that the beam will only set at the lowest fibre first, so long as

$$Q/P < \frac{1+\eta}{1-\eta} \kappa^2/hl.$$

Here $\eta$ is the stretch-squeeze ratio; $Q, P$ are respectively the shearing and contractive loads at the free end, and like Weisbach I neglect the buckling action of $P$; $\kappa$ is the sectional radius of gyration about a line through the centroid of the section per-

longitudinal traction (p. 256); Weisbach's results agree with those of Tredgold and Brix: see our Arts. 832 and 1249. (iv) Non-central contraction (p. 258). Here for the same reason as in (ii) I doubt the accuracy of Weisbach's results. (v) and (vi) contain other cases of combined traction and flexure. (vii) Combination of flexure and torsion. I do not feel satisfied with Weisbach's treatment of this case, for one reason, because he appears to have neglected the flexure which would be produced by his force $P$ (p. 260). (viii) Combination of traction and torsion; the same remark applies as in case (vii) (p. 261). (ix) Case of a strut subjected also to a deflecting force (p 263). The method is only approximate and I do not believe it would in practice lead even to approximately accurate results.

I may note that Weisbach in all these cases neglects the sliding strain, and so does not really fulfil the conditions he has himself laid down for compound strain.

[1379.] James Thomson. *On the Strength of Materials, as influenced by the existence or non-existence of certain mutual strains among the particles composing them. Cambridge and Dublin Mathematical Journal*, Vol. III. 1848, pp. 252—258.

It is well known that Cauchy and Poisson introduced into the expressions for the stresses $\widehat{xx}$, $\widehat{yz}$... six terms $\widehat{xx}_0$, $\widehat{yz}_0$... dependent on the initial state of stress. These extended formulae were not however applied to explain any of the phenomena of set. As a rule of course $\widehat{xx}_0$, $\widehat{yz}_0$... would be functions of position, but Cauchy and Poisson appear to treat them as constants: see our Arts. 598 and 616. The above paper of James Thomson seems to be among the first theoretical attempts to explain an initial state of strain and its bearing on set: see Art. 1207. The set here treated of is fore-set, not after-set, the time-element being disregarded. The paper is reproduced *in extenso* with a few additional notes by Sir William Thomson in his article on *Elasticity* in the

pendicular to the plane of flexure, $l$ is the length of the beam and $2h$ the diameter in the plane of flexure, which is supposed to contain a principal axis of the section: see Appendix, Note A. Weisbach's services to technical elasticity are, I think, slightly overrated by Rühlmann (*Vorträge über Geschichte der technischen Mechanik*, p. 421, Leipzig, 1885). This 'Pietät gegen seinen unvergesslichen Lehrer' is, however, very excusable.

*Encycl. Brit.* §§ 10—20. The object of this paper is stated in the following words:

To shew that the absolute strength of any material composed of a substance possessing ductility (and few substances, if any, are entirely devoid of the property) may vary to a great extent, according to the state of tension or relaxation in which the particles have been made to exist when the material as a whole is subject to no external strain (p. 252).

The term strain is here used in the modern sense of stress. What the paper is occupied with is the possibility of varying the elastic limits (i.e. extending the state of ease), thus the term *absolute strength* seems to be wrongly applied, for it usually denotes the limits of cohesion. Whether the absolute strength of a body can be increased by initial stress seems a doubtful matter: see our Arts. 1353 and 1524. In two sets of experiments recorded in Clark's work on the Britannia and Conway Bridges, we find opposite conclusions are reached: see our Arts. 1473 and 1486.

[1380.] I have found the reasoning contained in the paper extremely difficult to follow. As however the paper is very readily accessible in the *Encyclopaedia*, it is the less necessary to analyse its contents. I may, however, remark on one or two points.

I do not see how any conclusions such as are obtained on pp. 252—254 can be reached without some physical statement as to the relative amounts of set and elastic strain in the strain produced by any stress. Nor does it seem to me that after set begins stress will remain constant. This is certainly not true for a bar under uniform terminal traction; it holds only after the bar has become, at least locally, plastic, i.e. after stricture has set in. Further, if it be necessary that two elements of surface be given a definite slide, say $\sigma$, before the shear reaches the elastic limit $\eta$, I do not understand how such slide could ever be reached even at a moderate distance from the centre of the section before rupture took place at the contour. Hence the footnote to p. 253 as well as the statement in the text as to the increased torsional resistance do not seem to me convincing. Again, it would appear that, in the state of strain supposed by the writer, where there is a line of no strain in the section, shears parallel to the axis of the bar of varying magnitude would be called into play and thus the primitively plane sections be distorted into curved surfaces; this, I think, would affect the results obtained on p. 254. I have suggested the above difficulties rather that the attention of the reader may be drawn to the memoir, than that a mere reference to the memoir should lead him to disregard it.

[1381.] James Thomson proceeds on p. 255 to shew that, what we may term the variability of the state of ease, has led to many discordant experimental results. On p. 257 he defines the *superior and inferior limits* of elasticity in the sense in which we have used them in the present work. He then remarks that:

These two limits are not *fixed* for any given material, but that, if the change of form be continued beyond either limit, two new limits will, by means of an alteration in the arrangement of the particles of the material, be given to it in place of those which it previously possessed; and lastly, that the processes employed in the manufacture of materials are usually such as to place the two limits in close contiguity with one another, thus causing the material to take in the first instance a set from any strain, however slight, while the interval which may afterwards exist between the two limits, and also, as was before stated, the actual position assumed by each of them is determined by the peculiar strains which are subsequently applied to the material (p. 257).

This is expressed concisely by the statement that: The state of ease depends on the worked state of the material: see our Appendix Note C (i).

The author of the memoir refers to Eaton Hodgkinson's experiments on cast-iron as evidence of the close approach of the limits of elasticity due to the process of manufacture: see our Art. 969.

[1382.] James Thomson *On the Elasticity and Strength of Spiral Springs and of Bars subjected to Torsion. Cambridge and Dublin Mathematical Journal,* Vol. III. 1848, pp. 258—266.

The writer, after a few preliminary explanations, remarks:

The elasticity and strength of spiral springs have not, so far as I am aware, been hitherto subjected to scientific investigation; and erroneous ideas are very prevalent on the subject, which are not unfrequently manifested in practice by the adoption of forms very different from those which would afford the greatest advantages.

The special case treated in this memoir is of a tractional load in the axis of the cylinder on which the spiral lies. More general cases had previously been considered by Giulio and Saint-Venant, to say nothing of the investigations of Binet and Mossotti: see our Arts. 1219 and 1608.

James Thomson, by neglecting the stretching and sliding effect of the tractive load, in fact by supposing the helix unwound and subjected merely to a torsional force, arrives at a result which coincides with the approximate result (vi) of Giulio's memoir of 1842: see our Art. 1220.

Let $s$ be the length of the spiral, $w$ the tractive load in the axis, $a$ the radius of the cylinder on which the central thread lies, $r$ the radius of the section supposed circular; the rest of the notation coinciding with Giulio's. Then

$$s = n \times l = n \times 2\pi a \text{ nearly, and } w = H;$$

but, if $\zeta$ be the elongation of the spiral spring,

$$\zeta = n\,(h - h_0).$$

Hence $$\zeta = n\,\frac{l^3 H}{4\pi^2 a}, \text{ from Giulio's Equation (vi),}$$

$$= \frac{swa^2}{\mu\pi r^2 \times r^2/2} = \theta\,\frac{swa^2}{r^4} \quad \text{......................(i),}$$

since in Thomson's notation $\theta = 2/(\mu\pi)$. This is the result, p. 261. We see that it is only an approximate method of treating a special case.

[1383.] The equation obtained above relates to the elasticity of the spring. Thomson now proceeds to consider the strength, or the space through which it can be elongated without set.

Let $W$ be the greatest weight, $Z$ the greatest elongation the spring will take without set, $\nu$ 'the utmost couple producing torsion which can be resisted by a bar whose radius is unity composed of the same substance as the spring, and having its particles at various distances from its centre free from mutual opposing strains when it, as a whole, is subject to no strain' (compare Arts. 1379—1380). Then Thomson, as in his previous paper, takes $\eta$ to denote the limit of shear per unit area, and finds for the limiting couple for a bar of sectional radius $r$,

$$Wa = \int_0^r 2\pi x dx \,.\, \eta x^2/r$$

$$= \tfrac{1}{2}\pi\eta r^3 = \nu r^3 \text{ ..........................(ii).}$$

Now he assumes that $Z$ and $W$ will be related as the $\zeta$ and $w$ of the equation of elasticity (i), or

$$Z = \theta \frac{sWa^2}{r^4} = \theta\nu \,.\, sa/r \text{ .................(iii).}$$

The justification of this assumption is based on the remark that: 'in ordinarily formed spiral springs, the elongations continue proportional to the weights added, even up to the very greatest that can be resisted.' Thomson cites an experiment of his own: see the conclusions of Hooke and Mossotti, Arts. 7 and 250. The equation (i) we must, however, remark would hardly remain sufficiently approximate for a spiral of which the stress was very considerable, when the strain, while remaining elastic, also became considerable.

[1384.] The *resilience* of the spring, being the total quantity of work which can be stored up in it without producing set, is expressed by $\tfrac{1}{2}WZ = \tfrac{1}{2}\theta\nu^2 \,.\, sr^2 = \frac{1}{2\pi}\theta\nu^2 \times$ volume. Thus the resilience of a spring is for the same material proportional to the volume of the coil or weight of metal contained in it. This is of course a special case of the theorem due to Young and extended by Tredgold, Poncelet and others: see Saint-Venant's *Historique Abrégé*, pp. ccxvii—ccxix and our Arts. 982, 999 and Appendix, Note A, (3). Thomson draws from this result the conclusion that the springs of railway buffers should not be made of rectangular section, a form frequently adopted. But it seems to me that in the case of a prismatic rod on a rectangular base of sides $b$ and $c$, the sections ceasing to remain plane the quantity $\theta$ would be entirely changed, and we could not argue that the resilience would be less without knowing the ratio of $b/c$ and without further investigation. In fact equations (i) and (ii) would be completely altered in character.

On p. 263 will be found a general statement of theoretical results, and on pp. 264—265 some experimental details.

1385. *Ueber die Gesetze der Biegung elastischer fester Körper.* Von Herrn v. Heim, Major in der königl. Würtembergischen Artillerie. This memoir is published in Crelle's *Journal für... Mathematik,* Vol. XXXVII. 1848; it occupies pages 305—344.

Suppose an elastic rod to be subjected to the action of forces; it is possible that in the state of equilibrium the axis of the rod should become a curve of *double curvature:* Heim considers that the conditions of equilibrium for this case have not yet been accurately investigated. He refers especially to what had been given on this subject by Poisson, in the following words:

Unter den neuern Schriftstellern, welche die Lehre von dem Gleichgewicht und der Bewegung elastischer fester Körper zum Gegenstande ihrer Forschungen gemacht haben, nimmt unstreitig *Poisson* eine der ersten Stellen ein.

Er hat theils mehrere besondere Abhandlungen über diesen Gegenstand in den *Mémoires de l'Académie des sciences* Tome VIII. und in den *Annales de chimie et de physique* 1829 · geliefert, theils denselben in seinem *Traité de Mécanique,* 2te Ausg. 1833 mit einiger Ausführlichkeit bearbeitet, und sich hierdurch wesentliche Verdienste um den genannten Zweig der mathematischen Physik erworben. Jedoch sind einige der Ergebnisse seiner Untersuchungen, hauptsächlich aus dem Grunde, weil er die Haupt-Axen der Querschnitte der Körper entweder nicht berücksichtigt, oder, was wahrscheinlich ist, nicht gekannt hat, nicht frei von Ungenauigkeiten oder Unrichtigkeiten.

Dass dieses namentlich bei den allgemeinen Gleichungen über das Gleichgewicht einer elastischen Ruthe, wie sie *Poisson* im ersten Bande seines *Traité de Mécanique* Nro. 316 u. folg. giebt, und woraus er die Beständigkeit des Torsionsmoments der Ruthe im gebogenen Zustande ableitet, der Fall ist, soll hier umständlicher gezeigt werden (pp. 316, 317).

1386. Heim's own investigation is long and obscure; he does not state clearly what the *body* or the *element* is of which he considers the equilibrium; he seems to think it sufficient to speak of the equilibrium of an imaginary section made by a plane, instead of the equilibrium of a slice bounded by two such planes. Moreover for some important formulae which he uses he refers to the work of his published in 1838 and referred to in our Art. 906. One consideration however will be sufficient, I think, to shew that they are not satisfactory. We know that Poisson obtained the result that the torsional moment is *constant* [see Saint-Venant's account of Poisson's mistake in Art. 1602].

Now this result has been shewn by other writers to be inaccurate, as resting on an inadmissible assumption with respect to the elastic forces. But Heim obtains a formula which makes the torsion constant under certain circumstances; for instance it is constant if the section of the rod is a circle or a regular polygon: so that practically Heim arrives at Poisson's result, which we know is inaccurate[1].

1387. *Beitrag zur Lehre von den Schwingungen elastischer fester Körper.* Von Herrn v. Heim...Crelle's *Journal für... Mathematik,* Vol. XL. 1850, pages 1—20.

This does not relate to our subject, and seems of no value; the author objects to the solution of various dynamical problems given by Poisson and other writers, and offers investigations of his own which are quite untenable. One example will suffice to indicate the nature of the memoir. Poisson in Art. 493 of his *Mécanique* discusses the longitudinal vibrations of an elastic rod. By cutting the rod into slices and considering a single slice Poisson obtains such an equation as

$$dT = \left(X - \frac{d^2u}{dt^2}\right) dm,$$

where $T$ is the traction, and $X$ the body force parallel to the axis. Now our author says that this equation is not admissible; he says that we have by D'Alembert's Principle the equation

$$\int\left(X - \frac{d^2u}{dt^2}\right) dm + \int dT + P = 0,$$

where the integration is to extend over the whole rod, and $P$ denotes the load applied at one terminal. This of course is quite

[1] [Heim's memoir is long and tedious, but this result is not inaccurate as Dr Todhunter holds. Heim appears to be quite ignorant of the memoirs of Bellavitis and Saint-Venant: see our Arts. 935 and 1597. As will be seen by referring to those memoirs, the torsional moment is constant when the section possesses inertial isotropy. The memoir is of no value because earlier writers had obtained in a far simpler fashion the like results. ED.]

true; but then as this equation alone does not give sufficient information Heim proceeds to make an arbitrary hypothesis to enable him to express $u$ the shift as a function of position on the rod. He seems to imagine that in treating a dynamical problem relating to a rigid body we must formally use D'Alembert's Principle for the *whole* body, and that we may not by special considerations resolve the body into elements, and treat each element separately.

The author refers to his book published in 1838: see our Arts. 906—916.

[1388.] C. C. Person: *Relation entre le coefficient d'élasticité des métaux et leur chaleur latente de fusion; chaleur latente du cadmium et de l'argent. Annales de chimie et de physique*, T. XXIV. 1848, pp. 265—277. This memoir was read to the Académie on September 4, 1848.

The author considers that the work done in separating the molecules of a substance mechanically ought to be related to the heat required to separate them by fusion. Thus he argues there ought to be a relation between the stretch-modulus and the latent heat of fusion in any given material. This is however rather a leap, because the stretch-modulus is not a quantity which is related to cohesion but, as far as we know, only to elastic stress. Further, if we do not assume uni-constant isotropy, the bulk-modulus and not the stretch-modulus would seem a quantity more likely to be related to latent heat of fusion.

Referring to Wertheim's first memoir in the *Annales* (see our Art. 1292) Person notices that the latent heats of fusion are nearly proportional (*très-peu près proportionnelles*) to the stretch-moduli. This would give a relation of the form

$$E/E' = L/L',$$

$L, L'$ being latent heats of fusion and $E, E'$ stretch-moduli. This relation is not however close enough, and the empirical formula

$$\frac{E}{E'}\frac{1+2/\sqrt{w}}{1+2/\sqrt{w'}}=\frac{L}{L'},$$

where $w, w'$ are the specific densities of the materials, is then given as in close accordance with experiment. The numerical values, however, considered on p. 270 do not seem to me so close that we can conclude that this relation must be the true one. In regard to cadmium and silver, Person calculates their latent heats of fusion by the above formula from Wertheim's results and from direct experiments of his own. He finds:

| Latent heat of Fusion | Cadmium | Silver |
|---|---|---|
| From formula | 13·52 | 20·38 |
| From experiment | 13·58—13·66 | 21·07 |

When the variation coefficient is determined from zinc we have

$$L = 0{\cdot}001669\, E\,(1 + 2/\sqrt{w}).$$

Compare *Annales de chimie*, T. XXVII. p. 266, where however the $E$ seems to have dropped out.

Person's formula is at least suggestive, and we shall have occasion again to refer to it.

[1389.] A. Kupffer. *Recherches expérimentales sur l'élasticité des métaux. Première Partie. Mémoires de l'Académie...de Saint-Pétersbourg. Sixième Série. Sciences mathématiques, physiques et naturelles*, T. VII. *Sciences mathématiques et physiques*, T. V. St Petersburg, 1853, pp. 231—302[1]. This is the first contribution of this physicist to our subject, and marks the beginning of a long and very important series of experimental researches on the elasticity of metals. Probably no more careful and exhaustive experiments than those of Kupffer have ever been made on the vibrational constants of elasticity and the temperature-effect. The important memoir of 1852 as well as the grand work of 1860 will be considered in our second volume. The present memoir was read December 1, 1848. The Russian government had founded an Institute of Weights and Measures, to which was also entrusted the duty of investigating those properties of metals which can

[1] The titles of the St Petersburg memoirs reach the height of complexity in this period. It would be a great blessing to science, if all scientific societies would either style their transactions after the year, or else adopt a continuous numbering of volumes.

affect the standards of measurement. Foremost among these properties are those of temperature and elasticity; Kupffer in the present memoir proposes to investigate the latter.

[1390.] Kupffer's experiments in this memoir are all based on the torsional vibrations of metal in the form of wire. He ascertains the moment of inertia of the body suspended from his wire by means of a method suggested by Gauss for obtaining the moment of inertia of magnets. The theory of the torsional vibration adopted is that of Coulomb. Thus, if $\mu$ be the slide-modulus, $r$ the radius and $l$ the length of a wire, $P$ the period of a semi-oscillation, and $I$ the moment of inertia of the suspended mass:

$$\mu \frac{\pi r^4}{2l} = \frac{\pi^2 I}{P^2},$$

or,

$$\mu = \frac{2\pi Il}{r^4 P^2}.$$

Kupffer makes use of a constant $\delta$, which is thus related to our $\mu$,

$$\delta = \frac{2}{5\pi\mu} = \frac{r^4 P^2}{5\pi^2 lI}.$$

[1391.] He discovered very early in the experiments that the duration of the oscillations increased with the amplitude, and it became necessary to deduce the value of $P$ for an infinitely small arc, when found from a finite arc. If $P_0$ be the value of $P$ for an infinitely small arc, and $P_s$ its value for an arc of amplitude $s$, Kupffer found that the reduction was proportional to the *square root of the amplitude*; or,

$$P_0 = P_s - \alpha\sqrt{s}, = P_s\left(1 - \frac{\alpha}{P_s}\sqrt{s}\right),$$

$\alpha/P_s$ being a quantity which is not the same for wires of different material, so that $\alpha$ is a constant which depends not only on the resistance of the air, but on the particular elastic nature of the material. Its value was ascertained in every experiment. This law of reduction only holds for durations of oscillation as large and for surfaces of resistance as small as those occurring in Kupffer's torsion experiments. The laws of correction for lesser duration and larger surfaces of resistance are given on p. 251.

[1392.] The diameters of the wires were carefully measured not only by means of a microscope, but also by finding the weight of a definite length of wire, by weighing in water. In all the experiments we have careful statements of the heights of the barometer and thermometer. The normal temperature being taken at $13^{\circ}\frac{1}{3}$ R. The reduction of a semi-oscillation $P_0'$ at the temperature $t$ is given by

$$P_0 = P_0' - \beta (t - 13{\cdot}3),$$

where $\beta$ is a constant found for each set of experiments, and termed by Kupffer the 'coefficient of the influence of heat on elasticity.' (p. 299).

The experiments were made upon iron, copper (*cuivre jaune*), platinum, silver, and gold wires. The requisite calculations were undertaken by Napiersky: see our Art. 1396.

[1393.] On p. 298 Kupffer gives the mean values of $\delta$ and $\log 1/(5\delta)$ for the above wires. To obtain the slide-modulus $\mu$ we must multiply the value of $1/\delta$ by the fraction $2/(5\pi)$. The value of $\mu$ thus obtained will be in Russian pounds per square Russian inch[1]. For *mean* values we have

| Material | $\delta$ | $\mu$ | 𝕰 |
|---|---|---|---|
| Iron, No. 1 | $10^{-7} \times {\cdot}1088$ | 10,565,308 | 18571 |
| Iron, No. 2 | $10^{-7} \times {\cdot}1132$ | 10,154,644 | 17850 |
| Copper | $10^{-7} \times {\cdot}2139$ | 5,374,033 | 9446 |
| Platinum | $10^{-7} \times {\cdot}1269$ | 9,058,358 | 15924 |
| Silver | $10^{-7} \times {\cdot}2854$ | 4,027,700 | 7080 |
| Gold | $10^{-7} \times {\cdot}2974$ | 3.865,183 | 6794 |

[1] The Russian pound contains 409·512 grammes, and the Russian inch is equal to the English inch and contains 25·3995 millimetres.

The values of $\delta$ are Kupffer's, those of $\mu$ I have calculated in English pounds per square inch. $\mathfrak{E}$ is the pseudo-stretch-modulus in French measure, or the number of kilogrammes which would double by traction the length of a wire of one square millimetre section. $\mathfrak{E}$ is calculated on the supposition of uni-constant isotropy. Kupffer's experiments probably give us very accurate values of the slide-modulus for the above materials.

[1394.] Kupffer however supposed wires to be *isotropic* bodies and isotropy to be marked by only one constant. These erroneous suppositions led to his adoption of the above value of $\delta$. For in uni-constant isotropy $E/\mu = 5/2$, thus $\delta = 1/(\pi E)$ or $\delta$ is the extension of a wire of unit length and unit radius under a traction of one pound (Russian units). The values of the stretch-modulus thus calculated differ of course considerably from those obtained by other experimenters from simple traction. Kupffer's experiments thus conclusively prove either that isotropy possesses two constants, or that wires possess a cylindrical arrangement of elasticity, i.e. are aeolotropic.

[1395.] On pages 299—300 will be found a calculation of the dilatation coefficients of various metals for change in temperature. As this is based on the values of the stretch-modulus given by $\delta$, the results cannot be considered of value. If $1/\delta$ be defined as the 'coefficient of elasticity,' Kupffer concludes that:

> Le coefficient d'élasticité augmente également avec la température; il est probable que le coefficient d'élasticité augmente avec la tension (p. 301).

In other words: The slide-modulus increases equally with the temperature, and is probably increased by an increased traction perpendicular to the plane of the slide for which the slide-modulus is measured: see our Art. 1300.

Kupffer holds that his experiments were not sufficient for him to form any conclusions as to the law of the latter variation.

[1396.] A. W. Napiersky. *Beobachtungen über die Elasticität der Metalle. Poggendorffs Annalen, Ergänzungsband* III. 1853, pp. 351—373.

This memoir is dated Mitau, October 15, 1850, and was communicated by Kupffer. The experimental investigations are a repetition of those instituted by Kupffer and referred to in the previous Articles. The experiments are here upon iron, zinc, and silver wires, and the method is that of torsional vibration. The results are given in Russian units, the wires are supposed to possess uni-constant isotropy (see p. 353), and we are not given any particulars as to the working they may have received (p. 361); accordingly the numerical results do not seem to be of any great value.

The experiments were apparently made with great exactitude and they confirmed Kupffer's law of reduction, which is here given in a slightly different form:

$$P_0 = P\,\{1 - \beta\,(t - 13{\cdot}3)\}.$$

For iron $\beta$ was found to be equal to ·0002501 (p. 358). It does not however appear very clearly from the experiments whether the change in the periodic time was due to the direct thermal effect of heat in increasing the length of the wire, or its effect in altering the elasticity.

[1397.] J. D. Forbes. *On an Instrument for Measuring the Extensibility of Elastic Solids. Philosophical Magazine*, Vol. XXXV. pp. 92—94, 1849. This paper appears also in the *Proceedings of the Royal Society of Edinburgh*, II., 1851, pp. 172—175. It contains a method by which an instrument similar to that used by s' Gravesande for verifying Hooke's Law may be applied to find the stretch-modulus. The method depends upon central flexure, and does not seem of any special importance. One numerical example on a steel pianoforte wire is given.

1398. *Ueber die Veränderungen, welche in den bisher gebräuchlichen Formeln für das Gleichgewicht und die Bewegung elastischer fester Körper durch neuere Beobachtungen nothwendig geworden sind:* von R. Clausius. *Poggendorffs Annalen*, Vol. 76, 1849, pages 46—67.

This is an interesting paper. Clausius considers that experiments do not accord well with the theory of uni-constant isotropy; for example, we have seen in Art. 368 of our account of Poisson's investigations, that if the original *length* of a cylinder is increased by traction in the ratio of $1+\delta$ to 1, then the *volume* is increased in the ratio of $1+\frac{1}{2}\delta$ to 1: but Wertheim found from numerous experiments that instead of $1+\frac{1}{2}\delta$ we have really very nearly $1+\frac{1}{3}\delta$: see our Art. 1319. Some explanation is therefore required of the discrepancy between theory and experiment.

[1399.] Clausius then proceeds to investigate on what basis the theory of uni-constant isotropy is founded. He does not apparently question that intermolecular force is *central*, nor does he suggest that the action of a molecule $A$ on a molecule $B$ may depend upon the position or motion of a third molecule $C$. His inquiry is shortly this: Do the equations of isotropic elasticity contain only one constant, if the molecular force is central? If this be granted, how are we to explain the divergence between theory and experiment?

[1400.] The paper opens with a consideration of Wertheim's memoir of 1848 (see our Arts. 1319—1326), and Clausius shews that Wertheim had no right to use Cauchy's equations, and that he had confused a constant, marking an initial state of stress, with the bi-constant isotropy of a body primitively unstrained. Clausius next proceeds to consider the method by which Poisson and Cauchy reduce their two constants to one. He points out that Poisson arrives at this result by *neglecting the irregular part* of the action of molecules in the immediate neighbourhood of the chosen molecules, and so is able to replace the summations of his constants by integrals. Cauchy on the other hand retains his constants as summations, but by *supposing a perfectly uniform molecular distribution* in the case of isotropy finds a relation between his two remaining constants. Both these assumptions Clausius holds to be contrary to what we know of the ultimate elements of bodies. But although these suppositions are wrong, Clausius holds that the conclusion drawn from them is correct, or that isotropy is uni-constant. Clausius comes to this conclusion on the following grounds. The summations which appear as the constants of Poisson and Cauchy ought not to be calculated for a single molecule, for in that case, owing to irregular arrangement, even in what appear to be the most homogeneous of bodies, they would vary from molecule to molecule. These summations must be calculated for a *mean or normal arrangement* of molecules based upon taking an immense number of irregular individual arrangements. For such a normal arrangement we may assume with Cauchy a uniform distribution, or we may replace Poisson's irregular part of the action around one molecule by a regular distribution, and so our summations by integrals. We are led in both cases to the equations of uni-constant isotropy. Clausius remarks that the lower limit, zero, of Poisson's integrals in the memoir of 1828, ought now to be replaced by some unknown constant, the value of which however there is no need, nor in fact possibility, of discovering. He thus justifies Navier's process: see our Arts. 443, 532 and 922.

[1401.] Clausius next asks: How is it that the equations of uni-constant isotropy give results by no means agreeing with Wertheim's experiments on materials, which we are compelled to suppose very nearly isotropic? He answers this question, not by a doubt as to molecular force being central, but by the supposition:

Dass die Körper unter der Einwirkung fremder Kräfte eine innere Veränderung erleiden, welche in etwas Anderem besteht als einer blossen Verschiebung der Moleküle, da diese in den Formeln schon berücksichtigt ist, und dass dadurch die Körper für die Dauer der Einwirkung jene als Bedingung gestellten Eigenschaften theilweise verlieren können (p. 59).

[1402.] This hypothesis Clausius holds to be confirmed by the experiments of Weber and Wertheim. He refers first to the elastic after-strain as noted by Weber in 1835 (see our Art. 707); next he mentions Weber's researches of 1830 on metal wires (see our Art. 701), and holds with Seebeck[1] that a part at least of the change in traction there noted was due to elastic after-strain. He supposes that elastic after-strain in metals is either not so great as in organic substances or disappears far more rapidly. As I have

[1] Clausius cites for Seebeck's opinion the *Programm zur öffentlichen Prüfung der technischen Bildungsanstalt und der Baugewerken-Schule zu Dresden*, 1846, S. 35, a work inaccessible to me.

remarked (see Art. 706), this assumption of elastic after-strain to explain the discordance between uni-constant isotropy and experiment does not seem to me entirely satisfactory.

[1403.] Clausius cites in support of his opinion the results of Wertheim's experiments of 1842. He shews (p. 63) that the formula of Duhamel adopted by Wertheim for the ratio of the specific heats, namely

$$\gamma = 1{\cdot}8v'^2/v^2 - 0{\cdot}8,$$

—where $v'$ is the real velocity of sound in an infinite solid and $v$ the velocity supposing the specific heats equal—is erroneous, as it does not relate to the *linear* propagation of sound waves: see our Art. 1297. Wertheim ought to have taken on the uni-constant hypothesis

$$\gamma = \frac{1}{6v^2/v'^2 - 5},$$

a formula easily obtainable from our Art. 888.

With this formula Wertheim's results give the following values for $\gamma$ (p. 64),

| | | | |
|---|---|---|---|
| Cast Steel | 1·150 | | |
| Brass | 2·588 | | |
| Silver (drawn) | 1·209, | (annealed) | 1·092 |
| Gold „ | 1·484, | „ | 3·875 |
| Copper „ | 1·044, | „ | 1·955 |

These values Clausius holds to be absurd; in fact for glass and lead $\gamma$ becomes negative! But it must be noted on the other hand that Weber's experiments, so far as they go, do not lead to absolutely impossible values of $\gamma$, and there is of course a chance of some experimental error running through Wertheim's results, even if we admit the uni-constant isotropy of his material.

[1404.] The difference between the stretch-modulus as found from traction and from sound experiments Clausius holds to be due to elastic after-strain produced by continuous load. He thus distinguishes the state of elastic equilibrium from that of motion. To the former he would apply the formulae of bi-constant isotropy, to the latter those of uni-constant isotropy. In the former case the two constants must each be determined experimentally. This reasoning does not seem to me very satisfactory, for Braun has shewn that elastic after-strain differs from ordinary elastic strain in not admitting of superposition. Hence it would be impossible for us to apply even the bi-constant equations of elasticity to a strain of this kind.

Clausius appears however, on the last page of his memoir, somewhat to modify this view by adopting like Weber the hypothesis that the molecules are not only displaced but rotated by a system of load, and that the direction and amount of rotation depends on the distribution of load. Thus he holds it possible that the distribution and duration of load may affect the strength of molecular attraction, and so render the existing mathematical theory quite inapplicable.

[1405.] The memoir is suggestive, but not entirely satisfactory. It concludes thus:

Jedenfalls sieht man aus den angeführten Thatsachen, dass die Theorie der Elasticität noch durchaus nicht als abgeschlossen zu betrachten ist, und es wäre zu wünschen, dass recht viel Physiker sich mit diesem Gegenstande beschäftigten, um durch vermehrte Beobachtungen die sichere Grundlage zu einer erweiterten Theorie zu schaffen. Dabei würde es von besonderem Interesse seyn, wenn nicht nur über den Gleichgewichtszustand ähnliche Versuche wie der des Hrn. Wertheim unter möglichst veränderten Umständen angestellt, sondern auch die Schwingungsgesetze entscheidenden Prüfungen unterworfen würden, indem es dem Obigen nach nicht ohne Weiteres angenommen werden darf, dass diese ebenso von den bisherigen Formeln abweichen, wie die Gleichgewichtsgesetze (p. 66).

[1406.] 1849. *Report of the Commissioners appointed to inquire into the application of Iron to Railway Structur es.* London 1849. The Commissioners were Lord Wrottesley, Robert Willis, Henry James, George Rennie, W. Cubitt, and Eaton Hodgkinson, with Douglas Galton as secretary.

[1407.] The Report contains, pp. 1—263, appendices of experimental and in part theoretical results; pp. 264—378, minutes of evidence taken before the Commissioners, comprising the opinion of nearly all the leading British engineers of that day; pp. 379—435, an appendix composed of letters and data sent by various experimenters on the strength of iron, together with facts communicated by various iron-masters in answer to a circular. The volume concludes with 77 plates. It forms the most valuable experimental contribution made during the period we have under consideration in this our first volume to our knowledge of the elasticity and cohesion of iron. The effect of continuous and intermittent loads, of long-continued impacts, of moving loads, etc., etc., are all considered in this Report, with a mass of experimental data and scientific opinion which it would be hard to excel even in more recent researches. We can only afford space here to note some of the results which have more important theoretical value.

[1408.] Appendix A (pp. 1—114) contains experiments of Eaton Hodgkinson of a kind similar to those referred to in our Arts. 939—971.

We draw attention to the following points as bearing upon theory:

(i) A perceptible although very small difference is found in the absolute tractive (tensile) strength of cast-iron for different forms of section (p. 11): see our Arts. 858 and 1216; also Saint-Venant's edition of Navier's *Leçons*, p. 116.

The absolute contractive (crushing) strength of cast-iron seems also to vary slightly with the form of the section. The mean ratio of the absolute tractive and contractive strengths for the simple irons of these experiments was 1 : 5·6603 (p. 101). In previous experiments the author found the ratio to be 1 : 6·595 (p. 15). Combining these values we have 1 : 6·1276. If the *absolute strengths were in the same ratio as the elastic strengths* and the castings were supposed isotropic, we should have on the uni-constant theory the ratio = 1 : 4.

[1409.] (ii) An extensive series of experiments (by means of a ball swung as a pendulum) on oft-repeated transverse impact on iron beams offers data for comparison with theory, when a theory has been found (pp. 16—19 and p. 103). This is followed by a series on considerable transverse impact with the object of determining the maximum or rupture blow (pp. 20—36, pp. 39—44, and pp. 104—105). They are similar to those we have considered in

Arts. 939 and 942. Young's theorem on resilience (the inertia of the beam being neglected) is confirmed: see our Appendix Notes A, (3) and B, (*b*).

The deflections in cast-iron beams were always found to be greater than in proportion to the velocity of impact, whilst in wrought-iron they were nearly constant with impacts of very different velocities (p. 105).

Hodgkinson attributes this fact to the 'defect of elasticity' in cast-iron; there seems to have been no attempt previous to the experiments to reduce the material to a state of ease.

He remarks that his formulae (see our Art. 943) apply only to wrought-iron, or to *small* impacts on cast-iron.

[1410.] (iii) A series of experiments (by means of a freely falling ball) as to the effect of transverse impact on loaded beams of cast-iron will be found on pp. 37—38, and a series as to the effect of transverse impact on wrought-iron beams on p. 45. The first series shew that to increase the inertia of a beam subjected to transverse impact, without increasing its strength, increases very considerably its power of resisting impact (p. 106). Hodgkinson terms this series 'vertical impact' to distinguish it from what he terms the 'horizontal impact' of Art. 1409.

[1411.] (iv) On pp. 47—67 we have a long and most valuable series of experiments on the stretch or squeeze produced by tractive or contractive load on cast-iron bars.

Hodgkinson found that the relation between the *elastic* stretch $s$ and the tractive load $L$ was of the form,

$$L = as - bs^2.$$

If the load $L$ be expressed in pounds per square inch the mean values of $a$ and $b$ are,

$$a = 13{,}934{,}040,$$
$$b = 2{,}907{,}432{,}000.$$

For the elastic squeeze $s'$ due to a contractive load $L'$ he found (pp. 107—109)

$$L' = a's' - a's'^2,$$

where, if $L'$ be measured in pounds per square inch,

$$a' = 12{,}931{,}560,$$
$$b' = 522{,}979{,}200.$$

The set was deducted in all the experiments, though it would have been more satisfactory had the bars been reduced in the first place to a state of ease. Hodgkinson found in the case of a tractive load that the set-stretch $S$ was given with a moderate degree of approximation in terms of the elastic-stretch by a relation of the form

$$S = ps + qs^2, \text{ (p. 60)}$$

where $p$ and $q$ are constants for the same material.

[1412.] We shall draw attention to a memoir of Homersham Cox's, wherein it is shewn that a hyperbolic law of elasticity gives better results than Hodgkinson's parabolic law, while to assume the load a cubic function of the elastic stretch gives almost exact results: see our Arts. 1438—1442. What, however, is of special importance is, that *within the elastic limits,* and even for comparatively small strains, the proportionality of stress and strain ceases to be true. The cubic terms at least must in the case of a great number of cast-metals be retained in the expression for the work: see our Appendix Note D.

[1413.] (v) The next set of experiments (pp. 68—94) are on the deflection and transverse strength of long bars of cast-iron. We may note that in these experiments the *set* was found to vary nearly as the second, or more accurately as the 1·92 power of the deflection.

On p. 110 Hodgkinson shews how the formulae of Art. 1411 may be applied to the case of transverse flexure and the calculation of the position of the neutral line. He points out that in the formulae of his *Experimental Researches on the Strength of Iron,* it is only necessary to put $v = 2$: see our Art. 971.

[1414.] (vi) The final set of experiments in this first Appendix (pp. 95—100, see also pp. 5—7) is on the crushing of short iron prisms. The rupture surfaces of a great number of prisms of square, rectangular and circular cross-sections will be found on Plate III. to this Appendix. The characteristic rupture by the sliding off, as it were, of a wedge, and the peculiar forms of this wedge, are shewn in great variety: see our Arts. 729 and 948. Some typical figures will be found on our frontispiece. Hodgkinson writes of these experiments:

In all these cases fracture took place by the specimen forming wedges which slide past one another, and cut it up in angles dependent on the nature of the material. When the length of the specimen is sufficiently great to allow the wedge to slide off in the direction of least resistance, then the height of the wedge, in a cylinder of cast-iron, will be 3/2 of the diameter nearly. If the height of the specimen is less than 3/2 of the diameter, then that of the wedge will necessarily be less and the resistance to crushing greater, since fracture will be constrained to take place otherwise than in the direction of least resistance (p. 7).

It is not quite obvious what is meant by the 'direction of least resistance.'

Hodgkinson remarks that the same wedge-like rupture-surface is found in short prisms of timber, stone, marble, glass, etc., the wedge-angles being different for each substance: see our Art. 950.

[1415.] (vii) On p. 113 will be found a calculation of the constants $a, b, c, d$ for a formula of the type

$$L = as + bs^2 + cs^3 + ds^4.$$

It gives less error than when only the first two terms are taken; the discrepancies would have been still less if $a, b, c, d$ had been calculated by the method of Least Squares and not from certain definite experiments. Hodgkinson observes that the first two terms are sufficient for the present state of our experimental knowledge: see our Arts. 1411 and 1439.

[1416.] Appendix A (pp. 115—180) contains Hodgkinson's experiments on tubes and cells for the Tubular Bridges of Robert Stephenson. It embraces a great variety of experimental data, as well as the calculation of the strength and deflection of the Conway Tube. We shall refer to these researches of Hodgkinson when discussing the works of Fairbairn and Clark: see our Arts. 1477 and 1494.

[1417.] Appendix B is entitled: *Experiments for determining the effects produced by causing weights to travel over Bars with different velocities, made in Portsmouth Dockyard and at Cambridge,* by the Rev. Robert Willis, F.R.S., Jacksonian Professor, etc.; Captain Henry James, R.E., F.R.S., and Lieutenant Douglas Galton, R.E.

This Appendix is divided into two parts, the first, a *Preliminary Essay* by Willis, occupies pp. 181—214, and treats of the nature of the apparatus and experiments made both at Portsmouth and at Cambridge, of the mathematical theory and specially of the effect of the inertia of the bridge; the second part, by Captain James and Lieutenant Galton, occupies pp. 215—250, and gives the tabulated results of the Portsmouth experiments. At the end of the *Report* will be found eleven interesting plates of apparatus and trajectories bearing upon this Appendix.

[1418.] Willis' essay ought to be read in conjunction with Stokes' memoir in the *Cambridge Transactions:* see our Art. 1276. We may note a few points with regard to it.

The Portsmouth experiments, although they were not sufficiently fine to give an accurate form to the trajectory of the travelling load, had yet shewn that the dynamical deflection of the bar could amount to more than twice, or even thrice, the statical deflection (pp. 184 and 203). Such deflections, however, occurred for values of a certain constant $\beta$, which were not likely to occur in practice: see Homersham Cox's paper discussed in our Art. 1433. These experiments thus failed to give a limit (except in the case of a very short pair of steel bars, etc.) to the maximum deflection. Summing up the results of the Portsmouth experiments Willis, after noting the radical defect of the apparatus when applied for the purpose of drawing a trajectory comparable with theory, writes:

> The principal excellence of the Portsmouth experiments consists in the determination of the effect of velocity upon the breaking weights on a large scale, for which purpose they will be found to give a most valuable and novel collection of facts (p. 193).

In other words they do not throw much light on the mathematical theory of the travelling load. In order to be able to compare theory and experiment Willis constructed an apparatus at Cambridge of an exceedingly ingenious character. This apparatus not only gave automatically the trajectory, but by the addition of an arrangement, which Willis terms an 'Inertial Balance,' enabled him to measure the effect of the inertia of the bridge on its deflection.

[1419.] After discussion of the apparatus follows the theoretical investigation. We reproduce that part of it which is due to Willis:

To simplify as much as possible the mathematical calculation the carriage must be considered as a heavy particle, and the inertia of the bar neglected. Let $x$, $y$ be the coordinates of the moving body, $x$ being measured horizontally from the beginning of the bar and $y$ vertically downwards, $M$ the mass of the body, $V$ its velocity on entering the bar, $2a$ the length of the bar, $g$ the force of gravity, $S$ the central statical deflection, that is to say the deflection that is produced in the bar by the body placed at rest upon its central point, $R$ the reaction between the body and the bar. The deflection is small, and therefore this reaction may be supposed to act vertically, for it must be recollected that the reaction is perpendicular to the curve of the *bar* and not to the *trajectory*, and therefore, in the case of such small deflections as we have to deal with, the horizontal component of the reaction will be insignificant. Thus the horizontal velocity $V$ will remain constant during the passage of the body along the bar. Now we have seen that a given weight $W$ suspended to the bar at a distance $x$ from its extremity will produce a deflection $y = cW(2ax - x^2)^2$, $c$ being a constant depending on the elasticity and transverse section of the bar. But as the inertia of the bar is neglected, its elastic reaction upon the travelling weight will be equal to a weight that would, if suspended to the bar at the point where the travelling weight touches it, depress that point to the same amount below the horizontal line. Therefore

$$R = W = y/\{c\,(2ax - x^2)^2\}.$$

The constant $c$ may be determined by observing that if $R = Mg$ and $x = a$, $y$ becomes $S$. Whence, substituting in the above equation, we obtain $c = S/(Mga^4)$.

The forces which act on the body are its gravity and the reaction of the bar. Whence we obtain the equation of motion,

$$\frac{d^2y}{dt^2} = g - \frac{ga^4}{S}\frac{y}{(2ax - x^2)^2},$$

which becomes, since $V = dx/dt$,

$$\frac{d^2y}{dx^2} = \frac{g}{V^2} - \frac{ga^4}{V^2S}\frac{y}{(2ax - x^2)^2}.$$

Putting $x = 2ax'$, $y = 16Sy'$, $\beta = ga^2/(4V^2S)$, we obtain the equation used by Stokes in our Art. 1278. Willis then continues:

Having proceeded thus far, however, I found the discussion of this equation involved in so much difficulty, that I was compelled to request my friend G. G. Stokes, Esq., Fellow of Pembroke College, to undertake the development of it. His kind and ready compliance with my wishes, and his well-known powers of analysis, have produced a most valuable and complete discussion of the equation in question (pp. 197—198).

[1420.] So far as theory is concerned the rest of the paper draws only from Stokes' memoir, but it involves some interesting additional comparison of experiment with theory. Thus Willis tells us that the value of $\beta$ for real bridges varies from 14 to 600, so that the dynamical increment of the deflection would be from ·0017 to ·1 only. Thus the great development of deflection which appeared at Portsmouth does not belong to real bridges but to cases in which $\beta$ had far too low a value (p. 203).

The remarks on p. 204 are also of considerable interest:

By comparing the experimental and calculated values of the dynamical deflection it will be seen that, with the exception of the last set (on bars of *steel*), the calculated values are smaller than the real values.

The excess, from its irregularity, is evidently due in part to some sources of error inseparable from the nature of the experiments, as, for example, the *set*, which shews itself by the greater difference exhibited in the case of cast-iron, for the mean value of the excess in the five experiments on cast-iron bars is three-tenths (·32) of the statical deflection, whereas in the fourteen cases where wrought-iron was employed, the mean value of the excess is one-tenth (·12) of the statical deflection. In the experiments on steel bars, on the other hand, the calculated deflections are greater than the actual deflections. But the values of $\beta$, in the latter case, are smaller than in the experiments on wrought and cast-iron, being, with one exception, less than unity.

In the next section I shall shew that the inertia of the bar will account for the greatest part of the discrepancies above stated between the theoretical and experimental deflections, for it will appear that it tends to increase the theoretical deflections when $\beta$ is greater than about 2 and to diminish them when less.

This last remark should be taken as qualifying Stokes' statement quoted in Art. 1289.

Similar discrepancies between computed and observed results were found in the cases of the Ewell and Godstone Bridges experimented on by the Commissioners.

[1421.] Willis' experiments with the 'Inertial Balance' led him to the following conclusions:

(1) For all values of $\beta$ less than about unity the least sensible inertia added to the bar will diminish the central deflection due to the theoretical trajectory, namely, that in which the bar is supposed to have no inertia.

(2) For all values of $\beta$ greater than about unity, inertia gradually added to the bar will at first increase the central deflection due to the theoretical trajectory, will then bring it to a maximum, and finally will diminish it.

(3) The ratio of the masses of the bar ($B$) to the load ($L$) that corresponds to this maximum will be very nearly unity for $\beta = 3$, and for the larger values of $\beta$ and of $B/L$ will be expressed by the equation $B = \cdot 823 \beta L$.

This later result is due to Stokes: see a footnote p. 210 of the paper.

It will thus be seen that experiments lead us to rather remarkable results as to the effect of the inertia of the bridge. Since for real bridges $\beta$ is usually $> 14$, it follows from these experiments that the inertia of a bridge tends to increase the deflections due to the theoretical trajectory of no inertia (p. 209).

[1422.] Willis concludes his paper by remarking that the alarming increase of deflection with increased velocity in the Portsmouth experiments having been found to be very much diminished in experiments on actual bridges,

It became, therefore, necessary to investigate the laws of these phenomena; and as analysis, even in the hands of so accomplished a mathematician as Mr Stokes, failed to give tangible results, excepting in cases limited by hypotheses that separated the problem from practical conditions, it became necessary to carry on also experiments directed to the express object of elucidating the theory and tracing its connection with practice......It has been shown that the phenomena in question exhibit themselves in a highly developed state when the apparatus is on a small scale, but that on the contrary with the large dimensions of real bridges their effects are so greatly diminished as to be comparatively of little importance, except in the cases of short and weak bridges traversed with excessive velocities. The theoretical and experimental investigation, which is the subject of the above essay, will, however imperfect, serve to show that such a diminution of effect in passing from the small scale to the large is completely accounted for (p. 214).

[1423.] Pp. 215—249 are occupied with the data of the experiments of James and Galton on moving loads. These experiments may have practical value, but they do not seem to have been so arranged as to settle points of theoretical interest.

[1424.] On pp. 250—258 of the same Appendix B will be found some experiments by James on the transverse strength of rectangular bars of cast-iron statically loaded. There is nothing of theoretical value to be noted.

[1425.] On pp. 259—262 we have probably the first experiments on reiterated strain. It would appear that the cast-iron bars subjected to a continuous depression and release, when the depression is less than that produced by 1/3 of the statical breaking load, gained no set after the first 150 depressions and were not weakened by the process; that is, showed no diminution of the statical breaking load. A few experiments on 'slowly moving loads' close this Appendix.

[1426.] The *Minutes of Evidence taken before the Commissioners* occupy pp. 264—378. An analysis of the evidence will be found on pp. 264—283. The material contained in these minutes will well repay perusal not only by the practical engineer, but by the pure theorist. We will note a few points in the order of the analysis.

[1427.] *Proportion of Load to Breaking Load.* 1°. Dead load on girders, should be not more than 1/3 to 1/5 breaking load. 2°. Travelling load causing vibrations, should not be more than 1/3 to 1/10 of the breaking load; the majority being in favour of 1/6.

There was thus considerable diversity of opinion, although it tended to shew that the maximum travelling load should not amount to more than 1/2 the maximum dead load (p. 266). This may be compared with a similar result of Poncelet's: see our Art. 988.

[1428.] *Deflection of Girders and Temperature-Effect.* Some interesting facts as to effect of continual change of temperature will be found in the evidence (p. 267). A curious idea, apparently supported by nearly all the practical men, was that a girder would be less deflected by a load moving at a high velocity across it, than by the same load at rest. Hawkshaw and P. W. Barlow, however, held the contrary view, the latter remarking that he had noticed that an express train passing over a timber-viaduct 'seemed to push the bridge like a wave before it.' This is interesting in the light of Stokes' and Willis' researches; see our Arts. 1282 and 1422. Some experiments of Hawkshaw's (pp. 411—412) referred to in a letter to the Commissioners confirm his view. Robert Stephenson also referred to a case where an engine and train had pushed a suspension bridge like a wave, said to be two feet high, in front of it (p. 340).

[1429.] *Change of molecular structure in Iron.* This is the point to which we find frequent reference at this time, namely, as to vibrations reducing 'tough and fibrous' metal to a 'crystalline and brittle state': see our Arts. 1463 and 1464. There was considerable evidence produced to shew that such a change takes place, although Robert Stephenson considered it highly improbable, see p. 335 and our Art. 1464. The evidence seems to me to point on the whole to a repeated *transverse* impulse producing a fracture which is crystalline in appearance; thus the connecting rods of engines in which the impulse must be principally longitudinal are not affected by years of work, while axles and shafting which must receive a good deal of transverse impulse are repeatedly referred to as 'altered in structure', that is to say, the fracture has a crystalline appearance: see the case of a cast-iron gun referred to on p. 374 of the *Minutes*.

[1430.] *Elastic Limit.* A good deal of the evidence as to the nature and even the existence of the 'elastic limit' is very contradictory. Thus one engineer held that the time element has no effect, so long as the 'elasticity is not destroyed' (p. 288). Another held that the smallest weight 'impairs the elasticity of a beam'; especially, if there be changes of temperature, the load does not permit of a return to the first molecular condition (p. 299): see our Art. 876. The last witness held that all railway girders 'would gradually swag down,' i.e. would have an increasing although very small set (p. 300). A third witness had observed no swagging in girders (p. 305). Robert Stephenson also mentions the limit of elasticity in cast-iron as a definite point (p. 337), and a witness called by him states that time has no influence on strain within the limit of elasticity (p. 348).

These replies are to some extent noteworthy, first because they do not clearly recognise the state of ease, which in many cases was probably produced by testing the girder, and secondly because so far as they concern cast-iron the elastic limit is not the mathematical elastic limit, the limit of the proportionality of stress and strain, but the limit to the state of ease, a very variable quantity.

[1431.] A few minor points in the evidence may be grouped in this Article. Thus, elastic after-strain had been observed by one witness in the testing of anchors, which 'crept back' to their primitive form in the course of a week (p. 300). This should be noted in regard to Saint-Venant's view of elastic-after-strain being of no sensible magnitude in metals: see his edition of Navier's *Leçons* p. 745. Set does not, at least in the early stages of loading, affect 'strength' (pp. 316 and 328). It is not quite clear whether the witnesses refer to elastic or absolute strength; in the first case this is a confirmation of the Coulomb-Gerstner Law: see our Art. 806. The term 'initial strain' used by Robert Stephenson and others (p. 348) must be distinguished from initial stress in the sense of Cauchy and the present work: it refers not to a molecular condition of the material produced by a process of preparation (as the sudden cooling of a casting), but to a permanent stress mechanically maintained in certain parts of a structure.

Nearly all the witnesses spoke favourably of Hodgkinson's formulae and form for beams: see our Arts. 244 and 971.

[1432.] The *Minutes of Evidence* are followed by Appendices (pp. 379—435) containing letters from various well-known engineers with data as to the structure of bridges, and as to the strength, elasticity, price and other properties of the material composing them.

[1433.] Homersham Cox. *The Dynamical Deflection and Strain of railway Girders. The Civil Engineer's and Architect's Journal.* Vol. XI. London, 1848, pp. 258—264. This forms No. X. of an interesting series of *Notes on Engineering* contributed by Mr Cox to this Journal. Others of these notes on the strength of suspension bridges, of the Menai Bridge, etc. will be found to be connected with the practical side of our subject.

The author applies Poncelet's method of treating resilience problems to investigate the problem considered by Stokes in his memoir of 1849: see our Art. 1276. He proceeds from the principle of the Conservation of Energy to calculate the work done and so the deflection produced. The whole calculation is based upon very elementary principles. Cox concludes that the statical strain and deflection cannot be more than *doubled* by the transit at any horizontal velocity of a weight travelling along the bridge (p. 260). He also shews how the influence of the 'centrifugal force' of the travelling load may be of use in diminishing the pressure at high speeds, if the girder be given a *camber* or curve upwards. Stokes has questioned the first of these results in his memoir in the following words:

My attention has recently been directed by Professor Willis to an article by Mr Cox......In this article the subject is treated in a very original and striking manner. There is, however, one conclusion at which Mr Cox has arrived which is so directly opposed to the conclusions to which I have been led, that I feel compelled to notice it. By reasoning founded on the principle of *vis viva*, Mr Cox has arrived at the result that the moving body cannot in any case produce a deflection greater than double the central statical deflection, the elasticity of the bridge being supposed perfect. But among the sources of labouring force which can be employed in deflecting the bridge, Mr Cox has omitted to consider the *vis viva* arising from the horizontal motion of the body. It is possible to conceive beforehand that a portion of this *vis viva* should be converted into labouring force, which is expended in deflecting the bridge. And this is, in fact, precisely what takes place. During the first part of the motion, the horizontal component of the reaction of the bridge against the body impels the body forwards and therefore increases the *vis viva* due to horizontal motion; and the labouring force which produces this increase being derived from the bridge, the bridge is less deflected than it would have been had the horizontal velocity of the body been unchanged. But during the latter part of the motion the horizontal component of the reaction acts backwards, and a portion of the *vis viva* due to the horizontal motion of the body is continually converted into labouring force, which is stored up in the bridge. Now, on account of the asymmetry of the motion, the direction of the motion is more inclined to the vertical when the body is moving over the second half of the bridge than when it is moving over the first half, and moreover the reaction itself is greater, and therefore, on both accounts, more *vis viva* depending upon the horizontal motion is destroyed in the latter portion of the body's course than is generated in the former portion; and therefore on the whole the bridge is more deflected than it would have been had the horizontal velocity of the body remained unchanged.

Stokes then shews that the change in the square of the horizontal velocity although small as compared with the square of the initial horizontal velocity is yet commensurable with the square of the vertical velocity gained (p. 725). He also shews that if it is the weight of the body which is small as compared with the weight of the bridge, then the deflection of the bridge will be sensibly the same although the bridge be *cambered* (p. 733).

Finally we may note that Willis's experiments shew that the deflection due to a travelling load may amount to *thrice* the central statical deflection, no limit in fact was reached for the deflection: see Stokes' memoir (p. 707) and our Art. 1418.

[1434.] Homersham Cox. *On Impacts on Elastic Beams. Camb. Phil. Trans.*, Vol. IX. Part I. pp. 73—78. This paper was read on Dec. 10, 1849.

Cox after referring to the experiments conducted for the Royal Commission appointed to inquire into the Application of Iron to Railway Structures (see our Art. 1409 and Eaton Hodgkinson's experiments described in Art. 942) thus states his problem:

An elastic beam of uniform density and section throughout its length, abuts at each extremity against a fixed vertical prop, and is impinged upon at its centre by a ball moving horizontally, with an assigned velocity in a direction perpendicular to the length of the beam before collision, and subsequently moving in contact with the beam throughout its deflection. It is required to determine the deflection of the beam produced by the impact.

[1435.] The problem may be divided into two parts, (i) to find the alteration in the velocity of the ball at the instant of collision, (ii) to measure the effect of the elastic stresses in the beam in destroying the kinetic energy which the system has immediately after collision.

Now the curve taken by the beam, Cox holds, will not differ considerably from the elastic curve of a beam deflected by statical pressure at its centre; this is in fact the assumption made by Willis and Stokes in a similar case: see our Art. 1419.

Hence, if $f$ be the central deflection of a beam of length $a$, $y$ the deflection at distance $x$ from one extremity,

$$y = f \, . \, (3a^2x - 4x^3)/a^3 \quad \ldots\ldots\ldots\ldots\ldots\ldots\ldots \text{(i)}.$$

Let $y$ now be considered as the displacement produced in an indefinitely small time $t$ by the impact, $M$ the mass of the beam, then since $y/t$ is the velocity at the point $x$, the virtual moment of the momentum generated by the blow

$$= \int_0^a \frac{M}{a} dx \frac{y}{t} \, . \, y = Pf \quad \ldots\ldots\ldots\ldots\ldots\ldots\ldots\ldots \text{(ii)},$$

where $Pf$ is the product of the blow and its virtual velocity.

For the legitimacy of the principle from which this equation is deduced Cox refers to Poisson's *Traité de Mécanique*, Chap. IX. § 535.

Substituting from (i) in (ii) and integrating, we find

$$\tfrac{17}{35} \frac{f^2 M}{t} = Pf,$$

or, if $v = f/t$ be the initial velocity of the centre of the beam,

$$P = \tfrac{17}{35} Mv.$$

Now if $m$ be the mass of the ball and $u$ its velocity before impact,

$$P = m(u - v),$$

since $v$ is supposed the same for the ball and beam after impact. Thus we find

$$v = u \cdot m/(m + \tfrac{17}{35}M) \text{..................(iii).}$$

[1436.] The next step is to calculate the kinetic energy of the beam immediately after impact. This is easily found to be

$$\tfrac{1}{2}(mv^2 + \tfrac{17}{35}Mv^2) = \tfrac{1}{2}\frac{m}{m + \tfrac{17}{35}M}mu^2 \text{..............(iv).}$$

Upon these results Cox remarks:

Adopting, then, the elastic curve to represent the initial velocities of the several parts of the system, effecting the integration and supplying the numerical calculations, we find ultimately that rather less than one half the inertia of the beam may be supposed to act initially to resist the ball; or, to speak more precisely, that at the instant after impact the impinging ball loses as much of its motion as it would have done if it had impinged on another free ball having 17/35 of the mass of the beam (p. 76).

This very nearly agrees with the gratuitous *assumption* of Tredgold and the experimental result of Hodgkinson: see our Arts. 943 and 999 and Appendix, Note E.

[1437.] The next stage is to find the deflection of the beam produced by the change of this kinetic into potential energy. Now if the beam be statically deflected at its centre through a certain distance, it is found that the pressure required to produce this deflection is in a nearly constant ratio to the deflection. Let $\alpha$ be the weight which will statically produce a deflection of unit length. Then, if $f$ be any deflection, $\alpha f$ is the corresponding pressure necessary to maintain it, and $\tfrac{1}{2}\alpha f^2$ the work necessary to produce it, we must have

$$\frac{m}{m + \tfrac{17}{35}M}\frac{mu^2}{2} = \frac{\alpha f^2}{2},$$

or,

$$f = \sqrt{\frac{m}{m + \tfrac{17}{35}M}\frac{m}{\alpha}} \cdot u.$$

We have thus the deflection produced by a ball of mass $m$ striking with velocity $u$ a beam of mass $M$, the extremities of which are supported.

This formula is confirmed by Hodgkinson's experiments, except when the beam is very flexible and subject to a great velocity of impact. In practice beams of great rigidity are always employed. Of course care must be taken that the ends of the beam do not recoil from their bearings after impact. A useful comparison of theory and experimental result will be found on p. 78.

[1438.] *The Deflection of imperfectly Elastic Beams and the Hyperbolic Law of Elasticity. Camb. Phil. Trans.*, Vol. IX. Part II. pp. 177—190. This paper was read in part March 11, 1850, and in part October, 1850.

Cox begins his paper by a reference to James Bernoulli's memoir of 1694, especially that part of it—referred to in our Art. 22—wherein Bernoulli cites an experiment of his own as confuting Hooke's law. Cox quotes this result as an early notice of the 'inexactness of Hooke's Law,' and adds:

> The real law of elasticity of any material can be known only by direct experiments on the material itself, and it seems nearly certain that even for two different specimens of the same metal, the laws would be in some measure different.

Hodgkinson's hypothesis:

$$\widehat{xx} = as_x - bs^2_x,$$

where $\widehat{xx}$ is the traction, $s_x$ the elastic stretch, and $a$, $b$ constants, is then referred to: see our Art. 1411. Cox terms this very fitly the *parabolic law*, to distinguish it from Hooke's law

$$\widehat{xx} = Es_x,$$

which we might in a similar fashion term the *linear law*.

[1439.] It is shewn (p. 178) that the parabolic law is not in close accordance with the experiments of Hodgkinson recorded in the Railway Commissioners' Report: see our Art. 1411. This is partly due to the fact that the constants $a$ and $b$ were deduced in those experiments from selected cases and not by the method of Least Squares. But Cox holds that the *parabolic law* is not the true law, because the differences between the theoretical and actual values of the stresses are not promiscuously positive and negative, but are all negative or all positive for several terms together. It is thus possible to find a law for the differences. He shews that if a positive traction be expressed by the quantity

$$\alpha s + \beta s^2 + \beta' s^3,$$

and a negative traction by the quantity

$$\gamma s + \delta s^2 + \delta' s^3,$$

it is possible to calculate the constants $\alpha$, $\beta$, $\beta'$, $\gamma$, $\delta$, $\delta'$ by the method of Least Squares, so that the experimental agree almost exactly with the calculated values. The value with the above law of traction-stretch for the deflection of a beam terminally supported and centrally loaded is stated. But to obtain the constants from deflection experiments would require most complex calculations.

[1440.] In order to avoid these calculations Cox suggests what he terms the *hyperbolic law* of elasticity, or

$$\widehat{xx} = \frac{\alpha s_x}{1 + \beta s_x},$$

$\alpha, \beta$ being constants which are supposed to be different for positive and negative tractions. This law, he states, gives far more accurate results than the *parabolic law*, besides leading to greater simplicity in computation. The hyperbolic law evidently gives a rectangular hyperbola, with its asymptotes parallel to the axes, for the traction-stretch relation.

[1441.] The writer compares results calculated on the hyperbolic and parabolic laws respectively with Hodgkinson's experiments and shews that the mean error of the latter is 3 or 4 times as great as that of the former. The hyperbolic law is then applied to the theory of beams and their deflection calculated. The process is one of approximation, although occasionally of an ingenious character (pp. 184—188). The memoir concludes with a calculation of the four constants of the hyperbolic law for cast-iron of the quality used in Hodgkinson's experiments. Cox remarks finally:

The great desideratum for perfecting the Hyperbolic or any other hypothetical law of elasticity, is the want of knowledge of these variations of the strength and elasticity of the material, which depend on the magnitude of the castings. It is greatly to be desired that this defect of experimental data may not long continue unsupplied.

[1442.] It must, however, be remarked that this hyperbolic law as stated is purely empirical[1]. It is not in some respects so satisfactory as the parabolic law. For that supposes an expansion of the work function in integer powers of the strain, and a retention of the first two terms only.

It does not very clearly appear from Hodgkinson's experiments whether his parabolic law holds after the body is reduced to a state of ease, for he kept on applying increasing loads and deducting the set. The traction-stretch relation is certainly not linear for a bar of cast-iron in its state of ease. Does, however, this relation coincide with that obtained by applying a series of increasing loads and deducting set? I do not see that Cox or Hodgkinson has noted this point. As I read their memoirs we have only approximate and purely empirical formulae suggested as relations between the load and the elastic-stretch in the case of a series of increasing loads applied to a cast-iron bar not reduced to a state of ease.

[1443.] Several brief papers on matters related to our subject will be found in the *British Association Reports*, 1842—1850. We will devote the following nine Articles to them.

[1444.] *Glasgow Meeting*, 1841. *Notices of Communications*, pp. 201—202. *Experimental Inquiry into the Strength of Iron.* This abstract gives the details of some of W. Fairbairn's experiments on the strength of iron plates, and the influence upon it of riveting. A fuller account will be found in his book on the Tubular Bridges: see our Arts. 1494 and 1495.

[1445.] *Cambridge Meeting*, 1845. *Notices of Communications*, p. 26. *On the strength of Stone Columns*, by Eaton Hodgkinson.

This paper contains some interesting facts. I am unaware, as in the case of a number of the following communications by the same writer, whether it was ever published *in extenso.*

The experiments were made on prisms of square section; the side of the bases being 1 inch and $1\frac{3}{4}$ inches, the heights varying from 1 inch to 40 inches. We may note the following remarks as closely bearing on the Bernoulli-Eulerian theory of columns, and especially on Lamarle's memoir: see our Art. 1255.

[1] On a certain fairly plausible hypothesis as to 'initial stress' we may, however, deduce the Hyperbolic Law for the elastic stretch of a bar *not* reduced to a state of ease.

From the experiments on the two series of pillars it appears that there is a falling off in strength in all columns from the shortest to the longest; but that the diminution is so small, when the height of the column is not greater than about 12 times the side of its square, that the strength may be considered as uniform; the mean being 10,000 lbs. per square inch or upwards.

From the experiments on the columns 1 inch square, it appears that when the height is 15 times the side of the square, the strength is slightly reduced; when the height is 24 times the base, the falling off is from 138 to 96 nearly; when it is 30 times the base, the strength is reduced from 138 to 75; and when it is 40 times the base, the strength is reduced to 52, or to little more than one-third.

The stone was from the Peel Delph, Littleborough, Lancashire.

It will thus be seen that the strength of a short column is nearly proportional to the area of the section, though the strength of a larger one is somewhat less than in that proportion. These results are in accordance with Lamarle's theory: see Arts. 1258 and 1259.

[1446.] Hodgkinson then goes on to remark that in all columns shorter than 30 times the side of the square, fracture took place by one of the ends failing, shewing the ends to be the weakest part. Theory shews that when flexure takes place, if the ends are *built-in*, the two ends and the middle are equally the weakest points. I should judge that Hodgkinson's ends were not truly built-in, and this may account for their rupturing *first* at an end[1].

The fashion in which such struts or columns tend to rupture at the terminals is, according to Hodgkinson, by a wedge being sheared out. A complete theory of the flexure of columns would deduce this result theoretically. Hodgkinson talks rather vaguely of the 'tendency of rigid materials to form wedges with sharp ends, these wedges splitting the body up in a manner which is always pretty

[1] A number of cast-iron struts experimented on by A. B. W. Kennedy in 1885 shewed the tendency to *simultaneous* fracture at the ends and the middle in a very marked fashion. These struts were of square section and their length 20 times the diameter. One such strut is figured on the frontispiece. The shaded portions shew where the strut retained its original skin, i.e. the parts in tension. The form of the 'neutral line' is thus very obviously marked, and agrees with theory. It would appear to coincide very nearly with the axis of the strut at both the ends and centre before rupture.

nearly the same.' He refers to Coulomb's attempt to explain the matter theoretically: see our Arts. 120 and 729.

[1447.] His final conclusions may be reproduced:

As long[1] columns always give way first at the ends, shewing that part to be the weakest, we might economise the material by making the areas of the ends larger than that of the middle, increasing the strength from the middle both ways towards the ends. If the areas of the ends be to the area in the middle as the strength of a short column is to that of a long one, we should have for a column, whose height was 24 times the breadth, the area of the ends and middle as 13,766 to 9,595 nearly. This however would make the ends somewhat too strong, since the weakness of long columns arises from their flexure, and increasing the ends would diminish that flexure.

Another mode of increasing the strength of the ends would be that of preventing flexure by increasing the dimensions of the middle.

From the experiments, it would appear that the Grecian columns, which seldom had their length more than about 10 times the diameter, were nearly of the form capable of bearing the greatest weight when their shafts were uniform; and that columns tapering from the bottom to the top, were only capable of bearing weights due to the smallest part of their section, though the larger end might serve to prevent lateral thrusts. This last remark applies too to the Egyptian columns, the strength of the column being only that of the smallest part of the section. (p. 27.)

[1448.] *Southampton Meeting*, 1846. *Notices of Communications*, pp. 107—109. These pages contain some account of experiments by Fairbairn and Hodgkinson on hollow tubes. They led to some controversy at the meeting. The papers are given in full in the works of Fairbairn and Clark: see our Arts. 1466 and 1494.

[1449.] *Oxford Meeting*, 1847. *Notices of Communications*, p. 43. *On the Defect of Elasticity in metals subject to Compression.* This is the title of a communication by Hodgkinson of which no particulars are given: see our Arts. 1411 and 1412.

*Ibid*, p. 132. *Experiments on the strength of Iron Columns.*

[1] Apparently Hodgkinson terms 'long' a column whose length is 30 times its diameter: see Art. 1446.

Another communication by the same author, with no particulars: see our Art. 1477.

[1450.] *Swansea Meeting*, 1848. *Notices of Communications*, p. 119. *On Investigations undertaken for the purpose of furnishing data for the construction of* Mr Stephenson's *Tubular Bridges at Conway and Menai Straits*, by Eaton Hodgkinson. There are no particulars: see our Arts. 1416 and 1466.

[1451.] *Birmingham Meeting*, 1849. *Notices of Communications*, p. 118. *On the Strength and Elasticity of stone and timber*, by Eaton Hodgkinson. This paper refers to an extensive series of experiments, which were in progress: see our Art. 1414.

[1452.] *Edinburgh Meeting*, 1850. *Notices of Communications*, p. 2. *On the Laws of the Elasticity of Solids*, by W. J. Macquorn Rankine. The immediate object of this paper is: 'to investigate the relations which must exist between the elasticities of different kinds possessed by a given substance, and between the different values of those elasticities in different directions.' Six general theorems are stated without proof. As these theorems recur in Rankine's other papers, which we shall treat at length in our second volume, we need not consider them here.

*Ibid*, p. 172. *On the Hyperbolic Law of Elasticity of Cast Iron*, by Homersham Cox. Stokes communicated Cox's results to the Meeting: see our Arts. 1438 and 1440.

[1453.] *Annales des ponts et chaussées.* A certain number of memoirs bearing on our subject will be found in the volumes of this Journal for the years 1840—1850. We devote the following seven Articles to their contents.

[1454.] 1842. 1^er^ semestre. E. Flachat and J. Petiet. *Sur les ponts suspendus avec câbles en rubans de fer laminé*, pp. 336—393. This is a purely practical paper, a variety of experimental data and comparative statistics will be found in Chapitre IV. (pp. 357—393). The authors advocate the use of *câbles en rubans de fer laminé* as more advantageous than wire or *chaînes en barres* in the construction of suspension bridges.

En résumé, par rapport aux fils, l'avantage réel, incontestable, c'est la durée.

Comparés aux chaînes en barres, on trouve : sécurité, économie, durée, élégance, légèreté (p. 393).

[1455.] 1844. 2e semestre. G. H. Dufour. *Nouvelles épreuves d'un pont suspendu en fil de fer construit à Genève, de* 1822 *à* 1823, pp. 89—98. These tests are interesting as bearing on the time-effect in elasticity. It was shewn that the cables of this bridge after 21 years' service had their absolute and elastic strengths unimpaired.

[1456.] 1845. 2e semestre, pp. 252—256. A *résumé* is given of the memoir of Hagen on the stretch-moduli of wood referred to in our Art. 1229.

[1457.] 1848. 1er semestre. Bresse. *Études théoriques sur la résistance des arcs employés dans les ponts en fonte ou en bois*, pp. 150—193. The following is the problem treated in this memoir:

Un arc circulaire, à section constante, est posé sur deux appuis de niveau ; il est ensuite chargé de poids répartis uniformément suivant une ligne horizontale, et peut avoir en outre à supporter accidentellement des poids concentrés en son milieu. Il s'agit de determiner : 1° la poussée de l'arc, ou, en d'autres termes, la composante horizontale de la réaction des appuis ; 2° le maximum d'effort auquel doit résister la matière de l'arc.

This problem had been considered previously by Navier, Heim and Saint-Venant although with slight variations: see our Arts. 278, 913 and 1373.

The author's problem is less general than Saint-Venant's, in that he nowhere introduces the consideration of slide; it is more general, in that he has also a continuous load.

[1458.] Let $2P$ be the weight placed upon the summit of the arc, supposed of chord $2a$ and central angle $2\psi$, let $p$ be the pressure on the arc per unit of *horizontal* length; then Bresse finds (p. 153) that $Q$ the horizontal component on one of the points of support is given by

$$Q=\frac{(\frac{3}{2}\sin^2\psi-\psi\sin\psi\cos\psi+\cos\psi-1)P+(-\frac{1}{4}+\frac{1}{4}\psi\cot\psi+\frac{7}{12}\sin^2\psi-\frac{1}{2}\sin\psi\cos\psi)pa}{\frac{1}{2}\psi+\psi\cos^2\psi-\frac{3}{2}\sin\psi\cos\psi}.$$

This is deduced from formulae given by Navier in his *Leçons*, 1re partie, p. 287: see also our Arts. 278 and 279.

If we compare it with Saint-Venant's result we find that they agree if we put $p = 0$ in Bresse's form, and strike out of Saint-Venant's the terms depending on compression and slide: see our account of Saint-Venant's *Cours* of 1839 in Art. 1573.

[1459.] Bresse then shews that if $P = 0$, and if $f$ be the subtense of the arc,

$$Q < \frac{pa^2}{2f}.$$

This leads him to criticise a result of Ardant given in the memoir referred to in our Art. 937. Ardant states that

$$Q > \frac{pa^2}{2f}.$$

But Bresse points out, that if we retain terms involving $\psi^2$,

$$Q = \left(1 - \frac{2\psi^2}{7 - \frac{4}{3}\psi^2}\right) \frac{p\,(a^2 + f^2)}{2f},$$

and is thus $\quad < \dfrac{pa^2}{2f}$ (p. 155).

[1460.] The next step is to consider the form taken by the loaded arc. Bresse shews that, beside the terminals, there are two points, either of which he terms *point moyen neutre*, where the curvature remains unchanged.

Finally we have some consideration of the greatest central load which a continuously loaded circular arc will bear. It is based upon the maximum-traction in a fibre, not upon the stretch limit. Probably the consideration of slide would considerably modify this result. The memoir concludes with approximate expressions for the constants involved, and a table of the values of these constants for a considerable variety of angles.

A very important book by Bresse on the same subject, entitled: *Recherches analytiques sur la flexion et la résistance des pièces courbes*, Paris, 1854, will be discussed in our second volume.

[1461.] 1850. 1er semestre. V. Chevallier. *Recherches expérimentales sur la construction des portes d'écluses*, pp. 309—356. This memoir contains a valuable comparison of the theoretical and experimental curves of flexure for beams subjected to transverse load. A very close accordance was found between theory and practice (p. 316).

The memoir is followed by certain *Notes* giving a few cases of the theory of flexure drawn from Navier's *Leçons* and some remarks on the memoir of Wertheim and Chevandier noted in our Arts. 1312—1314. We may cite the following:

Dans les expériences présentées à l'Institut en octobre 1846, MM. Wertheim et Chevandier ont trouvé que les flèches des pièces chargées au milieu se composaient d'une partie sensiblement proportionnelle à la charge, et d'une partie permanente dont la grandeur variait avec l'intensité et la durée de la charge. Dans les expériences que j'ai faites, je n'ai trouvé des résultats semblables que pour des tringles peu homogènes et des charges trop fortes ou trop longtemps appliquées. Souvent même des pièces qui, déchargées, semblaient conserver une flèche permanente, reprenaient, à très-peu près, leur forme première au bout de plusieurs heures. Du reste, la charge ne restait guère au-delà du temps nécessaire pour observer les flèches: en un mot, elle n'était pas permanente (p. 350).

This would seem to shew that the after-strain noted by Wertheim and Chevandier was really not set but elastic after-strain.

[1462.] 1850. 2e semestre. Leclerc and Noyon. *Notice sur la construction du pont suspendu de Saint-Christophe, près Lorient*, pp. 265—336. This contains a great deal of valuable experimental matter. See especially p. 295 and the tabulated results pp. 333—336. The experiments were made with a hydraulic press in the naval dockyard at Lorient. We may note the following results:

1°. Maximum resistance of iron wire of good quality may be taken as 75 kilogs. per square millimètre.

2°. Maximum resistance of iron bar of good quality may be taken to be 34 kilogs. per square millimètre. (This is 20 % too low for good iron bar of to-day.)

3°. Loads of 10 kilogs. per square millimètre do not produce sensible extension in iron bars.

4°. In general loads up to 22 kilogs. or 2/3 of the rupture-load do not produce set in iron bars, and the stretches do not exceed 1/1000, i.e. stretch-modulus = 22,000 kilogs. per square millimètre (?).

[1463.] *Minutes of Proceedings of the Institution of Civil Engineers.* A number of papers referring to our subject will be found in Volumes I. to X. for 1840—1850. We may note a few related more closely to the theory of our subject:

(*a*) Vol. II. Session, 1842. pp. 180—184. *On some peculiar Changes in the Internal Structure of Iron, independent of, and subsequent to, the several processes of its manufacture,* by Charles Hood. This paper notes the changes in direction of crystallisation produced by impact, heat and magnetism, and thus the change in cohesion. The paper and the discussion contain only general statements, no theoretical investigation.

(*b*) Session, 1843. pp. 89—94. *Account of a series of experiments on the comparative strength of Solid and Hollow Axles,* by J. O. York. Experimental detail, with further discussion on the effect of vibration as supposed to produce crystallisation and so to affect the cohesive strength of iron.

(*c*) Session, 1843. pp. 126—133. *Experiments on Cast and Malleable Iron, at the Milton Iron Works, Yorkshire,* by David Mushet. No theoretical value.

(*d*) Vol. III. Session, 1844. p. 202. *On the causes of fracture of the Axles of Railway Carriages,* by Joseph Glynn. Further consideration of the influence of vibration on cohesive strength.

(*e*) p. 248. *Remarks on the position of the Neutral Axis of Beams,* by Charles Schafhaeutl. The author draws the attention of the Institution to Brewster's method of determining the neutral axis of beams by the transmission of polarised light through annealed glass beams under similar load: see my footnote, p. 640.

(*f*) Vol. IX. Session, 1849—1850. pp. 233—287. *On Tubular Girder Bridges,* by William Fairbairn. This is a discussion of the Torksey Bridge, with a certain amount of experimental detail. The theory introduced by the writer of the paper, as well as by Hodgkinson in the discussion which follows, involves only the ordinary formulae for the deflection of beams subjected to transverse load. There is frequent reference to the Report of the Iron Commissioners: see Art. 1406.

(*g*) pp. 294—302. *On the manufacture of Malleable Iron; with the results of experiments on the strength of Railway Axles*, by G. B. Thorneycroft. Interesting, as giving further particulars as to the effect of vibration on cohesive strength, and as figuring a number of surfaces of rupture.

Several of these papers bring into importance the need for a full theory of the influence of vibration upon molecular arrangement, and of the influence of molecular arrangement upon cohesion. Any mathematician attempting to work out such a theory will find statements suggesting various directions for research in the papers (*a*), (*b*), (*d*) and (*g*) referred to above.

[1464.] *Institution of Mechanical Engineers.* October, 1849, to October, 1850. Several papers as to the effect of continuous vibration on the strength of railway axles will be found in the *Proceedings* for these months, and may be profitably compared with the corresponding papers read before the *Institution of Civil Engineers:* see our Art. 1463.

The first paper by J. F. McConnell is principally of practical value; the author, however, lays stress on the change from the 'fibrous to the crystalline character' owing to vibration (October, 1849, pp. 13—21). The Chairman, Robert Stephenson, in the discussion said he had been unable to detect any change in molecular structure due to vibration (p. 22).

Another speaker said he had cold-hammered fibrous iron till it became crystalline; but in this first discussion the general opinion seems to have been against a molecular change due to vibration. An additional paper by J. F. McConnell will be found in the *Proceedings*, January, 1850 (pp. 5—19), with further evidence of a strong kind in favour of a molecular change being produced by vibration. The discussion was continued at the April meeting (*Proceedings*, April, 1850, pp. 3—14). Robert Stephenson was still doubtful as to the change in molecular condition, but his reasoning (p. 6) does not seem peculiarly convincing. The writer of the original paper considered that 'fibrous structure' was a misnomer; a molecular change took place which changed the iron from tough to brittle. Robert Stephenson referred to some interesting experiments made during the building of the Britannia Bridge, in which plates which were to have been delivered as fibrous, were found to be crystalline. By direct experiment, however, it was found that their tractive strength was above rather than below the average. He drew attention to a distinction which he thought ought to be made between a steady tractive stress and a vibrational stress, and apparently held that iron of crystalline structure might have greater resistance than fibrous iron to the former stress, and less resistance than fibrous iron to the latter stress.

In the *Proceedings*, July, 1850, pp. 35—41, will be found a paper by Thomas Thorneycroft entitled: *On the Form of Shafts and Axles.* An additional paper by the same engineer with the discussion on both will be found in the *Proceedings*, 1850, pp. 4—15. The only point of theoretical value which seems brought out in either paper or in the discussion is that by bending iron frequently backwards and forwards it may be rendered crystalline (pp. 8 and 15). That is, the rupture *appears* crystalline.

The papers referred to above do not seem strictly scientific either in their substance or their method. Nor do they achieve more than to draw attention to the need existing at that time for experiments on the influence of intermittent load and continuous impact on molecular structure and the elastic strength of materials.

[1465.] Edwin Clark and Robert Stephenson. *The Britannia and Conway Tubular Bridges with general inquiries on Beams and on the Properties of Materials used in Construction.* 2 vols. London, 1850. This book was written by Edwin Clark under the supervision of Robert Stephenson.

It contains, beside an elementary theory of beams subjected to transverse load on the old Bernoulli-Eulerian hypothesis, a large amount of experimental material, and not a few new physical facts which it will be needful for us to notice. The results of some of the experiments given in this work had already been published in the *Report of the Iron Commissioners*: see our Art. 1406.

[1466.] Vol. I. Section I. contains the early history of the design with an account of parliamentary evidence, etc.

Section II. contains the details of the preliminary experiments made by Stephenson, Fairbairn and Hodgkinson. On pp. 83—115 we have an account of some experiments on cylindrical and elliptical tubes. There is little in these experiments to which it is necessary to draw attention for our present purpose. We may however note a remark on p. 111, to the effect that the value of each horizontal layer of a beam in resisting transverse stress is directly proportional to its distance from the neutral axis and independent of its lateral position. This remark is of course a theoretical result of the Bernoulli-Eulerian theory of beams, it does not, however, hold if slide be taken into account. Probably it is not generally true, owing to the change of the stretch-modulus at the epidermis, or in the case of rupture, which is that of the book, owing to a weakness produced by a tendency to buckle, when the material, instead of being arranged in a tube form, is gathered about the vertical diameter of the section. Experiments are wanted on this point.

[1467.] Pp. 116—154 are occupied with details of experiments on rectangular tubes and beams of cross-section. Then follow Fairbairn and Hodgkinson's reports of their experiments on tubes, especially on rectangular tubes with 'cellular tops.' These reports were also published in Fairbairn's work: see our Arts. 1416 and 1494.

[1468.] Chapters III. and IV. of this section (pp. 155—205) contain an account of experiments on a large model tube (75 feet long, 4 ft. 6 in. deep and 2 ft. 8 in. wide) just one-sixth in length of one of those used for the Britannia Bridge, and general deductions from all the preliminary experiments. They embrace much interesting information, although of little importance for theory. We may remark, however, that though some results as to deflection and set are given, the main object sought after seems to have been the *absolute strength.* This appears to me rather a doubtful measure for the stability of a bridge. One or two isolated experiments will be found on *time-effect,* none on repeated load : see, however, pp. 458—460 on repeated *impact.*

[1469.] Section III. (pp. 207—292) is entitled: *General Principles of Beams.* The analytical investigations of this section as well as much of Section VIII. are due to Dr William Pole. The writer remarks:

The complete theory of a beam, in the present state of mechanical science, is involved in difficulties. *The comparative amount of strain at the centre of the beam, where the strain is greatest,* or at any other section, is easily determined, but the exact nature of the resistance of any given material almost defies mathematical investigation (p. 207).

The words in italics are clumsy, as they suppose the beam to have a distribution of load symmetrical about the centre. The difficulties mentioned arise apparently from the change in the nature of the elasticity of a beam from the surface inwards, and from the want of proportionality between traction and stretch. Throughout this section, as in other parts of the work, the term *strain* is used for what we now term *stress.*

[1470.] Chapter II. (pp. 239—257) of the section gives formulae for various kinds of beams on the suppositions of the proportionality of traction and stretch, and the equality of the stretch- and squeeze-moduli. Hooke's principle,—*ut tensio, sic vis,*—is here identified with *perfect elasticity*, but this seems to me a wrong use of the term; a material may be perfectly elastic, and yet stress not be proportional to strain: see our Appendix, Note D.

[1471.] Chapter III. commences with a geometrical theory of beams (pp. 258—267), which is due to C. H. Wild, and described in Clark's preface as 'extremely elegant and novel' in method. It can hardly be considered as a valuable contribution to our subject. The remainder of the chapter is occupied with the usual formulae for the deflection of beams.

[1472.] The following chapter is devoted to continuous beams, and treats the cases of beams supported at three, four, five or an indefinite number of regular points of support. Some portions of this theory of continuous beams seem to me ingenious and possibly appear here for the first time; thus we may note the elementary determination of the points of contra-flexure (pp. 275—

277), and the final proposition of the chapter on a beam subject only to its own weight:

The deflection of each span of a perfectly continuous beam is one-fifth that of an independent beam spanning the same opening.

Some of the results given in this chapter could be obtained with less analysis by the use of Clapeyron's Theorem.

[1473.] Section IV. is entitled: *Specific Experimental Inquiries.* The remarks as to set are interesting. It is pointed out that to reduce a body to a state of ease up to a certain load does not increase its absolute strength, and that when workmen term such a body stronger, this only arises from their measuring strength by the amount of strain (p. 305). Set in wrought iron is not proportional to the load, and the writer suggests that it probably varies as the square of the load 'as is exactly the case with cast-iron.' He is evidently referring to Hodgkinson's experiments on cast-iron: see our Arts. 969 and 1411, but it should be noted that Hodgkinson's later formula makes the set vary in part as the first, in part as the second power of the *elastic stretch,* not as the square of the load.

[1474.] On pp. 305 and 311 will be found an account of two interesting set-phenomena noted by Messrs Easton and Amos. In the one case a wrought-iron cylinder was subjected to enormous internal pressure; it received a set *altering its internal but not its external diameter;* additional set of the same kind was obtained with each reloading, but at last equilibrium was reached and the cylinder ceased to expand under the pressure. In the second case a very similar fact was noted with regard to cast-iron. In connexion with cast-iron cylinders it is noted on p. 305 that with great internal pressure they begin to open on the inside, the fracture gradually extending to the outside, increased thickness beyond moderate limits giving no increased strength. This should be compared with Lamé's results: see our Art. 1013, footnote.

[1475.] With regard to set and the plastic stage for iron the following remarks are of great interest, and should be noted by any one attempting to reduce these stages in the elastic life of a material to theory:

With a wrought-iron inch cube the set becomes so great with 12 tons that its shape and proportions begin to suffer; and where these are of any consequence, as in most practical cases they are, we come to the limit of its utility. It is not, however, yet destroyed until the load is about 16 tons. It then oozes away beneath additional strain, as a lump of lead would do in a vice, or like a red-hot rivet under the pressure of the riveting machine, and to some extent obeys the laws of liquids under pressure. If prevented from bulging or oozing away, the softest metal would bear an infinite weight, like water in a hydraulic press (p. 306).

[1476.] Turning to the phenomena exhibited by cast-iron, Clark points out that for similar weights it yields twice as much as wrought-iron, yet that its absolute strength is three times as great. Under high pressures cast-iron also appears like liquids to exert an equal pressure in every direction in which its motion is opposed (p. 311). Notwithstanding this, however, a cast-iron cube ultimately fails in an entirely different manner to wrought-iron, namely by wedge-shaped fragments being, as it were, sheared off: see our Arts. 949 and 1414, and the reproduction of some of the rupture-surfaces on our frontispiece.

[1477.] Pp. 314—318 are occupied with some account of Hodgkinson's memoir on pillars: see our Art. 954. This is followed by some very important experiments of Hodgkinson's on the resistance of plates of wrought-iron to a crushing load (pp. 318—335). The strength of such plates is shewn theoretically and experimentally to be directly as their width and the cube of their thickness, but inversely as the square of their length. The theoretical considerations on p. 320 do not seem to me quite satisfactory, but the result is easily seen to be true. As we have noted, the top of the Britannia tube is of a cellular form, and thus cannot be treated as a plate subjected to crushing load. We accordingly have given on pp. 335—364 a great number of experiments on the crushing of cellular and hollow struts. These experiments are due to Hodgkinson, and show very markedly the importance of *form*. Thus in the case of circular and double rectangular sections we have the following noteworthy results for 10 feet tubes (pp. 343, *et seq.*):

| | Form | External Dimensions Inches | Area of Section Sq. in. | Thickness of Plates Inches | Weight of Tube lbs. | Total Crushing Load lbs. |
|---|---|---|---|---|---|---|
| Rectangular (riveted) | ▭ | 8·1 × 4·1 | 1·885 | ·059 | 82 | 43673 |
| Circular | ○ | 4·05 | 1·7078 | ·278 | 59 | 47212 |

Notwithstanding the much better results *apparently* given by the circular form and Hodgkinson's strong recommendations, the rectangular form, chiefly, I expect, from the comparative ease of construction, was ultimately adopted for the 'cellular plate' at the top of the tube.

We shall state reasons later for questioning the application of these experiments to the 'cellular plate': see our Art. 1493.

[1478.] The concluding pages (365—370) of the second chapter of Section IV. are devoted to experiments on the crushing of brickwork, single bricks, sandstone and limestone. In the case of sandstone the fracture was by a wedge, as in Hodgkinson's experiments on cast-iron: see our Arts. 949 and 950. Limestone on the other hand formed *perpendicular* cracks and splinters in much the same way as glass struts have been observed to fracture. It would be interesting to explain theoretically by a consideration of their aeolotropic or isotropic formation why sandstone and limestone thus differ when subjected to a crushing load, the one apparently showing a rupture due to slide and the other to lateral stretch.

[1479.] Chapter III. (pp. 370—388) is entitled: *On Tension and Tensile Strength of Materials.* The valuable part of it is a reproduction of Hodgkinson's results published in the *Report of the Iron Commissioners*: see our Art. 1408.

[1480.] Chapter IV. (pp. 389—396) is occupied with experiments on the strength of rivets and on the shearing of iron. The terms shear and shearing strain are here used as accepted terms and in the sense of our shearing stress. On p. 517 (Vol. II.) we read 'the strain called by Mr Stephenson "the shearing strain".' It would thus appear that the term shear in this sense was first used by

Stephenson; I have not noticed the word before. It of course corresponds to the *force transverse* of Vicat: see our Art. 726.

The experiments on the strength of rivets seem to be among the earliest of the kind. The following conclusions were obtained when the rivet was treated merely as a body subjected to shear:

(i) The ultimate resistance to shear is proportional to the sectional area of the bar torn asunder.

(ii) The ultimate resistance of any bar to a shear is nearly the same as the ultimate resistance to a direct longitudinal traction.

This is not in accordance with more recent experiment; even on the hypothesis of uni-constant isotropy holding up to the instant of rupture the ratio of these two resistances should be 4/5.

[1481.] The frictional force produced by the cooling of red-hot rivets was then investigated. As the rivet cools it draws the plates together, and the friction produced by this pressure has to be overcome before the rivet can be affected by shearing force. The frictional force produced by a rivet of 7/8 inch diameter was found to vary from $4\frac{1}{2}$ to nearly 8 tons. Thus it would appear that in cases of *elastic* strain rivets may possibly act only through a tractive and not a shearing stress. In the case of the Britannia tube the stress on any rivet never amounts to $4\frac{1}{2}$ tons, which explains why the deflection is 'the same as that indicated by theory for a tube formed of one welded piece of iron without joints.'

[1482.] The fact is noted that long rivets break at the head sooner than short ones; there appears to be no theoretical reason why they should, as the stress produced by cooling ought to be the same in both. It is possibly due to irregular cooling, the shank in the neighbourhood of the head and tail cooling more slowly than towards the centre (p. 395).

Clark concludes his investigation of riveting with the remark,

Thus, also, by judicious riveting the friction may in many cases be nearly sufficient to counterbalance the weakening of the plate from the punching of the holes, so that a riveted joint may be nearly equal in strength to the solid plates united (p. 396).

This is certainly not true of *absolute* strength; does it refer to the *elastic* strength of a butt joint?

[1483.] Chapter v (pp. 397—460) is devoted to experiments on the transverse strength of beams and tubes. Although many of the experiments in this chapter are due to Hodgkinson and will be found in the *Report of the Iron Commissioners* (see our Art. 1413), it yet contains several additional facts.

We may draw special attention to some experiments on the transverse strength of cast-iron tubes due to Stephenson. The results are obtained for equal length, sectional area, and thickness. On p. 433 it is remarked:

Now, with a constant length and sectional area, if there were no advantage in any particular form, the strengths should be precisely in the ratio of the depths.

A constant $c$ giving 'the relative advantage of each particular form of section' is then calculated from the formula $W = c\omega a/l$, where $W$ is the central breaking load, $\omega$ the sectional area, $2a$ the vertical diameter, and $l$ the length of the tube. I think this formula is wrong.

If we suppose $T_0$ to be the traction per unit area which will just rupture a bar of cast-iron in tension, and suppose the breaking load to be in a constant ratio to the load which just produces set, we have:

$$W = 4T_0 \frac{\kappa^2 \omega}{la},$$

where $\kappa$ is the radius of gyration of the cross-section about the trace of the neutral surface on the plane of the section. Hence, if the tube be cylindrical and of radii $a$ and $a_1$, and $W_c$ the breaking load,

$$W_c = 4T_0 \frac{\pi (a^4 - a_1^4)}{4la};$$

if square with sides $2a$ and $2a_2$, and $W_s$ the breaking load,

$$W_s = 4T_0 \frac{4 (a^4 - a_2^4)}{3la}.$$

But the section being the same, $\omega = \pi (a^2 - a_1^2) = 4 (a^2 - a_2^2)$.

Now let $a_1 = n_1 a, \quad a_2 = n_2 a,$ then

$$W_c = 4T_0 \frac{\omega a}{l} \cdot \frac{1 + n_1^2}{4}, \qquad W_s = 4T_0 \frac{\omega a}{l} \cdot \frac{1 + n_2^2}{3}.$$

Thus
$$\begin{cases} \text{for a cylindrical form,} & c_1 = 4T_0 \cdot \dfrac{1+n_1^2}{4}; \\ \quad \text{,, \quad square \quad form,} & c_2 = 4T_0 \cdot \dfrac{1+n_2^2}{3} \end{cases}$$

If $n_1$, $n_2$ were mere numbers $c$ would certainly be a constant giving the relative advantage of each form of section. But we have

$$n_1^2 = 1 - \omega/(\pi a^2), \quad n_2^2 = 1 - \omega/(4a^2),$$

or, $n_1$ and $n_2$ depend upon the dimensions of the beam, i.e. on the ratio of the sectional area to the area enclosed by the external contour. We note of course that $n_2 > n_1$, and thus $c_2$ is always greater than $c_1$. Stephenson makes the order of strength square, elliptical, circular, rectangular tubes; but the values he gives to the constant $c$ have no value except for tubes having the same ratio of section to contour area as those in his experiments.

[1484.] On pp. 442—445 we have some experiments of Captain James on the ultimate strength of cast-iron bars, which bring out in a marked fashion the loss of strength in large bars, owing to the different elastic characters of the core and epidermis due to the process of casting. Thus 'bars planed down from the centre of larger bars are comparatively very weak': see our Arts. 858, 1216 and 1408.

[1485.] A very noteworthy result is given on p. 446, the first of the kind I have come across, namely, the change produced by flexure in the shape of the square cross-section of a wrought-iron bar. The *set* shape of the section pictured bears a striking resemblance to the *elastic* shape of the same section obtained from Saint-Venant's theory of flexure: see his *Leçons de Navier*, p. 34.

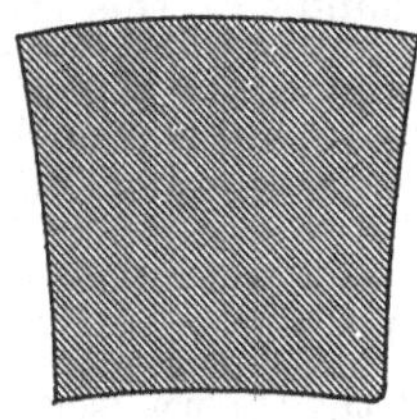

[1486.] On pp. 449—451 we have again a remarkable experiment, namely, on the increase of strength in a bar due to a primitive strain. Four equal bars of wrought-iron were placed in a furnace, and two out of them when they had attained a dull-red heat were arched or bent 'with a wooden mallet, so that the metal was not upset by hammering'; the other two remained straight. The curved bars when cold were straightened, so that they were given an initial stress. The four bars were then subjected to transverse load, and the deflections for the bars with initial stress were found much less for like loads than the deflections of those without initial stress.

In fact as regards deflection the strained beam may be considered a new material, of which the elasticity is quite different from that of the original beam (p. 451).

The result deserves attention, as being of constant practical application. It furnishes, moreover, a confirmation of the views explained in a previous chapter on the nature of permanent set arising from strain, and satisfactorily accounts for the many anomalies which characterise the conclusions arrived at by different authors from experiments on the elasticity of materials, in which the effect of previous strain has been overlooked (p. 450).

Compare our Arts. 1379 and 1473.

According to Clark not only is the deflection decreased, but the beam gains in absolute strength by an initial stress.

[1487.] Some remarks on resilience and repeated impact (pp. 454—460) are only quotations from Hodgkinson in the *Report of the Iron Commissioners.* The Young-Tredgold theorem shows that the resilience of a beam is as its volume, and Stephenson thus rightly argued that the motion or impact of trains would have no sensible effect on the enormous mass of his tube.

[1488.] In the last chapter of this section (pp. 461—466) we have a comparison between the theoretical and experimental deflections for continuous beams supported at four and five points. The experiments are due to Brunel and Clark. In the former case Brunel has compared the observed and computed reactions at the points of support. The elastic lines are carefully plotted out and figured. Considering that the theory used neglects slide, remarking the difficulty of getting a really uniform bar, and then of determining exactly its elastic constants and linear dimensions, I think that the computed deflections accord remarkably with experiment.

[1489.] Volume II. (pp. 467—821) is principally occupied with the description of the tubes and the history of their erection, and does not concern us as theorists. Section VIII. (pp. 726—787), however, is of interest; it contains calculations of the deflection and strength of the Conway Tube by Eaton Hodgkinson, and of the deflection and strength of the Britannia Tube presumably by Dr William Pole. These are probably the most important problems to which the Bernoulli-Eulerian theory of beams was ever, or ever will be, applied. For the Conway Tube the correspondence between the calculated and actual deflections may be described as the best proof ever given of the close approximation of that theory to fact when it is applied to beams under transverse load. The mean error is only ·01 inches, the central deflection being more than 8 inches. The theory is based upon a uniform moment of inertia, but as a matter of fact the moment of inertia of the tube decreased from the centre towards the ends. Allowing for this the most marked differences between actual and calculated deflection are at the ends, a result which we should naturally expect from the neglect of slide in the calculations.

The comparison of the like results for the Britannia Tube shews a far greater discrepancy, but this is natural, as the tube can hardly be considered as a continuous beam, the junction of the two large tubes at the centre or Britannia Tower having been made rather with the aim of equalizing the pressures on the three points of support than of producing continuity (pp. 780—781).

We have given sufficient evidence to shew the important contributions to physical knowledge made by these great engineering works. These contributions have considerable bearing on theory. They are frequently referred to by Saint-Venant in his edition of Navier's *Leçons*.

[1490.] Thomas Tate. *On the Strength of Materials; containing various original and useful formulae, specially applied to Tubular Bridges, Wrought-iron and Cast-iron Beams, &c.* London, 1850.

This is a work of 96 pages. It contains an application of the old Bernoulli-Eulerian Theory to beams of varied cross-section allied to those employed in the tubular bridges. Thus we have the moments of inertia and the relative strengths calculated for square circular and elliptic tubes, and also for rectangular tubes with 'cellular' tops.

[1491.] The writer practically assumes the proportionality of stress and strain up to the limit of cohesion.

If $M$ be the maximum bending moment for a beam terminally supported and centrally loaded, we have with the notation of Art. 1483, since $M = Wl/4$,

$$M = T_0 \,.\, \omega\kappa^2/a.$$

Tate terms $T_0$ the *modulus of rupture*, and supposes it constant for beams of the same material (p. 16). As a matter of fact this formula may possibly be true when $T_0$ is taken as the tractive load which just produces set (i.e. the elastic limit) in a material for which there is no defect of Hooke's Law: see our Appendix, Note D. But it is certainly not true as a formula for rupture, (i) when $T_0$ is taken as the tractive load which produces rupture in a bar, for the rupture-load has occasionally double the value given by this formula (see Saint-Venant's edition of Navier's *Leçons*, p. 90); or (ii) when the material has defect of Hooke's Law; or (iii) different moduli for stretch and squeeze, for in this as in the preceding case the neutral axis does not pass through the centroids of the cross-sections; or (iv) when the form of the cross-section is considerably modified; it is found in this case that $T_0$ is not independent of the shape of the cross-section.

Tate throughout his work calculates the strength of his beams, even of the Conway Tube, on the supposition that they are to be loaded till rupture. For permanent structures, it is difficult to understand how this can be any safe criterion of strength. It was, however, adopted also by Fairbairn and Hodgkinson.

It cannot be too often reiterated that the theoretically best form of section for a beam to be loaded only below the elastic limit by no means necessarily coincides with the section of the beam of greatest strength.

[1492.] Pp. 24—41 are devoted to the consideration of the

relative strength of beams of similar section. In the preface (p. vii) Tate lays some stress upon the originality of the formulae obtained for such beams; they are however very easy deductions from the above somewhat doubtful rupture-formula.

[1493.] Tate questions, and I think justly, some of the formulae given by Hodgkinson, which bear too often a purely empirical character, and therefore cannot be trusted beyond the range of material and form employed in the experiments on which they are based. Thus (p. 59) Tate objects to Hodgkinson's treatment of the cellular top structure of the Conway Tube as a strut, remarking that the whole tube must be considered as a beam under transverse load. For this reason he doubts Hodgkinson's recommendation of the circular cell in preference to the rectangular cell for the cellular top structure of the tubes. By an analysis (pp. 74—76) very similar to that of our Art. 1483 it is shewn that the hollow square section of the same height and area as the hollow circular section is the stronger of the two. Since the neutral axis in this case lies outside the section, we have to replace $\kappa^2$ of that Article by $\kappa^2 + d^2$, where $d$ is the distance of the centre of either section from the neutral axis. This addition, however, of the same quantity to $W_c$ and $W_s$ will not affect the question as to which is the greater, and the conclusions of that Article still hold.

It seems to me that even supposing the cellular top structure to act as a strut, Hodgkinson's experiments are hardly conclusive, because his rectangular cells with the same amount of material occupy *more space* than the circular—so he would really be using less material for a top structure of given size. If he intended to divide his circular section into two, as was proposed in the 'corrugated' top, and to compare the cellular tops of type

then since these contain the same amount of material and occupy about the same space, his recommendation was theoretically right, always supposing the top *might* be treated as a strut. His experiments only shew, as it is, the advantage of *form* in hollow pillars or struts, not in cellular tops; see our Art. 1477.

Tate, after endeavouring to account for Hodgkinson's results, remarks,

Be this as it may, we are not disposed to give up a principle established by theory until some *direct experiments* should prove the contrary (p. 76).

On the whole there is little in Tate's work of present value; the special problems discussed in it being of an elementary nature.

[1494.] Sir William Fairbairn. *An account of the Construction of the Britannia and Conway Tubular Bridges.* London, 1849.

This work, although marred by the unnecessary details of a very painful controversy between Robert Stephenson, Eaton Hodgkinson and the author, still contains a large amount of experimental matter relating to the strength of cylindrical and rectangular tubes used as girders. Thus we have (pp. 37—42) an abstract of Fairbairn's experiments on cylindrical, elliptic and rectangular tubes, together with tubes with 'cellular' structures at the top to strengthen the resistance to 'puckering' or to crushing force. These are followed by a report of Eaton Hodgkinson's (pp. 42—47) containing a comparison of the formulae for tubes of circular, elliptic or rectangular section with experimental results The 'cellular' structure for strengthening the tube seems to have been the idea which Hodgkinson and Fairbairn each claim to have originated (p. 105). The matter was discussed at the British Association Meeting for 1846 in Southampton: see our Art. 1448.

The most valuable experimental part of the work, however, will be found in the Appendix, pp. 209—288, being a detailed account of Fairbairn's experiments entitled: *Experimental Inquiry into the Strength of Malleable Iron Tubes.* These contain experiments on a vast variety of tubes, including those strengthened at the top by a cellular structure. On pp. 273—281 will be found formulae, adapted from the ordinary theory of flexure to tubes of this character. The inquiry concludes with a few experiments on rivets and the strength of iron plates: see Art. 1495.

Much of the matter in this book will also be found in Clark's work on these Bridges: see our Art. 1465.

[1495.] 1850. Sir William Fairbairn. *An Experimental Inquiry into the Strength of Wrought-Iron Plates and their Riveted Joints as applied to Ship-building and Vessels exposed to severe strains. Phil. Trans.* Part II. 1850, pp. 677—725. This paper is also reproduced in Fairbairn's *Useful Information for Engineers,* First Series, 3rd Ed., Appendix, pp. 251—320.

[1496.] The experiments described in the paper were made in the years 1838—39, but professional duties hindered the preparation of the paper till the above date. Eaton Hodgkinson assisted Fairbairn in the experiments, which were designed with the special view of testing the suitability of iron in the construction of ships. The distinct points under investigation were:

1°. The strength of plates when torn asunder by a direct tractive load in the direction of the fibre, and when torn asunder by a load applied across it.

2°. On the strength of joints in plates united by rivets as compared with the strength of the plates themselves.

3°. On the resistance of plates to a contractive load whether applied as a dead weight or by impact.

4°. On the transverse strength of wrought-iron ribs and frames.

[1497.] In the first section (Part I. pp. 678—687) Fairbairn comes to the following conclusions:

(i) There is very slight difference between the absolute strength of plate and of bar iron.

This by no means holds for plates and bars such as are now manufactured.

(ii) In whatever direction the plates are torn asunder their absolute strengths are nearly the same, being in the mean about 1/45 in favour of those torn *across* the fibres.

The 'direction of the fibre' is used to signify the direction of the axes of the rough bars when piled together before being rolled into the plate. The conclusion (ii) seems to me extremely improbable and suggests some experimental error.

[1498.] On p. 686 will be found a reference to some experiments made by Hodgkinson on iron wire and to others made by the Admiralty at Woolwich on iron bars, which shew that the absolute strength of a material is not reduced by repeated rupture. Some experiments apparently shew it to be increased: see our Art. 1503.

[1499.] In the second section on riveting (Part II. pp. 688—702), Fairbairn gives the preference to machine over hand-riveting. He does not appear to consider that any great influence is exercised by the frictional adhesion of the plates at the riveted joint. Thus he writes:

On comparing the strength of plates with their riveted joints, it will be necessary to examine the sectional areas taken in a line through the rivet-holes with the section of the plates themselves. It is perfectly obvious that, in perforating a line of holes along the edge of a plate, we must reduce its strength; and it is also clear, that the plate so perforated will be to the plate itself nearly, as the areas of their respective sections, with a small deduction for the irregularities of the pressure of the rivets upon the plate; or, in other words, the joint will be reduced in strength somewhat more than the ratio of its section through that line to the solid section of the plate......It may be said that the pressure or adhesion of the two surfaces of the plates would add to the strength; but this is not found to be the case, to any great extent, as in almost every instance the experiments indicate the resistance to be in the ratio of their sectional area, or nearly so (p. 691).

It does not seem to me that Fairbairn's experiments are at all conclusive as to the part played by friction in a riveted joint. I think that till the elastic limit of the rivet is passed friction must be an element in the resistance at least of butt joints: see our Art. 1482.

Fairbairn has also omitted to notice that the remaining material is weakened if the rivet-holes be punched.

[1500.] Fairbairn concludes that a double line of rivets is far superior to a single line, and gives the following *round* numbers as shewing loss of strength (p. 700):

Strength of plate = 100.
Strength of double-riveted lap joint = 70.
Strength of single-riveted lap joint = 56.

I do not understand the process by which these numbers are deduced, nor do they seem very satisfactory, for no distinction is made between the elastic and absolute strengths of the joint. The strain of a rivet is of a very complex nature, yet theoretically of a most interesting kind. We shall refer to the important memoirs of recent years in our second volume. Fairbairn can however lay claim to being the earliest and for many years the only experimenter in this field.

[1501.] Part III. (pp. 702—705) is occupied with the absolute strength of wrought-iron plates subjected to a normal load produced by the pressure of a blunt point at the centre of the plate. The results obtained are compared with similar experiments on oak planks. It would have been interesting to compare the forms assumed by loaded plates with those given by theory for circular plates: see our Arts. 494—504.

[1502.] Part IV. (pp. 705—713) is occupied with experiments on the absolute strength of the frames and ribs of vessels, i.e. beams formed by riveting angle-irons and plates together. These experiments have practical rather than theoretical value. They are followed (pp. 713—719) by simple analytical formulae for finding the neutral axis, moment of rupture, etc. for beams of various cross-sections on the Bernoulli-Eulerian hypothesis.

[1503.] The memoir concludes with an account of experiments by Loyd on the tractive strength of bars of wrought-iron and some, suggested by Kennedy, on the transverse strength of malleable iron beams of diverse section (pp. 719—725). All that it seems necessary to notice in these experiments are the remarkable statements:

(i) That the absolute strength of iron bars allowing for reduced sectional area is *increased* by successive breakages (p. 720): see our Arts. 1353, 1379 and 1524. Thus it increased on the average in four successive breakages from 23·94 to 29·20 tons per square inch.

(ii) The rupture-stretch of wrought-iron bars of the same section increases as their length decreases, although the absolute strength remains the same (p. 723).

These results would be suggestive as regards the changes set produces in molecular condition were we certain how far the first may not be due to the want of homogeneity in the bar and so to its rupturing successively at its weakest sections, and how far the second may not be due to the fact that the short lengths always include the *local* stricture.

[1504.] E. Phillips. *Sur les ressorts formés de plusieurs feuilles d'acier employés dans la construction des voitures et wagons. Comptes rendus,* T. XXXI. 1850, pp. 712—715. This is an extract of a memoir, which does not appear to have been ever published in full. The memoir contains a theoretical and experimental investigation of the best form of carriage springs. In the extract no analysis is given, but an interesting statement of theoretical and experimental results. These will be best taken in connection with later memoirs on the same subject by Phillips which will be considered in our second volume.

1505. Samuel Haughton. *On the Laws of Equilibrium and Motion of Solid and Fluid Bodies.* This is printed in the *Camb. and Dublin Math. Jour.*, Vol. I. 1846, pp. 173—182, and Vol. II. 1847, pp. 100—108. The whole is in effect reproduced with considerable additions in a memoir entitled: *On the Equilibrium and Motion of Solid and Fluid Bodies,* and published in the *Transactions of the Royal Irish Academy,* Vol. XXI., 1848, pp. 151—198; the memoir was read May 25, 1846. An abstract of the memoir is given in the *Proceedings of the Royal Irish Academy,* Vol. III. 1847, pp. 252—258.

1506. An account of this memoir would find its appropriate place in a history of the wave theory of light; we shall only briefly notice what strictly belongs to our subject.

The author says at the beginning of the memoir:

The object of the present paper is to deduce from simple physical considerations the laws of equilibrium and motion of elastic solid and fluid bodies, by the method followed in the *Mécanique Analytique* of Lagrange, which possesses the remarkable advantage of giving, by the same analysis, the *general* equations of any system, together with the *particular* conditions to be fulfilled at the limits......The present paper is an attempt to apply the same method of Lagrange to the case of material substances, whether fluid or solid, homogeneous or heterogeneous, and whether possessed of a crystalline structure or not; and more particularly to investigate the general dynamical laws of solid elastic bodies, and the conditions at the limiting surfaces which bound the solid.

[1507.] The following remarks, which will be found on p. 173

of the *Journal*, Vol. I., may be quoted as marking Haughton's position:

The most general conception of solids and fluids, is that of 'an immense assemblage of molecules separated from each other by indefinitely small distances'; if we add to this general notion, the assertion, that '*these molecules act on each other only in the line joining them,*' we shall have a definition of the medium whose laws I propose to investigate.

This statement stamps the writer as a rari-constant elastician[1].

[1508.] The distinction drawn between solids and fluids is

That in solid bodies the resultant of all the forces exerted by all the surrounding molecules on any molecule ($m$) is zero. That in fluids, whether liquid or gaseous, this is not the case, and that consequently the fluid (no external pressures or forces acting) would be dissipated (p. 174.)

[1509.] It may be noted that Haughton, notwithstanding the Navier-Poisson controversy, does not hesitate to replace his summations by definite integrals and this without any remark: see our Arts. 421, 532, and 615. The same objections which Poisson and Cauchy raised against Navier apply to Haughton: see Haughton's integrals on pp. 178 and 179 of the *Journal*, Vol. I.

1510. The general equation of equilibrium from which the author starts is

$$\iiint (X\delta u + Y\delta v + Z\delta w)\, dm = \iiint \delta V\, dx\, dy\, dz,$$

where $\delta V$ denotes, in the language of Lagrange, the sum of the moments of the internal forces. The essential part of the process is that which relates to the nature of the function $V$. The action between two molecules is assumed to be represented by an expression of the form $f(\rho, \alpha, \beta, \gamma, \rho')$, where $\rho$ is the distance

[1] Of course in the case of isotropy as a uni-constant elastician. Thus Sir William Thomson said in his Baltimore lectures:

'The pressural wave has been the difficulty. Cauchy starved the animal, M'Cullagh and Neumann didn't know of its existence, Haughton put it in an Irish car and it wouldn't go, Green and Rayleigh treated it according to its merits.' (*Nature*, March 1885, p. 462.)

I understand this to mean that the *bête noir* of those, who syllogize: ether = jelly = elastic solid, is quite unmanageable with one constant: see our Art. 930.

between the two molecules in their original state, and $\alpha, \beta, \gamma$ the direction cosines of this distance; $\rho'$ is the increment of $\rho$ produced by the operation of the forces.

By expanding and neglecting squares and higher powers of $\rho'$ we may represent $f$ thus,

$$f = F_0 + 2F_1\rho'$$

The *moment* of this, in Lagrange's sense of the word, is $fd\rho'$, that is

$$F_0 d\rho' + F_1 d\rho'^2;$$

and thus we have for $\delta V$ the value

$$\delta V = \Sigma\{F_0 \delta\rho' + F_1 \delta (\rho'^2)\},$$

and finally

$$V = \iiint (F_0\rho' + F_1\rho'^2)\, dm.$$

The author holds that the first part of the integral vanishes in the case of solids but not in the case of fluids.

1511. The general equations of equilibrium restricted to this approximate value of $V$ are fully worked out. Denote by $V_0$ and $V_1$ the parts of $V$ which depend on $\rho'$ and $\rho'^2$ respectively; then it is found that

$$V_0 = \frac{4\pi}{3}\left(\frac{du}{dx} + \frac{dv}{dy} + \frac{dw}{dz}\right)\int_0^\infty F_0\rho^3 d\rho,$$

where $u, v, w$ are the shifts. It is found also that $V_1$ contains 21 terms, but involves only 15 distinct coefficients.

1512. I will not delay on the applications made by the author of his formulae to the wave theory of light, but on account of the interesting nature of the results I will reproduce a portion of the abstract given in the *Proceedings of the Royal Irish Academy*. Haughton constructs six fixed ellipsoids, and from these derives a certain variable ellipsoid.

The axes of this ellipsoid will be the *three* possible directions of molecular vibrations, and the corresponding velocities of waves will be inversely as the lengths of these axes.

To this the following footnote is appended:

After Mr Haughton had obtained this construction, he found that M. Cauchy has given analytically, and for a particular case, a solution which involves an analogous ellipsoid; but M. Cauchy has not followed out the consequences of his analysis in the right direction, and has been misled in his attempt to apply his equations to the problem of light.

The abstract then proceeds thus:

The six ellipsoids just mentioned perform a very important part in the problem of elastic solids, as they reappear in the conditions at the limits, and afford a geometrical meaning for many of the results.

1513. Mr Haughton then determines from simple considerations the equation of the Sphaero-Reciprocal-Polar of the Wave-surface, or the *Surface of wave-slowness of elastic solids*, which occupies a position in this subject analogous to that held by the index-surface in Light. This surface, and the important results it leads to, are as far as Mr Haughton is aware, given by him for the first time; it is of the sixth degree, and has three sheets, and by means of it the direction of a vibration passing from one medium into another may be determined.

1514. The paper then proceeds to the discussion of three particular cases of elastic solids: (i) The case where the molecules are arranged symmetrically round three rectangular planes; (ii) Round one axis; (iii) The case of a homogeneous isotropic body.

In the first case the following results are deduced:

The traces of the surface of wave-slowness on the planes of symmetry consist of an ellipse and a curve of the fourth degree. The surface possesses four nodes in one of its principal planes, where the tangent plane becomes a cone of the second degree, and the existence of these points will give rise to a conical refraction in acoustics, similar to what has been established in physical optics.

In general, for a given direction of wave-plane, three waves will be possible, the corresponding vibrations of the molecules being in three directions, at right angles to each other, though not, in general, parallel or normal to the wave-plane. Mr Haughton investigates the possibility of the vibrations being normal and transversal, and finds that for particular directions of wave (given in the paper) the vibrations are, two in the wave-plane, and the third perpendicular to it. He discusses also at length the other two cases, together with the equations of condition which hold in general at the limits, and the geometrical interpretations of these conditions by means of the fixed ellipsoids.

1515. The part of the memoir which really concerns us may be considered to be pp. 151—168, corresponding to what is given in Vol. I. of the *Camb. and Dub. Math. Journal.* Out of these the following pages of the memoir are not found in the Journal, namely 159—163 which treat of the reduction of the constants from 12 to 9, and 166—168 which treat of the conditions at the limits.

In connection with this memoir by Haughton and another by him in Vol. XXII. of the *Transactions of the Irish Academy*, the reader may consult a paper by W. J. M. Rankine in the *Camb. and Dublin Math. Jour.*, Vol. VII., 1852, pp. 217—234.

1516. Samuel Haughton. *On a Classification of Elastic Media, and the Laws of Plane Waves propagated through them.* This is published in the *Transactions of the Royal Irish Academy*, Vol. XXII. 1855, pages 97—138; it was read January 8, 1849[1].

1517. So far as we are concerned with this memoir it may be said to consist in the establishment of the body-stress-equations of an elastic solid. They are given in the form

$$\rho\left(\frac{d^2u}{dt^2}-X\right)=\frac{d\widehat{xx}}{dx}+\frac{d\widehat{xy}}{dy}+\frac{d\widehat{xz}}{dz},$$

$$\rho\left(\frac{d^2v}{dt^2}-Y\right)=\frac{d\widehat{yx}}{dx}+\frac{d\widehat{yy}}{dy}+\frac{d\widehat{yz}}{dz},$$

$$\rho\left(\frac{d^2w}{dt^2}-Z\right)=\frac{d\widehat{zx}}{dx}+\frac{d\widehat{zy}}{dy}+\frac{d\widehat{zz}}{dz}.$$

The mode of obtaining them resembles that of which an account will be found in our notice of Haughton's memoir of 1850: see our Art. 1520.

The conditions

$$\widehat{zy}=\widehat{yz},\quad \widehat{xz}=\widehat{zx},\quad \widehat{yx}=\widehat{xy}$$

are obtained as by Cauchy and Poisson, but Haughton adds

[1] [In this memoir Haughton treats the rari-constant hypothesis as applying only to a particular class of bodies, and so to some extent retires from the position assumed by him in the memoir of May, 1846. But he reaches a very curious conclusion (pp. 126, 127), namely, that if the body consist of repelling and attracting molecules the function $V$ can contain *twenty-one* independent constants. He supposes in fact the reduction to *fifteen* constants in his former paper to depend on the representation of the constants by definite integrals, and not on the assumption of *molecular force being central.* He thus apparently does not understand why Green's equations are more general than Cauchy's. It is obvious that Haughton is in this section a rari-constant elastician in disguise, for if he accepts the hypothesis that the function $V$ depends only on change in molecular distance he ought to reduce his constants to *fifteen,* and in the equations of p. 133 to *one:* see Cauchy's and Clausius' memoirs noted in our Arts. 615, 616, 922 and 1400. ED.]

on p. 107, "These writers seem to have considered them as necessary for all systems; but this is not true...". The author proceeds to develope this remark, but I do not agree with him.

1518. The most important bearing of the memoir is on the wave theory of light; the author institutes comparisons between the different methods used by Fresnel, Cauchy, Green, and MacCullagh: see p. 274 of the *Report on Double Refraction* by Stokes in the *British Association Report* for 1862.

The following remark as to the principle of Virtual Velocities occurs on page 116 of the memoir:

The equation of virtual velocities, as stated by M. Poisson and other writers, supposes no virtual displacements but those for which equal and opposite displacements are possible; the correct statement of the principle is, that the sum of the virtual moments of the system can never become positive for possible displacements. For the full development of this important correction of the equation of virtual velocities, as given by Lagrange, I shall refer to a memoir of M. Ostrogradsky, contained in the *Mémoires de l'Académie de St Pétersbourg*, Tom. III. p. 130[1].

1519. It will be convenient to take together three papers published by Haughton in various volumes of the *Camb. and Dublin Math. Jour.* and entitled: *Notes on Molecular Mechanics.*

I. *On the general equations of Motion.* Vol. V., 1850, pp. 172—176.

II. *Propagation of Plane Waves.* Vol. VIII., 1853, pp. 159—165.

III. *Normal and Transverse Vibrations.* Vol. IX., 1854, pp. 129—137.

1520. The second and the third of these papers belong to the wave theory of light, and so do not fall within our range; they cite a memoir by Jellett in Vol. XXII. of the *Trans. R. Irish Acad.*: see our Art. 1526. We have then only to notice the first paper; the object of this is to establish the general equations of

[1] The following errata in Haughton's memoir may be noted:

Page 111. In equation (14) in many terms 2 is omitted.

Page 114. In equations (16) supply − on the left-hand side; this mistake extends its influence further. The fact is that in this memoir $V$ has the contrary sign to the $V$ of the first memoir.

equilibrium and motion. Two extracts from it will be of interest; they are from the commencement of the paper.

The equations which express the conditions of equilibrium and motion resulting from molecular forces have been investigated by many writers, and although the results arrived at have for the most part agreed with each other, yet the principles from which they have been derived have been so different, as to render this branch of mechanics less complete than many others on which less labour has been bestowed. In France, this subject has occupied the attention of Navier, Poisson, Cauchy, and St. Venant, while at home it has been cultivated with success by Mr Green and Prof. MacCullagh in the case of Light, and by Mr Stokes in the case of Hydrodynamics and Elastic Solids......

I shall adopt the method of the *Mécanique Analytique* of Lagrange, which is well adapted to such investigations as the present. In order to express by means of it the conditions of equilibrium and motion of a continuous body, it is necessary to distinguish the forces acting at each point into two classes, viz. molecular and external forces; including among the latter the resultants of the attractions of the points of the body which are situated at a finite distance, as these attractions result from gravitation, and should not be confounded with molecular forces. The forces being thus considered as divided into two groups, the equation of virtual velocities is the following:

$$\Sigma (P\delta p + P'\delta p' + \&c.) + \Sigma (Q\delta q + Q'\delta q' + \&c.) = 0,$$

$P$, $P'$, &c. denoting the external forces, and $Q$, $Q'$, &c. the molecular forces. The hypotheses which I shall make as to the nature of the molecular forces are two in number: first, '*that the virtual moments of the molecular forces may be represented by the variation of a single function,*' i.e.

$$Q\delta q + Q'\delta q' + \&c. = \delta V \quad\text{.......................}(1);$$

and secondly, '*that if $u$, $v$, $w$ represent the small displacements of any molecule from its position of rest, $x$, $y$, $z$, the function $V$ depends on the differential coefficients of the first order of $u$, $v$, $w$,* i.e.

$$V = F(\alpha_1, \alpha_2, \alpha_3, \beta_1, \beta_2, \beta_3, \gamma_1, \gamma_2, \gamma_3) \quad\text{..............}(2),$$

where

$$\alpha_1 = \frac{du}{dx}, \quad \alpha_2 = \frac{du}{dy}, \quad \alpha_3 = \frac{du}{dz},$$

$$\beta_1 = \frac{dv}{dx}, \quad \beta_2 = \frac{dv}{dy}, \quad \beta_3 = \frac{dv}{dz},$$

$$\gamma_1 = \frac{dw}{dx}, \quad \gamma_2 = \frac{dw}{dy}, \quad \gamma_3 = \frac{dw}{dz}.$$

1521. It will be seen from this quotation that the assumptions on which the process rests are very clearly stated. By the ordinary methods of the Calculus of Variations the following equations are obtained which must hold at every point of the body in the case of motion:

$$-\rho\frac{d^2u}{dt^2}=\frac{d}{dx}\cdot\frac{dV}{d\alpha_1}+\frac{d}{dy}\cdot\frac{dV}{d\alpha_2}+\frac{d}{dz}\cdot\frac{dV}{d\alpha_3},$$

$$-\rho\frac{d^2v}{dt^2}=\frac{d}{dx}\cdot\frac{dV}{d\beta_1}+\frac{d}{dy}\cdot\frac{dV}{d\beta_2}+\frac{d}{dz}\cdot\frac{dV}{d\beta_3},$$

$$-\rho\frac{d^2w}{dt^2}=\frac{d}{dx}\cdot\frac{dV}{d\gamma_1}+\frac{d}{dy}\cdot\frac{dV}{d\gamma_2}+\frac{d}{dz}\cdot\frac{dV}{d\gamma_3},$$

where $\rho$ denotes the density of the body.

[1522.] Paper III. may be fairly neglected for our present purpose as its applications are purely to the undulatory theory of light, but there are one or two points in the second paper which may properly be referred to here. The function $V$ has *forty-five* coefficients. These are only reduced in III. to *twenty-one* and afterwards to *nine* by a theorem of Jellett's; Haughton in III. (p. 134) finds two constants in the equations for wave-motion in an isotropic homogeneous elastic solid, but it seems to me that, if he adopts Jellett's theorem to reduce his constants, he ought only to have *one*, for that theorem contains the hypothesis of uni-constant isotropy. The equations based upon the function $V$ are investigated in II. in a manner very similar to that of Lamé and other writers for the case of a plane wave, but there are some additional points which deserve notice. Haughton investigates the possibility of:

(i) A change in the amplitude so that the type of wave-shift takes the form

$$u=p\cos\alpha\,.\,e^{-\frac{2\pi}{\lambda}q\,(lx+my+nz)}\sin\frac{2\pi}{\lambda}(lx+my+nz-vt),$$

where $lx+my+nz=0$ is a plane parallel to the wave front. It is shewn that the constant $q$ must be zero, or motion is not possible.

(ii) A change in the amplitude so that the type of wave-shift takes the form

$$u = p\cos\alpha \,.\, e^{-\frac{2\pi}{\lambda}\{q(lx+my+nz)-Kvt\}} \sin\frac{2\pi}{\lambda}(lx+my+nz-vt).$$

It is shewn that in this case $K$ must be equal to $q$. The amplitude is then a function of the phase, but its form does not indicate a diminished intensity of vibration. Haughton remarks that this agrees with a statement of Rankine's and refers to the *Camb. and Dublin Math. Jour.* Vol. VII. p. 218. I suppose the statement referred to is that contained in the footnote. It is to the effect that the function $V$, since it expresses the work, does not permit of a loss of energy by conversion into heat etc. Thus of course we might *à priori* have recognised the impossibility of such a shift-type as is suggested in (i).

(iii) A change in the density of the medium being taken into account. In this case Haughton deduces that there will be a relation between the altered and primitive density of the form

$$\rho = \rho_0 e^{-\theta} \dots\dots\dots\dots (a),$$

where $\theta$ is the dilatation and he obtains equations of the type

$$-\rho_0 e^{-\theta}\frac{d^2u}{dt^2} = \frac{d}{dx}\frac{dV}{d\alpha_1} + \frac{d}{dy}\frac{dV}{d\alpha_2} + \frac{d}{dz}\frac{dV}{d\alpha_3} \dots\dots\dots (b),$$

where $\alpha_1 = \frac{du}{dx}, \quad \alpha_2 = \frac{du}{dy}, \quad \alpha_3 = \frac{du}{dz}.$

But in obtaining equation $(a)$ from the equation of continuity Haughton neglects such terms as $\frac{d\rho}{dx}\frac{du}{dt}$, hence it seems to me he ought to neglect such terms as $\theta\frac{d^2u}{dt^2}$ in $(b)$, which would reduce that equation to the usual form.

He applies equation $(b)$ to the case of plane wave motion and takes

$$\rho = \rho_0 e^{-\frac{2\pi}{\lambda}p\cos\chi\cos\frac{2\pi}{\lambda}(lx+my+nz-vt)},$$

where $\chi$ is the angle between the normal to the wave-face and the direction of vibration. Thus in order that $\rho$ may not differ much from

$\rho_0, \chi$ must be nearly equal to a right angle, i.e. the vibration nearly in the wave-face; or, if the vibration be nearly normal, $\lambda$ must be very great, i.e. the velocity of normal vibration must be very great. As I have said, however, I am not satisfied with equation ($b$).

[1523.] Samuel Haughton. *On the Dynamical Coefficient of Elasticity of Steel, Iron, Brass, Oak and Teak. Royal Irish Academy, Proceedings,* Vol. VIII, 1864, p. 86. This communication was made on February 10, 1862. It refers to some experiments the full results of which the writer proposes to lay later before the Academy. I am unaware whether any such later communication was made.

The noteworthy result of these experiments is that, according to them, the dynamical coefficient of elasticity or the ratio of the velocity of separation to the velocity of approach of the two colliding spherical bodies is not a constant, but diminishes according to some unknown law as the velocity of collision increases.

The reader should, however, note the opposite conclusions of Morin and Hodgkinson, not to mention Newton: see Arts. 1183 and 941.

[1524.] A. Baudrimont. *Experiences sur la ténacité des métaux malléables faites aux températures* 0, 100 et 200 *degrés. Annales de chimie et physique,* T. XXX. pp. 304—311, 1850. This writer does little more than confirm Wertheim's results; at the same time, he claims a priority of investigation, although not of publication. He points out that the tenacity of metals varies with the temperature, generally decreasing as it increases; the rate of change in the tenacity is not in all metals proportional to the rate of change of the temperature. The peculiar nature of the tenacity of iron is noted: see our Arts. 1308 and 1353.

In the same volume of the *Annales,* p. 507, are some *Remarques à l'occasion du mémoire de M. Baudrimont sur la ténacité des métaux* by Wertheim. He vindicates his claim to priority in the determination of the influence of temperature on cohesion. A note of Baudrimont's in reply will be found in the *Annales,* Tome XXXI. p. 508. The controversy is of no interest.

Other memoirs of Baudrimont, including one presented to the

Académie on December 30, 1850, will be considered in our second volume.

[1525.] J. Lissajous. *Mémoire sur la position des noeuds dans les lames qui vibrent transversalement: Annales de chimie et physique*, T. xxx, 1850, pp. 385—410.

This memoir may be said to complete the labours of Euler and Strehlke (see our Arts. 49, 64, and 356). We have six possible cases for the lamina:

1°. Both ends free.

2°. Both ends built-in.

3°. One end free, one end built-in.

4°. One end free, one end fixed.

5°. One end built-in, one end fixed.

6°. Both ends fixed.

By 'fixed' we denote the terminal fixed in position, but not in direction. Strehlke had considered case 1°, Lissajous considers all the cases calculating the roots of the transcendental equations which give the nodes, and verifying his results experimentally. The memoir is very important for the theory of sound, but it lies beyond the scope of our present enterprise to do more than refer to it.

1526. John H. Jellett. *On the Equilibrium and Motion of an Elastic Solid.* This is published in the *Transactions of the Royal Irish Academy*, Vol. XXII, 1855, pp. 179—217; it was read January 28, 1850.

In this memoir we have a very elaborate investigation of the equations for the equilibrium and motion of an elastic solid, the solid being regarded as an aggregate of molecules, and Lagrange's method employed. Other writers who treat the problem in this way have assumed that the sum of the 'internal moments' of a medium may be represented by the variation of a single function; but Jellett does not make this assumption.

1527. The following extract gives an account of the basis of the memoir:

General Classification of Bodies. I. Hypothesis of Independent Action. II. Hypothesis of Modified Action.

The classification which I propose here to adopt, and which forms the basis of the present Memoir, is founded upon the following very obvious principle. The force, or influence, which one particle or molecule exerts on another, may show its effect either by causing a change in its *state*, or by causing a change in its *position*. Either or both of these changes may affect the influence which this particle in its turn exerts upon any of those around it. Thus, for example if $m$, $m'$, $m''$ be three particles acting upon each other by the ordinary attraction of gravitation, the action of $m'$ upon $m''$ will be modified by the action of $m$ only so far as their distance from each other is changed by it. The attraction of $m$ has no power to change the attraction of $m'$ upon any other particle, except by altering its distance from that particle. But the case would be altogether different if we supposed $m$, $m'$, $m''$, to be *electrified* particles. In this case the action of $m$ upon $m'$ would modify the action of that particle upon $m''$, not only by changing the distance between them, but also by changing their electrical state, and, therefore, the force which each exerts upon the other. In the former case, if $m'$ and $m''$ maintain the same relative position, the force which they mutually exert remains unchanged. In the second, even though the relative position of the two particles remains unaltered, their mutual action will be modified by the presence of a third particle. From this distinction an obvious classification follows. In the first class we place all bodies whose particles exert upon each other a force which is *independent* of the surrounding particles; a force, therefore, which can be changed only by a displacement of one or both of the particles under consideration. In the second class, which includes all other bodies, the mutual action of two particles is supposed to be affected by that of the surrounding particles (p. 181.)

1528. The equations are then investigated on the *Hypothesis of Independent Action:* see pp. 182—193 of the memoir. They are worked out without assuming the body to be *homogeneous.* This assumption is afterwards introduced: it is found that the equations now involve 54 coefficients.

1529. The memoir proceeds on pp. 193—199 to discuss the assumption that a single function $V$ exists by the variation of which the sum of the internal moments of the body may be represented. It is found that the function $V$ must involve 36 constants. The form obtained for $V$ is compared with forms used for the Theory of Light by MacCullagh and Green respectively; it is shewn that MacCullagh's form is inadmissible, and that Green's form is admissible only under conditions which

would render the body uncrystalline, and therefore incapable of being generally identical with the luminous ether[1].

The memoir then says:

Hence we infer, that if the supposed luminous ether be a medium such as either of these writers assume it to be, the mutual action of its particles cannot be independent. In other words, we must suppose that in such a medium the capacity which each particle possesses of exerting force on any other particle, is *modified* by the action of the surrounding particles[2] (p. 199).

1530. The memoir, on pp. 200 and 201, considers the case in which bodies are composed of attracting and repelling molecules. The conclusion obtained is this: *The equations of equilibrium or motion in a system of attracting or repelling molecules will in general contain thirty distinct constants.* Then the additional supposition is taken that the sum of the internal moments may be represented by the variation of a single function; the inference now is this: *The equations of equilibrium or motion of a body*

[1] [This argument against Green, it should be noted, is not that raised by Saint-Venant: see our Art. 1364 and Glazebrook's *Report on Physical Optics, British Association Reports*, 1885, p. 171. It is only the result of accepting the 'Hypothesis of Independent Action.' Ed.]

[2] [This result apart from optical theories is of peculiar interest as showing that the multi-constant theory urges us to assume that intermolecular force is a function not only of intermolecular distance, but of the arrangement of surrounding molecules. Yet this indirect action of a molecule $C$ on a molecule $B$, by its influence on the action of $A$ on $B$, would in most conceivable cases be either of a higher order than the direct action of $A$ upon $B$ or require time to produce its full effect. Thus suppose $A$, $B$ and $C$ were pulsating spheres in a fluid medium, the direction action of $A$ on $B$ would depend on their distance and on the amplitudes of their pulsations, these amplitudes would, it is true, be influenced by the pulsating of $C$ and by the position of $C$, but the effect would be of a much smaller order than the direct influence of $A$ and $B$ on each other's pulsations and position. As a first approximation at least we should have rari-constant elasticity. Perhaps we have here the cause of the divergence in value between the elastic constants as determined by vibrational and statical methods: see our Arts. 931 and 1404. Ed.]

*in which the molecular force acts in the line joining the molecules, and is represented by a function of the distance, will contain fifteen distinct constants.*

1531. On pp. 202—205 the memoir considers the *Hypothesis of Modified Action.* The discussion is summed up thus:

It is unnecessary to pursue the consequences of this principle further; for, as we have already seen, all the varieties of the general equations of motion, to the consideration of which the present Memoir is specially devoted, may be obtained from the more limited principle of independent action (p. 205).

1532. Pp. 206—209 are occupied in pointing out the serious difficulties which must occur in applying the equations, as many writers on physical Optics do, to the transmission of undulations from one body to another with which it is in contact. If we take a point in one body nearer to the common surface than the radius of molecular activity, the integrals on which the coefficients depend cannot be justly assumed to retain the same values as they have for a point remote from the common surface; they will not be constant but functions of the distance from the common surface The form of the equations of motion will therefore be completely altered, by the change of constant into variable coefficients, and by the introduction of new terms.

[1533.] Jellett sums up his conclusions on this point in five statements (pp. 207—208) which we reproduce here as they seem worthy of more consideration than has apparently been given to them:

(1) That in the case of a single medium of limited extent, the molecules which are situated at a distance from the bounding surface less than the radius of molecular activity, move according to a law altogether different from that which regulates the motion of the particles in the interior.

(2) That it is impossible to assign this law without formulating one or more hypotheses as to the nature of the medium.

(3) That if a plane wave pass through a homogeneous medium, it will not in general reach the surface; that is to say, the motion of the

particles in and immediately adjoining the surface will not be a wave motion composed of rectilinear vibrations.

(4) That if two media be in contact there will be a stratum of particles extending on each side of the surface of separation to a distance equal to the greatest radius of molecular activity; and that the motion of the particles in this stratum is altogether different from that of the particles in the interior of either medium.

(5) That, therefore, two media which are thus in contact, may be each perfectly capable of transmitting plane waves through them in all directions, and yet incapable of transmitting such a motion from one to the other; and that even in the case of reflection, in which the motion is transmitted back again through the same medium, the vibrations may cease to be rectilinear. The phenomenon of total reflection affords an instance of this.

Jellett can see no satisfactory answer to these objections and he thus finds difficulty in accepting the theories of reflection and refraction propounded by Cauchy, MacCullagh, Green, etc.

1534. Throughout the memoir the coefficients are represented by *integrals*, and pp. 209—215 are devoted to examining how far this is admissible; special reference is made to Poisson's opinion, expressed on p. 399 of his memoir of April, 1828, that *summation* could not safely be changed into *integration*. The conclusion of the discussion is thus stated:

*The methods of the integral calculus are applicable to questions of molecular mechanics, provided that the molecular force varies continuously within its sphere of action; and provided also that the sphere of molecular action is of such a magnitude as to admit of being subdivided into an indefinite number of elements, each element containing an indefinite number of molecules.*

[We may remark, however, that precisely similar results to those of Jellett may be obtained without replacing the summations by definite integrals, *i.e.* by a method similar to that of Cauchy: see our Arts. 615 and 1400. Ed.]

1535. The whole memoir deserves attentive study; it is the clearest and most satisfactory of the numerous discussions of the problem by the aid of Lagrange's method. The author refers more than once to the memoir by Haughton published in the same volume: see our Art. 1516[1].

1536. James Clerk-Maxwell. *On the Equilibrium of Elastic Solids.* The author of this paper is the present distinguished Professor of Experimental Physics in the University of Cambridge[2]; it is published in the *Transactions of the Royal Society of Edinburgh,* Vol. xx. 1853, pages 87—120; it was read on February 18, 1850. In the Table of Contents prefixed to the volume instead of *solids* we have by mistake *fluids* in the title of the paper.

1537. The author alludes to the mathematical theories of elasticity given by Navier, Poisson, and Lamé and Clapeyron; he holds that these are inconsistent with experimental results, citing those of Oersted: see our Art. 690, and considers that formulae involving only *one* coefficient are insufficient[3]. Accordingly he proposes to

[1] The following slip may be noted:

Page 193 first line: such terms as $P-1$ should be $P+1$.

[2] [I have left these words as a record of the date when Dr Todhunter wrote them. Clerk-Maxwell died in 1879. Ed.]

[3] [The arguments used are of a nature similar to those of Stokes:

The insufficiency of one coefficient may be proved from the existence of bodies of different degrees of solidity.

No effort is required to retain a liquid in any form, if its volume remain unchanged; but when the form of a solid is changed, a force is called into action which tends to restore its former figure; and this constitutes the difference between elastic solids and fluids. Both tend to recover their *volume*, but fluids do not tend to recover their *shape*.

Now, since there are in nature bodies which are in every intermediate state from perfect solidity to perfect liquidity, these two elastic powers cannot exist in every body in the same proportion, and therefore all theories which assign to them an invariable ratio must be erroneous (p. 87).

My objection to these arguments may be thus stated: Given a stress which exceeds the elastic limit, the strain will consist of two parts elastic strain and set, every such stress produces a definite amount of set. The uni-constant elasticians assert that their hypothesis holds only for the elastic strain and *not for the set.* Now in the case of a fluid the elastic limits approach nearer and nearer, and the least stress resulting in a *positive stretch* produces set, barely any elastic strain. The bi-constant hypothesis which is needful for viscous fluids is a hypothesis as to set and not as to elastic strain. When we say that there are bodies in every intermediate state between solid and fluid, do we not mean that for any given stress we can find bodies which can have for this stress every conceivable relative amount of elastic strain and of set, *i.e.* bodies ranging from the perfectly elastic solid to the perfectly setting fluid? Ed.]

adopt as the foundation of his theory two axioms, of which he gives the following account:

If three pressures in three rectangular axes be applied at a point in an elastic solid:

1. *The sum of the three pressures is proportional to the sum of the compressions which they produce.*

2. *The difference between two of the pressures is proportional to the difference of the compressions which they produce.*

The equations deduced from these axioms contain two coefficients, and differ from those of Navier only in not assuming any invariable ratio between the cubical and linear elasticity. They are the same as those obtained by Professor Stokes from his equations of fluid motion, and they agree with all the laws of elasticity which have been deduced from experiments (pp. 87, 88).

1538. The author refers to the researches of Cauchy contained in his *Exercices d'Analyse*, Vol. III. p. 180, but the reference should be to the *Exercices de mathématiques:* see Art. 614 of my account of Cauchy. The author says, with respect to Cauchy:

Instead of supposing each pressure proportional to the linear compression which it produces, he supposes it to consist of two parts, one of which is proportional to the linear compression in the direction of the pressure, while the other is proportional to the diminution of volume. As this hypothesis admits two coefficients, it differs from that of this paper only in the values of the coefficients selected. They are denoted by $K$ and $k$, and $K = n - \frac{1}{3}m$, $k = m$ (pp. 88, 89).

1539. The following reproduction of Maxwell's pages 90—92 will exhibit his method of considering strain.

The laws of elasticity express the relation between the changes of the dimensions of a body and the forces which produce them.

These forces are called *Pressures*, and their effects *Compressions.* Pressures are estimated in pounds on the square inch, and compressions in fractions of the dimensions compressed.

Let the position of material points in space be expressed by their coordinates $x$, $y$, $z$, then any change in a system of such points is expressed by giving to these coordinates the variations $\delta x$, $\delta y$, $\delta z$, these variations being functions of $x$, $y$, $z$.

The quantities $\delta x$, $\delta y$, $\delta z$ represent the absolute motion of each point in the directions of the three co-ordinates; but as compression depends not on absolute, but on relative displacement, we have to consider only the nine quantities:

$$\frac{d\delta x}{dx},\ \frac{d\delta x}{dy},\ \frac{d\delta x}{dz},\ \frac{d\delta y}{dx},\ \frac{d\delta y}{dy},\ \frac{d\delta y}{dz},\ \frac{d\delta z}{dx},\ \frac{d\delta z}{dy},\ \frac{d\delta z}{dz}.$$

Since the number of these quantities is nine, if nine other independent quantities of the same kind can be found, the one set may be found in terms of the other. The quantities which we shall assume for this purpose are—

1. Three compressions $\frac{d\alpha}{\alpha}, \frac{d\beta}{\beta}, \frac{d\gamma}{\gamma}$ in the directions of three principal axes $\alpha$, $\beta$, $\gamma$;

2. The nine *direction-cosines* of these axes, with the *six connecting equations*, leaving three independent quantities;

3. The small angles of rotation of this system of axes about the axes of $x$, $y$, $z$.

The cosines of the angles which the axes of $x$, $y$, $z$ make with those of $\alpha$, $\beta$, $\gamma$ are—

$$\begin{array}{lll} \cos(\alpha Ox) = a_1, & \cos(\beta Ox) = b_1, & \cos(\gamma Ox) = c_1, \\ \cos(\alpha Oy) = a_2, & \cos(\beta Oy) = b_2, & \cos(\gamma Oy) = c_2, \\ \cos(\alpha Oz) = a_3, & \cos(\beta Oz) = b_3, & \cos(\gamma Oz) = c_3. \end{array}$$

These *direction-cosines* are connected by the six equations,

$$a_1^2 + b_1^2 + c_1^2 = 1, \quad a_2^2 + b_2^2 + c_2^2 = 1, \quad a_3^2 + b_3^2 + c_3^2 = 1,$$

$$a_1a_2 + b_1b_2 + c_1c_2 = 0, \quad a_2a_3 + b_2b_3 + c_2c_3 = 0, \quad a_3a_1 + b_3b_1 + c_3c_1 = 0.$$

The rotation of the system of axes $\alpha$, $\beta$, $\gamma$, round the axis of

$$\begin{array}{l} x, \text{ from } y \text{ to } z, = \delta\chi_1, \\ y, \text{ from } z \text{ to } x, = \delta\chi_2, \\ z, \text{ from } x \text{ to } y, = \delta\chi_3. \end{array}$$

By resolving the displacements $\delta\alpha$, $\delta\beta$, $\delta\gamma$, $\delta\chi_1$, $\delta\chi_2$, $\delta\chi_3$ in the directions of the axes $x$, $y$, $z$, the displacements in these axes are found to be

$$\begin{aligned} \delta x &= a_1\delta\alpha + b_1\delta\beta + c_1\delta\gamma - \delta\chi_2 z + \delta\chi_3 y, \\ \delta y &= a_2\delta\alpha + b_2\delta\beta + c_2\delta\gamma - \delta\chi_3 x + \delta\chi_1 z, \\ \delta z &= a_3\delta\alpha + b_3\delta\beta + c_3\delta\gamma - \delta\chi_1 y + \delta\chi_2 x. \end{aligned}$$

But $\delta\alpha = \alpha\frac{\delta\alpha}{\alpha}$, $\delta\beta = \beta\frac{\delta\beta}{\beta}$, and $\delta\gamma = \gamma\frac{\delta\gamma}{\gamma}$;

and

$$\alpha = a_1x + a_2y + a_3z,\quad \beta = b_1x + b_2y + b_3z,\ \text{and}\ \gamma = c_1x + c_2y + c_3z.$$

Substituting these values $\delta\alpha$, $\delta\beta$, and $\delta\gamma$ in the expressions for $\delta x$, $\delta y$, $\delta z$, and differentiating with respect to $x$, $y$, and $z$, in each equation, we obtain the equations

$$\left.\begin{aligned}
\frac{d\delta x}{dx} &= \frac{\delta\alpha}{\alpha}a_1^2 + \frac{\delta\beta}{\beta}b_1^2 + \frac{\delta\gamma}{\gamma}c_1^2\\
\frac{d\delta y}{dy} &= \frac{\delta\alpha}{\alpha}a_2^2 + \frac{\delta\beta}{\beta}b_2^2 + \frac{\delta\gamma}{\gamma}c_2^2\\
\frac{d\delta z}{dz} &= \frac{\delta\alpha}{\alpha}a_3^2 + \frac{\delta\beta}{\beta}b_3^2 + \frac{\delta\gamma}{\gamma}c_3^2
\end{aligned}\right\}\ \ldots\ldots\ldots\ldots\ldots(1).$$

$$\left.\begin{aligned}
\frac{d\delta x}{dy} &= \frac{\delta\alpha}{\alpha}a_1a_2 + \frac{\delta\beta}{\beta}b_1b_2 + \frac{\delta\gamma}{\gamma}c_1c_2 + \delta\chi_3\\
\frac{d\delta x}{dz} &= \frac{\delta\alpha}{\alpha}a_1a_3 + \frac{\delta\beta}{\beta}b_1b_3 + \frac{\delta\gamma}{\gamma}c_1c_3 - \delta\chi_2\\
\frac{d\delta y}{dz} &= \frac{\delta\alpha}{\alpha}a_2a_3 + \frac{\delta\beta}{\beta}b_2b_3 + \frac{\delta\gamma}{\gamma}c_2c_3 + \delta\chi_1\\
\frac{d\delta y}{dx} &= \frac{\delta\alpha}{\alpha}a_2a_1 + \frac{\delta\beta}{\beta}b_2b_1 + \frac{\delta\gamma}{\gamma}c_2c_1 - \delta\chi_3\\
\frac{d\delta z}{dx} &= \frac{\delta\alpha}{\alpha}a_3a_1 + \frac{\delta\beta}{\beta}b_3b_1 + \frac{\delta\gamma}{\gamma}c_3c_1 + \delta\chi_2\\
\frac{d\delta z}{dy} &= \frac{\delta\alpha}{\alpha}a_3a_2 + \frac{\delta\beta}{\beta}b_3b_2 + \frac{\delta\gamma}{\gamma}c_3c_2 - \delta\chi_1
\end{aligned}\right\}\ \ldots\ldots\ldots(2).$$

Equations (1) and (2) are purely geometrical results; equations (1) coincide with the formula of my Art. 547; for the slides vanish for the principal axes. The other six equations I have not seen elsewhere; by adding so as to form three from them by elimination of $\delta\chi_1$, $\delta\chi_2$, $\delta\chi_3$ we get expressions for the slides very similar to that quoted in my Art. 1368, when we put in the latter the slides for the principal axes zero. But is Maxwell's mode satisfactory? Where does he use his hypothesis that $\alpha$, $\beta$, $\gamma$ refer to principal axes? Implicitly in treating these compressions as the only *relative* displacements.

It will be seen that neither the terminology nor the notation of Maxwell is peculiarly inviting.

1540. When we arrive at the mathematical investigations of the memoir concerning stress we see that something less is assumed than the verbal statement of the axioms would suggest: it is really assumed that the axioms hold with respect to *principal tractions* only.

Let $\widehat{xx}$, $\widehat{yy}$, $\widehat{zz}$ denote *principal tractions;* and assume that

$$\widehat{xx}+\widehat{yy}+\widehat{zz}=3n\left(\frac{du}{dx}+\frac{dv}{dy}+\frac{dw}{dz}\right),$$

and also that

$$\frac{\widehat{xx}-\widehat{yy}}{\dfrac{du}{dx}-\dfrac{dv}{dy}}=\frac{\widehat{yy}-\widehat{zz}}{\dfrac{dv}{dy}-\dfrac{dw}{dz}}=\frac{\widehat{zz}-\widehat{xx}}{\dfrac{dw}{dz}-\dfrac{du}{dx}}=m.$$

These give

$$\widehat{xx}=(n-\tfrac{1}{3}m)\,\theta+m\frac{du}{dx},$$

$$\widehat{yy}=(n-\tfrac{1}{3}m)\,\theta+m\frac{dv}{dy},$$

$$\widehat{zz}=(n-\tfrac{1}{3}m)\,\theta+m\frac{dw}{dz}.$$

Then we easily deduce from equation (3) of Art. 659 that

$$\widehat{x'x'}=\widehat{xx}\cos^2(xx')+\widehat{yy}\cos^2(yx')+\widehat{zz}\cos^2(zx')$$

$$=(n-\tfrac{1}{3}m)\,\theta+m\left\{\frac{du}{dx}\cos^2(xx')+\frac{dv}{dy}\cos^2(yx')+\frac{dw}{dz}\cos^2(zx')\right\},$$

and by using equation (2) of Art. 663 this becomes

$$\widehat{x'x'}=(n-\tfrac{1}{3}m)\,\theta+m\frac{du'}{dx'}.$$

Also we may write $\theta'$ instead of $\theta$ since this quantity is an invariant; so that for $\widehat{x'x'}$ we obtain an equation of precisely the same form as that for $\widehat{xx}$. Similarly we have for $\widehat{y'y'}$ and $\widehat{z'z'}$ equations of precisely the same form as those for $\widehat{yy}$ and $\widehat{zz}$ respectively. Again as in Art. 1368,

$$\widehat{y'z'}=\widehat{xx}\cos(y'x)\cos(z'x)+\widehat{yy}\cos(y'y)\cos(z'y)+\widehat{zz}\cos(y'z)\cos(z'z)$$

$$=m\left\{\frac{du}{dx}\cos(y'x)\cos(z'x)+\frac{dv}{dy}\cos(y'y)\cos(z'y)+\frac{dw}{dz}\cos(y'z)\cos(z'z)\right\}$$

$$=\frac{m}{2}\left(\frac{dv'}{dz'}+\frac{dw'}{dy'}\right)=\frac{m}{2}\,\sigma_{y'z'}.$$

This easily follows by adding the fourth and sixth equations of (2) Art. 1539.

Similarly $$\widehat{z'x'} = \frac{m}{2}\sigma_{z'x'}, \quad \widehat{x'y'} = \frac{m}{2}\sigma_{x'y'}.$$

The expressions thus found for the three tractions and the three shears practically agree with those of Maxwell. He assumes moreover that "$n$ is the coefficient of cubical elasticity, and $m$ that of linear elasticity[1]." To obtain these equations Maxwell relies partly on results demonstrated by Lamé and Clapeyron, and partly on formulae which he obtains himself, so that on the whole it is somewhat difficult to apprehend his method. He obtains equations of bi-constant form for the stresses like those first given by Cauchy: see our Art. 614; these lead to the ordinary equations of equilibrium of elasticity with two constants.

[1541.] On p. 95 Maxwell remarks that the shift-fluxions $\frac{du}{dx}$, $\frac{dv}{dy}$, $\frac{dw}{dz}$ must, if the temperature be variable, be diminished by a quantity proportional to the temperature. Thus he writes equations of the type:

$$\frac{du}{dx} = \left(\frac{1}{9n} - \frac{1}{3m}\right)(\widehat{xx} + \widehat{yy} + \widehat{zz}) + \frac{1}{m}\widehat{xx} + cq.$$

Which give for the tractions, equations of the type:

$$\widehat{xx} = (n - \tfrac{1}{3}m)\left(\frac{du}{dx} + \frac{dv}{dy} + \frac{dw}{dz} - 3cq\right) + m\left(\frac{du}{dx} - cq\right).$$

The shears remain unchanged. Here $cq$ is the stretch for a rise of temperature $q$, and $3nc$ equals the $\beta$ of our notation. This result agrees with that of Stokes. Maxwell does not, like Duhamel, give the equation for the flow of heat: see our Arts. 869, 883 and 1266.

Maxwell applies his equations to various interesting cases; some of these had already been discussed by Lamé and Clapeyron, but the present investigations are more general as involving *two*

[1] Maxwell uses $\mu$ for the $n$ of our account. It is equivalent to our $(3\lambda + 2\mu)/3$ (or to the dilatation-modulus $F$ of our Appendix, Note B); his $m$ is our $2\mu$.

constants instead of *one:* see Art. 1010 of my account of Lamé and Clapeyron.

1542. For example Maxwell's Case III. is that which we have in Lamé's *Leçons,* p. 188, though not quite so fully developed: see my Arts. 1012 and 1087.

Taking expressions for the stresses in cylindrical coordinates, such as equations (7) of Lamé's p. 184, we easily get by addition,

$$\widehat{rr} + \widehat{\phi\phi} + \widehat{zz} = (3\lambda + 2\mu)\,\theta\text{; so that}$$

$$\widehat{rr} = \frac{\lambda}{3\lambda + 2\mu}(\widehat{rr} + \widehat{\phi\phi} + \widehat{zz}) + 2\mu\frac{du}{dr},$$

$$\widehat{\phi\phi} = \frac{\lambda}{3\lambda + 2\mu}(\widehat{rr} + \widehat{\phi\phi} + \widehat{zz}) + 2\mu\left(\frac{u}{r} + \frac{1}{r}\frac{dv}{d\phi}\right),$$

$$\widehat{zz} = \frac{\lambda}{3\lambda + 2\mu}(\widehat{rr} + \widehat{\phi\phi} + \widehat{zz}) + 2\mu\frac{dw}{dz}.$$

Maxwell's equations (18), (19), (20) practically coincide with these three, allowing for difference of notation. He assumes that $\widehat{zz}$ is constant, that $v$ is zero, and that the expressions are independent of the angle $\phi$. Further, Maxwell supposes a uniform distribution of load over the inner and outer surfaces of his hollow cylinder so that $dw/dz$ is constant; hence $\theta$ and $\widehat{\phi\phi} + \widehat{rr}$ are constants. Thus $\widehat{r\phi}$ is zero and the first of the body-stress equations (Lamé, p. 182, (5)) reduces to

$$\frac{\widehat{\phi\phi} - \widehat{rr}}{r} = \frac{d\widehat{rr}}{dr}.$$

which coincides with Maxwell's equation (21).

1543. In these examples Maxwell calculates what he terms the 'optical effect of the pressure of any point.' He does this on the following basis:

I have found no account of any experiments on the relation between the doubly refracting power communicated to glass and other elastic solids by compression, and the pressure which produces it; but the phenomena of bent glass seem to prove, that in homogeneous singly-refracting substances exposed to pressures, the principal axes of pressure coincide with the principal axes of double refraction; and that the difference of pressures in any two axes is proportional to the difference of the velocities of the oppositely polarised rays whose directions are parallel to the third axis. On this principle I have calculated the phenomena seen by polarised light in the cases where the solid is bounded by parallel planes (p. 90).

[1544.] Thus if $I$ be the optical effect' at distance $r$ from the axis in Case III. of a hollow cylinder of height $b$, in which the ray is transmitted parallel to the axis, Maxwell puts

$$I = b\omega\,(\widehat{rr} - \widehat{\phi\phi}),$$
$$= b\omega\,\frac{a_1^2 a_2^2}{r^2}\,\frac{p_1 - p_2}{a_1^2 - a_2^2},$$

where $a_1$ and $a_2$ are the radii of the two bounding surfaces, $p_1$ and $p_2$ the contractive loads upon them respectively, and $\omega$ is a coefficient depending on the substance and probably a function of the linear elasticity: see p. 100 and Note C, p. 120 of the memoir.

The words difference of pressures in any two axes' are understood by the author in rather a large sense; for example, suppose that all the stresses vanish except the shear $\widehat{yz}$, then $\widehat{yz}$ is taken for the difference of pressure: see p. 96, last line[1].

The quotation in the preceding article shows that Maxwell was ignorant of Neumann's great memoir of 1841: see our Art. 1185. Neumann is much more definite than Maxwell in his measurement of 'optical effect:' see Arts. 1189, 1200.

1545. Maxwell's Case IV. relates to the equilibrium of a hollow sphere: see Lamé's *Leçons*, page 198, and my Arts. 1016, 1093. The elaborate equations which are obtained for a complete solution of the problem in all its generality are not required here, as it is assumed that $v = 0$, $w = 0$ and $u$ is independent of $\phi$, $\psi$. Thus in our notation,

$$\widehat{rr} = \lambda\theta + 2\mu\,\frac{du}{dr},$$

$$\widehat{\phi\phi} = \widehat{\psi\psi} = \lambda\theta + 2\mu\,\frac{u}{r};$$

so that
$$\theta\,(3\lambda + 2\mu) = \widehat{rr} + 2\widehat{\phi\phi}.$$

[1] What Maxwell practically uses is the *maximum-shear*. Thus on p. 96, $\widehat{yz}$ is the maximum-shear of the torsional strain, and by Hopkins' theorem (see Art. 1368) the difference of the principal pressures is equal to twice the maximum-shear. To make the optical effect proportional to the maximum-shear, that is, to the maximum-slide, seems to me to give a better physical meaning as well as to introduce consistency: see my remarks, Art. 1189.

Hence the first of the body stress-equations (Lamé, p. 198, (5)) reduces to

$$\frac{d\widehat{rr}}{dr} + \frac{2}{r}(\widehat{rr} - \widehat{\phi\phi}) = 0,$$

for there is no body-force. The equation just given agrees with Maxwell's equation (36).

[1546.] Under Case IV. Maxwell refers to the researches of Canton, Oersted, Perkins, Aimé and Regnault: see our Arts. 687—691, 1013 and 1227. He does not however notice the extraordinary statement of Oersted to which we have referred in Art. 689. Oersted based his attack on the theory of elasticity upon the assumption that lead was 18 times as compressible as glass, which he had deduced from some experimental results of Tredgold, but as Wertheim has remarked the compressibility of lead is only 3 to $3\frac{1}{3}$ times as great as that of glass[1]. When we note this the difference between the theory of uni-constant isotropy and Oersted's experiments presents nothing like the divergence that Oersted supposed. It is a divergence which might readily be accounted for by the want of isotropy in a glass bottle and the readiness with which lead takes a set: see our Art. 1326 on Wertheim's memoir of 1848 and Saint-Venant's edition of Navier's *Leçons*, pp. 666—667.

1547. Maxwell's Case V. relates to the equilibrium of an elastic beam of rectangular section uniformly bent. This is connected with Case III. in the manner thus suggested: "By supposing the bent beam to be produced till it returns into itself, we may

[1] This result may be deduced from the Tables given in Thomson's Article on *Elasticity* (*Encycl. Brit.*) If $F$ be the dilatation-modulus, on the theory of uni-constant isotropy $F = 2/3\,E$, where $E$ is the stretch-modulus. In Art. 77 I find quoted on the authority of Everett that $F$ for a certain glass specimen $= 354 \times 10^6$ in the metric system. In the same article $E$ for *cast* lead, such as Oersted's bottle must have been formed out of, is given as equal to $177 \times 10^6$ on the authority of Wertheim, hence for cast lead $F = \frac{1}{3}(354 \times 10^6)$, or in this case the compressibilities of lead and glass are as 3 : 1. Looking at such results it is hard for us to lay the same stress upon Oersted's experiments as Maxwell and Stokes seem to have done: see Arts. 1266 and 1326. We hold that these experiments are absolutely inconclusive in regard to the constant-controversy: see our Art. 925.

treat it as a hollow cylinder." An equation, numbered (46), is obtained involving two constants $c$ and $C_1$; the determination of the former amounts to fixing the situation of the *neutral line* in the beam. Approximations are given to some exact expressions; but the work of obtaining these approximations is not developed.

[I do not follow Maxwell's reasoning in this case at all. The section of a rectangular beam under flexure does not remain rectangular, and so I do not see how we can treat the beam as a portion of a hollow circular cylinder. For this reason I do not think Maxwell's equation (46) is more exact than that given by the ordinary Bernoulli-Eulerian hypothesis, possibly less so. The section after flexure becomes an equiangular trapezium, of which however the sides originally perpendicular to the plane of flexure are curved, their curvature being opposite to that of flexure: see our figure on p. 793. How can this be obtained from a portion of a circular cylinder? ED.]

1548. Case VI. is an example of the equilibrium of an elastic plate; it is treated in a practical approximative way. The reader will see that equation (51) is found by integrating that which immediately precedes it; then the equations of elasticity are obtained by assuming that we may use equations (19) and (20), and by taking a quantity $\frac{1}{2}(h_1 + h_2)$ as an approximate value of an unknown traction.

[1549.] Case VII. deals with the following problem: To find the conditions of torsion of a cylinder composed of a great number of parallel wires bound together without adhering to one another.

This is apparently an attempt to explain the anomalies which occasionally arise in applying Coulomb's theory of torsion to experiments on cylindrical rods of a 'fibrous texture.' It is questionable whether the supposition of a cylindrical elasticity due to the working of the materials and causing the plane sections to become curved on torsion would not give a better result.

Maxwell's investigation does not seem to me very satisfactory. He says nothing as to how the terminals of his wires are treated. I judge that, if they were fixed, the torsion of his bundle might be due as much to the sum of the individual torsions of the wires as to the effect of the tractions which have been produced by stretching them. There is a general reference to Young's *Natural Philosophy.* I suppose the particular passage meant is Vol. I. p. 139.

1550. Case VIII. is interesting; it is thus stated:

It is well known that grindstones and fly-wheels are often broken by the centrifugal force produced by their rapid rotation. I have therefore calculated the strains and pressure acting on an elastic cylinder revolving round its axis, and acted on by the centrifugal force alone[1].

[1551.] Maxwell's work seems to me of very doubtful value, as he treats the traction perpendicular to the plane of rotation as negligible at every point of the stone. The result is an unbalanced shearing stress on every element of the free surface. This shear is comparable with the traction which he supposes to burst the stone. According to Maxwell's equations the stone would burst first at the rim. His mechanical and optical results on p. 112 are hardly acceptable even as approximations without strong physical confirmation[2]. In addition he has taken a stress and not a strain maximum as the limit of elasticity (or, as he assumes, of cohesion.)

1552. Maxwell now adverts to the influence of a change of temperature on the equations; and he illustrates this point in some of his Cases: see his pp. 95 and 112—114. We have

[1] In the second equation of (57) for $\frac{1}{r^2}$ read $\frac{2}{r^2}$, and for $t$ read $t^2$.

[2] The problem of the grindstone has also been attempted by J. Hopkinson in a paper entitled: *On the stresses produced in an elastic disc by rapid rotation: Messenger of Mathematics*, Vol. II. 1873, pp. 53—54. He does not refer to Maxwell's discussion of the same problem, and so I suppose was ignorant of it. He also uses the stress and not the strain condition of rupture. Further, he neglects like Maxwell the traction perpendicular to the plane of rotation, but he arrives at different results because an equation from which he starts is erroneous. He has dropped an $r$ in the term $\sigma r\omega^2$, of his first equation. It is strange that he did not remark that his final values for the stresses are of wrong dimensions.

The problem of the grindstone when treated without approximation involves four series of Bessel's functions.

already referred to the memoirs of Duhamel on this matter: see Arts. 868—904.

[1553.] Of Cases IX. to XI., the first and last had been previously treated by Duhamel. Maxwell in Case IX. applies his equations to the optical phenomena described by Brewster, when polarised light is passed longitudinally through the material of a hollow glass cylinder the exterior and interior surfaces of which are maintained at constant temperatures. Case X. is an application of Maxwell's theory to the fact, noted by Brewster, that a polarizing force is developed when a solid cylinder of glass is suddenly heated at the cylindrical surface. I do not clearly understand how the temperature given by equation (63) is physically procurable or maintainable (see p. 114). The case is, however, interesting as shewing how stress may be determined from optical phenomena.

[1554.] Case XII. is an attempt to take into account the slide in a horizontal beam loaded at the centre. It is practically identical with Saint-Venant's first treatment of the problem in his *Cours lithographié* of 1837; the slide is considered uniform over the section: see our Art. 1571.

[1555.] Case XIII. is an example of the superposition of strain, and what may be termed the 'photography' of the resultant strain by means of polarised light. (pp. 115—117.)

[1556.] Case XIV. (p. 117) is a very valuable and interesting example of a method suggested by Brewster of determining the nature of a strain in an elastic solid by means of the colour-fringes exhibited by polarised light—what we have termed the photography of strain. We reproduce this case as it forms an excellent sample of Maxwell's method and of a mode of analysing strain which has hardly been given sufficient prominence notwithstanding the memoirs of Neumann and Maxwell.

Sir David Brewster has pointed out the method by which polarised light might be made to indicate the strains in elastic solids; and his experiments on bent glass confirm the theories of the bending of beams. (See our Art. 698, and footnote, p. 640.)

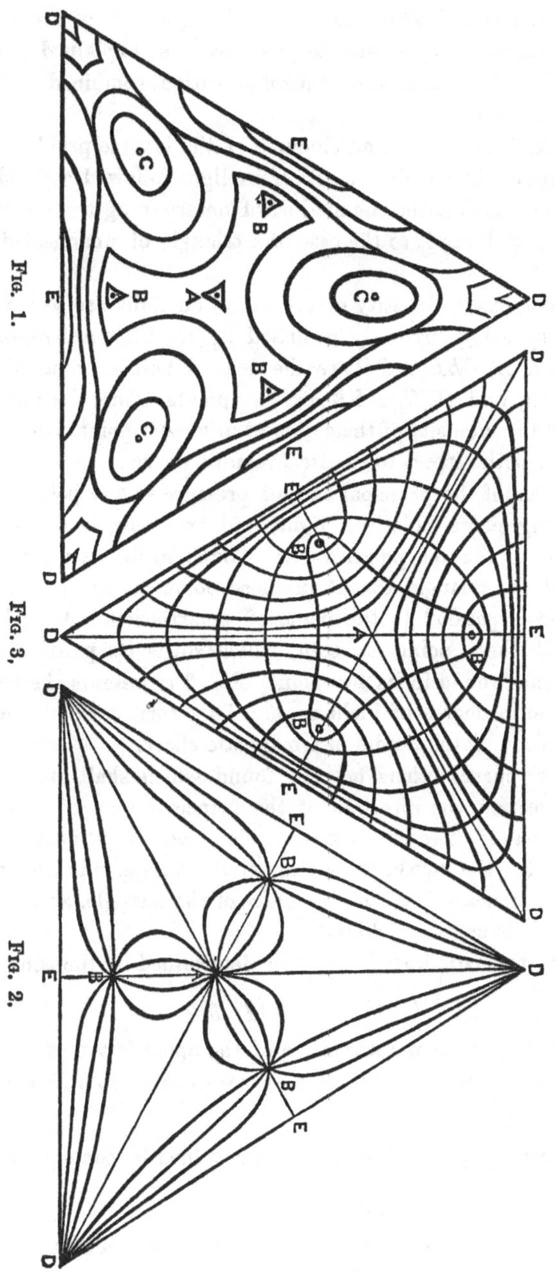


FIG. 1.

FIG. 3.

FIG. 2.

The phenomena of heated and unannealed glass are of a much more complex nature, and they cannot be predicted and explained without a knowledge of the laws of cooling and solidification, combined with those of elastic equilibrium.

In Case X. I have given an example of the inverse problem, in the case of a cylinder in which the action on light followed a simple law; and I now go on to describe the method of determining the pressures in a general case, applying it to the case of a triangle of unannealed plate-glass.

The lines of equal intensity of the action on light are seen without interruption by using circularly polarised light. They are represented in fig. 1, where *A*, *BBB*, *DDD* are the neutral points, or points of no action on light, and *CCC*, *EEE* are the points where that action is greatest; and the intensity of the action at any other point is determined by its position with respect to the isochromatic curves.

The direction of the principal axes of pressure at any point is found by transmitting plane polarised light, and analysing it in the plane perpendicular to that of polarisation. The light is then restored in every part of the triangle, except in those points at which one of the principal axes is parallel to the plane of polarization. A dark band formed of all these points is seen, which shifts its position as the triangle is turned round in its own plane. Fig. 2 represents these curves for every fifteenth degree of inclination. They correspond to the lines of equal variation of the needle in a magnetic chart.

From these curves others may be found which shall indicate, by their own direction, the direction of the principal axes at any point.

These curves of direction of compression and dilatation are represented in fig. 3; the curves whose direction corresponds to that of *compression* are concave towards the centre of the triangle, and intersect at right angles the curves of dilatation.

Let the isochromatic lines in fig. 1 be determined by the equation

$$\phi\ (x, y) = I/z = \omega\,(q-p)/z$$

where $I$ is the difference of retardation of the oppositely polarised rays, and $q$ and $p$ the pressures in the principal axes at any point, $z$ being the thickness of the plate.

Let the lines of equal inclination be determined by the equation

$$\phi_2\,(x, y) = \tan\phi$$

$\phi$ being the angle of inclination of the principal axes; then the differential equation of the curves of direction of compression and dilatation (fig. 3) is

$$\phi_5(x, y) = dy/dx.$$

By considering any particle of the plate as a portion of a cylinder whose axis passes through the centre of curvature of compression we find

$$(q-p)/r = dp/dr.$$

Let $\rho$ denote the radius of curvature of the curve of compression at any point, and let $s$ denote the length of the curve of dilatation at the same point,

$$\phi_3(x, y) = \rho, \quad \phi_4(x, y) = s$$

$$q - p = \rho dp/ds,$$

and since $q-p$, $\rho$ and $s$ are known, and since at the surface, where $\phi_4(x, y) = 0$, $p = 0$, all the data are given for determining the absolute value of $p$ by integration.

[1557.] We have thus the graphical solution of the problem. Maxwell remarks in conclusion that his curves shew the correctness of Sir John Herschell's ingenious explanation of the phenomena of heated and unannealed glass. He gives no reference. Herschell discusses the optical effect of strained and heated glass in his *Treatise on Light*, London, 1854, §§ 1083—1108. Maxwell of course refers to an earlier work, but I do not know what explanation he has in view.

The memoir concludes with a few notes of which the first recapitulates some results of Lamé and Clapeyron, and the second states the experiments on which Maxwell bases his somewhat artificial fundamental axioms: see Art. 1537.

1558. The memoir of Maxwell is noticed in the following places: *Cambridge and Dublin Mathematical Journal*, Vol. VI. p. 185; Moigno's *Statique*, p. 657; *Fortschritte der Physik in den Jahren* 1850 *und* 1851, p. 255. The last of these refers with respect to the optical phenomena noticed in Arts. 1544 and 1556 to the memoir by Neumann of 1841: see our Art. 1185.

[1559.] *Summary*. The activity of this decade is remarkable, not only on the physical and technical side but in pure theory. On the technical side it is no longer suspension bridges, but practical railway needs, which call forth not only innumerable experiments on iron, but corresponding developements of theory. The investigation of resilience problems by Stokes, Willis and Cox is a noteworthy example of theory directly produced by technical needs. The technical papers of the period abound with material suggesting profitable lines of mathematical, or physical investigation. We see the theory of elasticity applied with success to such vast structures as the tubular bridges of Stephenson, and at the same time the construction of those bridges reacting upon theory, by shewing the need of rectification and of developement. In this period also, following up the researches of Brewster, Neumann and Maxwell throw open the immense new field of photo-elasticity; while Wertheim and others explore the wholly unbroken ground upon which electricity, magnetism and elasticity meet. The vibratory motion of elastic solids is considered by Blanchet, Stokes and Haughton, although with these scientists the needs of the undulatory theory of light are predominant. The controversy over the rari- or multi-constancy of the elastic equations receives new light from memoirs of Stokes, Wertheim, Clausius and Jellett; while a host of physical and technical investigators enlarge our knowledge of set, of after-strain, and of the molecular conditions which influence elasticity as well as cohesion. Not in one country alone, but throughout the length and breadth of Europe we find men foremost in three of the great divisions of science (theoretical, physical and technical) labouring to extend our knowledge of elasticity and of the subjects akin to it. This, apparently spontaneous, labouring to a common end is not only of interest to the historian of modern science, it is full of meaning for the historian of human developement.

# CHAPTER IX.

## SAINT-VENANT'S RESEARCHES BEFORE 1850.

[1560.] We now come to one of the most eminent of living[1] elasticians; in his earliest writings and in the *Royal Society Catalogue of Scientific Papers* his name is given as Barré de Saint-Venant, but he is usually quoted as Saint-Venant. We confine ourselves in this chapter and volume to his earlier researches—those before 1850.

[1561.] *Leçons de mécanique appliquée faites par intérim par M. de St-Venant, Ingénieur des ponts et chaussées*[2]. 1837 à 1838.

This is the *Cours lithographié* frequently referred to by Saint-Venant himself, and constitutes the first contribution of our author to the subject of elasticity. It consists of lithographed sheets on the topic of the lectures given to the students. It is interesting to note that these lectures were delivered by Saint-Venant as deputy for the then professor of mechanics, Coriolis, at the *École des ponts et chaussées*.

[1562.] Remarks on the contents of these lectures by Saint-Venant himself will be found in: *Notice sur les travaux et titres scientifiques de M. de Saint-Venant*, Paris, 1858, pp. 3—6, and *Ibid.* Paris, 1864, pp. 3, 4, with several further references. These works were presented on successive candidatures for vacancies in the mechanical section of the Académie des Sciences. We may also cite the references in the *Historique Abrégé* (pp. cxxiii, ccxii, and else-

[1] M. de Saint-Venant died while these pages were in type: January 6, 1886.

[2] This work has of course never been for sale, and I owe the possibility of giving some account of it here to the extreme kindness of M. de Saint-Venant, who very readily lent me a copy as well as pointed out those portions which presented novelty of treatment—a kindness which I hope my readers will appreciate as I do.

where in the same work). The preliminary observations of the *Cours* are characteristic of the time and of the writer. We must remember that notwithstanding the splendid theoretical discoveries of Navier, Poisson and Cauchy, the only 'practical theory' which was still to be found in mechanical text-books for the ever-recurring beam problems was the Bernoulli-Eulerian hypothesis in more or less modified forms. The application of the general equations of elasticity to the problems of the flexure and torsion of beams had yet to be made. Practical engineers like Robison and Vicat (see Arts. 145, 735) were disgusted with mathematical theories and advocated what Saint-Venant here appropriately terms *l'appréciation par sentiment.* To reinstate theory in its true place, to make the theory of elasticity of practical value has been the life-work of Saint-Venant. Much of what he writes in the *Cours* of the relation of theory to practice deserves to be printed; we regret that our space only permits us to cite the following passages, which are so suggestive for the direction taken by the author's after-work.

L'usage des mathématiques cessera de s'attirer des reproches si on le referme dans ses vraies limites. Le calcul pur est simplement un instrument logique tirant des conséquences rigoureuses de prémisses posées et souvent contestables. La mécanique y joint bien quelques principes physiques que l'expérience a mis hors de contestation, mais elle laisse aux expériences particulières le soin de déterminer quelles forces sont en jeu dans chaque cas, et il règne toujours à cet égard plus ou moins d'incertitude qui affecte nécessairement les résultats. Ces résultats ne doivent point être considérés comme les oracles, dictant infailliblement ce que l'on doit décider; ce sont de simples renseignements, comme les dépositions de témoins ou les rapports d'experts dans les affaires judiciaires, mais des renseignements extrêmement précieux et dont on ne doit jamais se priver, car il est extrêmement utile à la détermination que l'on a à prendre, de connaître la solution exacte d'un problème fort rapproché de celui qui est proposé, et de pouvoir se dire, par exemple, "si les efforts étaient exactement tels ou tels, les dimensions à donner seraient telles ou telles." De cette manière le champ de *l'appréciation instinctive* se trouvera réduit aux différences qui ne peuvent pas être le sujet du calcul théorique; et l'on voit que ces deux méthodes, loin de s'exclure, peuvent concourir ensemble, se suppléer et s'aider mutuellement, se contrôler même quelquefois,—enfin contracter sous les auspices du bon sens, une alliance féconde en résultats utiles sous le double rapport de la convenance et de l'économie.

Speaking of the imperfections of the then existing theory, Saint-Venant says:

Si ces imperfections sont malheureusement nombreuses, cela vient de ce que la science appliquée est jeune et encore pauvre; avec ses ressources actuelles, elle peut déjà rendre de grands services, mais ses destinées sont bien plus hautes: elle offre un champ immense au zèle de ceux qui voudront l'enrichir, et beaucoup de parties de son domaine semblent même n'attendre que des efforts légers pour produire des résultats d'une grande utilité (p. 2).

[1563.] We will now note the novel points of the *Cours*. Beginning with some account of the labours of the great French elasticians, Saint-Venant corrects their definition, based on the molecular theory, of *stress across an elementary plane* at a point in a body. He gives sufficient reasons for his own definition, shewing that the old definition, although agreeing with his in its results on certain suppositions as to the distance between molecules and the radius of the sphere of molecular activity in relation to the dimensions of the elementary plane, is yet likely to lead to difficulties—such even as Poisson had met with[1] Saint-Venant's own definition of stress across an elementary plane is—'the resultant of the actions, whether attractive or repulsive, which the molecules situated on one side of the plane exercise upon the molecules upon the other side, when the direction of these actions traverse the plane.' The older elasticians define this stress as 'the resultant of the actions of all the molecules situated on one side of the elementary plane—considered as indefinitely produced—upon all the molecules contained in the interior of a right cylinder situated upon the other side of the plane which is taken as base of the cylinder.'

Compare our Articles 426, 440, 546, 616 and 678—679.

[1564.] On p. 9 begins an interesting dissection of strain as

[1] Saint-Venant refers to the *Journal de l'École polytechnique*, 20e *Cahier*. Arts. 49, 50, 51, 53, etc.

appearing in the stress-strain relations. We here find the term *glissement* introduced and defined. This is probably its first accurate treatment in the history of our subject: see Appendix, Note A (6). We reproduce the original definition:

*Glissement* des molécules d'un corps, sur une petite face prise à l'intérieur,—la tangente du petit angle formé par une perpendiculaire à cette face après qu'elle s'est déplacée avec les molécules adjacentes et par la droite matérielle qui y était primitivement perpendiculaire et qui s'est aussi déplacée.

Glissement estimé suivant la direction d'une droite tracée sur la face,—la tangente du même angle projeté sur le plan normal à la face passant par la droite donnée.

To the first paragraph we have cited Saint-Venant puts the foot-note:

Dans le mouvement des faces et des lignes entraînées avec les molécules primitivement adjacentes à ces faces ou à ces lignes, nous supposons que les faces restent planes et que les lignes restent droites: cela est permis à cause de leur étendue supposée très petite, et de la régularité qu'on suppose exister, si ce n'est dans les déplacements des molécules elles-mêmes, au moins dans les déplacements des points qui occupent des positions moyennes entre des molécules qui les environnent (p. 11).

In the section the *double-suffix* notation is used, possibly for the first time: see Art. 610, footnote.

[1565.] With regard to the general question of slides and the corresponding shears, we may remark that Coulomb had considered the effect of shear in producing rupture in his: *Essai sur une application des règles de maximis et minimis à quelques Problèmes de Statique, relatifs à l'Architecture* (*Savants étrangers* 1773, page 348 *et seq.*: see also our Art. 120). His theory however is not tenable. On the whole a more scientific view was presented by Young who, in his Lectures on Natural Philosophy, devoted some space to what he termed 'lateral adhesion,' or considering the corresponding strain 'detrusion': see our Art. 143. Young however gave no mathematical theory of the subject. Slides of course appear, although not under the name of *glisse-*

*ments*, in the investigations of Poisson and Cauchy, but their neglect in the ordinary theory of beams does not seem to have been regarded, and enabled Vicat in 1833 to make his vigorous protest against the mathematicians: see Art. 735. This probably induced Saint-Venant to consider the matter more closely, and we have the first-fruits in this *Cours*: see Appendix, Note A (6).

[1566.] We may note a remark of Saint-Venant's on p. 12 that it would be better to term *tension or traction* what is usually termed *pressure*, although he retains the latter word as sanctioned by usage. It is interesting therefore to find him in his edition of Clebsch writing:

> C'est une heureuse innovation de Clebsch, que d'appeler ces forces ou résultantes d'actions moléculaires *tensions* ou *tractions* et non pas *pressions*. (Foot-note, p. 18.)

The word used by Clebsch is *Zugkraft.*

[1567.] Pp. 16—17. Saint-Venant states that the true method of ascertaining the strength of a given body is to calculate the greatest stretch produced by loading it in the required fashion. This stretch must be less than a definite quantity, to be determined experimentally. He shews that the calculation cannot be made on the basis of the greatest traction not exceeding a certain amount, for this only agrees with the former in certain cases. We have here clearly pointed out an error made by innumerable English and German engineers and even perpetuated by such theoretical authorities as Clebsch (*Theorie der Elasticität*, S. 134—138 and elsewhere) and Lamé: see our Arts. 1013, 1016, footnotes.

[1568.] The now well-known theorem that the superposition of small strains is productive of the sum of the corresponding stresses is here probably stated distinctly for the first time:

> Les pressions répondant à divers petits déplacements ont pour résultante la pression qui proviendrait de déplacements équivalents à tous ceux-ci ensemble (p. 15 : see its use on p. 31).

The word displacement here must be taken as *relative* displacement, the theorem holds for the strains, not necessarily for the shifts. See however on this theorem Art. 929.

[1569.] On p. 18 Vicat's protest (see Arts. 721, 725, 735) is mentioned and also the disagreement of his results with Coulomb's theory of shear (see Art. 729); the latter however is not here criticised.

[1570.] A geometrical proof (now to be found in some text-books) is given that a slide consists of two stretches—one positive and the other negative and both equal in magnitude to half the slide—in directions making angles of 45° with that of the slide (p. 20). This is again, I believe, the first appearance of this result. It supposes that the squares of the shift-fluxions may be neglected.

[1571.] We now reach a very suggestive, if not quite satisfactory, attempt (p. 23) to introduce the consideration of slide into the ordinary (Bernoulli-Eulerian) theory of beams. While the sections are supposed, owing to the slides, no longer to be normal to the axis, they are still treated as *plane;* the slide is also treated as constant over a section. The maximum stretch is then obtained at any point when the slide is taken into account. The well-known formula for the maximum stress involving a radical (see Art. 995 and Saint-Venant's *Leçons de Navier*, p. 227) then makes its appearance for the first time in the history of our subject (p. 27). In the following year it was approved by Poncelet and has since (England excepted!) been universally adopted.

[1572.] On p. 44 we find some general remarks on a method for solving the problem of a beam subjected to any system of loads (this includes the 'statically' indeterminate problem of a body resting on more than three supports). We shall refer again to this method when mentioning its more complete development in the memoir of 1843. Here however we may quote a paragraph (§ 47) which contains a fact still repeatedly neglected in the elementary text-books[1]:

[1] An exception ought to be made in favour of Cotterill's *Applied Mechanics:* see p. 368. Still even in this excellent work a qualifying remark might well have been introduced in using the old theory (e.g. p. 329).

On doit faire attention, en mettant en pratique cette méthode générale, qu'il n'y a que les éléments encastrés ou assujétis d'une manière quelconque qui soient astreints, ou à rester dans la même direction, ou à faire constamment des angles donnés avec d'autres éléments. Ces conditions ne doivent pas être appliquées aux éléments immédiatement adjacens à ceux assujétis, car nous avons vu (§ 27) que si les glissements transversaux ne sont pas négligeables, deux éléments contigus d'une même pièce qui faisaient un angle infiniment petit avant les déplacements peuvent, après les déplacements, faire un angle fini quoique fort petit. Si donc on pose les équations, comme il est commode de le faire, seulement pour les parties des pièces qui sont en dehors des encastrements ou des assemblages, il ne faudra pas exprimer, dans les équations de condition, que les premiers éléments de ces parties ont une tangente commune avec les éléments assujétis, mais qu'ils font avec ceux-ci les petits angles déterminés en grandeur et en direction par les quantités $P_t/\mu\omega$ y relatives.

Here $P_t$ represents the total shear parallel to the section $\omega$, and $\mu$ is the slide-modulus.

[1573.] The first part of the *Cours* fills 48 pages, and we have noted the more interesting points in it. The second part, entitled: *Sommaire destiné provisoirement à faire suite à la partie rédigée*, contains 28 pages, upon which we will make a few remarks. The third part, on steam-engines, does not concern us here.

The second part commences with a considerable number of examples in which the formula of Saint-Venant for rupture is applied, occasionally in a form where the slide is neglected. We may draw attention to pp. 11—14, where curved pieces of constant section fixed at the extremities and placed in a vertical plane, so that they are symmetrical with regard to a vertical force applied at their summit, are treated of. The peculiarity of the treatment is the consideration of the slides. As the slides are considered constant over the section, the results even in the following case of the circular arc must be looked upon as approximations.

If, $2P$ be the vertical force,

$Q$ the horizontal component at either point of support,

$r$ the unstrained radius of the arc (supposed circular),

$\omega$ the section of the piece,

$\phi$ the inclination of the tangents at its terminals to the horizontal,

$E$ the stretch-modulus and $G$ the slide-modulus,

$\omega\kappa^2$ the moment of inertia of the cross section about a horizontal line in its plane through the central axis of the arc,

then Saint-Venant finds for the value of $Q$ the expression

$$P\frac{\frac{r}{2}\sin^2\phi\left(\frac{1}{\mu\omega}-\frac{1}{E\omega}\right)+\frac{r^3}{E\omega\kappa^2}\left(\frac{3}{2}\sin^2\phi-\phi\sin\phi\cos\phi+\cos\phi-1\right)}{\frac{r}{E\omega}\left(\frac{1}{2}\phi+\frac{1}{2}\sin\phi\cos\phi\right)+\frac{r}{\mu\omega}\left(\frac{1}{2}\phi-\frac{1}{2}\sin\phi\cos\phi\right)+\frac{r^3}{E\omega\kappa^2}\left(\frac{1}{2}\phi-\frac{3}{2}\sin\phi\cos\phi+\phi\cos^2\phi\right)}.$$

Compare the results cited in our Arts. 278 and 1458.

[1574.] If we suppose $\phi$ very small, or the length of the chord $2a$ very great as compared with the sagitta $b$, this reduces to

$$Q=\frac{25}{32}P\frac{b}{a}\left(1-\frac{15}{8}\frac{\kappa^2}{b^2}\right) \text{ (p. 12).}$$

The depression is given by

$$\frac{Pa^3}{128E\omega\kappa^2}+\frac{25}{16}\frac{Pa}{E\omega}\left(1+\frac{25}{64}\frac{a^2}{b^2}\right) \text{ (p. 13).}$$

Saint-Venant remarks that, as the first term is generally the most important, the depression is only 3/128 of that which it would be for a beam of length $2a$ loaded in the middle with a weight $2P$.

[1575.] The next problem treated (pp. 13—14) is that of a circular ring of radius $r$ placed vertically upon a horizontal plane and loaded on the top with a weight $2P$. Saint-Venant finds the depression[1]

$$=\left(\frac{\pi}{2}-\frac{4}{\pi}\right)\frac{Pr^3}{E\omega\kappa^2};$$

and the extension of the horizontal diameter $=\left(\frac{4}{\pi}-1\right)\frac{Pr^3}{E\omega\kappa^2}$.

[1576.] Then follow results similar to those of Euler, Lagrange and Heim (see Arts. 65, 85, 106, 910), and the part concludes with

[1] In the text the halves of these values are given, but the values cited in the *Comptes rendus* t. XVII. p. 1024 seem to me the correct ones.

some interesting examples of problems in which several struts support one load, or one beam is loaded at several points.

[1577.] We have not thought it needful to criticise certain portions of the work which are now-a-days superseded, for the very good reasons that the work has never been published, and that it is to the author himself that we owe the correction of the old theories. He would be the last to consider it without faults[1].

Nevertheless this *Cours* is peculiarly interesting, as it clearly suggests the lines of thought which were opening up before the great elastician and bears in it the germs of much of his later work. For this reason, as well as for the important points of our science which it first elucidates, we have not hesitated to consider it at some length.

[1578.] The next work of Saint-Venant which we have to notice is entitled: *Mémoires sur la Résistance des Solides suivis de deux notes sur la flexion des pièces à double courbure.* Paris, 1844. This is an offprint, with preface, of the following papers:

I. *Sur le calcul de la résistance et de la flexion des pièces solides à simple ou à double courbure, en prenant simultanément en considération les divers efforts auxquels elles peuvent être soumises dans tous les sens.* Extrait des *Comptes rendus*, t. XVII. pp. 942—954, Oct. 30, 1843.

II. *Intégration d'une équation différentielle qui se présente dans la Théorie de la flexion des verges élastiques.* Extrait du *Journal de mathématiques* (Liouville), t. IX. 1844, pp. 191—192.

III. *Suite au Mémoire lu le* 30 *octobre* 1843, (I) *Comptes rendus*, t. XVII. pp. 1020—1031.

IV. *Sur la torsion des prismes à base rectangle* {*et à base losange et sur une petite correction numérique à faire subir, en*

[1] Saint-Venant himself writes in his characteristic fashion to the Editor: "Vous verrez que mes Leçons de 1837 ont été un tâtonnement rempli de fautes." We may add that considering the haste in which they were prepared, their value is surprising.

*général, aux moments de torsion}* (*traitée à la manière de Cauchy*)[1]. Extrait des *Comptes rendus*, Nov. 20, 1843, t. XVII. pp. 1180—1190.

V. *Sur l'état d'équilibre d'une verge élastique à double courbure lorsque les déplacements éprouvés par ses points, par suite de l'action des forces qui la sollicitent, ne sont pas très-petits.* Extrait des *Comptes rendus*, July 1, 1844, t. XIX. pp. 36—44.

VI. A *Deuxième note*, with the same title as V., appears as an extrait des *Comptes rendus*, July 15, 1844, t. XIX. pp. 181—187.

In the preface Saint-Venant states that these six memoirs, printed only in extract, have for object the completion, correction and simplification of several points in the usual theory of the resistance of the solid bodies used in constructions. He further points out that proofs of various results merely stated in one memoir will be found in the others. We proceed to consider these memoirs individually.

I., III. and IV. were referred to a committee consisting of Poncelet, Piobert, Lamé and Cauchy; their report was drawn up by Cauchy and is printed on pp. 1234—1236 of the *Comptes rendus*, t. XVII. The report speaks very highly respecting the communications, concluding thus:

Les perfectionnements que les formules de M. de Saint-Venant ont apportés à la mécanique pratique, ainsi qu'à la mécanique rationnelle, ont été tellement sentis, que plusieurs d'entre elles sont déjà passées dans l'enseignement, et ont été données, en particulier, dans le cours fait par notre confrère M. Poncelet à la Faculté des Sciences.

En résumé, les divers Mémoires de M. de Saint-Venant nous paraissent justifier pleinement de la réputation que cet habile ingénieur, qui a toujours occupé les premiers rangs dans les promotions à l'École Polytechnique, s'est acquise depuis longtemps. Nous les croyons très-dignes d'être approuvés par l'Académie et insérés dans le *Recueil des Mémoires des Savants étrangers*.

[1579.] I do not think however that any further publication took place. IV. seems to be completely given in the *Comptes*

[1] In the corrected copy of these offprints with which M. de Saint-Venant favoured the Editor, the title is altered by the substitution of the words in round brackets for those in looped brackets. Sections $1^0$, $2^0$, $3^0$ and $14^0$ were also erased.

*rendus;* but of the communications I. and III. all we have in the *Comptes rendus* is the author's abstract. These communications may be considered to be to some extent superseded by the later researches of Saint-Venant, namely the memoir on Torsion of June, 1853, and the memoir on Flexure of 1856.

[1580.] § 1 of I. entitled: *Exposition,* criticises the old Bernoulli-Eulerian theory of flexure and contains the following remarks:

Je cherche, dans mon Mémoire, à combler ces lacunes, à réparer ces inexactitudes et à faire disparaître toute complication inutile. Je fais entrer dans le calcul les effets de glissement latéral dus à ces composantes transversales dont l'omission a été l'objet principal d'une sorte d'accusation portée par M. Vicat contre toute la théorie de la résistance des solides (see our Art. 725). Je montre comment, à l'aide d'une seconde équation de moments transversaux, on résout très-simplement ce cas général signalé par M. Persy (see our Art. 811), où l'équilibre posé comme à l'ordinaire ne saurait exister, et où la flexion de la pièce se fait nécessairement dans une autre direction que celle où elle est sollicitée à fléchir. J'étends les calculs de résistance aux cas de flexion et torsion simultanées qui doivent s'offrir souvent si l'on considère qu'une pièce tordue ne l'est presque jamais par ce qu'on appelle un couple. Je tiens compte de ce que les sections planes deviennent gauches, de ce qu'elles s'inclinent un peu sur la fibre centrale, et de ce que les fibres exercent les unes sur les autres une action qui n'est pas tout à fait à négliger. Je donne des équations différentielles nouvelles pour les petits déplacements des points des pièces courbes à double courbure, et les intégrales, d'une forme très-simple, que j'ai tirées de ces trois équations simultanées du troisième ordre à coefficients non-constants.

Saint-Venant also remarks that his memoir contains several practical applications and that some of his results had already been given in his *Cours lithographié.*

[1581.] § 2 is entitled: *Équations d'équilibre des forces intérieures et extérieures* (pp. 944—946). It involves a treatment of the equilibrium of a prism subject to flexure, when account is taken of the slides and of the forces producing flexure not lying in the same plane as that of flexure. A formula of Cauchy's for the torsion of a rectangular prism is assumed, which Saint-Venant at that time thought applicable to a prism of any section, but as he has shewn later it is only applicable to an elliptic section. The noteworthy part of the paragraph is the taking moments about the principal axes of the section to obtain the plane of flexure. This is the first consideration of this point, and it may perhaps have been suggested by a remark of M. Persy: see our Art. 811.

[1582.] § 3 is entitled: *Condition de résistance à la rupture ou à l'altération de l'elasticité* (pp. 947—949). This is the consideration of the formula involving a radical, which gives the maximum-stretch in a prism subject to flexure, when the slides are not neglected. As we have already seen, this formula was first given in the *Cours lithographié* (Art. 1571). A more complete discussion of this formula as well as the matter of the preceding section will be found in the memoir on Torsion of 1853 (pp. 316—320) and in Saint-Venant's edition of Navier's *Leçons* (pp. 220, 371 etc.), to both of which we shall return later. There is a footnote to the effect that Poncelet, in his *Cours de Mécanique industrielle de la Faculté* of 1839, had insisted on the importance of introducing the slides and had adopted Saint-Venant's formula (see our Art. 994). An elegant geometrical proof of the formula given by Poncelet in 1839 will be found in a footnote to Saint-Venant's *Leçons de Navier*, p. 374.

[1583.] § 4 is entitled: *Application à quelques exemples—Différences avec les résultats de l'ancienne théorie* (pp. 949—951). This contains practical applications, the principal of which are reproduced in the edition of Navier's *Leçons*. Thus the paradox of the 'solids of equal resistance' having no sectional area at the points of support is cleared up by the consideration of the shear.

A result may be quoted from p. 951, which I believe has not been reproduced. It is the case of a spring in the form of a vertical helix on a cylinder of radius $a$, sectional area $\pi r^2$, and of which the thread makes an angle $\beta$ with the horizon. If $P$ be the weight supposed to act in the axis of the helix, $R_0$ the traction per unit area, which would produce set in a prism of the same material as the helix, then the greatest load $P$ is given by the equation

$$R_0 = \frac{4P}{\pi r^3}\left[\tfrac{3}{8}\left(a+\frac{r}{4}\right)\sin\beta + \tfrac{5}{8}\sqrt{\left(a+\frac{r}{4}\right)^2\sin^2\beta + \left(a+\frac{r}{2}\right)^2\cos^2\beta}\right]$$

[1584.] § 5, entitled : *Détermination des déplacements des points des pièces solides ou des changements de forme qu'elles éprouvent* (pp. 951—953). This is the consideration of the equilibrium of a wire the mean thread of which is a curve of double curvature when subjected to any system of force. The main point to be noticed is the introduction of an angle $e$ and its variation $\epsilon$. This angle is that between the radius of curvature at any point and a principal axis of the corresponding section of the wire. Saint-Venant shews how the stretch at any point of the section depends on the variation of this angle. Paragraph 17 (p. 952) may be quoted as explaining this point:

On peut s'étonner de voir, dans mes équations, une certaine quantité toute nouvelle $\epsilon$, dont personne n'a encore tenu compte, et qui s'y trouve en quelque sorte sur le même pied que les angles de contingence et d'osculation plane $ds/\rho$ et $ds/r$. Un exemple montrera facilement, je pense, que ce *déplacement angulaire du rayon de courbure sur la section* devait entrer nécessairement dans notre analyse.

Qu'on se figure une verge élastique à double courbure serrée de toutes parts dans un canal fixe et rigide, mais où on puisse la faire tourner elle-même, car on suppose sa section circulaire ainsi que celle du canal. Dans ce mouvement, les fibres les plus longues se seront forcément raccourcies, les plus courtes se seront allongées, et il y aura eu aussi des torsions si les rotations imprimées à toutes les sections n'ont pas été les mêmes: l'élasticité de la pièce aura résisté énergiquement, dans tous les cas, à de pareils déplacements de ses points.

Cependant ni les rayons de courbure, ni les plans osculateurs de l'axe n'auront changé en aucune manière.

Donc les résistances dites à la flexion et à la torsion ne dépendent pas uniquement du changement des angles de contingence, et de ces angles que forment les plans osculateurs entre eux : elles dépendent, au même degré, *d'autre chose*, savoir, du genre de déplacement qui a eu lieu dans l'exemple cité; or c'est précisément, sur chaque section, ce déplacement angulaire que j'ai appelé $\epsilon$.

On voit donc que l'on chercherait vainement la solution du problème des changements de forme des pièces élastiques à double courbure en se bornant à considérer les points de leur axe. Il faut s'inquiéter aussi de ce qui se passe hors de l'axe. Cette observation explique, ce me semble, une erreur de Lagrange que Poisson n'a pas évitée (*Mécanique*, 2e éd. Nos. 317, 318), quoiqu'elle eût été signalée [en partie][1] par M. Binet dès 1814.

See our Articles 157—161, 173—175, 215—222, 423 and 935. It will be seen by an examination of these articles that Binet, Bordoni and Poisson considered only the mean thread of the wire, while introducing the angle between osculating planes (badly termed the angle of torsion) which had been omitted by Lagrange, that in addition Binet seems to have had some idea of the twist of the section round the thread; but it appears to have been first introduced into the calculations by Bellavitis and first practically treated by Saint-Venant.

[1585.] The final section of I. (p. 953) contains a statement of the terminal conditions to be satisfied when a system is composed of several solid pieces which are united together, pass over points of support, etc. The conditions are similar to those mentioned in Art. 1572, with an additional one in the case of pieces the mean thread of which is a curve of double curvature.

1586. II. In the penultimate section of I. (p. 952) Saint-Venant obtained for the case of double curvature the following equations:

$$\frac{\delta ds}{ds} = D, \quad \frac{1}{ds}\,\delta\frac{ds}{\rho} = F, \quad \frac{1}{ds}\,\delta\frac{ds}{\tau} = T \quad \ldots\ldots\ldots\ldots(1).$$

In these equations $D, F, T$ represent known functions of the coordinates $x, y, z$ of any point $m$ of the curved axis of the rod, $s$ the length measured up to the point $m$ of the arc of this axis, $ds/\rho$ the angle of contingence, and $ds/\tau$ the angle between two consecutive osculating planes of the curve at this point. Finally $\delta$ denotes the variation arising from the *very small shifts* of the points of the axis produced by the action of the given forces. Put $u$ for $\delta x$, $v$ for $\delta y$, and $w$ for $\delta z$; then $u, v$, and $w$ are to be found from equations (1).

[1] Addition of Saint-Venant in corrected memoir.

On the page cited of the *Comptes rendus* Saint-Venant gave without demonstration the integrals of the three equations, namely,

$$\left.\begin{aligned} du &= Ddx - dy\int\left(Tdz + \frac{\rho Z}{ds^2}F\right) + dz\int\left(Tdy + \frac{\rho Y}{ds^2}F\right) \\ dv &= Ddy - dz\int\left(Tdx + \frac{\rho X}{ds^2}F\right) + dx\int\left(Tdz + \frac{\rho Z}{ds^2}F\right) \\ dw &= Ddz - dx\int\left(Tdy + \frac{\rho Y}{ds^2}F\right) + dy\int\left(Tdx + \frac{\rho X}{ds^2}F\right) \end{aligned}\right\} \text{(2)},$$

where

$$X = dy\,d^2z - dz\,d^2y,\quad Y = dz\,d^2x - dx\,d^2z,\quad Z = dx\,d^2y - dy\,d^2x.$$

To verify the statement that these values do satisfy the equations (1) we must develop these. The last of them is somewhat more tedious than the others; we must observe that

$$\frac{ds}{\tau} = (L + M + N)\frac{\rho^2}{ds^5},$$

where $\qquad L = dx\,(d^2y\,d^3z - d^3y\,d^2z),$

and $M$ and $N$ can be derived from $L$ by symmetry: see Frost's *Solid Geometry*, 3rd ed. Art. 640.

Thus the equations (1) become when $s$ is taken as independent variable:

$$\left.\begin{aligned} dx\,du + dy\,dv + dz\,dw &= D(ds)^2 \\ d^2x\,d^2u + d^2y\,d^2v + d^2z\,d^2w &= D\frac{(ds)^4}{\rho^2} + \frac{F}{\rho}(ds)^4 \\ \frac{\rho^2}{ds^5}(\delta L + \delta M + \delta N) + (L + M + N)\,\delta\,\frac{\rho^2}{ds^5} &= Tds \end{aligned}\right\} \text{(3)}.$$

The last equation might be developed still further since we easily find that

$$\delta\,\frac{\rho^2}{ds^5} = -\frac{2F\rho^3}{ds^5} - \frac{3D\rho^2}{ds^5};$$

$$\begin{aligned} \delta L + \delta M + \delta N = du\,(d^2y\,d^3z - d^3y\,d^2z) &+ d^2u\,(dz\,d^3y - dy\,d^3z) \\ &+ d^3u\,(dy\,d^2z - dz\,d^2y), \\ &+ \text{similar terms in } v \text{ and } w. \end{aligned}$$

1587. It will be found on trial that the expressions in (2) do satisfy the equations (3); the verification is tedious, but may be performed with patience. It is to be observed, that if we take the arc $s$ for the independent variable we have from (1)

$$d^2u = dDdx + Dd^2x + \rho Fd^2x - d^2y\int\left(Tdz + \frac{\rho Z}{ds^2}F\right) + d^2z\int\left(Tdy + \frac{\rho Y}{ds^2}F\right),$$

and similar expressions for $d^2v$ and $d^2w$.

1588. We now return to the article in Liouville's *Journal.* Here Saint-Venant adopts a different process of integration. He eliminates $v$ and $w$ and obtains the following result:

$$du\, d^2y\, d^3z - du\, d^2z\, d^3y + dy\, d^2z\, d^3u - dy\, d^2u\, d^3z + dz\, d^2u\, d^3y - dz\, d^2y\, d^3u = Sds^6 \ldots (4).$$

Here $S$ represents a certain known function of the primitive coordinates of the point $m$, or of the arc $s$, which we may take for the independent variable. I have verified (4) by means of (2); here Saint-Venant derives (4) immediately from (1): the following is his statement of the process, which I do not understand.

Pour cela, effectuons les différentiations par $\delta$ des premiers membres des équations précédentes; après avoir remplacé $ds$, $ds/\rho$, $ds/r$ par leurs expressions générales connues en $x$, $y$, $z$, nous aurons trois équations différentielles simultanées du premier, du deuxième et du troisième ordre en $u$, $v$, $w$. Si nous éliminons deux de ces inconnues, par exemple $v$ et $w$ (ce qui se fait facilement en différentiant la première équation et en tirant les valeurs de $d^2v$, $d^2w$, d'où, à l'aide de la seconde équation, celles de $dv$ et $dw$ que l'on substitue dans la troisième) les intégrales disparaissent et nous obtenons une équation en $u$,...

Then follows equation (4).

1589. Saint-Venant then proceeds thus: if we replace in (4) $y$ and $z$ by their values in terms of $s$ derived from the equations to the curve we have only $u$ and $s$ left in the equation. The differential equation is of the third order and is linear, but the coefficients of $\frac{du}{ds}, \frac{d^2u}{ds^2}$, and $\frac{d^3u}{ds^3}$ are not constant, so that there is no general method for solving the equation. However the equation may be treated as follows: put $du = U\,dz$, and $dy = V\,dz$, and substitute in (4); we obtain 22 terms on the left-hand side, of which 20 disappear by cancelling, leaving $dz^3(dU\,d^2V - dV\,d^2U)$, that is $-dz^3\,dV^2\,d\dfrac{dU}{dV}$. Hence (4) becomes

$$-dz^3\left(d\,\frac{dy}{dz}\right)^2 d\,.\,\frac{d\,\dfrac{du}{dz}}{d\,\dfrac{dy}{dz}} = Sds^6;$$

this is integrable and gives

$$\frac{du}{dz} = -\int \lambda\, d\,\frac{dy}{dz},$$

where $\lambda$ stands for $$\int \frac{Sds^6}{dz^3\left(d\,\dfrac{dy}{dz}\right)^2}.$$

Now $\displaystyle\int \lambda d\,\frac{dy}{dz} = \lambda\,\frac{dy}{dz} - \int d\lambda\,\frac{dy}{dz}$; so that finally

$$du = dz\int \frac{Sds^6}{(dy\,d^2z - dz\,d^2y)^2}\,dy - dy\int \frac{Sds^6}{(dy\,d^2z - dz\,d^2y)^2}\,dz.$$

Saint-Venant finishes thus[1]:

La recherche des petits déplacements des points de l'axe d'une pièce courbe à double courbure est ainsi ramenée aux quadratures, et l'on voit que le polynôme différentiel qui forme le premier membre de l'équation (4) est intégrable quand, après l'avoir multiplié par $dz$ ou $dy$, on le divise par $(dy\,d^2z - dz\,d^2y)^2$.

[1590.] III. This memoir consists of an application of § 5 and § 6 of I. to the calculation of various cases of pieces of single or double curvature subject to systems of force. Saint-Venant again draws attention to the great divergencies of his theory from the usual (Bernoulli-Eulerian) hypothesis.

[1591.] Pp. 1021—1022 are occupied with the general equations when the mean thread has double curvature. In this case equation (19) depends upon Cauchy's erroneous theory of torsion; thus Saint-Venant's $\mu''$ (the $2\nu$ of our Art. 1594) is really a coefficient which he has taught us in later memoirs varies with the shape of the section.

[1592.] On p. 1022 we have the consideration of the case when

[1] Saint-Venant uses $\xi$, $\eta$, $\zeta$ for our shifts $u$, $v$, $w$ and $u$, $v$ for our $U$, $V$.

the mean thread is a plane curve and remains in the same plane when its form is changed. Let $u$, $v$ be the shifts of the point $x$, $y$ of the mean thread, $E$ the stretch-modulus, $\mu$ the slide-modulus, $\omega$ the section, $P_t$, $P_n$ the load components parallel to the tangent and to the normal at the point $x$, $y$ of the mean thread, $M$ the moment of these forces about the point $x$, $y$, $\omega\kappa^2$ the moment of inertia of the section $\omega$ about a line in its plane perpendicular to the plane of flexure, and $ds$ an element of arc of the mean thread; then Saint-Venant obtains the following equations for the shifts:

$$du = \frac{P_t}{E\omega}\,dx + \frac{P_n}{\mu\omega}\,dy - dy\int\frac{M}{E\omega\kappa^2}\,ds,$$

$$dv = \frac{P_t}{E\omega}\,dy - \frac{P_n}{\mu\omega}\,dx + dx\int\frac{M}{E\omega\kappa^2}\,ds.$$

The equations obtained by Navier (see our Art. 257) omit the first two terms on the right-hand side of both equations. They were in fact omitted by all writers before Saint-Venant, and the text-books of the present day often omit them still.

[1593.] Then follow various practical examples (pp. 1023—1031). We may note that of the vertical circular ring considered in our Art. 1575, the case of a horizontal circular ring acted upon by vertical forces, and especially the consideration of the helix subject to a load in its axis. We will briefly consider this last.

Let $x, y, z$ be the coordinates of a point on the mean fibre of the helix, which is supposed to lie on a cylinder of radius $a$ and to cut its generators at an angle $\pi/2 - \beta$. In cylindrical coordinates we may write

$$x = a\cos\theta, \qquad y = a\sin\theta, \qquad z = a\theta\tan\beta.$$

Let $w$ be the shift in $z$ and $\alpha$ in $a$, then, if $P$ be the axial load, Saint-Venant finds:

$$w = C^{\text{v}} + \frac{Pa\sin\beta\tan\beta}{E\omega}\,\theta + \frac{Pa^3}{\cos\beta}\left(\frac{\cos^2\beta}{2\mu\nu} + \frac{\sin^2\beta}{E\omega\kappa^2}\right)\times$$
$$[\theta + C\sin\theta + C'(1-\cos\theta)],$$

$$\alpha = -\frac{Pa\sin\beta}{E\omega} + Pa^3\sin\beta\left(\frac{1}{2\mu\nu} - \frac{1}{E\omega\kappa^2}\right)$$
$$+\frac{Pa^3\tan\beta}{\cos\beta}\left(\frac{\cos^2\beta}{2\mu\nu} + \frac{\sin^2\beta}{E\omega\kappa^2}\right)[1 + C\theta\sin\theta + C'\theta(1-\cos\theta)]$$
$$+ C'''\cos\theta + C^{\text{IV}}\sin\theta,$$

where $C$, $C'$, $C''$, $C'''$, $C^{IV}$, $C^{V}$ are constants to be determined by the conditions imposed on the terminals of the spiral, and the rest of the notation is that used in the earlier articles of this chapter.

[1594.] Saint-Venant then remarks that if the terminals are so attached that the helix does not lose any of its regularity in changing its form all these constants ought to be suppressed. Further, since the longitudinal stretch has in general only a feeble influence, he omits the second term in $w$ and the first in $\alpha$. His values now reduce to

$$w = \frac{Pa^3}{\cos\beta}\left(\frac{\cos^2\beta}{2\mu\nu} + \frac{\sin^2\beta}{E\omega\kappa^2}\right)\theta,$$

$$\alpha = \frac{w}{\theta}\tan\beta + Pa^3\sin\beta\left(\frac{1}{2\mu\nu} - \frac{1}{E\omega\kappa^2}\right).$$

In these equations $2\nu$ (Saint-Venant uses $\mu''$ for our $\nu$) is the quantity by which it is necessary to multiply the slide-modulus in order to obtain the 'torsional rigidity.' Thus, if in prism of unit length and any section the couple $M$ produce a torsion $\tau$,

$$M = 2\nu\mu\tau.$$

On the old theory of torsion $2\nu = \omega\,(\kappa^2 + \kappa'^2)$, and for a circular section $\nu = \omega\kappa^2$.

On the assumption that $\nu = \omega\kappa^2$ and $\mu = \frac{2}{5}E$, Saint-Venant (p. 1030) reduces his value of $w$ to

$$\frac{Pa^3}{E\omega\kappa^2\cos\beta}\left(\tfrac{5}{4}\cos^2\beta + \sin^2\beta\right)\theta.$$

[1595.] Referring to Giulio's memoir considered in our Art. 1218, Saint-Venant then concludes:

C'est pour ce dernier cas, ou pour des hélices d'un *pas* très-faible, que M. Giulio a fait ses intéressantes expériences, consignées dans un Mémoire lu, le 11 juillet 1841, à l'Académie de Turin. Sa formule, dressée à priori par des considérations particulières à l'hélice et non applicables à des cas où l'angle $\epsilon$ (see our Art. 1584) n'est pas nul, revient à

$$\omega = \frac{Pa^3}{E\omega\kappa^2\cos\beta}\,\frac{5}{4}\,\frac{\theta}{\cos^2\beta},$$

elle n'est identique avec la mienne que pour $\cos\beta = 1$.

Si l'on ne supprime que la quatrième puissance de $\sin\beta$ la formule de M. Giulio est

$$\frac{Pa^3}{E\omega\kappa^2}\left(\frac{5}{4}+\frac{15}{8}\sin^2\beta\right)\theta,$$

et la mienne

$$\frac{Pa^3}{E\omega\kappa^2}\left(\frac{5}{4}+\frac{3}{8}\sin^2\beta\right)\theta.$$

Ma formule représente donc un peu mieux que la sienne les résultats des experiences de ce savant, car il prend constamment $\cos\beta=1$ ou

$$w=\frac{5}{4}\frac{Pa^3}{E\omega\kappa^2}\theta,$$

pour y satisfaire.

[1596.] Let us now advert to the communication IV. of this set of memoirs. It relates to Torsion, and is founded on an approximate method given by Cauchy, a method which Saint-Venant here much simplifies, but which he finally abandoned: see Art. 661 of our chapter on Cauchy. The process assumes only one constant of elasticity. Wrong values are given to two moments of inertia on p. 1184. Section 4 seems quite unsatisfactory, and section 5, which extends Cauchy's results to a lozenge-shaped section, has been discarded by Saint-Venant himself.

Saint-Venant however brings out one point in this memoir for the first time, a point which is the basis of his later investigations, namely, that Cauchy's formula differs from the old formula, because his theory really involves the distortion of the plane sections. The old theory is false for all prisms, of which the section is not circular, because torsion distorts the plane sections. This point is here first noticed. Saint-Venant again emphasises it in the memoir of February 22, 1847. See I. of our Art. 1617. The distortion is termed *gauchissement* in both these memoirs.

See the memoir on *Torsion*, p. 361; Navier's *Leçons*, 3[e] éd. pp. clxxv—clxxxv, 620—627; Moigno's *Statique*, p. 640.

[1597.] The communications V. and VI. may be taken together: they relate to the equations of equilibrium of a piece the axis of which is a curve of double curvature, but they suppose the shifts not to be very small as was the case in § 5 of I. These notes are interesting as presenting the first general investigation of the problem, that is to say, they first take account of the angle $\epsilon$ referred to in our Art. 1584.

1598. The object of the first note is thus expressed at the beginning:

M. Binet, et ensuite M. Wantzel, viennent de donner (*Comptes rendus* des 17 et 24 juin) les intégrales des équations de la courbe élastique à double courbure provenant de la flexion et de la torsion d'une verge ou d'une portion de verge cylindrique et primitivement droite, sollicitée à ses extrémités seulement. Ces intégrales s'appliquent à des déplacements des points aussi grands qu'on veut, pourvu, bien entendu, qu'ils n'aillent pas jusqu'à altérer l'élasticité de la matière. Elles supposent admis ce théorème de Poisson "que le moment qui tend à produire la torsion (ou le moment opposé qui y résiste dans l'état d'équilibre) est constant dans toute l'étendue de la verge."

D'un autre côté, j'ai donné, le 30 octobre et le 6 novembre 1843 (*Comptes rendus*, tome XVII), des équations et leurs intégrales, pour une verge élastique dont la forme primitive et le mode de sollicitation sont absolument quelconques et en tenant compte de plusieurs éléments nouveaux, mais seulement lorsque *les déplacements restent très-petits*, ce qui est le cas le plus ordinaire des applications.

Je me propose dans cette Note:

1°. De donner les équations différentielles de l'état d'équilibre d'une verge élastique dans le cas le plus général et pour des déplacements quelconques de ses points.

2°. De montrer dans quelles limites le théorème de Poisson est applicable, ainsi que les équations dont il l'a tiré.

[1599.] On p. 40 a value is given to the constant $\nu$ (see our Art. 1594) which depends on the erroneous torsion theory of Cauchy, but, as Saint-Venant remarks in the corrected copy of the memoir before us, the analysis remains the same if this value be not given to $\nu$: see our Arts. 661 and 1630. If $\omega\kappa^2$, $\omega\kappa'^2$ be the moments of inertia about the principal axes of the section at a point, where $\rho$ is the radius of curvature, and $\tau$ the radius of torsion, $M$, $M'$ the moment of the forces about these principal axes and $M_t$ about the tangent, $e$ and $\epsilon$ being as in Art. 1584, then Saint-Venant finds:

$$M = E\omega\kappa^2\left(\frac{\cos(e+\epsilon)}{\rho} - \frac{\cos e}{\rho_0}\right),$$

$$M' = E\omega\kappa'^2\left(\frac{\sin(e+\epsilon)}{\rho} - \frac{\sin e}{\rho_0}\right),$$

$$M_t = 2\mu\nu\left(\frac{d\epsilon}{ds} + \frac{1}{\tau} - \frac{1}{\tau_0}\right).$$

These equations are obtained by a straightforward bit of analysis on the assumption that the sections are not distorted and that the mean fibre is not stretched.

[1600.] Saint-Venant eliminates $\epsilon$ between these equations and finds:

$$\frac{1}{\rho^2} = \frac{1}{\rho_0^2} + \frac{2}{\rho_0}\left(\frac{M}{E\omega\kappa^2}\cos e + \frac{M'}{E\omega\kappa'^2}\sin e\right) + \left(\frac{M}{E\omega\kappa^2}\right)^2 + \left(\frac{M'}{E\omega\kappa'^2}\right)^2,$$

$$\frac{1}{\tau} = \frac{1}{\tau_0} + \frac{M_t}{2\mu\nu} - \frac{\dfrac{d}{ds}\left[\rho\left(\dfrac{M'}{E\omega\kappa'^2}\cos e - \dfrac{M}{E\omega\kappa^2}\sin e\right)\right]}{\rho\left(\dfrac{1}{\rho_0} + \dfrac{M}{E\omega\kappa^2}\cos e + \dfrac{M'}{E\omega\kappa'^2}\sin e\right)}.$$

Also, $$ds = ds_0.$$

Various cases are mentioned in which these equations take a simpler form, but in several of these cases it seems impossible to integrate them. The exceptional case of our Art. 1586 is referred to.

[1601.] In particular, however, if the rod be primitively straight, and the moments of inertia $\omega\kappa^2$ and $\omega\kappa'^2$ equal (of 'isotropic section' as Saint-Venant elsewhere terms it), the equations reduce to the simple forms:

$$\left.\begin{aligned} 2\mu\nu\left(\frac{d\epsilon}{ds} + \frac{1}{\tau}\right) &= M_t \\ \frac{E\omega\kappa^2}{\rho} = M_n, \quad 0 &= M_\rho \end{aligned}\right\} \quad \text{......................(i)},$$

where $M_\rho$ is the moment about the line in each section which is the trace upon it of the osculating plane of the curve produced by the strain and $M_n$ about the normal to this osculating plane.

[1602.] These equations can be easily thrown into the form

$$E\omega\kappa^2\frac{dy\,d^2z - dz\,d^2y}{ds^3} + \chi\frac{dx}{ds} = M_x \text{ ..............(ii)},$$

where $M_x$ is the moment about a line parallel to the axis of $x$, and $\chi$ the left-hand side of the first of equations (i). We have similar equations for moments about lines parallel to the axes of $y$ and $z$.

These are the equations of Lagrange modified by Binet, who added the second term. They have been integrated by Binet and Wantzel on the supposition that $\omega\kappa^2$ is constant and that the forces which enter on the right-hand sides act only on the terminals of the piece. Saint-Venant remarks, however, that, except in this particular case of initial straightness and inertial isotropy of section, $M_\rho$ will never be zero.

Le théorème [de Poisson] $M_t =$ constante, [qui est] lié à $M_\rho = 0$, *n'aura donc lieu que dans ce cas-là* (où la pièce était primitivement droite, et où l'on avait $\omega\kappa^2 = \omega\kappa'^2$), et les équations (i), (ii) sont incomplètes dans tout autre cas.

See our Arts. 159, 175, 423, 571—572, 935, 1215, 1239.

[1603.] The concluding paragraph defines so exactly the relative merit of the various investigators on the subject of pieces of double curvature that we reproduce it here:

Si Poisson semble établir ce théorème et les équations (ii) d'une manière générale, c'est qu'il omet, dans son analyse, ce troisième moment $M^\rho$, qui tend à fléchir une verge courbe transversalement à son plan osculateur actuel si elle était déjà courbe, et, par conséquent, à changer le plan de sa courbure. Lagrange n'avait fait attention qu'au moment $M_n$, qui tend à augmenter ou à diminuer la courbure dans son plan actuel, ce qui suffit *pour les courbes planes restant planes*. M. Binet y a ajouté le moment $M_t$, tendant à tordre, et cela suffit dans le cas particulier que nous venons d'énoncer, lorsqu'on ne cherche que les équations générales de *l'axe* de la verge; mais, dans le cas général où la verge à double courbure était primitivement courbe, ou bien où l'axe étant rectiligne, la section n'a pas une des formes donnant $\omega\kappa^2 = \omega\kappa'^2$, il est indispensable d'introduire aussi dans le calcul ce troisième moment $M_\rho$ perpendiculaire aux deux autres, et qui tend à plier la verge droite obliquement aux axes principaux de ses sections, ou à faire tourner le rayon de courbure sur le plan des sections de la verge courbe. Mais cela exige impérieusement que l'on introduise aussi l'angle $\epsilon$ qui mesure cette rotation et dont la prise en considération est nécessaire, en tous cas, pour déterminer les déplacements des points hors de l'axe et pour fixer même la valeur de certaines constantes des équations définitives de l'axe.

[1604.] Near the beginning of the second Note (our communication VI.) Saint-Venant says:

Je me propose, dans cette Note, d'ajouter plusieurs observations à celles que contient la Note précédente, et de considérer divers cas où l'on peut déterminer facilement l'état d'équilibre de la verge pour des déplacements d'une grandeur quelconque.

1605. The reader will find stated here distinctly the difference between Poisson and Saint-Venant as to the problem of the equilibrium of an elastic rod of double curvature.

Consider the following three straight lines at any point of the curve formed by the axis of the rod; the tangent, the principal normal, and the normal to the osculating plane. Let the moments round these straight lines of all the forces acting on the rod between a normal section at the point $x$, $y$, $z$ and the free end be denoted by $M_t$, $M_\rho$, and $M_n$ respectively. Let the moments of the same forces round the fixed rectangular axes be denoted by $M_x$, $M_y$, and $M_z$ respectively. Let $X$, $Y$, $Z$ denote the following binomials respectively;

$$dy\,d^2z - dz\,d^2y,\quad dz\,d^2x - dx\,d^2z,\quad dx\,d^2y - dy\,d^2x.$$

Let $\rho$ be the radius of curvature at the point $(x, y, z)$. Then by the ordinary principles of Statics we have

$$\left.\begin{aligned} M_n &= \frac{\rho X}{ds^3} M_x + \frac{\rho Y}{ds^3} M_y + \frac{\rho Z}{ds^3} M_z \\ M_\rho &= M_x \frac{\rho}{ds} d\frac{dx}{ds} + M_y \frac{\rho}{ds} d\frac{dy}{ds} + M_z \frac{\rho}{ds} d\frac{dz}{ds} \\ M_t &= \frac{dx}{ds} M_x + \frac{dy}{ds} M_y + \frac{dz}{ds} M_z \end{aligned}\right\} \text{......(i).}$$

Differentiate the last of these and use the second; thus

$$\frac{dM_t}{ds} = \frac{M_\rho}{\rho} + \frac{dx}{ds}\frac{dM_x}{ds} + \frac{dy}{ds}\frac{dM_y}{ds} + \frac{dz}{ds}\frac{dM_z}{ds}.$$

Now in Poisson's *Mécanique*, Vol. I. page 627, it is shewn that

$$\frac{dx}{ds}\frac{dM_x}{ds} + \frac{dy}{ds}\frac{dM_y}{ds} + \frac{dz}{ds}\frac{dM_z}{ds} = 0 \quad \text{............(ii);}$$

this is in fact Poisson's contribution to the subject under consideration. Hence we have

$$\frac{dM_t}{ds} = \frac{M_o}{\rho} \quad \text{.........................(iii).}$$

Therefore $\frac{dM_t}{ds}$ vanishes, and $M_t$ is constant, *provided we assume* that $M_\rho$ is zero; and this assumption Poisson really makes. The point had been previously noticed by Bordoni: see our account of his memoir of 1821 in Art. 222.

Saint-Venant remarks that the equation (iii) is due to Wantzel. It was given by him in the memoir we have referred to in Art. 1239.

[1606.] The Note then proceeds to discuss various cases in which the equations for the elastic rod will be integrable. For example, for the interesting case when a primitively straight rod is acted upon only by a couple, the form is shewn to be a helix, a result previously obtained by Wantzel: see our Art. 1239. Saint-Venant remarks:

C'est une généralisation du résultat d'Euler consistant en ce que lorsque la courbe provenant de la verge, ainsi sollicitée, est plane, elle ne peut être qu'un arc de cercle.

See our Art. 58.

[1607.] A paragraph on p. 181 is interesting:

Observons à ce sujet que lorsque la verge primitivement droite et à section régulière, est assujettie à une de ses deux extrémités seulement, et libre ou simplement appuyée à l'autre, le calcul de $\epsilon$ n'est pas nécessaire pour déterminer les constantes de l'équation de l'*axe*. La forme de cet axe et sa position peuvent être déduites complétement, alors, des équations différentielles de Lagrange, complétées et intégrées par M. Binet.

[1608.] Finally the memoir concludes with a further reference to Giulio's memoir of 1841: see our Art. 1219. Saint-Venant now sees that with a proper distribution of force the distorted form of the elastic helix might still be a helix without the vanishing of the five constants of our Art. 1593.

He obtains equations which are easily deduced from those of our Art. 1599, on the supposition that $\epsilon = 0$ and $\omega\kappa^2 = \omega\kappa'^2$, namely,

$$Pr\sin\beta + N\cos\beta = E\omega\kappa^2\left(\frac{\cos^2\beta}{r} - \frac{\cos^2\beta_0}{r_0}\right),$$

$$-Pr\cos\beta + N\sin\beta = 2\mu\nu\left(\frac{\sin\beta\cos\beta}{r} - \frac{\sin\beta_0\cos\beta_0}{r_0}\right),$$

where $P$ is the force along the axis, $N$ the couple about it, which constitute the proper distribution; $r$, $\beta$ define the helix formed by the distorted rod and $r_0$, $\beta_0$ that formed by the primitive rod; and $\nu$ has the value given to it in our Art. 1594.

These equations are given by Thomson and Tait in Art. 605 of the second edition of their *Treatise on Natural Philosophy*. They have, however, chosen the opposite direction for the force $P$.

Saint-Venant remarks with regard to these equations:

Ces formules s'accordent avec celles du Mémoire de M. Giulio, quand $P$ et $N$ ont entre eux une relation telle que l'hélice s'allonge ou se raccourcisse sans se tordre ou se détordre, ou réciproquement: elles s'accordent avec celles que j'ai données le 6 novembre lorsque $N$ est nul, et que $r - r_0$, $\beta - \beta_0$ sont très-petits. La circonstance $\epsilon = 0$, et la supposition que les termes altérant la forme hélicoïdale s'évanouissent, rendent semblables les résultats de nos deux analyses, et les différences que j'avais cru y apercevoir n'étaient qu'apparentes.

See our Arts. 1219—1222 and 1382.

1609. As an illustration I will place here an example taken from the *Dublin University Calendar* for 1846; it is the only notice of the subject which I have found in English works.

The problem is thus stated:

Point out the error in Poisson's analysis which leads to the conclusion that the moment of torsion of an elastic rod is constant, whatever be the forces which act on it; and prove that if the rod be a horizontal plane curve of very small thickness, fixed at one end and acted on by its own weight only,

$$\rho\frac{d^2\tau}{ds^2} + \frac{d\rho}{ds}\frac{d\tau}{ds} + \rho^2\tau\frac{d^2x\,d^3y - d^3x\,d^2y}{ds^5} + W = 0,$$

where $\tau$ is the moment of torsion, $\rho$ the radius of curvature, $W$ the weight of the rod between the point $x$, $y$ and the free extremity.

Since the rod is of very small thickness the equations (i) of Art. 1605, which belong to the *axis*, may be taken as applying to the rod itself. We shall have

$$M_x = g\sigma\omega \int_s^l (y' - y)\, ds', \quad M_y = - g\sigma\omega \int_s^l (x' - x)\, ds',$$

where $\sigma$ is the density, and $\omega$ the area of a section of the rod.

Put $\tau$ for the $M_t$ of those equations (i): thus

$$\tau = M_x \frac{dx}{ds} + M_y \frac{dy}{ds} \quad \ldots\ldots\ldots\ldots\ldots\ldots\ldots\ldots(1),$$

$$\frac{d\tau}{ds} = M_x \frac{d^2x}{ds^2} + M_y \frac{d^2y}{ds^2} + \frac{dx}{ds}\frac{dM_x}{ds} + \frac{dy}{ds}\frac{dM_y}{ds};$$

but $$\frac{dM_x}{ds} = - g\sigma\omega \int_s^l \frac{dy}{ds}\, ds' = - g\sigma\omega\, (l - s)\, \frac{dy}{ds},$$

$$\frac{dM_y}{ds} = g\sigma\omega \int_s^l \frac{dx}{ds}\, ds' = g\sigma\omega\, (l - s)\, \frac{dx}{ds};$$

therefore $$\frac{d\tau}{ds} = M_x \frac{d^2x}{ds^2} + M_y \frac{d^2y}{ds^2} \ldots\ldots\ldots\ldots\ldots\ldots\ldots\ldots(2).$$

From this we get

$$\frac{d^2\tau}{ds^2} = M_x \frac{d^3x}{ds^3} + M_y \frac{d^3y}{ds^3} + \left(\frac{d^2y}{ds^2}\frac{dx}{ds} - \frac{d^2x}{ds^2}\frac{dy}{ds}\right) g\sigma\omega\, (l - s) \;\ldots(3).$$

From (1) and (2) we obtain

$$M_x = \frac{1}{\gamma}\left(\tau \frac{d^2y}{ds^2} - \frac{d\tau}{ds}\frac{dy}{ds}\right), \quad M_y = - \frac{1}{\gamma}\left(\tau \frac{d^2x}{ds^2} - \frac{d\tau}{ds}\frac{dx}{ds}\right)$$

where $\gamma$ stands for $$\frac{d^2y}{ds^2}\frac{dx}{ds} - \frac{d^2x}{ds^2}\frac{dy}{ds}.$$

Substitute in (3); thus we get

$$\left.\begin{aligned} \gamma \frac{d^2\tau}{ds^2} = \tau \left(\frac{d^3x}{ds^3}\frac{d^2y}{ds^2} - \frac{d^3y}{ds^3}\frac{d^2x}{ds^2}\right) + \frac{d\tau}{ds}\left(\frac{d^3y}{ds^3}\frac{dx}{ds} - \frac{d^3x}{ds^3}\frac{dy}{ds}\right) \\ + W\gamma^2, \end{aligned}\right\} \ldots(4),$$

where $W$ is put for $g\sigma\omega\,(l - s)$.

Now we have $$\gamma = \frac{d^2y}{ds^2}\frac{dx}{ds} - \frac{d^2x}{ds^2}\frac{dy}{ds} = \frac{1}{\rho};$$

and thus $$\frac{d^3y}{ds^3}\frac{dx}{ds} - \frac{d^3x}{ds^3}\frac{dy}{ds} = - \frac{1}{\rho^2}\frac{d\rho}{ds}.$$

Hence (4) becomes

$$\frac{1}{\rho}\frac{d^2\tau}{ds^2}+\frac{1}{\rho^2}\frac{d\rho}{ds}\frac{d\tau}{ds}+\tau\left(\frac{d^3y}{ds^3}\frac{dx^2}{ds^2}-\frac{d^3x}{ds^3}\frac{d^2y}{ds^2}\right)=\frac{W}{\rho^2}.$$

This agrees with the proposed formula, except that $\tau$ has a different sign, which merely amounts to changing the direction in which the moment is estimated.

1610. The volume XVII. of the *Comptes rendus* which we have had under notice contains other communications by Saint-Venant; but these do not relate to our subject. To one of them concerning fluid motion Saint-Venant alludes in Moigno's *Statique*, p. 694. From pp. 1310 and 1327 we find that Saint-Venant was one of six candidates for a vacancy in the Paris Academy, and by the committee of recommendation he was bracketed with Morin as superior to the other four; at the ballot Morin was elected, but Saint-Venant gained only the fourth place.

[1611.] Another memoir on the subject of the resistance of materials will be found in the same volume of the *Comptes rendus* (XVII.), p. 1275. It is entitled: *Mémoire sur le calcul de la résistance d'un pont en charpente, et sur la détermination, ou moyen de l'analyse des efforts supportés dans les constructions existantes, des grandeurs des nombres constants qui entrent dans les formules de résistance des matériaux; par MM. de Saint-Venant et P. Michelot.*

This memoir which is entirely of a practical character, relates to a bridge over the Creuse. We need not stay to consider it here, merely remarking that the graphical would now-a-days be preferred to the analytical calculus in questions of this kind.

1612. The next publication by Saint-Venant on our subject is thus entered in the Royal Society Catalogue of Scientific Papers: *Sur la définition de la pression dans les corps fluides ou solides en repos ou en mouvement. Soc. Philom. Proc. Verb.* Paris, 1843, pp. 134—138. I have not seen this volume, but the note appears to be reprinted in full in *L'Institut*, No. 524, January 10, 1844, where it occupies about two columns and a half.

The definition of stress across an elementary plane given by various authors for twenty years is stated to be this: *La résultante des actions exercées sur les molécules d'un cylindre indéfini, élevé sur cette face comme base, par toutes les molécules situées du côté opposé de la face et de son prolongement.* The inconvenience of this definition is pointed out in a manner of which a brief indication is given in Moigno's *Statique*, p. 619; and then the note furnishes the following historical information:

Je pense donc qu'il faut renoncer à la définition des pressions rapportée plus haut. J'ai proposé, en 1834, dans un mémoire, et ensuite, en 1837, dans un cours lithographié [see our Art. 1563], d'en adopter une autre, analogue à celle qui a été donnée du *flux de chaleur* à travers une petite face, par Fourier (ch. I., 96), et par Poisson (Mémoire de 1815, publié en 1821, *Journal de l'École polytechnique*, article 56). Cette définition consiste à appeler pression, sur une petite face plane quelconque, imaginée à l'intérieur d'un corps, ou à la limite de séparation de deux corps, *la résultante de toutes les actions attractives ou répulsives qu'exercent les molécules situées d'un côté de cette face sur les molécules situées de l'autre côté, et dont les directions traversent cette face.*

Déjà M. Duhamel avait reconnu la possibilité de définir ainsi la pression, car, dans un mémoire présenté en 1828, il la calculait, dans les corps solides élastiques, absolument comme il a calculé le flux de chaleur dans un autre mémoire daté de la même année et inséré au *Journal de l'École polytechnique* (21e cahier, p. 213): mais il n'y attachait qu'une faible importance, comme on peut le voir à un autre mémoire (t. v. des Mémoires des savants étrangers) où, pour en faciliter la lecture, il revient à la définition la plus connue de la pression.

In the 19th Cahier of the *Journal de l'École Polytechnique*, Art. 11, Poisson had shewn that the two definitions would lead to the same result with respect to the flow of heat, and Saint-Venant says that they will also with respect to pressure if certain quantities are neglected. He adds:

Mais rien ne dit que l'approximation dont on s'est contenté jusqu'à présent suffise dans des questions à examiner ultérieurement: il me semble même que déjà la difficulté des arêtes vives qui s'est présentée à M. Poisson (20e cahier du *Journal de l'École polytechnique*, nos. 25, 49, 50, 53 du mémoire du 12 octobre 1829) tient en partie à la définition de la pression sur le cylindre.

1613. The next paper by Saint-Venant is thus entered in the Royal Society Catalogue of Scientific Papers: *Sur la question, "Si la matière est continue ou discontinue."* *Soc. Philom. Proc. Verb.* Paris, 1844, pp. 3—16. This I have not seen; there is an allusion to it on p. 694 of Moigno's *Statique.* There is a brief abstract of the memoir in *L'Institut,* No. 528, February 7, 1844. The following statement occurs:

L'auteur en conclut qu'il n'existe pas dans la nature, de masse continue, grande ou petite, et qu'il convient de ne regarder les dernières particules des corps que comme des points sans étendue, *non contigus,* centres d'action de forces attractives et répulsives. C'est le système de Boscovich (*Theoria philosophiae naturalis reducta ad unicam legem virium in natura existentium,* 1763).

A note says:

Ce mémoire a été imprimé à part et se trouve chez Carilian-Gœury et Victor Dalmont, libraires, quai des Augustins, 39 et 41.

1614. The next paper by Saint-Venant is thus entered in the Royal Society Catalogue of Scientific Papers: *Sur les pressions qui se développent à l'intérieur des corps solides, lorsque les déplacements de leurs points, sans altérer l'élasticité, ne peuvent cependant pas être considérés comme très-petits.* *Soc. Philom. Proc. Verb.* Paris, 1844, pp. 26—28. This I have not seen; there is a brief abstract of the memoir in *L'Institut,* No. 537, April 10, 1844; here we have the two formulae given in Moigno's *Statique,* p. 670 [and reproduced in our Art. 1618. A full account of the memoir will be found in the *Notice sur les travaux...de M. de Saint-Venant,* Paris, 1858, referred to in our Art. 1562. The important point is the statement of the formulae cited in our Art. 1618].

1615. *Note sur les flexions considérables des verges élastiques;* this occupies pp. 275—284 of Liouville's *Journal de mathématiques,* Vol. IX., 1844.

The theory of elasticity is very slightly involved in this note; an equation is assumed for the equilibrium of a *primitively* curved elastic rod coincident with that, given by the Bernoulli-Eulerian theory, such as we find in Poisson's *Mécanique*, Art. 315. This leads to the differential equation

$$\frac{\frac{d^2y}{dx^2}}{\left\{1+\left(\frac{dy}{dx}\right)^2\right\}^{\frac{3}{2}}} = ax + by + c.$$

The note really consists of the integration of this differential equation, which is accomplished by the aid of Elliptic Functions.

1616. *Note sur la pression dans l'intérieur des corps ou à leurs surfaces de séparation. Comptes rendus*, Vol. XXI., pp. 24—26. The object of the note is thus explained at the beginning of it:

M. Cauchy, dans des Notes relatives à la Mécanique rationnelle, insérées au *Compte rendu* du 23 juillet, page 1765, veut bien citer, comme plus exacte que la définition la plus connue de la pression, celle que j'en ai donnée en 1834 et en 1837, et qui consiste à regarder la pression, sur un élément très-petit, comme *la résultante des actions de toutes les molécules situées d'un côté sur toutes les molécules situées de l'autre côté, et dont les directions traversent cet élément.* Déjà M. Duhamel, dans un Mémoire présenté en 1828, avait reconnu la possibilité de substituer une pareille définition, analogue à celle que Fourier donne du *flux de chaleur*, à celle par laquelle on considère la pression comme l'action totale des molécules contenues dans un cylindre droit indéfini ayant l'élément pour base, sur toutes les molécules situées de l'autre côté du plan de l'élément. J'ai prouvé, surtout dans une Note du 30 décembre 1843, insérée au no. 524 du journal *l'Institut*, que la définition nouvelle n'avait pas les inconvénients de l'autre ; car, en l'adoptant, les pressions sur la surface de séparation de deux portions de corps peuvent toujours être substituées identiquement à l'action totale des molécules de l'une de ces portions sur les molécules de l'autre, tandis qu'avec la définition par le cylindre indéfini, cette substitution entraîne ordinairement l'omission d'un certain nombre d'actions moléculaires, l'emploi multiple de certaines autres, et l'introduction d'actions étrangères.

Je me propose, dans cette Note, de montrer que la nouvelle définition de la pression se prête très-facilement à l'établissement de la formule fondamentale, dont on tire ensuite, par le calcul, toutes celles de la mécanique moléculaire.

A note says that the demonstration was communicated verbally to the *Société philomatique* on March 26, 1844. The demonstration is the same as that in Moigno's *Statique*, pp. 674 and 675.

See Moigno's *Statique*, pp. 619, 675. Saint-Venant on *Torsion*, pp. 249, 262.

1617. Three communications by Saint-Venant occur in the *Comptes rendus*, Vol. XXIV., 1847.

I. *Mémoire sur l'équilibre des corps solides, dans les limites de leur élasticité, et sur les conditions de leur résistance, quand les déplacements éprouvés par leurs points ne sont pas très-petits*, pp. 260—263.

II. *Mémoire sur la torsion des prismes et sur la forme affectée par leurs sections transversales primitivement planes*, pp. 485—488.

III. *Suite au Mémoire sur la torsion des prismes*, pp. 847—849.

In I. we have the two formulae to which I refer in Art. 1614.

[1618.] These formulae deserve notice. Saint-Venant had treated large *elastic* shifts in his memoir on rods of double curvature: see Art. 1598. He here remarks that the complete values for the stretches and slides, whatever be the magnitude of the shifts, $u$, $v$, $w$ will be given by formulae of the type

$$s_x = \frac{du}{dx} + \tfrac{1}{2}\left[\left(\frac{du}{dx}\right)^2 + \left(\frac{dv}{dx}\right)^2 + \left(\frac{dw}{dx}\right)^2\right],$$

$$\sigma_{yz} = \frac{dv}{dz} + \frac{dw}{dy} + \left[\frac{du}{dy}\frac{du}{dz} + \frac{dv}{dy}\frac{dv}{dz} + \frac{dw}{dy}\frac{dw}{dz}\right]$$

Further, the stretch $s_r$ in direction $\alpha$, $\beta$, $\gamma$ will still be given by a formula of the form

$$s_r = s_x \cos^2\alpha + s_y \cos^2\beta + s_z \cos^2\gamma + \sigma_{yz}\cos\beta\cos\gamma + \sigma_{zx}\cos\gamma\cos\alpha + \sigma_{xy}\cos\alpha\cos\beta \ldots\ldots (\text{i}).$$

Le problème de la recherche de déplacements de grandeur quelconque des points de corps élastiques sollicités par des forces données, est donc posé en équation[1].

[1] The equations have been given in a general form by Sir W. Thomson. See the *Treatise on Natural Philosophy*, Part II. p. 463. So far as we know the only special application of these formulae is that to plane waves in an isotropic medium communicated by the Editor to the *Cambridge Phil. Soc. Proceedings*, Vol. V., Part IV. p. 296.

[1619.] Let us investigate this point and note upon what assumptions Saint-Venant obtains the above formulae. For brevity let the shift-fluxions be denoted by subscripts. Let the distance $r$ between two adjacent particles $A$ and $B$ become $r'$; since $r$ is at our choice we take it so small that the cubes of its projections upon the axes may be neglected as compared with the squares. Then in a manner similar to that of Art. 612 we find:

$$r'^2/r^2 = \{l(1+u_x) + mu_y + nu_z\}^2 + \{lv_x + m(1+v_y) + nv_z\}^2 + \{lw_x + mw_y + n(1+w_z)\}^2$$

$$= 1 + l^2(2u_x + u_x^2 + v_x^2 + w_x^2) + m^2(2v_y + u_y^2 + v_y^2 + w_y^2) + n^2(2w_z + u_z^2 + v_z^2 + w_z^2) + 2nm(w_y + v_z + u_yu_z + v_yv_z + w_yw_z) + 2ln(u_z + w_x + u_zu_x + v_zv_x + w_zw_x) + 2ml(v_x + u_y + u_xu_y + v_xv_y + w_xw_y),$$

where for brevity we have put $l, m, n$ for $\cos\alpha, \cos\beta, \cos\gamma$. We have thus a relation of the following kind:

$$r'^2/r^2 = 1 + 2\{l^2\epsilon_x + m^2\epsilon_y + n^2\epsilon_z + nm\eta_{yz} + ln\eta_{zx} + ml\eta_{xy}\} \text{.........(ii).}$$

The strained value of $r$ is accordingly fully determined *whatever be the ratio of* $r'/r$ by the six quantities[1]

$$\epsilon_x, \epsilon_y, \epsilon_z, \eta_{yz}, \eta_{zx}, \eta_{xy}.$$

These quantities thus fully characterise the strain at a point and the work must be represented by some function of them. We can speak of them as the six components of strain, when the shift-fluxions and the strain are not necessarily very small quantities. Are these components, however, those which correspond physically to the three stretches and the three slides?

Now $r' = r(1+s_r)$, hence we find

$$s_r + s_r^2/2 = l^2\epsilon_x + m^2\epsilon_y + n^2\epsilon_z + nm\eta_{yz} + ln\eta_{zx} + ml\eta_{xy}.$$

Put $m = n = 0$, $l = 1$, and we find

$$s_x + s_x^2/2 = \epsilon_x.$$

It follows that $\epsilon_x$ is *not the stretch* in direction of the axis of $x$, *unless we may neglect* $s_x^2$. This is practically what Saint-Venant does; he is really treating cases in which any or all of the shift-fluxions may be great, but the strain at any point is small.

[1] Their values were first obtained by Green in 1839 (*Math. Papers* p. 297), but he nowhere clearly insists on their physical meaning and practical importance.

Les formules de la mécanique dite *moléculaire* ont été basées, jusqu'à présent, sur la supposition que les déplacements des divers points des corps élastiques auxquels on les applique sont extrêmement petits, de manière que la ligne de jonction de deux points quelconques ne change jamais que très-peu, non-seulement de longueur, mais encore de direction dans l'espace.

Or il s'en faut bien que cette condition soit remplie en général : une lame mince peut être ployée de manière que ses deux bouts arrivent à se toucher, et un cylindre d'un faible diamètre peut être tordu de plusieurs circonférences sans que l'élasticité, ni de cette lame, ni de ce cylindre, ait subi d'altération.

Il convient donc d'avoir des formules qui s'appliquent à des grandeurs absolument quelconques des déplacements, avec cette seule restriction, *que les distances mutuelles de points très-rapprochés ne varient que dans une petite proportion, afin que la cohésion et l'élasticité naturelle subsistent.* (p. 260.)

By displacements in this quotation as in Art. 1568 Saint-Venant denotes *relative* shifts, i.e. the shift-fluxions.

[1620.] Saint-Venant's argument may be thus applied to our equation (ii): $r'/r$ is very nearly unity, hence the quantity in brackets on the right-hand side of that equation must be very small. Taking the root of the right-hand side and neglecting the square of this quantity we have:

$$r'/r = 1 + l^2\epsilon_x + m^2\epsilon_y + n^2\epsilon_z + nm\eta_{yz} + ln\eta_{zx} + ml\eta_{xy},$$

or

$$s_r = l^2\epsilon_x + m^2\epsilon_y + n^2\epsilon_z + nm\eta_{yz} + ln\eta_{zx} + ml\eta_{xy}.$$

This is Saint-Venant's equation marked (i) above. We see therefore that when *the shift-fluxions are large, but the strain small,* the quantities $\epsilon_x, \epsilon_y, \epsilon_z, \eta_{yz}, \eta_{zx}, \eta_{xy}$ which are always strain-components represent the physical stretches and slides. These quantities are therefore the proper expressions for the stretches and slides for large shift-fluxions within the elastic limit, for in this case $r'/r$ is always small.

[1621.] We can however find physical meanings for the quantities $\epsilon_x \ldots \eta_{yz}, \ldots$ when the strain is not small. Thus $2\epsilon_x$ represents the rate of change in the *square of a small line* initially drawn parallel to the axis of $x$. Further it may be shewn that if $\delta x'$, $\delta y'$, $\delta z'$ denote the magnitude and position of the projections $\delta x$, $\delta y$, $\delta z$ of $r$ after strain, then $\eta_{zx}$ is the ratio of the area of the rectangle under $\delta z'$ and the projection upon $\delta z'$ of $\delta x'$ to the area of the rectangle under $\delta z$ and $\delta x$ the unstrained values of $\delta z'$, $\delta x'$. If the strain at any point be such that its square may be neglected this becomes the slide, or, $\eta_{zx} = \sigma_{zx}$.

Generally, $\eta_{zx}/\sqrt{(1+2\epsilon_x)(1+2\epsilon_z)} = \cos(z'x')$;
where (Art. 1564) $\cot(z'x') = \sigma_{zx}$.

[1622.] Further we may ask: If the strain be not small, but the shift-fluxions such that we need only retain their squares, do $\epsilon_x \ldots \eta_{yz} \ldots$ represent the stretches and slides?

We have in this case:

$$s_x = \sqrt{1+2\epsilon_x} - 1 = u_x + \tfrac{1}{2}(v^2_x + w^2_x)$$
$$= \epsilon_x - u^2_x.$$

Similarly,
$$\sigma_{zx} = \eta_{zx}/\sqrt{1 + 2(\epsilon_x + \epsilon_z) + 4\epsilon_x\epsilon_z - \eta^2_{zx}}$$
$$= \eta_{zx} - w_x(u_x + w_z) - u_z(w_z + u_x).$$

Hence $s_x$ and $\sigma_{zx}$ are not identical with $\epsilon_x$ and $\eta_{zx}$ to the second power of the shift-fluxions, except under very special conditions of strain (e.g. $u_x = u_y = u_z = 0$, or a case of pure sliding strain, etc.).

We conclude therefore that Saint-Venant's expressions, if treated as the physical stretches and slides, apply only for large shift-fluxions and small strain (e.g. elastic strain), but that treated as components of strain not identical with the stretches and slides, they may be used for a strain of any magnitude.

Thus we cannot apply the ordinary body-shift elastic equations, if we have merely shewn that the usual expressions ($u_x$, $w_y + v_z$, etc.) for the stretches and slides are small. It is necessary that *all* the shift-fluxions should be small. This is, perhaps, best expressed by saying that the *twists* must also be small. For example, these equations do not apply to small strain accompanied by large rotations as in the cases referred to by Saint-Venant: see Art. 1619.

1623. Then Saint-Venant proceeds to indicate the improvement which he had effected in the ancient theory of torsion, and this subject is continued in II. and III. All that these communications contain with respect to torsion is comprised in the extensive memoir by Saint-Venant on Torsion which was read to the Academy in June, 1853. The language underwent a slight change however; for at this date Saint-Venant separated what he calls the *gauchissement* into two elements, whereas in the later memoir, with the exception of a slight allusion on p. 362, he makes no such separation; but there is no change as to facts.

[1624.] These memoirs are epoch-making in the history of the theory of elasticity; they mark the transition of Saint-Venant from Cauchy's theory of torsion to that which we must call after Saint-Venant himself. The later great memoir on Torsion is really only an expansion of these three papers, and it may be well here to note what are the important new facts they contain.

[1625.] I., so far as it treats of torsion, contains only the statement based upon general physical principles, that the plane sections of a prism, if not circular, are distorted by torsion. This is shewn to accord with the experiments of Duleau on the torsion of bars of square and circular section of the same moment of inertia and of the same material: see our Art. 226.

Further, Saint-Venant quotes experiments of his own (p. 263):

J'ai vérifié expérimentalement cette théorie d'une autre manière; j'ai soumis à la torsion deux prismes de caoutchouc, de 20 centimètres de longueur, l'un à base carrée, de 3 centimètres de côté, l'autre à base rectangle, de 4 centimètres sur 2. Les lignes droites, tracées transversalement sur leurs faces latérales avant la torsion, seraient restées droites et perpendiculaires à l'axe s'il n'y avait eu aucun gauchissement des sections; elles n'auraient fait que s'incliner sur l'axe du prisme à base rectangle, en restant droites, s'il n'y avait eu que le premier gauchissement. Au lieu de cela, ces lignes, par la torsion, se sont courbées en doucine ou en *S*, de manière que les extrémités restaient normales aux arêtes, ce qui prouve bien le deuxième gauchissement dont on vient de parler.

[1626.] In II. we have the solution of the problem of torsion for a prism on rectangular base. This solution is identical with that afterwards given in the memoir on *Torsion*, p. 370, except that the coefficient in the case of the square is not calculated to the same degree of approximation (·841 instead of ·843462, *Torsion*, p. 382). Saint-Venant states that Wantzel had suggested the form of solution required by the differential equations.

It is also shewn that Cauchy's result for a prism of rectangular section is only true when the ratio of the sides of the rectangle is a very small quantity. Further, experiments of Duleau and Savart are quoted as confirming the new theory: see our Arts. 226 and 333.

[1627·] In III. the case of a prism of elliptic section is worked out. Saint-Venant points out that we require a solution of the differential equation,

$$\frac{d^2u}{dy^2}+\frac{d^2u}{dz^2}=0,$$

subject to the contour-condition

$$\left(\frac{du}{dy}+\tau z\right)dz-\left(\frac{du}{dz}-\tau y\right)dy=0.$$

This is precisely the form which the statement of Saint-Venant's problem has taken in later memoirs and works.

He gives the solution of an elliptic section of semi-axes $a$, $b$,

$$u=\tau\frac{a^2-b^2}{a^2+b^2}yz,$$

and remarks that the primitively plane sections become hyperbolic paraboloids, and that Cauchy's approximate solution for a rectangle becomes exact in the case of an elliptic section.

See Moigno's *Statique*, p. 670. Saint-Venant on *Torsion*, pp. 235, 248, 362, 363.

1628. *Mémoire sur les vibrations tournantes des verges élastiques* (Extrait par l'auteur). *Comptes rendus*, Vol. 28, 1849, pp. 69—72. Let $n$ denote the number of longitudinal vibrations, and $n'$ the number of torsional vibrations, corresponding to the lowest note for a rod built-in (*encastrée*) at one end. Then, as we have seen in Art. 470, Poisson obtained $\frac{n}{n'}=\sqrt{\frac{5}{2}}$. Wertheim, in the *Comptes rendus*, Vol. 27, 1848, p. 650, gave as the result of some investigations of his own $\frac{n}{n'}=\sqrt{\frac{8}{3}}=1\cdot6330$; he says that Savart by experiment obtained 1·6668, and that he had himself obtained by experiment 1·6309. In the present communication Saint-Venant adverts to the same point; he combines some experimental results with his own theory, and concludes that for rectangular prisms the evidence is in favour of Poisson's number, and for cylindrical rods in favour of Wertheim's. On pp. 126—128 of the volume Wertheim replies to the remarks of Saint-Venant: see our Arts. 333, 398, 1339—1343.

Saint-Venant seems to imply that Poisson's result had been obtained by other writers, but he does not give references[1].

1629. Saint-Venant gives only statements without demonstration; I extract a few passages:

Lorsqu'un prisme élastique est maintenu dans un état de torsion par deux *couples* agissant à ses extrémités, les rotations angulaires $\psi$ de ses diverses sections transversales sont données par l'équation

$$\mu\nu_1 \frac{d\psi}{dx} = M_x \quad \text{.........................(1),}$$

qui exprime l'égalité du moment $M_x$ de l'un des deux couples autour de l'axe du prisme (pris pour celui de $x$) avec le moment des réactions intérieures qui s'exercent à travers l'une quelconque de ces sections $\omega$; $\mu$ représentant le coefficient d'élasticité dit de *glissement* transversal;

$\nu_1$ une quantité qui est moindre que le moment d'inertie de la section autour de son centre de gravité, en raison de ce que cette section, primitivement plane, devient légèrement courbe......

L'équation (1) peut encore être posée pour une longue tige sollicitée par des forces agissant sur tous les points de sa masse et ne produisant qu'une faible torsion, $M_x$ représentant alors la somme des moments de celles appliquées depuis $\omega$ jusqu'à une extrémité.

On passe au cas du mouvement en remplaçant ces forces par les inerties

$$-\rho d\omega dx \frac{d^2(r\psi)}{dt^2},$$

$\rho$ étant la densité, $r$ le rayon vecteur d'un élément, et $t$ le temps.

Il en résulte, en différentiant les deux membres par rapport à $x$, celle

$$\frac{\mu}{\rho} \frac{\nu_1}{\omega k^2} \frac{d^2\psi}{dx^2} = \frac{d^2\psi}{dt^2} \quad \text{.......................(5).}$$

Elle donne, pour le nombre de vibrations tournantes du son le plus grave, en une seconde, $l$ étant la longueur de la tige,

$$n' = \frac{1}{2l}\sqrt{\frac{\mu}{\rho}\frac{\nu_1}{\omega k^2}}.$$

......

Le nombre des vibrations longitudinales de la même tige est exprimé, comme on sait, par

$$n = \frac{1}{2l}\sqrt{\frac{E}{\rho}}.$$

......

[1] See however our Arts. 470 and 661.

Ces formules sont propres à fournir des valeurs du rapport, en ce moment controversé, des coefficients d'élasticité *d'allongement* et de torsion $E$ et $\mu$. On sait que, d'après les formules de Navier, Poisson et MM. Cauchy, Lamé et Clapeyron, on aurait, dans les corps homogènes et isotropes, ou d'égale élasticité en tous sens,

$$E/\mu = 5/2 = 2{\cdot}5.$$

Mais Wertheim, en adaptant des résultats d'expériences d'équilibre à un commencement d'analyse de M. Cauchy, dont il rejette le complément, modifie profondément toutes ces formules, et prend

$$E/\mu = 8/3 = 2{\cdot}6666...$$

$\omega k^2$ represents moment of inertia of the section about the axis of the prism.

[1630.] Thus, according to

Poisson, Weber and Cauchy $\frac{n}{n'} = \sqrt{\frac{5}{2}}$,

to Wertheim $\frac{n}{n'} = \sqrt{\frac{8}{3}}$,

to Saint-Venant

$$\frac{n}{n'} = \sqrt{\frac{E}{\mu} \cdot \frac{\omega k^2}{\nu}} = \sqrt{\frac{5}{2} \frac{\omega k^2}{\nu}},$$

on the uni-constant hypothesis.

For a circular section $\nu = \omega k^2$; for a square section $\nu/(\omega k^2) = {\cdot}843462$, and other values of this ratio have been calculated by Saint-Venant in his memoir on *Torsion*. The present memoir is thus an essential addition to those enumerated in Art. 1617. It is practically reproduced with additions in the *Appendice* IV. to Saint-Venant's edition of Navier's *Leçons*, pp. 631—645. He there shews that his theory is in accordance with experimental result. See also our remarks on Wertheim's memoirs in the preceding chapter, Arts. 1339—1343.

[1631.] *Summary*. The important contributions of Saint-Venant to the theory of elasticity even before the date of the two classical memoirs on Torsion and Flexure, by which he is best known, will be rendered manifest by a perusal of this chapter. We find in this elastician a keen appreciation of practical needs, combined with a wide theoretical grasp, of which it would be hard to discover another example in the history of our subject. Had Saint-Venant done nothing more than correct the theory of flexure by the consideration of slide, the theory of elastic rods of double curvature by the introduction of the third moment, and the theory of torsion by the discovery of the distortion of the primitively plane section,—corrections all embraced in the researches considered in this chapter—he would deservedly have ranked among the foremost scientists who have contributed to our subject. We shall see in the course of our second volume, that these form but a small part of his services to elasticity.

With Saint-Venant and 1850 we bring our first volume to a close. We have traced the growth of the theory of elasticity from its childhood with Galilei to its maturity in Saint-Venant. We have seen it become an important and all-powerful instrument in the hands alike of the most practical of engineers and of the most theoretical of mathematicians. There is scarcely a branch of physical investigation, from the planning of a gigantic bridge to the most delicate fringes of colour exhibited by a crystal, wherein it does not play its part. The manifoldness of its applications increases with every advance of our mechanical and physical knowledge. To trace these applications from 1850 to the present day will be the task of our second volume. But when we consider the immense number not only of memoirs but of treatises which these years have produced, both reader and editor may perhaps be pardoned for a certain feeling of dismay mixing with their pleasure at the rapid progress of the science of elasticity.

# APPENDIX.

## NOTE A. ADDENDA.

(1) *The Stress-Strain Curve.* Arts. 18 and 979.

1692. James Bernoulli. *Curvatura Laminae Elasticae. Acta Eruditorum*, Leipzig, 1694, pp. 262—276.

On p. 265 Bernoulli writes:

> Esto Spatium rectilineum sive curvilineum quodvis *ABC*, cujus abscissae *AE* vires tendentes, ordinatae *EF* tensiones repraesentent etc.

This curve he terms the *curva tensionum* or *linea tensionum*. Bernoulli might thus be considered to have introduced a graphical method of representing the stress-strain relation. At the same time it will be seen by consulting the original memoir that Bernoulli's *linea tensionum* does not represent the curve obtained by measuring the strains produced in the same rod by a continually increasing stress. This seems to me to have been first done by Poncelet.

(2) *The Coulomb-Gerstner Law.* Arts. 119 and 806, footnote.

1784. Coulomb. *Recherches théoriques et expérimentales sur la force de torsion et sur l'élasticité des fils de métal. Histoire de l'Académie des Sciences, année* 1784, Paris, 1787, pp. 229—269. This memoir is reprinted Tome I. pp. 63—103 of the *Collection de Mémoires relatifs à la Physique publiés par la Société Française de Physique*, Paris, 1884. Dr Todhunter has referred to it in Art. 119. We may note the second section (pp. 255—269) of the memoir in this place. It occupies pp. 90—103 of the reprint and is entitled: *De l'altération de la force élastique dans les torsions des fils de métal. Théorie de la cohérence et de l'élasticité.* In this section Coulomb brings out clearly (i) that the absolute strength of a material depends upon the working or treatment it has received (*la force des métaux varie suivant le degré d'écrouissement et de recuit*);

(ii) that in the case of set, the set-slide produced by torsion is at first for very small sets proportional to the total slide and thus to the elastic slide, but that it very soon begins to increase in a much greater ratio; (iii) that the slide-modulus (Coulomb speaks of the *réaction de torsion*) remains almost the same after any slide-sets; (iv) that the elastic limit (at least for the case of torsion) can be extended by giving the material a set, thus by subjecting a wire to great torsional set a state of ease can be produced almost as extended as if the wire had been annealed.

Coulomb also notes that the resistance of the air had very little effect in diminishing the amplitude of the oscillations of his apparatus, and he apparently attributes the decrease in amplitude to something akin to 'fatigue of elasticity'.

Coulomb distinguishes elasticity and cohesion as absolutely different properties. Thus the cohesion can be much altered by working or other treatment. It would appear that the elastic limit is altered when the cohesion or absolute strength is altered[1]; scarcely however, as Coulomb apparently suggests, in the *same* ratio (§ XXXI.). The elastic constants however remain the same: see our Art. 806. This independence of the elasticity of the cohesion was confirmed by flexure experiments for the stretch-modulus (§ XXXIII.).

It may be noted that Coulomb uses the terms *glissement* and *glisser* for set-slide and not for elastic slide. He appears to hold that the difference in the cohesion of the same material in different states depends upon its capacity for receiving set-slide. If its parts cannot slide on each other, it is brittle; if they can, it is ductile or malleable.

It will be noted that the several suggestive points of this memoir remained for many years unregarded, till they were again rediscovered by Gerstner and Hodgkinson.

(3) *Resilience.* Arts. 136, 993 and 999.

(*a*) J. A. Borellus. *Liber de vi percussionis*, Bologna, 1667. Another edition of this work entitled: *De vi percussionis et motionibus naturalibus a gravitate pendentibus* (*Editio prima Belgica, priori Italicâ multo correctior et auctior*, etc.), appeared 1686, *Lugduni Batavorum.* There are two chapters in this work (Caput XVIII. and Caput XIX.) entitled:

[1] The ratio $\frac{\text{elastic limit}}{\text{cohesive limit}}$ for most materials in their *unworked* state diminishes as the cohesive limit increases, but with many materials in a *worked* condition (e.g. tempered steel) this is not true.

*Quomodò in flexibilibus corporibus impetus impressus retardetur aut extinguatur* (pp. 106—112 of the later edition).

*Qua ratione in corporibus flexibilibus resilientibus motus contrarii se mutuò destruant, renoventurque* (pp. 112—116 of the later edition).

These chapters at least by their titles suggest a consideration of the problem of resilience, and the examination of the figures on Plate III., opposite p. 106, suggests still more strongly that something of value might be found in them. Beyond giving, however, the name *resilience*, probably for the first time, to a number of problems now classed under that name, the work really contributes nothing to our subject, being composed of a number of general and extremely vague propositions.

(*b*) 1807. Young. *A Course of Lectures on Natural Philosophy and the Mechanical Arts.*

Young was, I believe, the first to introduce into English the term *resilience*, and to state the general theorem that: *The resilience of a prismatic beam resisting a transverse impulse is simply proportional to the bulk or weight of the beam.* The statement of this general proposition occurs on p. 147, Vol. I. of his *Lectures on Natural Philosophy.* On p. 50 of Vol. II. he returns to the matter with the following definition and theorem:

> The resilience of a beam may be considered as proportional to the height from which a given body must fall to break it.
>
> The resilience of prismatic beams is simply as their bulk.

This theorem he proves in the following characteristic fashion:

> The space through which the force or stiffness of a beam acts, in generating or destroying motion, is determined by the curvature that it will bear without breaking; and this curvature is inversely as the depth; consequently, the depression will be as the square of the length directly, and as the depth inversely: but the force in similar parts of the spaces to be described is everywhere as the strength, or as the square of the depth directly, and as the length inversely: therefore the joint ratio of the spaces and the forces is the ratio of the products of the length by the depth; but this ratio is that of the squares of the velocities generated or destroyed, or of the heights from which a body must fall to acquire these velocities. And if the breadth vary, the force will obviously vary in the same ratio; therefore the resilience will be in the joint ratio of the length, breadth and depth.

It will be observed that Young is here speaking of *cohesive* resilience, which must be distinguished from *elastic* resilience. The latter has of course greater practical importance. Compare Note E, (*b*).

(4) Fourier. Art. 207.

An account of a memoir by Fourier upon the vibrations of flexible and extensible surfaces, and of elastic plates, will be found on pp. 258—264 of the *Histoire de l'Académie des Sciences* in the *Analyse des Travaux...pendant l'Année* 1822. The *Analyse* will be found attached to Tome v. of the *Mémoires* (1821 and 1822) published in 1826. Delambre seems to have analysed this memoir, which never appears to have been published as a whole. It apparently contained solutions of the linear partial differential equations satisfied by the above forms of vibrations of the kind given for similar equations in Fourier's *Théorie de la Chaleur*, i.e. solutions in the form of periodic series and of definite integrals.

(5) Addendum to Chapter III.

H. F. Eisenbach. *Versuch einer neuen Theorie der Kohäsionskraft und der damit zusammenhängenden Erscheinungen*, Tübingen, 1827.

I have added a reference in the Addenda to this book as its title might lead a reader to believe something of value was contained in Eisenbach's theory. This would undoubtedly be the case were we to accept the author's own estimate of his discovery, the history of which he narrates in some 18 pages. The 'new theory of cohesion' consists in the hypothesis that the law of cohesion is based on a central intermolecular force which can be expanded in inverse powers of the square of the molecular distance. On this hypothesis Eisenbach attempts, with the crudest mixture of mathematical and physical absurdities, worthy of the Père Mazière, to explain cohesive, elastic and chemical phenomena. He speaks of the memoirs of Euler, Lagrange and Laplace as *vortreffliche Vorarbeiten* for his own great principle, although he criticises somewhat severely Euler's memoir of 1778: see our Art. 74. Those who are interested in the history of pseudo-science, in the paradoxes and self-admiration of circle-squarers, perpetual-motion seekers, gold-resolvers, and the innumerable 'Grübler die so lange über einem Trugschlusse brüten, bis er zur fixen Idee wird und als Wahrheit erscheint,' will find much amusement and instruction in Eisenbach's sections: *Eine Widerlegung aller bisherigen Kohäsionstheorien* and *Kurze Geschichte des Ganges meiner Erfindung.* Englishmen should remark that it was on board the 'Emilie' in the London Docks that the possibility of this great principle 'stepped like a flash of lightning before the soul' of this second, but sadly disregarded Newton.

(6) *Slide—Glissement.* Arts. 120, 143, 279 and 726.

I have shewn in Art. 120 that Coulomb had formed the important conceptions of 'lateral adhesion' and of 'sliding strain.' In the paragraph there quoted Coulomb talks of a stress tending *à faire couler la partie supérieure du pilier sur le plan incliné par lequel il touche la partie inférieure.* In Coulomb's memoir of 1773 (see our Art. 115) we have a section, pp. 348—349, on *Cohésion.* Here Coulomb describes an experiment on shearing force :

J'ai voulu voir si en rompant un solide de pierre, par une force dirigée suivant le plan de rupture, il fallait employer le même poids que pour le rompre, comme dans l'expérience précédente, par un effort perpendiculaire à ce plan.

He found that the shearing load must be slightly greater than the tractive load, but the difference was so little ($\frac{1}{44}$ of the total load) that he neglects it in the theory which follows. On the uni-constant hypothesis the limit of shearing load should be $\frac{4}{5}$ of the limit of tractive load. A little later on in the memoir (p. 353) Coulomb uses the expression *tendre à glisser;* for exactly the same conception as in the former paragraph he used *tendre à faire couler.* It will thus be seen that although Coulomb's theory is unsatisfactory he had still formed a clear conception of slide and shear.

In Art. 143 we have noted that Young in 1807 drew attention to the phenomena of 'lateral adhesion,' or as he terms it *detrusion.* It was however Vicat who first insisted on the mechanical importance of this form of strain and the accompanying stress. In his memoir of 1831 he defines shear (*force transverse*) in terms of the strain which it tends to produce. He uses the word *glisser:* see our Art. 726.

Vicat's remarks did not escape Navier, who in the second edition of his *Leçons* (see our Art. 279), which appeared in 1833, after describing the nature of shear thus defines the slide-modulus :

Un coefficient spécifique représentant la résistance du corps à un glissement d'une partie sur l'autre dans le plan de la section transversale (§ 152).

The merit of practically introducing slide and shear into the ordinary theory of beams rests, as we have seen in Arts. 1564—1582, with Saint-Venant.

(7) *Notation for the six stresses.* Art. 610, footnote.

In the table of notations I have attributed to Kirchhoff the notation:

| | $x$ | $y$ | $z$ |
|---|---|---|---|
| $x$ | $X_x$ | $X_y$ | $X_z$ |
| $y$ | $Y_x$ | $Y_y$ | $Y_z$ |
| $z$ | $Z_x$ | $Z_y$ | $Z$ |

I have since noticed that this convenient notation had been previously used by F. E. Neumann in his memoir of 1841. This is the important memoir on photo-elasticity discussed in our Chapter VIII.: see Art. 1185.

(8) *Wire-Drawing.* Art. 748.

On re-examining Karmarsch's paper, I find that, while he supposed the drawing of different metal wires through the same hole would ensure their having the same diameter, he still recognised that this diameter was not that of the hole (pp. 323 and 324). He thus really puts in evidence the existence of elastic after-strain in metals: see our Arts. 1402 and 1431. The whole subject is of primary physical importance as throwing light on the effect of 'working' (initial stress) on the elastic and cohesive properties of metals. It is worthy of an accurate experimental investigation in which due account would be taken of the time-effect, not only due to the rate of drawing, but also to the interval between successive drawings through the same or different holes. Such an investigation would give valuable data on the relation of initial stress to cohesion.

(9) *Becker's Formula.* Arts. 1291 and 1376.

According to Stokes the ratio of the dynamical deflection to the central statical deflection is given by

$$D/S = 1 + \frac{4V^2S}{gc^2},$$

nearly, if $\beta$ be $> 100$, and the mass of the bridge small as compared with that of the moving load. On the other hand if the mass of the bridge be great as compared with that of the moving load

$$D/S = 1 + \frac{25}{8}\sqrt{\frac{31}{252}}\frac{V\sqrt{S_1}}{\sqrt{gc^2}}.$$

See our Arts. 1278, 1287 and 1290.

Let us combine these and endeavour to form an empirical formula. We have

$$D/S = 1 + \frac{4S}{gc^2}V^2 + \frac{25}{8}\sqrt{\frac{31}{252}}\frac{V\sqrt{S_1}}{\sqrt{gc^2}}.$$

Now let $L$ be the mass of the travelling load, then $S$ is a quantity which varies as $L$, i.e. $= CL$, where $C$ is a constant which depends only on the material and the dimensions of the bridge.

Thus we may write

$$D - S = ALV + BL^2V^2 \text{ ......................(i)},$$

where $A$ and $B$ are constants depending only on the bridge not on the travelling load.

If we compare this empirical formula with Becker's in Art. 1376, we see that he has introduced into $A$ a factor depending on what he terms the adhesion of the driving wheels and rails; in addition in the second term he has $L$ for our $L^2$. It does not seem improbable that (i) would give as good results as Becker's formula, and it has a certain, although not very considerable, theoretical weight.

(10) *On Cores and Whorls.* Arts. 815 and 1378, footnote.

From Art. 815 we see that the equation to the neutral line for a section loaded at the point $a$, $b$, is given by

$$1 + \frac{ax}{\kappa'^2} + \frac{by}{\kappa^2} = 0.$$

This is obviously the polar of the point $(-a, -b)$ with regard to the momental-ellipse of the section. But the point is the central image of the load-point $(a, b)$. The load-point and the neutral axis are accordingly spoken of as *antipole* and *antipolar* with regard to the momental ellipse. If the load-point describe a curve, the neutral axis will envelope another curve which is termed the antipolar reciprocal of the first. If the neutral-axis envelope the contour of the section, the load-point describes a curve termed the *core*. The core is thus the antipolar reciprocal of the contour of the section with regard to the momental ellipse. So long as the load-point is within the core, every point of the section is subjected only to contraction. Thus the core plays an important part in those structures, which are to be so loaded that they are not subjected to a positive traction.

Another curve analogous to the core is frequently useful in strut and beam problems; it is the curve within which the load point must lie in order that the material may rupture first by compression (i.e. lateral extension). This curve may be termed from a botanic similarity the *whorl*. Let $\eta$ be the stretch-squeeze ratio, then the stretch $s_1$ at $x$, $y$ produced by a contractive load $P/\omega$ at $a$, $b$ is, by Art. 815,

$$= \eta \frac{P}{E\omega}\left(1 + \frac{ax}{\kappa'^2} + \frac{by}{\kappa^2}\right),$$

where $x$, $y$ are to be given values which make $\frac{ax}{\kappa'^2} + \frac{by}{\kappa^2}$ *positive.*

This is the stretch produced by a longitudinal squeeze; the direct stretch $s_2$, if one exists, will be obtained by giving $x$, $y$ such values that $\frac{ax}{\kappa'^2}+\frac{by}{\kappa^2}$ is a negative quantity greater than 1, or if we denote these values by $x'$, $y'$,

$$s_2=\frac{P}{E\omega}\left(-\frac{ax'}{\kappa'^2}-\frac{by'}{\kappa^2}-1\right).$$

Hence we must give to $x$, $y$ and $x'$, $y'$ such values that $s_1$ and $s_2$ are maxima, but for any given position of the neutral axis this will be where the tangents to the contour are parallel to the neutral axis, or are conjugate in direction to the line joining the load-point to the centroid of the section. For points on the whorl we must have

$$s_1=s_2,$$

or,

$$\eta\left(\frac{ax}{\kappa'^2}+\frac{by}{\kappa^2}+1\right)=-\frac{ax'}{\kappa'^2}-\frac{by'}{\kappa^2}-1.$$

Hence,

$$1+\frac{a}{\kappa'^2}\frac{\eta x+x'}{1+\eta}+\frac{b}{\kappa^2}\frac{\eta y+y'}{1+\eta}=0 \text{ ..................(i)}$$

is one relation to be satisfied by $a$, $b$.

Let $y=f(x)$ be the equation to the contour of the section, then since the tangents to the contour are to be parallel to the antipolar of $a$, $b$ we must have

$$f'(x)=f'(x')=-\frac{\kappa^2 a}{\kappa'^2 b} \text{ .....................(ii).}$$

Thus the construction is as follows: Choose any point $x$, $y$ on the contour of the section, find the point $x'$, $y'$ at which the tangent is parallel to that at $x$, $y$; then equations (i) and (ii) will give $a$, $b$. That is to say the load-point is the intersection of the diameter of the momental ellipse conjugate to the tangent at $x$, $y$ and the antipolar of the point $\frac{\eta x+x'}{1+\eta}, \frac{\eta y+y'}{1+\eta}$.

We are thus able to construct the whorl.

An important and interesting case is where the contour of the section possesses central symmetry. In this case $x'=-x$ and $y'=-y$. Hence the point $a$, $b$ on the whorl is the intersection of the line

$$\frac{1+\eta}{1-\eta}+\frac{ax}{\kappa'^2}+\frac{by}{\kappa^2}=0,$$

with the diameter conjugate to it. But the corresponding point on the core is the intersection of the line

$$1 + \frac{ax}{\kappa'^2} + \frac{by}{\kappa^2} = 0,$$

with the same diameter. Hence the whorl is in this case a figure similar and similarly situated to the core, with all its linear dimensions $(1+\eta)/(1-\eta)$ times the corresponding dimensions of the core.

Whorls may be easily constructed for the section of a beam or strut subjected at the same time to contractive and transverse loads; they are of value in determining conditions for set similar to that occurring in the footnote to Art. 1378.

Cores and whorls form an interesting feature in the graphic (or drawing-board) treatment of elastic problems.

## NOTE B.

### *Terminology and Notation.*

The following terms are used in the course of the present volume, and are here collected for the purpose of reference. I believe that after the chapter devoted to Poisson they have been used fairly consistently.

*Shifts.* The component-displacements of a point parallel to three rectangular directions are termed the *three shifts* and denoted by $u$, $v$, $w$. The first differential coefficients of the shifts with regard to these directions are termed the *shift-fluxions.*

*Strain.* This word is retained, as first suggested by Rankine, for the purely geometrical consideration of distortion. The *strain at a point* is determined by six strain-components. These are respectively the *three stretches* and the *three slides.* They are defined in Arts. 617, 612 and 1564. We denote them in the case of elastic strain by the symbols $s_x$, $s_y$, $s_z$, and $\sigma_{yz}$, $\sigma_{zx}$, $\sigma_{xy}$. For small shift-fluxions

$$s_x = \frac{du}{dx}, \quad \sigma_{yz} = \frac{dw}{dy} + \frac{dv}{dz}.$$

It would have been more symmetrical to have given the slides *half* the above values, but it seemed too great an interference with a nearly general custom.

For shift-fluxions of any magnitude the full values of the stretches and slides for a small strain will be found in our Art. 1618.

When neither the strain nor the shift-fluxions are small, the strain is expressed by the six components given in our Art. 1619. These components are not the stretches and slides, but the latter can be expressed in terms of these strain-components.

*Strain-ellipsoid:* see Art. 617. *Inverse Strain-Ellipsoid*, the ellipsoid in the unstrained body which becomes a sphere in the strained body: see Thomson and Tait, *Nat. Phil.* 2nd ed. I. p. 130. *Stretch-quadric:* see Art. 612. *Principal stretches:* see Art. 1539, (1).

The stretch in direction $r$ is denoted by $s_r$; the slide parallel to the direction $r'$, of a small plane at any point, the normal to which has the direction $r$, is denoted by $\sigma_{rr'}$

Stretch and slide may be positive or negative, thus a negative stretch denotes a *squeeze*.

*Dilatation* is used only for *cubical* dilatation, which may be positive or negative. It is denoted by the letter $\theta$, so that $\theta = s_x + s_y + s_z$, for small shift-fluxions.

The word *spread* is used for areal dilatation: see Art. 595.

The three *twists* are denoted by $\tau_{yz}, \tau_{zx}, \tau_{xy}$, so that

$$\tau_{yz} = \frac{1}{2}\left(\frac{dw}{dy} - \frac{dv}{dz}\right).$$

The advantage which would arise from introducing the $\frac{1}{2}$ into the slides is thus obvious.

Strain may be of various kinds—*elastic strain* or *set*. That part of a strain which does not disappear on the removal of the load is termed *set*. Set is measured by six components—the three *stretch-sets* and the three *slide-sets;* these are represented in the present work by $S_x$, $S_y$, $S_z$, and $\Sigma_{yz}, \Sigma_{zx}, \Sigma_{xy}$.

*Elastic strain* is of two kinds, one which disappears at once on the removal of the load, and the other which requires time to disappear. The latter is termed *elastic after-strain*, or the Weber-effect: see Art. 708; the former *elastic fore-strain*, or more shortly elastic strain. The influence of time may also be sensible in the matter of set, so that *fore-set* and *after-set* have obvious meanings, denoting the set produced by immediate or by long-continued load. The whole relation of time to strain is spoken of as the *time-effect*.

*Stress.* This word is reserved for the dynamic aspect of distortion. Stress may be of two kinds:

*Stress across a plane at a point.* This is defined in Art. 1563. The stress across a plane of which the normal is $r$ when resolved in direction $r'$ is denoted by the symbol $\widehat{rr'}$: see Art. 610, (ii). The component, perpendicular to the plane, of stress across a plane is termed *traction;* the component, in the plane, of stress across a plane is termed *shear*.

*Stress at a point* is determined by taking three rectangular planes at the point; it is found that the stress across these three planes can be completely represented by *six component stresses.* If the directions $x$, $y$, $z$ be normal to these planes, the six stresses are represented by the symbols

$$\widehat{xx},\ \widehat{yy},\ \widehat{zz},\ \widehat{yz},\ \widehat{zx},\ \widehat{xy}.$$

Of these the first three are termed the *three tractions*, the latter three the *three shears.*

*Stress across a plane* thus replaces the *pressure* or *tension* of some writers. The word shear (*effort tranchant ou cisaillant*) is retained entirely for stress; the confusion which frequently arises from its use for strain being obviated by the introduction of the word slide.

Traction may be either positive or negative (like acceleration in kinematic). It corresponds to the normal pressure or normal tension of some writers. It would be convenient occasionally to speak of a negative traction as a *contraction;* this word would thus refer to stress and not to strain. The three tractions at a point across the three planes for which the shears vanish are termed the *three principal tractions.*

All stresses are supposed, unless otherwise stated, to be measured per unit of area. The words tension and pressure have been used in such a variety of senses that it has been thought better to avoid them as far as possible in the present work.

*Stress-quadric:* see Art. 610, (iii). *Cauchy's Stress-ellipsoid:* see Art. 610, (iv). *Lamé's Stress-ellipsoid:* see Art. 1059. *Stress-director quadric:* see Art. 1059. *Shear-cone:* see Art. 1059.

Stress may be of two kinds; it may be the result of a given elastic strain, or may be produced by some treatment to which the body has been subjected before the strain which we propose to consider was applied. We speak of this latter form of stress as *initial stress,* and determine it by the three initial tractions and the three initial shears. These are denoted by the symbols, $\widehat{xx}_0$, $\widehat{yy}_0$, $\widehat{zz}_0$, $\widehat{yz}_0$, $\widehat{zx}_0$, $\widehat{xy}_0$: see Arts. 616 and 1210.

The influence of temperature on stress is spoken of as the *thermal* or *temperature effect.* The decrease in traction produced by unit increase of temperature in an isotropic body is represented by the symbol $\beta$, which is termed the *thermo-elastic constant:* see our Arts. 869 and 1541.

External forces applied to the mass of a material, such for instance as gravitating force, are spoken of as *body-forces* and denoted by the three components $X$, $Y$, $Z$. Body-force, unless otherwise stated, denotes in itself a measurement per unit of mass and is thus an acceleration.

External forces applied to the surface of a mass of material are all embraced under the term *load.* Load denotes in itself a measurement per unit of area. Load is of two kinds; *tractive load,* which is normal to the surface of a body, and may be either positive or negative (occasionally termed *contractive load,* the context however generally shews its sign), and *shearing load,* the direction of which lies in the tangent plane to the surface at the point of application. Load may be determined by its components parallel to three rectangular directions, or by a tractive and two shearing loads. In either case we speak of the *three load-components* at a point of the surface.

The equations of elasticity for a given body are of two kinds, *body-equations* and *surface-equations.* These equations may be expressed in terms of the shifts, or the strains or the stresses. Thus such terminology as *body-shift-equations, surface-stress-equations,* etc. is readily intelligible as well as convenient.

Stress is linked to strain by the aid of certain constants which are termed the *elastic-constants* or *set-constants* as the case may be. The number of these constants depends on the nature of the material considered. The terms homogeneous, heterogeneous, isotropic and aeolotropic are used in the senses adopted by Thomson and Tait: see their *Treatise on Natural Philosophy,* 2nd ed. Arts. 675—679. A heterogeneous aeolotropic body may have still some form of symmetrical heterogeneity. If a system of similar and similarly situated surfaces of the *n*th degree can be found, upon every one of which the elastic-constants remain the same for every point of the surface, the body may be said to have an, '*n-ic distribution of elasticity.*' We speak also of planar, cylindrical, spherical, etc. distributions of elasticity.

The hypothesis which makes stress a linear function of strain is termed the *generalized Hooke's law.* Supposing the generalized Hooke's law to hold, one molecular theory leads to an isotropic body having only one, a second molecular theory to its having two constants. These theories are spoken of respectively as the *uni-constant* and *bi-constant theories.* The same theories when applied to aeolotropic bodies are spoken of as the *rari-constant* and *multi-constant* theories: see our Arts. 921—932.

The constants which connect elastic stress with strain are termed *elastic coefficients.* For an isotropic body we use equations of the type

$$\widehat{xx} = \lambda\theta + 2\mu s_x,$$
$$\widehat{yz} = \mu\sigma_{yz}.$$

Here $\mu$ is termed the *slide-coefficient,* and $\lambda$ the *dilatation-coefficient.*

Certain combinations of these coefficients are termed *elastic moduli.* They are the forms under which the coefficients are reached in experiment. Thus:

$\dfrac{\mu(3\lambda+2\mu)}{\lambda+\mu}$ is termed the *stretch-modulus* and denoted by $E$.

$\dfrac{3\lambda+2\mu}{3}$ is termed the *dilatation-modulus* and may be denoted by $F$.

$\dfrac{\mu(3\lambda+2\mu)}{\lambda+2\mu}$ is termed the *spread-modulus* and may be denoted by $G$.

$\mu$ is termed the *slide-modulus,* and as it is thus the same as the *slide-coefficient* the latter name may be dispensed with.

If a terminal tractive load produce a longitudinal stretch $s$ in a bar, then there is a corresponding lateral stretch $= -\dfrac{\lambda}{2(\lambda+\mu)}s$. The quantity $\dfrac{\lambda}{2(\lambda+\mu)}$ is termed the *stretch-squeeze ratio* and denoted by $\eta$.

For an aeolotropic body, when we express stress in terms of strain, we use the following notation:

(i) $\widehat{xx} = |xxxx|\, s_x + |xxyy|\, s_y + |xxzz|\, s_z + |xxyz|\, \sigma_{yz} + |xxzx|\, \sigma_{zx} + |xxxy|\, \sigma_{xy}$,

(ii) $\widehat{yz} = |yzxx|\, s_x + |yzyy|\, s_y + |yzzz|\, s_z + |yzyz|\, \sigma_{yz} + |yzzx|\, \sigma_{zx} + |yzxy|\, \sigma_{xy}$.

This notation for the coefficients is easily intelligible and possesses considerable advantages in theoretical investigations, especially those which can be conducted by symbolic methods. We adopt Rankine's terminology for these coefficients; the reader will find an account of that terminology in our second volume.

## NOTE C.

### *On the Limits of Elasticity and the Elastic Life of a Material.*

(1) It is of primary practical importance to fix the limits within which the ordinary mathematical theory of elasticity holds. We shall accordingly in this note consider the various stages in the elastic life of a material and point out to which of them the theory applies. For the present we exclude the consideration of after-strain and time-effect. We commence with the following definitions:

The strain produced in a mass of material by any system of load and body-force is said to be elastic when the whole strain disappears on the removal of the load and body-force.

So long as the whole strain disappears the body is said to be in a *state of ease up to* that strain.

(2) Given an aeolotropic body with any amount of initial stress, let at each point of it a surface be described in the following fashion: Upon a radius vector revolving in every direction about the point let a length be marked off such that its inverse square is proportional to the greatest stretch which can be given to the material in that direction without set, the end of the radius-vector will thus trace out a surface, which may be termed the *surface of perfect elasticity* at the point[1]. If the body and the initial stress be homogeneous, the surface of perfect elasticity will be the same for all points. This surface marks the limit of elastic strain for the substance. Any strain which produces a stretch in any direction exceeding that measured by the inverse square of the corresponding radius vector of the surface of perfect elasticity, produces set or alters the state of ease. We may thus state the condition for a strain being perfectly elastic: A strain is wholly elastic, when the stretch sheets (as distinguished from the squeeze sheets, if there be any) of the stretch-quadric lie entirely outside the surface of perfect elasticity.

We will term a stretch sheet a positive sheet of the stretch-quadric.

(3) It has been usual among writers who adopt a stress limit of elasticity to speak of *two* elastic limits, *a superior and an inferior limit* of elasticity, instead of a limiting surface. All that is meant by these superior and inferior limits is that by reversing the direction of a load we interchange the stretch and squeeze sheets of the stretch-quadric, and thus by altering the magnitude of the load can make a positive sheet ultimately touch the surface of perfect elasticity approaching it from a different direction. The greatest positive and the greatest negative loads are then termed the superior and inferior limits of elasticity. This terminology however is not very scientific.

(4) If a surface be drawn in precisely the same fashion as the surface of perfect elasticity, but so that the inverse square of its radius vector in any direction is taken proportional to the stretch in that direction which would produce rupture, we have a second surface which we may term the *surface of cohesion*. The surface of cohesion obviously lies entirely inside the surface of perfect elasticity, and the

[1] The constant which determines the ratio of the stretch to the inverse square of the radius-vector is supposed the same for all the surfaces considered in this note; it may be conveniently taken as an unit of area.

limit of cohesion is marked by a positive sheet of the stretch-quadric touching the surface of cohesion. Strains which bring any part of a positive sheet of the stretch-quadric between the surface of perfect elasticity and that of cohesion produce set.

The surface of perfect elasticity will continually change its form as the material receives set, and this whether the elastic stretch be calculated on the basis of the primitive or set-dimensions of the material in question. Since it is the stress corresponding to the elastic strain and not to the set which at all times supports the load, and since according to the Coulomb-Gerstner law, the elastic constants alter but little with the set, it is obvious that as we increase the load, we must increase the strain, and as a rule the stretch. Hence the effect of an increasing load is not only to produce set but to *contract* the surface of perfect elasticity.

(5) The form and relative magnitude of the surfaces of perfect elasticity and of cohesion depend largely on the treatment or the working which a material has previously received: see our Arts. 692 (4°), 853–56 and 1003. In particular every strain which produces set appears to alter the surface of perfect elasticity. Outside the surface of perfect elasticity (which from its nature must be a *closed* surface) we have the state of ease, between the surface of perfect elasticity and the surface of cohesion we have the state of set. It is important to discover how far the surface of perfect elasticity can be contracted and made to approach in any or all directions the surface of cohesion. If we may in any way argue from the result of experiments on bars as to the physical conditions which hold for masses of material subjected to any form of strain, there is a practical limit up to which it is possible to contract the surface of elasticity. When a bar is subjected to increasing traction, a certain stretch is reached after which there is a sudden and rapid increase of stretch, during which the traction so far from increasing appears to diminish; only after a very great increase of set-stretch is the bar again in a condition to sustain an increase of traction. The physical nature of the material appears after this to have changed; for still larger tractions, strain, nearly all set, increases much faster than stress, and the material rises in temperature. The point at which this change takes place is very marked, and various names have been suggested for it, as the *limit of fatigue*, the *limit of stability*, and the *break-down point*. The latter name brings out the character of the phenomenon, but at the same time suggests a point related to absolute strength or cohesion; I have therefore spoken of this point in the present work as the *yield-*

*point.* Arguing from the analogy of a bar, there would be in every material in any definite direction a definite stretch at which the material would suddenly yield as a bar does. Supposing such to exist we may form a surface from these stretches in exactly the same manner as we formed the surface of cohesion, and we may term it the *yield-surface.* As the state of ease for a bar may be extended up to the yield-point, so we may suppose that the surface of perfect elasticity can be contracted till it coincides with the yield-surface. In this case no set would be manifest, till it appeared suddenly and with great rapidity on the stretch-quadric touching the yield-surface; an obviously dangerous state of affairs for practical purposes.

(6) The phenomenon of the yield-point is probably closely connected with the treatment or working (rolling, hammering, hardening, annealing, etc.) a body has received; and the sudden set it gains in the yield-stage immediately following the yield-point is not improbably a removal of all or part of the elastic and cohesive influence of that treatment or working. The yield-stage may be in fact a destruction of the record of the past life of the material. In particular it would seem that all 'initial stress' may disappear at this stage.

(7) It would be convenient to speak of the elastic life of the material up to the yield-stage as the *worked stage*, and after the yield-stage as the *raw* or *unworked stage.* The raw stage includes the stage of plasticity. A body may be in a state of ease when it is either in the worked stage or the raw stage, but there is this physical difference, that in the worked stage a stress which produces a strain beyond the elastic limit produces (always supposing the state of ease not limited by the yield-stage) only a small amount of set, while in the raw stage a stress producing a strain beyond the elastic limit will produce also a large amount of set. It is thus apparent that the surface of perfect elasticity, while it can be contracted across the yield-surface and possibly almost up to the surface of cohesion, is for practical purposes limited by the yield-surface, and indeed should lie well outside that surface, if the material is to shew signs of the limit of its practical capacity by exhibiting a small set.

(8) We have next to inquire whether a body strained within its state of ease (be it in the worked or the unworked stage) obeys the ordinary equations of the mathematical theory. This depends on whether we can assume that the generalized Hooke's law holds for all strains within the limits of perfect elasticity. We shall shew in the next note that for a

considerable number of materials this is not the case. For a variety of ductile materials, however, such as wrought iron, steel, copper, etc. it appears to be true that stress is proportional to strain for all strains within the state of ease. Further, between the limit of elasticity (the limit to the state of ease) and the yield-point the elastic part of the strain according to the Coulomb-Gerstner law is also proportional to the stress. If the state of ease be narrow, set appears at first to be proportional to stress; hence we see that the ordinary equations of elasticity hold for so much of the strain as is elastic, while Neumann's equations hold for so much of the strain as is set: see our Arts. 1207 and 1211. However before we reach the yield-point set although small is no longer proportional to stress, and so Neumann's equations cease to hold. Professor Kennedy tells me that the elastic part of the strain even in the unworked stage, for a bar under terminal traction apparently follows the same stretch-modulus as in the worked stage, so that we may perhaps consider the equations of elasticity to hold with the *same constants* for the elastic strain of a material in this stage. This is in fact a very important extension of the Coulomb-Gerstner law.

(9) It will be noted that the usual definition of the limit of elasticity as the beginning of set, as well as the limitation of perfect elasticity to materials for which the generalized Hooke's law holds, is not entirely satisfactory. I have endeavoured to indicate in this note the exact portions of the elastic life of a material for which the mathematical theory is true, and what are the proper strains to be taken in that theory for calculating the strength of a material.

(10) I owe to Professor Kennedy the following automatically drawn diagram of the elastic life of a bar of soft rivet steel subjected to a continually increasing tractive load. It may serve to bring out clearly the stages we have considered above.

The tractions are calculated per unit area of *primitive* cross-section and are measured along the horizontal circular arc. Stretches are measured parallel to the vertical axis.

$A$ to $B$ is the state of ease, traction is proportional to stretch, which is entirely elastic. $B$ is the limit of perfect elasticity.

$B$ to $C$ is the set-stage (*écrouissage*) which lies between the limit of perfect elasticity and the yield-point $C$. Set, although small, as we approach the yield-point increases more rapidly than the load.

$CC'$ is the yield-stage, and Professor Kennedy's researches shew a diminution of traction during this stage. He considers that the

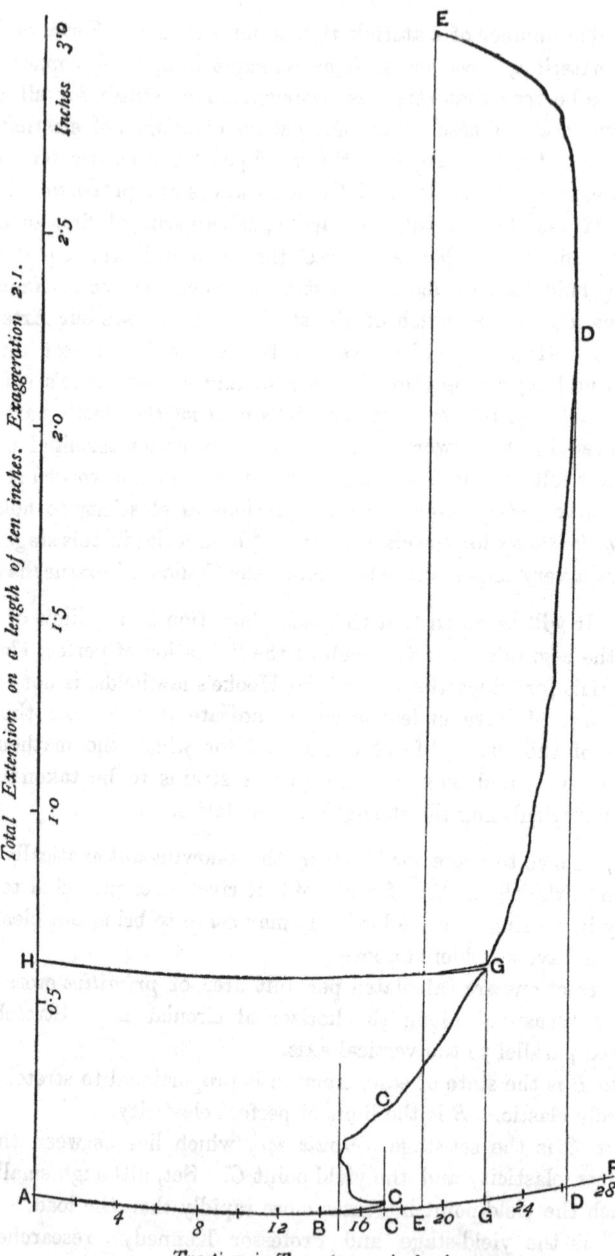
Total Extension on a length of ten inches. Exaggeration 2:1.
Inches 3·0
2·5
2·0
1·5
1·0
0·5
E
D
H
G
C
C
A
F
4
8
12
B
16
C
E
20
G
24
D
28
Traction in Tons per sq. inch of original area.

great increase of stretch takes place at different parts of the bar successively and not simultaneously during the stage.

$C'D$ is the first portion of the rough stage, here the elastic strain apparently obeys the Coulomb-Gerstner law, but there is a very large set. The bar has a uniform but decreasing cross-section and rises in temperature. The bar is not perfectly plastic in this stage for the traction is still increasing. Professor Kennedy terms $C'D$ the *stage of uniform flow*[1]. $HG$ shews how when the load is removed, and the bar gradually reloaded, its state of ease practically extends up to the maximum traction which it has previously sustained. The limit of elasticity has thus been carried beyond the yield-point. At $D$ a maximum traction is reached if calculated on the primitive cross-section, but from $D$ to $E$ the traction really increases if calculated on the reduced section.

About the point $D$ the bar begins to 'thin down' or 'flow' *locally*, a portion becoming more reduced in cross-section than the rest. I have termed this phenomenon in the present work *stricture* (*étranglement*), and the cross-section which thins down the *section of striction*. Professor Kennedy terms the stage which follows this the *stage of local flow* (*énervation*). I have preferred the name stricture to local flow, because the material does not appear to be in a condition of genuine flow or plasticity, for in that case the traction-stretch curve ought to be convex and not concave to the stretch-axis. $E$ is the *limit of cohesion.*

$D$ corresponds to the maximum *total load* and $E$ to the *terminal load.* The traction has in general at $E$ its maximum value if calculated for the section of striction.

## NOTE D.

### *On the Defect of Hooke's Law.*

We have had occasion several times to draw attention to the fact, that in certain materials, even for very small strains which are entirely elastic, stress is not proportional to strain, or the stress-strain (*e.g.* the traction-stretch) curve is not a straight line. We have seen that Hodgkinson

[1] See *Nature*, April 2, 1885, p. 504.

termed this peculiarity in the case of cast-iron the *defect of elasticity*[1] The name appears to be unfortunately chosen, and we prefer to term the phenomenon the *defect of Hooke's Law*. The elasticity indeed remains perfect but the *ut tensio, sic vis* principle is defective. The materials which are defective in this respect are of great importance in technical elasticity, and for the elastic strains which occur in practice it is not possible to suppose the stress-strain relation linear. It will be obvious that, in the ordinary sense of the term, these materials possess no stretch-modulus. What are we to understand then by the stretch-moduli which will be found tabulated for these materials in many works on physical and technical elasticity? Either the modulus has been obtained as the mean of a number of corresponding values of traction and stretch, in which case its value will depend entirely on the range of values chosen, or else it must be taken to represent the tangent of the angle which the tangent at the origin to the traction-stretch curve makes with the stretch axis[2]. This latter is the value suggested by the mathematical theory as the limit. But it is extremely difficult to determine this angle, unless we know the *form* of the stress-strain curve, because of the delicacy of the testing machine required to measure the ratio of vanishingly small stress and strain. Possibly sound-experiments might be the best method of ascertaining this ratio. This leads us to a very important remark, namely that the apparent isochronism of sound vibrations in materials of this nature does not permit of our assuming the principle that superposition of strains means superposition of stresses for such strains as occur in technical elasticity. In other words the ordinary mathematical theory of elasticity may for certain materials possibly give very correct results for such infinitely small strains as occur in sound vibrations, but we cannot argue from this that these same equations hold for even perfectly elastic strains such as we require to consider in technical elasticity: see our Arts. 928, 929, and 1404.

The accompanying plate represents the traction-stretch curves for

[1] It seems to me that Hodgkinson in his *Experimental Researches* (see our Art. 969) terms set 'defect of elasticity', but that in the *Report of the Iron Commissioners* he means by 'defect of elasticity' a defect in Hooke's Law, see our Art. 1411. As appears from that Article the two are not the same, because set has a term linear in the stretch. Saint-Venant speaks of 'une partie persistante sensible—ce que les Anglais appellent *set* ou *defect of elasticity*' (*Leçons de Navier*, p. 104). He does not appear to have remarked Hodgkinson's double use of the term.

[2] A third course would be that suggested by Wertheim's memoir on the human tissues: see our Art. 1318.

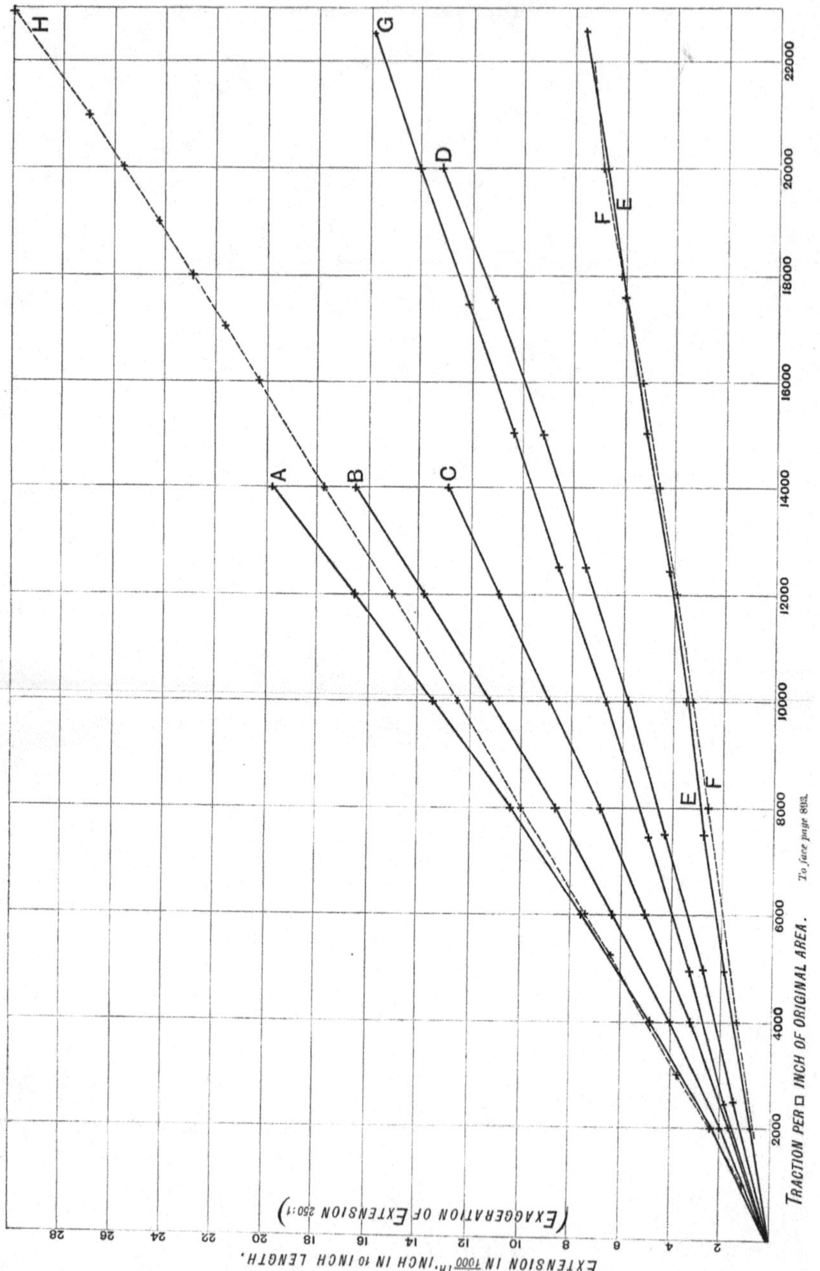


The material originally positioned here is too large for reproduction in this reissue. A PDF can be downloaded from the web address given on page iv of this book, by clicking on 'Resources Available'.

bars of various materials. These curves are the result of a series of very careful experiments made by my colleague, Professor Kennedy, who has kindly permitted me to publish them in this Note. In all cases the material had been reduced to a state of ease limited by a considerably higher traction than that applied, so that the bars in every case returned to their initial length and the whole strain was elastic. Tractions are measured along the horizontal, and stretches along the vertical axis. By joining the origin to the terminal of the stretch-traction curve the reader will be able to form an idea of its curvature, or the defect of Hooke's law.

A and B. Cast-Iron (Blaenavon).

C. Cast-Iron (Carron).

D. Cast-Iron (very tough mixture, tenacity 12·5 tons per square inch).

E and F. Wrought Iron and Steel. These curves are within the limits of experimental error *straight* lines. They shew the equality of the stretch-moduli for wrought-iron and steel.

G. "Manganese Bronze." Cold rolled.

H. Extremely hard steel used for dies at the Royal Mint. Tenacity 62·3 tons per square inch, and elasticity scarcely impaired up to 95 p. c. of that load. The scale of tractions for this curve is *four times as great* as for all the others, so that the curve is carried to a traction of over 88,000 lbs. per square inch. It will be seen to be practically a straight line throughout its whole length. The stretch-modulus for this steel is slightly greater than that of the wrought-iron and steel in E and F.

These curves will suggest to the mathematician what a field of investigation lies open in the mathematical theory of perfectly elastic bodies, when for certain materials in frequent technical use we recognise the defect of Hooke's law.

The 'defect of Hooke's Law' may possibly throw some light on the inconsistencies in Wertheim's experiments: see our Arts. 1297 and 1403.

## NOTE E.

*The Resilience of Beams subject to transverse Impact.* Arts. 939, 942.

(*a*) A full account of Hodgkinson's various experiments on impact and the resilience of beams will be found in *Der Ingenieur*, Bd. I. Freiberg, 1848, pp. 403—432. The account is by Rühlmann and is entitled: *Festigkeit prismatischer Körper gegen Stoss und insbesondere die Arbeiten und Versuche von* Eaton Hodgkinson *über diesen Gegenstand.* It adds little to Hodgkinson's investigations, but is a useful *résumé* for German engineers.

(*b*) It may not be without interest to compare Hodgkinson's results in Arts. 939 and 942 with the results of a theory based upon the hypothesis, that during impact the beam takes the same form at each instant as it would if the deflection at the point of impact were produced by a statical load. This hypothesis is probably nearly the truth when the velocity of impact is not very great. I have extended the method of Homersham Cox to a few more general cases. Let $v$ be the velocity of impact, $M$ the mass of the striking body, $m$ that of the beam, $l$ the length of the beam, $E$ its stretch-modulus, $\omega$ its section and $\kappa$ the radius of gyration of the section about a diameter perpendicular to the plane of flexure, $f$ the greatest deflection at the point of impact, which we will suppose at distances $a$ and $b$ from the ends of the beam $(a+b=l)$. We will first suppose the ends of the beam supported as in Hodgkinson's experiments and Cox's theory, the impact being horizontal. If the impact were vertical the statical deflection due to the weight of the falling body must be added. I find:

$$\left.\begin{aligned} f &= \frac{ab\,Mv}{\sqrt{3E\omega\kappa^2 l\,(M+\gamma m)}}, \\ \text{where} \quad \gamma &= \tfrac{1}{105}\left\{1+2\left(\frac{a}{b}+\frac{b}{a}+3\right)^2\right\} \end{aligned}\right\} \text{..............(i).}$$

Here $\gamma$ is a coefficient which measures, as it were, the *effective mass* of the beam. It may be termed the *mass-coefficient of resilience.*

Hence $\gamma$ is least for central impact and increases to infinity as we approach the ends. At the centre $\gamma=\frac{17}{35}$, which agrees with Cox's result; at 3/4 span $\gamma=\frac{731}{945}=\frac{27}{35}\left(1+\frac{2}{729}\right)=\frac{27}{35}$ nearly. Thus, if $f_1, f_2$ be the deflections at 1/2 and 3/4 span, produced by the same blow, $Mv$, we have:

$$\frac{f_2}{f_1}=\frac{3}{4}\left\{\frac{1+\frac{17}{35}\frac{m}{M}}{1+\frac{27}{35}\frac{m}{M}}\right\}^{\frac{1}{2}}, \text{ nearly....................(ii).}$$

Further, if $T_0$ be the traction which will, if applied longitudinally, rupture a bar of the same material, and $2h$ be the diameter of the beam in the plane of flexure, then on the ordinary theory the rupture deflection is given by

$$f' = \frac{ab}{3h}\frac{T_0}{E}.$$

Hence the blow which will just break the beam is given by

$$B = Mv = T_0 \sqrt{\frac{\omega l}{3E}\left(\frac{\kappa}{h}\right)^2 (M + \gamma m)} \dots\dots\dots\dots\dots \text{(iii)}.$$

If $B_1$ be the blow which will break the beam at 1/2 and $B_2$ the blow which will break the beam at 3/4 span, we have:

$$\frac{B_1}{B_2} = \left\{\frac{1 + \frac{17}{35}\frac{m}{M}}{1 + \frac{27}{35}\frac{m}{M}}\right\}^{\frac{1}{2}}, \text{ nearly} \dots\dots\dots\dots\dots\dots\dots \text{(iv)}.$$

We see at once that Hodgkinson's results (2) and (3) of Art. 939 hold only when the ratio $m/M$ may be neglected in equations (ii) and (iv). Similarly we note that (v), (vi) and (vii) of Art. 942 are only very rough approximations to the truth.

Turning to Young's definition of the resilience of a beam as proportional to the height from which a given body must be dropped to break it, we have from equation (iii),

$$\text{Resilience} \propto v^2 \propto \frac{T_0^2}{3EM}\,\omega l \left(\frac{\kappa}{h}\right)^2 \left(1 + \gamma\frac{m}{M}\right),$$

that is proportional to $\omega l\left(1 + \gamma\frac{m}{M}\right)$.

Hence Young's Theorem (see Note A, (3), (b)) that the resilience is proportional to the volume ($\omega l$) is accurately true only when we neglect $m/M$ as small.

(c) I may note three additional results:

(i) If the beam instead of being supported at the two ends be built-in at both ends, the mass-coefficient $\gamma = \frac{13}{35}$ for a central impact, and the deflection is given by

$$f = \frac{l^2 Mv}{8\sqrt{3E\omega\kappa^2 l\,(M + \frac{13}{35}m)}}.$$

(ii) If the beam be built-in at one end only and the impact be applied horizontally to the free end, the mass-coefficient $\gamma = \frac{33}{140} = \frac{1}{4} - \frac{1}{70} = \frac{1}{4}$ nearly, and

$$f = \frac{l^2 Mv}{\sqrt{3E\omega\kappa^2 l\,(M + \frac{33}{140}m)}}.$$

(iii) If the beam be built-in at one end, supported at the other and struck in the middle, $\gamma = \frac{764}{1715} = \cdot 445$ nearly, and

$$f = \frac{l^2 Mv}{16\sqrt{\frac{3}{7}E\omega\kappa^2 l}\,(M + \cdot 445m)}.$$

The calculation of $\gamma$ for a variety of other cases presents no difficulties.

# INDEX.

*The numbers refer to the articles of the book and not to the pages unless preceded by p. ftn.=footnote.*


Absolute strength: see *Cohesion*

Adherence: see *Cohesion*

Adhesion, lateral: see *Slide*

Aether. Mazière, 29; Descartes, 34; Bernoulli, 35; Euler, 94; Sanden, p. 78; Mossotti, 841; Kelland and Earnshaw, 1230

After-strain: see *Strain*

Aimé. Compression of liquids, 691

Airy. Flexure of a bar supported at several points and application to standards of length, 1247

Alloys. Elasticity of, Wertheim, 1302—3, Bevan, 751 *d*

Ampère. Molecules, atoms, 738

Annealing. Effect on strength wires, 692, 830, 856—8; on extension of wires, 692; on stretch-modulus of metals, 1295, 1301, of glass, 1311

Arcs. Circular, resistance of, 278, 937, 1457—60, 1573—4

Arches. Expansion by heat, Rennie, 838; wooden circular, experiments on, Ardant, 937; Poncelet, review of theories of, 996

Ardant. Experiments, beams, arches, 937; Poncelet on experiments of, 983—5

Arosenius on columns, 819

Atoms. Sexual characteristics, Petty, 6; attraction of, Newton, 26, John Bernoulli, 35, Euler, 94; polar properties, Desaguiliers, 29 $\gamma$, Ritchie, 408; Frankenheim, Kirwan, 829; definition of, Ampère, 738

Attraction. Elective, query on, Newton, 26; to explain cohesion, s' Gravesande, 42; is gravity applicable to cohesion or another law varying with distance, 164, 165, Belli, 361, 752—8, Mossotti, 842, Faraday, 843; of drop of water to earth, 166; of two thin plates, 168; of two spheres, 169; capillary, 171; law for single particles, 172, 439, 543, 977; history of theories of, Belli, 165—6, 211, 308; Nobili, 211, Voute, 212—4, Eisenbach, p. 876; and repulsion of spheres, Laplace, 307; laws of, Laplace, 308, 312—5; theory of, Laplace, 313; Kelland, Ellis, Earnshaw, controversy on law of, 1230—32; Brünnow, 1235—7

See also *Molecule*, *Atom*

Avogadro. *Fisica de' corpi ponderabili*, value as a text-book, 847; influence of caloric on molecular force, 844—5; molecular forces and experimental results, 845; vibrations of bodies, 846

Axis. Of equilibrium (=neutral axis), J. Bernoulli, 21; of elasticity, bodies with three axes, Savart, 340—5, Cauchy, 615; of gypsum and other crystals, elastic, optic, thermal, 788—93, 919

See also *Crystal*

Axles. Solid and hollow, strength of, 1463 *b*; fracture of, 1463 *d*; effect of vibrations on, 1463 *g*, 1464; form of, 1464

Bacon. Experiments on compression of water in lead spheres, p. 78, ftn.

Balance. Flexural, Weber, 718; inertial, Willis, 1418, 1421; torsion, Coulomb, 119, Biot, 183, Hill, 398, Ritchie, 408, Mainardi, 1214

Ballistic Machine. Bernoulli, 246; Mossotti, 251

Banks. Experiments on wood and iron beams, 147
Bar. Non-central tractive loads, effect on, Tredgold, 832—3; Brix, 1249
Cylindrical, movement of points of, temperature varying, Duhamel, 898; equation for, sp. heats equal, 899; initial temp. uniform and allowed to cool, 900; initial temp. varying allowed to cool under traction, 901; cooling, ends fixed to moveable masses, 902—3; made to stretch another bar, 904
Resilience of, Poncelet, 988—993.
Uniform, flexure of, Airy, 1247
Long cast-iron experiments, on flexure and strength of, 1413
Deflections of, by travelling loads, 1417—22
Sounds produced in metal, by electric currents, 1333 and ftn.
See also *Beam*, *Rod*
Barbé. Experiments on traction and strain in iron, 817
Barlow. Strength of Timber, 188; review of beam theories, 189; position of neutral axis, 189—90, 192—4; on Girard's results, 190—1; deflection of beams, 191; Hodgkinson on, 233; report on strength of materials, 945
Base of fracture. Definition and resistance of, Galilei, 3; Varignon's formulae for resistance of, 14
Baudrimont. Experiments on annealing wires, 830; effect of worked state on elastic constants, 831; effect of temperature on cohesion of metals, 1524
Beams. Continuous, Navier, 281; theory of, Clark, 1472; experiments and theory, do., 1488; Saint-Venant, 1572
Tractive loads. (See *Struts* and *Columns*.) Rupture of Bülfinger, 29 β; curvature of fixed, Euler, 67; of varying density and section, Euler, 68; of uniform section weight considered, do., 68; strength of, Young, 134; stiffness of, do., 138; buckling and flexure of, Tredgold, 196, Heim, 907—10; Saint-Venant's problem, what value are we to take for load? 911; Euler's theory, Lamarle, 1255—62; core and whorl of section, 815, pp. 879—81; experiments on, Musschenbroek, 28; iron, Duleau, 229; neutral line for, p. 44 ftn., Young, 138, Robison, 146, Duleau, 229, Whewell, 737, Heim, 907—10, Lamarle, 1255—62
Transverse loads. *Theory*. Hypothesis of inextensibility, Galilei, 2; base of fracture, do., 3; forms for equal resistance, do., 4; absolute and relative resistance, Varignon, 14; points of rupture, Parent, 17; lemmas on strained fibres, Bernoulli, 20—21; rupture of, Bülfinger, 29 β, Euler, 76; flexure of, Coulomb, 117; compressibility in, Eytelwein, 150; deflection of, Barlow, 191—2; general formulae for, Hodgkinson, 231; stress-strain, laws for, do., 234, 240; theories and extensions of, Navier, 279; built-in one end, do., 281; flexure of uniform section, 816; equilibrium of built-in, Heim, 914; Emerson's Paradox, 187, p. 74 ftn., 952 (ii); tendency to break under, A. Smith, 1181; Moseley on, 1233; Lamarle on theories, 1253—4; deflection and imperfect elasticity of, Hodgkinson, 1438—9; hyperbolic law for cast-iron, Cox, 1440—2; comparison of flexure by theory and experiments, Chevallier, 1461; general principles and formulae for, Clark, 1469—71; formulae for various sections, Tate, 1490—3; equilibrium of elastic beam of square section uniformly loaded, Maxwell, 1547; introduction of slide into theory of, St Venant, 1571, 1582—3, Maxwell, 1554; problem for any system of loads, St Venant, 1572; criticism of old theory, do., 1580
Transverse loads. *Experiments on Iron*. Buffon, Musschenbroek, 28, Banks, 147, Rennie, 187, Barlow, 190; best form of, Hodgkinson, 237, 243—4; neutral axis, &c., Hodgkinson, 238—40; T shape, Hodgkinson, 241; malleable, Hodgkinson, 242; non-fibrous, Vicat, 732—3; wrought and cast, Gerstner, 809; cast, Bramah, 837; cast, fracture of, set in, Hodgkinson, 952—3; effect of skin, 952; effect of long-continued load, Hodgkinson, 967; on temperature-effect, Fairbairn, 953, 968; flexure and rupture, Giulio, 1216; cast, Cubitt, 970; neutral axis, &c., Colthurst, 998; long cast, Hodgkinson, 1413; wrought, change of section of, Clark, 1485; cast tubes, Stephenson, 1483; increase of strength by initial strain, Clark, 1486; cast, weaker at core, James, 1484
*Wood*. Rupture of, Mariotte, 10, 28, Euler, 76, Girard, 131, Banks, 147; flexure of, Dupin, 162, Hodgkinson,

233; flexure and set in, Gerstner, 807—8; neutral axis, &c., Colthurst, 998; flexure of (to determine stretch-modulus), Hagen, 1229, Pacinotti and Peri, 1250—2, Wertheim, 1312—13; Lamarle, 1253—4

Neutral axis. Galilei, 3, Varignon, 14; Bernoulli, 20—21; Coulomb, 117; Riccati, 121; Young, 139; Eytelwein, 149; Barlow, 190—4; Duleau, 227; 228; Hodgkinson, 231, 232, 233, 238, 239, 240, 279, 971; Hill, 403; Persy, 811; Colthurst, 998; Giulio, 1216; Schafhaeutl, 1463; Clark, 1469—71

Resilience of. Hodgkinson, 939—43, 1409—10; comparison of Hodgkinson's experiments with theory, pp. 894—96; Young, p. 875, H. Cox, 1434—37; Clark, 1487; Poncelet, 988—94

Theory of. Higher Approximations. Poisson, 570 and ftn., Heim, 907, Lamarle, 1253, Saint-Venant, 1615

Vibrations of. D. Bernoulli, 46, 49; Euler, 64; Riccati, 121; Strehlke, 356; Poisson, 466—471; Heim, 916; Bell, 1372; Lissajous, 1525; Savart, 347—351; Cauchy, 660. See also *Vibrations*, *Rods*, *Nodes*

Travelling Load etc. see *Bridges*

Beatson, electro-elastic properties, p. 674, ftn., 1333, p. 720, ftn.

Becker. Experiments on deflections of cast-iron bridges, 1373; method of measuring do., 1374, results, 1375, empirical formula 1376, p. 876

Belgrado. Position of neutral line, on s'Gravesande's theory, scale of elastic strengths, 29 $\delta$

Bélidor, de. Experiments on rupture of beams, 28 $\beta$

Bellavitis. Equilibrium of elastic rods, Poisson's error, 935

Belli, G. Molecular forces, 163—4; Newton and Laplace's theories, 164; solid of greatest attraction, 165; problem of hanging drop of water, 166; law of molecular force unknown, 167; attraction of thin plates, 168, of spheres, 169; refraction of light, 170; capillary attraction, 171; law of attraction between particles, 172, 361; of molecular attraction, 752—3; of cohesion, 754; ordinary hypotheses of molecular attraction, 755; is gravitational attraction applicable? 756; hypothesis of attraction, 757—8

Bells. Tones of, Euler, 96 $\beta$; his theory applied, p. 55, ftn.; to Harmonicum, 93 $\gamma$

Bell. Stretch-modulus from musical notes, 1372

Bergeron. Resistance of elastic solids, 1238

Bernoulli, Daniel. Letters to Euler on vibrations of elastic laminae, 44—8; equation for transverse vibrations of a bar, 45, 49; modes of vibration, 50

Bernoulli, Jas. (the older). Papers on elastic lamina, 18; as originator of stress-strain curve, p. 871; notices of Galilei, Leibniz, and Mariotte's work, 19; lemmas on fibres of strained bodies, 20; idea as to position of axis of equilibrium, 21; rejects Hooke's law, 22; Girard on this, 127; curvature of elastic line, and equation to it, 24; Poisson on his methods, 25; Riccati on do., 32

Bernoulli, Jas. (the younger). Vibrations of elastic plates, 122; on Chladni's experiments, 122; on Euler's hypothesis, 122; spiral springs, 246

Bernoulli, John. On elasticity, 35; his *Discours*, 42; hardness and elasticity, 42; letters to Euler, 43

Bernoullis. Genealogy of family, p. 72, ftn.

Bevan. Experiments on elasticity of ice, 372; adhesion of glue, 374; strength of bones, 375; cohesion of wood, 376, of cast-iron, 377; modulus of torsional elasticity, 378—9; importance of time element, 751 *a*; stretch-modulus of gold and alloys, 751 *d*; ring of a gold coin, 751 *d*

Bresse. Strength of circular arcs, 1457—60

Binet. Elastic curve of double curvature, 173; radius of torsion in do., 174; equilibrium of forces on polygon of rigid rods, 174—5; on Lagrange's treatment of rods of double curvature, 175; spiral springs acting by torsion, 175; Thomson and Tait and Saint-Venant on, 175; integral of equation for elastic curve of double curvature, 1240

Biot. His *Traité*, 181; theory of molecular forces, 182; limit of elasticity, 182; set, 182; elasticity of crystals, 182, of threads, 183; torsions of do., 183; elastic laminae, 183; torsion balance, 183; vibrations of elastic bodies, 184; Galilei as discoverer of nodal figures, 358; Pictet on his expts. 876

Blanchet. Propagation and polarisation of motion in crystallised media, 1166

—8; Poisson, report on, 1166—7; nature of vibrations when initial disturbance small, 1169; summary of results for velocities of propagation, 1170; delimitation of wave in propagation of vibratory motion, 1172—4; surface of wave, 1175 *a*, *b*; limits of wave, 1175; discussion with Cauchy on, 1175—6; result of his investigation of waves in a 36-constant medium, 1177; continuous disturbance of a crystallised medium, 1178

Blondel. Controversy on solids of equal resistance, 5

Boilers. Calculation of thickness of shell of, 694; cause of explosions of, Lamé, 1035; thickness and form of shell, 1038; Piobert on, 1039

Bone. Strength of mutton, Bevan, 375; stretch-modulus of beef, Bevan, 375

Bonnet. Elastic-equations in curvilinear coordinates, 1241—6

Bornet. Experiments on iron, 817

Bourget. Vibrations of flexible cords, 576

Boussinesq. Isostatic surfaces, 1152; Thomson and Tait's reconciliation of Poisson and Kirchhoff, attributed to, p. 252, ftn.

Bordoni. Equilibrium of elastic rods of double curvature, 215; includes effect of traction, 216; angle of torsion, 217; equation to elastic line, 218—222; on Poisson's results, 223—224

Borrellus. Resilience, p. 874

Bramah. Experiments on cast-iron beams, 837

Brass. Stretch-traction curve, Poncelet, 985

Braun. Experiments on elastic after-strain, 717

Brewster. Teinometer, 698, p. 640, ftn.; researches on polarising effect of strain in glass, strain produced by heat; application to glass models of arches, laws of tints, superposition of small strains, p. 640, ftn.

Bricks. Adhesion of, to mortar, Morin, 905; crushing strength of brick-work and, 1478

Bridge. Impulsive effect of travelling load on, Stokes, 1276—1277, Willis, 1419; solution of problem, inertia neglected, 1277—84; do., inertia considered, 1288—91; experiments on effect of travelling load, Willis, 1417—22; proportion of dead and travelling to breaking load, 1427; dynamical deflection in girders of, Cox, 1433; cast-iron, deflections of, Becker's experiments, 1373—76, p. 878

Tubular. Experiment on tubes, Hodgkinson, 1416; description of, Torksey, 1463 f.; Britannia and Conway, description of, 1489; deflection of do., 1489; Tate, 1490

Suspension. Navier, 272; Dufour, 692; Von Mitis, use of steel for, 693; Vicat, 721, experiments on wire and bar-iron for, 723; Brix, do., 849; Leblanc, do., 936; application of theory of resilience to, 992; experiments on iron for, Lamé, 1001—3; use of cables in, 1454; effect of time on elasticity of cables, 1455; at Lorient, Leclerc, 1462

over Creuse. Calculations for, Saint-Venant, 1611

Brix. Elasticity and strength of iron wires used in suspension bridges, 848; comparison of French and German wire, 849; description of testing machine, 849; Hooke's law true till set begins, 850; after-strain, and set, 852; relation of set to elastic limit, 853; set in wires, and effect of loads, 854; yield-point, 855; set in annealed wires, 856; beginning of set, 857; elastic properties of annealed wire, 858; effect of skin on elasticity of wires, 858; non-axial tractive loads, 1249

Brünnow. Molecular attraction, 1235; disputes inverse power as law of cohesion, 1236; problems on attraction, 1237

Buffon. Experiments on rupture of iron and wooden beams, 28 $\epsilon$

Bülfinger. *De solidorum resistentiâ*, reference to previous authors, breaking loads of beams, consideration of Galilei and Mariotte-Leibniz theories, compression of fibres of beams, Hooke's law, position of neutral line, 29 $\beta$

Burg. Strength of materials, 866—7

Byrne. New theory of strength of materials, 1352

Caignard Latour. Experiments on compression of liquids, 318; on longitudinal vibrations of cords, 465; on strength of boiler plates, 694; on notes of metal wires, 802

Caloric, 313, 543, 844, 976, 1230

Camber of bridges, effect on dynamical deflections, 1433

Caoutchouc. Properties of, C. G. Page, 1370, Gough, p. 386, ftn.: see also *India Rubber*

Capillary attraction. Belli, 171; Laplace, 311; Poisson, 567
Carbonate of lime. Nodal systems in plates of, Savart, 344
Catenary. Euler, 71
Cauchy. On Navier's theory of elastic solids, 271; Poisson on his theorem, 553; Saint-Venant on do., 554, 555; equilibrium and motion in interior of solids and fluids, 602; theory of six stress-components, principal tractions, 603—6; his theorem, stress across a plane at a point, 606, 659; stress in a fluid, 609; elastic stresses, stress-quadric and ellipsoid, 610—11; stretch and dilatation, stretch-quadric, 612; body-stress-equations, 613; equation of equilibrium and motion of a solid or fluid, 614; of a motion of a system of material points, initial stress and bi-constant isotropy, 615-16; theorems relating to dilatation, strain-ellipsoid, 617; elastic laminae, bi-constant equations for, 618; definitions of lamina and plate, 619; plane lamina of constant thickness, 620—4, of variable thickness, 625; curved lamina, constant thickness, 626—7; elastic plate, 628; plane and constant thickness, Poisson's problem, 629—37; isotropic, 638—49; non-elastic, variable thickness, 650; contour-conditions of Poisson, 651—2; elastic plate, aeolotropic, 656; elastic rod, of rectangular cross section, sound-vibrations, 653—5; aeolotropic rod, 657; relation between stresses for two sets of rectangular axes of a body, 658; longitudinal vibrations of a cylindrical or rectangular rod, 660; torsional do., 661, 669; Saint-Venant on results, 661; equation of equilibrium and motion of system of particles under attraction, 662; general expression for stretch, 663—5; general expressions for stress in terms of strain for an aeolotropic body, 666—7; on Savart's law of vibrations, 668; elastic equations deduced from consideration of bodies as molecular or continuous structures, 670—1; stress in a double system of particles under attraction, 672—3; analysis of stress, 674—677; stresses within a solid and near surface, 676; definition of stress at a point, 678—9; report on Wertheim's experiments, 682—3; torsion of prisms, Saint-Venant on Cauchy's view, 684; report on Blanchet's paper on waves in crystalline media, 1172; controversy with Blanchet on wave surfaces, 1175—6; Maxwell on Cauchy's researches, 1538
Chevallier. Flexure and after-strain of beams subjected to transverse load, 1461
Chevandier. Elasticity of glass, 1310—11, of wood, 1312—14. See *Wertheim*
Chladni. His experiments on sound, Jas. Bernoulli, 122, Biot, 184; his books, p. 411, ftn.; his experiments on vibrations of cylinders, 470; his figures in crystals and wood, Savart, 337; Savart on, 352; Strehlke on, 354—360; Faraday on, 745; Wheatstone on, 746
Circular. Ring, vibrations of, Euler, 93, Lexell and Hoppe, 95 $\beta$; Germain, 290—99; fixed vertically, depression of, St Venant, 1575—6, 1593; arc, flexure of, Navier, 278, Ardant, 937, Bresse, 1457—60
Clapeyron. Equilibrium of elastic solids 277, 1001, etc. See Lamé. His theorem, Lamé on, 1067—8; example of triangular frame loaded vertically, 1069; modification of his theorem by introduction of body forces, 1070; example of, heavy elastic prism hung up and loaded, 1070
Clark. *The Britannia and Conway Tubular bridges*, etc., 1465; experiments on cylindrical and elliptical tubes, 1466; on rectangular tubes, &c., 1467; on very large model tube, 1468; on cellular tubes of wrought iron, 1477; calculations for the two tubes, 1489; beams, continuous, 1472, 1488; wrought iron, state of ease and set of, 1473; transverse strength of cast-iron, 1483—4; wrought iron, change of shape under flexure, 1485; increase of strength from initial stress, 1486; cylinders subjected to internal loads (large), 1474; set and plastic stage in wrought iron, 1475; extension and ultimate strength of cast-iron, 1476; experiments on crushing of brickwork, 1478; on shearing strength of rivets, 1480; on friction set up by cooling of rivets, 1481; on strain in rivet, 1482; resilience, 1487
Clausius. On Weber's experiments on temperature-effect, 706; relation of uni-constant theory to experiment, 1398; number of elastic constants, 1399; concludes there is only one constant, 1400; failure of experiment to agree with theory due to after-strain, 1401—2; application to Wertheim's results, 1402—3; effect of

after-strain on stretch-modulus, 1404—5
Clebsch. Theory of Elastic Bodies (translation by Saint-Venant), 8; on vibration of elastic plate, Germain, 291; vibration of elastic sphere, 463; strain-ellipsoid, 605
Coefficient, see under *Elastic*
Cohesion. Newton, 26, 27; Musschenbroek, 28 γ; Coulomb, 120, pp. 873—74, p. 876; Young, 134, 142; Voute, 212—3; Rondelet, of mortar, 696; due to Newtonian attraction? 164—5, 361, 754—757; Frankenheim, 821; of gases and fluids, 822; of solids, 823; surface of, p. 886; relation to magnetism, 829 and ftn.; attractive force necessary to cause, Mossotti, 842; Faraday on his ideas, 843; molecular force of, 976; bricks and mortar, Morin, 905; Lamé, 1034; Poncelet, 976—77; and heat, Wilhelmy, 1356; affected by temperature, 334, 1353: see *Temperature Effect*; Eisenbach, theory of, p. 876; see also *Attraction, Atom, Molecule, Strength, Rupture*
Cohesive Limit, 365, 852, 986, 1301 (9°), p. 874 ftn., p. 891
Colladon. Experiments on compression of liquids, effect on vessel, 688
Collision of elastic bodies, history of, p. 26, ftn.; experiments on, Hodgkinson, 941, Morin, 1183, Haughton, 1523
Colour-fringes. Use of to measure strains, Brewster, p. 640, ftn.; Neumann, 1185—1213; Maxwell, 1556—7
  Homogeneous strains. Strained 6-face, 1191; rectangular glass rod transversely loaded, 1191; use of to measure elastic constants, 1193
  Heterogeneous strains. Plate with parallel faces, 1195; varied distribution of temperature, 1196; case of sphere, 1197; thin plate, temperature constant only in direction of thickness, 1198—1202; long thin rectangular plates cemented at edges, 1203; hot plate edge on cold surface, 1204; comparison of theory with experiment, 1205
Colthurst. Experiments to find neutral axis and stress-strain relation in wooden and cast-iron beams, 998
Column. Euler. Curvature of, 65; moment of stiffness of, 65; load partially transverse, 66; buckling and rupture of, under vertical load, 74; moment of stiffness in terms of cross-section, 75; position of neutral line for, 75 and ftn.; equations for buckling under its own weight, 77—85, 910; greatest possible height of, under own weight, 80—85, 910
  Fuss. Application to framework of Euler's formulae, 96 *a*
  Lagrange. Form of (enlarged at centre), 106; on Vitruvius' theory, 106; reference to Euler, 107; equation for bent cylindrical, 108, 110; as a surface of revolution, 111; generated by conic, 112; cylindrical determined to be best form; 113
  Robison on Euler's results and on neglect of compression, 145—6, 833
  Tredgold. Strength of timber, 196; double parabola as form for, 195—6; theory of, 833; position of neutral axis, 198, 834
  Duleau. Euler's theory and experiments, 226—229
  Navier on Euler's theory, 281
  Pagani. Equilibrium of, 397
  Thury. Experiments on hollow cast-iron, 750
  Arosenius. Theory of, 819
  Heim. Theory of, 907—10; height and stability, 910
  Hodgkinson. Experiments on cast-iron, 954; ratio of strength of built-in to pivoted, 955—56; point of fracture with different ends, 957; load for flexure, 958; Euler's formulae, modification to include compression, 959; effect of dimensions in long, 960—62; breaking loads of short, 962; hollow cylindrical strength of, 963; time-element in regard to, 964; stone, strength in regard to length, 1445—7; iron, experiments on, 1449
  Poncelet. Best form for, 987
  See also *Strut, Beam, Tractive Load, Pillar*
Commission, Royal, to enquire into use of iron for railway purposes, 1406—32
Compression. Mariotte's idea of, in beam, 10; Jas. Bernoulli, introduction of, in strained beam problem, 19; of beam by longitudinal loads, Musschenbroek, 28 δ; laws of, Bülfinger, 29 β; strains of beams, Hodgkinson, 232—3; ratio of to pressure, Hodgkinson, 234; effect on position of neutral axis, do., 235; modulus of for cast-iron, do., 240; ratio to extension for cast-iron, do., 239, for wrought iron, do., 242; of liquids, p. 78 ftn.,

318, 686—691, 1011, 1357—8, 1359; of a sphere, Poisson, 535; of water in glass vessels, effect on vessel, Oersted, 687, Lamé, 1011; of rollers and spheres by tangent planes, experiments, Vicat, 729—30; heat liberated in solids by sudden, Duhamel, 881—2; sudden, of spheres, do., 888; for different bodies, values of, Lamé, 1034; of solids, Regnault, 1357—8; of metals, defect of elasticity in, 1449; see also *Stretch* (squeeze), *Traction* (contraction)

Conical Refraction, 1105, 1514

Conservation of energy, first conception of, 39; Euler, 54

Constants, see under *Elastic*

Cooling. Of free sphere, rate of, Duhamel, 895—6; of bars, ends fixed to moveable masses, 902—3, to other bars, 904

Coordinates. Trilinear, 1076; spherical, 1092—5, 1112—48, 1154—63; cylindrical, 1022, 1087—9; curvilinear, 1037, 1096, 1149—53, 1241—6

Copper. Stretch-modulus of, 751 *d*, 1292, 1294, 1299, 1309, 1393

Cords. Flexible, equilibrium of, Euler, 71, Poisson, 568, De Schultén, 818; flexible, vibrations of, Euler, 69—70, Poisson, Bourget, 576; do. history of, p. 42 ftn.; flexible or elastic, equilibrium of, Euler, 72

Elastic, equilibrium and motion of, Poisson, 464; explanation of tension in, 464; longitudinal vibrations of, 465; experiment of Caignard-Latour, 465

Elastic and non-elastic, vibrations of, Savart, Duhamel, 1228

See also *Strings*

Cores. Theory of applied to beams, 815, p. 879

Coriolis. Experiments on strength of lead cylinders, after-strain, skin influence, 720

Coulomb. Saint-Venant on, 115; position of neutral line, 117; Young and Saint-Venant on, 117—118; torsion of metal threads, 119, p. 69, ftn.; cohesion, 120; Girard on, 120; Duleau on, 227; Vicat on, 729; effect of worked state on absolute strength, extension of elastic limit and sets, pp. 873—874; lateral adhesion, sliding strain, p. 876; Coulomb-Gerstner Law, p. 441 ftn., p. 872, pp. 887—891, 852, 1224, 1431

Cox. Dynamical deflections &c. in railway girders, Stokes' criticism, 1433; impact on elastic beams, 1434; calculation of deflection, 1435—38, pp. 894—6; hyperbolic law of elasticity for cast iron, 1439—42 and 1452

Crushing. Coulomb, 120; Rennie, 186; Vicat, 729; experiments on short prisms, cast-iron, 948, 1414—5, 1476, bone &c., 950; laws for, 948—50; surface of rupture, 951; of brickwork and stone, 1478

Crystals. Elasticity of, Biot, 182, (regular) Savart, 335; analysed by Chladni's figures, Savart, 337, do. (rock-crystal), 341—3, do. (carbonate of lime), 344; rock-crystal, nodal lines of, Savart, 342; position of axes of, 343; of three rectangular axes, body-shift equation for, 615

Cohesion and elastic properties of, Frankenheim, 825; Voute, 213

Homoedric. Six elastic constants for, 794; stretch-moduli of, 795—6; stretch-modulus quartic for, 797—800; application to measure stretch-modulus, 800—801

Inner structure of, cleavage &c., 336

Systems similar to gypsum. Neumann, 788; comparison of different axes, 788—89; elastic and optic axes, 790; thermal axes, 791; identity of all these with crystalline, 792; application to case of gypsum, 793

Crystalline. Stress in bodies, Poisson, 582—6; equilibrium and motion in, do., 582-6, 594, 600, 615, 666; medium, propagation of waves, Blanchet, 1166—71; medium, vibratory motion in, O'Brien, 1360—65; state of iron after vibration, 1463—1464.

Curvilinear coordinates. Lamé's theory of, and application to elasticity, 1037, 1096, 1149—1153; change of three body stress-equations to, 1150; of equation for homogeneous isotropic solids, 1153; Bonnet, application to elastic equations, 1241—6.

Cylinder. Rupture of hollow, Galilei, 2; best form for column, Lagrange, 112—113; equilibrium of indefinite, Lamé, 1012, 1022, 1088; hollow cast iron, strength of as columns, Thury, 750, Hodgkinson, 963; wrought and cast iron, strength of, under great internal pressure, 1474; of wires, torsion of, Maxwell, 1549; wooden, torsion of, Gerstner, 810. See *Column*, *Bar*, *Wire*

Cylindrical coordinates. Change to, from rectangular, 1087; application to theory of elasticity, 1089, 1199, 1542, 1551.

Defect of Hooke's Law, 969, 1411, pp. 891—93
Deflection. Of Bridges. Experiments on, 1373; by travelling loads, experiments, 1417—22; Britannia, 1489. Of girders by travelling loads, experiments, 1433. Of elastic beams by impact, 1434—37; 1438—42. See also *Beam, Flexure*
Delanges. Equilibrium of body supported at more than three points, 155
Denison. Experiments on American timber, 836
Density. Of earth, mean, 166; of water compared with earth, 166; of solids before and after deformation, 776, 1522 (iii)
Desaguliers. *Thoughts on Elasticity*, 29 γ
Descartes. Theory of Elasticity (subtile aether), 29 α, 34
Dilatation. Poisson, 586, Poncelet, 980, Cauchy, 612, 663—5; effect of movement of body as a whole on, Poisson, 587; Piezometer, Colladon, 688 (see under *Compression*); Poisson on results of, 688; due to heat, coefficient of, for various metals, Kupffer, 1395; of glass, experiments on, Regnault, 1227, 1357; definition of, p. 882. See also *Compression*
Dufour. Wire suspension bridges, 692; experiments on wire of Geneva bridge, 692; elasticity and strength of cables of suspension bridges after twenty years' use, 1455
Duhamel. Specific heats of metals, 705; effect of change in temperature on elastic equations, 868—70; equilibrium of a hollow sphere, variable temperature, 871; vibrations of sphere of variable temperature, 872—5; thermal variations in elastic constants, 878; ratio of specific heats, 879—80; heat liberated by sudden compression, 881—2; general thermo-elastic body and surface equations, 883—5; applied to following five cases, wave motion in an infinite solid, 886—7, equilibrium of a body suddenly compressed, 888, wire suddenly loaded, 889—90, equilibrium of a spherical shell, 891—4; rate of cooling of a free sphere, 895—6; sound problems, fiddle bow, 897; vibrations of a cylindrical rod under varying temperature, 898—901; cooling of bars with ends fixed to moveable masses, 902—3, and to another bar, 904; on N. Savart's experiments on vibrating cords, 1228
Duleau. Strength of forged iron, 226; neutral line of beams, 227; beam problems, various load systems, 228; neutral line of columns, 228; torsion experiments, 229, cited by Wertheim, 1340, by Saint-Venant, 1625; Hodgkinson on, 239
Dupin. Experiments on flexure of wooden beams, 162

Earnshaw. Nature of molecular forces, 1230, 1232
Eisenbach. New theory of cohesion, p. 876
Elastic coefficients, distinguished from moduli, p. 884; dynamic, 941, 1523
Elastic constants. Thermal variations in, Duhamel, 878; controversy on, multi- and rari-constancy, 921—33; assumption as to, and relation between, Lamé, 1034, 1040, 1065, 1163; optical methods to measure, Neumann, 1193; number of, Stokes, 1265; unaltered by working, Coulomb, p. 874; ratio of to one another, Wertheim, 682; six for crystals with three rectangular axes, Poisson, 794; relation between for uni-constant isotropy, Wertheim, 1320—26, 1342—51; relation between in crystalline medium, Fresnel and O'Brien, 1363—4; number of independent, Clausius, 1400; values of by vibrational and statical experiments disagree, 931, 1404, p. 702 ftn., p. 814 ftn.; definition and terminology for, pp. 884—885. See also *Modulus*
Elastic curve or line. Jas. Bernoulli, 18, equation to, do., 24; variation of elasticity along, Euler, 52—57; *vis potentialis* of, do., 55; equation to, fixed at one or two points, do., 56, for single force, do., 58, for two or more forces, do., 58, Saint-Venant on, 58; nine species of, Euler, 59; the rectangular, do., 59; equation for elasticity variable, do., 60—61, for forces at every point, do., 62; in same plane with loads, Plana, 151—4; of circular arc, 278, 1458, 1573; memoir on, Piola, 362; classification of, Hill, 399—401; for a heavy rod, do., 401; Lagrange's error in, do., 402; equation to, do., 403—4; family of these curves, do., 405; forms of equations for, do., 406; *characteristica* of, do., 407
of double curvature. Lagrange, 159; Binet, 173—175; Bordoni, 215—222; difference between Poisson and Bordoni, 223—4; moment of

torsion, Poisson, 423; third moment, Bellavitis, 935; Mainardi, 1215; integration of equations to, 1239—1240; complete discussion of problem and its history, Saint-Venant, 1584—1609

Elastic fluids. See *Fluids*

Elastic life of a material, p. 885—891

Elastic limits, 240, 854, p. 886, p. 874, 1381. See also *Iron*, *Wood*, etc.

Elastic solids, equilibrium of. Young, 136—41; Navier, 253, 265—71; Clapeyron, 277; Pagani, 395; Poisson, 431—3, 540—551; Cauchy, 602—3, 614—5; Haughton, 1505—15, 1520—21; Jellett, 1526—35; Maxwell, 1536—41

Elasticity. Birth of modern theory of, Navier, 1; hypothesis of, Petty, 6; law of, Hooke, 7, 34, Newton, 26, Musschenbroek, 28 $\gamma$; hypothesis of atoms being loadstones, Desagulier, 29 $\gamma$; *elastica vis*, Belgrado, 29 $\delta$; Cartesian theories of, 34; due to captive ether, John Bernoulli, 35; attraction as cause of, s'Gravesande, John Bernoulli, 42; theory of, Riccati, 33, 38—40, Euler, 94, Sanden, Hannelius, Krafft, pp. 77—79, Young, 134, 142, 143, Biot, 182, Fresnel, 320, of cast-iron, Hodgkinson, 240; of bodies of three rectangular axes, Savart, 340; glass threads, Ritchie, 408; theory of torsional, Ritchie, 408; history of, Girard, 125, Poisson, 435; effects of temperature on, Duhamel, 868—70; defect of, in cast-iron, Hodgkinson, 969; laws of, Maxwell, 1536—39; theory of, distinct from cohesion, Coulomb, p. 874; surface of perfect, pp. 885—891

Greater and less, meaning of, 522

Elective attractions, Newton's query on, 26

Electric currents in vibrating rods and wires, 1248; influence on elastic properties, Wertheim, 1304—6, De la Rive, 1332, 1336; effect on flexure of bar, Guillemin, 1327—30, Wartmann, 1331; sounds in bars produced by, history of, p. 719 ftn., experiments on, Wertheim, 1333—5, Sullivan, 1248, Beatson, p. 720 ftn.

Ellipsoids of strain, stress, stretch. See *Strain*, *Stress*, *Stretch*

Ellis. Criticism of Mossotti's hypothesis, 1231

Emerson, his *Mechanics*, p. 74, ftn.; his paradox, 187, 952 (ii)

Energy. Conservation of, 39, 54: see also *Work*

Ether. Theories of structure of, 930; on refractive power of, 1187; fluid, laws of equilibrium, Lamé, 1025—31; do. in a diaphanous body, 1032—3; existence of, 1108; constants of, Stokes, 1266. See also *Aether*, *Crystalline Medium*, etc.

Explosion in boilers, cause of, Lamé, 1035

Euler. Letters of Bernoulli to, 43, 44—47; small oscillations of rigid and flexible bodies, 51; elastic curve, equation for, 53, 56—62; Calculus of Variations, 53; least action, 54—55; oscillations of elastic lamina, 63, 93; of rods, four cases, 64; on columns, 65—8, 74—85; Lagrange on his methods, 107; vibrations of cords, 69, 70, of laminae and elastic rods, 71—2, 86—92; oscillations of elastic rings, 93; nature of elasticity, 94; miscellaneous memoirs, 95; pupils, 96 $\alpha$—$\gamma$; Giordano Riccati on, 121; Jas. Bernoulli on (vibration of membranes), 122, p. 73 ftn.; Robison on his theory of columns, 145—6; Barlow, do., 189; Duleau do., 226—229; Hodgkinson, do., 954—64; Lamarle, do., 1255

Extension of fibres of strained beam, Mariotte, 10, Leibniz, 11, Varignon, 14, Bernoulli, 20, 24, Musschenbroek, 28 $\delta$, Bülfinger, 29 $\beta$; relation of sum of, to sum of compression on beams, Hodgkinson, 232; laws of, do., 234; effect on modulus and neutral axis, do., 235; ratio to compression for cast iron, do., 239; of wrought-iron, do., 242; elastic, Navier, 273; by longitudinal vibrations in rods, Savart, 351; of threads and elastic plates, Poisson, 430; of annealed and unannealed wires, Dufour, 692; ratio to compression in wrought-iron, cast-iron and fir beams, Colthurst, 998

Eytelwein. On beam problem, 149—150

Fairbairn. Controversy with Hodgkinson and Stephenson, 1448, 1494
Experiments on cast-iron made by hot and cold blast, 953; wrought-iron pillars, strength of, 1354; iron plates and riveting, 1444; do., for ships, 1495; do., strength of joints, 1496; do., strength with and across fibre, 1497; iron wire, 1498; riveted joints, 1499—1500; plates under transverse load, 1501; strength of frames and

ribs of ships, 1502; absolute strength of wrought-iron, 1503; hollow tubes, 1448; tubes of various shapes, 1494; tubular bridges, 1463 *f*; 1494

Faraday. Acoustical figures, Savart's theory, 745; theory of attractive and repulsive molecular forces, 843

Fiddle. Vibration of produced by bow, Duhamel, 897

Flachat. Use of cables in suspension bridges, 1454

Fluids, Elastic. Sound-waves in Euler, 95; motion and equilibrium of, Laplace, 307; laws of molecular action in, Laplace, 309, 312—315; velocity of sound in, Laplace, 316—317; theory of, Pouillet, 319; motion and equilibrium of, Poisson, 540—558, 559—63; pressure at a point in, Cauchy, 609; equilibrium of, do., 673; equation of continuity in, Piola, 776; motion of, do., 778—9; theory of, do., 782—7; cohesion of, 822; in motion, friction of, Stokes, 1264—66; difference from solids, Haughton, 1508

Forbes. Compression of glass vessels, p. 379 ftn.; instrument to measure stretch-modulus, 1397

Fourier. Vibrations of elastic membranes, 207; waves in do., 207; vibrations of simple plates, 208, of a lamina, 209—10, of flexible and extensible surfaces and plates, p. 876

*Forza viva e morta.* 29 δ, 35, 38, 40

Frankenheim. *Die Lehre von der Cohäsion*, 821; cohesion of gases and fluids, 822, of solids, 823; elasticity, 824; crystallography, 825; structure of solids, 826; after-strain in glass, lead, iron, 827; influence of temperature on elastic properties, 828; absolute strength, 829; relation of cohesion to magnetism, 829 and ftn.

Fresnel. Double refraction of light, 320; theories of molecular motion and vibrations, 320; St Venant on do., 320; Cauchy on do., 604, on formulae of, 611; elastic axes of crystals, 790; O'Brien on, 1363

Friction. Young, 143; effect on solids, Morin, 905; in riveted joints by cooling of rivets, 1481

Funicular (or link) polygon. Pagani, 381; Poisson, 568

Fusinieri, 396

Fuss. Application of Euler's formulae for struts to framework, 96 *a*

Galilei. *Discorsi*, 2; fracture of rods, beams, hollow cylinders, 2; solids of equal resistance, 4; motion of bristles on sounding boards, 746

*Galilei's Problem.* Statement of, 3; application of Hooke's law to, 10; Leibniz, 11; de la Hire, 12; Varignon's treatment of, 13, 15; Jas. Bernoulli, 19; Bülfinger on his hypothesis, 29 β; Euler on, 95; Banks, 147; Gregory, 148; Hodgkinson, 231

Galton. Experiments on effect of travelling loads, 1417—22. (See *Willis*)

Gehler. *Physikalisches Wörterbuch* (historical notices), 695

Germain, Sophie. Theory of elastic surfaces, 283; account of life, p. 147 ftn.; prize essay, account of, 284; hypotheses for elastic plate, 285; on Poisson's results, 286—7; general equation for equilibrium of surface, 288; thickness of surface, Poisson and Clebsch, 289; vibrations of cylinder and circular rings, 290, 294—8, of lamina, 290—91, 292—3; comparisons of theory with experiment, 299; vibrating surfaces, 300—305; general remarks on her own work, 306

Gerstner. *Handbuch der Mechanik*, 803; experiments on piano-wires, 804, on ratio of traction to stretch, 805, on flexure of wooden beams, 807—8, on flexure of cast and wrought-iron beams, 809, on torsion of wooden cylinders, 810

Gerstner's Law, 806; Coulomb-Gerstner Law, p. 441 ftn. 852, pp. 871—2, pp. 887—91, 1224, 1431; Gerstner-Hodgkinson Law, p. 441 ftn. 852, 969, 984, 1217

Girard. On solids of equal resistance, 5, 123, 125, 131; experiments on beams at Havre, oak and deal, 28 η, 123, 131; on Coulomb, 120; history of elasticity, 125; experiments and formulae for beams, 125; on resistance to slide, 126; Galilei's and Mariotte-Leibniz theories, 127—28; Bernoulli's position of neutral axis, 128; on Varignon, 129; elastic curves, beams and columns, 130; on Euler and Lagrange, 130; Barlow on, 189—90; report on Vicat's papers, 724—35

Giulio. Experiments on iron used in Italy, 1216, 1217; rupture, flexure and elastic line of beams, 1216; Gerstner's Law, 1217; torsion of iron wire, 1218; spiral springs, 1219—1223; set and elastic after-strain, 1224—6; fore-set and after-set, 1226, Saint-Venant on, 1595, 1608

Glass. Threads, torsion of, 408—9; do., how to draw, 410; vessels, compression of, 686—91; polarisation of light by strained glass, 698, 1191—1205, p. 640 ftn., 1543, 1553—1557; compression of, experiments on by Parrot, 747; deflection of bar to measure elastic constants, Neumann, 1193; set in spheres and cylinders of, by rapid cooling, do., 1212; dilatation of, experiments on, Regnault, 1227; cohesion and elasticity of, experiments on, Wertheim, 1310—11

Glue. Cohesion of, in a thin coat and in mass, Bevan, 374

Glynn. Fracture of axles, 1463

Gold. Stretch-modulus of, note of coin, 751 *d*

Golovin. Notes of harmonicum, 96 $\gamma$

Grandi, Guido. Controversy on solids of equal resistance, 5

Grassi. Experiments on compression of liquids, 1359; application of Wertheim's formulae, 1359

s'Gravesande. Composition of elastic bodies, Belgrado on, 29 $\delta$; stretch-traction relation, 37; laws of elasticity, 42; Forbes' modification of s'G.'s apparatus, 1397

Green. Refraction &c. of light at common surface of two non-crystallised media, 917; propagation of light in crystallised media, 917; body-shift equations for free vibrations of elastic solids, 918—19; values of strain-components for large shift-fluxions, p. 865, ftn.; McCullagh on his methods, 920; controversy on elastic constants, 921—33; Saint-Venant on his methods, 934, Jellett on do., 1529 and ftn.

Gregory. On Galilei's results, 148

Gresy, Cisa de. Equilibrium of flexible surfaces, 199; application of Lagrange's method to Poisson's theories, 200—6.

Griffin. Motion of a rigid body, 765

Grindstones. Effect of centrifugal force on, 1551, p. 827 ftn.

Guillemin. Electric currents, effect on elasticity, 1327; Wertheim on his results, 1328; letter to Arago on do., 1330

Gut. Experiments on, Karmarsch, 749

Gypsum crystals. Relation between various axes of, 788—93

Hagen. Experiments on moduli of wood, 1229

Hannelius. Cause of elasticity, p. 79

Harmonicum. Notes of, Golovin, 96 $\gamma$

Haughton. Equilibrium and motion of solids and fluids, 1505—6; definition of molecular structure, 1507; distinction between solids and fluids, 1508; general equations of equilibrium, 1509—11, 1519—1521; do., application to wave-theory of light, 1512; surface of wave-slowness in elastic solids, 1513; application to various elastic bodies, 1514; classification of elastic media and laws of plane waves, 1516; body-stress-equations for solids, 1517; do., application to wave-theory of light, 1518; propagation of plane waves, 1510; normal and transverse vibrations, 1522; dynamical coefficients of elasticity of iron teak, &c., 1523

Hausmann. Magneto-elastic properties of stricture in bar-iron, 1179

Heim. Flexure of beams, 906—9; maximum height of conical and other columns, 910; equilibrium of elastic rods with curved central axis, 913; solids of equal resistance, 915; vibrations of elastic beams, 916, 1387; best form of rails, 916; equation of curve of double curvature, 1385

Hill, C. J. Torsion of metallic threads, 225, 398; elastic curves, 399—407; on osculating parabola, 406

Hire, de la. *Traité de Mécanique*, 12

Hodgkinson. Beam, strained state of fibres in, 230; position of neutral axis for glass, etc. 231; calculation of neutral axis for beam, 232; on Barlow's errors in do., 233; two moduli used in do., 234; position of neutral axis, 235, in wood, experiments on, 235, in iron, experiments on, 237; position of neutral axis alters with moduli, 238; cast-iron, experiments to determine best form of beam, 243; do., of equal top and bottom ribs, experiments on, 244; effect of lateral impact on, 939, 942, 943; experiments on cast-iron, 952; on Cubitt's do., 970; cast-iron, position of neutral axis, 971; horizontal long-continued impact, effect of, 1409, p. 895; vertical impact on loaded cast-iron beams, 1410; deflection and strength of long cast-iron beams, 1413

Cast-iron, experiments on extension and compression of, 239; on stretch-moduli, 240; on limits of elasticity, 240; on absolute strength, 240; on strength of T bar, 241; comparison with elasticity of wrought iron, 242,

245; tractive strength of, 940, 966, 1408; made by hot and cold blast, strength of, 947; time element in beam of, 967; temperature-effect on, 968; defect of elasticity in, 969, 1449; torsional resistance of, 972—3; contractive strength, 1408; stress-strain relation for bars of, 1411—12

Wrought-iron, experiments on, stress-strain relation for, 234; extension and compression of, 242; strength of, 1354

Columns. Cast-iron, experiments, 954; ratio of strength of flat and round ends, 955; form of round ends, 956; points of fracture, 957; general results, 958; Euler's formulae modified to include compression, 959; effect of dimensions on strength of, 960—2; hollow, 963; time-effect, 964; physical constants of, 965

Wrought-iron, strength of, 1354

Stone, experiments on, 1445; variation of strength with length, 1445—7

Crushing. Experiments on iron short prisms, 948—9, 1408, 1414, 1415; on bone and marble, 950; on stone and timber, 1451; on theory of surfaces of rupture, 951

Collision of imperfectly elastic bodies, 941

Experiments on tubes etc. for tubular bridges, 1416, 1448, 1450, 1477, 1493, 1494

Hood. Internal changes in iron during manufacture, 1463 *a*

Hooke. *De Potentiâ restitutiva*, 7; theory of springs and springy bodies, 7; congruity etc. of bodies, 9, 34; suggestion as to gravitation, p. 6 ftn.

Hooke's law. Statement of, 7; generalised, 8, 928, p. 884, p. 888; Mariotte on, 10; Leibniz on, 11; Bernoulli on, 22; Bülfinger on, 29 *β*; Riccati on, 31; near limit of elasticity, 367; defect of, p. 891, curves showing defect, pp. 892—893

Hooks, strength of, Brix, 1249

Hoppe, 96 *β*

Hopkins, internal stresses in rock-masses, 1366—9; his theorem as to maximum shear, 1368

Hyperbolic law of elasticity for cast-iron, 1440—2, 1452

Ice. Stretch-modulus of, Bevan, 372, compared with that of water by Young, 372, Thomson on, 372 ftn.

Ignaz von Mitis. Experiments on relative strengths etc. of iron and steel, 693

Impact. Of elastic bodies, history of, p. 26 ftn.; Riccati on, 40; Poisson, 579; Hodgkinson (relation of dynamic coefficient to stretch-modulus), 941; Morin, 1183; Haughton, 1523

Transverse, on beams, Hodgkinson, 939, 1409—10; experiment and theory, 942—43; Cox, 1434—37, pp. 893—896

Longitudinal, on beams, Poncelet, 988—993

See also *Resilience*, *Beam* (Travelling Load)

India-Rubber. Experiments on, Gough, p. 386 ftn.: Page, 1370; Wertheim, 1322; Saint-Venant, 1625. Is it an 'elastic' solid? 924, p. 504 ftn.

Iron. Cast. Strength of, Tredgold, 197, 999, Hodgkinson, 237—44, 952, 940 and ftn., 955—70, 1408, James, 1484, Bramah, 837, Giulio, 1216, Bevan, 377, Lagerhjelm, 363—4; crushing strength of, Rennie, 186—7, Hodgkinson, 948—51, 1414, (of pillars) 954—65, (of struts) 1477; elastic limit in, 240, 1430; set in, Hodgkinson, 969, 1411; extension and compression in, 239—41; effect of hot and cold blast on, Hodgkinson, 946—47, Fairbairn, 953; hollow cylinders as pillars, Thury, 750; beams, Musschenbroek and Buffon, 28 *ε*; do. best form for, Hodgkinson, 243—4, 1431; general properties of, 1423—1432

Difference in properties of cast and wrought, Hodgkinson, 245, Clark, 1473—1477

Wrought. Strength of, Duleau, 226, 229, Lagerhjelm, 363—4; (of plates) 1495—1502; (bars after repeated fracture) 1502; (of bars) 1462; (effect of heat, "thermo-tension") 1353; Giulio, 1216—7; (stricture of round and square) 1003; set in, crushing, 1473—6; magnetisation by stricture, 1179; extension and compression of, 242; stretch-modulus (from sound experiments), 370; (same for all kinds) 366; (variation in neighbourhood of elastic limit) 367; elastic limit, relation of to cohesive limit, 365, 852, 986, 1301; set, 817; after-strain, 817

Wire. Strength of, 692—3; do., and stretch at elastic limit, 693; time-effect, 736; effect of annealing, 858; experiments on piano-wires, 803; set, yield-point, effect of annealing in, 850—858 (see also *Wire*); state of ease in, want of isotropy, 1001—3; kinds useful for ship-cables, 817; cables of suspension bridges, 692, 723, 817, 936

use of in railway structures, 1406—32; molecular changes in, due to vibration, 1429, 1463, 1464

Isostatic Surfaces. Discovered by Lamé, 1036—37; definition of, 1152; Boussinesq on, 1152

Isotropy. First appearance of bi-constant, formulae for, 614, 1267; initial stress, as leading to bi-constant formulae for, 615—6; controversy over the uni- and bi-constant theories, 921—33; number of constants of, Lamé's views, 1007, 1034, 1051, 1073, 1164, Stokes', 1266, Maxwell's, 1537 and ftn., Clausius', 1399—1405, p. 893, Wertheim's, 1319—1326, 1338, 1339—1351; of glass vessels, 691; of wires, 831, 858, 924, 1002; of bars and plates, 858, 924

James. Experiments on travelling loads on beams, 1417, 1423, on transverse strength of cast-iron beams, 1424, on reiterated stress on beams, 1425

Jellett. Hypothesis of intermolecular action, 1526—7; attracting and repelling molecules, number of constants, 1530; remarks on Green, 1529; objection to usual surface-equations, 1532—3; coefficients represented by integrals, 1534—5

Johnson. Effect of thermal treatment on tenacity of iron, 1353

Jurin. On springs, p. 58, ftn.

Karmarsch. Experiments on metal wires, 748, on gut, 749

Kelland. On molecular force, 1230; on Mossotti's hypothesis, 1230—32

Kennedy, A. B. W. Experiments on position of neutral line in columns, p. 44 ftn.; stretch-modulus of steel, p. 702 ftn.; strength of cast-iron, p. 512 ftn.,; effect of temperature on strength, p. 730 ftn.; stress-strain curves different metals, pp. 889—893

Kirchhoff. On Poisson's contour conditions for elastic plates, 487—8, 652; his results cited by Wertheim as supporting the hypothesis $\lambda=2\mu$, 1342, 1348

Kirwan. Relation of magnetism and cohesion, p. 452 ftn.

Krafft on elasticity, p. 79

Kupffer. Torsional vibrations of metal wires, 1389—92; values of slide- and stretch-moduli for metal wires, 1393—4; supposes wires to have uni-constant isotropy, 1394; dilatation coefficients for change of temperature, 1395

Lagrange. Equations for bent springs, 97—104; spiral springs, 105; columns, equation for, 106—110, as surface of revolution, 111, various forms of, 112, of greatest efficiency, 113; equilibrium of flexible strings, 158, of membranes, 158, of elastic wire or lamina, 159; error of, Plana on, 153; Bertrand, Binet and Poisson on, 159, 174—5; supposed error of, Schultén and Poisson on, 536—9, Ostrogradsky on, 744; Piola on his methods, 760—64

Lagerhjelm. Experiments on density and elasticity of iron, cast and wrought, 363; method of testing, 369; relation of limit of elasticity to cohesive limit, 365; modulus of elasticity same for all kinds of wrought-iron and steel, 366; limit of elasticity and breaking load vary with character, 367; raising of limit of elasticity, 367; specific weight of ruptured bar, 368—9; comparison of moduli deduced from extension and sound experiments, 370

Lama, Francesco de. Experiments on beams, 28 δ

Lamarle. Experiments on flexure of wooden beams, 1253; relation of stretch and squeeze-moduli, 1254; on Euler-Lagrangian theory of struts, 1255; struts doubly pivoted, 1256—7; do., ratio of diameter to length for set or ruptured preceding buckling, 1258—9; comparison of theory with Duleau's experiments, 1260; report on, by Liouville and Poncelet, 1261; general results, 1262

Lamé. Navier's report on Lamé and Clapeyron's memoir, 277, 1023—4; experiments on iron for suspension bridges, 1001, 1002—3; yield-point and state of ease, aeolotropic character of wires, 1002—3; and Clapeyron, their memoir of 1828 on elastic homogeneous bodies, 1005—6; uni-constant surface and body equations, 1007; stress-ellipsoid, stress-director-quadric, shear-cone, 1008, 1059; uniform surface traction, 1010; compression of water, 1011; cylindrical shell with surface tractions, 1012; cylinder under torsional strain, 1014; gravitating sphere, 1015; spherical shell under surface tractions, 1016; solid bounded by an infinite plane, 1018—19; by two parallel planes, 1020—21; indefinite cylinder, 1022; experimental values of slide-modulus, 1034; on boiler explosions, 1035; thickness and form of boilers, 1038—9; isostatic surfaces, 1036—7; the *Leçons*, 1040—4; definition of stress across a plane, 1045; stress-equations of equilibrium, 1046—9; formulae for stresses, 36; coefficients, 1050—51; bi-constant formulae for isotropic body and general solution of equations, 1052—63; for special body-forces, 1064—66; Clapeyron's theorem, 1067—8; application to triangular frame, 1069, modification of by introducing body-forces, 1070, case of heavy elastic wire, 1070; equilibrium and motion of elastic strings, 1071, of lamina, 1072—3; vibrations and nodal lines of plane membrane, 1074; application of trilinear coordinates, 1076—7; analysis of plane waves, 1078; equilibrium of a right six-face, 1079—81; vibration of a rectangular prism, 1082; application to Savart's Experiments, 1083—6; equations in cylindrical co-ordinates, 1087—8; vibrations of a cylindrical body, 1089—91, equations in spherical co-ordinates, 1092—3; elastic equilibrium of a spherical shell, 1094—5; spherical envelope subjected to *any* surface-loads, 1110—48, 1154—61; application of curvilinear coordinates, 1096, 1149; transformation of body stress-equations to do., 1150—1153; isostatic surfaces, 1152; on bi-constant isotropy, 1151; general principles and fundamental lemmas, 1163—1165

Light. Equilibrium and vibrations of ether, 1025—33; refraction of, in homogeneous aeolotropic medium, 1097; number of elastic coefficients and relations between, 1098, 1100; equation to wave-surface, 1101—2; explanation of refraction &c. on wave theory, 1103—4; conical refraction, 1105; conditions for propagation of wave in elastic medium, 1106—7; general remarks on undulatory theory, 1108—9

Death of, 1165

Lamé's Problem (see also *Shell, spherical*). 1154—62

Lamina, Elastic. Equilibrium of, Jas. Bernoulli, 18, Plana, 151—3, Lindquist, 114, Lagrange, 159, Navier, 262, Duleau, 228, Poisson, 570, Cauchy, 618—27, Lamé, 1072; vibrations of, D. Bernoulli, 44—50; nodes of, 50, 121, 356, 1525; Euler, 63, 64, 86 and ftn., 93; Biot, 184, Plana, 176—80; Fourier, 207—10, Germain, 290—9, Lamé, 1072; flexure of, used in balances, Weber, 718; see also *Bar*, *Rod*, and *Plate*

Laplace. Law of attraction of particles, Belli on, 164; *Mécanique Céleste*, 307—19; attractions of shells and spheres, 308, 312; molecular attraction in fluids, 309, 311; velocity of sound in air, 310, 316, 317; matter and caloric, 313—5; Belli on his molecular hypothesis, 755

Latent heat of fusion. Relation to stretch-modulus, 1388

Lateral Adhesion. Girard, 126; Young, 143, 1565, p. 877

Lead. Compression of, 689, 1326, 1546 and ftn.; crushing strength of, 720; effect of skin, 720; stretch-modulus of, 1292; (effect of heat on) 1307; time-effect on, 978; velocity of sound in, 1184 ($8^0$)

Least Action, Principle of, Euler, 54

Least Constraint, Euler, 54

Leblanc. Experiments for suspension bridges, 936

Leclerc. Experiments on strength and stretch of bar- and wire-iron, 1462

Leibniz, 11: see *Mariotte-Leibniz theory*

Leslie. Gerstner's law attributed to, p. 441 ftn., 984

Lexell. Notes of bells, 96 $\beta$

Light. Propagation of in crystallised and non-crystallised media, Green, 917—20; theory of double refraction in aeolotropic homogeneous media, Lamé, 1097—1109; differential equation for equilibrium and motion of ether, 1025—33. See also *O'Brien*, *Haughton*, *Stokes*, etc.

Limestone. Crushing strength, 186, 1478

Limit of elasticity. Definition of, p. 886 (3); Biot, 182; of iron and steel, 692—3; cast-iron in tension and compression, Hodgkinson, 240—241; definitions of, Lagerhjelm, 365; relation to cohesive limit, 365; alteration of with character of iron, do., 367; raising of, do., 367; beginning of set called, Brix, 853; determination of by colour fringes, 1192; determination for the metals, Wertheim, 1293—6; relation to rupture-stretch, do., 1296; of alloys, relation to that of constituents, 1303; possibility of varying, 1379—81, p. 874 and ftn.; evidence on, before Royal Commission on Iron, 1430; general remarks on, p. 886, p. 889

Lindquist. Flexure of elastic lamina, 114

Liouville. Integration of differential equation for dilatation, 524, p. 290 ftn.; report on Blanchet's papers on wave motion, 1172

Liquids. Compression of, 318, 686—91, 1227, 1357, 1359

Lissajous. Nodes for vibrating rods, 1525

Liebherr. Testing machine, 697

McConnel. Molecular change of iron due to vibration, 1464

MacCullagh on Green's method, 920—21

Magnetism. Relation to cohesion, 829 and ftn.; produced by stricture in iron, 1179; effect on stretch of a bar, 1301, 1329—1335; De la Rive, 1336

Mainardi. Vibrations of an elastic sphere, 860—4; torsion balance, 1214; elastic rod, equilibrium of, 1215

Manfredi. On impact, 29 ε

Marchetti. Controversy on solids of equal resistance, 5

Mariotte. Application of Hooke's law to Galilei's problem, 10

Mariotte-Leibniz theory. Statement of, 11; De la Hire on, 12; Varignon on, 13; Jas. Bernoulli on, 19—22; Musschenbroek on, 28 δ; Bülfinger on, 29 β; Belgrado on, 29 δ; Girard on, 127—30; Hodgkinson on, 231

Martin. Set and time-effect in iron bars, 817

Masson. Experiments on metal rods, 1184

Maxwell. Memoir on elastic solids, 1536—1556; uses bi-constant equations, 1537; analysis of strain, 1539, of stress, 1540—41; application to simple cases of cylindrical coordinates, 1542; optical effect of stress at any point, 1543—4; equilibrium of hollow sphere, 1545; on Oersted, 1546; rectangular beam uniformly bent, 1547; approximate treatment of plate, 1548; torsion of a cylinder formed of wires, 1549, of grindstones and fly-wheels, 1550—51, p. 827 ftn.; influence of change of temperature on elastic equations, 1552; optical effects of strain produced by heating, 1553; slide in beam-problem, 1554; analysis (photography) of strain by colour fringes, 1555—7

Mazière. Theory of "subtile matter," application of Cartesian doctrine to aether, 29 α

Membrane. Elastic. Equilibrium of Lagrange, 158, Cisa de Gresy, 199—206, Pagani, 382—4, Poisson, 472—3; vibrations of, Euler, 122, p. 73 ftn., Fourier, 207, Savart (nodal systems) 329, Pagani (circular) 385—394, Lamé, 1074—75, Lamé (triangular) 1076

Mersenne. Experiments on beams, 28 δ

Metals. Crystalline structure of, Savart, 332

Möbius. Equilibrium of flexible and elastic strings and elastic rods, 865

Modulus. Stretch-. (Modulus of elasticity). Definition of, Young, 137, Frankenheim, 824, p. 885, Euler, 75, Lamé, 1066. Of the metals, Wertheim, 1292—4, 1299, 1301, Kupffer, 1389—95, Napiersky, 1396, Masson, 1184, Bevan, 751 *d*; iron, cast, Hodgkinson, 234, 240; iron, Giulio, 1216, Leclerc, 1462, Lagerhjelm, 366—7, Brix; 849; steel, 986; cast-iron at surface, 821; gold, copper, 751 *d*; alloys, 751 *d*, 1303; relation of slide- to stretch-moduli for metals, 379, 972—3; wood, Hagen, 1229, 1456, Pacinotti and Peri, 1250—52, Wertheim, 1313—4; ice, Bevan, 372; glass, 1310—11; human tissues, 1315—1318; determined by sound experiments, 370, 1293, 1297, 1301, 1344—8, 1372; differs from traction-results and why, 1404; determined from wire in form of helix, 1221, by torsional experiments, 1339—1345, by flexure experiments, 800, 1216, 1254; effect of annealing on, 853, 1293, 1301, of change in density on, 1293, 1301, of heat on, 1295, 1298—9, 1307—10, of electric current on, 1304—6; relation of to molecular distances, 1299—1300, to latent heat, 1388, to slide-modulus, 972—3, 1319—26; instrument to measure, 1397; in homoedric crystals, Neumann, 794—7; quartic, 798—801; effect of set on, 852, p. 441 ftn., p. 887, p. 889; proportional to specific gravity in case of wood, 1252 (cf. 379)

Modulus. Slide-. For wood, 378 (proportional to specific gravity), 379; relation to stretch-modulus, 379, 972—3, 1319—1326, 1339—1343; for metals, 1034, 1393—6; unaltered by slide-set, p. 874

Molecular Force. Law of, Belli, 163—72, 752—8; Biot, 182; Nobili, 211; Mossotti, 840—3; Franklin, 841; Avogadro, 844—5; Kelland, 1230; Earnshaw, Ellis, 1231; Brünnow, 1235—7; Eisenbach, p. 876; Wertheim, 1325

Stress deduced from, Poisson, 436—44, 616; use of integrals for, 436—444, 527—32; Clausius, 1398—1401; Haughton, 1507—9; Jellett, 1526—35; Saint-Venant, 1563; influence on number of elastic constants, 555, 615, 922, 923, 931—2, 1007, 1163, p. 814 ftn.; various assumptions as to law applied to stress, Poisson, 543—544, 585, 592, 597; Poncelet, 975—6, (his graphical method) 977, p. 873

Structure, change of, by vibration, 1429, 1464

Molecules. Ampère, 738; sphere of activity of, Piola, 777, Jellett, 1533; size of, Avogadro, 846; distances of, related to stretch-modulus, 1299—1300; Wilhelmy, 1356

Moment of stiffness. Euler, 65, 66, 74, 75, Lagrange, 108, Young, 138, Poncelet, 979, Hodgkinson, 939

Monochord. Construction of, Weber, 700

Morin. Experiments on cohesion of bricks and mortar, 905; effect of friction on slide, 905; resistance to bodies rolling on one another, 1183; elastic impact, 1183

Mortar. Cohesion and adherence of, Rondelet, 696; adhesion to bricks and stones, Morin, 905

Moseley. Flexure of beams, 1233; impact and resilience, 1233; *Illustrations of Mechanics*, 1234

Mossotti. Spiral springs, 246—51, equations for motion of, 248, hypothesis with regard to, 250, experiments on, 249; the Ballistic machine, 251; forces which regulate internal constitution of bodies, 840; Franklin's hypothesis, 841; two systems of particles, ether and matter used to explain cohesion, 842—3; controversy on his molecular theory, 1230—3

Mushet. Experiments on cast and malleable iron, 1463*c*

Musschenbroek. *Introductio ad coherentiam*, 28 δ; explanation of elasticity by *internae vires*, experiments etc., on extensions and flexures of beams of wood, strength of struts, Euler on M.'s experiments on struts, 28 δ, 76

Multi-constant. See *Constant*, *Molecular Force*

Napiersky. Torsional vibration of wires, 1396

Navier. Originator of modern theory of elasticity, 1, 253; flexure of rods, straight and curved, 254; equation for curvature of elastic rod built-in at one end, 255; length of curve for do., 256; equation for curvature of rods originally bent, 257; flexure of elastic plate, 258; references to Poisson, Germain and Fourier, 259; equilibrium of elastic plates, 260—61; of elastic lamina, 262; rupture of plate, 263; influence of thickness of lamina on equations, 264, 528; general equations for elastic solids, 265—8; Saint-Venant on do., 269; report on Cauchy's memoir, 271; suspension bridges, 272; longitudinal vibrations of rods terminally loaded, 272—273; experiments on tensile strength of various materials, 275, on rupture of shells, 275; controversy with Poisson on equilibrium of elastic solids, 276, 527—34; contour-conditions of elastic plate, 529—32; molecular actions, 531—2, 922, 1400; report on Lamé and Clapeyron, 277; flexure of circular arc, 278; beams, summary of experiments and theories on, 279; torsion of rectangular prisms, 280; limits of safety, 281; continuous beam theory, 281; columns, experiments on, 281; Poisson on his results for equations of solids, 448; report on Lamé's memoir on equilibrium in interior solids, 1023—4

Neumann, F. E. Elastic, thermal, optic, crystalline axes identified in gypsum and allied crystals, 788—93; homoedric crystals, elastic properties of, 794, stretch-modulus for, 795, stretch-modulus quartic, 799—801

Analysis of strain due to load, temperature, set, by effect on light, 1185 *et seq.*

*Homogeneous strain*, general treatment and photo-elastic equations, 1185—90; glass bar transversely loaded, 1191; elastic limit from colour-fringes, 1192; elastic constants from photo-elastic measurements, 1193; relation of refractive index to density, 1194

*Heterogeneous strain*, plate, 1195; thermo-elastic equations, 1196; case of sphere, 1197; thin plates, 1198—1200; circular plates and annuli, 1201; circular arc and "spokes," 1202; two

plates cemented at edges, 1203; hot plate on cold surface, 1204; comparison with experiment, 1205
*Theory of set*, 1206—7, p. 889; relation of set to elastic strain, 1208—9; initial stress, 1210—11; application to glass sphere and cylinder suddenly cooled, 1212
Neutral axis, *Beams transversely loaded*, Mariotte, 10; Varignon, 15; Bülfinger, 29 $\beta$; Bernoulli, 21; Coulomb, 117; Riccati, 121; Young, 134, 138—9, 144; Eytelwein, 149; Barlow, 189—194; Duleau, 227—9; Hodgkinson, 231—35, 238—242, 971; Giulio, 1216; Colthurst, 998; Persy, 811
*Struts*, Euler, 75; Kennedy, 75 ftn.; Robison, 145—6; Tredgold, 198; Barlow, 189—194; Duleau, 227—9; Whewell, 737; Navier, 255; Hodgkinson, 954, 965
Newton. *Optics*, 26; impact of elastic bodies, 37; Musschenbroek on, 28; Belli on, 164; Laplace on, 310; his molecular theory, 755, 1230—2
Nobili. Molecular attraction, 211
Nodal Points, Lines and Surfaces
*For rod*, Euler, 50; Riccati, 121; Strehlke, 356—7; Lissajous, 1525; Savart, 322; Seebeck, 1180
*For prismatic bars*, Savart, 327, 347—9; Peyré, 820
*For Plates*, Savart, 329, 337—40, 352; Strehlke, 353—5, 359; Poisson, 513—20; Wertheim, Kirchhoff, 1344—8; Faraday, 745; Wheatstone, 746
*For sphere*, Poisson, 460
*For membrane*, Lamé (rectangular) 1074—5, (triangular) 1076; Poisson, 472 (ii); Pagani (circular), 391—4
Systems, possible in a body, Savart, 329; rotation of system, 331
Used to determine aeolotropic structure, 332, 337—9, 340—344
Notes. Of a rod, 49, p. 33 ftn., 86 and ftn., 88—91, 96, 429, 470—1, 655, 699, 704; of circular plates, 510—20, 511—21, 1342; of a gold coin when rung, 751 *d*; of hard-hammered wires and tuning-forks, 366, 802; of a sphere, 459—60; of the harmonicum, 96 $\gamma$; of reed-pipes, 700; of a cord, 465, 1228, p. 42 ftn. See also *Vibrations*, *Sound*, *Nodal lines*
Noyon. Experiments on iron, 1462

O'Brien. Symbolic forms for equations of elasticity, 1360—65
Oersted. Compression of water, experiment on, 687; do., in lead vessels, 689; do., in glass vessels, effect on vessel, 690; on Poisson and Colladon's results, 690; remarks on his results, 925, 1326, 1357, 1546 and ftn.
Optics. Newton's treatise on, 26
Optical effects of stress. Brewster, 698, p. 640, ftn.; Neumann, 1185—1212; Maxwell, 1543—4
Oscillations. See *Vibrations, Note, Nodal Line*
Ostrogradsky. Shifts for vibration in elastic medium expressed by definite integrals, 739—41; on Legendre's coefficients, 742—3; on supposed error of Lagrange, 744

Pacinotti and Peri. Experiments on wood, 1250; stretch-modulus from flexure, torsion and traction, 1250—2; ratio of elastic strain to stress before and after set, 1252; values of height stretch-modulus for all wood constant, 1252
Pagani. On rods and membranes, 380; funicular polygon, 381; flexible membranes, 382—4; vibrations of circular elastic membranes, 385—94; on Poisson's and Navier's treatment of elasticity, 395; on bodies supported by more than three struts, 396
Page. Elastic properties of Caoutchouc, 1370
Paoli. Molecular motion and life, 361; Belli on, 361
Parabola. As generating curve of solids of equal resistance, Galilei, 4; do. for column, Lagrange, 112; osculating, of a curve, Hill, 406
Parabolic law of elasticity in cast iron, 1411, 1438—39
Parent. Controversy on solids of equal resistance, 5; rupture of beams, point of, 17; do. experiments on, 28 *a*; Bülfinger on, 29 $\beta$
Parrot. Compressibility of glass, 747; effect on thermometer bulbs in deep sea, 747
Parseval's Theorem, 178
Particles, material under mutual attraction and repulsion, Cauchy, 615, 662; stress in a system of material, 616; of a double system under mutual attraction, 672—3; stretch, dilatation etc. under change of form, 674—5; definition of, Ampère, 738. See *Molecule, Attraction*
Peri. See *Pacinotti*
Perkins. Experiments to determine compressibility of water, 691, 1011
Permanent load. Defin tion of, Vicat,

726; importance of in structure, Vicat, 727
Perronet. Experiments on beams, 28 ζ
Person. Relation of stretch-modulus to latent heat of fusion, 1388
Persy. Neutral surface of beam in uniplanar flexure, 811
Petiet. Use of cables in suspension bridges, 1454
Petty. *Duplicate Proportion and Elastique Motions*, 6
Peyré. Nodal surfaces of solid prisms and tubes vibrating longitudinally, 820
Pfaff. Translator of Lagerhjelm's work, 363
Phillips. Impulsive effect of travelling loads on bridges, 1277; best form of carriage springs, 1504
Photo-elasticity: see *Polarisation*, *Brewster*, *Neumann*, *Maxwell*
Pictet. Permanent effect of temperature on iron bars, 876
Piobert. Lamé's rules for boiler thicknesses, 1039
Piola. Applies Lagrange's method to theory of elasticity, 362, 759—764; applies Poisson's method and finite differences, 767—75; equation of continuity of fluid motion, 776; analysis of stress, 777; erroneous theory of fluid motion, 777—87
Plane. Solid bounded by indefinite, Lamé, 1018—19; by two indefinite and parallel places, 1020--21
Plana. Treatment of elastic curves, 151—4; correction of Lagrange, 153—4; elastic lamina, oscillation of, 176—80; Fourier on his equations, 208—9
Plaster. Cohesion and adhesion of, 696
Plastic Stage. Specific weight at, Lagerhjelm, 369; Gerstner's law not true at, Giulio, 1217; in crushing wrought-iron cubes, Clark. 1475; in elastic life, p. 891. See *Stricture*
Plate. General discussion of elastic, Navier, 258—261; Germain, 283—306; Poisson, 474—493; Cauchy, 629—652
- Contour-conditions, Free or loaded, Navier, 260, 527—34; Poisson, 486—7; Cauchy, 641, 646—8, 651—2; Kirchhoff, Clebsch, Thomson and Tait, Saint-Venant, Boussinesq, 488; built-in, 491, 642, 652
- Equilibrium of circular-, heavy and normally loaded, free, supported, or built-in contour, 494—500; centre and contour supported, 501—504
- Equilibrium of rectangular, Navier, 261—4; Cauchy, 636, 643—4, 647; Maxwell, 1548
- Vibrations of circular, notes and nodes, Poisson, 505—21; Savart on copper, 520; Wertheim, Kirchhoff, 1344—8; Savart, 326, (metal) 332, (wooden, crystalline, rock-crystal) 337—342; Poisson and Savart, 432; Strehlke, 359
- Vibrations of rectangular, Jas. Bernoulli the younger, 122; Germain, 290, 298; Savart, 326, 329; Strehlke, 354—5, 359
- Vibrations of, nodal lines for all kinds of contours, Savart, 352; Strehlke, 353—5; Wheatstone, 746. Integration of equation for, Poisson, 424—5; Fourier, p. 876; Cauchy, 627, 645—6
- Stretch in, when subject to uniform contour-traction, 430, 483
- of aeolotropic structure, Cauchy, 649
- without elasticity, variable thickness, Cauchy, 650
- Velocity of sound in, Cauchy, 649
- Rupture by steady pressure, Vicat, 734; Navier, 263
- Iron, Fairbairn, Experiments on strength of, 1444; resistance to crushing, 1477; strength of, 1496; with and across fibre, 1497; riveted, strength of, 1499—1502; normal load, 1501
- Photo-elastic properties of, Neumann, when strained, 1195, when heated, 1198—1202; circular, and circular annulus, 1201; rectangular, cemented together, 1203; hot, on cold surface, 1204
- Attraction of two thin circular, Belli, 168

Poisson. Equilibrium of elastic surfaces, 412—15; on Jas. Bernoulli's work, 413; equilibrium of surface flexible and non-elastic, 416; remarks on Lagrange's method, 417; vibrations of elastic plate, 418; elastic stress near boundary, 419; equation to surface of constant area, 420; use of integrals in questions of molecular force, 421; elastic curve of double curvature, 423; waves in water and in infinite elastic plate, relation of, 424; vibrations of elastic plates, integral of equations for, 425; distribution of heat in a solid body, Saint-Venant on, 426; vibrations various, of elastic rod, 427—8; frequencies of different kinds of, 429; stretch in wires, 430; *first memoir* on general theory of elastic solids, 431—3, 434; history of theories of elasticity, 435; use of definite integrals, 436, 441; stress in terms of strain, 438, 442; laws of molecular force, 439, 443—4. Definition of stress—Saint-Venant on, 440, 1563; objection to calculus of variations, 446; equilibrium under normal pressure, 447—8; elastic sphere, 449—463, 535; elastic cords, 464—5; elastic rods, 466—71; elastic membrane, 472—3; elastic plate, 474—493; application to circular plate, 494—521; greater and less elasticity, 522; integral of equations for vibrations in indefinite medium, 523—6; controversy with Navier, 527—34; on a supposed error of Lagrange, 536—39; *second memoir* on general theory of elastic bodies, 540—2; molecular force and stress, 542—6; analysis of stretch, 547, 549, 550; stress (initial), 548, (in crystalline body), 552—558; propagation of motion in an elastic medium, fluid and solid, 560—6; capillary action, new theory of, 567; *Traité de mécanique*, 568—81; *third memoir*, elasticity of crystalline bodies, 582—601; on his position as rari-constant elastician, 615—6, 922—4, 925, 927, 980, 1026, 1034, 1050, 1073, 1264—6, 1267, 1320—4, 1346, 1400—2, 1520, 1534, 1537, 1563; compression of vessel in compression of water experiments, 688—90; Weber on his theory of vibrations in rods, 699; Bellavitis on his theory of equilibrium of rods, 935; report on and relation to Blanchet's paper on wave motion, 1166—67; death of, 601

Polarisation of light. Application of, to investigate strains, Brewster, 698, p. 640 ftn., Neumann, 1185—1213, Maxwell, 1543—4, 1553—57
 of wave motion in an indefinite crystalline elastic medium, Blanchet, 1166—70

Polariscope. Use in analysing strain, Neumann, 1190

Poncelet. *Mécanique Industrielle*, 974; definition of elasticity, 975; cohesion and molecular forces, 976—7; want of isotropy in wires and bars, 978; stress-strain curve, 979; condition of rupture, 980; work, 981—2; resilience, 982; time-effect, 983; traction-stretch curves, 985; moduli for iron and steel, Lagerhjelm's experiments, 986; best form of column, 987. *Resilience*, sudden loading with and without initial velocity, 988—90, bar loaded, receiving blow, 991, application to bridges, 992—3; lectures at Sorbonne, 994; stretch condition of rupture, 995; on theory of arches, 996; report on Lamarle's strut-theory, 1261; on Saint-Venant's memoir on resistance of solids, 1578—9

Pouillet, elastic fluids, 319

*Potentialis, Vis*, Euler, 55—6

Pressure. See *Stress*

Prism. Flexure of, 21; torsion of rectangular, Navier, 280; vibrations of, nodal surfaces of, Savart, 327; under terminal tractive load, Lamé, 1011; vibrations of rectangular, Lamé, 1082—6; crushing of short, Hodgkinson, 948—51, 1414; equilibrium of, subject to flexure, 1581—2; torsion of, 1625—7. See also *Beam, Rod, Bar, Wire*

Prony. Report on Vicat's papers, 724—735

Quartic. Stretch-modulus, Neumann, 799

Rails. Best form for, Heim, 906, 916

Rankine. Laws of Elasticity of Solids, 1452

Rari-constant theory, 921—31, p. 884. See *Constants, Elastic*

Rayleigh, Lord. *Theory of Sound*, 96 β, 470

Réamur. Experiments on cords and beams, 28 γ

Reed-pipes. Experiments on notes of, Weber, 700

Refraction of Light. On corpuscular theory, 170; Fresnel, 320; formulae for, Cauchy, 611; O'Brien, 1363—5; in homogeneous aeolotropic medium, Lamé, 1097—1108; caused by strain, Neumann, 1185—1203, Maxwell, 1543—58

Refraction, Conical, of Light, Lamé, 1105; of Sound, Haughton, 1514

Refractive Index, decrease of with increase of density, 1194

Regnault. Dilatation of, and want of isotropy in glass, 1227; compression of liquids and solids, 1357

Renaudot. Travelling loads on railway bridges, 1277

Rennie. Experiments on strength of materials, 185, on crushing of metal blocks, 186, on tensile strength of metal bars, 186, on crushing of wood and stone, 186, on transverse strength of cast iron, 187, on Emerson's paradox, 187; expansion of arches by heat, 838

Resilience. Young's definition and theorem, p. 875; Tredgold, 195, (modulus of) 999; Borellus, p. 874; Poncelet, 981
Beams subject to longitudinal impact, with or without initial velocity, 988—90; with permanent load, 991; application to bridges, 992, to pile driving, 1233; Duhamel, 813; Sonnet, 938 (5)
Beams subject to transverse impact, experiments on, Hodgkinson, 939—942, 1409—10, 1487; theory of, 943, 1434—7, pp. 893—6
Beams subject to impact from travelling load, Stokes, 1276—1291; Willis, 1418—1422; experiments, 1417, 1423; Becker, 1376, pp. 878—9
of spiral springs, Thomson, 1384
Resistance, relative and absolute, Varignon, 14. See also *Strength*, *Elastic Limit*
Solids of. See *Solids*
Riccati, Jacopo. Experiment on laws of elastic bodies, 30; remarks on Bernoulli, 30—32; use of acoustic properties of bodies, 31; relation of traction to stretch, 31, 32; general explanation of elasticity, 33; review of earlier theories and writers, 34—36; ignorance of Hooke's law, 37; ideas of *forza viva* and *morta*, 38; non-destructibility of *forza viva*, 39; value of his work as replacing Metaphysic by Dynamic, 41.
Riccati, Giordano. Vibrations of beams and position of neutral axis, 121; vibration of membranes, p. 73 ftn.
Rings. Elastic, oscillations of, Euler, 93; Lexell, 96 β; Germain, 290, 294—6
Ritchie. Elasticity of glass threads, 408; atomic theory of torsion, 408; torsion of glass threads and their application to torsion balance, 409; method of drawing glass threads, 410
Rive, De la. Vibrations in bar iron due to elastic currents, 1329; Wartmann on his experiments, 1331; Wertheim on do., 1332; experiments on vibrations caused by elasticity and magnetism, 1336
Rivets. Strength of, Clarke, 1480; frictional effect of, 1481; stress in long, 1482; Fairbairn's experiments on, 1499—1500
Rizzetti. Experiments on impact of elastic bodies, 37
Robison. Strength of materials, 145; on Euler's formulae for columns, 145, 198, 833, 954; position of neutral axis in struts, 146; Barlow on, 190
Rock masses. Stress in and lamination of, Hopkins, 1366—7
Rods, Equilibrium of. *Axis of single curvature*, Jas. Bernoulli, 18—25; Euler, 71—3; Girard, 125; Poisson, 466, 568—72, 574—5; Cauchy, 653, (aeolotropic), 657; Heim, 907; Navier, 254—7, 278; Lagrange, 97—105, 159; Saint-Venant, 1615; Young, 138—140; Hill, 399—407
*Axis of double curvature*, Lagrange, 159—61; Binet, 173—5; Bordoni, 215—24; Poisson, 423, 571; Bellavitis, 935; Mainardi, 1215; Binet and Wantzel, 1240; Heim, 1385—6; Saint-Venant, 1584—1608
Vibrations of. *Longitudinal*, Biot, 184; Navier, 274; Savart, (nodes) 322, 329, (thick) 347—9; Poisson, 427, 577; Cauchy, 660; Sonnet, 938; Lamé, 1089—91; Heim, 1387
*Transverse*, D. Bernoulli, 43—49, 50; Euler, 51—64, 86—92; G. Riccati, 121; Biot, 184; Strehlke, 356; Poisson, 428, 467—9, 581; Cauchy, 657; Plana, 178—180; Fourier, 209—210; F. Savart, 323—5
*Torsional*, Coulomb, 119, p. 873; Duleau, 229; F. Savart, 323, 327, 333; Cauchy, 661, 669; Wertheim, 1339—43; Saint-Venant, 1628—30; Lamé, 1085, 1091; Giulio, 1218
*Ratio of transverse to longitudinal frequencies*, Poisson, 429, 471; Cauchy and Savart, 655; Weber, 699
*Of torsional to longitudinal*, Poisson, Savart, Chladni, Rayleigh, 470; Wertheim, 1341; Saint-Venant, 1629—30
Torsion of, Bevan, 378—9, 972—3; Hill, 398
Velocity of sound in, Cauchy, 654; (stretch-modulus from), 370, 704, 1184, 1293, 1301, 1311, 1350—1, 1404, (of steel) p. 702 ftn., 1372. See also *Specific Heat*
Electric currents set up by vibrations in, 1248, p. 720 ftn. See also *Electric currents*
See also *Bar*, *Beam*, *Wire*
Rollers. Experiments on compression of, Vicat, 729—30
Rolling of bodies on one another, Morin, 1183
Rondelet. Experiments on plaster, mortar etc., 696

Saint-Venant. Remarks on Hooke's Law,

8, on Jas. Bernoulli's idea as to position of "axis of equilibrium," 21, on case of elastic curve under a couple, 58, on Coulomb, 115, 117, 119, on Young, 139, on Poisson's definition of pressure, 426, 440, on P.'s use of definite integrals, 542, on P.'s treatment of elastic lamina, 571, on Cauchy's theorem, and number of elastic constants, 554—5, on Cauchy's memoir on torsion of prisms, 661, 683, 684, on Vicat's attack on theory of elasticity, 722, 725, 1569, on Green's methods, 934, on Stokes' dynamical deflection of bridges, 1277, on Regnault's experiments on compression, 1358; controversy with Wertheim, 1343; lithographed lectures, 1560—77; on relation of theory to practice, 1562, definition of stress, 1563, 1612, 1614, 1616; of slide, 1564; on use of word "traction," 1566; stretch-condition for strength, 1567; superposition of small strains, 1568; discussion of slide, 1570—1572; circular elastic arc and ring, 1573—75, 1593; struts, 1576; memoirs on elastic rods, 1578, Poncelet's report on memoir, 1578; criticism of old theory of flexure, 1580; equilibrium and rupture of a prism subjected to flexure, 1581—2; set of helicoidal springs, 1583; introduction of shear for solids of equal resistance, 1583; equations for elastic rod of double curvature, 1584—92; application to helix and discussion of Guilio's results, 1593—5, 1608; history of problem of rods of double curvature, 1598, 1602—1607; torsion, distortion of plain sections, 1596, 1623—25; do., case of prism, rectangular base, 1626, of prism, elliptic section, 1627; torsional vibrations of elastic rods, 1628—30; fluid motion, 1610; bridge over the Creuse, 1611; is matter continuous or not? 1613; flexure of elastic rod, large displacements, 1615; form of expressions for strain-components, where shift-fluxions are large, 1617—22.

Sanden, H. von, Dissertation on Elasticity, pp. 77, 78.

Savart, F. Experiments on sound, 321; nodes of rod vibrating longitudinally, 322; analysis of vibrations in rods, plates, etc., 323—328; experiments on nodal surfaces of plates, membranes and flat rods, 329, on rotation of nodal lines of plates, 331; analysis of structure of metals from nodal lines of circular plates, 332; experiments on torsion of rods of various cross-section, 333—4, 398, 1218, 1340, 1628; relation of heat to torsional resistance, 334; analysis of crystalline bodies by the nodal lines of plates, 335—338; structure of wood by same method, 339; analysis of bodies with three rectangular axes of elasticity, 340, of rock-crystal, 341—343, of carbonate of lime, 344; nodal surfaces for longitudinal vibrations of large rods and bars, 347—50; comparison of extension in a bar by load and vibration, 351; experiments on nodal lines of plates of different forms, 352; Poisson on S.'s experiments on rods, 427—429, 470—1; measurement of nodal circles in copper plates, 520; Cauchy on S.'s experiments, 626, 668; Lamé on S.'s experiments on rectangular prisms, 1053—6

Savart, N. and Duhamel on effect of elasticity in vibrating cords, 1228

Schafhaeutl. Neutral axis of beams, by Brewster's method, 1463 ε

Schultén. Equilibrium of elastic wires, 536—9, of flexible cords, 818

Seebeck on vibrating rods, nodal lines and loops, 1180

Séguin, on law of molecular force, 1371

Set. Fore- and after-set, 708, p. 882; 1226, 1379, pp. 885—91

Slide-set, p. 874 (including effect of set on slide-modulus)

Coulomb-Gerstner Law, p. 874, p. 441 ftn., 806—810, p. 889, p. 891, 852, 1224, 1252

Gerstner-Hodgkinson Law. Stretch of piano-wires, 806; flexure of wooden beams, 807—8; torsion of wooden cylinders, 810; Giulio rejects, 1217, 1224; Hodgkinson, 969, (qualified) 1411; Neumann's hypothesis does not agree with, 1208

Neumann's theory of, 1206—12, p. 889; applied to glass sphere and cylinder rapidly cooled, 1212

Biot on, 182; Poisson on effect on density, 368

Cast-iron, Hodgkinson, 240, (produced by smallest load) 952, (time-effect) 953, (bars) 1413, (hollow cylinder) 1474

Wrought-iron, Hodgkinson, (bars) 1473, (hollow cylinder) 1474; Lamé, 1003

Iron-wire, Giulio (produced by very small load), 1217; Brix, (relation to state of ease, elastic limit, yield-

point) 850—5, (how affected by annealing) 856, (beginning of) 857.
In helical springs, Guilio, 1224—6; J. Thomson, 1383—4
Metal wires, Wertheim (begins at smallest loads, and proceeds continuously?), 1296, 1301
Wood, Gerstner, 807, 810; Wertheim, 1314; Pacinotti and Peri, 1252
In Human tissues, 1318
Shear. Definition of, p. 883. History of, p. 877, 1565. Vicat's protest, 721, 725; word due to Stephenson, 1480
Hopkins' theorem, on maximum, 1368
Omission of in old beam theories, Vicat, 722, 725; effect in case of short beams, 723, 1569
Experiments on shear of rivets, Clark, 1480
Experiments on ratio of shearing to tractive strength, Coulomb, p. 877
See also *Slide*
Shear-cone, Lamé's, 1008, 1059
Shells, Spherical. Experiments on rupture of iron, Navier, 275; equilibrium of, under sudden uniform tractive load, Duhamel, 891—3, Maxwell, 1553; equilibrium of, Lamé, 1016, 1094, 1110—48, 1154—62; condition of rupture, p. 552; vibration of, Lamé, 1093; application to case of earth, Lamé, 1095
See also *Spherical Envelope, Sphere (hollow)*
Cylindrical, equilibrium of, Lamé, 1012, 1022, 1088; condition of rupture, 1013
Shift, Shift-fluxion, Body shift-equations, &c., definitions of, p. 881, p. 884
Silk threads, torsion of, Coulomb, p. 69 ftn.
Elastic after-strain in, Weber, 707—9, 711—19
Absolute strength of, do., 710
Silvabelle's problem, 165
Skin changes, influence of, on lead cylinders, Coriolis, 720; on iron wires, Brix, 858; on iron beams, Hodgkinson, 952; on iron (oxide effect), Lamé, 1002; on wrought iron bars, Giulio, 1216; on cast do., Hodgkinson, Clark, 1466, 1484
Slide. Definition of, 1564, p. 881; history of, 1565, pp. 876—7; value of for large shift-fluxions, 1618—22
Effect of friction on, Morin, 729, 905, absence of in older beam theories, 1565
Geometrical proof that small slide consists of two stretches, Saint-Venant, 1570
Introduction into beam theory (sections still *plane*) 1571—2, Maxwell, 1554; example of circular arc in a vertical plane, Saint-Venant, 1573—4, of circular ring, do., 1575
Equilibrium of prism under flexure, 1581
Saint-Venant's formula for rupture, including effect of, 1582
Smith, A. Analytical expression for tendency of continuous beams to break, 1181
Propagation of wave in an elastic medium, 1181
Solids of equal resistance. How generated, 4; controversy on, 5; Varignon on, 16; Parent on, 17; Girard on, 125—131; Heim on, 915; Saint-Venant (shear considered), 1583
Solids. General equations for elastic, Navier, 253, 265—70; Poisson, 434—48, 548—53, 589; Cauchy, 613—16, 666—7, 677; Piola, 761, 768—70, 777, 780; Duhamel (including temperature-effect), 869, 883; Green, 917—9; Lamé and Clapeyron, 1007; Lamé, 1054—60, (in curvilinear co-ordinates), 1150—3; Neumann (thermo-elastic equations), 1196; Bonnet (in curvilinear co-ordinates), 1241—6; Stokes (including temperature-effect), 1266; O'Brien (in symbolic form), 1360—5; Haughton, 1510—11, 1516—17, 1521; Jellett, 1526—34; Maxwell, 1540—1; Avogadro (first text-book, appearance of), 845
in what differ from fluids, Haughton, 1508
Sonnet. Longitudinal vibrations, nodes etc. of rods, 938
Sound. Waves in elastic fluids, Euler, 95; velocity of in air, Laplace, 310, 316, 318; caused in a bar by electric currents, memoirs on, p. 719 ftn.; Wertheim, 1333—5; De la Rive, 1336
Specific weight. Near rupture-surface of iron, Lagerhjelm, 368; Poisson, 368—9
heats of metals, Weber, 701—6; Duhamel, 874, 877—880, 899; Wertheim, 1297, 1350; Clausius and Seebeck, 1402—4; Bell, 1372
Sphere. Attraction of two, Belli, 165, 169; attraction and repulsion, Laplace, 307, 308, 312, 315, 319; equilibrium of elastic sphere of attractive matter,

Lamé, 1015; vibrations of elastic, Poisson (no load), 449—460; (contractive load), 461—3; Mainardi, 860—4; equilibrium of hollow sphere, Maxwell, 1545; Poisson, 535; Lamé, 1358; do., temperature considered, Duhamel, 871; vibrations of hollow, temperature variable, Duhamel, 872—4; colour fringes for, Neumann, 1197; equilibrium of hollow, under sudden uniform tractive load, 891; do., temperature of each point a radial function, 894, 1197; rate of cooling of free, heat liberated by contraction considered, 895; do., surface at uniform and at varying temperature, 896; experiments on compression of, Navier, 275; Vicat, 730; Wertheim, 1326; Regnault, 1358; set in, due to rapid cooling, Neumann, 1212

Spherical envelope. *Lamé's Problem.* Equilibrium of, 1110—15; given distribution of load, 1110—15; acted on by constant force, 1116; rotating round fixed axis, 1117; force varying as distance from centre, 1118—48; application of spherical co-ordinates to, 1154—62. See also *Shell, Spherical*

Spread, definition of, 595, p. 882

Springs, Laws of. Hooke, 7—8; Jurin, Camus, Deschamps, p. 58 ftn. Lagrange, 97. Plane springs, Lagrange, 98—105; 1182; (for carriages), 1504. See also *Bar, Rod,* etc. Helical or spiral, Hooke, 7—8; Binet, 173—5; Bernoulli, 246; Mossotti, 247—51; Giulio, 1219—26; Wantzel, 1240; Thomson, 1382—4; Saint-Venant, 1583, 1593—5, 1606—8

Squeeze. Definition of as negative stretch, p. 882; stretch-squeeze ratio, p. 885, 368—9, 1321—5

Stages. Of elastic life of material, p. 888 (7), p. 889 (10). See also *Worked Stage*

Standard of Length. Method of supporting, 1247

State of Ease. Explanation of, pp. 885—91; importance of, recognized by Weber, 700, Lamé, 1003; disregarded by Hodgkinson, 952, by Wertheim, 1296, by Cox and Hodgkinson, 1442; relation to safe load, Gerstner, 808; Leblanc, 936; as regards rari-constant theory, 924, 926, p. 817 ftn.; effect on elastic after-strain, Giulio, 1224—1226; possibility of extending, Thomson, 1379; explains discordant experiments, do., 1381; effect on absolute strength, Clark, 1473; in wires under torsion, Coulomb, pp. 873—4; depends upon "worked condition" of material, p. 888 (7)

Steel. Modulus as compared with iron, 366, 986, 1293, p. 702 ftn.; strength compared with iron, 693; stretch at limit of elasticity, 693

Stephenson, R. Tubular Bridges, 1465—89; opinion as to change in molecular structure by vibration, 1429, 1464, as to limit of elasticity in cast-iron, 1430; as to effect of travelling loads, 1428; introduces expression "shearing strain," 1480

Stokes. Views on the constant-controversy, 922, 925—30, 1266—7, p. 817 ftn.; friction of fluids in motion, 1264; bi-constant and thermo-elastic general equations, 1265—6; constitution of luminiferous ether, 930, 1266; propagation of arbitrary disturbances in elastic media (waves of twist and dilatation), 1268—75; dynamical effects of travelling loads on girders, 1276; of loads small compared with weight of girders, 1277, Saint-Venant on his results, 1277, solution of equation, 1278—87; inertia of bridge considered, 1288—91, p. 878

Stone. Experiments on, Vicat, 728; Hodgkinson, 950, 1445—7, 1451; Clarke, 1478; Rennie (arches), 838, (crushing of), 186

Strain. General, and elastic fore-. Definition of, p. 882; analysis of, Poisson, 547, 586, (effect of rotation) 587, 595, Cauchy, 603-5, 612, 617, 663-5, Maxwell, 1539, Saint-Venant, 1564; Superposition of small, Stokes, 928, 1266, Maxwell, 1555, Saint-Venant, 1568, Weisbach, 1377-8, Wheatstone, 746; components of, for large shift fluxions, Saint-Venant and Green, 1614, 1617—1622; analysis by optical methods, Brewster, 698, p. 640 ftn., 1463 (*e*), Neumann 1185-1212, Maxwell, 1556—7

Elastic after-. Definition of, p. 882; discovery of, Weber, 708; his theory of, 711—715; relation to fore-strain, 716; Clausius on W.'s theory, 1401—5; non-superposable, Braun, 717; Coriolis, in lead, 720; History of, Frankenheim, 827; in wires, 736, p. 878 (8); in anchors, 1431; in human tissues, Wertheim, 1317—8; time-influence in after-strain, 720, 817; Pictet, 876; in helical springs, 1224—6

Strain-Ellipsoid. Definition, p. 882;

Cauchy, 605, 612, 617; Piola, 776. *Inverse*, p. 882; relation to Fresnel's ellipsoid of elasticity, Neumann, 1188—9

Strehlke. Nodal lines of plates, 354—5, 359, Chladni on, 355; nodes of rods free at the ends, 356—7; on Galilei's claim to have discovered the Chladni-figures, 358; correction of Tyndall and others, 360

Strength, Absolute. See *Cohesion*

Stress. Definition of, pp. 882—3, 1563; notation for, 610 (ii); old use of word strain for, 1469

Across a plane, Poisson, 426, 440, 546, 589, 598; Cauchy, 603, (his theorem) 606, 659, 610 (ii), 616, 678—9; (effect of temperature on), Duhamel, 869, 875, Stokes, 1266, Neumann, 1196, Maxwell, 1541; Lamé, 1007, 1045, 1163—4; Saint-Venant, 1563, 1612, 1614, 1616

Six components of, p. 883; notation for, 610 (ii); Poisson, 538, 548, (aeolotropy) 553, 591—4; Cauchy, 610, 613, 615, 616, (resolution of) 658—9, (aeolotropy with initial stress) 666—7; Lamé, (analysis of) 1007—9, (including aeolotropy) 1046—1059; Stokes, 1266; Duhamel, 869, 883; Maxwell, 1541

In spherical coordinates, 1092, 1113

In cylindrical coordinates, 1022, 1087

In curvilinear coordinates, 1150, 1153

Stress. Maximum, as condition of set (or rupture) not the true one, Poncelet, 995; Saint-Venant, 1567; Lamé's errors, p. 550 ftn. and p. 552 ftn., 1038 and ftn., 1551 and ftn., 1378 and ftn.

Stress-ellipsoid. Cauchy's, properties of, 610 (iv), p. 883. Lamé's, 1008, 1059, p. 883

Stress-quadric. Definition of, 610, (iii), p. 883, 614

Stress-director-quadric, 1008, 1059, p. 883

Stress-strain relation, its linearity, 8, (Stokes) 928, 1266, p. 884, (Pacinotti) 1252, (Colthurst) 998; its want of linearity, Riccati, 31, Bernoulli, 22, Gerstner, 803—7; Hodgkinson, 231, 969, 1411, Poncelet, 985, (defect of Hooke's Law) pp. 891—3; its graphical representation, Jas. Bernoulli, p. 873, Poncelet, 979, (for brass &c.) 985, (area of curve represents work) 981; Hodgkinson (parabolic law), 1411, 1438—9; Cox (hyperbolic law), 1318, 1412, 1440—2; its form in elastic after-strain, Weber, 714, Wertheim, 1318; remarks on, 929, with set, (iron cables) 817, (iron wire) 852

Stretch. Definition of, p. 881; value in any direction, 547, 612, 617, 663; of glass in relation to its dilatation, 688—690; of iron-bars before set, 1462; of iron and steel at elastic limit, 693; maximum, of wires (Gerstner), 804; of a wire by sudden traction (temperature effect), Duhamel, 889—890; ratio of, in cases of solid, membrane and thread under uniform tractive-load, Lamé, 1703, cf. 368, 483

Of a bar by electric currents, Joule, p. 719 ftn.; Wertheim, 1333—35; value of, for shift-fluxions of any magnitude, 1617—22; maximum, is true condition of set for bodies, 995, 1567, Lamé's errors, 1013, 1015, 1038, and ftns.; for a prism under flexure, slide considered, Saint-Venant, 1582

Stretch-quadric. Equation and properties of, Cauchy, 612, 614, p. 882, pp. 885—9

Stretch-squeeze-ratio, definition, p. 885; theoretical and experimental values of, 369, 1321—5

Striction. Section of, definition, p. 891

Stricture. Definition, p. 891; in iron wires, Dufour, 692; neglected by Gerstner, 804; noted by Poncelet, 979; by Lamé, 1002; difference of for square and round bars, 1003; Masson, 1184; magnetisation at point of, Hausmann, 1179

Strings. Notes of stretched, used to investigate laws of elasticity (Riccati), 31, 40, Euler, 69, 70, N. Savart, 1180, 1228; bibliography of, p. 42 ftn.; elastic, equilibrium of, Lagrange, 158; elastic, extensions of under load, Navier, 273; motion of two points connected by, Navier, 274; equilibrium and motion of, Poisson, 464, 465, Lamé, 1071; equilibrium of heavy and loaded, 1070. See also *Cords*, *Bar*, *Rod* (longitudinal vibrations of)

Struts. Euler, 65—68, 74—86; Lagrange, 106—113; Robison, 145—6, Tredgold, 198, 833—4; Duleau, 228; Whewell, 737; Arosenius, 819; Heim, Greenhill, 910; possibility of applying Saint-Venant's flexure method to, 911; crushing force of short, and form of rupture-surface, 948—51; cast-iron, Hodgkinson, (on Euler) 954, (flat and rounded ends) 955—6; position of fracture and modification of Euler's theory, 957—9;

empirical formulae for strength of, 960—1; short, Gordon's formula, 962; hollow, 963; time-element, 964; Lamarle's treatment of Euler-Lagrangian theory, 1255; conditions for set before or at buckling, 1256—62; form of neutral line for strut, Kennedy, p. 44 ftn., p. 521 ftn., p. 777 ftn. See also *Column*, *Beam* (longitudinal load)
Sturm. Compression of liquids, 688
Sullivan, experiment on electric currents set up in wires by vibration, 1248, p. 720 ftn.
Surface. Isostatic, 1036—7, 1152; of perfect elasticity, p. 886 (2); of cohesion, p. 886 (4); yield, p. 887 (5); of directions (=stress-director-quadric), 227, 1008, 1059.
Elastic equilibrium and motion of, Poisson, 412—22, (flexible and not elastic) 416; Germain, 283—305; Fourier, p. 876. See also *Membrane*, *Plate*
Surface-Equations, Navier, 267; Poisson, 448; Cauchy, 614; Piola, 780; Duhamel, 883; Lamé, 1046; Jellett, 1532—3

Tait, *Properties of Matter*, p. 325 ftn.
Tate. Formulae for flexure of beams and tubes of various sections, 1490—1; relative strength of beams of similar section, 1492; on Hodgkinson's results for cellular structures, 1493
Teinometer, Chromatic. Brewster, 698, p. 640 ftn.
Temperature. Deep-sea measurements of, 747, 827
Temperature-Effect, defined, p. 883
Experiments on, strength of iron-wires, 692 (8); on metal bars, 876; on strength of cast-iron beams, 953, 968; on iron-bar, 1353 and ftn.; on metal wires, 1293—1301, 1307—9, 1524; on slide-modulus, 1395, 1396; on girders, 1428; on torsional resistance, 334
Theory of, on elastic strain, Weber, 701—3; Neumann, 790—2, (change of sphere into ellipsoid, thermo-elastic ellipsoid) 791; Frankenheim, 828; general equations, (Duhamel) 869, 883, (Neumann) 1196, (Stokes) 1266, (Maxwell) 1541; effect of variation in temperature producing strain and colour-fringes, (Brewster) 698, p. 640 ftn., (Neumann) 1185—1205, (Maxwell) 1553; temperature producing strain (Duhamel), sphere, 871—4, 894, propagation of sound, (infinite medium) 886—7, (bar) 1403; sudden compression, 888—91; cooling of sphere, 895—6; stress in bar due to varying temperature, 899—903
Theory for set in heated and unannealed glass, (Neumann) 1209—12, (Maxwell) 1556-7
Thermo-elastic constant (for isotropy), p. 883, 869, 1541
Testing machines, 364, 697—8, 709, 1001, 1294, 1374, 1417, 1419
Thermo-elastic constant (isotropy), p. 883, 869, 1541
Thermo-elastic equations, p. 883, 701, 869, 883—5, 1196, 1266, 1541
Thermo-tension, 1353
Thomson and Tait. *Natural Philosophy*, p. 82 ftn., 175, p. 189 ftn.; on Poisson and Kirchhoff's contour-conditions for plates, 488
Thomson, Sir W. Torsion of glass fibres, p. 206 ftn.; time-element in torsional tests, p. 390 ftn.; on pressural wave, p. 803 ftn.
Thomson, Thomas. Effect of hot and cold blast in making cast-iron, 946
Thomson, Jas. Possibility of raising elastic limit, 1379—80; state of ease, 1381; elasticity, strength and resilience of spiral springs, 1382—4
Thorneycroft. Experiments on axles, 1463 *g*; form of axles and shafts, 1464
Thread. Torsion of metal, 119, p. 873, 398; silk, horse-hair, p. 69 ftn.; Biot, 183; glass, 408—9
Equilibrium of, Poisson, 430, 464—5; Ostrogradsky, 744. See *String*, *Cord*, *Wire*
Glass, how to draw, 410; silk, elasticity and strength of, 707—19
Thury, de. Experiments on hollow cast-iron struts and columns, 750
Time-Element. Defined, p. 882; relation to after-strain, Weber, 708, Thomson, p. 390 ftn.; effect on strain in lead cylinder experiments, Coriolis, 720; importance of, Bevan, 751*a*; as cause of after-strain, Martin, 817, Frankenheim, 827; Experiments on, for wires, Leblanc, 936, Vicat, 736, Poncelet, 983, Masson, 1184, Giulio, 1217; effect on struts, Hodgkinson, 964, on beams, 967; Fairbairn, 953; effect on cables of suspension bridges, Dufour, 1455; on helical springs, Giulio, 1224—6; on wood, Wertheim, 1314. See also *Strain*, *Elastic after-* and *Set*
Tissues, Human. Elasticity and after-strain of, Wertheim, 1315—18
Torsion. Theory, Coulomb, 119, p. 873;

Cauchy, 661, 669, 684; Navier, 280; Maxwell, 1549; Saint-Venant, (distortion of plane sections) 1596, 1623—5, (rectangular bar) 1626, (elliptical bar) 1626
Experiments, silk and horse-hair, 119; Biot, 183; rods of various section, Savart, 333—4; wires, Hill, 398; glass threads, Ritchie, 408—9; wood, Gerstner, 810, Pacinotti, 1251, Bevan, 378—9, 972, 973; Vicat, 731
Spiral springs act principally by, Binet, 175; Giulio, 1220; J. Thomson, 1382
Angle of, Bordoni, 217
Radius of, in elastic curve of double curvature, Binet, 174; Bordoni, 216—24; Saint-Venant, 1599
See also *Vibration*, *Modulus*, *Slide-*
Torsion Balance. Biot, 183; Hill, 398; Ritchie, 408, 410; Mainardi, 1214
Torsional Vibration. See *Vibration* and *Torsion*
Traction. Defined, pp. 882—3; use of wood for tension, Saint-Venant, 1566; excentric, effect on strength of bar, Tredgold, 832—4, Brix, 1249. Sudden, see *Temperature-Effect* (Duhamel)
Tractions, principal. Defined, p. 883; properties of, Cauchy, 603, 614, 1008, 1540
Travelling load. Impulsive effect of, on girders, Stokes, 1276—1291; effect on bars, experiment and theory, Willis, 1417—25; proportion of, to breaking load, 1427; deflection of girders by, 1428; dynamical effect of, Cox, 1433
Tredgold. Transverse strength of timber, 195; resilience of materials, 195; flexure and compression of beams, 196; strength of columns and form of, 196; on resilience, 999; strength of cast-iron, 197, 999; neutral lines for columns, 198, 833—4; strength of a bar under excentric tractive load, 832—3, 1249
Tubes. Cylindrical, rectangular and elliptical. Experiments on strength of, 1466, 1467, 1494; calculation for, 1490—3; large model of Britannia Tube, experiment on, 1468; cellular, experiments on strength of, 1477, 1494; transverse strength of, experiment on, and theory of, 1483; malleable iron, 1494
Twists. Defined, p. 882; used in expression for work, 933; propagation of twist-wave in an infinite medium, 1268—75; must be small if ordinary elastic equations are to hold, 1622
Variations, Calculus of, 53, 113, 203, 420
Varignon. On "solids of equal resistance," 5, 16; applications of Galilei's and Mariotte's hypotheses, 13; absolute and relative resistance of beams, 14; hypothesis as to neutral surface, 15
Velaria. Euler on, 71
Velocity of sound. In air, Laplace, 310, 316, 318; in bars, to determine their moduli, Lagerhjelm, 370; Weber, 702, Bell, 1372; Clausius, 1403; Masson, 1184; Wertheim, 1297, 1350—51, in lamina of indefinite size, Cauchy, 649; in rods, do., 655; in lead, Masson, 1184; in wood, Wertheim, 1252; in liquids, do., 1337—8
Vibrations. Of iron produce molecular change (?), 1429, 1463—4; produced by electric current, 1330—6; propagation of, in an infinite crystalline medium, Blanchet, 1166—78; Green, 917; Lamé, 1097—1108; in an isotropic medium, Poisson, 523—6, p. 299 ftn., 564—5, Ostrogradsky, 739—41, 743, Duhamel, 886, Stokes, 1268—75, Green, 917—21, Haughton, 1511—15, 1518, 1519—22, Lamé, 1033
of cords, 69, 70, p. 42 ftn., 576, 464—5, 1228, (by fiddle-bow) 897. See *Cords*, *Strings*
of beams, bars, rods, prisms, wires (*transverse*), 44—50, 64, 87—92, 209—10, p. 876, 121, 184, 290—9, 327—8, 356—7, 467—9, 581, 655, 657, 1382—6, 1180; (*large, nodal lines of*), 820; (*longitudinal*), 184, 272—3, 322, 329, 347—51, 577, 655, 660, 938, 1387, 1083—4; (*torsional*), 119, 184, 229, 333, 661, 669, 1085, 1628—30, 1389—96; classification of, 323—8, 427—9, 846, 1086; ratio of different frequencies of, 429, 470—1, 576, 654, 655, 699, 1630; set up electric currents (?), 1248, p. 720 ftn.; of tempered and hard-hammered iron and steel, 366, 802. See also *Bar*, *Rod*, *Prism*, *Wire*
of membranes, p. 73 ftn., 207, 329, 386—394, 472—3, 1074—7
of plates, 122, 184, 208—10, 290, 298, 326, 329—30, 332, 352—5, 425, 432, 485, 505—22, 645, 745—6, 1344—48
of rings, 93 and ftn., 96 β, 290, 294—6
of spheres and spherical shells, 449—463, 860—4, (with variable temperature) 871—6, 1093. See *Sphere*
of cylinders, 1089—91
of spiral spring, Mossotti, 246—51
Vicat. Suspension bridge over the Dor-

dogne, 721; effect of slide on theory of elasticity, 721—2; suspension bridges of wire or bar-iron, 723; phenomena preceding rupture in certain solids, 724; controversy with Prony and Girard, 724—5; shearing force, 726; instantaneous and permanent loads, 727—8; experiments on rollers and effect of lateral extension, 729—30; torsion experiments and their bearing on theory, 731; experiments on non-fibrous built-in beams, 732—3, on rupture of plates, 734, on long-continued loads on iron wires, 736; value of his work, 735

Vitruvius. Best form of columns, 106

Viviani. Controversy on solids of equal resistance, 5

Voute. Attraction, cohesion and crystallisation, 212; action at a distance, 213; cohesion, 213; forms of particles, 213

Wantzel. Integral for equation of elastic curve of double curvature, 1240; helical form of rod subject only to terminal couples, 1240; on equations for torsion of rectangular prism, 1626

Wartmann. Effect of electric currents on elasticity of wires, 1331

Water. Compression of, Bacon, p. 78 ftn., Canton, 372, Oersted, Colladon and Sturm, Aimé, 687—91, Perkins and Lamé, 1011, Regnault, 1227, 1357, 1546; modulus of (!), 372

Waves. Propagation of in elastic medium, Stokes, 1268—75, A. Smith, 1181, Poisson, 523—6, 564—5, p. 299 ftn., Ostrogradsky, 739—41, 743, Duhamel (including temperature effect), 886—7, Green, 917—21, Haughton, 1511—5, 1518, 1519—22, Lamé, (in undulatory theory of light, including discussion of wave-surface, refraction, etc.) 1097—1108, (analysis of plane-) 1078, Blanchet, (in crystalline medium, solution by integrals) 1166—78, (initial disturbance limited) 1169—71; (discussion with Cauchy on form of surface of) 1175, successive waves in crystalline medium) 1178

Velocity of. Wertheim (in solids and fluids), 1349; Poisson, (in water and elastic plates) 424, (in fluids) 560—3

Sound, in elastic fluid, Euler, 95

See also *Sound*, *Vibrations*

Weber. Experiments on ratio of frequencies of transverse and longitudinal vibrations, 699; investigations of notes of reed-pipes, 700; reduces wires to state of ease for his monochord, 700; two specific heats of metals, 701—6; elastic force of silk threads, 707, 719; elastic after-strain discovered by, 708, hypothesis as to, 711—5, relation to fore-strain, 716; description of testing machine and methods of experiments, 709—10; compensating balance, 718; Clausius on, 1401—5

Weisbach. Theory of superposed strains, 1377—8

Wertheim. Cauchy on memoirs of, 682; velocity of sound in different woods, 1252; researches in elasticity, 1292; methods to determine stretch-modulus for metals, 1293—4; relation of temperature to elasticity, 1295; elastic limits, 1296; stretch-modulus, relation to velocity of sound, 1297, temperature, effect on, 1298, relation to molecular distances, 1299, variation of, with molecular distances and with temperature, 1300, influence of low temperature on, 1307—9, wood, &c., 1313—14, relation to slide-modulus, 1319—22; alloys, elastic properties and cohesion of, 1302—3; relation between electromagnetic and elastic properties, 1304—6; glass, different kinds, elastic properties of, 1310—11; wood, elastic properties of, 1312—14; tissues of the human body, elastic fore- and after-strain in, 1315—18; on value of stretch-squeeze ratio ($\lambda = 2\mu$), 1319—26; experiments on hollow glass and metal tubes, 1323; electric current, effect on flexure of bars, 1328; do., controversy with Guillemin, De la Rive and Wartmann, 1328—32; do. sound produced by, history of, 1333, p. 719 ftn.; do. influence on bars of iron, 1334—5; do., De la Rive on W.'s experiments, 1336; velocity of sound in liquids, 1337—8; torsion of rods, 1339—42, controversy with Saint-Venant, 1343; vibrations of circular plates, 1344—8; ratio of velocity of transverse and longitudinal waves, solids and liquids ($\lambda = 2\mu$), 1349; velocity of sound in rods (error of earlier memoir), 1350—51; Clausius on, 1398—1405

Weyrauch. On strain-ellipsoid, 605; molecular action central but not a function of distance, 922

Wheatstone. Acoustical figures, 746, history of, 746; theory of superposition, 746

Whewell. *Mechanics*, 25, 71, 141, 737; on Young's theory of absolute strength, 141; neutral axis of struts, 737; letter of Oersted to, 690

Whorls, pp. 879—81
Wilhelmy. Common unit for heat and cohesion (?), 1356
Willis. Dynamical effect of travelling loads, 1417; description of Portsmouth experiments, 1418; Cambridge experiments, inertial balance (results), 1418, 1421; differential equation of problem, 1419; Stokes' calculations, 1420; on increase of deflection with velocity, 1422; Stokes on W., 1276—91, *passim*
Wires. Equilibrium of, *of single curvature*, Lagrange, 159; Poisson, 536; Schultén, 536—7, 539; Ostrogradsky, 744; heavy and loaded, 1070
*of double curvature*, Lagrange, 157—61; Binet, 173—5; Bordoni, 215—22; Mossotti, 246—51; Poisson, 423, 571—2; Bellavitis, 935; Mainardi, 1215; Heim, 1385; Binet and Wantzel, 1240; Saint-Venant, 1583—1609
See also *Rod, Spring*
Experiments on, Dufour (cohesion, stretch of annealed and unannealed), 692; (temperature-effect) 692; Karmarsch (cohesion, after-strain, and drawing of), 748, p. 878 (8); Vicat (long-continued load), 736; Weber (state of ease for), 700, (cohesion and temperature-effect), 701—6; Leblanc (time-element), 936; Giulio (time-element and set), 1217; Wertheim (elastic constants for, effect of annealing, temperature-effect), 1292—1301, 1307; (vibrations in, due to electric currents) 1333—6; Sullivan (electric current set up by vibrations?), 1248, p. 720 ftn.; Kupffer (torsional vibrations, temperature-effect on elastic constants), 1389—96; Giulio (torsion), 1218. (See *Torsion* and *Vibration*.) *Iron*, Caignard-Latour (tempered and hard-hammered), 802; Gerstner (set and elastic-strain), 803—6; Baudrimont (cohesion, effect of worked state, annealing, change in temperature), 830—1, 1524; Brix (cohesion, after-strain, set, elastic-limit, yield-point, effect of annealing, of working, of skin), 848—58
Sudden tractive load on, (temperature-effect), Duhamel, 889—90
Isotropy of: see *Worked stage, Skin influence*
See *Vibrations*
Wood. Experiments on beams, Mariotte, 10; Parent, 28 $\alpha$; Bélidor, 28 $\beta$; Réaumur, 28 $\gamma$; Musschenbroek, 28 $\delta$; Buffon, 28 $\epsilon$; Perronet, 28 $\zeta$; Girard, 28 $\eta$, 131; Banks, 147; Dupin, 162; Barlow, 188—194; Tredgold, 195; Hodgkinson, 235; Gerstner (set), 807—9; Colthurst (neutral axis), 998
Torsion of, Bevan, 378, 972—3; Gerstner, 810; Pacinotti and Peri, 1250—52
General experiments for elastic constants of, etc., Hagen, 1229; Pacinotti and Peri, 1250—2; Lamarle, 1253—4; Wertheim (including effect of humidity), 1312—4
Plates of, Savart, 339, Wheatstone, 746
Circular arcs of, Ardant, 937
Work of elastic body. Analytical expressions for, Green, 918—9, 933, Clapeyron, 1067—70, Haughton, 1510, 1520, Jellett, 1527, Poncelet, 981—91, Cox, 1433, 1436—7; graphical representation of, Poncelet, 981
Worked stage or condition. Defined, p. 888 (6) and (7); cohesion altered by, p. 874; material in worked stage rarely isotropic, 830—1, 858, 924, (influence on yield-point), 855; possibly affects after-strain, 736; influence on state of ease, 1381. See also *Skin influence* and *Yield-point*
Wurtz. Controversy on solids of equal resistance, 5

Yield-point. Defined, p. 887 (5) and (6), p. 889; in iron, 692, 855—6, 979, 1184, Wertheim (ignorant of), 1296
Yield-surface defined, pp. 887—9
York. Strength of solid and hollow axles, 1463, *b*
Young. Life of Coulomb, 118—119; cohesion and elasticity, 134, 142, 136; transverse force and strength of beams, 134; position of neutral axis of beams, 134, 139, 144; list of works on elasticity, 135; his modulus, 137; stiffness of beams, 138; strength and deflection of beams, 139; Saint-Venant on, 139; elastic rods, 140; Whewell on his results, 141; passive strength and friction, 143; lateral adhesion, 143; modulus of ice and water compared (!), 372; theorem in resilience of beams, p. 875, p. 895, 982, 1384
Young's Modulus. See *Modulus, Stretch-*

CAMBRIDGE: PRINTED BY C. J. CLAY, M.A. & SONS, AT THE UNIVERSITY PRESS.